D1476049

# Engineering Geology
## An Environmental Approach

# Engineering Geology
## An Environmental Approach

Perry H. Rahn
South Dakota School of Mines and Technology

Elsevier
New York • Amsterdam • Oxford

Elsevier Science Publishing Company, Inc.
52 Vanderbilt Avenue, New York, NY 10017

Sole distributors outside the United States and Canada:

Elsevier Science Publishers B.V.
P.O. Box 211, 1000 AE Amsterdam, The Netherlands



Library of Congress Cataloging in Publication Data

Rahn, Perry H., 1936–
Engineering geology.

Bibliography: p. 539
Includes index.
1. Engineering geology. I. Title.
TA705.R28 1985 624.1′51 85-10190
ISBN 0-444-00942-6

Current printing (last digit):
10 9 8 7 6 5 4 3 2 1

Manufactured in the United States of America

# Contents

# Preface

The Association of Engineering Geologists defines engineering geology as "... the application of geologic data, techniques and principles to the study of naturally occurring rock and soil materials or subsurface fluids. The purpose is to assure that geologic factors affecting the planning, design, construction, operation, and maintenance of engineering structures and the development of groundwater resources are recognized, adequately interpreted, and presented for use in engineering practice." Some areas of practice of the engineering geologist include: 1) the investigation of foundations for all types of structures, such as dams, bridges, high-rise buildings, and residences; 2) the evaluation of geologic conditions along tunnel, pipeline, and highway routes; 3) the evaluation of earth materials for construction purposes; 4) the investigation and development of surface and ground-water resources; and 5) the evaluation of natural geologic hazards, such as floods, landslides, and earthquakes.

Engineering geology is the oldest branch of geology. William Smith, one of the founders of geology, was a practicing engineering geologist who, during the early part of the nineteenth century, described the stratigraphy and construction problems along large canals in southern England. Although not rigorously defined, a geological engineer differs from an engineering geologist in background and emphasis. Typically, a geological engineer has an engineering background, including an undergraduate degree in engineering with geologic emphasis in upperclass studies; an engineer geologist is a geologist who has later received education and practice in engineering disciplines.

In the past two decades, it has become vogue to use the term environmental geology for many of the same subjects formerly covered by engineering geology. Both terms mean essentially the same thing, i.e., the study of applied geology: how man is affected by geological phenomena, and how man himself can trigger geologic processes. Engineering geology is quantitative (to the degree feasible) to allow for the assessment and prediction of the effects of various processes. Environmental geology is a much broader field, encompassing related sciences such as biology and meteorology. Environmental geology represents that part of ecology that relates earth science disciplines to man. Generally, environmental geology texts have less technical treatment of subjects than engineering geology. In the past ten years, numerous excellent environmental geology books have been published. The intended audience for these books is the nongeology student who is seeking an applied science course. While these books serve to expose a large audience to environmental geology or geologic hazards, for the most part they are not at a technical level sufficient for geologists who may pursue a career in this field.

Collections of papers in environmental or engineering geology share these shortcomings and have no continuous development of topics at a consistent level of difficulty.

This book is intended as a text for upperclass college students majoring in geological engineering. The book may also be useful to geology students who are interested in environmental geology or the application of geology to engineering. Geology students who do not have an engineering background (including prerequisite courses such as mechanics of materials) should be able to use this book if they independently pursue engineering subjects. Practicing engineering geologists may find this book a useful summary of engineering geology subject material and literature. However, the book is not a complete state of the art or synthesis of all of the diverse subjects encompassing engineering geology. Rather, material was selected so that it would be educational and interesting for the student who is first encountering the subject.

Every teacher has his own preferred approach, and may select different portions of this book for presentation in a different sequence. I have developed what I consider to be a logical sequence of subject matter, and to some degree subject material presented in one chapter depends upon material described in previous chapters. Some material is developed from one chapter to the next; for example, slope stability is discussed in Chapter 4 (Rock Mechanics) and further developed in Chapter 6 (Mass Wasting). Engineering geology includes applied aspects of hydrology. For this reason, two chapters on hydrology (Chapter 7, Ground Water, and Chapter 8, Fluvial Processes) are included in this book.

Emphasis is given in this book to geomorphological processes. A firm knowledge of processes which act on the surface of the earth is a requisite to understanding engineering geology. Today, unfortunately, geomorphology is rarely taught to undergraduates. The increasing awareness and need for engineering geologists in today's technical world may rekindle future interest in quantitative geomorphology.

An engineer must solve problems, using quantitative methods where applicable. Examples of problem-solving are given throughout this text, and problems are given at the end of the chapters. Solving problems is an excellent way to learn analytical techniques. The problems range from simple ones which mimic existing solutions, to complex problems involving more analysis and synthesis by the student, requiring in some cases consultation from outside references. SI (metric) units are the primary dimension mode in this book. English units are used where such units are quoted by others, or where knowledge of both units is deemed particularly useful.

This book contains a preponderance of examples (good and bad) from the United States in areas where I have lived and worked such as South Dakota. This is not because South Dakota has all the problems, but simply because I am more familiar with this area.

I would like to thank Arvid M. Johnson for making many helpful suggestions to this book. I appreciate the encouragement given by Richard J. Gowen and Alvis L. Lisenbee of South Dakota School of Mines and Technology. Sheryl Eddy and Lenora Hudson are to be thanked for their help in typing. David L. Royster, Michael P. Kennedy, and James G. Rosenbaum kindly sent photographs for inclusion in this book. John E. Eberlin, Kevin T. Brady, and other students in my classes at South Dakota School of Mines and Technology have made a major contribution refining this book, and they deserve a special thanks. I would like to thank Allan Ross of Elsevier Science Publishing Co. for his editorial assistance.

One of the rewards that comes from writing is the increased correspondence and knowledge provided by the readers. I look forward to this correspondence.

Perry H. Rahn
Rapid City, South Dakota
*December 1985*

# 1
# Population

> The power of population is infinitely greater than the power in the earth to produce subsistence for man.
>
> Thomas Malthus, 1766–1834

> We come now to New York, which enjoys the double advantages of an excellent harbor and a large navigable river, which opens communication with the interior parts of the country; and here we find a flourishing city, containing forty thousand inhabitants, and increasing beyond every calculation.
>
> Isaac Weld, Jr., 1807

> Unlike plagues of the dark ages or contemporary diseases we do not understand, the modern plague of overpopulation is soluble by means we have discovered and with resources we possess. What is lacking is not sufficient knowledge of the solution but universal consciousness of the gravity of the problem and education of the billions who are its victim.
>
> Martin Luther King, Jr., 1929–1968

> I feel like I'm really contributing something. Even after I'm dead, I'll go on.
>
> Mother of 17 children, Rapid City, SD, 1980

## 1.1 Introduction

A main concern of the engineering geologist is to allow for the improvement in man's condition. It is important, therefore, to understand what man's condition is and to grasp the magnitude of population growth—not only to obtain a perspective on future urban pressures and related geotechnical problems, but also, as an educated person, to be cognizant of the population explosion, its impact on the environment, and the limits of growth.

Ancient civilizations have crumbled because they fouled their own environment. Several thousand years ago the irrigation-based civilization in the Tigris and Euphrates Valley in Mesopotamia collapsed as the soil became salinized. The Romans succumbed not only because of the invading Huns but because the Romans destroyed their own forests and contaminated their own water. Ancient Indian tribes in the New World such as the Maya in Mexico and Hohokam in Arizona perished from drought and irrigation malpractice.

Many people in today's world seem to be enjoying growth and material prosperity unmatched in historic times, and yet there are signs that the planet earth is in trouble. A special report to the President of the United States, compiled by 13 U.S. government agencies (Council on Environmental Quality, 1980), warned that only international cooperation could arrest degradation of the world environment, exhaustion of resources, and overpopulation. "If present trends continue," the report stated, "the world in 2000 will be more crowded, more polluted, less stable ecologically, and more vulnerable to disruption than the world we live in now. Serious stresses involving population, resources, and environment are clearly visible ahead. Despite greater material output, the world's people will be poorer in many ways than they are today."

Nobody can tell what the future will bring, and it is often difficult to separate facts from propaganda spread by either the doomsday environmentalists or the optimistic promoters of growth. This chapter presents some data on population. Chapters 13 and 14 deal with environmental questions related to the future availability of mineral, energy, water, and land resources.

## 1.2 World Population

Man is believed to have evolved about 2 million years ago. Ehrlich and Ehrlich (1970) estimate that the total population of earth 10,000 years ago (8,000 B.C.) was approximately 5 million people. By the time of Christ, the population was around 250 million, and it increased to 500 million (doubled) by the year 1650. It then doubled to 1 billion around 1850, doubled again to 2 billion by 1930, and doubled again to 4 billion by 1975. The 1983 world population was about 4.72 billion.

The growth of human population is exponential (Fig. 1.1). In other words, the world's population has not only increased continuously (with minor irregularities such as the fourteenth century plague), but the rate of increase has also grown. Viewed in terms of geologic time, the "explosion" of human population is truly an exceptional biologic event, probably without parallel in the entire history of the earth.

Figure 1.1 World population growth (from van der Tak et al., 1979).

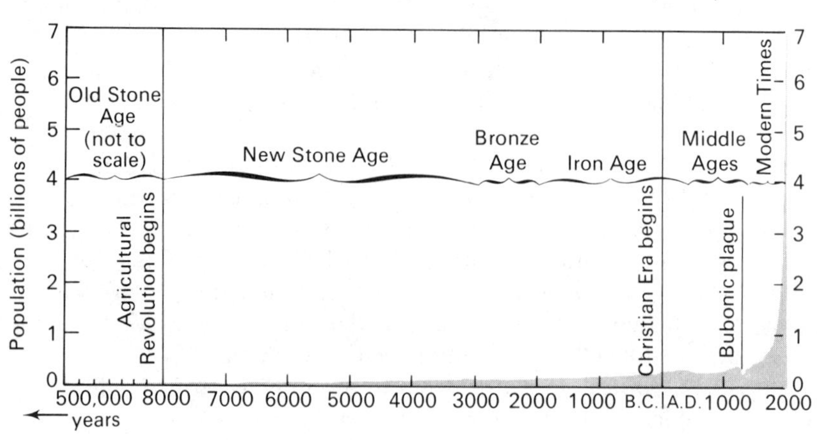

Table 1.1 Population Data and Projections for Selected Countries

| | 1975 (million) | 2000 | Percent increase by 2000 | Average annual percent increase | Percent of world population in 2000 |
|---|---|---|---|---|---|
| World | 4090 | 6351 | 55 | 1.8 | 100 |
| More developed regions | 1131 | 1323 | 17 | 0.6 | 21 |
| Less developed regions | 2959 | 5028 | 70 | 2.1 | 79 |
| Major regions | | | | | |
| Africa | 399 | 814 | 104 | 2.9 | 13 |
| Asia and Oceania | 2274 | 3630 | 60 | 1.9 | 57 |
| Latin America | 325 | 637 | 96 | 2.7 | 10 |
| U.S.S.R. and Eastern Europe | 384 | 460 | 20 | 0.7 | 7 |
| North America, Western Europe, Japan, Australia, and New Zealand | 708 | 809 | 14 | 0.5 | 13 |
| Selected countries and regions | | | | | |
| People's Republic of China | 935 | 1329 | 42 | 1.4 | 21 |
| India | 618 | 1021 | 65 | 2.0 | 16 |
| Indonesia | 135 | 226 | 68 | 2.1 | 4 |
| Bangladesh | 79 | 159 | 100 | 2.8 | 2 |
| Pakistan | 71 | 149 | 111 | 3.0 | 2 |
| Philippines | 43 | 73 | 71 | 2.1 | 1 |
| Thailand | 42 | 75 | 77 | 2.3 | 1 |
| South Korea | 37 | 57 | 55 | 1.7 | 1 |
| Egypt | 37 | 65 | 77 | 2.3 | 1 |
| Nigeria | 63 | 135 | 114 | 3.0 | 2 |
| Brazil | 109 | 226 | 108 | 2.9 | 4 |
| Mexico | 60 | 131 | 119 | 3.1 | 2 |
| United States | 214 | 248 | 16 | 0.6 | 4 |
| U.S.S.R. | 254 | 309 | 21 | 0.8 | 5 |
| Japan | 112 | 133 | 19 | 0.7 | 2 |
| Eastern Europe | 130 | 152 | 17 | 0.6 | 2 |
| Western Europe | 344 | 378 | 10 | 0.4 | 6 |

From Council on Environmental Quality, 1980.

The population in some countries is increasing faster than others. Table 1.1 shows that the people of the world, as a whole, are presently increasing at an annual rate of about 1.8%; but people in Central American countries are increasing at 3.2%/yr, whereas northern Europe is increasing at only 0.5%/yr. Sociological reasons for the larger growth rates generally involve religious factors (birth control taboos) as well as economic factors (more children help support the parents).

Figure 1.2 shows the distribution of people according to age in Mexico, the United States, and Sweden. The contrast is remarkable and illustrates the severity of population growth in a country such as Mexico or India (Population Reference Bureau, 1970). It is particularly disconcerting to realize that the countries with rapid growth rates are also the poorer countries (the "developing countries"), where food and other resources are inadequate or barely adequate. Population and environmental problems in many countries in Africa and Latin America are

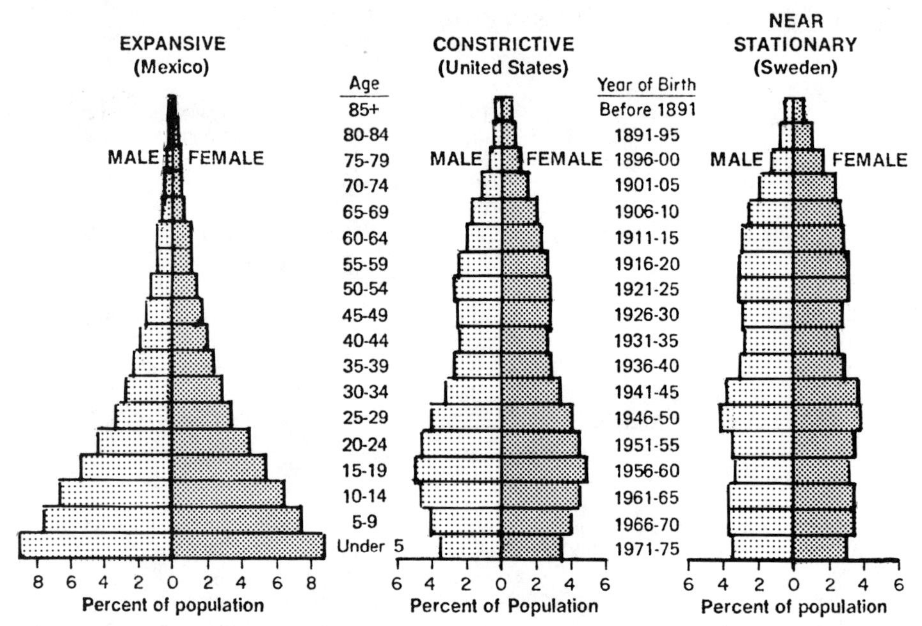


Figure 1.2 Age structure population pyramids (year 1976) for Mexico, United States, and Sweden (from Haupt and Kane, 1982).

critical, yet the education of the people and amelioration of problems of these countries is difficult to achieve where the majority of people are not only uneducated but are under 15 years of age.

## 1.3 Exponential Functions

One of the most useful equations in engineering is the relationship of the future value of some parameter. If its growth rate is constant, this equation can be simply derived as follows. The increase in number of items present ($dN$) is proportional to the original number of items ($N$) times some increment of time ($dt$). Then:

$$dN = \lambda N dt,$$

where $\lambda$ is a constant. Arranging terms, we have:

$$\frac{dN}{N} = \lambda dt.$$

Integrating both sides, we obtain:

$$\int \frac{dN}{N} = \int \lambda dt,$$

$$\ln N = \lambda t + c.$$

If the number of items present at $t = 0$ is $N_0$, then:

$$\ln (N_0) = \lambda(0) + c,$$

$$c = \ln N_0.$$

Substituting, we obtain:

$$\ln N = \lambda t + \ln N_0,$$

$$\ln N - \ln N_0 = \lambda t,$$

$$\ln \frac{N}{N_0} = \lambda t.$$

Raising both sides of the above equation to the $e$th power, we obtain:

$$\frac{N}{N_0} = e^{\lambda t}, \tag{1.1}$$

where $e$ = the base of the natural logarithm, approximately 2.718.

Equation (1.1) can be used to show how money increases in a savings account if the interest is compounded. For example, suppose \$1,000 is deposited for 5 yr at 12%/yr interest. The interest accrued after 1 yr at 12% is \$120, so $N$ after $t$ = 1 yr is \$1,120. From Eq. (1.1),

$$\frac{\$1{,}120}{\$1{,}000} = e^{\lambda(1\ \mathrm{yr})},$$

$$1.12 = e^{\lambda}$$

$$\ln 1.12 = \lambda,$$

$$\lambda = 0.11333.$$

Substituting $\lambda$ into Eq. (1.1) and using $t$ = 5 years, we have:

$$\frac{N}{\$1{,}000} = e^{(0.11333)(5)},$$

$$N = \$1{,}000(e^{0.56665}) = \$1{,}000(1.76235) = \$1{,}762.35.$$

Equation (1.1) can be used to demonstrate the exponential decay of a radioactive element. Consider strontium-90, which has a half-life of 29 yr. It may be desirable to determine what percent of a given amount of strontium-90 would be present after 100 yr. The "half-life" means that the ratio of radioactive atoms present at $t_0$ to the number present after some half-life time ($t_{1/2}$) is 0.5. Therefore,

$$\frac{N}{N_0} = 0.5.$$

The exponential formula can be used to solve for the "disintegration constant" ($\lambda$):

$$\frac{N}{N_0} = e^{-\lambda t}. \tag{1.2}$$

(*Note:* in this case, the disintegration constant has a negative sign because, unlike interest or population growth, the number of items is decreasing with time.) For strontium-90 with a half-life of 29 yr,

$$0.5 = e^{-\lambda(29\ \mathrm{yr})}$$

Taking the natural logarithm of both sides, we obtain:

$$\begin{aligned} \ln 0.5 &= \ln e^{-29\lambda}, \\ -.693 &= -29\lambda, \\ \lambda &= 0.0239 \text{ yr}^{-1}. \end{aligned}$$

The ratio of $N/N_0$ can then be solved using $t = 100$ yr:

$$\begin{aligned} \frac{N}{N_0} &= e^{-\lambda t}, \\ &= e^{-(0.02389 \text{ yr}-1)(100 \text{ yr})}, \\ &= e^{-2.39}, \\ &= 0.0916 \text{ or about 9.2\% of the original number of strontium-90 atoms.} \end{aligned}$$

(*Note:* Unlike the calculation of interest accumulated in a savings account, where money is transacted in exact amounts, the answer for this strontium-90 calculation (9.2%) has only two significant figures. This is because the original data (29 yr half-life) has only two significant figures. It is not sound engineering to indiscriminately give answers which contain all the numbers displayed on a hand calculator.)

In terms of population predictions, the exponential formula can be used to relate annual growth to the "doubling time" of a given population. Consider the population of China, which in 1980 was 1 billion and which doubles every 25 years:

$$\begin{aligned} \frac{N}{N_0} &= e^{\lambda t}, \\ 2 &= e^{\lambda(25 \text{ yr})}, \\ \ln 2 &= \ln e^{25\lambda}, \\ 0.693 &= 25\lambda, \\ \lambda &= 0.0278 \text{ yr}^{-1} \\ \frac{N}{N_0} &= e^{(0.0278 \text{ yr}-1)(1 \text{ yr})}, \\ \frac{N}{N_0} &= 1.0282. \end{aligned}$$

This equals a 2.82% increase per year. Thus, every year there are 28.2 million more Chinese. This large number can be put into perspective by considering the following situation. In 1978 one of the world's largest disasters, a magnitude 7.6 earthquake, killed 600,000 people in China (Table 7.1). (This is approximately the same number of people as the total population of South Dakota.) It can be seen that China replaced the loss of human life from this disaster in less than 8 days!

The underlying assumption of using the exponential function formula to predict future conditions requires that λ remain constant. Of course, no one can forecast the future with complete accuracy. Many unpredictable circumstances can develop—consider the New Yorker who, before the invention of cars in 1880, predicted the city would eventually be buried under horse manure.

Urban migration, the migration from farms to cities, is also occurring worldwide. The population of Mexico City is expected to grow to 32 million by the year 2000, which will create the largest city in the world. The population of the world's major cities is increasing at a faster rate than world population. In 1800, London was the only city in the world with 1 million inhabitants. In 1940, there were about 40 cities of over 1 million and by 1970 there were over 100 of these major centers in the world. Population density ranges from Edinburgh, Scotland, for example, with 35 inhabitants per hectare, to parts of Delhi and Calcutta, India, where the density reaches an incredible 0.16 people per square meter (Legget, 1973).

## 1.4 United States Population

There were probably less than 100,000 native American Indians in the area now known as the United States when the first Europeans stepped onto Plymouth Rock in 1620. Since then, the population has increased rapidly to a 1983 population of 234 million (Fig. 1.3). This is about 5% of the people in the world. Much of the growth came about by immigration, particularly near the turn of the twentieth century, rather than by births from within. The United States, like the rest of the New World, is rapidly filling up, and overcrowded conditions are present, particularly near the larger cities. There is a trend of migration in the United States from farms to the cities. In 1800 only 6% of the U.S. population lived in urban areas, but today more than 70% live in cities or their suburbs.

The amount of land consumed by urbanization is increasing as cities grow in population. About 3% of the United States and 15% of the United Kingdom has been "developed" into houses, highways, railways, airports, etc. The amount of new land which will be needed for future urban growth is of concern to many people who care for the natural landscape or who are concerned about the ability

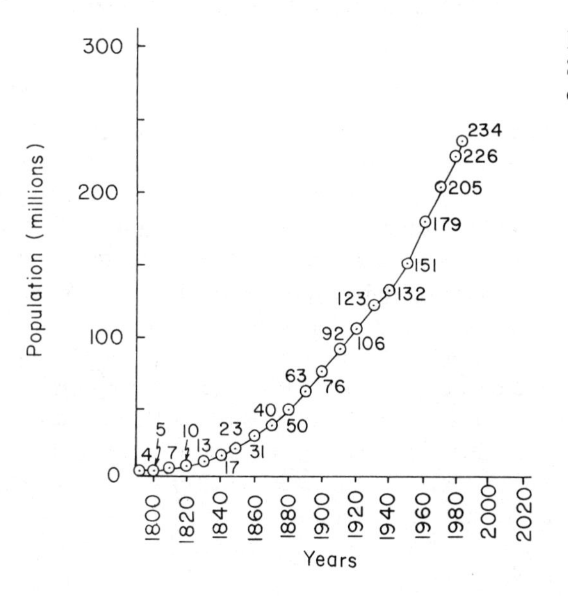


Figure 1.3 Population of the United States from 1790 to 1983, based on data from the Bureau of the Census.

of the remaining agricultural land to support the growing populace. The growth of urban centers also poses special challenges to the engineering geologist, as it is with the aid of his skills that mitigation of geologic hazards in these areas will occur. The many geotechnical problems associated with the rapid rates of construction in urban areas has given way to the term "urban geology" for engineering geology in cities.

In the United States, the annual increase (net immigration plus the excess of births over deaths) adds about 1.5 million to our population. The annual growth rate for the United States is presently about 0.6%/yr (Table 1.1). Legal immigration, including refugees, was 0.8 million in 1980, but illegal immigration probably exceeded 1 million. (Legal immigration to the United States, which was very large during the nineteenth and the beginning of the twentieth centuries, has slowed down, partly because many Americans feel that the country cannot continue to be the safety valve for the world's burgeoning population. There are those who feel it is time to change the philosophy illustrated by the inscription on the Statue of Liberty: "Give me your tired, your poor, your huddled masses yearning to breathe free. . . .")

Less than two centuries ago the New World offered such seemingly unlimited potential for resources that even Thomas Jefferson, one of the most learned of our founding fathers, was able to speak in his First Inaugural Address of a nation ". . . with room enough for our descendants to the hundreth and thousandth generation." But now the land is already taken up and no one speaks of unlimited land for future generations, certainly not for thousands of generations. While growth is still promoted by many developers in this country, anyone who has commuted to work in Chicago or New York traffic cannot help but realize that growth is not necessarily what constitutes a good quality life.

The United States, like any country, has a limited area. The good old days of "manifest destiny," which led to the rapid settlement and unhampered exploitation of this nation's resources, are over. It is obvious to most people that this country, which originally seemed to have infinite resources, will have to begin to adapt to conservation practices such as reuse of resources and land-use planning.

## 1.5 Future Population

The exponential equation can be used to make some chilling population predictions. In two doublings a population quadruples, in three doublings it will grow by a factor of $2^3 = 8$, in four doublings it will grow by a factor of $2^4 = 16$, etc. Rapid exponential growth can be illustrated by consideration of the example where one grain of wheat is placed on the first square of a chessboard, two on the second, four on the third, etc.; then the 64th square will have $2^{63}$ grains, or more than all the wheat ever harvested on the planet earth.

If the earth's population continues to double every 35 yr, then in only 1000 years there will be $10^{18}$ people on earth, or about 1600 people per square meter of the earth's surface, land and sea! Even more preposterous figures can be generated. In a few more hundred years the weight of people would equal the weight of the earth, and later the ball of people would be expanding outward at the speed of light. These absurd situations are given to illustrate that there can be no question that world population growth must end. It is physically impossible to continue at the present rate much longer.

Just how the termination of population growth will occur remains to be seen.

Figure 1.4 Moskito Indian woman drawing water from a contaminated well, eastern Nicaragua. There is no potable water, no sanitation facilities, and no hope for improvement for many people in third world countries. The average Moskito Indian couple has ten children, four of whom die of childhood diseases.

There is some reason for hope that nations can become enlightened to the point that each country can solve its own problems. For example, reports from Communist China indicate that financial rewards are being offered to couples for having only one child. Some optimistic demographers have estimated that global population may level off somewhere around 10 billion people by the year 2050.

The decline in the standard of living throughout many countries is a warning signal that the world is approaching ecological saturation. Affluence evades underdeveloped countries where population increases outpace increases in resource development (Fig. 1.4). Multitudes are entombed in the slums of Cairo and Calcutta, the barrios of Mexico City, and the ghettos of New York, condemned to the prison of their own burgeoning numbers. In much of Latin America and Asia, meadow and forest disappear before the surging tide of humanity. Starvation haunts millions in Africa. In the United States, farms and shady streets succumb to shopping malls and condominiums, and quiet suburban towns become part of a giant urban sprawl.

The photographs of planet earth taken from space offer a sobering realization of the finiteness of this globe. Growth is a term which everyone seems to promote, be it on a scale of car sales, or the gross national product. And yet, viewed from space, one cannot escape the conclusion that this planet is finite and can only support a limited population. In this context, there can be only one question to be answered concerning population growth: Not *will* it end, but *how* will it end?

Either the birth rate must go down or the death rate must go up (Meadows et al., 1972). The prestigious "Club of Rome" stated in 1972 that "the most probable result will be a rather sudden and uncontrollable decline in both population and industrial capacity." This may be true, but there are signs of optimism. In the developed countries, at least, the birth rate is dropping; that is, the population is still increasing, but the *rate* of increase is no longer increasing.

The issue of population must be squarely faced by everyone on this planet. We can't keep bulldozing our way to a solution.

## Problems

1. The land area in the world is about $1.5 \times 10^8$ km$^2$. Using present world population growth rates, determine how many years will elapse before one person will have only 1 m$^2$ of land ("standing room only").

2. (From Bartlett, 1980.) Bacteria multiply by division so that 1 bacterium becomes 2, the 2 divide to give 4, the 4 divide to give 8, etc. For a certain strain of bacteria, the time for this division is 1 min. Assume one bacterium is put in a bottle at 11:00 A.M., and it is observed that the bottle is full at 12:00 noon. Here is a simple example of exponential growth in a finite environment. This is mathematically identical to the case of the exponentially growing consumption of our finite resources of fossil fuels. Keep this in mind as you ponder two questions about the bacteria:
   **a.** When was the bottle half-full?
   **b.** If you were an average bacterium in the bottle, at what time would you first realize that you were running out of space? There is no unique answer to this question, so consider: at 11:55 A.M., when the bottle is only 3% full and is 97% empty, how many of you would perceive that there was a problem?

3. In what year will Mexico have the same population as the United States?

4. If gas sold for $.39/gal in 1978 and $1.24/gal in 1983, how much will it sell for in 1990?

5. If the present rate of world population growth had occurred since the time of Christ (when the population was about 250 million), how many people would there now be per square meter of earth? (Earth's total land area = $1.5 \times 10^{14}$ m$^2$.)

6. How many additional people are on earth every minute?

7. The half-life of radium is about 1620 yr. What is its decay constant? How many years are required to reduce a given mass of radium to 1%?

8. Westing (1981) estimates that since the evolution of *Homo sapiens,* approximately 50 billion people have lived on earth. The 1983 population is what percent of all the people who have ever lived?

9. Rapid City, SD, grew from 6000 to 46,000 people between 1910 and 1980. In what year will Rapid City be as big as Denver's present population of 500,000?

10. It has been proposed to colonize space with earth's excess population. In the year 1986, how many spaceships would have to lift off every day in order to evacuate the increase in the world's population? (Assume 100 people/spaceship.)

# 2

# Maps and Aerial Photographs

And so geology, once considered mostly a descriptive and historical science, has in recent years taken on the aspect of an applied science. Instead of being largely speculative as perhaps it used to be, geology has become factual, quantitative, and immensely practical. It became so first in mining as an aid in the search for metals; then in the recovery of fuels and the search for oil; and now in engineering in the search for more perfect adjustment of man's structures to nature's limitations and for greater safety in public works.

Charles P. Berkey, 1939

We must get geology off the bookshelves and into the streets. We must get out of our ivory towers and into the action. . . . Our published works, learned though they may be, are of little value if they aren't used, or can't be used, by people at whom they are aimed.

Hollis M. Dale, Assistant Secretary for Mineral Resources,
U.S. Department of Interior, 1979

## 2.1 Topographic Maps

An engineering geologist should investigate all possible sources of published information prior to extensive field work. This includes topographic and geologic maps, soils maps, aerial photographs, and remote sensing data. This chapter presents an introduction to sources of information for maps and remote sensing techniques that are available to the engineering geologist.

Topographic maps are a fundamental tool of geologists, mining, civil, and geological engineers. Topographic maps present a detailed record of a land area and show geographic positions and elevations of natural and man-made features. These maps show the shape of the land by means of brown contour lines (lines of equal elevation). Topographic maps are an inventory of the physical features of the earth's surface, including the names of many features (Fig. 2.1).

The art of map construction is called "cartography." Maps prepared by the U.S. Geological Survey (USGS) have a standard color code. The network of streams, lakes, springs, and other water features is shown in blue; woodland features are shown in green; and the principal works of man, such as roads, buildings, railroads, and power lines, are shown in black. Urbanized areas have a red tint. Revisions of maps are overprinted in purple color.

Large scale topographic maps such as 1:10,000 to 1:1000 are desirable for most engineering geology work. Unfortunately, these detailed maps are rarely avail-

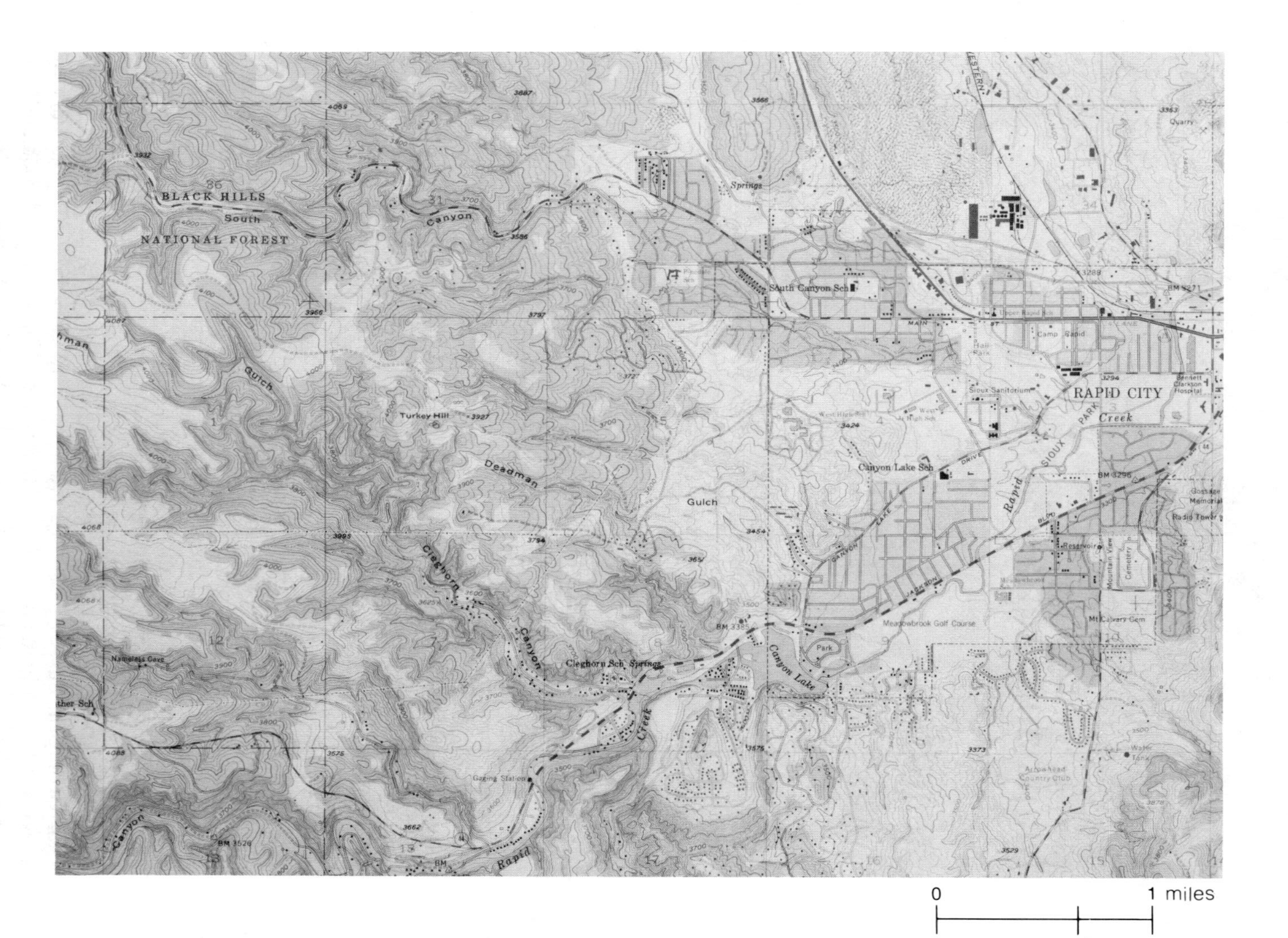


Figure 2.1 Topographic map of part of the Rapid City West 7½′ Quadrangle, SD.

able, and recourse must be made to the less detailed 1:24,000 (7.5-min quadrangle) scale maps published by the USGS.

In the United States, the USGS is responsible for the preparation and sale of topographic maps. Requests for map indexes for areas east of the Mississippi River, including Minnesota, should be sent to U.S. Geological Survey, Branch of Distribution, 1200 South Eads St., Arlington, VA 22202. Requests for indexes or maps for areas west of the Mississippi River, including Louisiana, should be sent to U.S. Geological Survey, Branch of Distribution, P.O. Box 25286, Federal Center, Denver, CO 80225. The USGS topographic maps are generally keyed to latitude and longitude lines which divide the earth into discrete areas called quadrangles. No flat map can portray a spheroidal surface with 100% accuracy, but the error becomes negligible if areas of less than 1° of latitude and longitude are used. If 1° (60 min) of latitude and longitude are each divided into eight segments, the resulting 7.5-min quadrangle map is produced. The 7.5-min qadrangle is the backbone of the USGS National Topographic Program. The maps are published at a scale of 1:24,000. The 7.5-min quadrangle map is nearly rectangular in shape. Because longitude lines converge on the North Pole in the northern hemisphere, the actual area covered in a 7.5-min quadrangle ranges from 181 $km^2$ in the southern United States to 127 $km^2$ along the Canadian border.

Topographic maps in the 7.5-min (1:24,000 scale) or 15-min (1:62,500 scale) series are available for all but the most remote areas of the continental United States. Also available, as part of the National Topographic Mapping System (NTMS), are 2° sheets (1° latitude and 2° longitude) at 1:250,000 scale. These smaller scale maps, while available for the entire United States, rarely depict features with sufficient detail to be of use to the engineering geologist, however. The U.S. Defense Mapping Agency has digitized all 1:250,000 scale topographic maps of the contiguous United States. These data are available through the USGS National Cartographic Information Service, Reston, VA 22092. The elevation data are interpolated from a 63.5-m grid spacing (Schowengerdt and Glass, 1983).

In the past, contour lines on USGS topographic maps are published in feet above sea level. The USGS has recently begun the arduous task of preparing maps in the metric system, requiring the revision of contour lines to meters above sea level. Some metric quadrangle maps at 1:100,000 scale have been published. In recent years computers have been used to store map information in digital form; the maps can be easily changed to take into account new geographic features or produce a map showing only certain features.

Cartographers rely on aerial photographs to make topographic maps (Meunier, 1980). In recent years "orthophotomaps" have been published; these maps have aerial photographs printed on top of a 7.5-min quadrangle base to realistically produce an authentic portrait of the earth's surface. In flat places such as the Florida Everglades, Prudhoe Bay of Alaska, or the lake region of northern Minnesota, a standard 7.5-min topographic map is considerably enhanced by the photographic additon of swamp, lake, and vegetation patterns.

Areas inundated by the "100-yr flood" have been outlined on some topographic maps as part of the National Program for Managing Flood Losses. These maps show quite clearly, even to the non-professional person, areas subject to flooding.

## 2.2 Geologic Maps

The importance of geological information to engineering geology applications cannot be overemphasized. Geologic maps with topographic overlay provide nearly all the information included in topographic maps as well as extensive data about the geology of an area (Fig. 2.2). Unfortunately, geologic maps are not always available, or are published at small scales lacking sufficient detail for engineering work.

Determining the availability of published geologic maps for any given area can be a time consuming job. To research the availability of maps in an unfamiliar area, the investigator should consider government publications, journals, and university files. In the United States, the USGS has begun to publish indices to geologic maps on a state-by-state basis (Fig. 2.3). Publications by the USGS which may include geologic maps, includes Professional Papers, Bulletins, Water-Supply Papers, and Map Series (Birdsall, 1973). University geology departments contain numerous graduate dissertations that may contain unpublished geologic maps. Publications useful for securing information on the status of geologic maps are the *Bibliography and Index of Geology,* and the Geological Society of America. Trautman and Kulhawy (1983) describe additional sources of engineering geology information pertaining to geologic maps, geophysical and hydrological data, and remote sensing.

Inquiries concerning the availability of geologic maps can also be sent to the state geologic surveys. Appendix C gives the names and addresses of the various state geologic surveys in the United States.

The domain of the engineering geologist is typically within the top few meters of the earth's surface. Therefore, surficial deposits are of critical importance in a geologic map. Many geologic maps, although prepared by competent geologists, do not show sufficient information concerning surficial deposits to be useful to the engineering geologist. (Some USGS geologic maps do not even show deposits of alluvium, although mapped at 1:24,000 scale in areas where alluvium is thick and abundant.) In some places, particularly glaciated areas, two different geologic maps can be prepared, showing either surficial geology or bedrock geology. The surficial geology is mapped by a glacial geologist and shows sand and gravel, till, eolian deposits, bedrock outcrops, and other information useful to the engineering geologist. The bedrock geology, mapped by a "hard-rock" geologist, shows the lithology and structure of the bedrock below the glacial drift. Special geologic maps may be designed to suit the needs of a specific terrain or engineering geology project. For example, in granitic areas of Colorado, alluvium and regolith thickness maps were prepared to assist in the siting of residential septic-system requirements (Schmidt and Pierce, 1976).

In 1980 the Engineering Geology Division of the Geological Society of America (GSA) and the Association of Engineering Geologists (AEG) formed a joint committee to further the development of standardized symbols for engineering geology maps. Presently, the direction seems to be moving toward integration of geomorphological mapping techniques and symbols that reflect the generalized

Figure 2.2 Geologic map of part of the Rapid City West 7½′ Quadrangle, SD (after Cattermole, 1969).

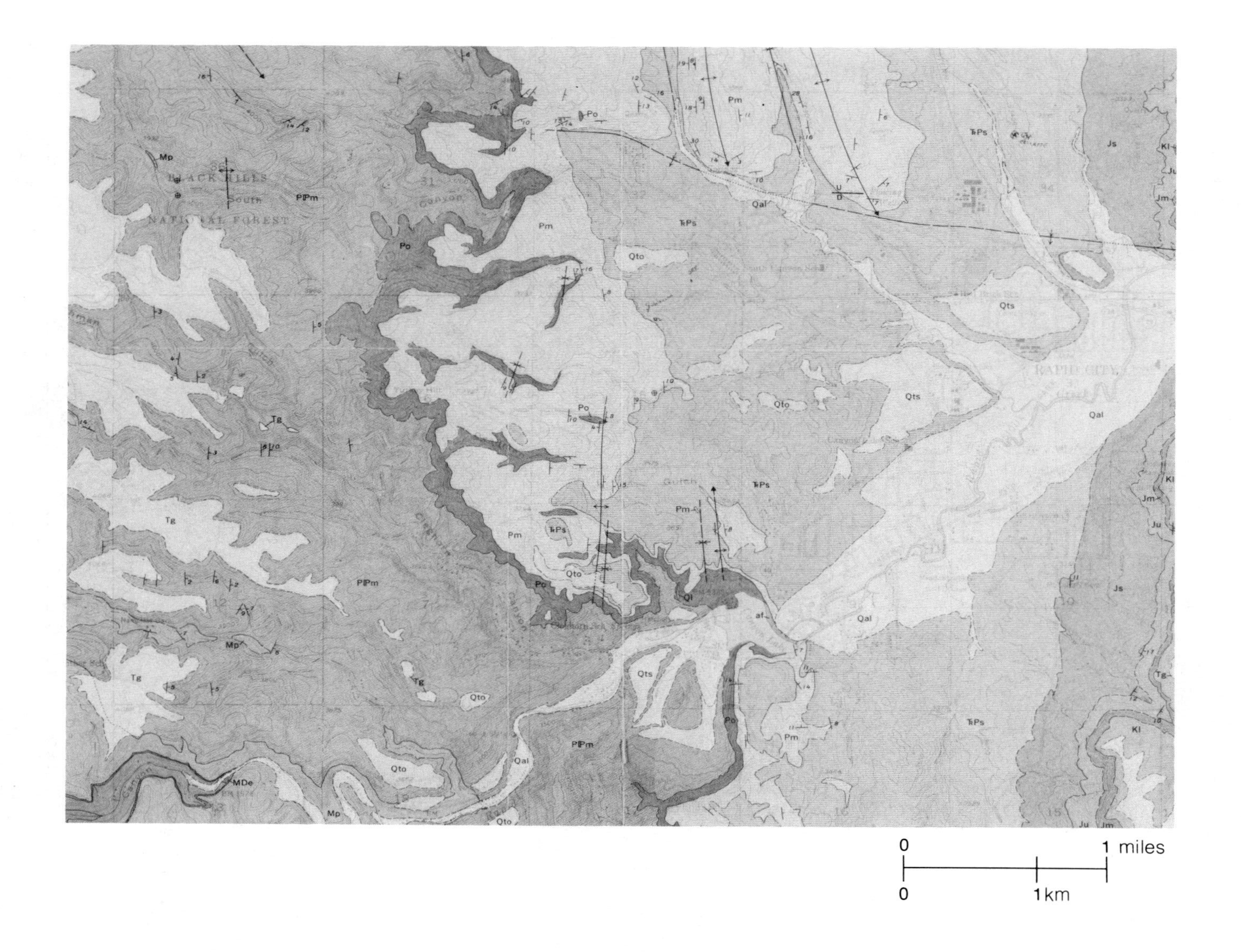

Mp
PIPm
Po
Pm
TRPs
Qto
Qal
Qts
Js
Kl
Ju
Jm
Tg
Ql
af
MDe
Canyon
0
1 miles
0
1km

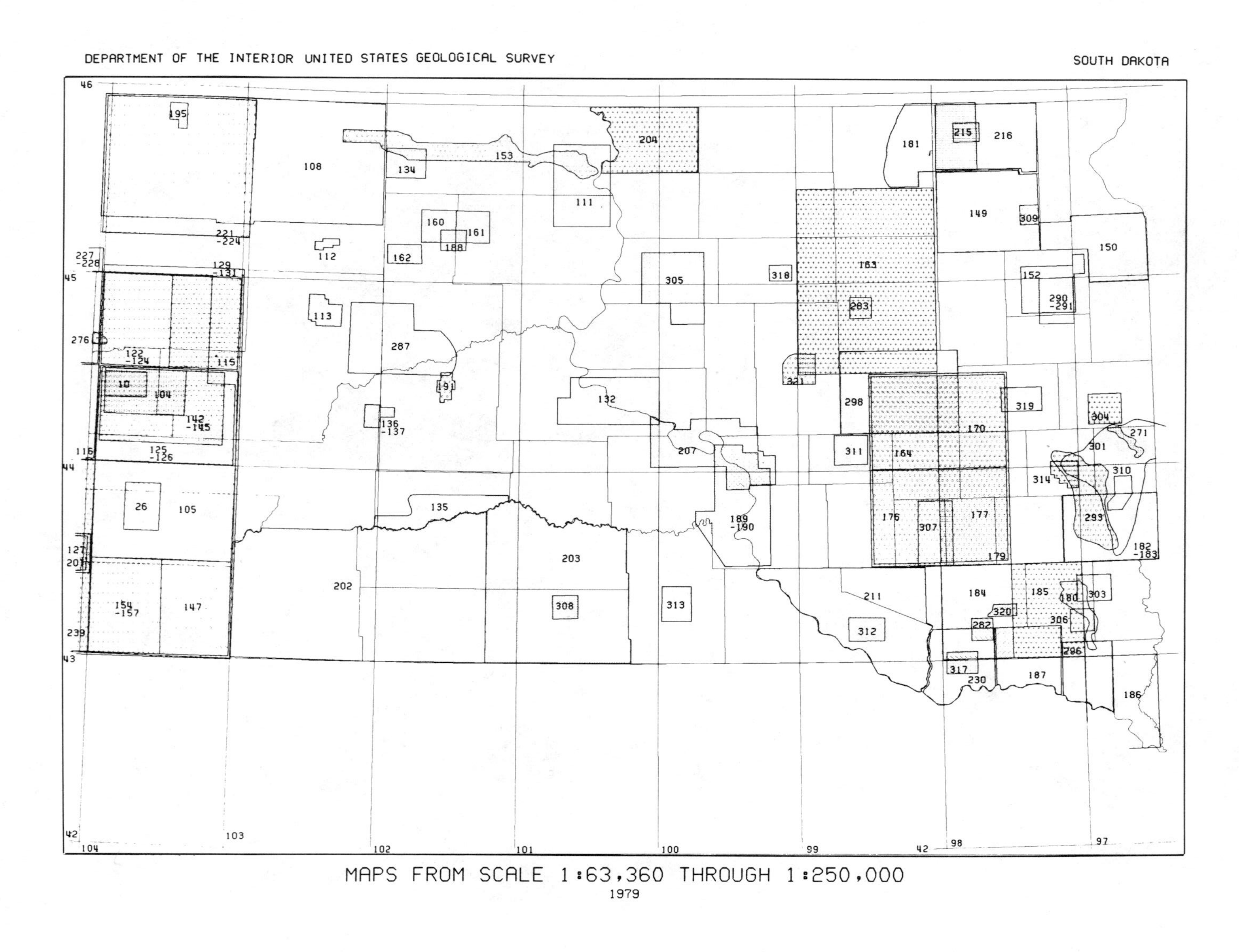
DEPARTMENT OF THE INTERIOR UNITED STATES GEOLOGICAL SURVEY
SOUTH DAKOTA
MAPS FROM SCALE 1:63,360 THROUGH 1:250,000
1979

nature of engineering geology map units such as the Unified Soil Classification System which is widely used in the geotechnical profession (Hatheway, 1981). The Engineering Group of the Geological Society of London has stressed the need for geomorphological maps for engineering geology purposes.

Galster (1977) proposed a simplified and universally applicable system of engineering geology map units. For example, instead of identifying a surficial deposit under a Quaternary terrace as $Q_t$, new designations are proposed, based on genesis and composition; for example $A_{sg}$ stands for alluvium composed of sand and gravel. The AEG is looking into the use of word descriptions such as "slope stability," "rippability," "expansive soils," etc., to visually communicate to a map user the actual result of an engineering geology analysis.

A recent trend of geological mapping in urban areas is the preparation of a series of practical maps, or folios, derived from a geologic map. For example, three classes of maps may be used for a land-planning study. The first class included maps showing topography, bedrock geology, and soils. Derivative maps showing slopes, drainage, ground water potential, and hazards may then be prepared. Finally, interpretative maps may be prepared by planners based on needs for specific resources and land-use priorities.

There are numerous examples of military use of geological conditions. For example, military maneuvers are affected by the depth of bedrock or "trafficability." The very outcome of critical battles may have been different had geological maps been available and widely used. Had British troops realized how well entrenched the American colonists were at Bunker Hill (a drumlin), they might not have been repulsed in 1775. In 1863, Union forces lost 23,000 troops at Cemetery Ridge near Gettysburg, PA, due in part to the lack of soils for digging foxholes because of shallow soils over diabase. The successful evacuation of British and French troops from Dunkirk, France, in 1940, was largely due to the inability of German Panzer tanks to traverse the marshy terrain and to the fact that the Allied troops were able to dig protective foxholes in sand dunes along the beach. Casualties were light because the sand tended to smother the bombs dropped by German Stuka dive bombers.

## 2.3 Aerial Photographs

### *2.3.1 Acquisition of Air Photos in the United States*

Aerial photographs supply another dimension to maps or field study. Surficial deposits, landslides, joint patterns, land use, and other phenomena not indicated on maps may be readily apparent on aerial photographs. Air photos are an essential tool for the field geologist, and should be used extensively for the preparation of any geologic map. Engineering geology applications of photogeology include a diversity of subjects such as landslide recognition, dam site evaluation, and the location of high-capacity well sites using fracture traces.

---

Figure 2.3 Geologic map index for South Dakota at scales 1:63,360 through 1:250,000 (from U.S. Geological Survey).

The entire United States has been photographed, and negatives at 1:20,000 scale are available from the federal government. Enlargements up to about 1:5000 can be obtained while maintaining sufficient clarity for mapping purposes. Two major problems in the use of aerial photographs are 1) the time required to purchase contact prints (perhaps 3 months) and 2) the fact that many photographs may have been taken up to 20 years previously.

The following three federal agencies in the United States supply aerial photographs:

1. U.S. Agricultural Stabilization and Conservation Service (ASCS), U.S. Dept. of Agriculture. The ASCS collects and files aerial photographs of farmlands, range lands, and non-government owned forest lands. In the eastern United States: Eastern Aerial Photograph Laboratory, 45 S. French Broad Ave., Asheville, NC 28801. In the western United States: Western Aerial Photograph Laboratory, 2505 Parley's Way, Salt Lake City, UT 84109.
2. The U.S. Forest Service (USFS) lands are mainly the mountainous areas of western United States. The general office is Building #85, Federal Center, Denver, CO 80225.
3. U.S. Geologic Survey is part of the U.S. Department of Interior. The USGS contracts for aerial photographs as part of its program making topographic

Figure 2.4 Stereoscopic viewing of two aerial photographs. Note the photo index sheet in the foreground.

maps. Three regional offices are: Eastern Mapping Center, 536 National Center, Reston, VA 22092; Western Mapping Center, 345 Middlefield Rd, Menlo Park, CA 94025; and Mid-Continent Mapping Center, 1400 Independence Rd, Rolla, MO 65401. Currently there is an attempt to consolidate USGS aerial photograph space imagery and photography data at EROS Data Center, Sioux Falls, SD 57109.

The USGS also attempts to maintain the status of all government air photos. The data is contained on 1:500,000 scale index maps. The main office of the USGS is Washington, DC 20244. Additional sources of remote sensing are given by Hunt (1983).

To acquire specific aerial photographs, it is recommended that a "Photo Index Sheet" be purchased prior to ordering individual air photos. These sheets are composite photographs of the individual air photos (Fig. 2.4). If a USFS office or a county ASCS office is located nearby, the interested party can visit the facility and expedite the selection process.

Commercial aerial photographs can also be obtained through private companies. The cost of acquiring these photographs may be expensive, but an advantage is that the photographs are timely and can be made at low altitudes over a specific target.

### *2.3.2 Photogrammetry*

Aerial photographs are typically taken at altitudes of about 3000 m above the ground. The camera is aimed vertically downward and triggered to provide about 60% overlap on successive photos. The plane covers the complete area required by successive flight lines (Fig. 2.5).

Two photographs can be used to acquire a stereoscopic (three-dimensional) view of the earth's surface. Stereoscopic viewing is attained by transferring photograph center marks from two successive prints onto each adjoining print, and lining up the four points (Fig. 2.5). A pocket lens stereoscope is recommended for enlarged viewing, separating the photographs equal to the distance between the viewer's eyes, roughly 7 cm.

There are many excellent texts available on the use and geological interpretation of aerial photographs. For example, see American Geological Institute, 1951; Lueder, 1959; Ray, 1960; Miller and Miller, 1961; Mollard, 1962; Lattman, 1965; Colwell, 1973; Sabins, 1978; Lillesand and Kieffer, 1979; El-Ashry, 1977; Siegal and Gillespie, 1980; Carter et al., 1980; Verstappen, 1983; and Colwell, 1983.

In recent years there have been many improvements in techniques and manipulation of data in the field of remote sensing. For engineering geology, however, the conventional black-and-white air photo contains almost all the information contained in the more expensive color prints (Fig. 2.6). Nevertheless, the extra money spent on color or color-infrared film is usually small in contrast to the viewers time, and it may be worthwhile to consider these films if available. There is no single film type which is most useful for all fields of endeavor used by the engineering geologist. I have found that the normal color photos best depict engineering geology features such as bedrock composition, surficial deposits, soil moisture, etc., whereas color-infrared photos best depict a high degree of vegetational contrasts which may be related to hydrogeologic phenomena.

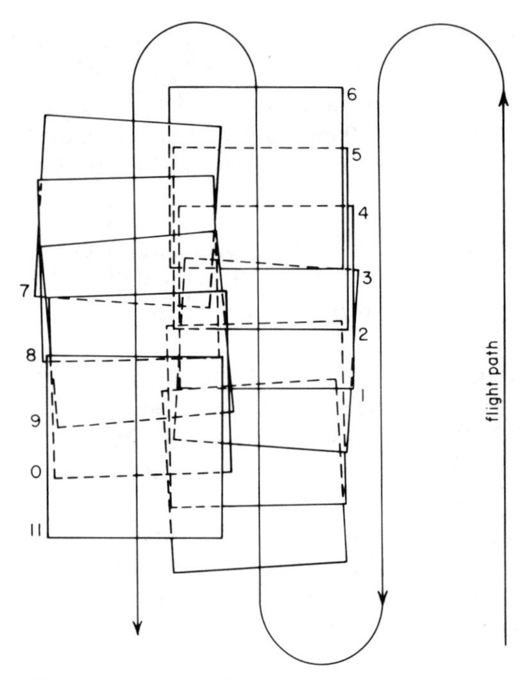


Diagram showing idealized flight plan. Vertical photographs are taken with 60 percent overlap along flight path and 25 to 30 percent sidelap between adjacent flight lines.

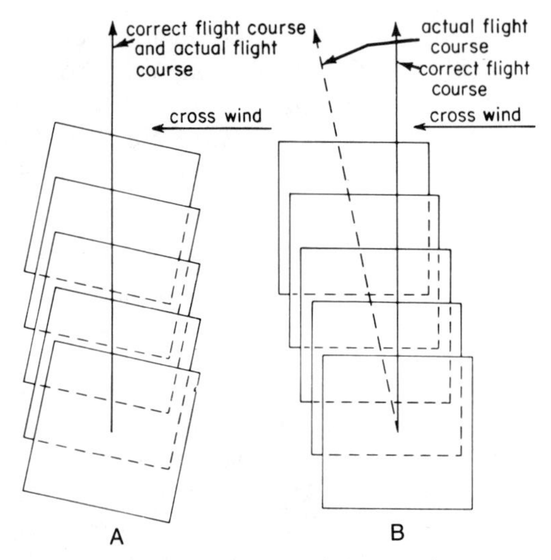


Diagram showing relationship of photographs to proper flight course under conditions of cross wind causing "crab" A or "drift" B. Note that, unless the camera is turned in its mount, all photographs will be skewed with respect to the actual course of the aircraft whether there has been crab or drift.

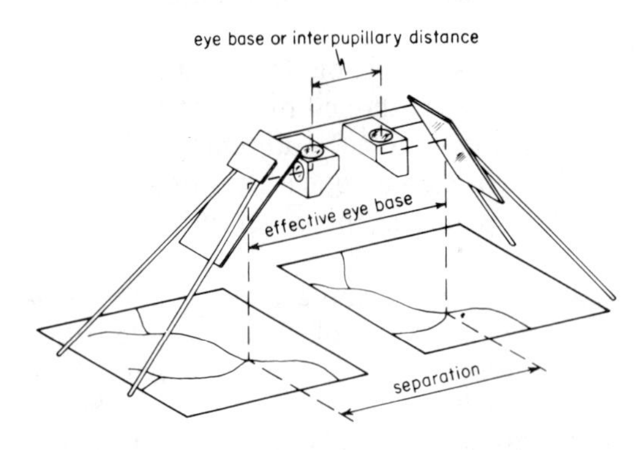


Diagram showing pair of photographs properly oriented for stereoscopic viewing. The stereoscope axis must be parallel to the straight-edge direction (flight-line direction) for proper viewing.

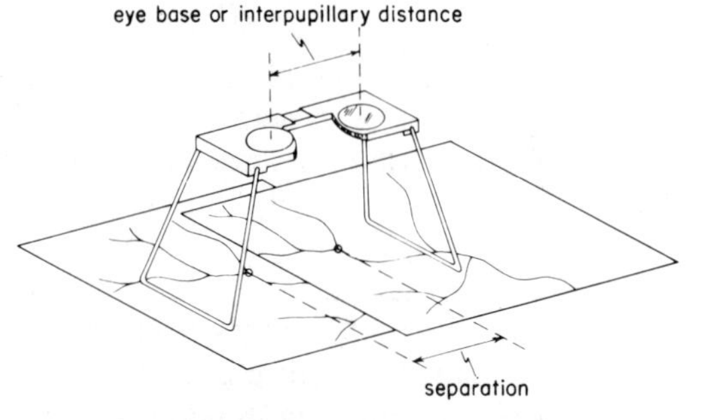


Diagram showing eye base (interpupillary distance) and the separation of photographs in stereoscopic viewing with a lens stereoscope.

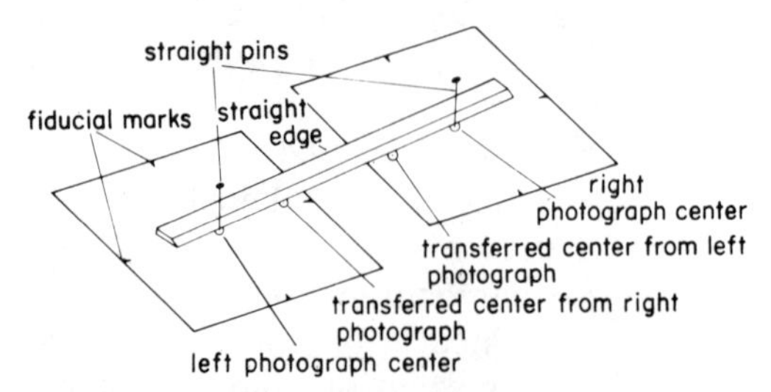


Diagram showing eye base (interpupillary distance), effective eye base, and separation of photographs in stereoscopic viewing with mirror stereoscope.

Figure 2.5 Elements of using aerial photographs (from Lattman, 1965).

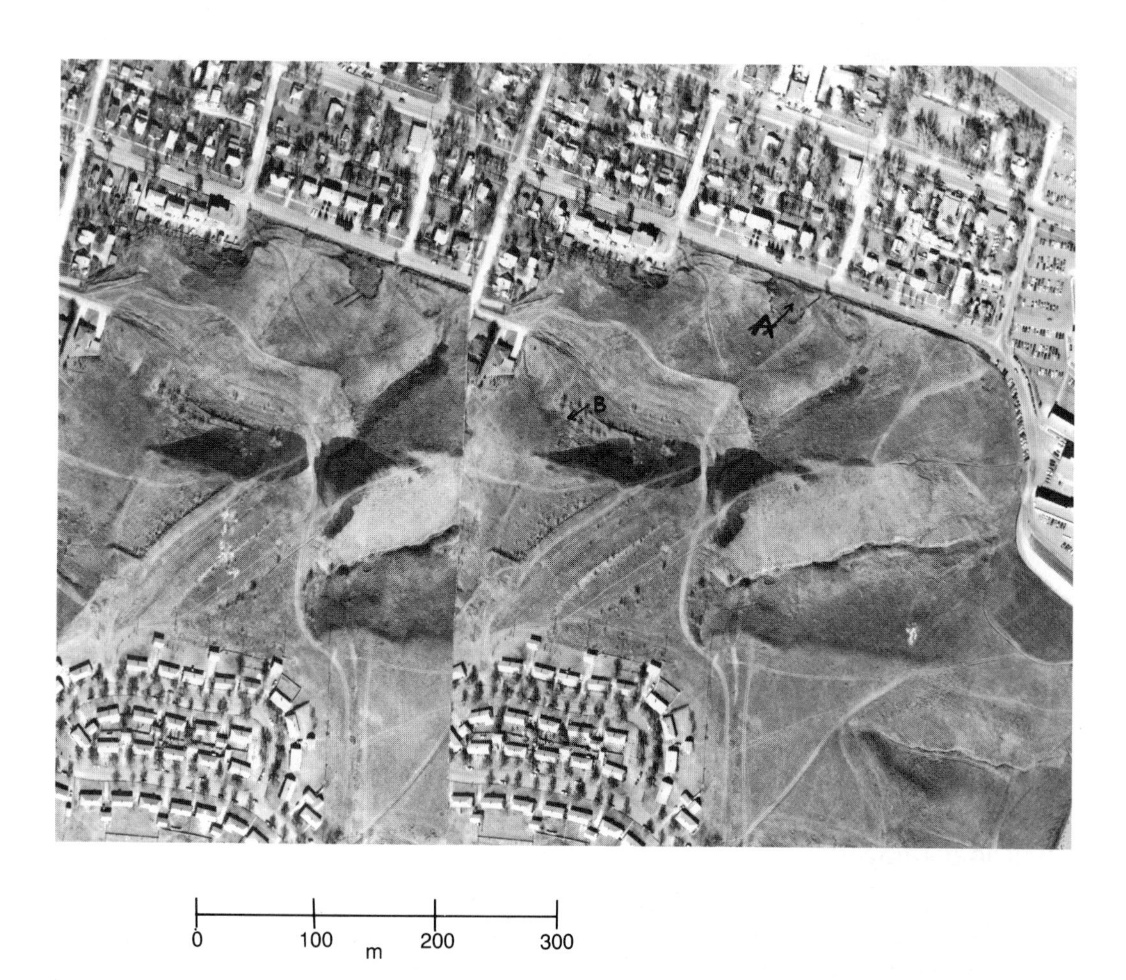


Figure 2.6 Aerial photographs of a portion of Rapid City, SD. This stereo pair shows a 50-m-wide active slump (**A**) and a 300-m-wide ancient landslide (**B**). House excavations at the base of the large landslide could initiate movement.

## 2.4 Thermal Imagery

Thermal imagery differs from conventional photography in that the camera employed responds to heat rather than to light. A thermal image on film looks like a photograph but shows temperature differences in terms of tone (a lighter tone for warmer objects and a darker tone for colder ones). For instance, cool submarine springs discharging near a shoreline into warmer sea water would show up as dark patches or streaks. The technique of mapping fresh ground-water discharges into the sea by means of aerial thermal imagery has been successfully applied in the Hawaiian Islands. Aerial photography and thermal imagery of the coastline of Jamaica were used to locate submarine fresh water springs (Kohout et al., 1981).

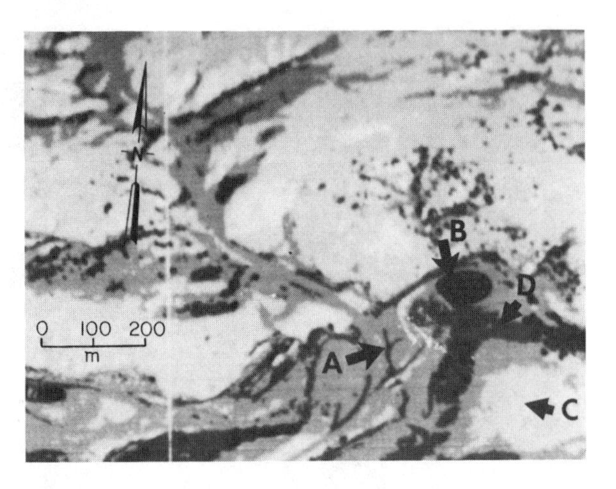


Figure 2.7 A daytime thermal image from an area near Wind Cave in the southern Black Hills, SD, on September 11, 1976, showing the intersection of two fracture traces (**A**) in the limestone terrain. Also shown are temperature differences between a stockpond (**B**), range vegetation on a hill top (**C**) and range vegetation at lower elevations (**D**). (From Dalstead et. al., 1977.)

Another application of thermal imagery is in the location of fracture traces, linear features indicative of subsurface joints or faults where productive water wells could be located (Fig. 2.7). (Chapter 7 treats this subject in more detail.)

Spectacular thermal imagery of active volcanoes in Iceland shows the pattern of hot spots changing in time (Friedman et al., 1969). Research on the usefulness of monitoring active and dormant volcanoes with thermal imagery in the United States is in progress.

## 2.5 Radar Imagery

Radar, developed during World War II to monitor aircraft, has in recent years been adapted to peaceful uses including rainfall analysis and terrain mapping. Side-looking aerial radar (SLAR) is used for mapping surface areas that could not be photographed using conventional methods. SLAR may be used in any weather, day or night. Radar operates at relatively long wave lengths, producing high-resolution imagery due to the reflectance characteristics of the earth's surface.

Radar imagery is a record of the interaction of electromagnetic radar waves transmitted to earth and returned in a non-uniform manner to produce a composite image of the area being mapped. The natural contour of the earth's terrain and amount of vegetation break up and return the radar waves to produce detailed imagery.

SLAR scans in strips. An on-board convertor changes the radar echoes into electron beams which are displayed on a cathode-ray tube. A special camera records each line of scanline, which moves at a speed proportional to aircraft ground speed, producing an image similar in appearance to an aerial photograph (Fig. 2.8).

The application of SLAR to engineering geology presently appears to be suited to the location of faults (Sabins, 1973) or to the preparation of maps in cloudy areas where conventional aerial photography is ineffective (Colwell, 1973).

A radar-imaging system was used on the space shuttle "Columbia" in 1981. The images, made from an altitude of 250 km, show physiographic features enhanced by terrain roughness such as contact between stoney alluvial fans and

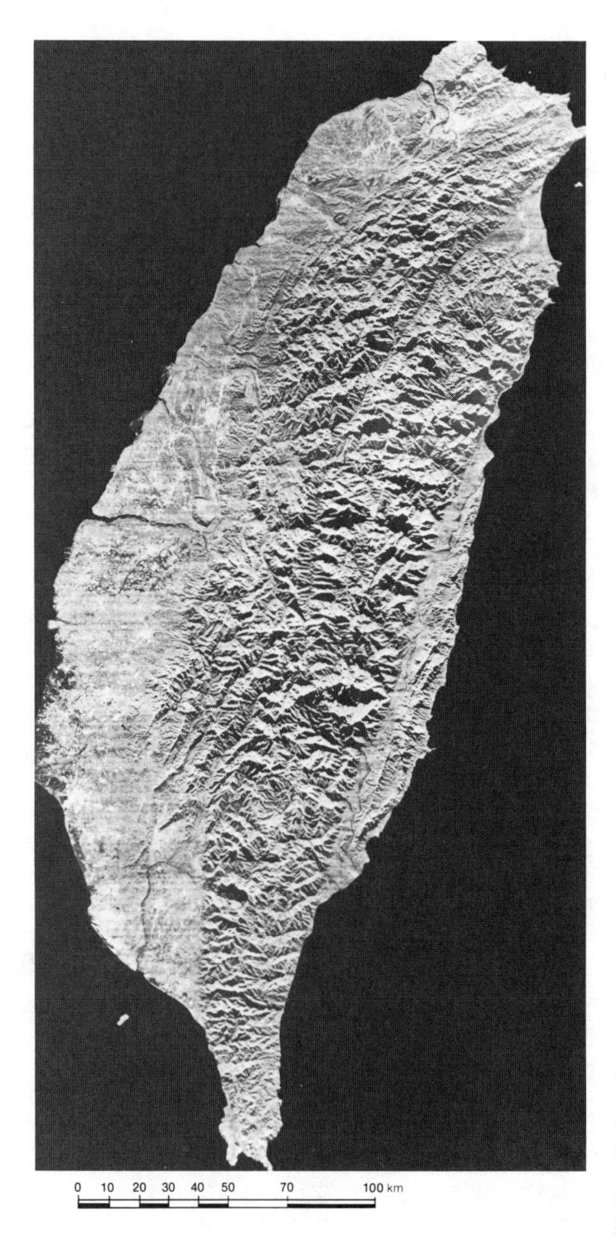


Figure 2.8 Radar imagery of Taiwan. This area is typically cloud-covered, and radar imagery offers a unique glimpse of the terrain. Large faults trending north–northeast are visible.

fine-grained playas (Elachi, 1982). Space shuttle radar images are available to the public through the following agency: National Space Science Data Center, Code 601.4, Goddard Space Flight Center, Greenbelt, MD 20771. Radar from the Venus probe "Pioneer" has mapped the topography of that planet beneath the steamy cloud layer.

A new development in radar imagery is known as the monopulse radar system. This technology offers the ability to penetrate soil depths of 10 m, or to the top of the water table if it is less than 10 m deep. "Subsurface interference radar" is a new commercial device which reportedly is capable of recording continuous, high-resolution data for surficial deposits for depths up to 30 m. In karst areas of

Florida, ground-penetrating radar profiles have been used to detect subsurface features to 5 m depths (Glaccum et al., 1982). Using ground-probing radar devices, Michigan scientists in 1983 located 300 drums of toxic waste buried 6 m deep at a sanitary landfill near Detroit.

## 2.6 Satellite Imagery

Landsat is the name of any of the automated satellites launched by the U.S. National Aeronautics and Space Administration (NASA). The satellites circle the earth at a polar orbit and continuously scan and transmit images to earth.

Landsat is one of the technological marvels of this century. Originally launched in 1972, the first satellite was called EROS (Earth Resources Orbiting Satellite), later ERTS (Earth Resources Technology Satellite), and presently Landsat. Landsat-2 was launched in 1975, Landsat-3 in 1978, and Landsat-4 in 1982.

Landsat satellites circle the earth 14 times a day at an altitude of 920 km (Landsat-4 has an altitude of 705 km). They have a sun-synchronous orbit that allows them to image the complete earth every 18 days. A multispectral scanner (MSS) is the primary imaging system on each Landsat. The images are sent to earth as millions of data bits and are printed in a vertical air photo format. Landsat images show a ground area 185 × 185 km on a side.

In contrast to conventional aerial photographs, Landsat images are generated from digital electronic signals which measure and record the intensity of reflected sunlight in four spectral bands of the electromagnetic spectrum. These are detected by a scanning telescope and an array of MSS photoelectric detectors on board the satellite. The four spectral bands cover the visible green and red (0.5–0.6 and 0.6–0.7 μm) and two near-infrared bands (0.7–0.8 and 0.8–1.1 μm). The sensor output is telemetered from the satellite to ground stations and recorded on tape. From the magnetic tapes, black-and-white images are generated for each of the four spectral bands. Landsat imagery can be obtained as individual black-and-white photographs of each spectral band or as a color composite, including the infrared bands, which gives it a "false color" appearance similar to color infrared.

The polychromatic or multispectral nature of Landsat images is useful in hydrology and agriculture, such as the study of the quality and composition of water, potential water content of snow, moisture content, composition of soils, types and state of vegetation cover, and factors relating to stresses on the environment. For example, clear water appears dark in all four spectral bands. The water is sediment-laden or contains high concentrations of nutrients where it appears bright in the short wavelengths and dark in the long wavelengths. Vigorous vegetation appears very dark in the visible spectrum but is very bright in the infrared, while less vigorous or diseased vegetation becomes brighter in the visible and darker in the infrared.

Images in the separate spectral bands can be overlain as color composites which display the spectral contrasts in varying shades of blue, green, and red. This permits the identification and measurement of botanical, hydrographic, environmental or man-made features on the earth's surface. In other more sophisticated techniques, data from the tapes is fed directly into computer-controlled displays to enhance measurements of certain desired characteristics. Landsat spectral data consists of 2340 × 3232 picture elements (pixels), each having a resolution of about 80 m. This data can be programmed as a 100 × 100 pixel alphanumeric printout (picture print), allowing for computer summation of land areas having certain spectral signatures (Howe and Patrick, 1979). Using certain manipulations and assumptions, it is possible, for instance, to almost instantaneously assess by Landsat computers how many hectares of wheat are growing in any country in the world.

Landsat images can be enhanced by a number of techniques, some of which are very new (Zall and Michael, 1980; Short, 1982). The MSS is designed to record a wide range of reflectance values, from those associated with low reflectance such as basalt, to those associated with high reflectance such as snow. Reflectance values, actually varying tones of gray, can be "stretched" so that tonal contrasts of a scene are redistributed over a full range of available values. Another process is "density slicing" whereby the continuous tonal image is converted into a series of digital steps, each representing a different increment in density (Ross, 1976). "Cibachrome" is a photographic material for reproducing multispectral and density-sliced images in color.

The USGS is currently experimenting with merging of digital topographic data combined with merged Landsat data to produce a stereoscopic pair. Parallax is introduced artificially by shifting picture elements—the left image is shifted as a function of elevation and desired relief displacement.

It is important to obtain Landsat images for the time of year which best suits the intent of the analysis. For example, late summer color images are best for irrigation or soil moisture studies. Low sun-angle snow-covered scenes are good for geomorphological analysis (Fig. 2.9).

The application of Landsat imagery to geology and water resources is well documented (Williams and Carter, 1976; Short et al., 1976; NASA, 1976). Small-scale engineering geology applications are available, particularly for work where repetitive coverage at a scale of about 1:500,000 is desired. This includes targeting mineral exploration (Taranik and Trautmein, 1976; Taranik, 1982), regional aquifer identification (Moore and Deutsch, 1975; Rahn and Moore, 1981), volcano activity (Williams et al., 1974), lineament and fault analysis (Zall and Michael, 1980), earthquake hazards (Carter and Rinker, 1976), flood hazard identification (Deutsch, 1976), and the extent of human degradation of air, land, and water resources (Williams and Carter, 1976; Short et al., 1976). The basics of satellite remote sensing techniques with emphasis on the use of computer-processed Landsat data for the recognition of geographic features in central Pennsylvania is presented in a workbook form by NASA (Short, 1982).

Small-scale water features are visible on Landsat; for example, on Lake Ontario, the Niagara River turbidity plume, or "gyre," has been observed on Landsat imagery to extend 10 km into the lake. Sources of pollution (such as the Welland Canal), movement of ice, and sediment patterns due to wave erosion and longshore drift can be monitored by repetitive coverage from Landsat (Falconer et al., 1981). The extent of devastation caused by the May 18, 1980, eruption of Mount St. Helens is clearly visible by comparing a pre-eruption (e.g., September 11, 1979) image with a post-eruption (e.g., August 19, 1980) image.

Digital image analysis equipment and techniques have been used effectively to enhance, display, and analyze Landsat data (Geotz et al., 1975). Original 70 mm negative transparencies can be reprocessed so as to provide improved contrast between features. For example, Deutsch and Ruggles (1978) enhanced images of

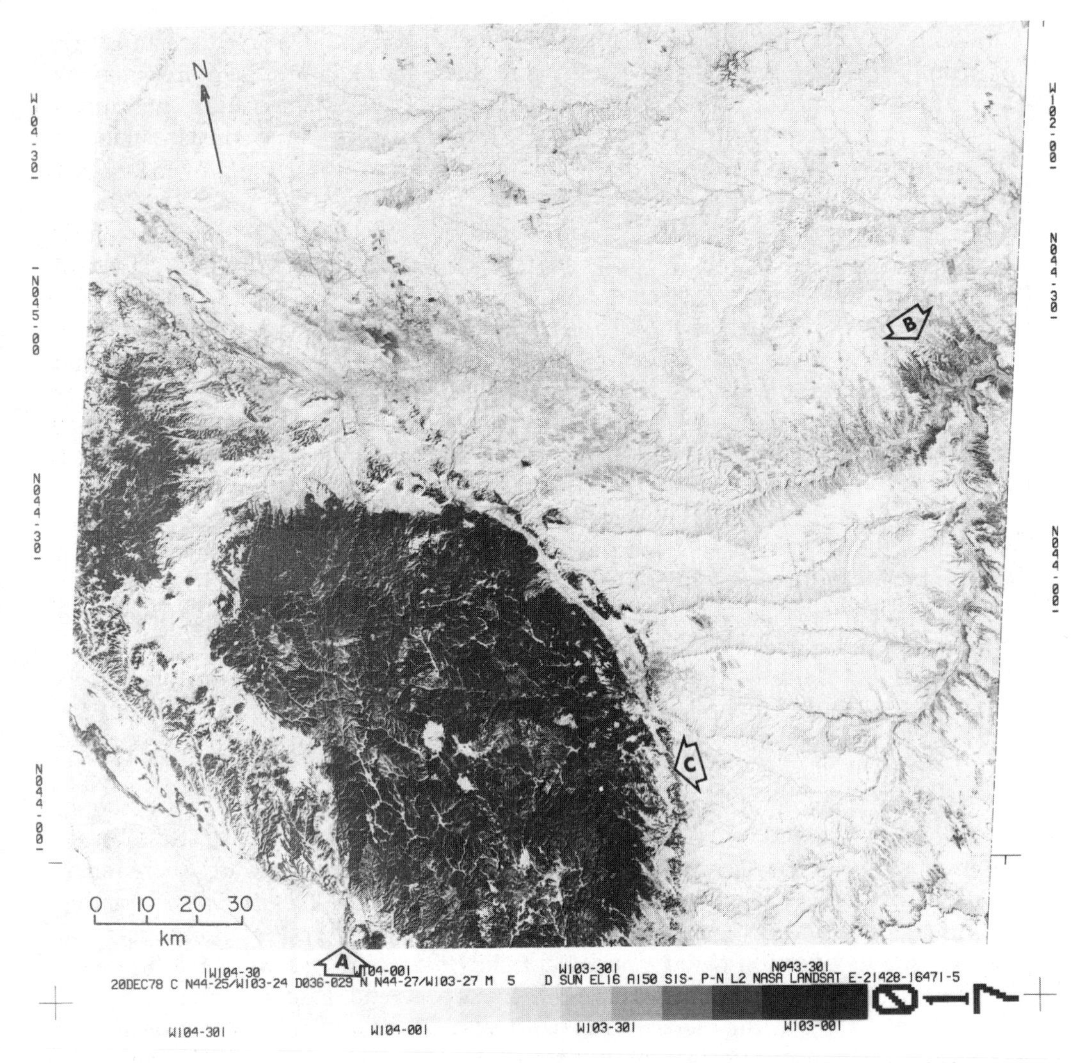


Figure 2.9 Landsat image of the northern Black Hills, SD, taken Dec. 20, 1978. In general, the dark areas are pine-covered hills; the white is snow-covered prairie. A. North-trending lineament. **B.** Badlands along Cheyenne River. C. Dakota Sandstone hogback.

the flooded areas of the Indus River in Pakistan by the use of "temporal composites," i.e., the use of different colored filters on Landsat negatives taken at different times to produce a temporal composite print.

The USGS has experimented with the use of satellite imagery of the southern Powder River Basin in Wyoming to help locate uranium deposits. The images are enhanced to show vegetational differences related to bedrock (sandstone, shale) composition, which, theoretically, relate to the favorability of uranium mineralization. Fussell (1980) reports that possible oil-bearing structures can be identified using lineaments which can be seen through a surficial cover of sand dunes in northwestern Nebraska.

The EROS Data Center, U.S. Geological Survey, Sioux Falls, SD 57198, is the main location where Landsat images can be obtained from the U.S. government. The EROS Data Center also handles earth-oriented imagery from Skylab and the earlier satellite missions. Information free to prospective users include "The EROS Program" (Reprint no. 167) and "Studying the Earth from Space." (A U.S. Geological Survey EROS program booklet is also available from the Supt. of Documents, Washington, DC 20402, Stock No. 2401–2060.) Landsat images are limited by cloud cover; however, the EROS Data Center will provide computer data on percent of cloud cover for various images for the latitude and longitude of the area in question. Products processed at the EROS Data Center include aerial photography obtained for the U.S. Department of Interior, photography and imagery acquired by NASA research aircraft and the Skylab, Apollo, and Gemini spacecraft, and imagery acquired by Landsat satellites.

Some of the newer satellites have heat sensors. Enhanced night-time thermal infrared imagery and digital data from the NOAA-5 polar orbiting satellite were used to map large scale drainage patterns and landforms in North and South Dakota (Schneider et al., 1981). NASA's Heat Capacity Mapping Mission Satellite (HCMM), launched on April 26, 1978, was used to delineate lands along the Big Sioux River, SD, which had been recently flooded. The imagery showed high moisture areas which appear cooler on daytime imagery as a result of thermal inertia and evapotranspiration differences (Heilman and Moore, 1981).

Radar altimeters in the GEOS-3 and SEASAT spacecrafts are able to make terrain measurements. The GEOS-3 satellite provided measurements from an altitude of 840 km with a precision of about 40 cm land elevation, and was used to study land subsidence in the San Joaquin Valley (Krabill and Brooks, 1981). Igneous dikes buried beneath 2 m of alluvium in the Mojave Desert were detected by SEASAT radar (Blom et al., 1984). In 1982, the "Shuttle Imaging Radar" discovered an ancient river channel system buried beneath a blanket of sand several meters thick in the Sahara Desert.

In 1985, the French government plans to launch the SPOT satellite. SPOT is an acronym for "Le Système Probatoire d'Observation de la Terre." SPOT will provide high (10 m) resolution images, better than the best instruments on Landsat, which can only see objects at least 30 m wide. The cameras on SPOT will be stereoscopic, and stereoscopic coverage will be obtained from successive orbits.

## Problems

**1.** Secure a 7.5-min topographic quadrangle map of the area where you live.

**2.** Secure a copy of the most detailed geologic map of the area where you live.

**3.** Purchase an aerial photograph stereo pair of the area where you live.

**4.** Order a 1:1,000,000 scale black-and-white Landsat image of the area where you live. (Order Band 5 for good water/land contrast.)

# 3

# Weathering and Soil-Forming Processes

---

No engineering structure is better than the material of which, and on which, it is built.

A. B. Brink, 1979

Let us not be very hopeful about our human conquest over nature. For each such victory, nature manages to take her revenge.

Friedrich Engels

---

## 3.1 Soil and Bedrock

Many engineering geologists have encountered a situation where it is necessary to differentiate between soil and bedrock. The contact may be gradational or very sharp. Contractor specifications which read "excavate to bedrock" in an area of gradational contact may lead to construction delays and litigation. An example of this sort occurred during the excavation of the Erie Canal when a contractor (Mr. Barrett) asked James Hall, the state geologist of New York, for his opinion as to whether he should get paid for excavating soil or bedrock in a certain calcareous shale bed.

Lockport
June 8th, 1839

Jas. Hall Esquir
State Geologist 4th

Dear Sir

I am very desirous of obtaining your opinion of the rock which occurs in the Excavation opposite the present Locks in this village upon the North Side of the Canal. That is its Geological Classification and the constituent parts of the different Classes as nearly as you can judge from their appearance as it presents its self to view Commencing with the Gray lime stone on the surface the rock appears to Change by incensible degrees from the under surface of the grey lime to the bottom of our Excavation Containing a greater proportion of Alumina as we descend the Strata. In Our Original Estimate this Material was called Slate rock & Shale we supposed it would all Come under the head of Shale below the gray lime stone. In observing the face from which the rock has been recently blasted I See that there are very heavy layers of this material with apparently little or no Seams, but on Exposure to the Atmosphere it soon yields and crumbles to pieces in small Cubes, and by Continued Exposure it becomes decomposed or disintegrates and forms a very tenatious clay. . . . by giving your opinion upon this Subject you will confer a very great favour. . . .

It is fair to state to you that this information is desired to Enable the Canal Commissioner and my self to decide a question raised by the Canal Contractors (unreadable name) in relation to their Contract for the Excavation of this Material. In the Contract they have a price for "Solid rock" and a price for "Slate rock & Shale."

They Claim that the whole is Solid rock. Therefore your professional opinion will be of great Service

I am very respectfully
Yours

A. Barrett
Chief Engineer

Alfred Barrett Esquir
Chief Engineer &c

Dear Sir,

I have received your favor of yesterday and hasten to give an answer to your inquiries in relation to the rock occurring in the excavation opposite the Locks, I understand you to require the names by which the several rocks are known geologically—

The face of the cliff presents the following rocks in the descending order—1st about 10 feet of gray, encrinal limestone of inches to 2 feet thick. Below the limestone are about 6 feet composed of layers of a few inches thickness and alternating with seams of shale. This rock may be termed an argillo-siliceous limestone and probably it contains magnesia or a trace of iron and perhaps manganese. It belongs to the variety which are termed hydraulic limestones, though this term is rather vague in its application, Below the hydraulic limestone are nearly 80 feet in thickness of "calcareous shale"—(The same rock was termed calciferous slate" by Prof. Eaton) with occasional layers of siliceous limestone and from one to four inches thickness. In a general description these would scarcely be mentioned as the amount is so small as not to affect the character of the mass as a whole which comes strictly within the denomination of shale, and no other name can properly be applied to it. Where it has never been exposed to the weather it separates into solid blocks which readily cleave into irregular laminae, this is a character common to all our shale rocks. The upper part of this rock is more compact and apparently contains a larger porportion of carbonate of lime than that below which has a more slaty structure. On exposure this shale decomposes into a tenaceous clay which is in the condition of much of that portion along the banks or sides of the cliffs below the locks. The change from the shale to a perfect state of decomposition is very gradual and there are so many intermediate stages that it may not be easy to decide the point when one begins or the other ends, yet for all practical purposes the distinction is sufficiently obvious.

I have here stated distinctly my opinions of the characters of these rocks, and the names are those by which they are known to all geologists. I make these statements impartially without reference to individuals or circumstances, regarding only truth. Should there be any points which are not satisfactorily explained I shall give any farther explanation with pleasure. You are aware that there can be nothing assumed or arbitrary on my part as rock terms are suited and in common usage among geologists of Europe and America. I cannot give an opinion of what constitutes "solid rock" in your contracts. It is evident that you have certain specifications attached to it in contradistinction to shale—if the term solid rock is applied to this then there is no further use of the term "slate rock & shale" except as applicable to the weathered edges of such strata which are partially decomposed. The circumstances of the mass crumbling into cubical or angular fragments denotes the presence of some scalene matter, which in this case is probably sulphate of iron and sulphate or magnesia arising from the decomposition of iron pyrites, the sulphuric acid uniting both with the iron of the pyrites and with the magnesia of the rock.

With regard to hardness this rock is far inferior to the limestone and is one of the softest rocks which occur in any series.

I am very respectfully
yours
James Hall
Geologist 4th Dist.

James Hall did not hedge the issue by avoiding a direct response, but simply tried to explain that the contact between firm unweathered shale bedrock and loose soil was gradational.

Real and semantical problems involving similar problems of soil/bedrock contacts have happened over and over during the past century. To a large degree, the argument of what constitutes soil and bedrock originates with people who do not understand geology.

## 3.2 Weathering

In order to understand soil/bedrock contacts, it is necessary to study weathering and soil-forming processes.

The gradational change of bedrock to soil is due to weathering (Reiche, 1950; Ollier, 1969; Carroll, 1970; Hunt, 1972; Birkeland, 1974, 1984). Weathering is the physical and chemical disintegration or decomposition of geologic deposits. All rocks exposed at the earth's surface have been buried to some degree in the geologic past and subjected to greater heat and pressure. At or near the earth's surface, the environment is different physically and chemically from condiitons at depth, and the rocks change to accomodate these new conditions.

For discussion purposes, it is possible to classify weathering processes into two categories, 1) physical weathering, and 2) chemical weathering.

### *3.2.1 Physical Weathering*

Physical weathering is the mechanical breakdown of rock, and includes processes such as pressure and temperature changes which tend to break up rocks. These include the following.

*Frost.* If water is available and the temperature drops below 0°C, ice forms and the resulting 9% increase in volume creates a pressure that can produce tremendous compressive forces. Frost is probably the chief reason for the break up of jointed bedrock outcrops in high mountains with cool climates. Thus frost action works best in jointed rock or where other agencies of weathering initiate cracks in rocks.

The spalling of concrete has been found to be due to the formation of ice in the pores of the aggregate (Iyer et al., 1975). Highly porous and absorptive particles of aggregate, such as some cherts, argillaceous limestones and dolomites, and shales, cause "popouts" or general surface scaling of concrete due to freezing and thawing (Mielenz, 1962).

Figure 3.1 Unloading joints (sheeting) in granite, Sierra Nevada Range, Oroville, CA.

*Unloading.* During erosion, underlying rocks experience reduced pressure as overburden is removed. The rocks expand as pressure is released, a process called unloading. Lithostatic stress release and accompanying strain (dilation) occurs slowly, during geologic time, but it is possible for cracks to suddenly propagate in quarries, or for wall rocks in mines to burst with explosive violence (Bain, 1931).

In certain rocks, particularly massive plutonic igneous rocks (e.g., granite) and massive sandstone, the process of unloading causes large joints (sheeting) to develop (Jahns, 1943). These joints tend to be oriented parallel to the general slope of the terrain. A granitic mountain may appear to be spalling off like layers of a giant onion. The magnificent domes of Yosemite Park, CA, and Stone Mountain, GA, are manifestations of the process of unloading (Gilbert, 1904; Mathes, 1930; Wahrhaftig, 1965).

Sheeting may have considerable engineering significance. For instance, in a vertical cut into a granite mountainside, joints caused by unloading that dip at an angle parallel to the slope of the mountainside form perfect avenues along which failure can occur (Fig. 3.1). Figure 3.2 shows a cross section of a spillway cut for a dam in New Hampshire. Widely spaced sheeting joints allowed a small rockslide to occur in 1943, 3 yr after construction. This was followed by a major rockslide. Rehabilitation costs totalled 70% of the initial cost of the entire dam.

The engineering geologist should be aware that any excavation can trigger unloading to occur. Natural erosion of overlying rocks have already induced

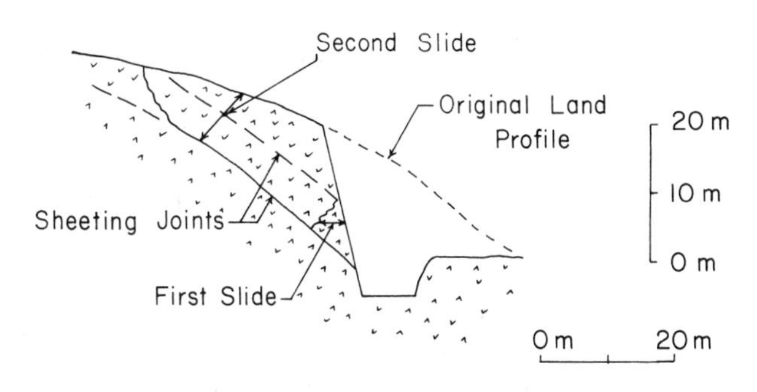


Figure 3.2 Rockslide at Surry Mt. Dam, NH (after Pariseau and Voight, 1979).

unloading stresses in any exposed rocks. Further removal of material by man can create rapid strain. Railroad cuts slowly heave together over the years, and quarries may shrink in size due to unloading. Rock quarry operations have been known to produce rapid strain release, which may be due to lithostatic stress but also include tectonic stresses (Block et al., 1979). (Chapter 4 contains quantitative examples of unloading phenomena.)

During the construction of tunnels, unloading may cause spalling rock or "popping rocks." For example, during the 1906 construction of the Catskill Aqueduct (for the New York City water supply), frequent occurrences of unloading phenomena were observed in tunnels in granite gneiss and the Shawangunk Conglomerate. When a deep shaft was put down in the Storm King granite on the west side of the Hudson River, the popping action was so pronounced that walls had to be repeatedly scaled to remove loosening spalls which caused danger to workmen below (Sanborn, 1950).

Rock bursts are dangerous manifestations of unloading. The walls of deep mines, particularly in quartz-rich rock such as granite or quartzite, can literally explode, killing the miners. In the quartzite host rock areas of the lead-silver mines in Coeur d'Alene, ID, microseismographs are installed to monitor for precursors of damaging rock bursts. Rock cores from deep drilling areas have blown up due to rapid unloading. "Pressure ridges" are areas of rock in quarries which suddenly pop up on floor excavations. In 1981, a magnitude 2.5 earthquake was generated by displacement along a 0.6-km fault in a California diatomite quarry. The fault on the quarry floor showed a maximum displacement of 23 cm dip-slip and 23 cm strike-slip. The rupture is believed due to unloading following the excavation of 44 m of diatomite (Yerkes et al., 1983).

Some shales exhibit a form of unloading. Shale has the capacity for rebound during an excavation or due to natural erosional processes. For example, the removal of overlying glacial ice on the Bearpaw Formation (a Cretaceous shale) in western Canada is believed to be responsible for the unloading and attendant gradual weakening of the rock, resulting in landslides along the South Saskatchewan River (Peterson, 1958). Erosional unloading may cause anomalous fluid pressure in shale (Neuzil and Pollack, 1983).

*Salts.* The combination of moisture and salts (halite, gypsum, etc.) has been observed to cause decay of building stones (Winkler, 1978). The growth of salt crystals can cause stresses in the rock which physically break it apart. Thermal expansion of salt crystals which have grown in former voids may also work to disintegrate rock.

*Miscellaneous physical weathering processes.* Rocks expand when they are heated. In the desert sun, rocks reach temperatures of 60°C (Kerr et al., 1984). Typical coefficients of thermal expansion are $10^{-6}$ m/m per °C. Temperature changes which result in small expansion and contraction of mineral grains were long thought to be a major cause of the disintegration of rocks. This theory probably has little credence, however, because Griggs (1936) experimentally heated and cooled rocks and found no disintegration. He exposed dry granite to the equivalent of 244 yr of diurnal temperature fluctuation between 0°C and 61°C, but found visible cracks formed only when water was added. While a campfire or forest fire may provide sufficient local heat and stress for the splitting of rocks, most geologists believe that daily and seasonal temperature changes are not of sufficient magnitude to weather rocks, even on desert outcrops (Blackwelder, 1933).

On the moon, temperature changes from +130°C during lunar day to −150°C during the lunar night have been observed. During a lunar eclipse, changes of more than 140°C in less than 1 hr have been determined. These temperature changes probably create sufficient stress to break up lunar rocks. Weathering by temperature-induced stress, along with micrometeorite impact and solar radiation, helps form the ubiquitous mantle of lunar dust.

Other physical weathering processes, such as colloid plucking by soil colloid formation (Reiche, 1950), evaporation of salt water accompanying wave splash (Mustoe, 1982), and organic forces such as tree roots, lichens, etc., also exert mechanical forces which help weather rocks. Primitive plants such as lichens may also cause weathering of rocks by forming chelate compounds which remove silica and metal ions.

### 3.2.2 Chemical Weathering

Chemical weathering reactions are exothermic and produce minerals of increased volume. The reactions cause a decomposition (chemical change within the minerals constituting the rock), and the expansion results in the physical disintegration or break up of the rock. Dissolved mineral ions (solutes) found in ground water are largely the result of chemical weathering of rock through which water has passed.

As a simplification, let us consider the following four kinds of chemical weathering reactions:

*Hydration.* Hydration is the process whereby a mineral combines with water to form a hydrated mineral. An example is the hydration of anhydrite to gypsum:

$$\underset{\text{anhydrite}}{CaSO_4} + 2H_2O = \underset{\text{gypsum}}{CaSO_4 \cdot 2H_2O}. \tag{3.1}$$

*Hydrolysis.* Hydrolysis is the reaction of a mineral with water to produce a new mineral or minerals. An example is the weathering of feldspar by reacting

Figure 3.3 Exfoliation of basalt and formation of grus, near Weiser, ID.

with water to form clay and some solutes. For example, consider the hydrolysis of orthoclase:

$$\underset{\text{microcline}}{4KAlSi_3O_8} + 22H_2O = \underset{\text{kaolinite}}{Al_4Si_4O_{10}(OH)_3} + 4K^+ + 9OH^- + 8H_4SiO_4. \quad (3.2)$$

Because of the expansion of feldspathic minerals undergoing hydrolysis, the rock increases in volume. The granular decay of coarse-grained igneous rocks such as granite is largely attributed to the hydrolysis of the feldspars (Leopold et al., 1964). Other granitic minerals such as biotite also contribute to weathering in granite.

Equation (3.2) and other representative hydrolysis reactions are oversimplified because these reactions undoubtedly require several sequential steps before the products are formed. Also, they are very slow processes and may require years to reach equilibrium. Many reactions have been studied in the laboratory but almost always at elevated temperatures, circa 200°C (Faust and Aly, 1981).

Because of dilational stresses due to chemical weathering, many massive igneous rocks tend to develop spherical shells and spheroidal boulders. This process is called exfoliation. The spheroidal boulders may be buried in the decayed products which include shell fragments as well as disaggregated particles called grus (Fig. 3.3).

*Solution.* Limestone ($CaCO_3$) dissolves due to its reaction with percolating water which contains dissolved carbon dioxide. The reaction can be written simply as:

$$H_2O + CO_2 + CaCO_3 = Ca^{++} + 2HCO_3^-. \tag{3.3}$$

Because carbonic acid, bicarbonate, and carbonate can exist in water, depending on pH (Freeze and Cherry, 1979), it is often more meaningful to express the solution or precipitation of calcite in terms of three reversible reactions:

$$H_2O + CO_2 = HCO_3^- + H^+ = H_2CO_3, \tag{3.4}$$

$$HCO_3^- = H^+ + CO_3^=, \tag{3.5}$$

$$CaCO_3 + H^+ = Ca^{++} + HCO_3^-. \tag{3.6}$$

Meteoric water (rain and snow) picks up carbon dioxide in the atmosphere and in the soil zone. The atmospheric $CO_2$ content of 0.03% is increased up to 3% in the soil zone due to root respiration and organic decay. The hydrogen ions thus produced (Eqs. (3.4) and (3.5)) dissolve the limestone (Eq. (3.6)).

When loss of dissolved $CO_2$ occurs, such as near a spring or pumped well, loss of the gas causes the equations to work in reverse, and calcium carbonate precipitates, causing incrustation around the spring or on well screens (Baron, 1982).

The effectiveness of chemical weathering of limestone and its relationship to ground water quality has been studied in the high plains (Back et al., 1983) and in Florida and the Yucatan Pennisula of Mexico (Back and Hanshaw, 1970). It was found that, in general, ground water is undersaturated with respect to calcite in the recharge areas and progressively reaches equilibrium and supersaturation downgradient.

Chemical weathering by solution is responsible for the rapid weathering of limestone and marble dimension stones and tombstones in humid regions. Compare, for example, the rates of weathering of the granite and marble tombstones shown in Figs. 3.4 and 3.5. Jennings (1983) reports that limestone tombstones weather at a rate of at least 5 mm/100 yr in England; and, based on weathering of glaciated bedrock surfaces, limestone outcrops have weathered at a rate of about 4 mm/100 yr over the past 12,000 yr.

The rapid deterioration of building stones in urban areas has been linked to the increase in pollutants, particularly "acid rain" associated with the release of sulfuric acid from coal-fired power plants (Winkler, 1966). The rate of chemical weathering depends on the hydrogen ion concentration (pH) as well as moisture availability. Delicately carved faces on the marble statues on the Arch of Constantine, the official entrance to Rome built in 315 A.D., have been almost completely destroyed in the past 20 years due to automobile exhausts. Many Italian monuments are now hidden behind cloth cages to help protect them.

*Oxidation.* Oxidation involves the ionic combination of an element with oxygen. By definition, a substance is oxidized when it loses electrons. For iron, the reaction is:

$$Fe^{+2} - e = Fe^{+3} \tag{3.7}$$

Mineralogical examples of iron oxidation include the weathering of pyrite to

Figure 3.4 Weathering of 110-year-old marble tombstone, Connecticut. The lettering is almost obliterated.

Figure 3.5 Granite tomestone from Connecticut cemetery shows no weathering despite exposure for 113 years in an area having 114 cm/year precipitation.

limonite, or the weathering of siderite to hematite:

$$\underset{\text{Siderite}}{4FeCO_3} + O_2 + 4H_2O = \underset{\text{Hematite}}{2Fe_2O_3} + 4H_2CO_3. \quad (3.8)$$

### 3.2.3 *Rates of Weathering*

Minerals that crystallize at high temperatures are generally the easiest to weather (Goldich, 1938). Bowen's reaction series (Bowen, 1922) shows the general order of crystallization of minerals in igneous rocks as a magma cools:

| | | |
|---|---|---|
| Hotter (~1000°C) | Olivine | |
| | Pyroxene | |
| | Hornblende | Ca Plagioclase |
| | Biotite | |
| | Orthoclase | Na Plagioclase |
| | Muscovite | |
| Cooler (~300°C) | Quartz | |

From the very nature of weathering, it follows that the minerals which have high crystallization temperatures are farthest from equilibrium with atmospheric conditions, and hence are the first to weather. Quartz, with a low crystallization temperature, is virtually immune from chemical weathering; this, and the abundance of silicon and oxygen, explain why quartz is the most common mineral in rocks exposed at the earth's surface.

Climate controls the rate of chemical reactions. Thus, in a humid area limestone weathers rapidly and forms a valley, but in arid regions limestone is a resistant rock and in many desert locales forms the cap-rock for buttes, mesas, and hogbacks. Ancient limestone buildings in the arid middle east are remarkably unweathered.

A general ranking of the rates of chemical weathering of common rocks is given in Table 3.1. (Freeze and Cherry (1979) present a quantitative analysis of the solubility of individual minerals based on the "equilibrium constant" ($K_{eq}$) for reversible reactions.) Table 3.1 gives an indication, at least in an engineering sense, why foundations should not be built on certain rocks such as gypsum. Gypsum is not only weak in terms of structural strength, but it dissolves readily. This problem is compounded by houses where roof drains empty near the house, allowing for rapid solution and differential settlement.

Dam construction problems associated with soluble foundations are discussed by Sherard et al. (1963). A noteworthy example is the Buena Vista dam in California, where considerable seepage through the foundation had been detected for many years. In 1938 a chemical analysis of reservoir water was compared to seepage water, and it was estimated that seepage was responsible for dissolving as much as 3 $m^3$ of solids (mainly gypsum) from the foundation each day.

Granite is a particularly resistant rock, and building stones show virtually no evidence of weathering even in moist climates (see Fig. 3.5). Rodgers and Holland (1979) report only minute changes of orthoclase to kaolinite along hairline cracks in Wisconsinan glacial boulders (~10,000 yr old). However, older (Illinoian ~100,000 yr) glacial granite boulders are decomposed.

Table 3.1 Rate of Weathering of Common Rocks

| Category | Rock | Rate |
|---|---|---|
| Primary minerals | Gypsum (halite)<br>Calcite (arragonite, dolomite) | Soluble, unweathered or as secondary deposits |
| | Olivine-hornblende (diopside)<br>Biotite (chlorite, glauconite)<br>Albite (microcline, anorthite) | Easily and rapidly weathered |
| Secondary minerals | Quartz<br>Illite (muscovite)<br>Hydrous mica intermediates<br>Montmorillonite | Slowly weathered |
| | Kaolinite<br>Gibbsite<br>Hematite (goethite, limonite)<br>Anatase (rutile, ilmenite, corundum) | Weathered extremely slowly |

Adapted from Leopold et al. 1964.

Birkland et al. (1979) discuss the use of weathering rinds and degree of grussification of granite stones as a means of relative dating of glacial deposits. Birkeland (1982) notes that weathering rind thickness, corner angularity, surface oxidation, and pitting of clasts in Holocene glacial deposits allow for the subdivision of five different ages of glacial and/or rock-glacier deposits.

Weathering reactions are also important for understanding the properties of dimension stone. Stone has long been one of man's primary means of construction. The Romans built enormous aqueducts and coliseums with stone. Today stone quarries are a major mining industry. Modern buildings rarely use stone as structural members as in the past. Rather, dimension stone is used as an ornamental facing on outside walls.

Tombstones provide vivid evidence of the rates and kinds of weathering which occur in different rocks (Figs. 3.4 and 3.5). Even in humid areas such as New England, the rates of weathering are slow. The different weathering rates of various tombstone lithologies are related to the general topography of New England in that rocks that weather fast form low topographic areas (Rahn, 1971a).

Practically speaking, most common rocks used in engineering structures, either as dimension (facing) stone or as a structural wall, will retain their integrity over the life of the structure. For example, exposed parapet capings of Jurassic-aged Portland Limestone on St. Paul's Cathedral in London were found to have weathered only 13 mm in 250 years (Schaffer, 1933).

The importance of weathering to engineering geology lies mainly in the recognition of soil/bedrock contacts and properties of sound rock, weathered rock, or soil. Because almost all engineering excavations are near the surface, they involve rocks weathered to some degree.

### *3.2.4 Concrete Aggregate*

Another engineering application of rock weathering relates to the use of rock as aggregate for concrete. For example, crushed limestone was found to be inferior to quartzite as concrete aggregate for interstate highways, not only because of the relative hardness of quartzite, but because its lower porosity precluded water entrance into the aggregate, and hence freezing and breaking up of the rock (Iyer et al., 1975).

Concrete is composed of coarse aggregate, fine aggregate, cement, and water. The ratio of the above (by volume) is approximately 4:2:1:1. When hardened, the cement (normally Portland variety) paste includes hydration products and a minute pore system. Hardened concrete has sufficiently low permeability and absorptivity to prevent rigid ingress and destructive agents, the most important of which are water, air, acids, and sulfates. The increase in hydrostatic pressure within the cement or the particles of aggregate as a result of the crystallization of ice and consequent increase in volume occupied by ice and water is noted to be particularly destructive. Highly porous aggregate, such as some limestones, cause popouts and surface sealing.

The reaction of alkalies (sodium and potassium) with certain forms of silica can cause the expansion and disintegration of concrete (McConnell et al., 1950; Erlin, 1969). Aggregate containing concentrations of about 0.5% alkalies with chalcedony, opaline chert, rhyolite, or cristobalite should be considered suspect. Chert is a hard rock, but chert and other cryptocrystalline silicates are some of the worst reactors in concrete. The hardening of concrete (a hydration reaction) produces heat and releases alkalies, particularly the hydroxides of calcium, sodium, and potassium. Alkalies can be derived from 1) release from aggregate particles such as water-soluble salts of sodium or potassium, including zeolites or clays, 2) alkaline mixing water such as sea water, or 3) alkaline waters that penetrate the concrete after hardening (Mielenz, 1962). These alkalies react with certain cryptocrystalline silicates to form silica gels which absorb water from the concrete paste and exert osmotic pressure, creating tensional cracks in the hardening concrete. The small cracks can act as avenues of subsequent water penetration and destruction of concrete by freeze-thaw. Chert is a common component in stream gravel, and presents a serious problem which can, fortunately, be reduced by the use of low alkali cement.

Ordinary Portland cement can be chemically attacked by ground water containing high sulfate. In addition, fragments of shale found in coarse aggregate may disintegrate in time. Shale fragments can be removed by separation prior to use as aggregate in a high specific gravity liquid.

### *3.2.5 Weathering and Landscape Reduction*

Over geologic time, the amount of weathering and subsequent removal of weathered products (erosion) in a drainage basin are equal. The two processes can be considered to be in "dynamic equilibrium," and the earth's surface is reduced accordingly.

Chemical weathering is most important in moist climates and in areas underlain by soluble rock (Judson and Ritter, 1964). Because limestone, dolomite, or evaporites dissolve, leaving essentially no residual products, the rate of weathering and landscape reduction can be determined by measuring the rate of removal of dissolved constituents in surface or ground water. For example, the total dissolved solids (TDS) of large springs draining a Paleozoic limestone terrain (aver-

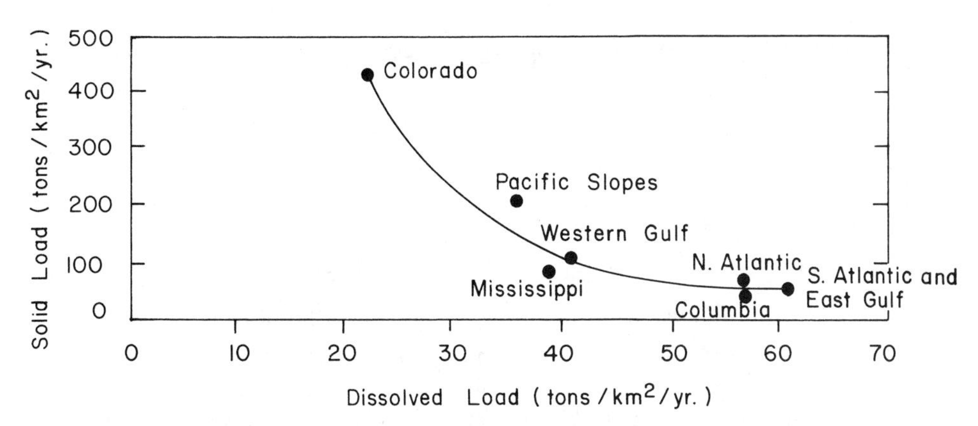


Figure 3.6 Relationship of dissolved and solid load in streams in various parts of the United States (after Judson and Ritter, 1964).

age annual precipitation = 43 cm/yr) circling the Black Hills in South Dakota ranges from 150 to 2280 mg/l. The solutes are mainly $Ca^{++}$, $Mg^{++}$, and $HCO_3^-$. Based on the product of discharge times the TDS of each of 23 springs, a total of $143 \times 10^6$ kg/yr of rock, is weathered and removed via springs draining into tributaries of the Cheyenne River. Assuming a rock specific gravity of 2.7 g/cc, the volume of limestone removed is:

$$\frac{143 \times 10^6 \text{ kg/yr}}{2.7 \times 10^3 \text{ kg/m}^3} = 53 \times 10^3 \text{ m}^3\text{/yr.}$$

Because the area of limestone exposed and drained by these springs is 3900 km2, then the rate of landscape reduction by chemical weathering is:

$$\frac{53 \times 10^3 \text{ m}^3 \text{ yr}}{3.9 \times 10^9 \text{ m}^2} = 14 \times 10^{-6} \text{ m/yr} = 14 \text{ mm/1000 yr.}$$

This chemical erosion rate is similar to other measurements reported by Walling (1977) and Jennings (1983). The total rate of landscape reduction of Black Hills limestone terrain would actually be somewhat larger than this, because occasionally a flash flood removes some clastic debris (limestone fragments and residual soil formed from clay and sand particles in the limestone). This debris constitutes the suspended and bedload of floods in the ephemeral streams draining the limestone area. Due to the karst topography, however, these floods are very rare, and are probably not as effective in reducing the limestone landscape as chemical weathering.

Figure 3.6 illustrates the amount of dissolved and solid (detrital) load in U.S. streams. Note that the Colorado River has an enormous solid load, derived from overland runoff and stream erosion in semiarid areas containing poorly vegetated slopes. In contrast, streams in the southeastern U.S. derive a much greater percent of their total load from dissolved constituents. Use of these data to calculate long-term landscape evolution may be suspect because of the increased erosion accompanying man's activities (Douglas, 1967).

Many rocks, such as granite, do not dissolve, but chemically change to form a residual soil upon weathering. Barth (1961) proposed a formula for the amount of rock weathering if weathering produces insoluble products as well as solutes:

$$W = \frac{D_i}{c_i - s_i}, \quad (3.9)$$

where:

$W$ = amount of rock weathered,

$D_i$ = amount of an element $i$ removed in solution,

$c_i$ = fractional concentration of $i$ in the original rock,

$s_i$ = fractional concentration of $i$ in the residual soil.

For a hilly granitic terrain in Rhodesia (average annual precipitation = 122 cm), Owens and Watson (1979) used calcium analyses in streams to determine the rate of granite weathering:

$$W = \frac{2.0 \text{ kg/ha/1000 yr}}{0.0005 - 0.0000} = 400 \text{ kg/ha/1000 yr.}$$

This corresponds to a depth of chemical weathering (and by inference landscape reduction) of about 15 mm/1000 yr. Because granite minerals such as quartz are insoluble, the 15 mm of chemical weathering was estimated to produce a residual soil of 11 mm thickness in 1000 yr. Presumably this residual soil is removed by stream erosion (suspended and bedload) in order to maintain a dynamic equilibrium of weathering and erosion.

Gibbs (1967) studied the salinity of the Amazon River, the world's largest river, which contributes 18% of the world's total runoff reaching the ocean. Of the dissolved salts reaching the ocean (36 mg/l), approximately 86% are supplied by only 12% of the total area of the Amazon Basin which is the mountainous region of the Andes. The reason for this is that chemical weathering proceeds rapidly in the mountainous areas where fresh rocks are exposed by erosion, whereas in the lowlands an enormous thickness of saprolite (mostly gibbsite and kaolinite minerals which are weathering end products) effectively isolates unweathered bedrock from the vast amounts of rain that characterizes the jungle area.

General surface lowering of the earth's landscape at 20–80 mm/1000 yr for gentle to moderate slopes have been reported by Young (1969). Judson (1968) reports that, based on average sediment loads for rivers in the United States, the landscape is being eroded at a rate of 60 mm/1000 yr. In a study of 70 rivers in the United States, approximately 20% of the total measured load is carried in solution (Leopold et al., 1964). This percentage ranges from 1% in the semiarid basin of the Little Colorado River in Arizona to 64% in the humid basin of the Juniata River in Pennsylvania. The worldwide ratio of mechanical to chemical denudation is about 5:1 (Birkeland, 1984). (Additional information of fluvial erosion and sedimentation is presented in Chapter 8.)

## 3.3 Soil and Bedrock Contacts

The origin and properties of soil are technically complex subjects. The word "soil" is typically used by engineers to indicate fragmented material below cobble size that can be excavated without blasting. Bedrock is the counterpart. The topsoil of agricultural soil scientists (pedologists) is more restricted, and includes soil horizon development near the surface which is capable of supporting plant life. The Glossary of Geology (American Geological Institute, 1972) defines bedrock as: "A general term for the rock, usually solid, that underlies soil or other unconsolidated surficial material." Soil (sometimes loosely referred to as saprolite, regolith, or mantle) is defined as "unconsolidated material above the bedrock. . . ."

The definition of soil and bedrock is an exercise in semantics; the definition largely depends on a person's area of interest. To a civil engineer, soil is simply unconsolidated material which typically disintegrates in water. Agricultural engineers and most geologists consider soil to be the weathered product of rock or surficial deposits containing varying proportions of organic material. A civil engineer could look at a 100-m-thick exposure of glacial till and call the entire exposure "soil," whereas a pedologist may recognize a soil (or soil "profile") in only the top 1 m of the till. To most geologists, soil is simply the product of in situ weathering processes. It is regrettable that semantical difficulties such as these exist. The problem is related to old scientific translations from German and Russian. In recent years soil terminology problems have been further compounded by the introduction of numerous cumbersome terms by American pedologists (Soil Survey Staff, 1975). [Olson (1981), Birkeland (1984), and Schlemon (1985) have practical explanations of these terms.]

What constitutes "firm bedrock" is also a moot point. To a geologist familiar with granite terrain, the Pierre Shale (a Cretaceous, semi-consolidated shale in the midwest) may not appear to be bedrock. But midwestern geologists refer to it as bedrock.

Hansmire (1981, p. 78) proposes the following useful quantitative classification for the definition of soil and rock:

> For very soft soils an unconfined compression strength of 100 psf (4800 N/m$^2$) would be typical for a fresh alluvial clay deposit that would not support a person standing on it. Lightly loaded buildings can be supported by shallow foundations on a medium to stiff clay with design loadings of 2000 psf (96,000 N/m$^2$) and would be difficult to cut with a knife. An unconfined compressive strength of about 20,000 psf (960,000 N/m$^2$) is normally the division between soil and rock. The transition from soft to higher strength rock is not as well established and may range from 200,000 psf (9,600,000 N/m$^2$) to 500,000 psf (24,000,000 N/m$^2$).

Other methods of classification of soil and rocks include weatherability, hardness, durability, and joint spacing (Goodman, 1976).

### *3.3.1 Residual Soils*

Soils and surficial deposits (referred to as soils by civil engineers) can be either residual or transported in origin (Fig. 3.7).

Residual soils develop in situ, and their characteristics depend on the kind of bedrock from which they are derived. Examples of residual soils are given below.

*Granite (and other massive igneous rocks).* The predominant physical weathering processes which act on granite rocks include unloading and frost action. Chemical weathering causes a decomposition of the rock mass. Wahrhaftig (1965) and Krank and Watters (1983) believe that the major process of weathering of

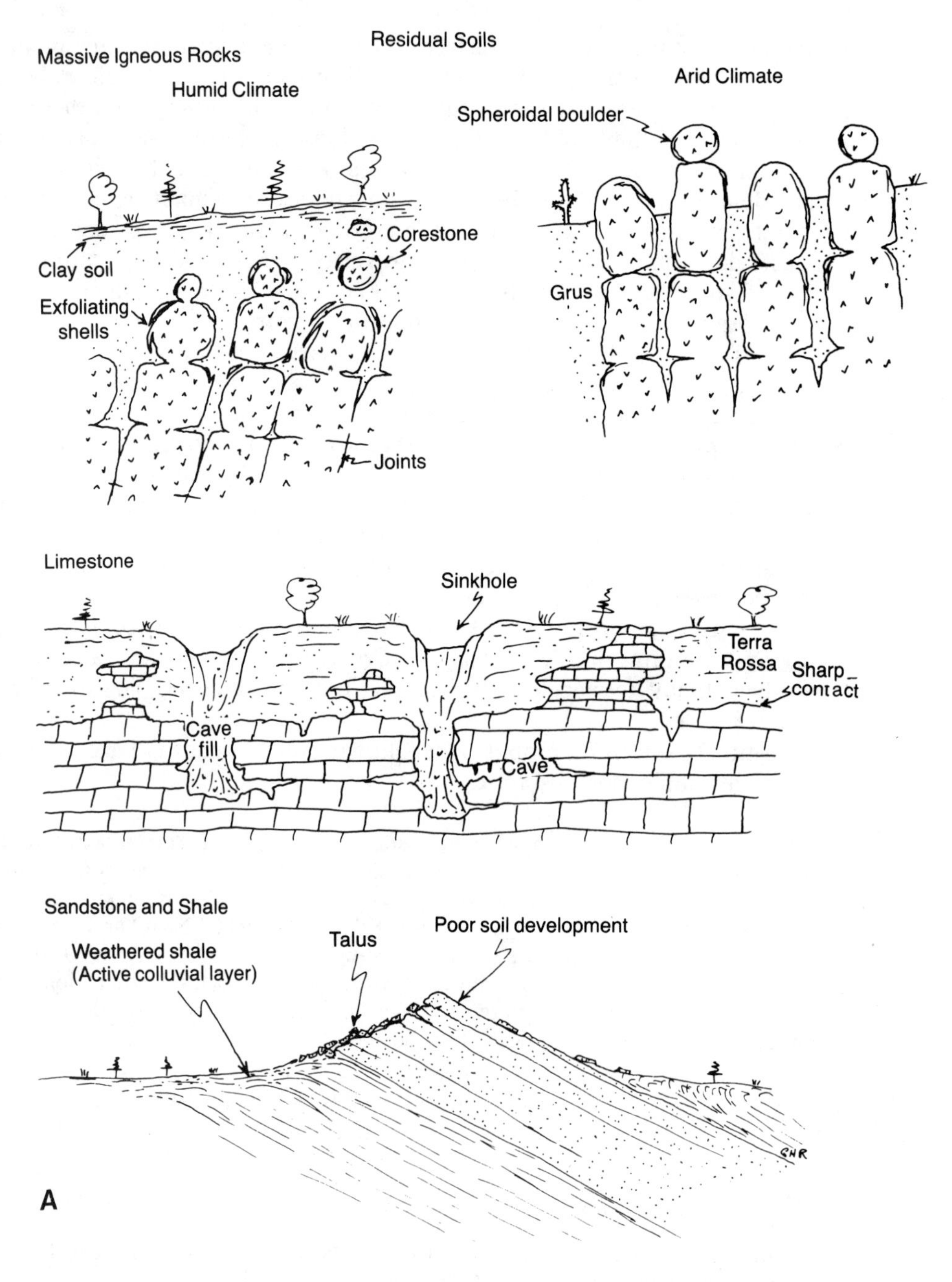


Figure 3.7 Idealized sketches showing (**A**) residual and (**B**) transported soils.

Sierra Nevada granodiorites is the expansion of biotite in contact with ground water. The expansion of biotite produces microfractures which progressively break down the original rock to a saprolite.

Figure 3.7A contains a sketch showing the appearance of igneous rock outcrops. Three states of weathering may occur: sound rock, exfoliation shells, and grus. Grus is disintegrated granite; the size of the fragments is typically coarse

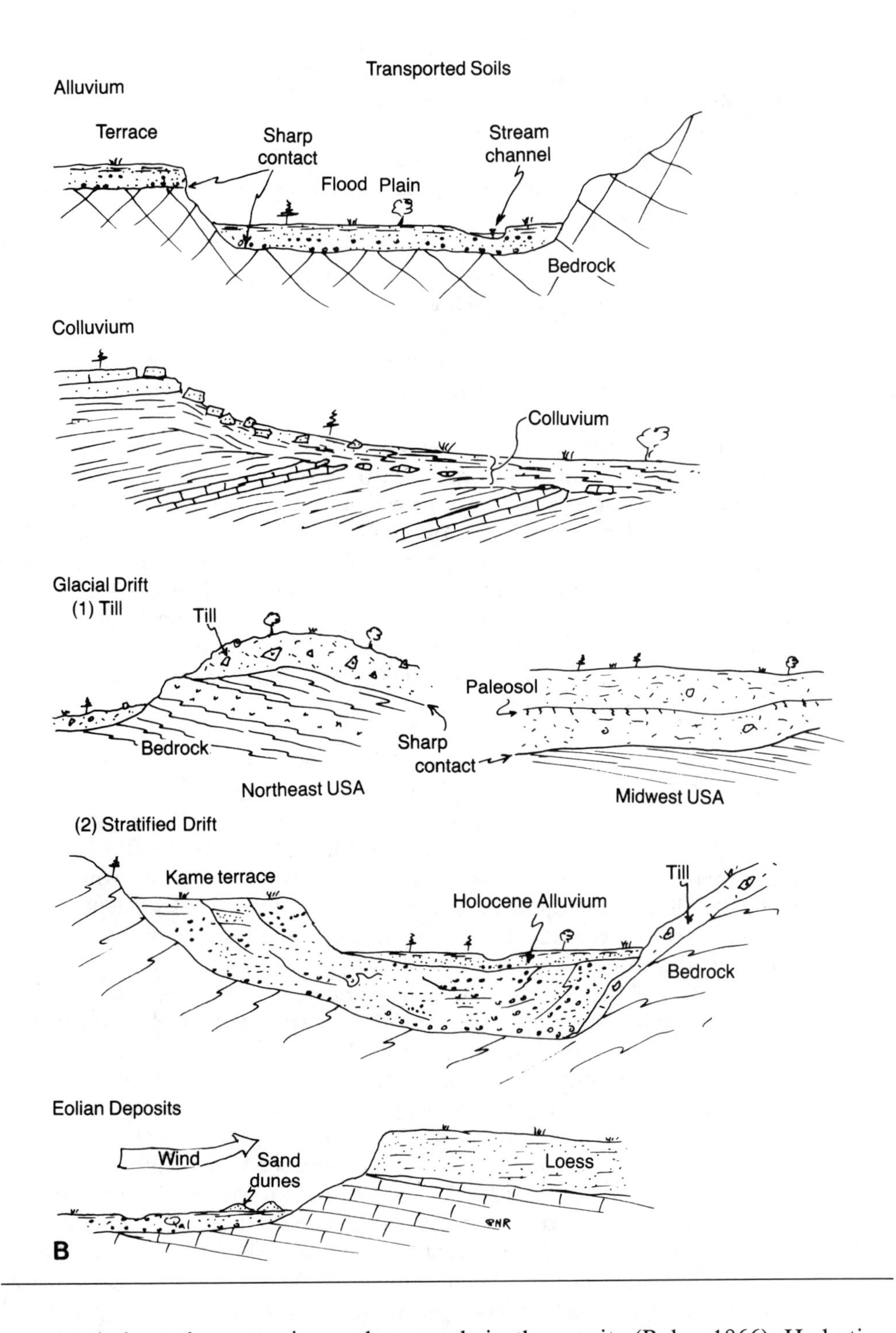


sand about the same size as the crystals in the granite (Rahn, 1966). Hydration and hydrolysis are largely responsible for the weathering and resulting dilation of the sound granite, although the actual change in chemical elements between the three states is small (Larsen, 1948).

Chemical weathering is slower in arid climates than in humid climates; thus the depth of "rotting" is less. Due to intense thunderstorms characteristic of deserts, grus tends to get washed away, causing the spheroidal boulders to protrude above the land surface. In humid climates, the depth of chemical weathering is great, and grus can be 100 m thick or more. Hatheway (1980) reports granite weathered to 20-m depths in tropical Singapore. Ruxton and Berry (1957) describe engineering problems associated with excavations in the decomposed granite of Hong Kong.

Weathering of crystalline rocks in the Piedmont area of the eastern United States has resulted in a saprolite averaging about 30 m thick in upland areas. The saprolite is decomposed rock characterized by the retention of parent rock structures and textures, but primary minerals are mostly replaced by clays and iron oxides. These saprolites can be excavated by a shovel. In the Piedmont, most water wells are located in saprolite because wells in the underlying bedrock are not very productive.

The main engineering problem in residual soils formed on igneous rock is the identification of a sound horizon on which to seat a foundation. Due to the gradational nature, there is no depth at which one can unequivocally say that soil ends and bedrock begins. The presence of individual boulders of sound rock (core stones) in residual soils presents special problems in foundation engineering. The problem is acute where the residual soil possesses a collapsible grain structure, but the core stones are too large to be excavated by earth-moving equipment or cannot be penetrated by an augered pile hole (Brink, 1979). Another problem is the misleading identification of sound bedrock in an excavation where the top of an isolated core stone is exhumed.

*Limestone.* In contrast to granite, the soil-bedrock contact in carbonate (limestone or dolomite) terrains is typically quite sharp. Figure 3.8 shows the sharp contact exposed in the wall of a limestone quarry near Philadelphia, PA. The thickness of soil in a carbonate terrain varies considearbly (~0–100 m) due to the formation of caves and sinkholes. After a sinkhole forms, slope wash will gradually fill in the sinkhole causing a thick soil of small areal extent.

Theoretically, because limestone weathers by solution, there are no weathered products; the entire rock is carried away by ground water. In reality, however, because of small beds of clay, chert, and sand, which are usually present in limestone, a residuum is left after solution of the carbonate rock. The red, clayey soil typically derived from limestone is called terra rossa. Over geologic time the thickness of the residual soil is maintained by the continued chemical weathering of the bedrock below and the stripping of the surface by erosion.

The irregular nature of a limestone/soil contact was dramatically evident in the excavations for the foundations of the Tennessee Valley Authority dams. Vertical solution channels over 60 m deep were discovered in Mississippian limestone. Although filled with terra rossa and cave-fill deposits, the solution zones leaked enormous amounts of water from the reservoirs. Following the appearance of downstream springs and "sand boils," which threatened the concrete dams, grouting programs were undertaken. At Hales Bar Dam, for example, from 1919 to 1921 an attempt was made to stop leakage by pumping hot asphalt into the cavities functioning as conduits. Sixty-eight holes were drilled and over 220 $m^3$ of asphalt was pumped into them (Burwell and Moneymaker, 1950). The Hales Bar Dam project, estimated at $3 million and scheduled for completion in 2 yr, cost $12 million and required 8 yr for completion. In 1940 the entire foundation had to be grouted again due to continued enormous leakage.

Figure 3.8 Sharp soil-bedrock contact on limestone at Conshohockon, PA. The terra rossa soil is 0–10 m thick. The water table in this abandoned quarry is at about 50 m in depth.

Differential settlement is common in thick residual soils overlying weathered carbonate rocks. Residual soil over solution cavities may be quite thick; building foundations or other man-made structures located in the residual soil should be rigorously analyzed (Lifrieri and Raghu, 1982). A similar problem occurs where a "floating boulder" (a core stone) is encountered and is assumed to be part of the bedrock. Even greater hazards occur in areas containing solution cavities at depth. Sowers (1975) attributes the 1969 bridge collapse in Tarpon Springs, FL, to failure of a roof of a cavern. The settlement of a junior college complex near Miami, FL, was believed due to the crushing of highly weathered porous limestone.

Sudden sinkhole collapse is a geologic hazard. Below the residual soil, solution selectively dissolves layers of limestone along joints, bedding planes, fossiliferous layers, etc., until a cave develops. The sudden collapse of the roof of the cave results in a sinkhole. This geologic process happens naturally, or can be accelerated by man's activities such as ground water withdrawal or the construction of parking lots which allow for abnormal discharge of rain water into the ground (Parizek, 1971a). (See Chapter 9 for further discussion of sinkholes.)

The sudden collapse of a sinkhole can cause more than the inconvenience of disrupted farmland, forest, or a broken highway. Sinkhole collapse resulted in the rupture of a gas main in Allentown, PA, which filled caves with gas, threatening to ignite. In the same region, near Saucon Valley, PA, sinkholes have developed due to the lowering of the water table due to a deep zinc mine (Fig. 3.9).

Figure 3.9 Sinkhole at Saucon Valley, PA, developed due to water table decline associated with underground zinc mining.

In an arid to semiarid climate, limestone typically forms a resistant unit, forming an obstruction to construction of pipelines, sewage systems, etc. One of the more sobering reminders of the engineering significance of soil-bedrock contacts to military combat is offered by Brooks (1920, p. 87) in his discussion of the trench warfare in the Western Front during World War I:

> During the great battle of Verdun a body of troops was ordered to intrench itself on the high plateau of the Côtes de Meuse. Even a casual examination of the geologic map would have shown that the plateau was underlain by hard limestone with less than a foot of soil. This material could not be excavated with the light tools furnished or even with proper equipment in the time available. As a consequence there was a large and needless loss of life.

*Shale—expansive soils.* Very little residual soil develops over sandstone or shale because the bulk of the minerals contained in the rocks (quartz, clay minerals) are relatively immune from weathering processes (Table 3.1). Thick transported soils can accumulate, however, as transported soils due to mass wasting processes, e.g., colluvium as described below. White (1962) describes the weathering of the Cretaceous Pierre Shale as first forming loose platey fragments and then forming a clay-rich soil. Shrinking and swelling of the clay as it dries and becomes wet causes mass movements, which gives a smoothly rounded topography to the western South Dakota prairie.

Clay is a very fine-grained deposit with plastic properties when moist, and composed largely of hydrous aluminum and magnesium silicates (Gillott, 1968). Clay exhibits considerable strength when air-dried. (The term clay has been used widely to designate the percentage of soil fragments finer than 0.002 mm (0.005

mm in some cases), but it is recommended that this usage be discontinued since, from an engineering point of view, its strength properties are more important than its size per se.) Clays which have resistance to deformation (low plasticity and compressibility, typical of the kaolinite group) are called lean clays. Less desirable engineering properties such as highly plastic and compressible clays (typical of the montmorillonite group) are called fat clays.

Shale is similar to clay, but it has a fissile or laminated structure parallel to the bedding. Slate is well indurated, possessing cleavage. Weathering processes tend to make shale and slate revert back to clay, so from an engineering point of view, weathered slate or shale possess properties of clay.

Engineering problems associated with shale or clayey soils are not due to their soluble nature, but to their low shear strengths, which make them hazardous for foundations or for any construction, particularly in landslide-prone areas. Shales used in highway fills often settle due to slaking of the shale pieces. Unfortunate past experiences dealing with the nonhomogeneous nature of shale and clay and their poor strength and weathering characteristics have led most engineers to treat shales as problem material.

At the time of deposition of subaqueous clay, the water content is very large. After burial the water content decreases. The process of void ratio reduction is termed compaction by geologists and consolidation by engineers. The importance of settlement of the land surface (especially differential settlement) due to loading of soils has led to detailed theoretical and empirical studies. Not only is water content important in the ability of soils to sustain loads, but it has been found that some deposits may become "quick" or completely fluid under rising ground water conditions, as in a seepage area. Other fine-grained soils are thixotropic in that they lose their strength when disturbed as by a vibrating load. Laboratory tests show that clay cores may support a considerable load in an undisturbed condition, but when remolded (at the same moisture content) behave like a fluid and may be literally poured from a beaker (Gillott, 1968). (See Section 5.5.4.)

The physical properties of shale vary considerably. Some shales have strengths comparable to concrete, and, as they have low permeability, make ideal dam sites. Other shales are not very consolidated, and have attained a degree of consolidation only slightly greater than clay soils. Low-grade shales that have settlement problems are also subject to rebound. If the superincumbent weight of materials removed in an excavation appreciably exceed the weight of the man-made structure placed in the excavation, rebound may occur. According to Burwell and Moneymaker (1950), "The rebound characteristics of shales like the Bear Paw, Pierre, and Fort Union of North Dakota and South Dakota are such that deep spillway cuts may cause buckling of spillway linings and sufficient differential rebound movements in the foundations of spillway structures to require special design provisions. For example, the Fort Peck Dam spillway, located on the Bear Paw shale, is anchored deeply into its foundation with a series of 60-inch diameter reinforced steel dowels."

Clayey soils are also hazardous due to expansion and contraction accompanying soil moisture changes. Unlike kaolinite clay minerals which form from the weathering of feldspars, shale and clayey soils typically include large percentages of illite or montmorillonite clay. The kaolinites form very stable clays because their tight, inexpandable crystal structure resists the introduction of water into their lattices and the consequent expansion or heaving when saturated. On the

Figure 3.10 Slaking of Cretaceous shale block which had been excavated 1 year earlier.

other hand, montmorillonite crystal sheets are bound rather loosely, allowing for space water molecules to insert themselves between the sheets, causing expansion or swelling (Tourtelot, 1974). When a saturated montmorillonite dries out, it is subject to shrinking or cracking. In pure form, montmorillonite clays may swell to over 15 times their dry volume, although most soils contain only small amounts of montmorillonite so that not many swell to more than 1.5 times their dry volume (Jones and Holtz, 1973). Expansive soils cause damage when they shrink upon drying or when they expand upon wetting, and the resulting soil movements disrupt houses, multistory buildings, sidewalks, streets, and utilities (Building Research Advisory Board, 1968). Leaking sewers or swimming pools, improper storm drains, or poorly built houses with inadequate foundations develop damage ranging from sticking doors and hairline plaster cracks to complete destruction.

The process of shrinking and swelling can result in expansive soil problems, or in slaking, the crumbling of sound outcrops or rock specimens into flakes or granular particles (Fig. 3.10). For example, because of slaking of the shale in the Miocene Castaic Formation, it was necessary to place earth fill within 24 hr of exposure of fresh bedrock surface during the construction of the Castaic dam in California (Hanegan, 1973). Poorly cemented shales may be quickly reduced to a pile of chips upon several cycles of wetting and drying, while well-cemented shales are quite resistant to such changes. (Note: Some soil engineers define slaking as the disintegration of dry soil when it is submerged in water, whereas other soil engineers define slaking as the disintegration of soil under alternating wetting and drying.)

A slaking test has been devised by highway engineers to evaluate slaking problems which are typical of calcareous shales. The usual procedure is to immerse rock samples in water and observe any disintegration that occurs. Slaking durability depends on 1) the ability of fluids to penetrate the rock, 2) the ability of the fluids to cause disruption by solution of cement or disruption of bonds or pre-pressure relationships, and 3) the inability of the rock skeleton to resist the disruptive forces. Clay-bearing rocks and sandstones cemented by soluble minerals are the most susceptible to slake deterioration. Among the several mechanisms that may account for slaking of shale, ion exchange appears to be dominant. Capillary effects, desiccation upon drying, and unloading probably also play an important role in slake deterioration.

Another example of soil problems was described by Baker (1975). Urban planning in the Boulder, CO, area is highly dependent upon bedrock and surficial deposits, and their derived soils. Three types of parent material exist: loess, a terrace gravel, and the Cretaceous Pierre Shale. The terrace gravel forms the best engineering substrate because it has good grain-to-grain strength properties and contains few swelling clays. Plastic clays are present in the Pierre Shale, however, and swelling and expansion exert differential pressure on overlying foundations which can break them up. Soils developed on loess are subject to solution of the calcite cement, causing collapse of structures; the chain of events culminating in loess failure may be initiated by something as innocuous as watering the lawn.

Expansive soils represent one of the costliest engineering geologic hazards, exceeding floods, landslides, and earthquakes in terms of average yearly destruction in the United States. Damage to family homes cost homeowners about $300 million annually. Still, each year over 250,000 new homes are built in expansive soil.

Expansive soils are found throughout the western United States, particularly those derived from Cretaceous shales. Some of these shales contain volcanic clay (bentonite) beds which are thick enough to be mined commercially as low permeability sealants for dams or house basements. Regions underlain by expansive soils can be recognized in some prairie rangeland areas by subtle but distinctive topography called gilgai. Gilgai are networks of small downslope ridges and grooves having wavelengths on the order of 2–3 m and amplitudes of about 0.1 m. Gustavson (1975) used aerial photographic interpretation of gilgai on undisturbed prairie lands as a means of mapping potential hazards from clay expansion.

Expansive soils are capable of developing tremendous pressures when moisture is introduced. Laboratory pressure devices have recorded $1.4 \times 10^6$ $N/m^2$ in montmorillonite shale soils from Dallas, TX (Krynine and Judd, 1957). These forces are capable of buckling concrete floor slabs. When expansive shale dries, it shrinks and can cause wall cracking or other forms of distress in structures. The wetting of expansive soils typically occurs every season when the frost leaves the ground and spring rains add a surge of moisture.

Moderately swelling soils are present throughout much of the Denver area. Swelling soils (typically possessing liquid limits of about 55% and plasticity indices of about 30%) swell 3–10% under normal loads of 500 $N/m^2$ (Costa and Bilodeau, 1982). Swelling pressures can be as great as 15,000 $N/m^2$. Structural damage can occur when swelling is as little as 1%. Both shrinking and swelling can occur with moisture variation, and either can cause damage to streets or structures.

Availability of water in expansive soils is controlled by climatic conditions; however, surface drainage, amount of vegetation, yard maintenance performed by the homeowner, and depth of the water table act to modify the moisture availability. Any of these factors which help to maintain a constant moisture content act to reduce the shrink-swell activity. One way to reduce structural damage in expansive soils is to dig the structural walls or piers below the zone of seasonal moisture change. Factors such as improper drainage, leaky swimming pools, etc., that make large fluctuations in the soil moisture content result in an increased shrink-swell activity. Hammer and Thompson (1966) found that large trees planted near houses could cause structural failure. Because tree roots may cause local soil moisture deficiencies and shrinkage, it was recommended that trees should be planted no closer to the foundation than one-half of their mature height. A study of expansive soil damage to homes in Texas, Mathewson et al. (1980) showed that yard maintenance and the thickness of an active soil zone (soil experiencing seasonal moisture changes) were the most significant parameters affecting house damage. In Rapid City, SD, houses most significantly damaged are ones which have roof drain spouts emptying near the house foundation; houses are less affected where drains extend 3 m or more away from the house.

*Sandstone.* Sandstone typically weathers slowly in any climate and forms highlands such as mesas, buttes, and hogbacks. For example, the silica-cemented Tuscurora Sandstone forms the most prominent ridges in the central Appalachian Mountains. There is virtually no residual soil on this formation, only large boulders of sandstone forming block fields or talus. Calcite-cemented sandstone, on the other hand, weathers relatively quickly to sand, and in many places forms commercial deposits of sand.

In general, sandstone is a good foundational material, although unweathered sandstone may be an impediment for foundation or service line excavation.

## 3.3.2 Transported Soils

Transported soils are surficial deposits which accumulate due to the erosion, transportation, and deposition of weathered residual soil or bedrock. Four common surficial processes and their resulting deposits are described below (Fig. 3.7B).

*Colluvium.* Colluvium results from the process of creep, whereby soil and weathered bedrock slowly move downslope due to gravity. The contact with the underlying bedrock can be sharp or gradational (Fig. 3.11). Colluvium resembles glacial till in appearance. Typically colluvium consists of unstratified, seemingly randomly oriented angular blocks of bedrock in a clayey matrix. In geologic time, colluvium is actively moving. Colluvium is very widespread and can reach 100 m thick in humid areas. In unglaciated humid areas, practically the entire land surface is covered by a mantle of colluvium, only interrupted by bedrock outcrops, or other surficial deposits such as glacial drift or alluvium. Thick accumulations of colluvium may have formed during past climatic regimes, such as in periglacial areas which may have had permafrost conditions.

Engineering geologists should view colluvium with suspect because any disturbance could cause accelerated movement. Although in the life of a structure

Figure 3.11 Colluvium exposed above limestone bedrock in a stream cut near Hermosa, SD.

($\sim$100 yr) the direct structural damage due to natural rates of movement may be small, man's activities such as excavations may accelerate creep, or cause a landslide, thereby leading to building distress or failure. Radbruch-Hall (1978) cites numerous examples of creep failure in the United States and Europe.

Creep differs from the process of landsliding in that creep operates slowly. Creep has more displacement near the surface of the land, typically causing slow rotation of telephone poles, tombstones, walls, etc. (The curved trunks of trees, attributed to creep by some geologists, may in fact be due to site-specific growing conditions unrelated to creep (Phipps, 1974).) Landslides also differ from creep in that landslides occur as massive displacements of soil or rock along a specific failure surface. (See Chapter 8 for a more detailed discussion of mass wasting.)

*Alluvium.* Alluvium includes all sediment deposited by streams. The deposits are stratified into layers of silt, sand, gravel, and clay. In mountainous areas, alluvium consists largely of boulders. In areas dominated by sluggish streams, alluvium typically is clayey silt. Figure 3.12 shows alluvium exposed in a stream terrace. Coarse-grained alluvium is an excellent aquifer where saturated due to the sorted, stratified nature of the debris. Along the edges of flood plains in humid areas, colluvium may constitute a large part of the valley fill (Lattman, 1960).

Two processes act to produce alluvial deposits. During intense floods, coarse-grained debris such as cobbles are transported as bedload. The stream may scour

Figure 3.12 Alluvium exposed in a kame terrace, Willimantic, CT. Brunton compass indicates scale.

down as far as bedrock, and upon diminution of discharge cobbles are deposited back to original grade. At the same time, overbank flooding occurs, and silt is deposited on the flood plain (Vanoni, 1975). Thus alluvium typically consists of two components: the lower part is a bedload-derived coarse fraction mantled by overbank silt above. It is fortuitous that nature provides this dual process. Alluvial valleys are the most arable places on earth, thanks to the fact that arable silt masks the sand and gravel below. The sand and gravel is also an aquifer which can supply ground water or can supply moisture to the flood-plain crops as "subirrigation."

The engineering properties of alluvium vary over a wide range. Oxbow lakes resulting from meander cutoffs eventually fill with fine-grained alluvium. The subsequent meander pattern which develops in the lower Mississippi River Valley is affected by the location of these clay plugs which resist erosion (Fisk, 1951). Uniform deposits of sand and gravel can support high footing loads and are not very susceptible to shrinkage or swelling. One problem in alluvial foundations is the delineation of specific sand, gravel, or clayey lenses; if slab or column loads are distributed onto heterogeneous layers, differential settlement can occur. Another problem of alluvial soils occurs in areas such as New Orleans, where very thick deposits of saturated clayey silt occur, giving rise to lack of bearing strength for large loads. In these areas, pilings are extensively used (see Chapter 5). (Note: by far the biggest environmental hazard associated with man's utilization of alluvial soils is floods; this subject is dicussed in Chapter 8.)

Terraces are abandoned flood plains no longer related to the present stream regime. Most terraces are Pleistocene or Holocene in age, probably formed by climatic and/or stream regime changes associated with Quaternary climate and

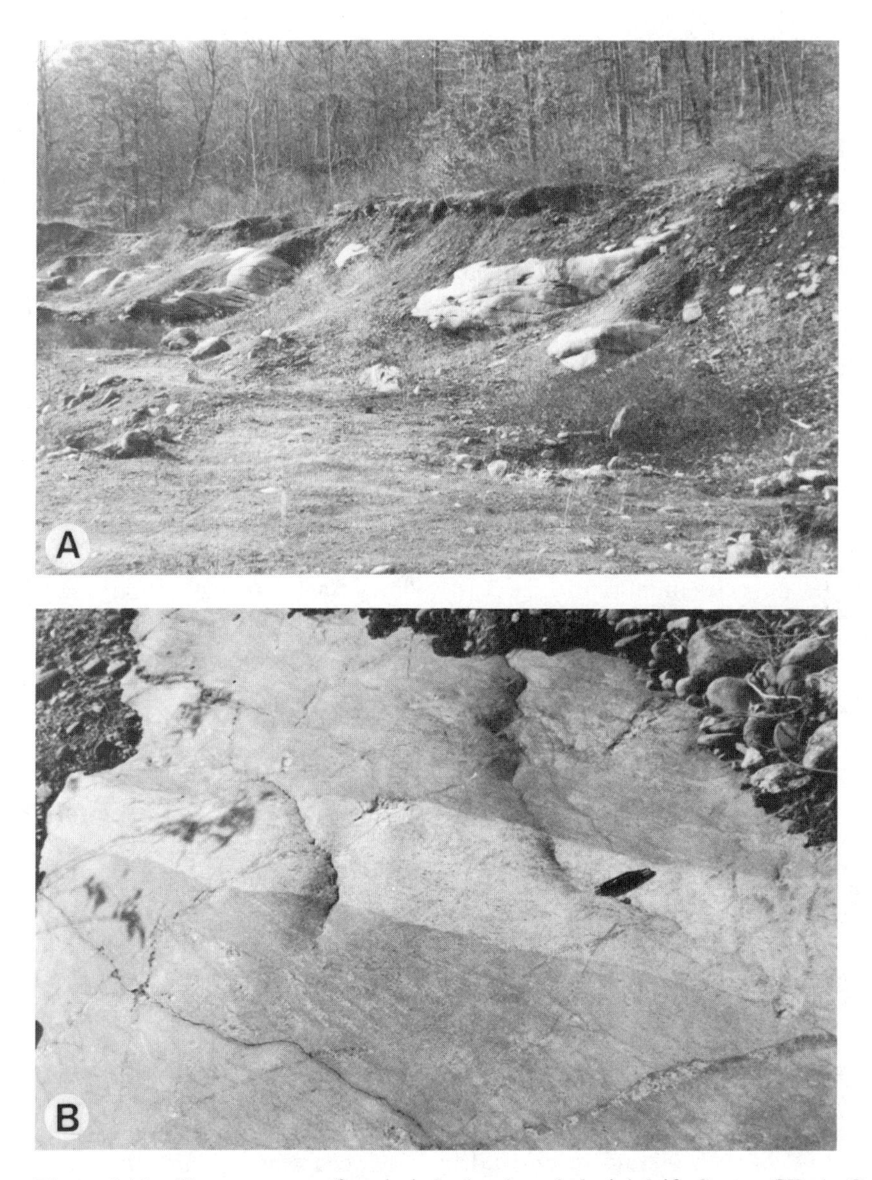


Figure 3.13 Sharp contact of gneissic bedrock and glacial drift, Storrs, CT. **A**. Contact exposed in an excavation. **B**. Close-up of glacially polished bedrock showing truncated pegmatite vein in gneiss.

glaciation. Terraces are ideal construction sites as they are relatively flat, not subject to floods, and are typically underlain by sand and gravel.

*Glacial drift.* Glacial drift includes all deposits formed by glaciers. Much of the land area of the northern hemisphere was covered by continental glaciers during the Pleistocene epoch. An engineering geologist working in these areas should acquire a sound knowledge of glacial geology. The contact between drift and the underlying bedrock is usually very sharp (see Fig. 3.7B and 3.13).

There are two types of glacial drift: till (nonstratified drift) and stratified drift. Till is deposited directly by glacier ice, either plastered down and compacted by an advancing glacier (lodgement till) or left as loose piles of debris as the ice melts at the terminus of a stationary or retreating glacier (ablation till). Lodgement (or "basal") till is found under drumlins, and consists of still, hard till possessing stones oriented with their long axes generally parallel to the direction of ice movement. Ablation till is typically cohesionless and sloughs naturally to the angle of repose (Hatheway, 1983). Lodgement till generally has better geotechnical properties in terms of strength, but may be several orders of magnitude less in permeability than ablation till, thus affecting the yields of water wells or the operation of sewage drain fields. The contrast in engineering properties of these two types of tills is exemplified by the construction of the St. Lawrence Seaway and Power project. On the basis of borings, ablation till was expected. Soon after excavation, a dense ($\rho$ = 2.7 g/cm$^3$) lodgement till was encountered, and costs increased by about 400%. The contractors involved either went bankrupt, defaulted, or entered litigation for extra payment (Legget, 1979).

Till is typically nonstratified, unsorted, and contains angular to sub-rounded rock particles of all sizes (Fig. 3.14). Platey fragments within lodgement till may be oriented roughly parallel to the ice movement, giving the till a texture referred to as fabric (Mills, 1977). The composition of till fragments reflects the type of bedrock over which the glacier passed. Thus in New England, due to the hard bedrock (e.g., gneiss and granite), the till typically contains large boulders of granite. These boulders pose a severe impediment to farmers plowing the land (Fig. 3.14A). Throughout much of the midwestern United States the bedrock is predominately shale, and hence the till is very clayey with only a few erratic boulders to bother the farmers. Till generally has favorable engineering characteristics. Lodgement till is uniform in nature, and is very compact, almost impenetrable to a hand shovel.

Sediment deposited by streams of water derived from melting glaciers is called stratified drift. The debris is reworked till; the particles have been sorted, rounded, and stratified by running water. Stratified drift is, in fact, alluvium and as such has the properties of alluvium. In New England the contrast in the engineering properties of till and stratified drift is exemplified by the location of cemeteries. Cemeteries which were used roughly 100–300 years ago are scattered rather randomly across the landscape, wherever small villages or a family plot happened to be, in both till and stratified drift. The occurrence of large quantities of boulders in the till is a severe obstacle to digging in till; as a consequence, modern cemeteries (as well as most landfills) are generally found only in stratified drift because the sand and gravel is easier to excavate (Rahn, 1971b).

*Lacustrine and marine deposits.* The fine-grained sediments typically deposited in lakes are limited in areal extent, but can be very troublesome for foundations. A dramatic example of the engineering influence of clayey lacustrine deposits (ancient glacial Lake Agassiz) occurred in 1911 when a newly constructed grain elevator near Winnipeg, Canada was filled for the first time. The entire concrete structure, placed on a foundation only 6 m into the underlying clay, sank and tilted 27°. Little damage occurred to the grain elevator, and so it was righted and placed on a good foundation. Lacustrine deposits are associated with the Great Lakes. Many of the foundation problems in the Chicago area originate from thin

Figure 3.14 Glacial till, near Storrs, CT (from Rahn, 1971b). **A**. Surface appearance of pasture land. **B**. Exposure showing poorly sorted texture.

lacustrine sediments deposited in Lake Michigan when the Great Lakes were more extensive than today.

Many shallow marine coastal deposits that accumulated during the Pleistocene are now above sea level because of isostatic rebound or sea level lowering. Examples include the Boston blue clay, 25 m of marine clay underlying the low parts of Boston. The sensitive Leda clay, found in the St. Lawrence Valley, is of marine origin. Other Pleistocene marine deposits, such as along the southern California coast, include sand and gravel, which generally make acceptable foundation material.

The engineering properties of beach sand were emphasized during World War II in the Pacific campaign. U.S. Marines were able to easily dig foxholes for protection upon landing on some beaches. But on some islands, such as Iwo Jima, Marines found it impossible to dig foxholes in volcanic ash deposits. The Japanese took advantage of local geology by building caves in coral limestone to shelter their men and cannons.

*Eolian deposits.* Eolian deposits include windblown sand (sand dunes) and windblown silt (loess).

The geomorphology and dynamics of sand dunes are described by Bagnold (1941) and Sharp (1966). Sand dunes invariably have a nearby source of sand. Extensive sand dune fields are usually related to a nearby source of unconsolidated to semiconsolidated sandstone. For example, the source of sand dunes for the Sand Hills of Nebraska is the Ogallala Formation of Pliocene age. The vast Sand Hills area (35,000 $km^2$) is covered by sand dunes averaging 8 m in thickness. The sand dunes of the Sahara Desert are derived from the Cretaceous Nubian Sandstone. Smaller areas of sand dunes are usually found within a short distance (1 km) of a source of sand. Examples include the southeastern shoreline along Lake Michigan; Cape Hatteras, NC, areas south of braided stream beds along the Platte River in Nebraska; and areas downwind of Playa lakes at White Sands National Monument, New Mexico.

Some dune fields have traveled considerable distance. The Algones dunes near Yuma, AZ, for example, are a nuisance for the operation of the All-American Canal, which carries water from the Colorado River to the Imperial Valley. The source of sand is believed to be reworked sand from shorelines of Pleistocene pluvial lakes some 5–30 km distant (Smith, 1978).

Sand dunes rarely serve as sites of construction due to the arid and/or inhospitable environment. Active sand dunes are subject to continual movement and are totally unsuitable for construction. Even inactive sand dunes have a delicate vegetal cover which can be broken down by man's activities causing them to become active (Péwé, 1982). In the Salton sea area of California, houses have been overwhelmed by sand dunes which migrate easterly at about 12 m/yr (Shelton, 1966). The Sahara Desert dunes move up to 100 m/yr.

Fine wind-blown sand grades into loess, which is much more widespread than sand dunes. Loess is ubiquitous throughout the midwest of the United States, capping bluffs along the major rivers (Fig. 3.15) and forming fertile soils over broad areas (Ruhe, 1969). In the upper Mississippi River Basin, loess was deposited mainly between 14,000 and 22,000 years ago (Ruhe, 1973).

A peculiar engineering property of loess is its ability to stand in vertical cuts. For example, vertical railroad cuts made over 100 years ago in Mississippi are still stable. At depths greater than 3 m, calcareous cement may occur in loess because infiltrating meteoric water has not dissolved the delicate calcareous internal skeleton. Clay minerals also act as cementing agents (Gibbs and Holland, 1960). This cementing material apparently develops along ancient root holes. The calcareous skeleton combined with evaporation of mineralized water on the surface of a man-made excavation helps bond the loess so that it is stable in vertical

Figure 3.15 Loess exposed at Sioux City, IA.

Figure 3.16 Geologic cross section showing loess at Vicksburg, MS (after Kolb and Steinriede, 1967).

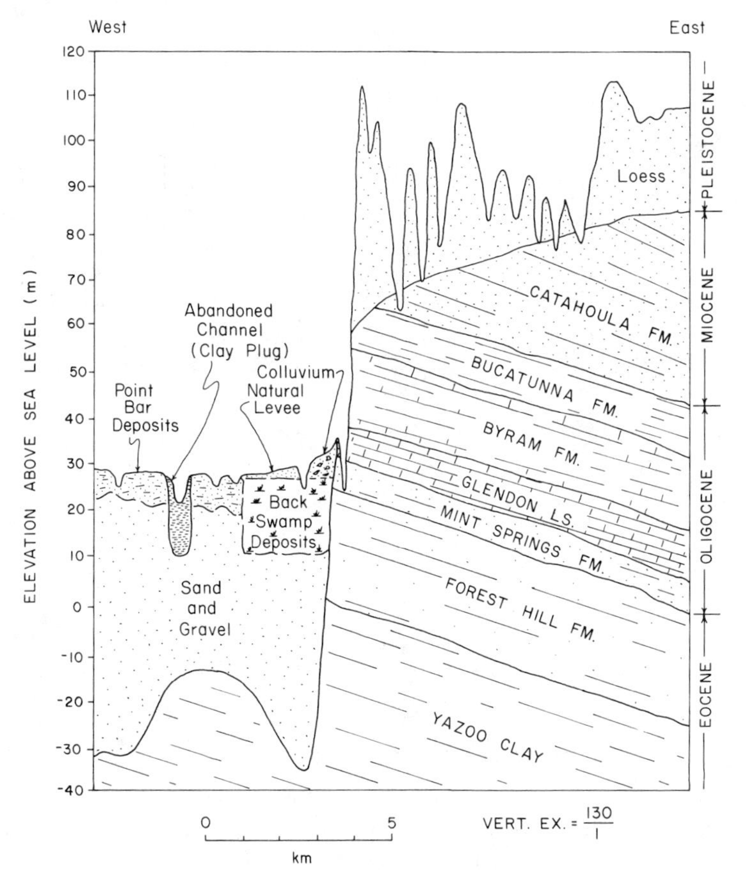

cuts. Continued evaporation of soil water precipitates minerals on the vertical walls, thus strengthening the exposure with time. Loess is susceptible to failure by landslide or gully erosion where gently sloping cuts are made.

Loess is generally acceptable media for foundations from the standpoint that it is uniform in composition and has very low swelling properties. Individual cases of land settlement after water has been added to loessal foundation soils are reported. Loess dry densities of less than 1.3 g/cm$^3$ are especially suspect for settlement (hydrocompaction) when they become wet.

At Vicksburg, MS, loess has been extensively studied by engineering geologists (Fig. 3.16). Loess blankets the eastern side of the Mississippi River for most of its length, deposited by northwestern prevailing winds during Pleistocene time. At Vicksburg the loess is 10–30 m thick at the bluffs to about 1 m thick 100 km to the east. The loess is composed of well-sorted clayey silt (Fig. 3.17), which becomes more fine-grained with increasing distance from its point of origin. The

Figure 3.17 Typical grain-size curves for loess east of Vicksburg, MS (from Krinitzsky and Turnbull, 1967).

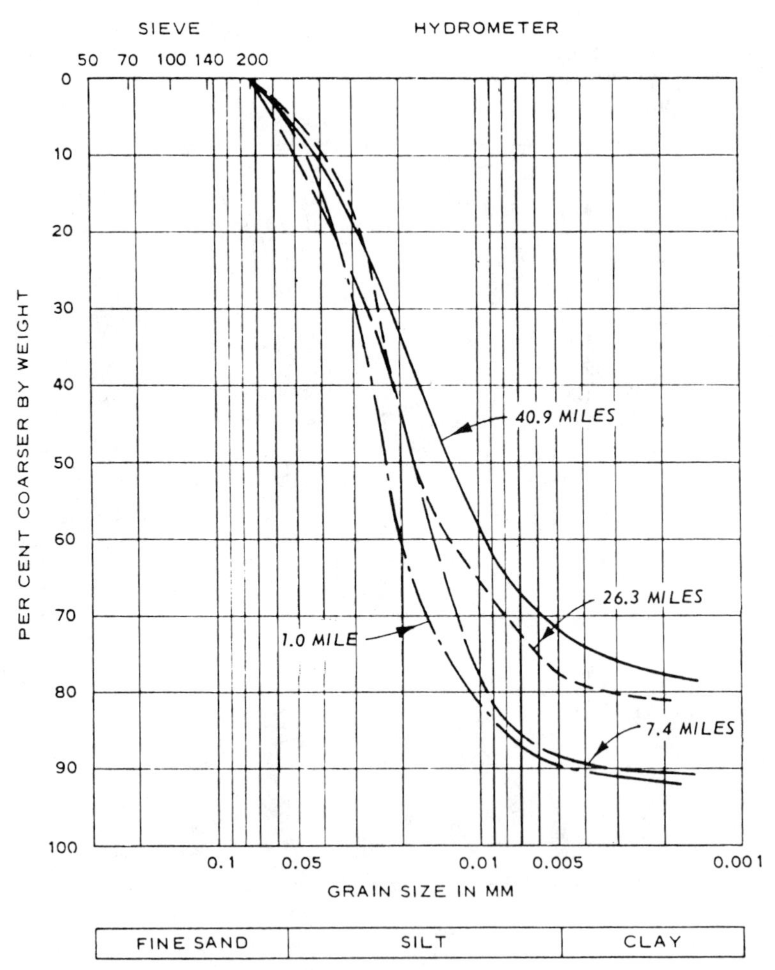

loess deposits at Vicksburg formed such an ideal media for construction of underground shelters that the Civil War General Ulyssess S. Grant was unable to dislodge the Confederate troops by bombardment and won the Battle of Vicksburg only by starving out the Confederate troops.

## 3.4 Soil-Forming Processes

The weathering and surficial geologic processes described above produce weathered rocks and surficial deposits that are commonly referred to by civil engineers and others as "soil." Nevertheless, soil scientists and most geologists think of soil as only the thin profile developed on weathered rocks or surficial deposits. The study of soil-forming processes and soil profiles is a science unto itself, whose scope is beyond the limits of this book. Nevertheless, there are lessons for the engineering geologist to learn from a cursory examination of soil-forming processes.

Soil scientists have developed a climatic classification of soils. It has long been recognized that topsoils tend to be thick and clayey, rich in red iron minerals in tropical areas, but tend to be thin and contain a lot of calcium carbonate and calcium sulfate in deserts. For this reason tropical soils used to be called "pedalfers" because of high aluminum and iron content, and desert soils were called "pedocals" because of high calcium content. This climate ("zonal") classification of soils has been expanded into "Great Soil Groups" (Table 3.2). Essentially, pedalfers are now called laterites, and pedocals are called caliches. More recent revision of soil classification by pedologists involves complex recognition of climatic as well as vegetative factors (Soil Survey Staff, 1975). The recognition of an ancient soil (paleosol) in Tertiary rocks (Retallack, 1983) or Quaternary deposits (Birkeland, 1984) can be used as a climatic indicator.

Table 3.2 Soil Classification According to the Great Soil Groups ("Zonal" Soils)

<table>
<tr><th rowspan="2">Zone</th><th colspan="9">Amount of moisture</th></tr>
<tr><th colspan="2">Arid</th><th>Semiarid</th><th colspan="2">Wet and dry</th><th colspan="2">Subhumid</th><th>Humid</th><th>Wet</th></tr>
<tr><td>Cold</td><td colspan="9">Tundra</td></tr>
<tr><td rowspan="2">Cool and temperate</td><td rowspan="4">Desert (caliche)</td><td rowspan="4">Gray desert (sierozem)</td><td rowspan="4">Brown soils</td><td rowspan="4">Chestnut</td><td rowspan="4">Chernozem</td><td rowspan="4">Prairie</td><td rowspan="4">Degraded chernozem</td><td colspan="2">Podzol</td></tr>
<tr><td colspan="2">Brown podzolic</td></tr>
<tr><td rowspan="2">Temperate</td><td colspan="2">Gray–brown podzolic</td></tr>
<tr><td colspan="2">Gray wooded soils</td></tr>
<tr><td>Warm and temperate</td><td colspan="2" rowspan="3">Red desert</td><td rowspan="2">Reddish brown</td><td colspan="2" rowspan="2">Reddish chestnut</td><td rowspan="2">Reddish prairie</td><td rowspan="2">Degraded Red–yellow podzolic</td><td colspan="2" rowspan="2">Red–yellow podzolic</td></tr>
<tr><td>Hot</td></tr>
<tr><td>Tropical</td><td colspan="7">Laterite</td></tr>
</table>

| | Zone | Horizon | Description |
|---|---|---|---|
| | Organic debris lodged on the soil; usually absent on soils developed by grasses. | $A_0$ | Organic debris. |
| The solum. (This portion includes the true soil developed by soil-building processes.) | Zone of eluviation. | $A_1$ | A dark-colored horizon containing a relatively high content of organic matter but mixed with mineral matter. Thick in chernozem and very thin in podzol. |
| | | $A_2$ | A light-colored horizon, representing the region of maximum leaching (or reduction) where podzolized* or solodized.** The bleicherde of the podzol. Absent in chernozem,† brown,† sierozem,† and some other soils. |
| | | $A_3$ | Transitional to **B** but more like **A** than **B**. Sometimes absent. |
| | Zone of illuviation. (Exclusive of carbonates or sulphates as in chernozem, brown, and sierozem soils. In such soils this horizon is to be considered as essentially transitional between **A** and **C**.) | $B_1$ | Transitional to **B** but more like **B** than **A**. Sometimes absent. |
| | | $B_2$ | A deeper-colored (usually) horizon representing the region of maximum illuviation where podzolized or solodized. The orstein of the podzol and the claypan of the solodized solonetz. In chernozem, brown, and sierozem soils, this region has a definite structural character, frequently prismatic, but does not have much if any illuvial materials; it represents a transition between **A** and **C**. Frequently absent in the intrazonal soils of the humid regions. |
| | | $B_3$ | Transitional to **C**. |
| | The parent material. | C | Parent material. |
| | Any stratum underneath the parent material, such as hard rock or a layer of clay or sand, that is not parent material but may have significance to the overlying soil. | D | Underlying stratum. |

*Process of water leaching downward through A and B horizons.
**Process of accumulating surface minerals through leaching upward, produced by evaporation in areas of low rainfall causing moisture movements to be toward the surface.
†Members of great soil groups (see Table 3.2).

Figure 3.18 A hypothetical soil profile having all the soil horizons.

The solum includes the soil horizons modified by soil-forming processes. Ideally these processes form a distinct A and B horizon. As shown in Fig. 3.18, D is the source material, which could be anything from sound bedrock to glacial till, and C is weathered parent material, which may include decayed granite, terra rossa over limestone, etc. Climatic controls on the development of the solum lead to the following threefold classification of soils (soil groups).

### 3.4.1 Podzol

Podzol soils develop in moist, cool climates such as the forested areas of northeastern United States. The soil horizons illustrated in Fig. 3.18 are those of a podzol soil. The process of forming a podzol soil is called podzolization. In this process, the most soluble ions (calcium, sodium, magnesium, as well as iron and aluminum organic compounds called micelles) are peptized (become mobile) in the A horizon. They are transported downward by percolating water, and flocculate (precipitate) in the B horizon. As a result, the A horizon is friable and is bleached to a grey color, and the B horizon is still, blocky, and clayey, often with a columnar structure. The A horizon is acidic due to organic acids from the forest litter; the B horizon begins where the pH increases to about 6.5.

Podzol soils have no special engineering significance. On many residual surficial deposits or surficial deposits of till, the podzol soil horizons may be unnoticed in excavations. In well-developed podzol soils, the B horizon may be very hard and troublesome to excavate.

Table 3.2 shows the range in climate soil names from podzol soils to the tropical laterite soils to the desert caliche soils. It is these latter two soil groups that deserve the most attention by engineering geologists because of their unusual properties.

### 3.4.2 Laterite

Residual red soils (loosely called laterites) form in tropical (wet and hot) climates. In North America, laterite soils form in Florida, Cuba, and Mexico. Laterites extend over large areas of Brazil, India, southeast Asia, Australia, and central Africa. They are best developed as residual soils developed on bedrock rich in iron and aluminum (such as basalt) and in tropical climates with alternating wet and dry periods (Hunt, 1983).

Laterites are products of intense chemical weathering, where the residual soil is higher in iron and aluminum and lower in silica than merely kaolinized rock. Typically organic accumulation is minimal, and silica and mobile ions (such as alkalines) are leached through the entire soil horizon. Laterization is the process whereby the precipitation of iron and aluminum compounds in the B horizon becomes excessive, leading to the formation of a brick-red, clayey B horizon which is called a laterite or more properly a cuirasse. The laterite horizon is typically composed of kaolinite, goethite, gibbsite, and quartz, but may contain up to 80% $Fe_2O_3$ in the form of goethite or hematite (Goudie, 1973). In some places the presence of a water table promotes a very tough cuirasse (iron and aluminum hydrates, called "hardpan" or "duricrust").

Laterite soils can be an impediment to construction. In Vietnam, excavation of airport runways during the Vietnam War was hampered by these tough soils.

Figure 3.19 Laterite exposed in a roadcut in northeastern Nicaragua. The soil, developed in alluvium, has a bright red color.

In Nicaragua (Fig. 3.19), I have seen attempts at hand-digging of water wells frustrated by laterite soils developed in clayey alluvium. In the Nicaraguan wet season, well-digging is impossible due to swampy conditions; in the dry season, digging is extremely difficult because the laterite dries to a brick-like consistency. In central Africa, saprolites approximately 30 m thick, containing laterite soils in the upper approximately 10 m, mask the bedrock lithologies below and make prospecting difficult.

Laterites are resistant to erosion and may cap hill tops or slopes where vigorous erosion has produced a truncated soil horizon (Chorley et al., 1984). Laterites have a long history of use as building materials such as the bricks used in the construction of walls in Thailand or the Ankor Wat temple complex of Cambodia (Banerji, 1982).

## 3.4.3 Caliche

In hot, arid places, weathering is slow, and a soil profile is virtually nonexistent. Nevertheless, calcium carbonate and calcium sulfate compounds accumulate in the top few meters of desert surfaces, particularly in alluvium. The formation of these desert soils is called calcification, and the resulting soil is called caliche.

Calcification is restricted to subarid and arid climates where precipitation is typically insufficient to allow for the movement of meteoric water downward to the water table. Caliche at or near the surface of alluvial fans decreases water infiltration rates (Cooley et al., 1973). In Nevada, Lattman (1973) describes the caliche near Las Vegas and its resulting impediment to the recharge of ground water. Caliche may act as a cap rock to preserve alluvial fan surfaces (Gardner, 1972) and affect arroyo geometry (Van Arsdale, 1982).

Gile et al. (1981) shows that caliche thickness is proportional to the age of the terrace along the Rio Grande River in New Mexico. Higher terraces are older and have thicker caliches. Due to excavation problems, caliche caused the abandonment of construction of a NASA facility airport on a high terrace of the Rio Grande where the caliche is about 5 m thick.

In parts of southern Arizona, caliche resembles concrete; I have found places where prospectors with bulldozers were unable to break through a caliche-encrusted colluvial slope where malachite float gave promise of copper veins in the bedrock.

In the Atacama Desert in Chile, nitrate-rich caliche has been mined as a source of nitrate used in explosives and fertilizers (Ericksen, 1983).

## 3.5 Soil Maps

Soil surveys are routinely published by the U.S. Department of Agriculture; these maps provide information on climate, soil-forming processes, and parent material. These reports can be used by engineering geologists to determine depth to bedrock, water table conditions, permeability, shrink-swell potential, and other soil properties. The soil units mapped generally are confined to the top 2 m. Soil survey maps are printed into an aerial photograph mosaic at 1:31,000, 1:24,000, 1:20,000, or 1:15,840 scale.

The soil map uses a soil "series" in the same manner that a geologist calls a rock unit a "group." All soil series are developed by the same genetic combination. Subdivision of the soil series is done into mappable soil units called soil type and soil phase, which includes characteristics such as erodibility, slope, and stoniness.

# 4

# Rock Mechanics

> If a builder builds a house for a man and does not make its construction firm, and the house which he has built collapses and causes the death of the owner of the house, that builder shall be put to death . . . if it destroys property, he shall restore whatever it destroyed, and because he did not make the house which he built firm and it collapsed, he shall rebuild the house which collapsed at his own expense.
>
> Hammurabi, of Babylon
> (2067–2025 B.C.)

## 4.1 Introduction

Mechanics of materials is a branch of engineering which deals with stresses and properties of materials. Over the past century, structural engineers have developed excellent techniques for determining stresses in steel and concrete, allowing for the construction of magnificent structures such as the steel Eiffel Tower (1889), the concrete Boulder Dam (1935), and the steel and concrete Sears Tower (1974). Man-made materials such as steel and concrete are homogeneous (same composition throughout) and isotropic (same directional properties throughout), and hence are easier to analyze than natural materials such as soil and rock. Many analytical methods used in rock mechanics utilize simplifying assumptions such as isotropic and homogeneous rock.

Due to man's need for knowledge of rock behavior for foundations, tunnels, mining, and utilization of underground space, there is a great need for future research in rock mechanics (National Research Council, 1981). Rock mechanics, a branch of mining engineering, is still in its infancy. In spite of the fact that rock has been used for thousands of years as a building material, it has only been in the past few decades that quantitative data or mathematical analyses (models) for rock mechanics have been attempted. Many of the techniques are borrowed from soil mechanics based on the simple law of Coulomb, which relates shear strength in elastic materials to the friction factor and the normal stress. Of course, any theory of rock mechanics that considers rocks to be no more than homogeneous mass, with no joints, faults, or directional properties of large masses in situ, is just a first approximation. It is important to know these first approximations, yet it is also important to recognize that rock mechanics applications in engineering geology require good geologic information, experience, and common sense.

In civil engineering techniques, uniform materials such as steel or concrete may be tested. Most rocks are anisotropic, however. It makes little sense, for example, to place much reliability on the laboratory testing of a 5 × 15 cm specimen of a rock, where field relationships show 1-m joint spacing. It is, nevertheless, necessary to understand the laboratory techniques as they form a background for engineering properties of models and formulas derived therefrom. There are a number of excellent textbooks available treating the subject of rock mechanics. These include Terzaghi and Peck (1948, 1967), Krynine and Judd (1957), Obert and Duvall (1967), Coates (1967), Farmer (1968), Zienkiewicz and Stagg (1968), Lambe and Whitman (1969), Johnson (1970), Jaeger (1972), Hoek and Bray (1974), Attewell and Farmer (1976), Lama and Vutukuri (1978), Goodman (1976, 1980), Huang (1983), and Brady and Brown (1985). For more in-depth treatment of rock mechanics, the reader may consult these references. In the past decade computer solutions to rock mechanics problems have been developed. Finite element analysis is a digital computer method for stress and other tensor field problems of large size. Finite elements are especially powerful for nonlinear rock mechanics problems having anisotropic or heterogeneous conditions.

## 4.2 Stress and Strain

The unit of force, called a newton (N), is defined as the force which gives an acceleration of 1 $m/s^2$ to a mass of 1 kg. Stress ($\sigma$) is the force per unit area that exists within a specified plane in a material. Actually stress cannot be measured, it can only be inferred. However, stress can be theoretically calculated if it is assumed that stress is uniformly distributed in a material. For instance, if a tensile force ($F$) of $2 \times 10^4$ N is applied over a steel rod having a circular cross-sectional area ($A$) equal to 4 $cm^2$, then the average force per unit area, or the average tensile stress is:

$$\sigma = \frac{F_t}{A},$$

$$= \frac{2 \times 10^4 \text{ N}}{0.0004 \text{ m}^2}, \tag{4.1}$$

$$= 5 \times 10^7 \text{ N/m}^2 .$$

Units of stress, $N/m^2$, are also known as Pascals (1 $N/m^2$ = 1 Pa).

Strain ($\epsilon$) is a measure of the deformation of a material when a load is applied. Since strain is a ratio of lengths, such as meters per meter, strain has dimensionless units. Consider a bar subject to axial tension having an original length $L_0$. As the axial load is applied, the original length increased to $L$, and the axial strain $\epsilon$ is:

$$\epsilon = \frac{L - L_0}{L_0}.$$

Within an elastic range, strain is proportional to stress (Hooke's Law). Thus:

$$E = \frac{\text{Stress}}{\text{Strain}} = \frac{\sigma}{\epsilon}, \tag{4.3}$$

where $E$ = the constant of proportionality, known as Young's modulus or the modulus of elasticity. The value of $E$ is determined experimentally from the slope

Table 4.1 Mechanical Properties of Rocks and Selected Materials

| Rock type | Locality | $\theta$ Density (g/cm$^3$) | $E$ Modulus of elasticity ($\times 10^9$ N/m$^2$) | Ultimate strength: $\sigma_c$ Compressive strength ($\times 10^6$ N/m$^2$) | Ultimate strength: $\sigma\epsilon$ Tensile strength ($\times 10^6$ N/m$^2$) |
|---|---|---|---|---|---|
| Amphibolite | California | 2.94 | 92.4 | 278 | 22.8 |
| Andesite | Nevada | 2.37 | 37.0 | 103 | 7.2 |
| Basalt | Michigan | 2.70 | 41 | 120 | 14.6 |
| Basalt | Colorado | 2.62 | 32.4 | 58 | 3.2 |
| Basalt | Nevada | 2.83 | 33.9 | 148 | 18.1 |
| Concrete | — | 2.7–3.2 | 2.1–1.0 | 0.41–0.21 | 0.04–0.02 |
| Conglomerate | Utah | 2.54 | 14.1 | 88 | 3.0 |
| Diabase | New York | 2.94 | 95.8 | 321 | 55.1 |
| Diorite | Arizona | 2.71 | 46.9 | 119 | 8.2 |
| Dolomite | Illinois | 2.58 | 51.0 | 90 | 3.0 |
| Gabbro | New York | 3.03 | 55.3 | 186 | 13.8 |
| Gneiss | Idaho | 2.79 | 53.6 | 162 | 6.9 |
| Gneiss | New Jersey | 2.71 | 55.16 | 223 | 15.5 |
| Granite | Georgia | 2.64 | 39.0 | 193 | 2.8 |
| Granite | Maryland | 2.65 | 25.4 | 251 | 20.7 |
| Granite | Colorado | 2.64 | 70.6 | 226 | 11.9 |
| Graywacke | Alaska | 2.77 | 68.4 | 221 | 5.5 |
| Gypsum | Canada | — | — | 22 | 2.4 |
| Limestone | Germany | 2.62 | 63.8 | 63.8 | 4.0 |
| Limestone | Indiana | 2.30 | 26.96 | 53.1 | 4.07 |
| Marble | New York | 2.72 | 54.0 | 126.9 | 11.7 |
| Marble | Tennessee | 2.70 | 48.3 | 106 | 6.5 |
| Phyllite | Michigan | 3.24 | 76.5 | 126 | 22.8 |
| Quartzite | Minnesota | 2.75 | 84.8 | 629 | 23.4 |
| Quartzite | Utah | 2.55 | 22.06 | 148 | 3.5 |
| Salt | Canada | 2.20 | 4.64 | 35.5 | 2.5 |
| Sandstone | Ohio | 2.17 | 10.52 | 38.9 | 5.17 |
| Sandstone | Utah | 2.20 | 21.37 | 107 | 11.0 |
| Schist | Colorado | 2.47 | 8.96 | 15.0 | — |
| Schist | Alaska | 2.89 | 39.3 | 129.6 | 5.5 |
| Shale | Utah | 2.81 | 58.19 | 215.8 | 17.2 |
| Shale | Pennsylvania | 2.72 | 31.2 | 101.4 | 1.38 |
| Siltstone | Pennsylvania | 2.76 | 30.6 | 113 | 2.76 |
| Slate | Michigan | 2.93 | 75.85 | 180 | 25.5 |
| Steel | — | 7.85 | 200 | 365 | 365 |
| Tuff | Nevada | 2.39 | 3.65 | 11.3 | 1.17 |
| Tuff | Japan | 1.91 | 76.0 | 36.0 | 4.31 |

From Lama and Vutukuri, 1978; Marin and Sauer, 1954; Am. Inst. Steel Constr. 1956.

of the straight line portion of the stress-strain curve. The units of $E$ are the same as stress, i.e., N/m$^2$. Typical values of $E$ are given in Table 4.1.

The effect of the tensile axial load in the steel rod described above is not only to increase the length of the section $L$ under consideration, but also to simultaneously decrease the lateral dimensions. The original transverse width, $B_0$, before loading will decrease to a new value $B$. Then the change in width per unit width is lateral strain $\epsilon_l$:

$$\epsilon_l = \frac{B - B_0}{B_0}. \tag{4.4}$$

The ratio of the numerical values of the lateral strain $\epsilon_l$ to the axial strain $\epsilon$ has been found experimentally to be a constant of an elastic material and is called Poisson's ratio $\nu$:

$$\nu = \frac{\epsilon_l}{\epsilon}. \tag{4.5}$$

Poisson's ratio can be visualized by considering the compression of a cylinder. Assume the original cylinder has 1-m height ($H$) and 1-m diameter ($D$), so that its volume is: $\pi/4\ \mathrm{m}^3$. If the cylinder is compressed to a new height, 0.99 m, its new diameter $D_2$ can be solved assuming no change in volume:

$$\frac{\pi D_1^2 H_1}{4} = \frac{\pi D_2^2 H_2}{4},$$

$$\frac{\pi(1\ \mathrm{m})^2(1\ \mathrm{m})}{4} = \frac{\pi\,(D_2)^2\,(0.99\ \mathrm{m})}{4},$$

$$D_2 = 1.005\ \mathrm{m}.$$

From Eq. (4.5),

$$\nu = \frac{\dfrac{(1.005 - 1)}{1}}{\dfrac{(1 - 0.99)}{1}} = 0.5.$$

When real rocks are subject to axial loading in the laboratory, some change in volume does occur because $\nu$ is less than 0.5. Experiments show that Poisson's ratio usually varies from 0.25 and 0.33 for rocks or metals, but may be 0.5 for rubber-like materials. Cork, oddly, has a Poisson's ratio of about zero.

Ultimate stress is defined as the maximum load resisted by a specimen divided by the cross-sectional area. Typical values are given in Table 4.1.

Figure 4.1 shows a sketch of an unconfined granite core being subjected to

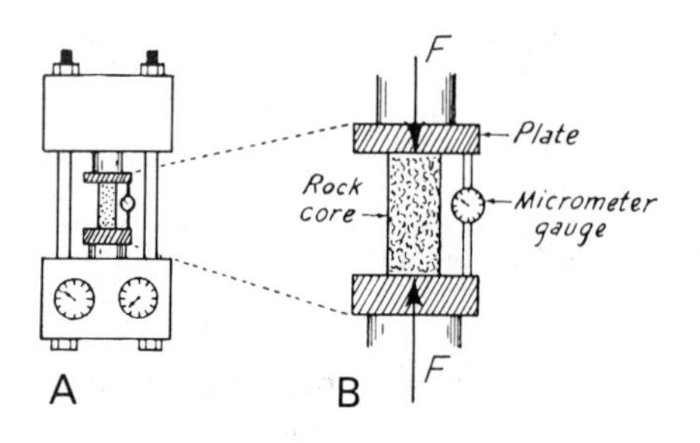


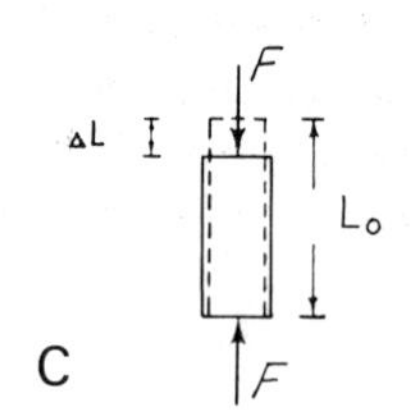


Figure 4.1 Compression of a rock core. **A.** Loading machine. **B.** Micrometer gauge used to measure displacements. **C.** Deformation of rock core (after Johnson, 1970).

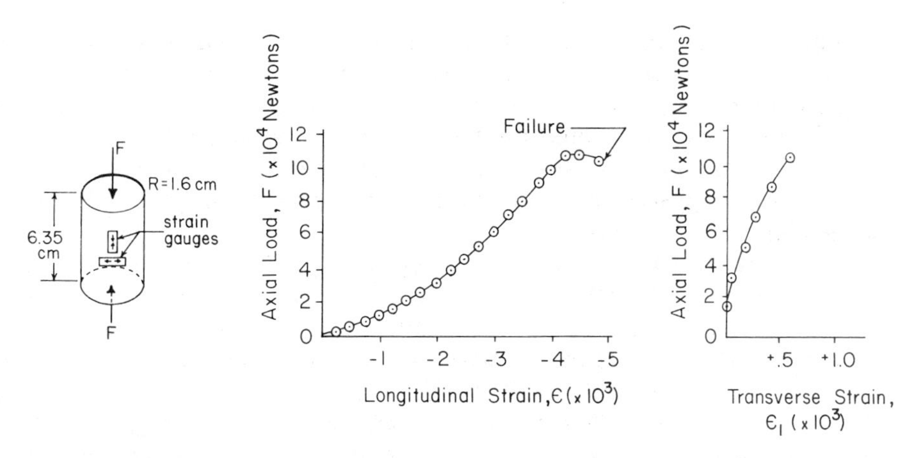


Figure 4.2 Relationship between axial load and longitudinal and transverse strain (after Johnson, 1970).

compression. The axial load versus longitudinal and transverse strain are plotted in Fig. 4.2. Note that the slope of the stress-strain curve is not exactly a straight line, indicating that the rock is not truly a linear elastic material. Nevertheless, in the range of 100,000 N, the value of Young's modulus can be calculated from Eq. (4.3):

$$E = \frac{\sigma}{\epsilon} = \frac{100{,}000 \text{ N}/\pi\,(0.016 \text{ m})^2}{4000 \times 10^{-6}},$$

$$= 3.11 \times 10^{10} \text{ N/m}^2.$$

The values of $E$ shown in Table 4.1 range over two orders of magnitude and are an indication of the range in strength, or competency of the rocks. In situ measurements or rock properties are desirable for practical rock mechanics, however. Research in this field is meager due to high costs of instrumentation. Jaeger (1972) and Pratt and Voegele (1984) describe several techniques, such as rosettes of strain gages, which are fixed to the wall of a tunnel or outcrop. A slot is then excavated, which relieves the stresses, and the strain is determined. A pressure-sensitive flat jack is then introduced into the slot, and the pressure observed for which the strain meter again reads the original condition.

Audible sound is an indication of the physical properties of a material. For example, an established method for detecting the presence of a crack in a ceramic vessel is by its ring. Rocks possessing great compressive strength, such as quartzite and diabase, typically are brittle and have little deformation under a load; they have high values of $E$, about $10^{11}$ N/m$^2$, and probably will ring if struck by a hammer. Rocks that respond with a dull thud to the blow of a hammer are weak and probably have low values of $E$, about $10^{10}$ N/m$^2$. Tillman (1982) describes an ultrasonic testing device that determines the strength of coal.

The rate of loading is a test variable that affects both the modulus of elasticity and the compressive strength. The effects of rate of loading on concrete cylinders is well known. For example, quickly loaded specimens have higher $E$ values and higher compressive strength than slowly loaded specimens. Some materials such as rock salt exhibit markedly different stress-strain curves due to varying rate of strain. The phenomena whereby rocks continue to deform after a specimen has been loaded is called creep. Compressive tests on rock specimens are normally carried out at a high rate of loading ($10^5$–$10^6$ N/m$^2$ per second), so that there are only limited published data on creep of rocks. The behavior of rock samples under slow loading rates seems to indicate that a fracture propagates along the most severely stressed crack or flaw in a specimen (see Jaeger, 1972, for further explanation).

Ideally, rock masses should be homogeneous and isotropic for their behavior to be predicted. However, even for a seemingly homogeneous and isotropic rock such as granite, field conditions are quite different than a single laboratory test specimen. Granite outcrops, for example, may contain joints or may be partially decomposed to grus. Even unweathered, unjointed granite possesses anisotropic properties; in New England granite quarries, rock cutters note crystal directional properties and refer to the hardest direction to saw granite as the hardway.

In 1981, the Panel on Rock-Mechanics Research Requirements of the U.S. National Research Council looked at rock mechanics research and gloomily concluded that " . . . there are no well-understood procedures by which laboratory data can be used to predict reliably the behavior of rock masses *in situ*." Let us consider the following application of Young's modulus to an in-situ rock mechanics problem (modified from Johnson, 1970).

> Granite is quarried in the Fletcher quarry near Westford, Massachusetts, by a combination of burning and wire-saw cutting. Vertical slots or channels are cut 10 to 15 m deep and about 10 m laterally into the sides of the quarry. The channels are bounded on each side by a burn cut, a sinuous slot about 15 cm wide, formed by moving a large torch up and down a side of the quarry, slowly forming the slot. The blocks of granite within the channels, bounded by the burn cuts, are removed to form the channel which is about 2.5 m wide.
>
> Suppose that three pins were placed in a 30 m wide sawblock of granite, before it was sawed, in the positions shown in Figure 4.3A and 4.3B. Two pins were cemented in holes drilled one meter apart at the western edge of the sawblock and one pin was centered in a small hole drilled near the eastern corner of the sawblock (Fig. 4.3C).

When the sawblock was wire-sawed along the northern boundary, the slab of granite containing the pins lengthened as shown in Fig. 4.3D. The strain in the $x$-direction is:

$$\epsilon = \frac{L - L_0}{L_0} = \frac{0.0093 \text{ m}}{30 \text{ m}} = 3.1 \times 10^{-4}.$$

If $E = 4 \times 10^{10}$ N/m$^2$ (typical value from Table 4.1), then:

$$\sigma = E \cdot \epsilon = (4 \times 10^{10} \text{ N/m}^2)(3.1 \times 10^{-4}),$$
$$= 1.24 \times 10^7 \text{ N/m}^2.$$

If lithostatic stresses (analogous to hydrostatic forces in water) were acting on the granite when it crystallized, and if it is assumed that the unloading of the sawed granite represents all of the strain release since the time of crystallization of the granite, it is possible to estimate the depth of emplacement of the granite as follows. Assume the density $= 2.6$ g/cm$^3$ $= 2.6 \times 10^3$ kg/m$^3$, so that each cubic meter block causes a force of $F = ma = (2.6 \times 10^3 \text{ kg})(9.8 \text{ m/s}^2) = 2.55 \times 10^4$

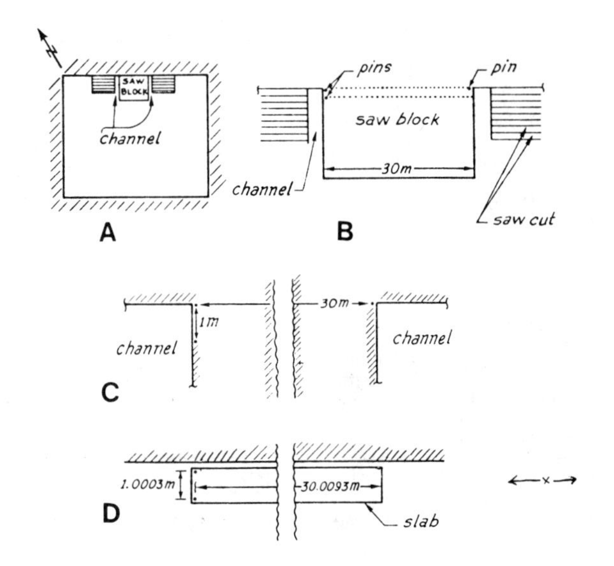


Figure 4.3 Plan views of a granite quarry showing deformation of a granite slab during quarrying (after Johnson, 1970). **A.** Sketch of quarry. **B.** Location of reference pins in saw block. **C.** Distance between pins prior to quarrying. **D.** Distance between pins after quarrying.

N on each $m^2$ of bottom surface. Thus the lithostatic stress per m depth $= 2.55 \times 10^4$ N/m$^2$/m. Therefore:

$$\text{depth of emplacement} = \frac{\text{stress release}}{\text{lithostatic stress per m depth}},$$

$$= \frac{1.24 \times 10^7 \text{ N/m}^2}{2.55 \times 10^4 \text{ N/m}^2\text{/m}},$$

$$= 486 \text{ m}.$$

Other factors need to be considered in order to realistically calculate depth of emplacement. These include 1) determination of strain released prior to quarrying, 2) determination of stress still in the quarried rock, 3) effect of tectonic stresses, and 4) effect of temperature-induced stresses.

## 4.3 Compressive Strength

Rock has high compressive strength. Typical values are given in Table 4.1. By using the average value of ultimate compressive strength of granite from Table 4.1, theoretically granite blocks could be stacked on top of each other to a great height.

$$\text{Height} = \frac{\text{ultimate compressive strength}}{\text{stress per m}},$$

$$= \frac{223 \times 10^6 \text{ N/m}^2}{2.55 \times 10^4 \text{ N/m}^2 \text{ per m}},$$

$$= 8700 \text{ m}.$$

There is a crude relationship between the hardness or consistency of soils and rock and their unconfined compressive strength. Based on a suite of simple crush-

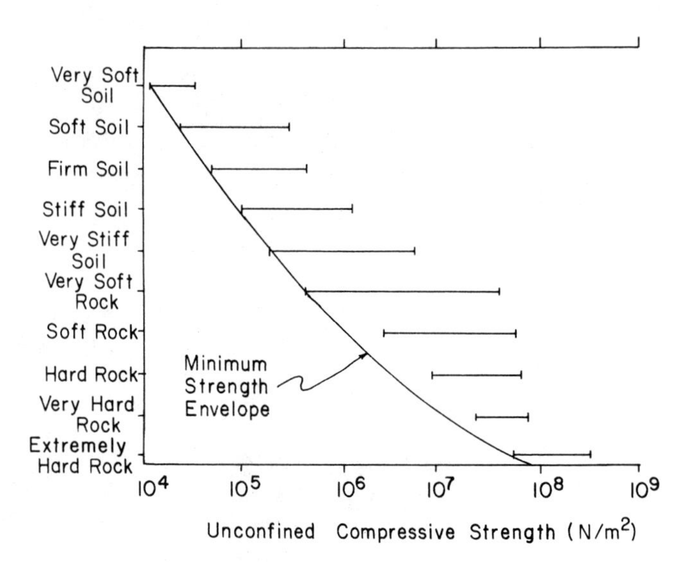


Figure 4.4 Relationship between consistency or hardness and the unconfined compressive strength of the material (after U.S. Department of Transportation, 1977).

ing strength tests, a lower strength envelope has been established (Fig. 4.4), and a classification scheme has been established. The hardness classification scheme is plotted as ordinate in Fig. 4.4.

There are many other soil and rock classification schemes, but Fig. 4.4 is as good as any, because it quantitatively evaluates, in an engineering sense, what distinguishes soil from rock.

## 4.4 Tensile Strength and Bending Moments

The tensile strength of rocks is important in terms of tunnels, archways in buildings, and the dynamic fracturing of rocks by explosive action. Table 4.1 lists typical tensile strengths. Uniaxial tension tests of sound granite specimens indicate that peak tensile strengths are about $7\text{–}18 \times 10^6$ N/m$^2$ (Segall and Pollard, 1983). Farmer (1968) reports that as a general rule, for brittle rocks, the tensile strength of rocks is between 0.4 and 0.1 that of the compressive strength. Theory of tensile fracture of rock originally developed for glass by Griffith, assumes that a fracture is initiated in a brittle material by tensile failure around microcracks or flaws present in the rock.

While a small rock specimen tested in the laboratory may possess tensile strength, an outcrop of the same rock in the field may have no tensile strength. In the field, almost all rock masses have joints, bedding planes, or cracks along which tensile forces could cause separation. Therefore, large masses of rock essentially cannot be relied on to possess any tensile strength.

The uniaxial tension test, similar to those used to pull steel bars apart, has been used for rock specimens, but with generally unsatisfactory results owing to the concept of the weak link. A beam test (described below) may also be used, but again the weak link concept suggests that as the length of the beam increases, more erratic results will be obtained.

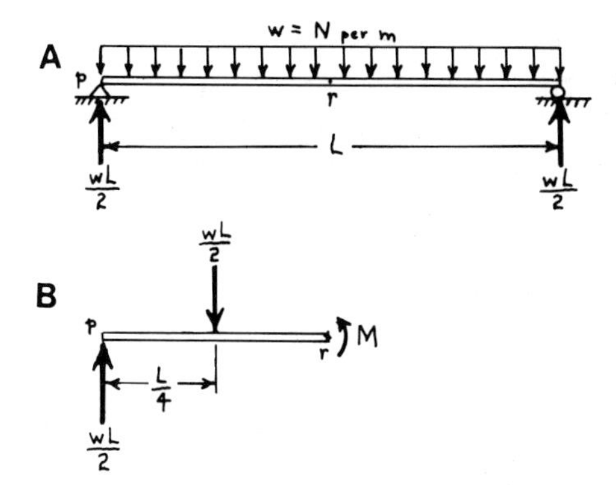


Figure 4.5 Bending stresses in a beam (after Marin and Sauer, 1954). **A**. Load diagram. **B**. Free-body diagram of the left half of the beam.

Other indirect tests have been devised, such as the Brazilian test. In this case, a cylinder of rock is oriented on its side, and a vertical load applied. The cylinder is crushed between horizontal plates. The result of the load produces a complicated distribution of stress. The formula for calculating the average tensile stress ($S_t$) in the vertical plane along the core axis is:

$$S_t = \frac{2P}{\pi DL}, \tag{4.6}$$

where $P$ is the external load, $D$ the diameter of the sample, and $L$ its length (Coates, 1967).

Perhaps the most severe impediment to the use of stone as a building material is its inability to serve as a beam due to the fact that bending imposes a tensile stress in the beam. From consideration of basic statics, let us analyze the stresses in a simply supported beam. A beam of rectangular cross section having length $L$ shown in Fig. 4.5A. Let the only load be the weight of the beam itself, or $w$ per unit length. The total load is $w \cdot L$. A free-body diagram can be constructed to show that a moment exists within the beam, and is a maximum at the center (point $r$). A free-body diagram can be drawn for half of the beam. The uniform load can be replaced by an equivalent load $wL/2$ located at the midpoint of this diagram (Fig. 4.5B). Because the sum of the moments about p is 0,

$$\mathrm{M} = \text{(force)(arm)}, \tag{4.7}$$

$$\mathrm{M} = \frac{wL}{2} \cdot \frac{L}{4} = \frac{wL^2}{8},$$

where M is the moment at the center of the beam.

The stress ($\sigma$) due to a bending moment ($M$) is related to the moment of inertia ($I$) and the distance ($c$) from the neutral axis to the point in question (Marin and Sauer, 1954):

$$\frac{M}{\sigma} = \frac{I}{c}, \tag{4.8}$$

where the moment of inertia ($I$) for a rectangular cross section of width $b$ and thickness $d$ is:

$$I = \frac{bd^3}{12}.$$

Consider a 2-m-long simply supported granite beam of density ($\rho$) = 2.6 g/cm$^3$ = 2600 kg/m$^3$, width ($b$) = 0.3 m, and thickness ($d$) = 0.1 m. Assume the rock has no joints and is able to sustain itself as a beam. From Eq. (4.8), the maximum tensile stress in this beam would be:

$$\sigma_t = \frac{Mc}{I} = \frac{\frac{wL^2}{8} \cdot \frac{d}{2}}{\frac{bd^3}{12}} = \frac{3wL^2}{4\,bd^2}.$$

In earth's gravity, the force exerted by 1 m$^3$ of granite would be 2.55 $\times$ 10$^4$ N. Because this beam has a volume of (0.3 m)(0.1)(1 m) = 0.03 m$^3$ per meter of length, then the unit weight is:

$$\begin{aligned} w &= (2.55 \times 10^4\ \text{N/m}^3)\ (0.03\ \text{m}^3\text{/m}), \\ &= 764\ \text{N/m}. \end{aligned}$$

Thus:

$$\begin{aligned} \sigma_t &= \frac{3wL^2}{4bd^2}, \\ &= \frac{3(764\ \text{N/m})(2\ \text{m})^2}{4(0.3\ \text{m})(0.1\ \text{m})^2}, \\ &= 7.64 \times 10^5\ \text{N/m}^2. \end{aligned}$$

This value of tensile stress is lower than published ultimate tensile strengths for selected granite specimens (Table 4.1), and indicates that solid granite beams even longer than 2 m could possibly be supported without failure. In reality, however, it is unlikely to find large pieces of completely homogeneous or isotropic rocks; in fact, the presence of a single joint will render the rock useless as a tensile member.

The above analysis shows that the highest tensile stress in a simply supported beam occurs in the center in the lower side. Simply supported beams were used for construction by the ancient Greeks. For example, the marble beams across the top of the magnificent columns of the 2500-year-old Parthenon (Fig. 4.6A) have approximately the same lengths as used in the example above. To increase the tensile strength of those beams, the Greeks installed iron brackets in the lower part. The clamps were lead-coated to resist rust. Modern reinforced concrete construction uses the same principle except that steel reinforcing rods are imbedded into the lower part of concrete beams to give adequate tensile strength (Dunham, 1953).

The general lack of tensile strength in rocks and the resulting inability to serve as beams was ingeniously circumvented by the Romans. The Romans spanned canyons and constructed large coliseums by use of the arch, whereby all blocks of rock are in compression (Fig. 4.6B). The magnificent Renaissance churches of Europe testify to the ingenuity of engineers using stone arches and domes (Mark, 1978).

Today, steel, possessing extraordinary tensile and compressive strength $\simeq$ 370 $\times$ 10$^{10}$ N/m$^2$), is used for cables and beams in bridges and buildings. When solid steel I-beams became available for construction in the last century, the modern

Figure 4.6 Examples of rock construction used by the ancients. **A**. The Parthenon, Athens, Greece. Built in 438 B.C., the marble beam across the top of the columns spans a length of about 2 m. **B**. A Roman arch provides compressive stress within all the rocks.

American city took shape. Prior to the turn of the century, large buildings in New York and Chicago, for example, were limited to about 16 stories; if taller walls were constructed, crushing forces due to static loads or wind loads could be produced in bricks or weak rocks in lower stories (Schneider and Dickey, 1980). Once steel came into its own, the skyscraper was born.

## 4.5 Shear Strength

Landslides, mine cave-ins, and rock failures of many kinds are commonly the result of sliding or failure due to shear stress. Except in the vertical walls of a building or other man-made configurations, it is unlikely that rock stresses are strictly axial. Indeed, even axial loads produce shear stresses along planes oriented at an angle to the axial load, and failure may occur on these planes. True compressive failure in a rock can only occur through internal collapse of the rock structure due to compression of pore space, thus resulting in grain fracture and movement along grain or crystal boundaries. The compressive strength of a rock specimen is, however, typically a reflection of its shear strength.

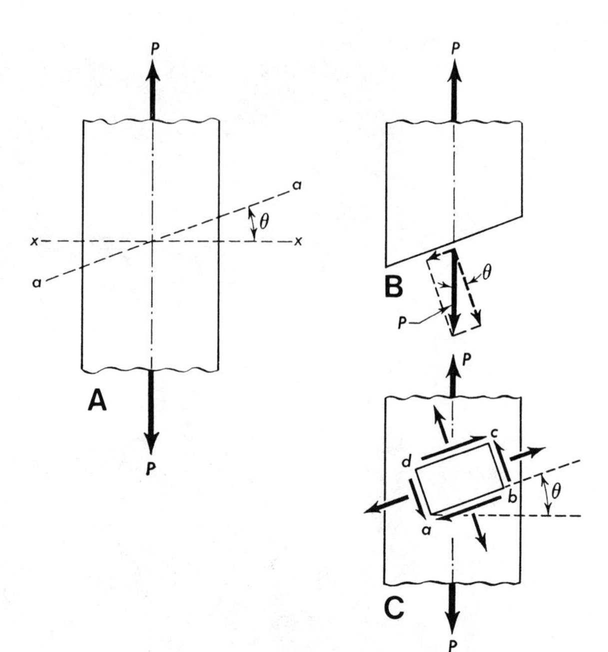


Figure 4.7 Internal forces in a member subject to tension (from Marin and Sauer, 1954). See text for explanation.

The shear strength of massive rocks tested in the laboratory are roughly equal to 20% of the value of the compressive strength. The shearing strength of a rock along a discontinuity is affected by a number of factors, including the infilling material (fault gouge, etc.) and "asperities" (roughness) of the discontinuity.

Figure 4.7 shows a structural member subject to an axial tensile force. A plane $a$–$a'$ is passed through the body, making an angle $\theta$ to the $x$–$x'$ axis. A free body diagram of the member above $a$–$a'$ is shown in Fig. 4.7B. The force $F$ can be divided into two components, ($F \sin \theta$) is parallel and ($F \cos \theta$) perpendicular to the plane $a$–$a'$. The component $F \sin \theta$ is the shearing force. The component $F \cos \theta$ is the normal force on the plane $a$–$a'$. If $A$ is the area of the member cut by the plane $a$–$a'$, and $A_0$ is the area cut by the transverse plane $x$–$x'$, then:

$$A_0 = A \cos \theta. \tag{4.9}$$

The shearing force per unit area (i.e., the shear stress) $\tau$ at any point on the inclined surface is equal to:

$$\tau = \frac{F \sin \theta}{\mathrm{A}_0/\cos \theta} = \frac{F}{A_0} \sin \theta \cos \theta = \frac{F}{2A_0} \sin 2\theta. \tag{4.10}$$

From this equation it can be seen that the shear stress is zero when $a$–$a'$ is inclined normal to the axis and approaches zero again as $\theta$ approaches 90°. The maximum value of the shear stress ($\sigma_s$ max) occurs at $\theta = 45°$, because $\sin 2\theta = \sin 2(45°) = \sin 90° =$ unity. Therefore,

$$\tau_{\max} = \frac{F}{2A_0}. \tag{4.11}$$

In summary, shear stresses at a point in a stressed body depend upon the particular plane passing through the point being considered. Thus, even for a body subject only to axial load, shear stresses are present on all planes except those given by $\theta = 0°$ and $\theta = 90°$. Figure 4.7C shows the resolution of forces on a small unit of rock along *a–a′*.

As expected from the above theory, isotropic and homogeneous rock specimens subject to compressive loads have been found experimentally to fail along planes roughly 45° to the axial load. Values of shear stress are generally in the range of tensile strengths for rocks. The likelihood of failure along a plane of weakness in anisotropic rocks will be affected by the relative inclination of the plane with respect to the direction of the major principal stress. Donath (1968) reports on the failure of Martinsburg Shale specimens having a prominent planar anisotropy (slatey cleaveage) oriented at varying confining stress and angles to the axial load. He found that the weakest rocks were ones having cleavage oriented 30° to the axial load.

The majority of joints found naturally in stratified rocks are believed to be the result of compression (Lahee, 1952). Compression along a single axis may produce two, four, or more joint sets, and there is a tendency for the fractures in intersecting sets to be perpendicular to one another and for all the sets to be approximately 45° to the direction of compression. Two joint sets thus mutually related are called conjugate sets. Close inspection may show small shearing displacements along these joint surfaces.

## 4.6 Triaxial Tests and Mohr Diagrams

A large axial load which causes a major principal stress may be supplemented by two mutually perpendicular stresses called the intermediate principal stress and minor principal stress. The axial loads described in the preceding sections are unconfined, meaning that during the laboratory testing no lateral support is given to the specimens, hence the intermediate and minor stresses are zero. If, however, specimens are jacketed to simulate confining loads, different stress-strain curves result with higher E values and higher ultimate strengths (Fig. 4.8). Most specimens tested in these triaxial compression tests fail along one or two oblique shear fractures. At higher confining pressures, specimens do not fail by fracturing, but exhibit ductile flow as the cylindrical test specimens deform to a barrel-shape (Fyfe et al., 1978).

Unconfined compressive strength tests on small rock cylinders (1-cm height

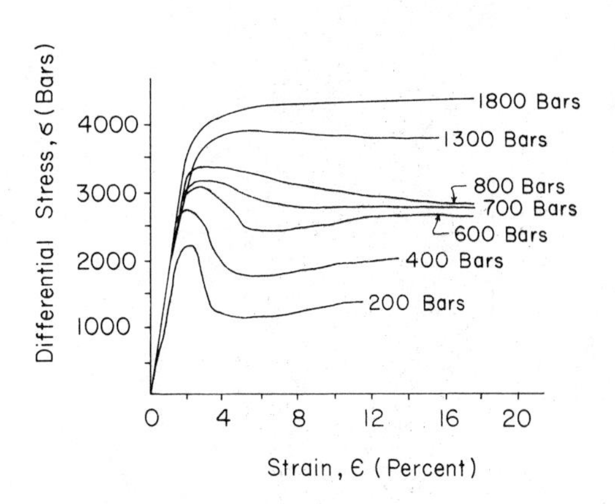


Figure 4.8 Stress-strain curves for the Crown Point Limestone showing varying confining pressures (after Donath, 1968). (Note: 1 bar = 1 atmosphere = 14.7 psi $\simeq 10^5$ N/m$^2$.)

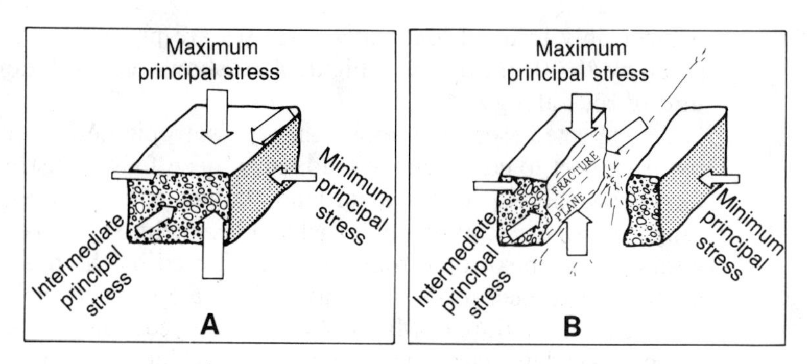


Figure 4.9 Rock broken by a differential stress system (from Miller and Miller, 1977). **A**. Differential compressive stress resolved into principal stress axes. **B**. Fracture plane develops parallel to the maximum and intermediate stress axes.

and 1-cm diameter) usually fail suddenly by shearing stresses acting along failure planes oriented 45° with the axial load, as explained above. Occasionally, a test specimen fissures or cracks vertically along the load axis. These resemble tension failures. In some older stone buildings, tensional failures of this sort may be observed in stone slab pedestals under heavily loaded stone marble columns (Krynine and Judd, 1957).

When rocks under differential compressive stress fail, it has been noted that the planes of fractures often propagate parallel to the axes of the maximum and intermediate principal stress (Fig. 4.9). These planes of fracture will, therefore, be normal to the axis of minimum principal stress.

The triaxial compression test (Fig. 4.10) consists of a heavy steel cylinder filled with a fluid (e.g., kerosene) which is subject to variable pressure in an attempt to duplicate the actual confining pressure of natural rock or soil. In a triaxial test, a cylindrical specimen is subjected to an equal all-around pressure, and, in addi-

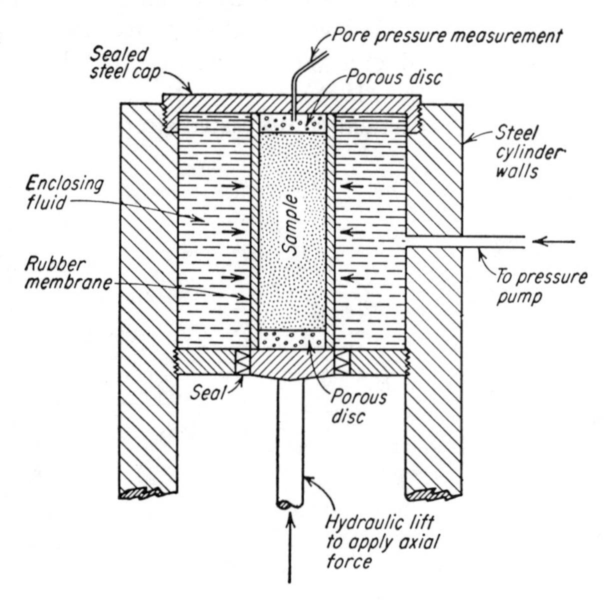


Figure 4.10 Diagramatic sketch of triaxial test on a rock sample (after Krynine and Judd, 1957).

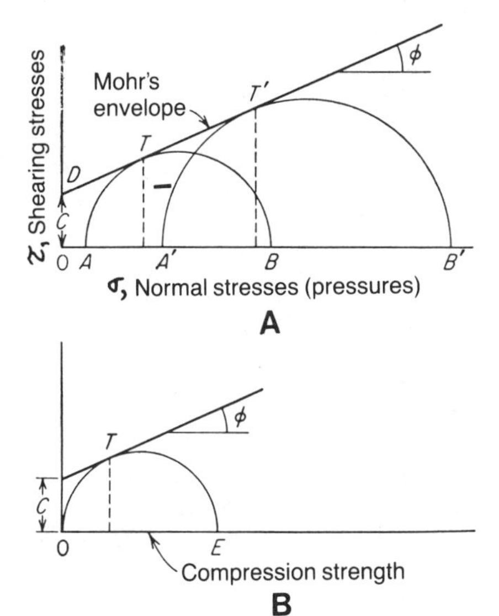


Figure 4.11 Mohr's diagram for two triaxial tests (**A**), and for a triaxial test without restraint or for an unconfined compression test (**B**). (From Krynine and Judd, 1957.)

tion, to an axial pressure that may be varied independently. Triaxial tests can be plotted graphically as Mohr circles or more commonly, Mohr half-circles, which are called Mohr diagrams (Fig. 4.11). The procedure was developed by Otto Mohr in 1882. Normal (confining) stress is plotted as the abscissa. Stresses are expressed as unit stresses, $N/m^2$. A cylindrical specimen is subjected to a known confining pressure that equally stresses all surfaces of the specimen. Then the axial stress is increased until the specimen fails. Two tests are run where the specimen is subjected to two confining pressures (the minor principal stresses, distance *OA* and *OA′* on Fig. 4.11A). The distances *OB* and *OB′* are the axial stresses (the major principal stresses) required to break the rock in these two tests, respectively. Circles (or, more conventionally, semicircles) are drawn using *AB* and *A′B′* as diameters. A common tangent to both circles can be drawn that will intersect the ordinate (shear stress) at some point *D*. The distance *OD* is the "unit cohesion" (*c*), and $\phi$ is the angle of internal friction. Values of $\phi$ typically range from about 30° to 40° and *c* from 1000 to 6000 $N/m^2$ for firm rock (Farmer, 1968; Hoek and Bray, 1974). (If more than two tests are performed on rocks, the Mohr envelope (area below tangent line) may be bounded above by a gradually flattening curve for rocks rather than a straight line as is usually the case for soils.) From trigonometry it can be shown that if perpendiculars from the points of tangency *T* and *T′* of Mohr envelope (dashed vertical lines in Fig. 4.11A) are dropped to the abscissa, these ordinate lengths are values of the shearing stress in the failure plane in the corresponding tests. The shearing stresses are read on the ordinate scale.

Triaxial tests are often performed on soil specimens as well as rock specimens (Peck et al., 1974; Holtz and Kovacs, 1981). Consider a sand where $c = 0$ and $\phi = 34°$. Suppose the confining pressure is $1.7 \times 10^5$ $N/m^2$, and it is desired to find the compressive stress needed to make this specimen fail. The answer can be determined graphically using Mohr's circle as follows. On arithmetic paper, a line is drawn through the origin at an angle of 34°. The value of $1.7 \times 10^5$ $N/m^2$ is located on the abscissa. A circle is drawn, passing through $1.7 \times 10^5$ $N/m^2$ and

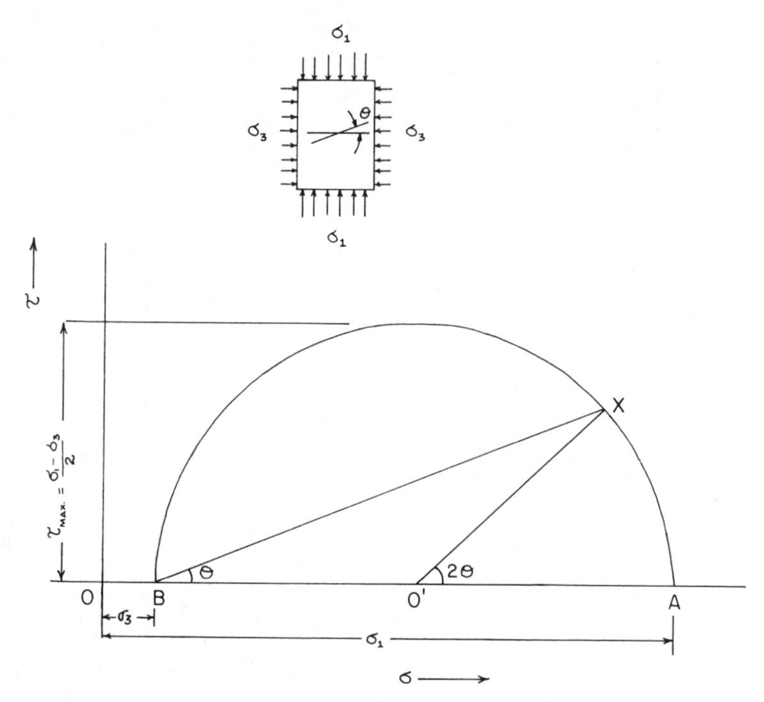


Figure 4.12 Mohr diagram for the state of stress at a point.

tangent to the 34° line. This circle intersects the abscissa at $6.8 \times 10^5$ N/m$^2$. Therefore, the specimen should fail at $6.8 \times 10^5$ less $1.7 \times 10^5 = 5.1 \times 10^5$ N/m$^2$ in excess of the confining pressure.

A benefit of the Mohr envelope is that it diagrammatically allows for determination of the stability of material under varying loads and confining conditions. If the envelope curve is known, then an axial and confining stress can be plotted, a circle drawn, and, if any part of the circle lies outside the envelope, failure should occur (Lambe and Whitman, 1969).

Another advantage of the Mohr circle technique is that it allows for solution of the stresses on all planes within a specimen. Consider Fig. 4.12 where a plane is oriented at angle $\theta$ to the direction of the minimum principal stress. The shearing stress on the planes having maximum axial stress is zero. This would be the horizontal plane and all vertical planes through the specimen. Thus the points $A$ and $B$ in Figure 4.12 lie on the horizontal axis, and have values of $OB$ minimum principal stress $\sigma_3$ (the confining stress) and $OA$ maximum principal stress $\sigma_1$ (the compressive stress). A Mohr circle is drawn. Note the maximum shear stress is equal to one-half the difference between the maximum and minimum principal stresses, as shown. The center of the circle is $O'$. An angle $2\theta$ is drawn from $O'$, intersecting the circle at point $X$. (Or $X$ can be derived from an angle $\theta$ drawn from point $B$.) The horizontal and vertical coordinates of point x are equal to the normal stress ($\sigma$) and the shearing stress ($\tau$). Thus it is possible to determine the stresses anywhere within a specimen. For example, if $\sigma_1 = 4 \times 10^7$ N/m$^2$ and $\sigma_3 = 2 \times 10^7$ N/m$^2$, a Mohr's circle solution for the normal and shear stresses on a

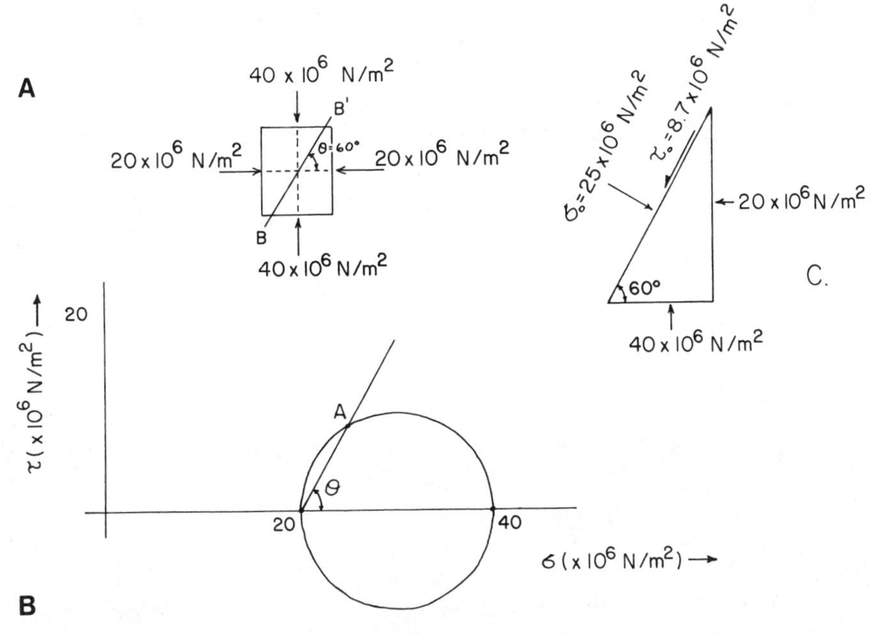


Figure 4.13 Example of Mohr solution (modified from Lambe and Whitman, 1969). **A.** Stress system on rock specimen. **B.** Mohr circle. **C.** Resulting schematic reorientation of stress.

plane oriented at $\theta = 60°$ will be $\sigma = 2.5 \times 10^7$ N/m² and $\tau = 0.87 \times 10^7$ N/m² (Fig. 4.13).

Fluid, if present and trapped in the interstices of a sample being tested, will change the stress distribution on a rock sample, and hence the Mohr diagram. Part of the axial stress as well as part of the confining stress is carried by the fluid. Thus the stress applied to the solid skeleton of the sample (effective stress) is the difference between the total and fluid stress. The following example (modified from Coates, 1967) shows the construction of two Mohr diagrams: 1) with, and 2) without consideration of effective stress. See Fig. 4.14.

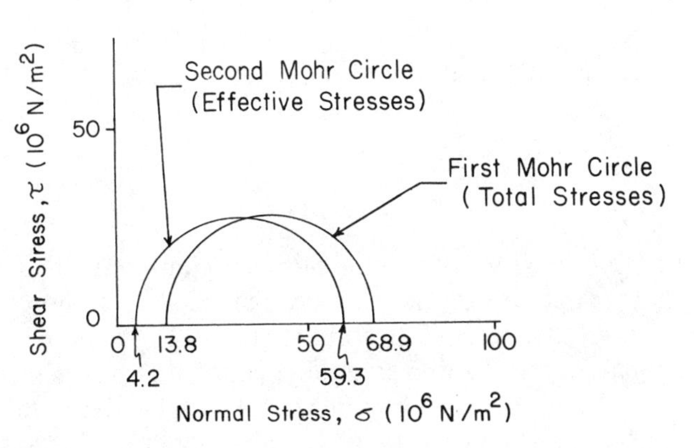


Figure 4.14 Mohr diagram plot (after Coates, 1967).

In a triaxial compression test the confining pressure, or the fluid pressure outside the membrane enclosing the sample, is $13.8 \times 10^6$ N/m$^2$. The total axial stress at failure is $68.9 \times 10^6$ N/m$^2$. The sample contains water in its interstices, and the pressure in this water at failure is measured as $9.6 \times 10^6$ N/m$^2$. Plot the Mohr Circles for the stress conditions at failure in terms of total stresses and effective stresses. The total stresses at failure are:

$$\sigma_3 = 13.8 \times 10^6 \text{ N/m}^2,$$
$$\sigma_2 = 68.9 \times 10^6 \text{ N/m}^2.$$

The first Mohr Circle is plotted as shown in Fig. 4.14. The effective stresses at failure are:

$$\sigma_{3'} = 13.8 \times 10^6 - 9.6 \times 10^6 = 4.2 \times 10^6 \text{ N/m}^2,$$
$$\sigma_{1'} = 68.9 \times 10^6 - 9.6 \times 10^6 = 59.3 \times 10^6 \text{ N/m}^2.$$

The second Mohr Circle is plotted as shown in Figure 4.14. Because the second Mohr's circle lies at different coordinates than the first Mohr's circle, it can be seen that $\tau$ and $\sigma$ on any plane in the rock is properly described by consideration of the *effective* stress on the rock skeleton itself if the rock has interstices filled with fluid.

To summarize, there exists at any stressed point three orthogonal planes, called the principal planes, on which there are zero shear stresses. The normal stresses that act on these planes are called the principal stresses. The largest is the major principal stress $\sigma_1$, the smallest the minor principal stress $\sigma_3$ and the third the intermediate principal stress $\sigma_2$. Considering two-dimensional analysis (which is acceptable for most engineering geology analyses), we are interested in the plane which contains the major and minor principal stress, $\sigma_1$ and $\sigma_3$. The Mohr circle can be used to find the stresses in any plane in any direction oriented $\theta$ to the direction of minor principal stress by graphical construction. Figure 4.13 shows the Mohr diagram for the state of stress at a point. Note that $\theta$ is drawn from $\sigma_3$. As derived in Eq. (4.10), the maximum shear stress exists at $\theta = 45°$. This is confirmed by inspection from Fig. 4.13. The normal stress ($\sigma_0$) and the shear stress ($\tau_0$) at any plane is defined as point *A*, whose solution is:

$$\sigma_0 = \sigma_1 \cos^2\theta + \sigma_3 \sin^2\theta = \frac{\sigma_1 + \sigma_3}{2} + \frac{\sigma_1 - \sigma_3}{2}\cos 2\theta, \tag{4.12}$$

$$\tau_0 = (\sigma_1 - \sigma_3)\sin\theta\cos\theta = \frac{\sigma_1 - \sigma_3}{2}\sin 2\theta. \tag{4.13}$$

These values can be read directly from the Mohr circle.

## 4.7 Coulomb's Law and Frictional Forces

### *4.7.1 Theory*

Figure 4.15A is a free-body diagram of a cubic block of granite resting on a horizontal table, where the weight ($W$) of the granite is balanced by an upward force ($R$) of the table. If a horizontal force $F$ is applied (Fig. 4.15B), movement will begin only if the frictional forces $F_f$ are overcome. Assume the granite has a mass of 1 kg. Then its weight ($W$) on earth = (1 kg)(9.81 m/s$^2$) = 9.81 N. In this example, the weight of 9.81 N supplies the normal force ($R$), which holds the surfaces

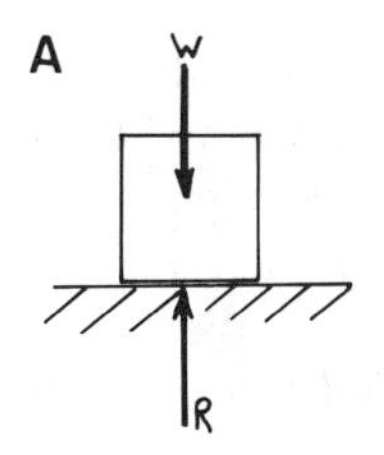


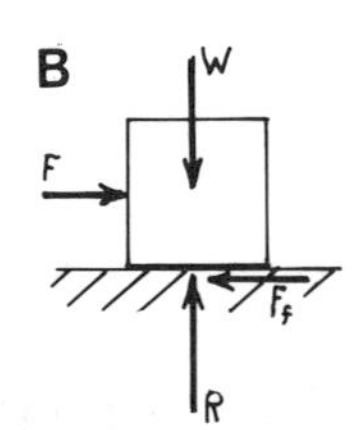


Figure 4.15 Frictional forces on a block on a flat table. **A.** At rest. **B.** With horizontal force applied.

together. Assume the horizontal force ($F_f$) necessary to cause movement of the block is 2.94 N. The ratio of the friction force to normal force is the coefficient of static friction ($u_s$):

$$u_s = \frac{F_f}{R}, \qquad (4.14)$$

$$= \frac{2.94 \text{ N/m}^2}{9.81 \text{ N/m}^2} = 0.30.$$

The force required to initiate sliding is usually slightly larger than the force necessary to maintain sliding at a constant speed. Thus the coefficient of static friction ($u_s$) is more than the coefficient of kinematic friction ($u_k$). Engineering geologists generally only concern themselves with $u_s$ values; however, the fact that $u_s > u_k$ suggests, for example, that once a landslide begins moving it can be kept in motion relatively easily.

From Eq. (4.14), the resistance to sliding is directly proportional to the frictional nature of the two surfaces and the normal force holding the surfaces together. Table 4.2 shows experimental values of the coefficient of friction between some common materials.

If the weight increases in the above model, the normal force $R$ increases pro-

Table 4.2 Approximate Values of Coefficient of Static Friction ($\mu_s$) for Dry Surfaces

| Materials | $\mu_s$ |
|---|---|
| Metal on metal | 0.15–0.60 |
| Metal on wood | 0.20–0.60 |
| Metal on stone | 0.30–0.70 |
| Metal on leather | 0.30–0.60 |
| Wood on wood | 0.25–0.50 |
| Wood on leather | 0.25–0.50 |
| Stone on stone | 0.40–0.70 |
| Earth on earth | 0.20–1.00 |
| Rubber on concrete | 0.60–0.90 |

From Beer and Johnston, 1977.

portionally. For example, if another 1-kg cube of granite in Fig. 4.15 is stacked on top of the first block, then the force required for movement will be 5.886 N, etc. The data can be plotted as stress (load/unit area). A straight line will be formed by a plot of normal stress as the abscissa versus frictional (shearing) stress necessary to overcome friction as the ordinate. This angle is called the angle of static friction. Thus the tangent of the friction angle is equal to the coefficient of friction:

$$u_s = \tan \phi = \frac{\tau}{\sigma},$$

or,

$$\tau = \sigma \tan \phi. \tag{4.15}$$

The relationship between shear and normal stresses was proposed by the French engineer Coulomb, and is sometimes called Coulomb's law. The term cohesion ($c$) is the shear stress ($\tau$) at zero normal stress ($\tau = 0$). In the example of the granite block resting on a table, $c = 0$ because the granite does not adhere to the table.

If cohesion is $>0$, the Coulomb equation takes the form of:

$$\tau = c + \sigma \tan \phi. \tag{4.16}$$

It should be noted that this equation is really the same equation as defined by

Table 4.3 Some Approximate Friction Angles and Cohesion Values for Rocks and Joint Infilling Materials

| Rock | $\phi°$ (intact rock) | $\phi°$ (discontinuity) | $\phi°$ (ultimate) | c (massive rock) kN $m^{-2}$ |
|---|---|---|---|---|
| Andesite | 45 | 31–35 | 28–30 | |
| Basalt | 48–50 | 47 | | |
| Chalk | | 35–41 | | |
| Diorite | 53–55 | | | |
| Granite | 50–64 | | 31–33 | 100–300 |
| Greywacke | 45–50 | | | |
| Limestone | 30–60 | | 33–37 | 50–150 |
| Monzonite | 48–65 | | 28–32 | |
| Porphyry | | 40 | 30–34 | 100–300 |
| Quartzite | 64 | 44 | 26–34 | |
| Sandstone | 45–50 | 27–38 | 25–34 | 50–150 |
| Schist | 26–70 | | | |
| Shale | 45–64 | 37 | 27–32 | 25–100 |
| Siltstone | 50 | 43 | | |
| Slate | 45–60 | | 25–34 | |

| Infilling material | $\phi°$ (approximate) |
|---|---|
| Remolded clay gouge | 10–20 |
| Calcitic shear zone material | 20–27 |
| Shale fault material | 14–22 |
| Hard rock Breccia | 22–30 |
| Compacted hard rock aggregate | 40 |
| Hard rock fill | 38 |

From Hoek and Bray 1974; U.S. Department of Transportation, 1977.

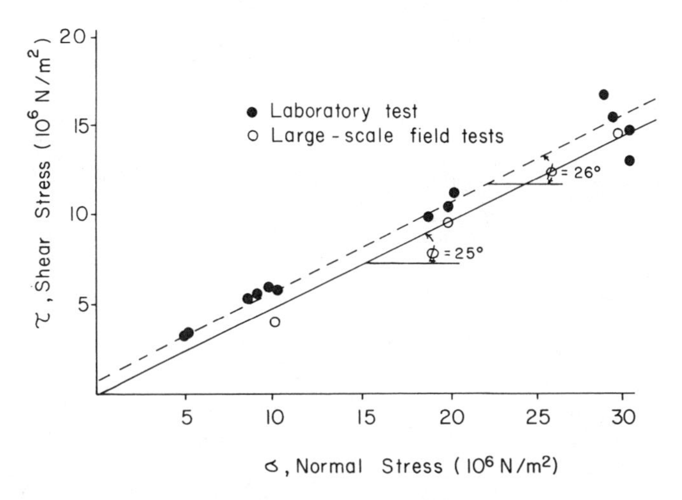


Figure 4.16 Comparison between the friction angle ($\phi$) measured in small-scale laboratory tests and in large-scale tests on joints in limestone (after Hoek and Bray, 1974).

line *DTT'* in the Mohr circle solution (Fig. 4.11A). The relationship of these variables is sometimes referred to as the Mohr-Coulomb theory.

In many engineering geologic problems, such as landslides (Chapter 6) and earthquakes (Chapter 11), *fluid* pressure is an important consideration because it affects the normal stress. Hubbert and Rubey's (1959) paper ("Role of fluid pressure in mechanics of overthrust faulting") documents the fact that hydrostatic uplift pressure ($\mu$) of entrapped fluids counteracts the normal weight. As a result, the force necessary to cause shear displacement with high fluid pressure is not as great as the force necessary in the absence of pore pressure. The resulting "effective stress" is ($\sigma - \mu$):

$$\tau = c + (\sigma - \mu) \tan \phi. \tag{4.17}$$

Table 4.3 and Fig. 4.16 show typical values of cohesion and friction angle determined for large slope masses. The values were determined from analysis of theoretical equations involving $c$ and $\phi$ mobilized at failure for massive slope failures.

Figure 4.16 shows the results of shear tests on joints in limestone which were tested in the laboratory and in situ. The lab samples were 8 cm × 30 cm in size and the in situ tests involved a joint surface 2 m × 2 m. The in situ test utilized the application of normal and shear loads with the aid of flat jacks packed between the rock and the walls of an underground chamber. The data shows residual shear stress, i.e., the value after large initial strains had essentially reduced the shear stress to a constant value. From Fig. 4.16, it can be seen that the friction angle $\phi$ is essentially the same for lab or in situ tests. Cohesion is nearly zero.

### *4.7.2 Examples*

*Example 1.* Stability analysis of rock masses in doubtful equilibrium follows the simple rules of statics (Jaeger, 1972). Consider an idealized sketch of a poten-

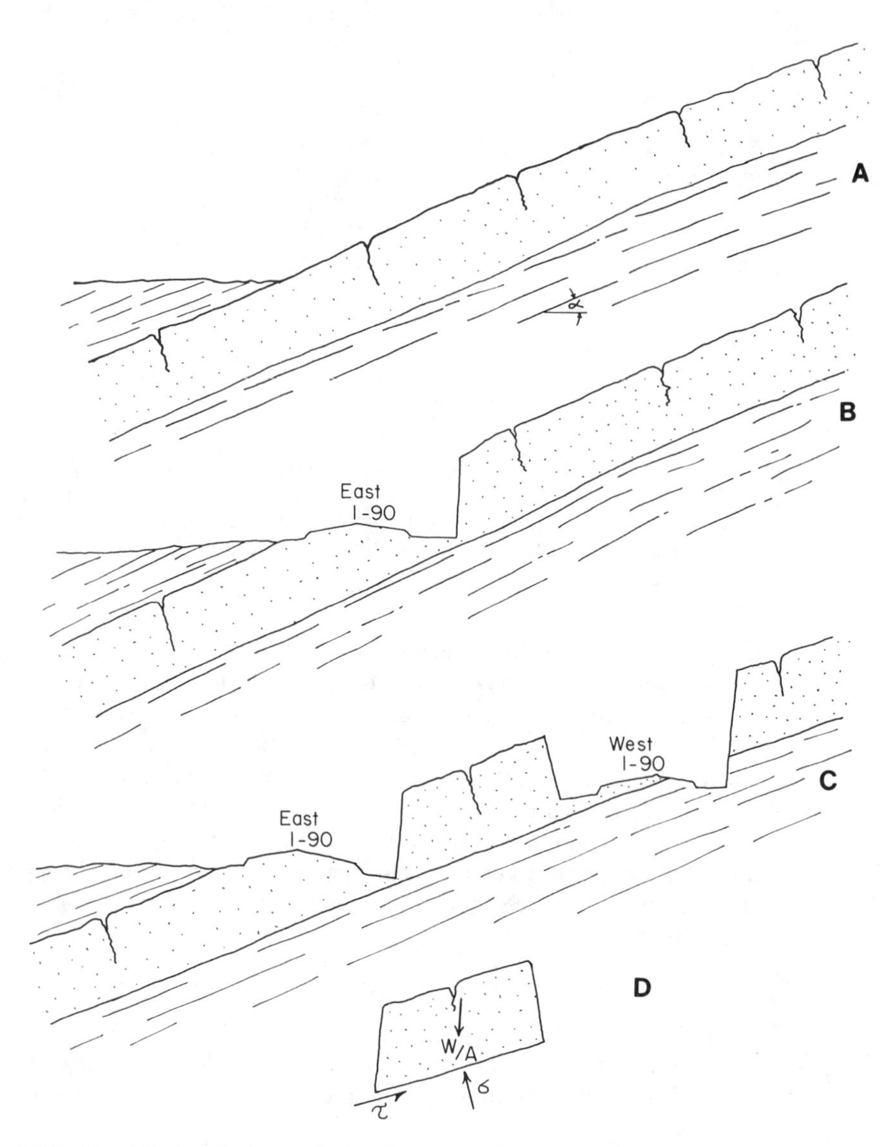


Figure 4.17 Simplified stability analysis of a gently dipping sandstone bed. **A**. Natural cross section. **B**. Road cut at base. **C**. Road cut at base and above. **D**. Resolution of stresses.

tial translational sliding of a dipping sandstone bed (Fig. 4.17). When the natural dip slope (Fig. 4.17A) is cut at the base (Fig. 4.17B), the potential for failure exists. In order for failure to occur, the frictional plus cohesive forces along the bedding plane (plus possible updip tension within the bed) must be overcome by the gravitation-induced component of shear. Assume no cohesion; i.e., the sandstone block is not attached to the shale below. If a joint occurs, cutting completely across the bed, tensional stress acting to hold the sandstone from above is completely destroyed; in fact water entering an updip joint could build up a hydrostatic stress which would help push the block down the slope. When an additional updip road cut is made above the block of sandstone (Fig. 4.17C), all possible

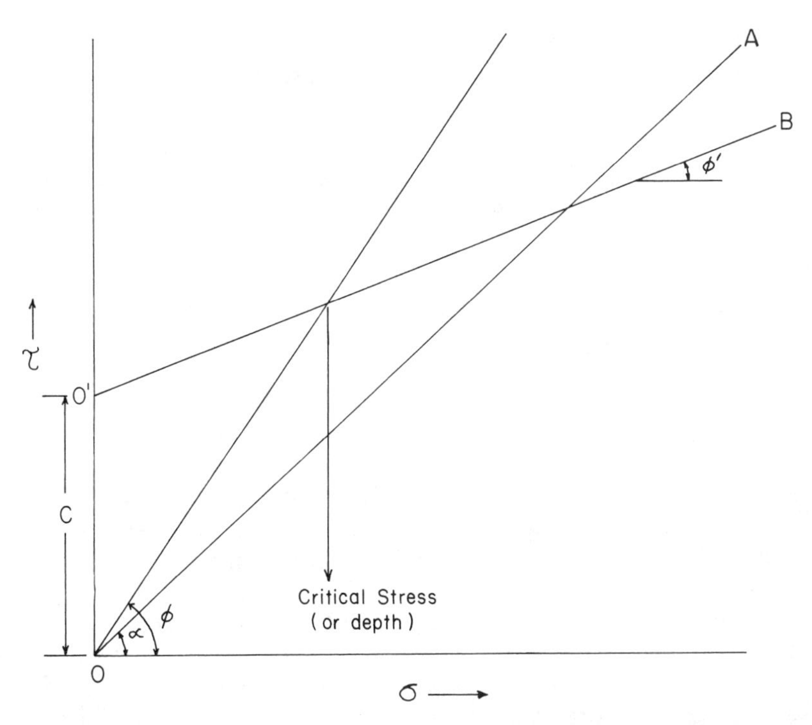


Figure 4.18 Theoretical relationship between shear stress ($\tau$) and normal stress ($\sigma$) for cohesive (line *O'B*) and cohesionless (line *OA*) material. The bedding angle is $\alpha$. See text for description.

tension in the sandstone is destroyed, and the block can be analyzed in terms of a free-body diagram (Fig. 4.17D). The frictional force per unit area ($\tau$) is related to the normal force per unit area ($\sigma$) on the detachment plane as shown by Eq. (4.17) where $c = 0$. Thus,

$$\tau = (\sigma - \mu) \tan \phi. \tag{4.18}$$

From Fig. 4.17D, the normal stress $\sigma$ is the component of the stress due to the landslide weight normal to the failure plane:

$$\sigma = W/A \cdot \cos \alpha. \tag{4.19}$$

For cohesionless material, the resisting force $\tau$ is simply equal to the tangent of the friction angle ($\tan \phi$) times $\alpha$:

$$\tau = (W/A \cdot \sin \alpha)(\tan \phi). \tag{4.20}$$

*Example 2.* For cohesionless conditions, the only requirement for stability is that the angle of internal friction $\phi$ of the material in the failure plane be greater than the angle of the bedding ($\alpha$). Consider line *AO* (Fig. 4.18) which plots below the tan $\phi$ line. If the material possesses cohesion ($c$) and an internal friction angle $\phi$, it will also be stable at all values of depth. Consider another material which possesses cohesion but has a lower friction angle $\phi'$ as shown by *O'B* on Fig. 4.18. Then a bedding angle $\alpha$ can be safely steeper than the internal friction angle only up to a certain stress. Thus, line *O'B* plots above the tan $\phi$ line up to a critical stress, which is in turn related to a critical depth by Eq. (4.18). Below this depth the shearing strength of the materials would be exceeded.

*Example 3.* The stability of a rock mass on a sloping surface such as shown in Fig. 4.17 is significantly affected by changes in pore pressure, but not in the weight. For example, consider the stacking of a 1-kg block on top of another, as discussed above. In a completely dry environment, then, if the weight of the block is increased (example: adding the weight of a building on top of the block), there should be no difference in the sliding stability of the block because an increase in the weight increases *both* the normal stress and the shear stress; hence the ability of the new shearing strength to withstand the increased downslope component of the weight is unchanged. According to this logic, then, the reason landslides occur during heavy rain is not because of the added weight, but because infiltrating rain causes instability either by 1) $\phi$ may decrease with increasing moisture, or, more likely, 2) the pore pressure $\mu$ increases, substracting a greater proportion of the normal stress, ultimately leading to a hypothetical condition whereby the block is "floating," and hence has no frictional restraining force at all [where $\sigma = \mu$ in Eq. (4.18)].

An excellent example of pore pressure uplift phenomena is a hovercraft, such as the one that operates on the English Channel. The effectiveness of fluid pressure in facilitating movement can be demonstrated in the laboratory using water poured into a can perforated at the bottom which rests on an inclined slope, or by using an air-cushion apparatus (Rahn, 1978). High fluid pressure is responsible for the movement of many landslides (see Chapter 6) and for the slumping of dipping sediments on deltaic slopes (Postma, 1984).

*Example 4.* Coulomb's equation (Eq. (4.17)) may be determined by a Mohr's circle analysis. If the principal stresses $\sigma_3$ and $\sigma_1$ correspond to a state of failure in the specimen, then at least one point on Mohr's circle of stress must represent a combination of normal and shearing stresses that caused failure on some plane through the specimen.

Figure 4.19A shows Mohr's circles plotted for two tests. As a number of tests increases, it is apparent that the envelope of failure circles (the rupture line) represents the locus of points associated with failure of the specimens. If the rupture line is straight, it may be represented by Eq. (4.17), the Coulomb equation.

The coordinates of the points made by intersection of the rupture line with a Mohr's circle (point $A$ on Fig. 4.19A) represents the combination of normal and shearing forces acting on the rupture plane. The angle $\alpha$ is the inclination of the plane on which failure took place.

From the geometry of Fig. 4.19B, it may be shown that for any failure circle:

$$2\alpha = 90^\circ + \phi. \tag{4.21}$$

*Example 5.* Figure 4.17D can also be used to analyze the angle of dip of beds ($\alpha$) which could produce sliding (example below modified from Hoek and Bray, 1974). The normal stress which acts across the sliding surface is:

$$\sigma = \frac{W \cos \alpha}{A},$$

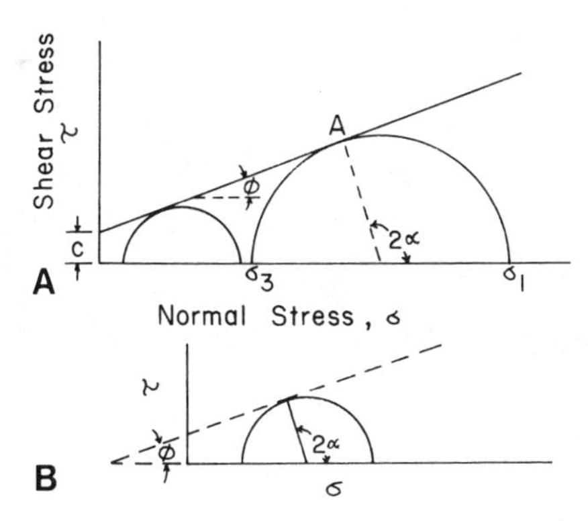


Figure 4.19 Mohr's rupture diagram (after Peck et al., 1974). **A.** Relationship to Coulomb equation. **B.** relationship between $\phi$ and $\alpha$.

where $A$ is the area of the base of the block. Substituting this into Eq. (4.16), we obtain:

$$\tau = c + \frac{W \cos \alpha}{A} \tan \phi.$$

Using total resisting shear force ($R$),

$$R = \tau \cdot A,$$

$$R = cA + W \cos \alpha \cdot \tan \phi.$$

The block will be just at the point of sliding when the gravitational (or disturbing) force acting down the plane ($W \sin \alpha$) is equal to the resisting shearing force:

$$W \sin \alpha = cA + W \cos \alpha \cdot \tan \phi. \quad (4.23)$$

If cohesion $c = 0$, this simplifies to

$$\alpha = \phi.$$

In other words, the angle of internal friction of a cohesionless substance (for example, sand) is the same as the angle of repose (for example, for lee face of a sand dune).

*Example 6.* The following example illustrates a slope stability model for an incipient landslide of certain geometric proportions which rests alongside a reservoir (modified from Lane, 1974):

> A growing body of evidence (e.g., Vaiont Reservoir, Italy) indicates the case of the rising reservoir as likely to be critical for analyzing stability of reservoir slopes. Fortunately there is a method of analysis available from soil mechanics which has been used for some years to check stability of the upstream slope of earth dams for the condition of a partial pool or submergence of the slope toe. To illustrate this, assume conditions, as shown in Figure 4.20. A rock slope with bedding planes (or joint sets) dipping toward the reservoir creates a tendency for sliding on the inclined bedding plane.

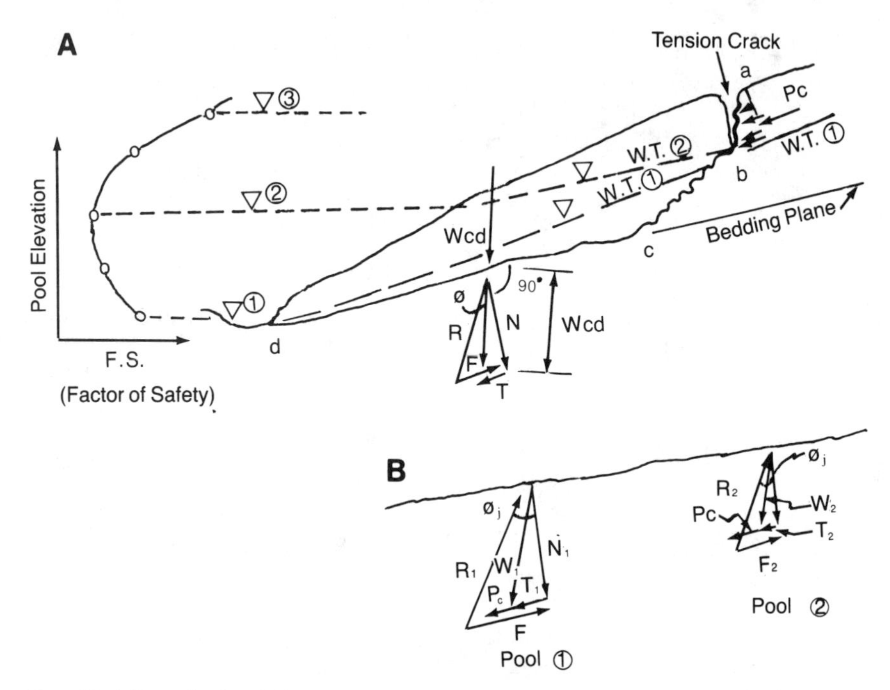


Figure 4.20 Stability of a landslide subject to uplift pressures (after Lane, 1974). **A**. Cross section. **B**. Force diagrams (neglect surface *bc*).

Consider an initial condition with a low pool (#1) and corresponding initial water table (W.T. #1); then as the reservoir rises to pool #2, the water table correspondingly rises to W.T. #2. In terms of effective stress (the intergranular stress creating frictional strength) the rise from pool #1 to #2 serves to reduce the effective unit weight from moist or nearly saturated weight (say 2670 N/m$^3$) to buoyant weight ($2670 - 1000 = 1670$ N/m$^3$). This is a major reduction in the weight $W_{cd}$ above the line *cd*. The loss of normal pressure results in a loss of frictional strength on the failure surface *cd*.

For simplicity, neglect the length *bc* beneath the wedge portion of the failure surface; the figure then shows the effective forces acting on plane *cd*. Here $W_1$ is computed from saturated unit weight above the water table and buoyant weight below, and then resolved into a normal force $N_1$ plus the tangential or sliding force $T_1$ (assuming a stress relief crack or joint *ab*, in which high rainfall has created a cleft pressure $P_c$, then the total sliding force is the sum of $T_1 + P_c$). The resisting force is $F_1 = N_1 \tan \phi_1$ where $\phi_1$ is the friction angle for the bedding plane (or joint). For pool #1, as shown in the force diagram in Figure 4.20B:

$$F_1 > T_1 + P_c,$$

so the slope is stable.

With the reservoir risen to pool #2, Figure 4.20B shows the force diagram. Due to the rise in reservoir and water table, $W_2$ is much less than for pool #1, as is also the case for $N_2$ and the resulting friction or resisting force $F_2$. Although the tangential or sliding force $T_2$ is also reduced, the total driving force remains significant, being represented as:

$$T_2 + P_c > F_2,$$

so the slope is unstable.

In interest of simplicity, the driving and resisting forces on the bc portion of the failure surface have been omitted; however, such forces can be handled by considering successive vertical slices. The method illustrated is adaptable to a sliding wedge and plane type of failure surface, or it can also be utilized with a circular rotation type of surface.

It should be noted that this particular analysis indicates that the stability improves at higher pools; the decrease in frictional strength F being more than offset by the greater reduction in the driving force T. In practice this type of analysis is conducted for different pool levels and the results plotted in the form of Figure 4.20A, which serves to locate the critical pool giving the minimum factor of safety, F.S. (pool 2 for the example shown). In recent analyses of some five earth dams, the critical pool was found in the range of 40 to 50 percent of the slope height. It seems a reasonable probability that the critical pool elevation for a rock slope could be expected in the same general range.

In the case of the catastrophic slope failure at Vaiont dam (see Chapter 6), a mechanism such as explained above may be what precipitated the landslide. Failure on reservoir slopes, however, is often caused by rapid withdrawal of reservoir water and consequent pool lowering. In this case, an adjacent incipient landslide could remain saturated, hence the sliding force $T$ is very large, but, within the failure plane, high water stresses still exist (relic conditions from inability of water to drain instantaneously). The resulting heavy mass is thus able to easily move due to the high pore pressure (Terzaghi and Peck, 1948; Blyth and de Freitas, 1984). Examples of this kind of slope failure were noticed immediately following the 1976 failure of the Teton Dam in Idaho, when hundreds of landslides developed on the reservoir slopes after the reservoir drained.

Another example of the possible practical use of pore pressure theory lies in the potential of earthquake modification by man's injection of fluid under pressure (see Chapter 11). Other analytical techniques for design of stable slopes are presented in Chapter 6.

## 4.8 Stereographic Projection

Problems in engineering geology that involve relationships between planes in space can be analyzed using stereographic projection. Applications include the determination of the plunge of the intersection of discontinuities, the analysis of a potential rock slide, as well as application in photogrammetry and assessment of the state of stress in rocks (Goodman, 1976).

Figure 4.21 is a Wulff stereonet, which is commonly used as the system to represent the spherical slope of the earth on a flat surface. The net is viewed as the lower great circle surface plus half of a hemispherical surface; the orientation of any plane in space can be viewed as its intersection with this surface (Fig. 4.22).

In order to project a plane onto the net, a transparent overlay is laid on top of the net. The strike of the plane in question is plotted as a line formed by the intersection of the plane with the great circle surface. The overlay is rotated to true north, the dip angle plotted according to increments on the net, and the half-circle traced in from the net. The transparency is then rotated back to its original proper orientation.

Figure 4.23 illustrates stereoplots for a man-made cut slope in four different geological environments. The plot of the cut slope is shown as the half-circle plot, plus strike line, and the plots of the rock planar anisotropics are shown as poles.

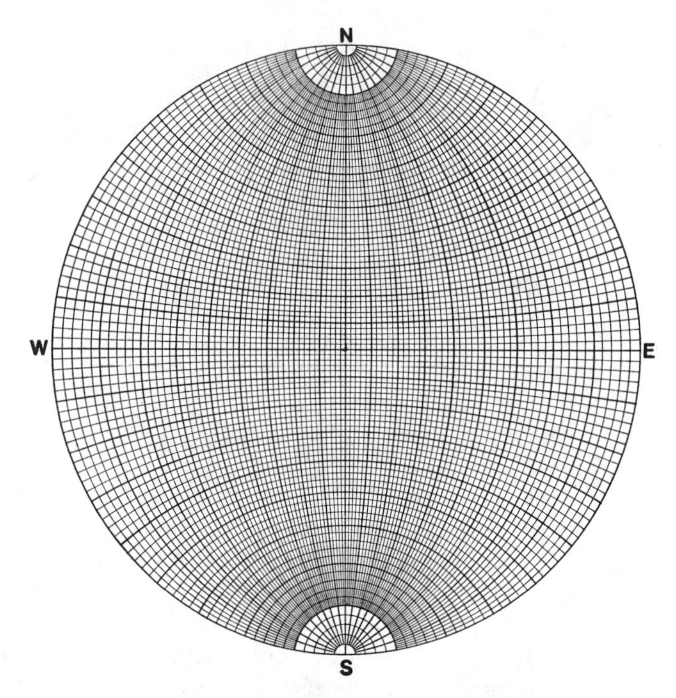


Figure 4.21 Wulff stereonet marked in 2° intervals. This net is most useful for the construction of great circles during the analysis of structural data.

(Poles are constructed as the point on the sphere made by projection of a radial line normal to the plane.) Polar plots can be contoured.

Figure 4.24 is an example of a stereographic plot of two faults in an area where an east–west road cut is proposed. The road cut is to be made at a 55°S dip. The intersection of these three surfaces can form a wedge-type landslide failure mass. (In highway and open-pit mines in rock, this type of failure is perhaps the most common, for example, where a fault or joint crosses bedding planes. See Fig. 4.25.)

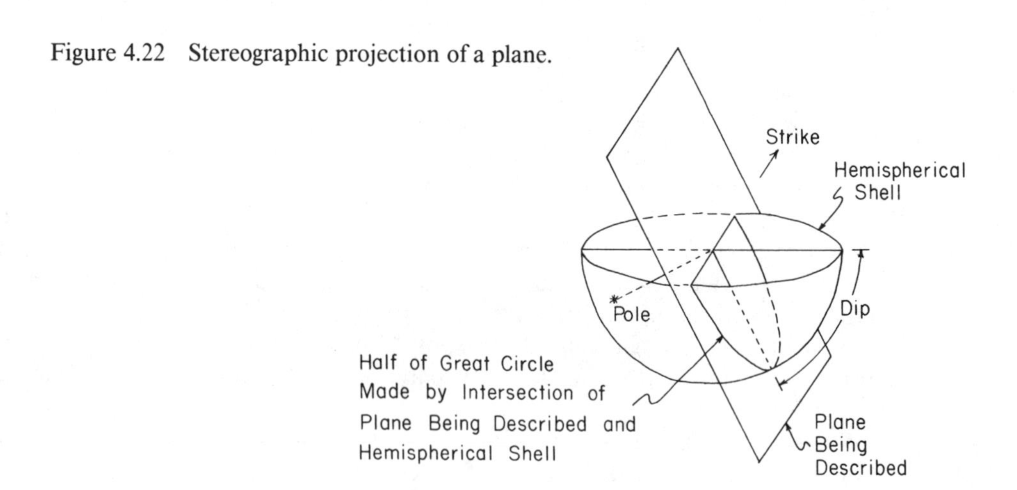


Figure 4.22 Stereographic projection of a plane.

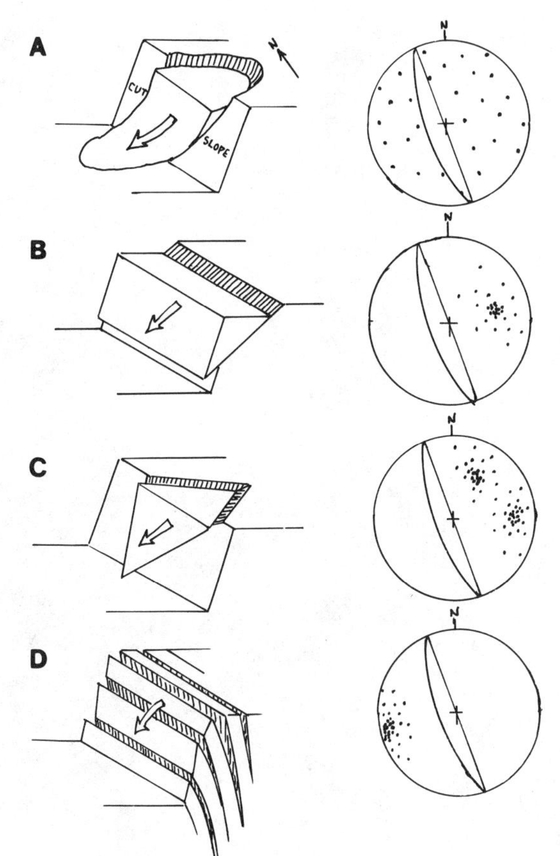


Figure 4.23 Main types of slope failure and appearance of stereoplots of structural conditions likely to give rise to these failures (after Hoek and Bray, 1974).

Figure 4.24 Example problem using a Wulff stereographic net (from Piteau and Peckover, 1978). Angle relations between faults, which form potential wedge failure, and proposed cut slope are shown.

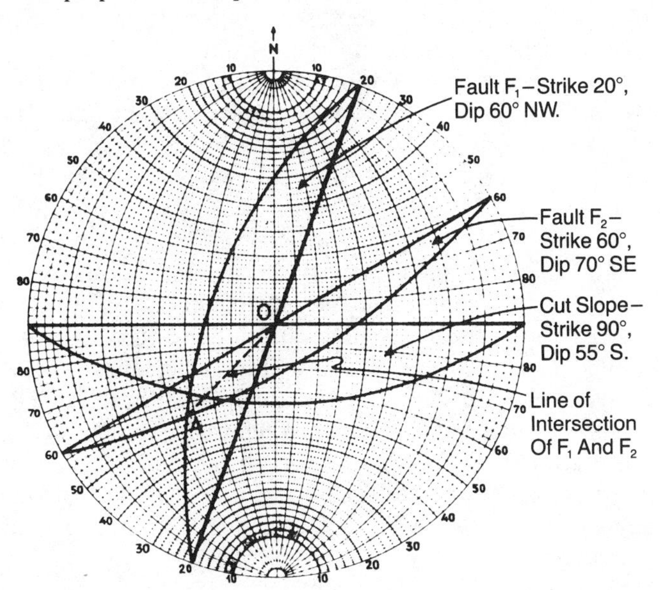

Figure 4.25 Wedge failure during 1984 excavation for spillway enlargement of Deerfield Reservoir, SD. The rock is a quartzose slate which has two prominent cleavage directions.

Figure 4.26 is a sketch showing Fig. 4.24 wedge in three dimensions. For this analysis it can be assumed the wedge formed by two faults "daylights" out at the base of the cut; if in fact the stereoplot represents two rock anisotropics such as general joint trends, then this assumption is quite valid.

The intersection of any two planes plotted on the stereonet is the line that passes through the center of the projection and intersects the surface of the sphere where the two planes intersect. The attitude of the line (i.e., strike and plunge) can be measured from the stereonet. For example, in Fig. 4.24, the line of intersection of fault ($F_1$) and fault ($F_2$) is the line *OA,* whose strike is N 45°E, and the plunge of the line is 36° SW.

The value of this technique is that a man-made cut can be designed to avoid a wedge-type failure. In this example, if a road cut slope is made which strikes exactly N 45°W (perpendicular to the wedge failure line), then the road cut (on the NE side) would be in jeopardy if it were made at 36° or greater. Since the strike of the proposed road cut in Fig. 4.24 is actually N 90°E, the safe design cut is established by graphical interpolation between the wedge intersection plunge of 36°SW and the fault $F_2$ dip of 70°SE, which results in a dip of approximately 55°S at the plane of the cut slope. (This interpolation can be achieved by rotating the stereonet until the highway is oriented N–S. Then the maximum roadcut dip can be read from the great circle.)

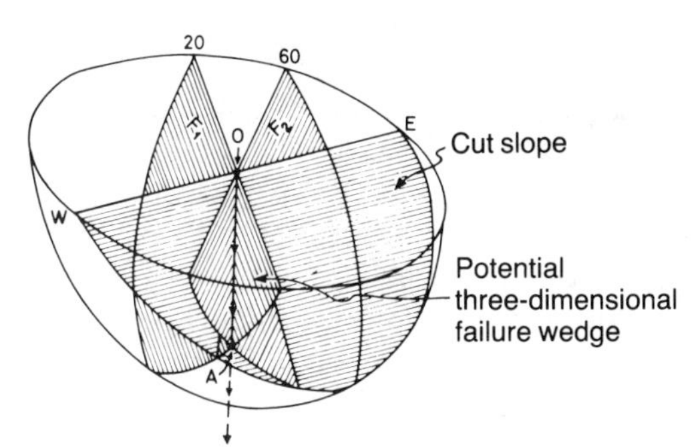


Figure 4.26 Three-dimensional diagram of spatial relations of salient features of rock model shown in Figure 4.24 (from Piteau and Peckover, 1978).

Rigorous analytical methods for predicting rock slope stability must be tempered by reason. Slopes have failed even where slope stability analyses indicate the slope was theoretically stable, using very high safety factors. Bryan (1981) describes the complex failures of a 10-m-high chalk bluff along the Tennessee-Tombighee Waterway in Alabama: "The described failure mechanism involving tensional stress and shrinkage cracking cannot be predicted or rationally analyzed using laboratory test data. One can only be forewarned on the basis of experiences. . . . Geologists are taught early to observe existing slopes, both natural and excavated, in order to set safe excavation slope heights and angles."

## 4.9 Tunnels

Tunnels play an important part in transportation and water systems in urban and mountainous areas. The Swiss engineers are famous for having developed ingenious railroad and highway tunnels in a very complex geologic setting. In the United States, over 4000 km of tunnels were driven during the period 1965–1975. Water and sewer tunnels accounted for 82% of this length, rapid transit 16%, and motor vehicle 2%. There are 44 cities in the world with subways; many of these have been excavated in complex geologic conditions. Immense tunnel transportation schemes including a London-to-Paris and a Washington-to-Boston railroad have been proposed.

This section describes basic geologic considerations for use in the design of tunnels. Many books and reports have been written on tunnel engineering, some of which discuss specific examples of tunnels. The reader may consult references by Krynine and Judd (1957); Proctor et al. (1966); Trefzger (1966); Jaeger (1972); Legget (1973); Proctor (1973); Richardson and Mayo (1975); and Bowen (1984). Tunneling in soft ground or rock is described by Terzaghi (1950a), Terzaghi and Peck (1967), Peck (1969), Pincus (1973), Knesel (1976), Hansmire (1981), and Fox and McHuron (1984).

### *4.9.1 Basic Tunnel Techniques*

Tunnels can either be excavated in soft ground or hard rock. In cases of soft ground (plastic clay, etc.) it is possible to drive a tunnel with an open front. Moles are ingenious boring machines that grind out the rock at the cutting face. Initially used in soft rocks such as shales, they have been adapted to use in soft limestones. The shield method utilizes a circular steel frame, equipped with steel cutters, which rotates and is jacked forward. The shield is pushed forward by hydraulic jacks that react against completed lining at the rear of the shield. A problem with shallow tunnels in weak soils is surface settlement or lateral movement can occur as tunneling proceeds (Clough, 1977; Hansmire, 1981). Soft-ground tunnels being driven under the water table may encounter difficulties with oozing ground as well as undergo a continuous fight with water. A 30-km subway is under preliminary engineering design for Los Angeles. Most of the route will be in soft ground, favorable to high-speed boring machines. Of concern is 5 km of "gassy" ground near the La Brea Fossil Tar Pits, and two active faults which must be crossed.

In shallow cover and soft ground, the tunnel may be constructed by the cut and cover method. In this case an open trench is made, preassembled tunnel sections (concrete, steel, or timber) are emplaced from above, and the tunnel is backfilled. Most of the London subway (the so-called underground), the Toronto subway, and part of the Buenos Aires subway were constructed in this manner. To a large degree, the Washington DC subway (the Metro) and the Metropolitan Atlanta Rapid Transit Authority (MARTA) are being built in this manner. Sections of the Bay Area Rapid Transit (BART) subway, built in 1960 in the San Francisco area, were built below sea level in the Bay Mud by the cut and cover system, where a trench was dredged in the Bay bottom, concrete sections were towed and lowered into place, and later covered. The same technique was used at the 2-km-long Hampton Roads Bridge Tunnel in Virginia (Meadors, 1977). An interesting aspect to the BART system is the earthquake hazard. The tube does not cross any faults known to be active, but the prefabricated tube sections were cushioned by laying in a trench dug into soft bay mud. To allow for possible movement during an earthquake, the tube's connections to its terminal buildings are flexible, permitting movement of several centimeters up or down, in or out, and sideways. Leveson (1980) gives an interesting account of subway history and geology in the United States, particularly in New York City, where there are 380 km of subways, 212 km of which are tunnels.

In most cases tunnels are driven through rock. Tunnel terms are largely borrowed from mining engineers. A tunnel is open at both ends, as opposed to a drift (or adit) which has one entrance. A raise (or stope) is an excavation that is inclined upward; if open to the surface, it is a shaft. Figure 4.27 shows a longitudinal cross section of a tunnel through a hill. The weathered rock at tunnel portals is excavated as an open quarry and lined before the tunnel excavation begins. Steel tunnel supports include posts (vertical members) and ribs (curved top portions above the "spring line") of many different designs. The ribs are usually curved in the shape of a semicircle in order to more evenly distribute the roof stresses.

Drilling in a hard rock tunnel is typically accomplished by drills mounted on the back of a large truck, called a jumbo. The jumbo is backed against the face (heading) and about 100–200 holes are drilled, about 3–4 m deep. The holes are loaded with explosive, jumbo and personnel removed from the tunnel, and the heading is blasted. (Blasting is a science unto itself; tunnel engineers have developed techniques to acquire adequate rock removal and yet prevent excessive overbreak (Richardson and Mayo, 1975). Overbreak (Fig. 4.27) is undesirable because the cavity between the steel ribs and the roof must be backpacked with railroad ties, concrete, or other media often at the contractor's expense.) The blasted rock (muck) is removed by "muckers" and dump trucks as soon as the

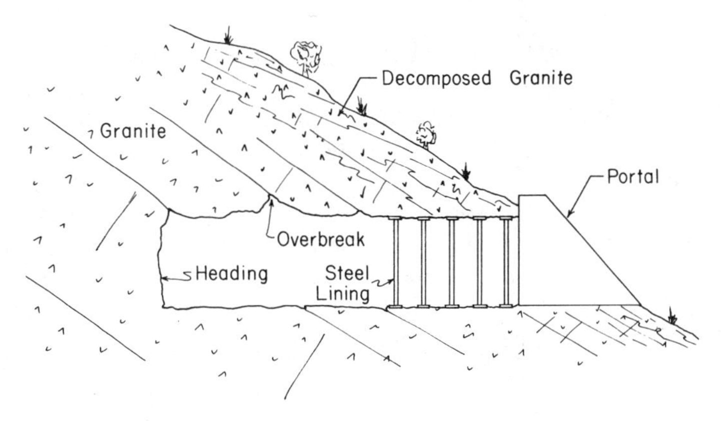


Figure 4.27 Tunnel drift showing use of steel supports at portal area.

tunnel has been ventilated. A load-haul-dump (LHD) machine, which digs into the muck and lifts it over it into a waiting truck, is very convenient where haulage distance is not great. Three cycles of drilling and mucking is considered a good 8-hr work shift.

Underground support in the form of rock bolts are commonly used in mines or tunnels where steel supports are deemed unnecessary. Figure 4.28 shows details of a typical roof bolt installation. Roof bolts are desirable in flat-lying sedimentary rocks because they bind several strata together and thus strengthen the rock into a form of arch. The flat, unsupported roof of a mine or tunnel may be likened to the unsupported beam (Fig. 4.5) so that tensional failure is likely directly overhead. By drilling from the tunnel roof deep up into firm rock above, and developing high tension in the rock bolt, the roof rock is effectively com-

Figure 4.28 Sketch of typical rock bolt installation in a tunnel (after Krynine and Judd, 1957).

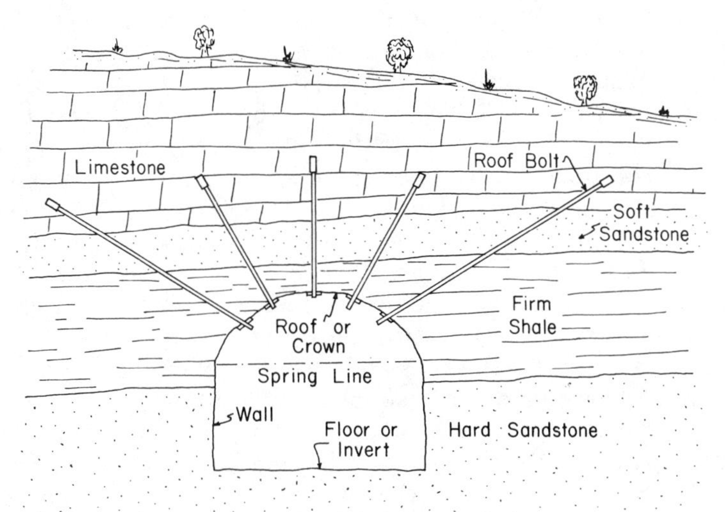

pressed, and thus failure by tension (or simply by loose slabs falling out) is significantly reduced. The effectiveness of a rock having an anchoring device at one end and a rigid bearing device at the other depends entirely upon the ability to maintain tension between the two ends of the bolt (Thomas, 1962).

Concrete is generally emplaced around the ribs and posts for water highway tunnels. In some cases, grout is sprayed onto wire mesh. The final engineering design for a tunnel depends, of course, on the purpose of the tunnel. Adequate permanent ventilation is required in long highway tunnels. For example, the two-lane Big Walker Tunnel on Route I-77 in Virginia has a larger ventilation system for the uphill lane due to increased vehicle exhaust.

### *4.9.2 Geological Factors in Tunnels*

The estimated cost for underground excavation depends more upon geologic conditions than any other type of civil engineering project. According to Proctor (1973), the most important parameters to be described in a geologic report are the soundness of the rock, water inflows, gas, joint spacing, weathering, and fault zones. He further states (p. 187):

> In the planning stage, the geologist should suggest routes that 1) do not cross large faults, or cross them at high angles, 2) pass through the most competent rocks possible, 3) avoid high ground-water areas where excessive inflow of water into the tunnel may be anticipated, 4) avoid areas that may contain oil or gas, and 5) avoid formations which contain 'squeezing rocks' such as bentonite and anhydrite.

There is a saying in the tunnel profession: "You pay for borings whether you get them or not." It is very important for an engineering geologist to accurately depict geologic conditions. Ground conditions are the big hazard in tunnel contracting, and the more complete the information, the less risk and the lower the bids. Test borings are necessary, as well as basic geologic knowledge. The Washington subway provides an example of a major engineering geology undertaking which draws successfully on geologic reports by Darton (1947) and other U.S. Geological Survey reports. When completed, about 40% of the subway will be in crystalline rock and 60% in alluvial and coastal sediments.

Peachtree Center Station, located in the heart of Atlanta's 100-km-long subway system, is to be finished by 1987. The originally proposed near-surface cut-and-cover design for the Peachtree Center Station has severely affected downtown traffic. A station constructed at greater depths was thought to be prohibitedly expensive, but geotechnical investigations showed unusually competent bedrock beneath the area. The station was mined out of massive gneiss, the bottom of the excavation being 38-m below the surface. It was designed similar to the Roman arch, in which the rock is laced together with resin-encapsulated steel rods (dowels) to form a solid mass that arches across the opening.

Figures 4.29 and 4.30 are sketches by an Italian tunnel engineer, Professor A. Desio. Of course, geology can never be simplified to "cartoons" such as this, but one hopes that they show concepts that the engineering geologist can use in real-life situations. Figure 4.29 shows how the strike and dip of a fault may play an important role in tunnel engineering. The intersection of a fault or fault zone in a tunnel can cause a myriad of problems, such as flowing ground water under high

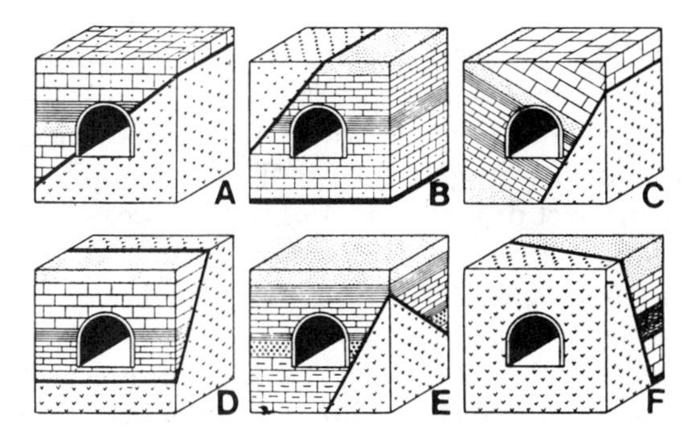


Figure 4.29 Different positions of a tunnel with respect to a fault. (After Prof. A. Desio, "Geologia applicata all' ingegneria," Hoepli, Milan, Italy.) (From Krynine and Judd, 1957.) Panel **A** shows a tunnel located within the faulted zone, whereas panels **B** and **C** correspond to the location of the tunnel in the hanging wall and in the foot wall, respectively. The tunnel in panels **D** and **F** crosses the fault, obliquely in the latter case. In panel **E** the tunnel is outside the fault.

pressure. Jaeger (1972) describes difficulties encountered with caving ground and inflow of sand and water at the Malgovert tunnel in France. The ground pressure was so great that steel ribs spaced 0.3 m apart twisted and reinforced concrete beams sheared through. Faults can cause additional expenses. The Eisenhower (Straight Creek) Tunnel, which penetrates the Continental Divide in Colorado, is 2.7 km long, and was bid at $50 million but cost $112 million to complete. The major problem was broken rock in a major fault zone. The final design consisted of grouting the fault zone with concrete and then excavating by conventional techniques.

The engineering significance of a fault was emphasized in the case of the construction of the Pennsylvania Turnpike. The tunnels in the Pennsylvania Turn-

Figure 4.30 Influence of rock stratification on the tunnel lining. (After Prof. A. Desio, "Geologia applicata all' ingegneria," Hoepli, Milan, Italy.) (From Krynine and Judd, 1957.) Panels **A**, **B**, and **C** show more or less uniform vertical pressure on the lining, whereas panels **D** and **F** show the pressure concentration at one side of the tunnel caused by oblique strata. Panel **E** shows a case of heavy pressure at the key of the arch. The locations shown in panels **A**, **D**, and **E** are favorable for office study of pressures, since in these cases, a two-dimensional stress distribution may be assumed.

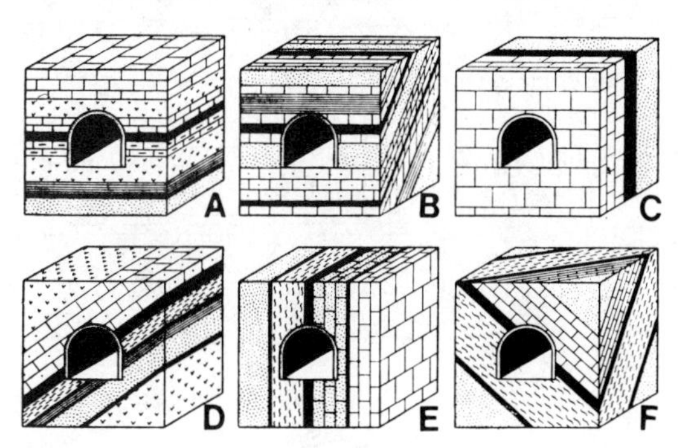

pike are located in the "Ridge and Valley" province of the Appalachian Mountains; the ridges are made of hard sandstone or quartzite and the valleys are shale or limestone. In 1885 the South Penn Railroad (owned by Andrew Carnegie) was built to bring anthracite coal to Pittsburgh's steel mills. Nine railroad tunnels were partially driven, but the project proved to be financially infeasible, and the tunnels were abandoned. In 1938, the Pennsylvania Turnpike was built. One tunnel (Allegheny #2) about 1.6 km long, was driven parallel to and only 25 m from the old railroad tunnel. In both of these tunnels it was noted that a fault was encountered ("soft, oozing material") a short distance in from the portal. Two famous geologists, Arthur Cleaves and Marlin Billings, had studied the area and had predicted the bad roof conditions in the 1938 turnpike tunnel based on the presence of the fault in the old railroad tunnel. The 1938 turnpike tunnel was completed but had only two lanes of traffic, thus it was decided in 1965 to widen the highway to four lanes by excavating a new tunnel parallel to and about 30 m from the original 1938 tunnel. Despite the presence of the fault and bad ground conditions known to have been present in both the adjacent 1885 and 1938 tunnels, the geologic information apparently remained in unused dusty files because there was no mention of the fault to the contractor in 1965 who suffered a $1.5 million loss due to caving ground, lost time, increased number of steel liners, and other costs. There was no "change of conditions" in the contract with the state, and the contractor eventually brought successfully a suit against the Commonwealth of Pennsylvania (Martin S. Michelson, personal communication, 1967).

Figure 4.31 Grouting used for the Washington, DC, Metropolitan rapid transit tunnels (from Clough, 1981; after Kuesel, 1976). The crowns of the upper two tunnels are in sand and gravel. The tunnels pass beneath the I-95 Freeway and the 7th Street Bridge, coming within 1.2 m of the foundation. A silicate grout was injected into the overlying sand and gravel as shown above (vertically hashured area). The grout successfully strengthened the overlying material, and only 2 cm of settlement was observed.

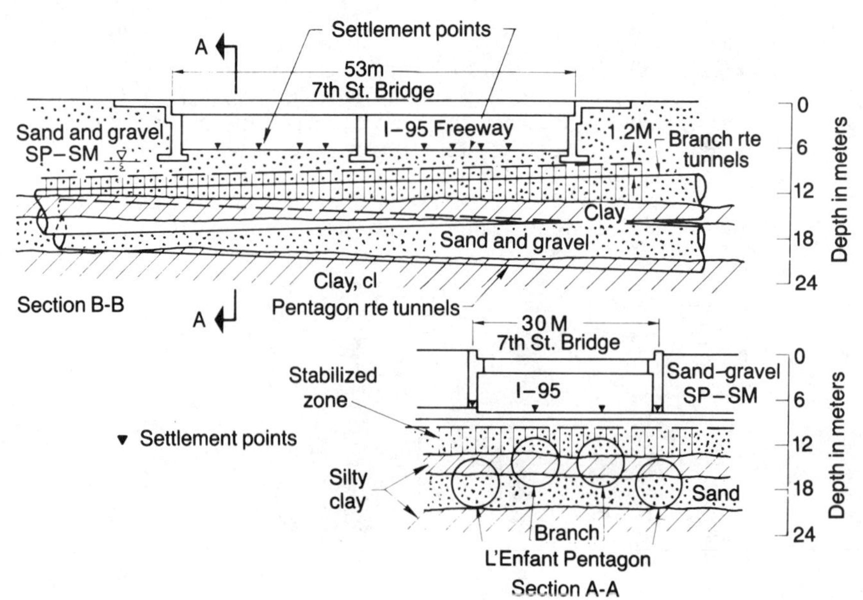

The lesson learned from this is not only that engineering geology is important, but that the data must be made available to the engineers and contractors who do the work.

Figure 4.30 shows how the dip of sedimentary rocks can effect the tunnel stability. Geologic cross sections, perpendicular to the tunnel centerline, would be the same everywhere in the tunnel in sketch A, D, and E. So a cross-sectional model could be made to predict stresses, water conditions, etc. However, conditions as shown by sketch B, C, and F would mean the geology would be constantly changing. The unstable nature of the excavation would be further increased by the presence of water which would tend to exert pressure on the tunnel, leading to a blowout. Other factors which can cause squeezing ground in tunnels include altered volcanic ash (bentonite), poorly consolidated mudstone, weathered schist, and altered veins (Richardson and Mayo, 1975).

In weak rocks it may be necessary to drill holes ahead of the tunnel excavation, and pump cement grout into the rock to render it more amenable to excavation. Figure 4.31 illustrates the extent of grouting used for construction of the Washington subway. Note the extensive use of grouting in the permeable sand and gravel.

Where tunnels encounter permeable ground that is also weak and subject to caving, it is possible to freeze the ground as excavation progresses. Freezing has been successfully employed in the construction of the different sections of the metro systems in London, Leningrad, Tokyo, and Helsinki, for sewer tunneling in New York City, and for railroad tunnels in Italy (Megaw and Bartlett, 1982).

## *4.9.3 Water in Tunnels*

The presence of ground water in underground workings can be more than a nuisance. Many early mining ventures succumbed because water flooded the workings. The invention of the modern high-capacity water pumps has reduced this problem, however.

If water is expected, the excavation for a long inclined tunnel should always being at the downhill portal. This allows the water to drain out by gravity. During tunneling, water may enter at faults, aquifers, or gouge zones. Usually a caving ground area is accompanied by excessive water. Figure 4.32 shows three gouge zones near a tunnel. Zone *c* may eventually cause uplift and failure in the floor due to water pressure.

Water can be expected anywhere that an excavation is below the water table. In a sense, a tunnel is like a giant horizontal well, draining the water table to a

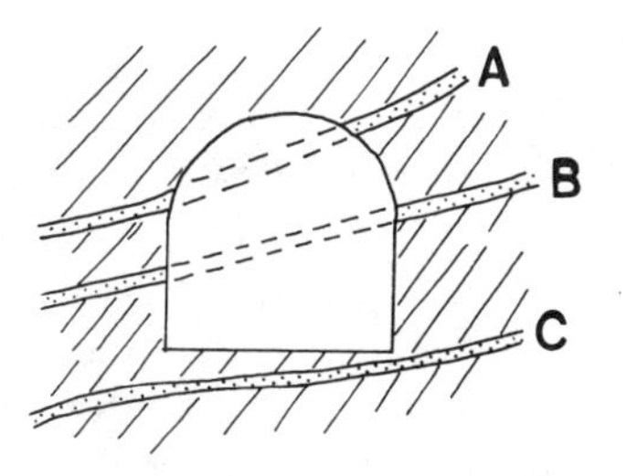


Figure 4.32 Gouge zones in a tunnel (after Krynine and Judd, 1957). The gouge zones **A** and **B** may allow for water ingress and cause cave-ins. Gouge zone **C** may cause uplift at the tunnel floor.

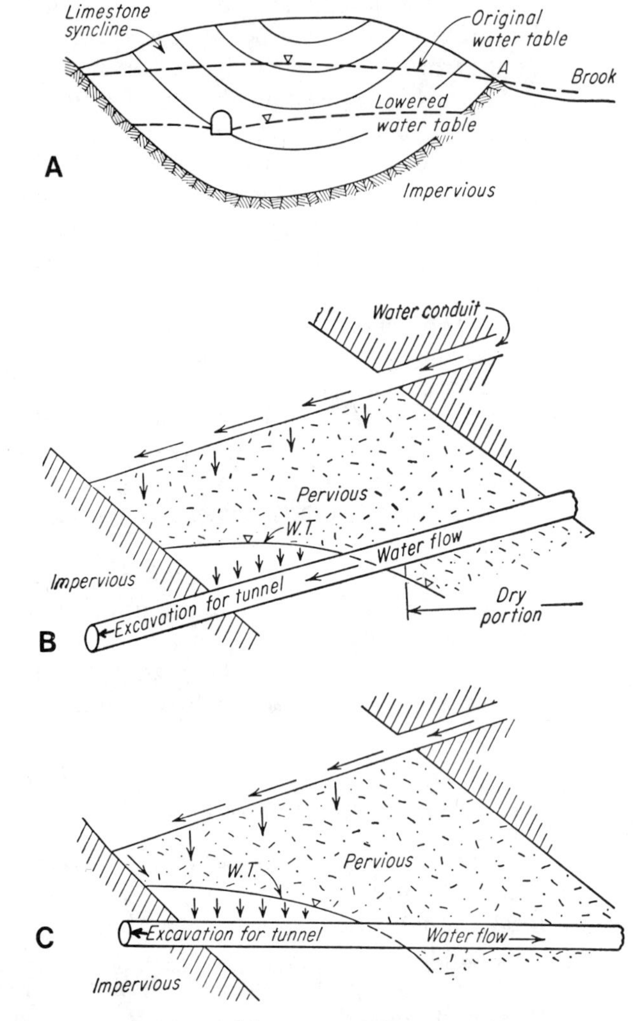


Figure 4.33 Change in ground water regime caused by a tunnel (from Krynine and Judd, 1957).

new position (Fig. 4.33A). If an inclined tunnel is bored downward, water may flood the head working area (Fig. 4.33B). If the tunnel is bored upward (Fig. 4.33C), water would drain out, but may require adequate drains throughout the construction and the operational life of the tunnel.

## *4.9.4 Examples of Tunnels*

Numerous tunnels have been excavated in California, mostly in connection with the needs of the Los Angeles area for water. Active faults, weak rocks, and natural gas seeps are encountered in the Los Angeles area. In 1971, 17 workers were killed by an explosion in a 9-km-long tunnel at San Fernando, despite warnings that natural gas existed in the area. The tragedy occurred because the contractor failed to provide monitoring or expert engineering geology advice; the contractor was found guilty of violating California Labor Code provisions.

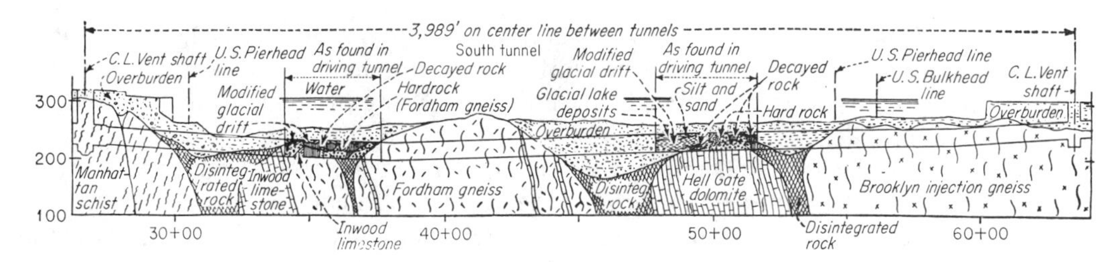


Figure 4.34 Generalized geological section across the East River along the line of the Queens Midtown Tunnel, New York City, showing the variable subsurface conditions encountered (from Legget, 1973).

The Tecolote Tunnel, through the Santa Ynez Mountains near Santa Barbara, CA, was one of the most difficult tunnels ever driven. The tunnel is part of the grandiose California Water Plan developed during the 1960s. Water was encountered in a fault in the Matillija Sandstone (Trefzger, 1966).

Preliminary surveys of the Tehachapi Tunnel in southern California, an enormous tunnel designed to carry 1000 $m^3$/sec of Feather River water from the San Joaquin Valley over the Tehachapi range to Los Angeles, revealed that the alignment would intercept the active San Andreas fault underground. Engineering geologists convinced the State Department of Water Resources officials that the risks associated with this situation exceeded the cost of a higher (and costlier pumpage) alignment, and the tunnel alignment was changed.

The history of New York City's water supply is fascinating. New York civic authorities built a gigantic water supply system starting in 1842 with the Croton Dam located 48 km north of the city. Following came the Catskill Mountain system, involving dams and tunnels bringing water over 160 km to the city. Assisting in the engineering geology was Dr. Charles P. Berkey, one of the leaders of the new field of engineering geology. Later came the extensive Delaware River project, bringing water from that distant watershed. An interesting hydrogeologic problem ensued. In the Brooklyn area, ground water pumpage had lowered the water table over 10 m below sea level due to heavy pumping in the early part of the twentieth century, but by 1947 the availability of surface water brought to New York City from the Delaware River watershed caused a cessation of ground water pumping (except for air condition pumping which returned water to the ground). Subways had been built during the years of heavy ground water pumpage. As the water table resumed its original position, flooding of basements and subways occurred until remedial measures were taken (Leveson, 1980). In 1970, New York City began an enormous water tunnel improvement project which is supposed to be completed by 2009 A.D. The concrete-lined water tunnels represent the costliest public-works project in American history.

New York City also has numerous highway and subway tunnels. For example, the underwater section of the Queens Midtown Tunnel (Fig. 4.34) in New York City is 1.3 km long. Compressed air was used as this tunnel was driven through a complex geology including crystalline bedrock and glacial deposits.

The I-95 Fort McHenry Tunnel across Baltimore Harbor, MD, under construction in 1983, is the largest single project for the U.S. interstate highway program. The 2.7-km tunnel will be built by dredging across the harbor and placing 16 pairs of 106-m-long prefabricated steel-shelled concrete tubes into the trench. The completed tunnel will have eight lanes of traffic. To help protect the Chesapeake Bay fishing and recreation area from turbidity during dredging operations, containment basins were constructed to allow for settling of solids from the dredged material (Siefring and Hart, 1982).

In Chicago, an underground storm Tunnel and Reservoir Plan (TARP) is in progress. The physiography around the southeastern edge of Lake Michigan, the result of glaciation, allows for surface water manipulation to the benefit of man. Because the Des Plaines River, a tributary of the Illinois and Mississippi Rivers, nearly reaches Lake Michigan, it was decided to dig a canal across the drainage divide and reverse the direction of flow of the South Branch of the Chicago River, thus allowing for canal linkage from the Mississippi to the Great Lakes. (At the same time, Chicago's sewage could be discharged into the Mississippi River. Because Lake Michigan is the source of Chicago's drinking water, the typhoid death rate dropped dramatically after completion of the Sanitary and Ship Canal in 1900 (Leveson, 1980).) Construction of the TARP facilities to handle storm water runoff and domestic and industrial wastes from 360 $km^2$ of the Greater Chicago area poses a tremendous challenge. Giant tunnels, excavated in the horizontally bedded Ordovician Niagara Dolomite, will divert storm runoff underground through giant sewer tunnels, allowing for removal of storm runoff to the Des Plaines River (and ultimately the Mississippi River). Thus storm water runoff contamination of Lake Michigan, the source of Chicago's drinking water, will be prevented (Grimes, 1977). When completed, TARP will include 211 km of tunnels ranging from 3 to 10.6 m in diameter, bored at depths of 45–90 m. It will also include a massive storage reservoir capable of holding 150 million $m^3$. The cost of TARP escalated from $1.2 billion in 1972 when the project began to $12.5 billion estimated in 1983, making this one of the most expensive construction projects in history. The main 10.6-m diameter tunnel is being bored by an immense mole, consisting of 69 rotating cutting discs chewing through rock at an average rate of 1.4 m/hr.

In addition to tunnels, rock mechanics expertise is needed in the construction of large underground excavations. Underground commercial warehouses are viable construction techniques where insulation is needed and energy costs are high. Underground mining of the Bethany Falls Limestone at Kansas City, MO, for example, has allowed for the construction of over 60 hectares of underground warehouses and offices. The naturally level floor, competent rock, and lack of ground water provide excellent construction characteristics (Stauffer, 1978). Rock-bolting to secure safer roof rock is occasionally necessary. Railroad spurs have been constructed into the warehouses. The limestone mines and tunnels are refrigerated and now ingeniously serve as giant freezers for french fries and other foods. Offices are also located in other underground limestone sites at Kansas City. The U.S. Postal Service and 100 other tenants enjoy the quiet and constant year-round temperature, requiring no heating or air conditioning.

Large underground facilities are used for defense purposes. Destroyers in the Swedish Navy can be hidden and rehabilitated in underground excavations. The North American Air Defense Command near Colorado Springs, CO, was built in 1962. The installation, capable of accommodating a staff of 700, was excavated in the Pike's Peak granite. Highly weathered rock and close jointing were encountered despite the 420 m of rock cover above, and a reinforced concrete dome was required (Blaschke, 1964).

As urban centers become more populous, they may begin to mimic Tokyo, where five million passengers are on the subways every day. The subways are crowded in a country where a population equal to half that of the United States is crammed into a state smaller than Montana. Land values are determined by proximity to subway stations because proximity to subways means proximity to work for the commuter. In 1982, tiny house plots ($\sim$10 m$^2$) sell for $200,000 because they are near the subway.

## 4.10 Dynamic Loads

Dynamic loads differ from static loads in that the stresses are induced over a very short period of time. Dynamic loads originate, for example, by striking an outcrop with a hammer, setting off explosives, or earthquakes and nuclear explosions.

A dynamic load is characterized by 1) the amount and speed of the loading and 2) its ability to produce shock or stress waves in the rock. A shock wave is produced where high pressure actually exceeds the brittle compressive strength of the rock for a split second; the rock behaves hydrodynamically, causing an oscillatory wave motion which propagates outward very rapidly, but carrying insufficient energy to permanently destroy the material in its path. These elastic waves consist of body waves, primary (P) waves, and secondary (S) waves and surface waves. Farmer (1968) and Roberts (1981) show theoretical formulas for analyzing dynamic loading. Seismic waves are discussed more fully in Chapters 11 and 12.

Serious problems may arise from rock blasting in urban areas. Quarry owners have to restrict the intensity of blast operations. Theoretical and practical information is available (Northwood and Crawford, 1965; Attewell and Farmer, 1976, and references contained therein), but it is difficult to specify tolerable vibration velocity or acceleration levels without considerable on-site experimentation, and even then any recommendation may be quite arbitrary.

On nearly every urban project involving blasting, claims are commonly submitted by nearby property owners for damages allegedly caused by flying rock, air concussion, and vibration induced damage to foundations or wells. Northwood and Crawford (1965) show a direct relationship between explosive charge and distance to the outermost limit of detectable damage. They show the conservative safe limit $d$ (m) related to charge $E$ (kg of dynamite):

$$d \cong 5.2E^{0.7} . \tag{4.24}$$

## Problems

1. The unconfined compressive strength of Tertiary welded tuff from the Nevada Test Site was found to be $51 \times 10^6$ N/m$^2$ (Parrish and Senseny, 1981).
   a. Suppose the flat-lying tuff bed is extensively mined in a checkerboard ("room and pillar") fashion for disposal of high-level radioactive wastes. What is the ratio of pillar/room area so that a 200-m thickness of overburden strata may safely be supported? Assume density of overburden = 2.5 g/cm$^3$ and the factor of safety = 2.
   b. What is the ratio so that the pillars would be able to withstand an additional load of $10^7$ N/m$^2$ caused by a nuclear bomb explosion just above the ground surface? (Keep F.S. = 2.)

2. Use data from the Westford, MA, granite quarry (Fig. 4.3 and text) to solve for the depth of emplacement of the granitic magma if thermal stresses as well as lithostatic stresses are considered. Assume the coefficient of thermal expansion of granite is $0.9 \times 10^{-6}$ m/m/°C. Assume that the mean annual temperature of Barre, VT, is 10°C, and the temperature at the time of final crystallization was 500°C.

3. (Data from Krynine and Judd, 1957.) Draw a Mohr envelope and determine c and $\phi$ using the following triaxial test data from granite at the Grand Coulee dam, Washington. The average unconfined compressive strength was $6.5 \times 10^7$ N/m$^2$. When the confining pressure was $0.69 \times 10^7$ N/m$^2$ the axial stress required to break the sample was about $12 \times 10^7$ N/m$^2$ and increased to $15 \times 10^7$ N/m$^2$ when the confining pressure was about $1.4 \times 10^7$ N/m$^2$. (Note: draw approximate best fit line to the three circles.)

4. A 100-N force acts on a 30.6-kg block placed on an inclined plane on the earth's surface as shown below. Determine if the block below is stationary or moving. If it is moving, calculate its acceleration. The coefficients of static and dynamic friction between the block and the plane are $u_s = 0.25$ and $u_k = 0.20$, respectively. (Hint: Make a free-body diagram assuming static conditions. If forces are not balanced, make a free-body diagram under dynamic conditions. If forces still are not balanced, solve for the net force along the inclined plane and solve for acceleration.)

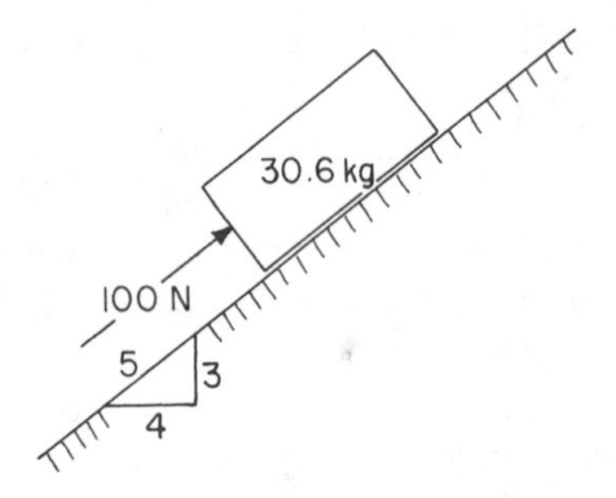


5. Refer to Fig. 4.7. On arithmetic paper plot seven points showing the peak stress (this occurs approximately at 3% strain in each case) as ordinate versus the confining pressure as the abscissa. Use values of N/m$^2$. The ordinate can be labeled maximum principal stress and the abscissa least principal stress.
   a. Assume a linear relationship (typical of strong rocks). What is the maximum unconfined strength?
   b. Assume this rock is subject to lithostatic loads (maximum principle stress) in a 100-m-deep mine. What confining stress should be exerted on the walls of the mine to keep the rock from crushing?

6. (Adapted from U.S. Department of Transportation, 1977.) Assume a 300-m-high cut is to be made in massive, argillaceous rocks. The rocks have three well defined joint sets.
   Joint Set *A:* strike N 78°E, dip 72°N.
   Joint Set *B:* strike N 78°E, dip 78°S.
   Joint Set *C:* strike N 17°W, dip 80°W.
   In addition, bedding planes strike N 45°E and dip into the slope at 26°SE. Assume the cut slope is over 1 km in length, is essentially straight, and trends N 10°E. The road will cut into the bluff of rock so that only the one side of the road cut need be analyzed; the dip of this cut will be on the east side of the road (i.e., the cut slope dips westerly).

**a.** Determine the strike and plunge of the line made by the intersection of the two critical planar anisotropies which could form a wedge failure.
**b.** Design a safe road cut for this highway.

**7.** Refer to the plot below showing shear stress versus displacement for joint surfaces in specimens of porphyry (Hoek and Bray, 1974). Based on the three curves, plot shear strength ($\tau$) versus normal stress ($\sigma$) for peak strength conditions. Derive an appropriate formula using Eq. (4.16) format.

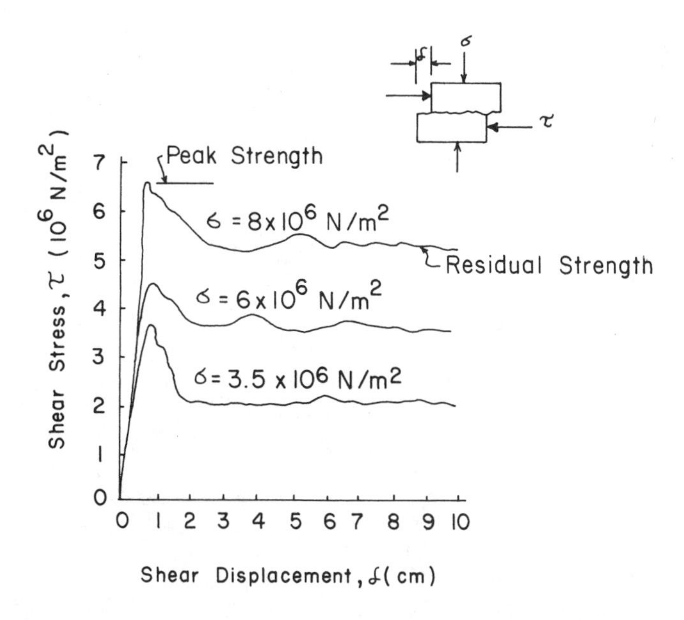


**8.** The world's deepest well, at Kola, USSR, is 12 km deep (Kozlousky, 1984). (The mass of the drill string, 900 tons, exerts enormous stresses on the drill rig; a bottom-hole turbine is used to turn the bit.) Core is very cautiously recovered because of rock bursts. Assuming a density of 2.9 g/cm$^3$, what is the pressure release from rock recovered from 12 km depth?

# 5

# Soil Mechanics

> Human history becomes more and more a race between education and catastrophe.
>
> H. G. Wells

> Nature, to be commanded, must be obeyed.
>
> Sir Francis Bacon

## 5.1 Introduction

Soil mechanics owes its origin to Karl Terzaghi, an Austrian Civil engineer, who published a book *Erdbaumechanik* in 1925. He emigrated to the United States and worked at M.I.T. and Harvard University. Geotechnical engineering developed in this century in response to immense construction projects such as airfields in the 1940s and the vast network of highways and bridges built in the 1950s and 1960s. Soil mechanics rapidly evolved as new techniques and instruments were developed. The basic definitions and soil testing procedures briefly described below are generally widely accepted. Many of the more esoteric theoretical soil mechanics formulas and techniques are not universally accepted; one need only consult soil mechanics texts (Hough, 1957; Terzaghi and Peck, 1967; Lambe and Whitman, 1969; Perloff and Baron, 1976; Holtz and Kovacs, 1981; Das, 1983) in the past several decades to see the rapid changes in development of equations and equipment used in soil mechanics.

Soil has diverse properties, and hence numerous assumptions are required in order to quantify soil parameters used for engineering studies. Nevertheless, many soil engineering formulas and testing techniques have been developed by civil engineers, and soils engineers have successfully designed construction projects in soil. Most predictive models stem from formulas based on empirical measurements; many seem quite crude at first glance, such as Atterberg limits.

Complete enumeration of soil mechanics techniques in this chapter is not practical. Only an introduction to basic concepts is presented. Many applied problems relating to soil or rock mechanics (permeability, pore pressure, slope stability) are covered in other chapters of this book.

It is important for engineering geologists to become familiar with soil mechanics techniques because engineering geologists frequently work with soils engineering (geotechnical) firms, and can help bridge the gap between the civil engineers and the real geologic environment. For example, there is a difference between the engineering classification of soils based on geotechnical measurements and a typical geologic map of the area. An engineering geologist can show a soils engineer how to interpret a geologic map; this can significantly enhance a geotechnical evaluation of an area. Peck (1968) points out that:

> It is my personal experience that engineering descriptions alone do not permit a sufficiently rational classification of subsurface materials for the design and construction of many engineering works in even the best known urban areas. Only if the data are organized on the basis of stratigraphic units does the mass of engineering test data become meaningful.

## 5.2 Soil Texture

Soils can be divided into soil types based on the texture of the surface soil (usually the *A* horizon). The texture of the *A* horizon is frequently determined by visual inspection but may require a sieve analysis. Once the percentages of sand, silt, and clay present in a given sample are known, reference can be made to a chart such as Fig. 5.1 to determine the soil type. Names such as sandy clay loam, clay loam, and silty clay are common.

Soils engineers have established definite limits for the different particle sizes found within a soil. The amount of each particle size group is determined by mechanical analysis. The gravel and sand fractions are determined by sieving; silt, clay and colloid contents are determined by sedimentation tests. There are slight differences in the way various agencies define the soil sizes (Fig. 5.2). The distribution of particle sizes that compose a soil is called the gradation of a soil. For instance, soil consisting of all sizes of debris is called *well graded* by the soils engineers. (Note: a geologist would call this same distribution *poorly sorted.*) Figure 5.3 illustrates a particle size distribution for eight different sediments. The proportional makeup of sediment 5, for example, is gravel 3%, sand 23%, silt 64%, clay 10%.

Figure 5.1 Textural classification of soil (from Asphalt Institute, 1964).

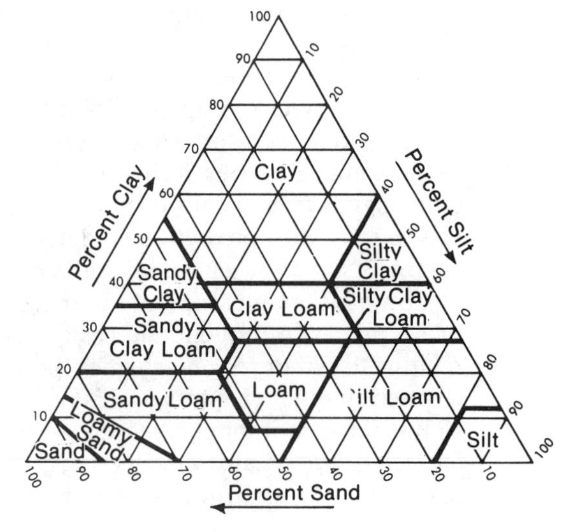

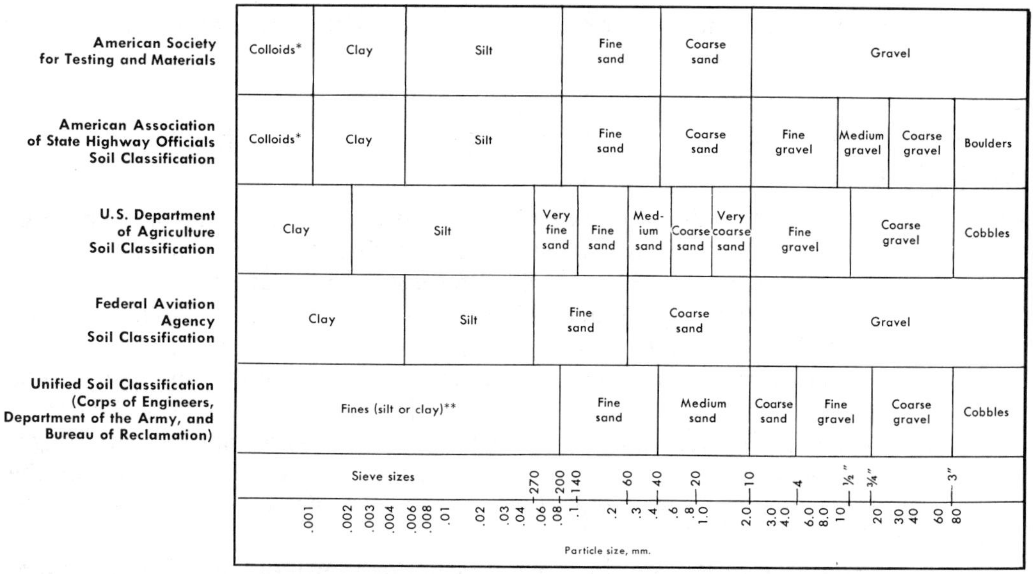

| Agency | Size classes (fine to coarse) |
|---|---|
| American Society for Testing and Materials | Colloids* · Clay · Silt · Fine sand · Coarse sand · Gravel |
| American Association of State Highway Officials Soil Classification | Colloids* · Clay · Silt · Fine sand · Coarse sand · Fine gravel · Medium gravel · Coarse gravel · Boulders |
| U.S. Department of Agriculture Soil Classification | Clay · Silt · Very fine sand · Fine sand · Med-ium sand · Coarse sand · Very coarse sand · Fine gravel · Coarse gravel · Cobbles |
| Federal Aviation Agency Soil Classification | Clay · Silt · Fine sand · Coarse sand · Gravel |
| Unified Soil Classification (Corps of Engineers, Department of the Army, and Bureau of Reclamation) | Fines (silt or clay)** · Fine sand · Medium sand · Coarse sand · Fine gravel · Coarse gravel · Cobbles |

Sieve sizes: 270 200 140 60 40 20 10 4 ½" ¾" 3"

Particle size, mm: .001 .002 .003 .004 .006 .008 .01 .02 .03 .04 .06 .08 .1 .2 .3 .4 .6 .8 1.0 2.0 3.0 4.0 6.0 8.0 10 20 30 40 60 80

*Colloids included in clay fraction in test reports.
**The L.L. and P.I. of "Silt" plot below the "A" line on the plasticity chart, Fig. 5.6. and the L.L. and P.I. for "Clay" plot above the "A" line.

Figure 5.2 Soil size as defined by various agencies (from Portland Cement Association, 1962).

A parameter commonly used by soils engineers is the uniformity coefficient, which is roughly analogous to the standard deviation. The uniformity coefficient ($C_u$) is the ratio of $D_{60}$ to $D_{10}$, where $D_{60}$ is the soil particle diameter at which 60% of the soil weight is finer and $D_{10}$ is the corresponding value at 10% finer (Lambe and Whitman, 1969). Sand is considered well graded if $C_u > 6$. The uniformity

Figure 5.3 Particle-size distribution curves for sediments in Czechoslovakia (Bazant, 1979). 1. Vltava River gravel. 2. "Gap-graded" gravel. 3. Letna terrace, uniform sand. 4. Pankvac terrace, gap-graded clayey sand. 5. Micovna loess. 6. Hodonin silt. 7. Ruzyne clay. 8. Branany bentonite.

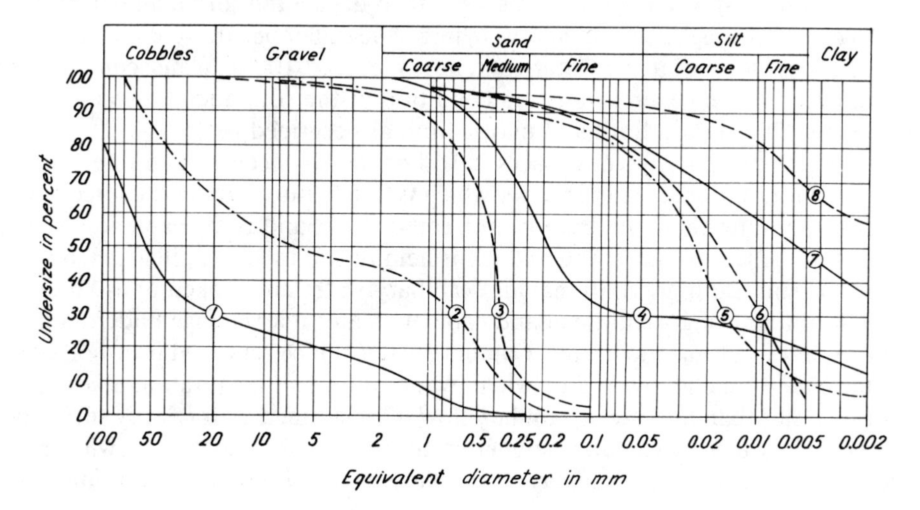

coefficient of sediment 5 shown in Fig. 5.3 is 6.4. This soil would be termed a well-graded silt loam.

## 5.3 Soil Moisture

Anyone who has walked across a dry mud flat knows that it may be impassible on a rainy day. Even small moisture content changes affect soil strength. The strength of London clay, for example, decreases from about $100 \times 10^6$ N/m$^2$ at a moisture content of 22% to about $6 \times 10^6$ N/m$^2$ at a moisture content of 34% (Poulos and Davis, 1980).

Soils engineers use the following criteria to describe soil moisture. Soil can be represented as consisting of three states of matter: solid, water, and air. Figure 5.4 is a schematic diagram for a soil in which all three states are present. Weight terms are shown on the right side of the diagram and volumetric terms on the left. The subscripts, a, w, s, and v are used to denote air, water, solids, and voids, respectively.

The ratios of certain weight and volume terms shown in Fig. 5.4 are employed in soil engineering to estimate settlement, compaction, and other engineering properties of soil. Some of these ratios, or proportions, are defined symbolically as follows:

$$\text{water content: w.c.} = \frac{W_w}{W_s} \times 100\% \,, \tag{5.1}$$

$$\text{void ratio: } e = \frac{V_v}{V_s} \,, \tag{5.2}$$

$$\text{porosity: } n = \frac{V_v}{V_t} \times 100\% \,. \tag{5.3}$$

The water content of a soil is defined as the weight of water in a soil divided by the oven-dry weight of the soil. It is usually expressed as a percentage. (Note: some geologists define moisture content as $W_w/W_t$.) The water content is determined by first weighing the moist soil, then drying the soil in an oven (at 110°C) to constant weight, and then reweighing. The difference in weight gives the moisture loss. Note that the water content can exceed 100%. Water content is useful to describe the general moisture condition of the soil, and is also used in other tests such as liquid limit and plastic limit as described below.

Water can occupy void spaces in soil or rock by occurring in three places:

*a.* Gravitational water is water which is free to move under the force of gravity. It is the water that will drain from a soil or rock, and is available to water wells.

*b.* Capillary water is water which is held in soil by the capillaries (small pores) in the soil. Both gravitational and capillary water are spoken of as "free water." The amount of gravitational and capillary water may change under natural conditions of wetting and drying. Capillary water can be removed from soil only after the water table is lowered or when evaporation or transpiration takes place at a rate faster than the rate of capillary flow. Fine-grained soils (clayey silt) may have a capillary fringe extending a meter or more above the water table, whereas coarse grained soils (gravel) may have a capillary fringe of only a few centimeters.

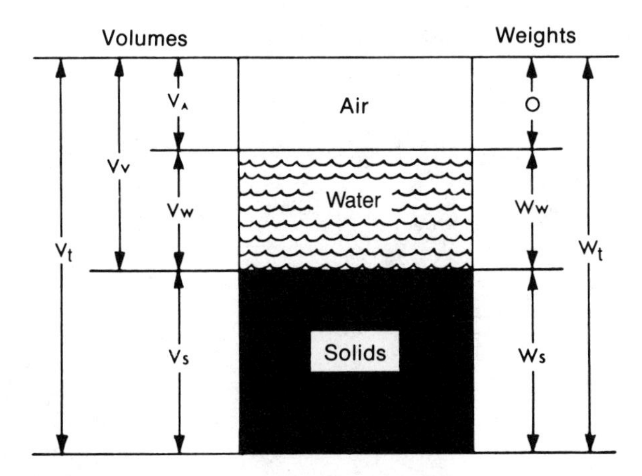


Figure 5.4 Schematic diagram showing three states of matter in a soil (Asphalt Institute, 1964).

*c.* Hygroscopic moisture (absorbed water) is water which is retained by soil after gravitational and capillary moisture are removed. Hygroscopic water is virtually immobile, and is held by soil grains in the form of a very thin film. It has both a physical (electric dipole effect) and chemical affinity for the mineral grain.

## 5.4 Atterberg Limits

Tests have been devised to determine the moisture content of a soil when it changes from one physical condition to another. The tests also show the water-holding capacity of different types of soils. These tests, conducted on the minus No. 40 sieve-size (medium sand or smaller) material, are widely used as parameters to classify soils for structural purposes. The tests, called Atterberg limits, seem crudely empirical when first introduced to a new student, but they are widely used by practicing soils engineers.

### *5.4.1 Liquid Limit*

The liquid limit (LL) is the moisture content at which a soil passes from a plastic to a liquid state. The test is made by determining, for a soil containing different moisture contents, the number of blows of a standard cup (Fig. 5.5) at which two halves of a soil cake will flow together for a distance of about 1.3 cm. Soil moisture versus number of blows are then plotted on arithmetic paper, and the moisture content at which the plotted line (called a flow curve) crosses the 25-blow line is the liquid limit.

The liquid limit can be used to describe the moisture content as a percentage of the oven-dry weight of the soil. For instance, if a soil has a liquid limit of 100%, then the weight of contained moisture equals the weight of dry soil, or the soil (at the liquid limit) is half water and half soil on a weight basis. A liquid limit of 50% shows that the soil (at the liquid limit) is two-thirds soil and one-third water by weight.

Soils with high liquid limits generally indicate high clay content and low load-carrying capacity. A cohesive soil such as silt or clay has liquid limits that may run as high as 100% or greater. Most clays in the United States have liquid limits

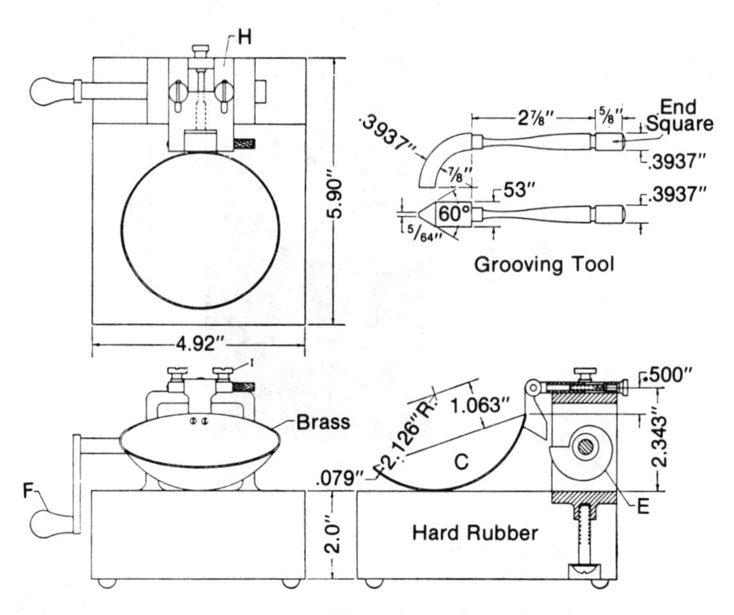


Figure 5.5 Liquid limit device (from Asphalt Institute, 1964).

between 40% and 60%, although clays containing the mineral montmorillonite have a profound influence on the liquid limit. Some "bentonite" (volcanic montmorillonite) clays from Wyoming can have liquid limits near 400% (Gillott, 1968). (In the narrowest sense of the name, bentonite is the name of a pure montmorillonite clay found near Fort Benton, MT. Bentonite is widely used as a general term to describe colloidal clays which disperse in suspension with water to form a colloidal gel.)

The cohesion of soil is a general indication of its strength in that a cohesive soil sticks together and does not deform. Therefore, the liquid limit test is a general index of cohesion because cohesion has been largely overcome at the liquid limit. Cohesionless soil, such as sandy soil, has low liquid limits, on the order of 20%. For sandy soils the liquid limit test is of little significance in judging load-carrying capacity.

## *5.4.2 Plastic Limit*

Plasticity is a characteristic of clayey soils which allows them to be deformed and remolded by hand without disintegration. The plastic limit (PL) of soils is the moisture content at which a soil changes from a semi-solid to a plastic state. This condition is said to prevail when the soil contains just enough moisture that it can be rolled into 3-mm diameter threads without breaking. The test is conducted by trial and error, starting with a soil sufficiently moist to roll into threads 3-mm diameter; the moisture content of the soil is gradually reduced until the thread crumbles.

The plastic limit is governed by clay content. Some silt and sand soils that cannot be rolled into these thin threads at any moisture content have no plastic limit and are termed nonplastic. The plastic limit test is of no value in judging the relative load-carrying capacity of nonplastic soils.

Soils having high plastic limits contain silt and clay, and the moisture content of these soils has a direct bearing on their load-carrying capacity. A very important change in load-carrying capacity of soils occurs at the plastic limit because soil at the plastic limit essentially changes from a solid to a plastic upon addition of water. Load-carrying capacity decreases very rapidly as the moisture content is increased above the plastic limit. On the other hand, load-carrying capacity increases very rapidly as the moisture content is decreased below the plastic limit.

Soils exposed at the earth's surface generally reach or exceed their plastic limit at some season of the year, particularly in humid climates. Therefore, the bearing capacity of the subgrade and pavement design requirements should be based on the lower load-carrying capacity condition rather than on drier and correspondingly higher load-carrying capacity conditions.

### *5.4.3 Plasticity Index*

The plasticity index (PI) is defined as the numerical difference between liquid limit and plastic limit (PI = LL − PL). The plasticity index gives the range in moisture content at which a soil is in a plastic condition. A small plasticity index, such as 5%, shows that a small change in moisture content will change the soil from a semisolid to a liquid condition; this is an undesirable condition for a foundational material. Such a soil is very sensitive to moisture (unless the silt and clay content combined is very low, on the order of less than 20%). A large plasticity index, such as 20%, shows that considerable water can be added before the soil becomes liquid, and the soil is a desirable foundational material. (On the other hand, soils with very high PI (> 35%) may have a high swell capacity (Soil Conservation Service, 1979).) When the liquid limit or plastic limit cannot be determined or when the plastic limit is nearly the same as the liquid limit, the plasticity index is reported as nonplastic.

The use of a weighted plasticity index is used as a design guide for building light foundations on expansive soils (Building Research Advisory Board, 1968). Weighting factors are used in the calculation of the effective plasticity index to emphasize the influence of the uppermost soil properties and minimize the influence of soil at deeper levels. Mathewson and Dobson (1982) have shown that, in addition to plasticity index, stratigraphy and the depth of the active zone (the depth at which soils no longer shrink or swell due to normal seasonal moisture change) also affect soil properties such as shrink-swell activity.

In summary, the liquid limit, plastic limit, and plasticity index—along with the shrinkage limit (rarely used by engineers)—are called the Atterberg limits. A practical use of Atterberg limits can be shown if the liquid limit is plotted as a function of plasticity index. Figure 5.6 shows the plasticity chart for soils developed by Casagrande. Soils possessing high PL and PI are said to be highly plastic or "fat" clays, whereas those with low values are described as slightly plastic or "lean." The *A* line drawn on the plasticity chart is used as a way of differentiating claylike materials (above the *A* line) from those which are siltlike (below the A line). In general, the poorer engineering soils are those which lie below the A line.

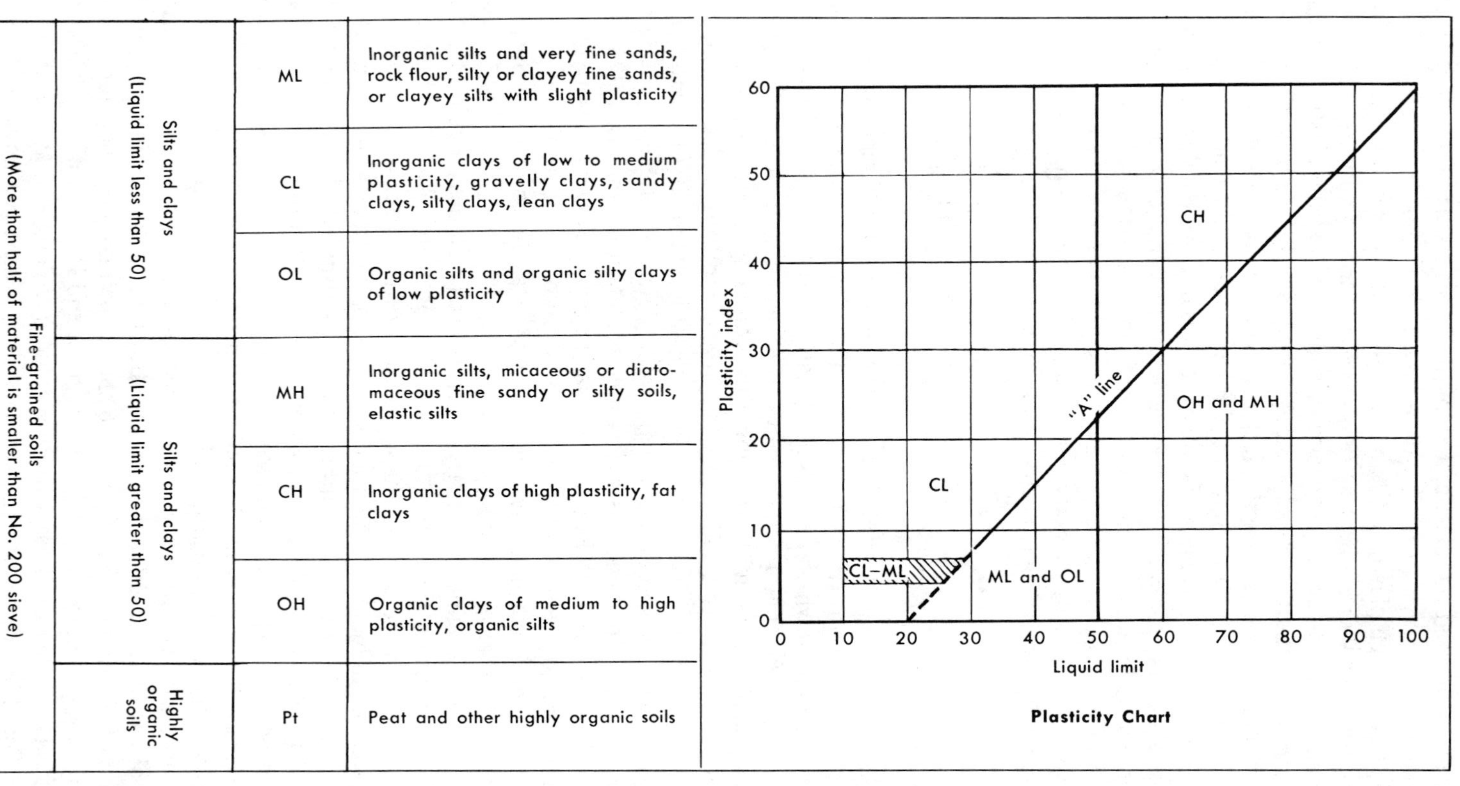

| Fine-grained soils (More than half of material is smaller than No. 200 sieve) | | | |
|---|---|---|---|
| Silts and clays (Liquid limit less than 50) | ML | Inorganic silts and very fine sands, rock flour, silty or clayey fine sands, or clayey silts with slight plasticity |
| | CL | Inorganic clays of low to medium plasticity, gravelly clays, sandy clays, silty clays, lean clays |
| | OL | Organic silts and organic silty clays of low plasticity |
| Silts and clays (Liquid limit greater than 50) | MH | Inorganic silts, micaceous or diatomaceous fine sandy or silty soils, elastic silts |
| | CH | Inorganic clays of high plasticity, fat clays |
| | OH | Organic clays of medium to high plasticity, organic silts |
| Highly organic soils | Pt | Peat and other highly organic soils |



*Division of GM and SM groups into subdivisions of d and u are for roads and airfields only. Subdivision is based on Atterburg limits; suffix d used when L.L. is 28 or less and the P.I. is 6 or less; the suffix u used when L.L. is greater than 28.

**Borderline classifications, used for soils possessing characteristics of two groups, are designated by combinations of group symbols. For example: GW-GC, well-graded gravel-sand mixture with clay binder.

Figure 5.6 Plasticity Chart and Unified Soil Classification for fine-grained soils (from Portland Cement Association, 1962).

Atterberg limits are affected by the clay composition. In general, among clayey soils, liquid and plastic limits are higher for montmorillonite clays than for kaolinites or illites, mainly because montmorillonites are able to disperse into very fine particles with large water-absorbing volume. Atterberg limits also increase with increasing total clay content, and hence reflect low shear values (Casagrande, 1948).

The term liquidity index (LI) is defined as the ratio of the moisture content of a particular soil sample divided by its liquid limit. If the LI $> 1$, the soil exceeds its liquid limit. Costa and Baker (1981) show analyses of moisture contents of mass wasting deposits taken just after movement had occured. Most landslides and slumps have LI $< 1$, but most debris flows and mud flows have LI $> 1$.

## 5.5 Engineering Properties

The basic engineering characteristics of soils—internal friction angle (similarly defined as in rock mechanics), cohesion, compressibility, elasticity, capillarity, and permeability—are interrelated and indicate properties that determine the suitability of soils for engineering use. The strength of the soil—its load-carrying capacity and resistance to movement or consolidation—is related to these parameters. Depending on the proposed use, other properties such as rippability and volume-change characteristics may be useful in evaluating soil properties.

The following sections describe some commonly used soil tests for describing density, compaction, settlement, and strength of soils.

### *5.5.1 Density*

The basic procedure used to determine in-situ moisture content and density consists of removing a sample of soil and 1) determining its wet weight, oven-dry weight, and moisture content, and 2) determining the volume of the cavity previously occupied by the soil. The bulk density is defined as the ratio of the oven-dry weight to the cavity volume. (In situ densities can also be determined by means of undisturbed samples, and there are several methods involving nuclear devices.)

In practice, a soil sample (approximately 13-cm diameter) extending for the full depth of the layer being tested is removed; the resulting cavity should be approximately cylindrical in shape. The moist soil is preserved in a plastic sack, taken to a lab, and weighed, and its moisture content determined. The volume of the cavity or hole is determined by accurately measuring the amount of material of known unit weight required to fill the hole. Sand (sand density-cone method, Fig. 5.7), water (water-balloon method), and oil have been used for this purpose.

The performance of soil underlying pavement structures depends to a great extent upon uniform compaction of the subgrade. Therefore, road-building agencies or other contracting agencies usually control compaction by specifying minimum requirements based on 1) soil density, 2) compactive effort, or 3) a combination of the two. Most agencies specify some minimum density and limit the range of moisture content used in the soil that is being compacted. In most instances the minimum acceptable density is usually specified as a percentage of maximum density.

Figure 5.7 In-situ field determination of soil density.

The in situ density of soil at depth is dependent to a large degree on the weight of the overlying soil. The geostatic or lithostatic loading of soil follows the same principles explained in Chapter 4. A typical dry density of soil is 1.53 g/cm$^3$; therefore, 1 m$^3$ cube of soil would have a mass of 1530 kg/m$^3$. Due to earth's gravity, this mass has a unit weight of 15,000 N/m$^3$. If unit weight is constant with depth, the relationship between depth ($Z$), vertical stress ($\sigma_v$), and unit weight ($\gamma$) is:

$$\sigma_v = Z\gamma. \tag{5.4}$$

If the unit weight of soil varies continuously with depth, the vertical stress can be evaluated by means of the integral:

$$\sigma_v = \int_0^Z \gamma \, dZ. \tag{5.5}$$

For example, suppose that the unit weight of ($\gamma$) soil at the surface is 15,000 N/m$^3$, and laboratory tests show that $\gamma$ increases at the rate of 0.003 times $\sigma$, where $\gamma$ is in N/m$^3$ and $\sigma_v$ is N/m$^2$:

$$\gamma = 15{,}000 \text{ N/m}^3 + 0.003\sigma_v.$$

To find the stress at 30 m, for example, a trial and error solution can be used, or, from calculus, using Eq. (5.5), we obtain:

$$\sigma_v = \int_0^Z (15{,}000 + 0.003\sigma_v)dZ,$$

$$\frac{d\sigma_v}{dZ} = 15{,}000 + 0.003\sigma_v\,.$$

The solution to this linear differential equation is

$$\sigma_v = 5 \times 10^6\,(e^{0.003Z} - 1).$$

For $Z = 30$ m,

$$\sigma_v = 5 \times 10^6\,(e^{0.003(30)} - 1) = 4.71 \times 10^5\ \mathrm{N/m^2}.$$

Some soils show the effects of having been squeezed by a great load in the geologic past. A preconsolidation curve can be determined from a consolidation test. The magnitude of preconsolidation in soils has been used, for example, to approximate the thickness of glacial ice that previously rested on soils (Harrison, 1958). Clayey shale may rebound upon excavation. This phenomenon is due to overconsolidation (Brooker, 1967). It is similar to the weathering process of unloading. Because shale is a rock that has had a history of sedimentation and burial, its stress history includes a time when a large overburden acted upon it. Thus, rebound due to lithostatic pressure release can be expected in shale excavations. According to Bowles (1978), part of the rebound is also due to an increase in the water content of the clay minerals.

### *5.5.2 Compaction*

The term compaction in engineering refers to the practice of artificially increasing the density of a soil mass by rolling, tamping, vibrating, or other means. The density of a soil is measured in terms of its mass per unit volume, expressed as kilograms of soil per cubic meter. Density is designated as either wet density or dry (bulk) density.

Several factors influence the maximum density which can be obtained by compaction. Of primary importance are 1) the moisture content of the soil, 2) the nature of the soil—that is, its texture, gradation, and physical properties—and 3) the type and amount of compactive effort. If soil is very wet, artificial compaction to densify a highway fill or earth dam will not achieve very good results because the soil shifts and bulges as the weight (sheep's foot roller, etc.) moves back and forth. On the other hand, if the soil is too dry, it cannot be compacted easily, either. It follows that there is an optimum moisture content at which maximum density can be achieved.

The American Society for Testing and Materials (ASTM) has standardized the method of testing a soil specimen in order to determine the optimum moisture content necessary to achieve optimum compaction. The Standard Proctor compaction test uses a 944-$\mathrm{cm^3}$ soil specimen which is compacted into five layers in a cylindrical mold. Each layer receives 25 blows by a 2.5-kg mass which free-falls 31 cm. Four or five specimens compacted at water contents within the range of

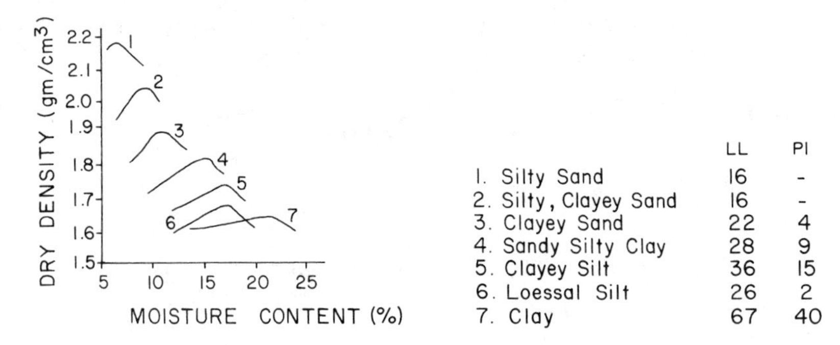


Figure 5.8 Compaction test results (from Das, 1983).

about 2% of optimum water content are usually enough to establish the optimum water content and the maximum density. The test results are plotted in the form of a moisture-density diagram, and a smooth curve is drawn through the points. Figure 5.8 shows typical moisture-density diagrams for compaction tests made with different soils showing coarse-grained soils require slightly less moisture for maximum density than fine-grained soils.

The shape of the moisture-density curve, which may vary from a sharply peaked parabolic curve to a flat curve, gives valuable information concerning the influence of moisture on the load-supporting value of the soil. For example, a flat curve indicates a soil that will have the same load supporting power over a wide range in moisture content.

The main purpose of the modified Proctor laboratory compaction test is to specify moisture content to be used for the field compaction of soil (for earth dams, highway fills, etc.). The moisture content of a fill can be specified so that during construction the engineer in charge can decide whether more or less water should be added during the compaction effort.

### *5.5.3 Consolidation and Settlement*

Consider the stress imposed onto a soil as a result of a load delivered to it from a spread footing of a large building. In order to calculate the settlement of the land expected as a result of the compression of a soil stratum under a given loading, it is necessary to determine the increase in stress caused by the loading at various depths. Land settlement is due to the consolidation (compression) of soil, and is principally controlled by the rate at which water can flow from the soil and the rate at which pore-water pressure is dissipated. Attewell and Farmer (1976) show theoretical and empirical treatments of consolidation based on Terzaghi's (1943) theory. The rate of consolidation can be estimated by certain formulas.

A famous example of settlement is the central domed building of the Massachusetts Institute of Technology. Built on the Boston Blue Clay, the building settled 25 cm. Most of the settlement occurred in the first 10 yr after construction as the weight of the building squeezed water out of the clay.

Differential settlement is of particular concern. An early example is the Washington Monument, 169 m high, which had test borings of only 5.4-m depth in

1876 when it was built. Although at one time it was found to be out of plumb by 44 mm, the tilt was corrected by an ingenious loading system. The Leaning Tower of Pisa has settled 3 m on the south side, but only 1 m on the north side. The tilt further aggravates the situation by increasing the bearing stress on the south side (see Section 5.8.1).

The allowable bearing capacity of a soil is the maximum pressure that can be applied to a soil without leading to excessive settlement (~5 cm). Bazant (1979) shows empirical data for Czechoslovakian standards; for footings for buildings up to five stories high, founded at depths of 1.5 m, and resting on a single geologic condition, the allowable bearing capacity ($N/m^2$) depends on the soil which Bazant has calculated as follows:

unweathered rock = 60,000,000
slightly weathered rock = 15,000,000
weathered rock = 1,000,000
gravelly soils = 400,000 to 1,000,000
clayey gravelly soils = 80,000 to 400,000
sand = 100,000 to 800,000
silt or clay = 50,000 to 400,000

To illustrate how compaction data are used in a settlement calculation, the two following illustrative examples are given.

*Example 1* (modified from Lambe and Whitman, 1969). For an elastic soil, the strain developed as a result of a load applied over a circular area can theoretically be computed by adding up all the stress increments along a vertical line below the load as determined by Eq. (5.4). Equation (5.4) can be used to determine settlement of soft rock as well as soil (Coates, 1967). If $Z$ is very deep, then an empirical relationship is:

$$p = \Delta q \cdot \frac{R}{E} \cdot 2\,(1 - \upsilon^2), \tag{5.6}$$

where:

$\Delta q$ = incremental surface stress,
$R$ = radius of loaded area,
$E$ = Young's modulus,
$\upsilon$ = Poisson's ratio.

For example, suppose a cylindrical water tank of radius 23 m is built on a soil, where $E = 96 \times 10^6\ N/m^2$ and $\upsilon = 0.45$. Assume the tank, when loaded, develops a stress of $2.6 \times 10^6\ N/m^2$ on the soil. The settlement at the center of the tank is:

$$p = \Delta q \cdot \frac{R}{E} \cdot 2\,(1 - \upsilon^2),$$

$$= (2.6 \times 10^6\ N/m^2) \cdot \frac{(23\ m)}{(96 \times 10^6\ N/m^2)} \cdot 2\,(1 - 0.45^2),$$

$$= 1.0\ m.$$

*Example 2* (after Hough, 1957; also see Peck, et al., 1974; Bowles, 1982). Suppose it is determined that a clay formation has an in-place void ratio of 1.4 and liquid limit of 60%. The clay has a thickness of 3 m. The effective overburden pressure ($p$) on the clay is 96,000 N/m$^2$. It is desirable to find the change in thickness of the clay if the stress is increased to 146,000 N/m$^2$.

If the clay is normally loaded, an approximation of the compression index (also called compaction index) $C_c$ may be obtained by the following empirical formula:

$$\begin{aligned} C_c &= 0.008\,(\mathrm{LL} - 10\%), \\ &= 0.008\,(60 - 10), \\ &= 0.40. \end{aligned} \tag{5.7}$$

The change in void ratio ($\Delta e$) is related to $C_c$ and the loading as follows:

$$\begin{aligned} \Delta e &= C_c \log\left(1 + \frac{\Delta p}{p}\right), \\ &= (0.40 \cdot \log\left(1 + \frac{50{,}000}{96{,}000}\right), \\ &= 0.0728. \end{aligned} \tag{5.8}$$

The original thickness ($H$) of a soil stratum will have a settlement ($\Delta H$) due to a change in void ratio as follows:

$$\begin{aligned} \Delta H &= H \cdot \frac{\Delta e}{1+e}, \\ &= 3\ \mathrm{m} \cdot \frac{0.0728}{1+1.4}, \\ &= 0.09\ \mathrm{m}. \end{aligned} \tag{5.9}$$

The settlement calculated by Eq. (5.6) is the final value attained after the termination of consolidation. For noncohesive soils the final value is attained in a few days, but, for cohesive soils, a longer time is required (Bazant, 1979). The rate of consolidation is related to the rate at which water can be squeezed out of the compacting sediment; therefore, the rate of consolidation is related to the permeability and length of drainage path (Holtz and Kovacs, 1981). If total settlement in a local area exceeds 0.1 m, there can be trouble with pipes (gas, water, or sewage). [Total settlement for some structures exceeds 5 m in places where ground water declines occur (Chapter 10).]

Some cities are located on firm bedrock, and have minimal foundation problems. For example, New York City's World Trade Center is founded on solid bedrock, the Manhattan schist. Bedrock at this site is about 20 m below street level, and excavation to this depth through miscellaneous old fills, timber docks, organic silt, and sand posed quite a problem. To prevent ground water from entering the pit, a trench was dug around the perimeter of the site, and a bentonite slurry emplaced to keep the trench wells from collapsing. Reinforcing rods and concrete were then emplaced, forming an impermeable stable wall so that the foundation for the building could be excavated (Kapp, 1969). The soil removed from the giant hole was trucked to a nearby shallow water area, creating a valuable piece of new real estate.

At some places where poor subsoils exist, foundations have been installed by allowing the weight of the building to push the foundation to a desired level. A special cutting edge is laid out around the perimeter of the building, and, as concrete is added to the walls and soil excavated from the interior, the entire structure sinks into the ground. This method of foundation siting has been widely used for the construction of bridge piers. The use of building foundation emplacement by the sinking caisson method has been used successfully in the poor soils in Tokyo and Geneva.

An example of the successful use of settlement is illustrated by the 1967 design of the Portsmouth, NH, interchange of U.S. Route 1-95, where five bridges and 16 km of highway were to be built on 10 m of quick clay. (At this site the quick clay would turn to "soup" if disturbed.) Sand drains were used as a means of speeding up the dewatering of a site to help stabilize the soils. Engineers designed a dewatering method whereby steel casing was very gently inserted into the clay so as not to disturb it, and sand was then poured into the casing and the casing extracted. The sand columns served as drains which allowed for settlement of the quick clay to occur, thus making the soil stronger and able to withstand the load of the new highway.

Differential settlement has been ameliorated at some places by preloading the foundation before construction. This loading can be accomplished by the temporary placement of earth fill over the soil to be compacted.

### *5.5.4 Strength*

The strength, or load-supporting capacity, of soils varies considerably. In addition, the strength of any specific soil can vary under different conditions of moisture and density. Some soils are "sensitive" in that a vibrating motion such as an earthquake causes them to lose their strength. Sensitive soils have natural water contents above the liquid limit. Their tenuous strength, due to a honeycomb soil particle structure in an undisturbed state, is lost due to a disturbance such as an earthquake.

A thixotropic condition in fine sand or silt is due to the rearrangement of grains caused by some dynamic disturbance. When shaken, the grains rearrange themselves, causing a loss of porosity. If the material is saturated so that the water is unable to quickly escape, the water prevents rearrangement of the grains and the material turns into a viscous fluid (Terzaghi, 1950b). Also known as spontaneous liquefaction, the thixotropic effect has caused numerous foundation and slope failures. The liquefaction of an unstable sediment may be caused by vibrations such as those produced by pile driving or quarry blasts.

Quick clay is a form of sensitive soil that is typically deposited in marine water and later leached by fresh ground water so that salt lost in the pores causes the soil to lose its strength when disturbed or loaded. When the strength of a soil becomes zero, a quick condition exists. In 1978, at Rissa, Norway, quick-clay landslides caused 33 hectares of farmland to vanish into a lake within 40 min. Over centuries the upward flow of fresh ground water had leached salt from the pores of the clay. The slight addition of the weight of a farmer's excavated soil on the edge of the lake triggered the soil failure.

Quicksand is loose, saturated sand which is affected by the upward flow of ground water, such as exists in a spring or ground water seep. The shear stress is

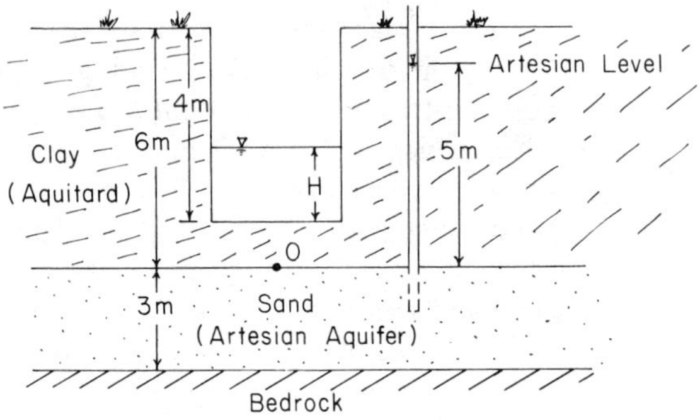

Figure 5.9 Example problem illustrating an artesian aquifer. Determine the depth of water ($H$) necessary to prevent a sand boil. Assume the clay has a density of 2 g/cm$^3$.

zero because upward-moving water forces the sand grains apart to the extent that no stress can be transmitted by the soil skeleton. Thus, the soil behaves as water, i.e., shear stress is zero. [Note: the nontechnical literature abounds with stories of quicksand sucking their victims down. A human body would not sink below the surface of quicksand because the density of the human body (~1 g/cm$^3$) is less than quicksand (~1.5 to 2 g/cm$^3$).]

The following example illustrates the role of water pressure which could create a quicksand (or boiling condition). Figure 5.9 shows an artesian aquifer 6 m below the land surface. The potentiometric surface rises to 1 m below the land surface. A canal is dug, and it is desired to know how deep to keep the water in the ditch so that a sand boil will not occur in the canal. At point 0, the total stress due to the weight of 2 m of clay plus $H$ m of water must be equal to or greater than the uplift pressure due to artesian water:

$$2(\gamma_{\text{clay}}) + H(\gamma_{\text{water}}) = 5(\gamma_{\text{water}}),$$
$$2(2 \text{ g/cm}^3) + H(1 \text{ g/cm}^3) = 5(1 \text{ g/cm}^3),$$
$$H = 1 \text{ m}.$$

The strength of soils often is determined by its ability to withstand shearing stresses. Equation (4.17) is the Coulomb equation relating normal stress, pore pressure, and friction angle to the shear strength of rock or soil:

$$\tau = c + (\sigma - \mu) \tan \phi.$$

Consider the following example of the use of this equation (adopted from Peck et al., 1974): Compute the shearing resistance against sliding along a horizontal plane at a depth of 20 m in a deposit of sand. Assume $\phi = 32°$ and $c = 0$, the water table is at 7 m depth, and the density of moist sand above the water table is 1.9 g/cm$^3$ and the saturated sand below the water table is 2.1 g/cm$^3$. To solve this problem, the total normal stress ($\sigma$) at 20 m depth can be determined as simply the lithostatic load, or (7 m) · (1900 kg/m$^3$) · (9.8 m/s$^2$) + (13 m) · (2100 kg/m$^3$) · (9.8 m/s$^2$) = 397,540 N/m$^2$. The hydrostatic stress ($\mu$) due to the water pressure at 20 m depth is (13 m) (1000 kg/m$^3$) (9.8 m/s$^2$) = 127,400 N/m$^2$. Substituting into Eq. (4.17), we have:

$$\tau = 0 + (397{,}540 \text{ N/m}^2 - 127{,}400 \text{ N/m}^2)(\tan 32°),$$
$$= 169{,}000 \text{ N/m}^2.$$

As an illustration of the complex nature of soil strength, and the historic poor use of geology in many past engineering projects, consider the sensitive marine clay (the Leda clay) found in the St. Lawrence Valley of Quebec. Cylindrical soil samples of this clay are capable of supporting heavy loads; yet the same sample *at the same moisture content* when remolded flows like a liquid. The town of Les Eboulements (eboulement is a French word for landslide) is located on the Leda clay. Legget (1976, p. 316) muses: "How strange and significant it is to think that the planners and builders of a modern housing project, about 80 miles northwest of Les Eboulements, who could have availed themselves of modern geological and geotechnical studies had they wished, placed their buildings on undisturbed Leda clay in a location so critical that on May 4, 1971, a massive landslide—that of St. Jean-Viannet—took place, carrying 40 new homes to destruction and 31 persons to their deaths."

Methods for determining soil strength range from use of complex formulas dependent on tests of cohesion, internal friction, and shear, to simple evaluation of field loading tests on projects in service. Many of the tests and concepts described are the same as those used in rock mechanics (see Chapter 4). A brief summary of some of these tests is given below (modified from Portland Cement Association, 1962).

*Internal friction and cohesion.* Internal friction is the resistance to sliding within the soil mass. Gravel and sand impart high internal friction, and the internal friction of a soil increases with sand and gravel content. For a sand, the internal friction is dependent on the gradation, density, and shape of the soil grain and is relatively independent of the moisture content. Clay has low internal friction that varies with the moisture content. A powder-dry, pulverized clay has a much higher internal friction than the same soil saturated with moisture, since each soil grain can slide on adjoining soil grains much more easily after it is lubricated with the water.

The angle of internal friction is the angle whose tangent is the ratio between the resistance offered to sliding along any plane in the soil and the component of the applied force acting normal to that plane. Values are given in degrees. Internal friction values range from 0° for clay just below the liquid limit to as high as 34° or more for a dry sand; a very stiff clay may have a value of 12°. (Note: according to Lambe and Whitman (1969), rough quartz surfaces have coefficients of friction ($u_s$) equal to 0.5 (friction angle, $\phi = 26°$). The friction angle for quartz sands varies from about 30° for coarse silt to 22° for sand. The friction angle is independent of normal load for nonsheet minerals but for clay particles the friction angle is typically about 13°, and cohesion may be significant.) The governing test should be based on the most unfavorable moisture conditions that will prevail when the soil is in service. This angle of internal friction is not the same as the natural angle of repose or degree of slope of the soil in fills since other factors, such as cohesion, are of influence.

Cohesion is the shearing strength of an unstressed soil. It is affected by the mutual attraction of particles due to molecular forces and the presence of moisture films. Hence, the cohesive force in a particular soil will vary with its moisture content. Cohesion is very high in clay but of little or no significance in silt and sand. Powder-dry, pulverized clay, for example, will probably have low cohesion. However, as the moisture content is increased, cohesion increases until the plastic limit is reached; then a further increased moisture will reduce cohesion. By oven-drying this wet clay, most free water is removed and the remaining moisture will hold the clay grains together so firmly and give the soil such high cohesion that a hammer is required to break the particles apart. These conditions are illustrated, respectively, by the dry dirt road that dusts easily but carries large loads in summer, the muddy, slippery road of spring and fall, and the hard-baked road immediately after light summer rains.

Results of cohesion tests may vary from 0 $N/m^2$ in dry sand or wet silt to $10^5$ $N/m^2$ in very stiff clays. Very soft clays may have a value of $10^4$ $N/m^2$. The governing test should be based on the most unfavorable moisture condition that will prevail during the life of the project.

The stability and hence the structural properties of soil are affected to a large extent by internal friction and cohesion. These combine, in most soils, to make up the shearing resistance. The combined effects are influenced by other basic factors such as capillary properties, elasticity, and compressibility. All these factors and the site on which the soil is located determine the moisture content that will prevail in the soil in service; they also govern the load-carrying capacity, which is the primary concern.

The clay-gravel road made up largely of gravel and sand, with a small amount of silt to fill voids and a small amount of clay to give cohesion, illustrates a soil of high bearing value produced by high internal friction (due to sand and gravel) and high cohesion (due to clay). Wet clay illustrates a soil of low bearing value because internal friction is negligible since no coarse grains are present, and cohesion is low since it has been destroyed by moisture. The same clay, air-dry, will have high bearing value due to high cohesion brought about by the removal of moisture.

Various laboratory tests have been devised to determine the shearing strength of soils; i.e., the direct shear test, the triaxial compression test, and the unconfined compression test. These are briefly discussed in the following sections.

*Direct shear test.* A soil specimen is placed in a split mold and shearing forces are applied to cause one portion of the specimen to slide in relation to the other portion. The test is conducted on specimens at several different loads normal to the shearing force. The unit normal forces applied and the shear stresses of failure are plotted to determine the internal friction and cohesion of the soil. The direct shear test is used for both cohesive and cohesionless soils.

There are vane shear devices on the market which can be used on lab specimens or in the field. There is, however, some disagreement among soil engineers concerning the interpretation of the results (Bowles, 1982).

*Triaxial compression test.* A soil specimen is encased in a rubber membrane and subjected to a constant lateral pressure through a liquid or gas around the specimen. A veritcal axial load is then applied and increased to failure of the specimen. The test is repeated with different lateral pressures. The test data are analyzed graphically by use of Mohr circles to determine the cohesion and internal friction of the soil. (See Sect. 4.6 for an example of a Mohr's circle solution for a soil test.) The results are used in various formulas to determine the load-carrying capacity of the soil for dams, buildings, pavements, etc. Several types of equipment and variations in test procedure have been developed.

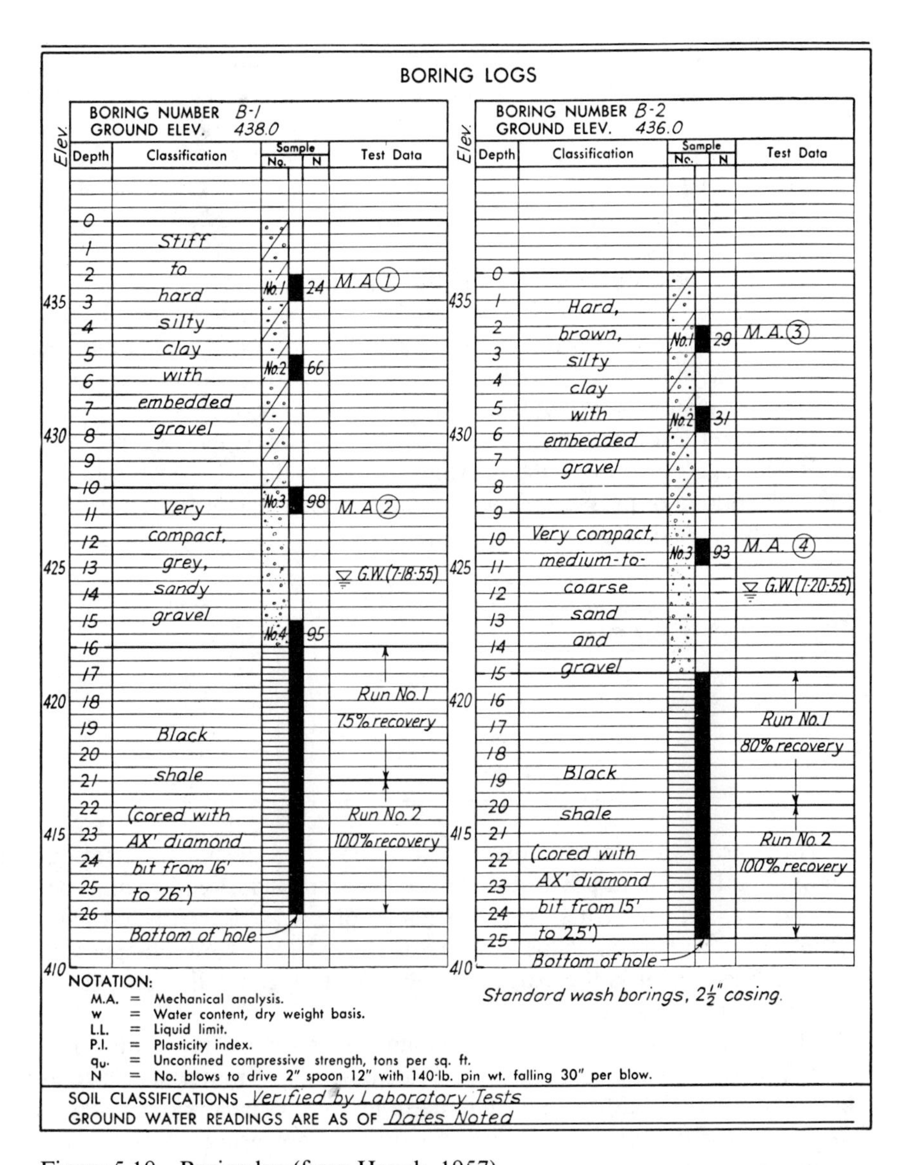


Figure 5.10 Boring log (from Hough, 1957).

*Unconfined compression test.* The unconfined compression test is similar to the triaxial compression test except that no lateral pressure is used. A vertical axial force is applied until the specimen fails along a shear plane or by bulging. The vertical strains or deformations are measured along with the applied load increments. The shear strength is usually assumed to be half of the compressive strength. The unconfined compression test is the simplest laboratory method commonly in use to determine the shear strength of cohesive soils.

*Punching shear test.* Prepared soil samples are placed in a cylinder with a hole in the bottom of less diameter than that of the cylinder. A plunger, varying in diameter from that of the cylinder to less, is forced into the sample at various specified rates, pushing the soil through the hole in the bottom. There are variations in the equipment and procedures used. Test results are used in various formulas to give the soil's bearing value and the thickness of flexible pavement that is required to carry a specified wheel load. Boring logs which describe the number of blows to drive a soil probe a given distance (Fig. 5.10) are a useful variation of the punching shear test.

*California bearing ratio.* A modified punching shear test was developed by the California Division of Highways in which a flat-ended piston of 4.96 cm end diameter is forced into two compacted sample molds of 15.2 cm diameter. There is no opening in the bottom of the specimen mold to permit extrusion of material. The rate of piston movement is controlled, and pressure readings are taken for various penetration depths. The standard of comparison for computing a material's bearing value is calculated using the following relationship between penetration and load or pressure on a "standard" well-graded crushed stone which has the following characteristics:

| Penetration (cm) | Standard load ($10^6$ N/m$^2$) |
|---|---|
| 0.25 | 6.9 |
| 0.51 | 10.3 |
| 0.76 | 13.1 |
| 1.02 | 15.9 |
| 1.27 | 17.9 |

The bearing value of a sample is determined for a specific penetration by dividing the load for that penetration by the standard load for the same penetration. For example, if a specimen requires a load of $3.1 \times 10^6$ N/m$^2$ to obtain 0.25 cm penetration, its bearing value will be $3.1/6.9 \times 100 = 45\%$. The percentage value has become known as the California bearing ratio, generally abbreviated to CBR, with the percent sign omitted.

The Corps of Engineers and some highway departments use the CBR principle in conducting tests to evaluate the bearing value of materials. Several agencies have their own modifications. There are numerous papers in the *Proceedings of the Highway Research Board* and in other engineering publications that give details on the various testing techniques and on data interpretation.

*Expansion tests.* Various test procedures have been developed for measuring the expansion properties of soils. Test results are used primarily to determine the depth (weight) of cover material required to prevent detrimental expansion of the soil when used in a pavement structure. The test specimen is compacted to a predetermined density, at proper moisture content, in a mold, and a supply of water is made available. Surcharges, equal in weight to the weight of the cover material that will overlay the soil in the ultimate pavement structure, are applied to the top of the specimen. The expansion that occurs during some given soaking period is measured at the actual change in length of the specimen, so that the pressure exerted by the expanding soil can be measured by means of a calibrated restraining gauge. The same specimen is then used for the CBR value determination.

*Consolidation test.* This test was devised by Terzaghi to determine the consolidation that would take place in a soil under specific loadings. Sometimes called a compression test and one of the first soil load-bearing value tests evolved, it is used to estimate the settlement that may take place in soil under large structures, such as buildings and bridge piers, and in very high earth embankments.

The test apparatus consists principally of a small, short cylinder that is filled with soil placed between two porous stones. The soil specimen is consolidated by a piston placed on the upper porous stone; any moisture forced from the specimen may escape through the porous stones. The piston is mounted on the short end of a lever arm, with weights on the opposite end. Dials are mounted to measure consolidation.

To conduct the test, the sample is loaded and deformations recorded at stated time intervals. The loads correspond to the anticipated field loads, and the time interval is plotted against the consolidation, in percent. Results are analyzed in terms of determined field conditions. Since the soil sample is completely confined, the test is applied only to field conditions of a similar nature such as building foundations and high fills.

Figure 5.11 Norwegian marine clay; results of a boring in Drammen (from Lambe and Whitman, 1969). The water content in the soil varies seasonally from $w_{min}$ to $w_{max}$.

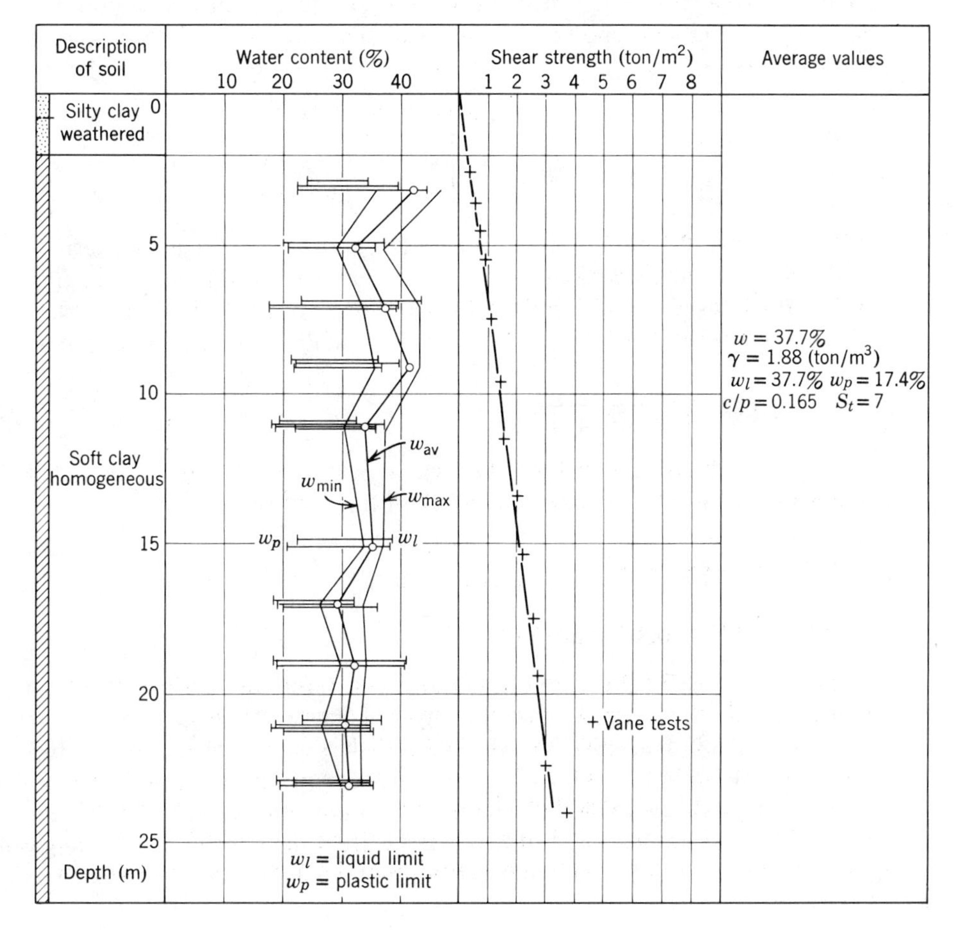

Figure 5.11 illustrates a subsoil profile and data collected from a boring of Pleistocene marine clay in Norway. Note that due to natural consolidation, both the void ratio and water content decrease with depth, and therefore strength increases.

## 5.6 Retaining Walls

Soil pressure on retaining walls and other structures can be roughly estimated by the following so-called Rankine equations. If the wall yields somewhat, the soil pressure reaches a minimum value (active pressure), but if the wall is forcibly restrained against the backfill, the pressure increases to a maximum value (passive pressure). As a rough estimate, it is assumed that the lateral pressure is lithostatic (behaves like fluid but has density of soil or rock), then the active pressure is approximately 0.25 and the passive pressure is approximately 4.0 times the lithostatic pressure, respectively.

For cohesionless soils, a more accurate analysis was developed by Coulomb (Wilbur and Norris, 1948; Bowles, 1982) wherein the soil pressure is resolved into a resultant force. Figure 5.12 shows the cross section of a retaining wall of unit length. The wall has height $H$ and the batter of the wall is defined by the angle $\theta$. Rupture along an inclined plane in the soil is assumed. The resultant force of the triangular area shown acts at one-third the wall height, in a direction making an angle $\phi'$ to a line perpendicular to the wall. The total force $F$ (Newtons) per unit length of wall is:

$$F = \frac{1}{2}\gamma H^2 \left[ \frac{\csc\theta \sin(\theta - \phi)}{[\sin(\theta + \phi')]^{0.5} + \left[ \frac{\sin(\phi + \phi')\sin(\phi - i)}{\sin(\theta - i)} \right]^{0.5}} \right]^2 , \tag{5.10}$$

where $\gamma$ is the unit weight of the soil (N/m$^3$) and $\phi$ is the angle of internal friction of the soil. The angle $\phi'$ is the angle of internal friction for the soil resting on the wall, which has about the same value as $\phi$ for a rough wall but is somewhat less than $\phi$ for a smooth wall. If $\phi = \phi'$, $i = 0°$, and $\theta = 90°$, Eq. (5.10) reduces to

$$F = \frac{1}{2}\gamma H^2 \frac{\cos\phi}{(1 + 1.414 \sin\phi)^2} . \tag{5.11}$$

For example, if we consider sand where $\phi = \phi' = 30°$, and $\gamma = 15{,}000$ N/m$^3$ acting on a 3-m vertical wall, the resultant force on the wall is:

$$F = \frac{1}{2}(15{,}000 \text{ N/m}^3) \frac{\cos 30°}{(1 + 1.414 \sin 30°)^2}$$

$$= 43{,}000 \text{ N per linear m of wall.}$$

Theoretical earth-pressure calculations can rarely be justified for a particular retaining wall, however, because the physical characteristics of the backfill or natural slope are not usually known. Nevertheless, a knowledge of earth-pressure theory permits recognition of the more important variables and their influence on retaining walls. Most soil mechanics textbooks (for example, Lambe and Whitman, 1969; Peck et al., 1974; Bowles, 1982) treat the subject of retaining walls in some detail. Section 6.1.3 has a description of the use of reinforced earth retaining walls.

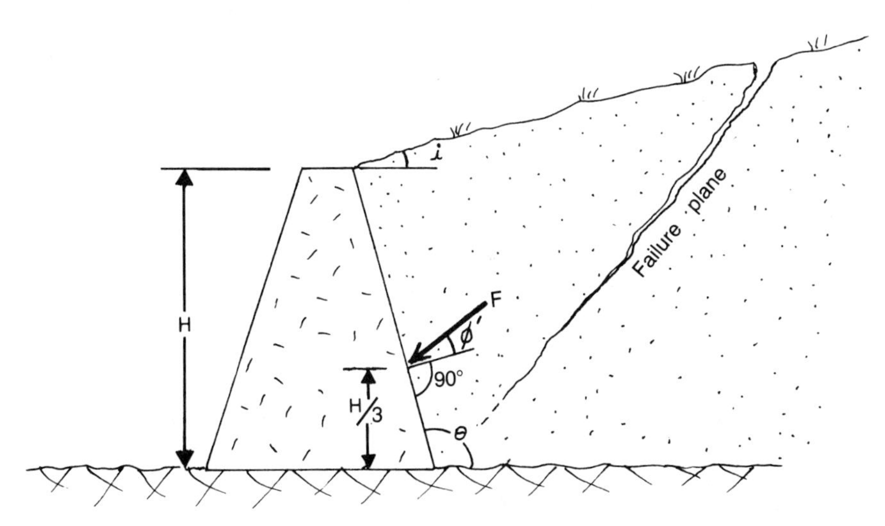


Figure 5.12 Resultant force acting on a retaining wall due to cohesionless soil.

## 5.7 Piles

The use of piles is one of man's oldest methods of making a foundation in soft soils. Where soil is unsuitable for transfer of a load by a slab or spread footing, bearing piles or caissons (Fig. 5.13) are used.

Piles are divided into two types, 1) friction and 2) end-bearing piles. A friction

Figure 5.13 Caisson (drilled pier) being drilled near the state capitol at Pierre, SD (photo by Vernon Bump, South Dakota Dept. of Transportation).

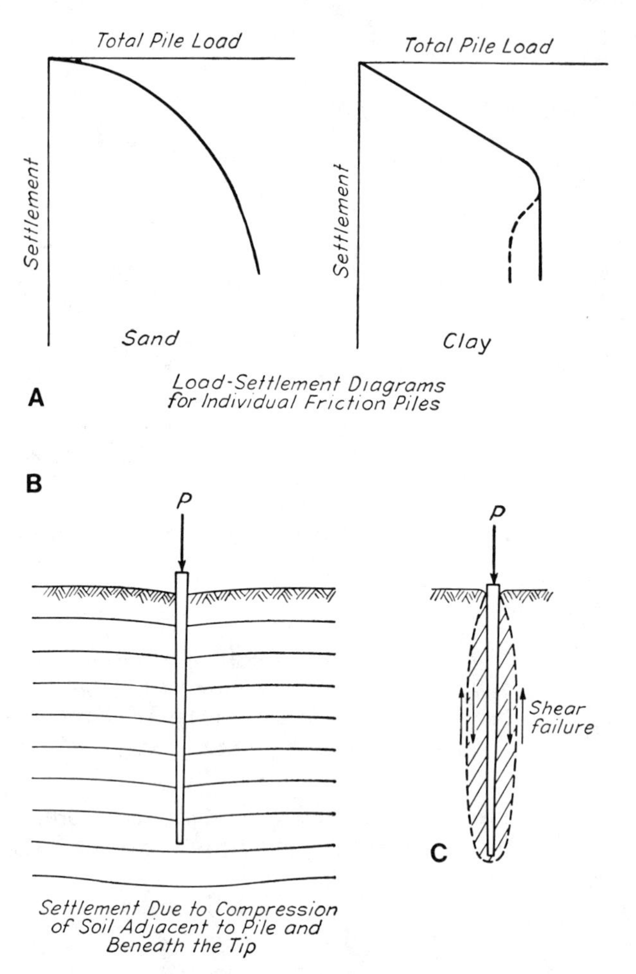


Figure 5.14 Load-settlement diagram for a friction pile (from Hough, 1957).

pile gains support by friction from the sides of the soil it is in contact with (Fig. 5.14), and the length of the pile is determined by the surface area necessary to hold up the load. Figure 5.14 shows how settlement increases in a regular fashion with increased load on a pile founded in sand due to gradual compression in the sand adjacent to the pile. In a clay, however, sudden shearing failure can occur above an ultimate load.

An end-bearing pile simply transmits the load to a firm strata; it is essentially an extension of a vertical column of the structure above, which also receives some lateral support from buckling by the soil. An end-bearing pile is assumed to receive no vertical support from the material in which it is embedded. Piles may be grouped together for additional support, and may be partially friction and partially end-bearing in nature (Fig. 5.15).

Piles may be made of wood, concrete, or steel. Timber piles are still widely used. Wood is remarkable in that if kept completely wet (or completely dry), it will last indefinitely. However, if placed near an area of fluctuating moisture such

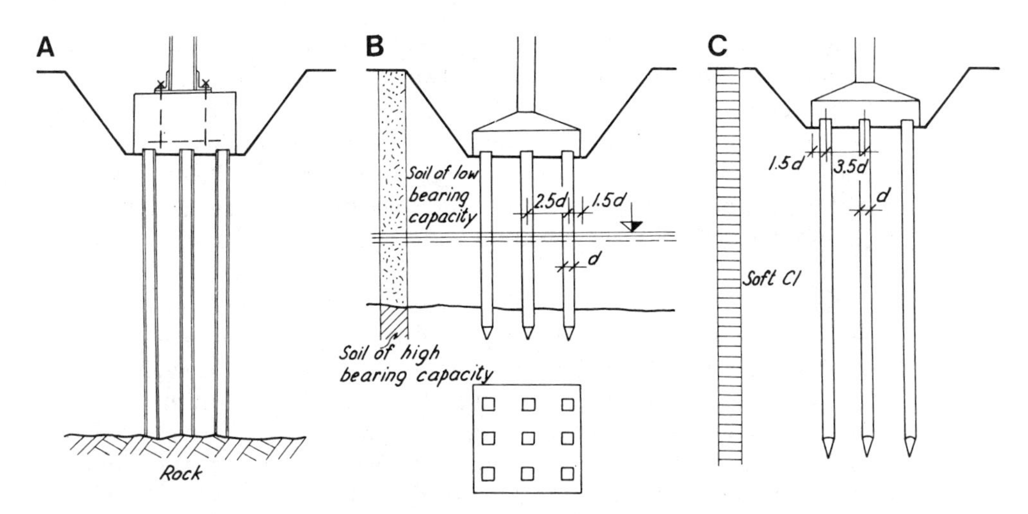


Figure 5.15 Transfer of load from pile to subsoil may be accomplished by (**A**) end-bearing piles, (**B**) partially embedded piles, or (**C**) friction piles (from Bazant, 1979).

as the water table, rotting may occur. Foundation problems in numerous ancient European cathedrals can be traced to rotting timber piles or rafts (flat-lying mats) of logs used as a foundation. Steel piles can be driven through obstructions such as boulders, and are generally used as end-bearing piles due to their great strength.

In New Orleans where the subsoil is largely deltaic mud many kilometers thick, there are few firm subsoil foundations; hence friction piles are utilized. New Orleans was initially built on the relatively narrow natural levees of the Mississippi River. Expansion onto adjacent drained swamps resulted in foundation problems. Early attempts to construct buildings on concrete slabs resulted in sinking or tilting. Present practice utilizes wooden piles which at some locations are driven to an underlying marine clay.

Large-diameter drilled caissons (also known as cast-in-place piles, or drilled piers) are becoming increasingly popular in weak foundation material such as shale where a large diameter end-bearing unit is needed (Fig. 5.13). Western South Dakota, for example, is underlain by Cretaceous shales. The upper several meters of shale bedrock typically weathers to a fissile material with low shear strength and is susceptible to swelling and mass wasting. Caisson holes are typically about 1 m in diameter, 10 m deep, and filled with reinforced concrete. In the 100-story John Hancock Center, built in Chicago in 1970, the footings rest on 239 caissons, 57 of which reach down through glacial till to limestone bedrock at depths of up to 60 m. The remaining 182 caissons rest on hardpan found at depths of up to 25 m. In another example, loads from the Gateway Arch in St. Louis, MO, are delivered by drilled piers through alluvium (variable sand, silt, and clay) to the Mississippian limestone at about 12 m depth. Grouting of the limestone was done to ensure sound bearing capacity (Reitz, 1965). A new wharf built in New Orleans utilizes caissons filled with stones (Munfakh, 1985).

There are two approaches to the calculation of the ultimate load that a pile is capable of supporting: the static approach, which utilizes equations derived from soil mechanics using soil-property parameters, and the dynamic approach, which estimates the ultimate load from analysis of the rate at which a pile is driven into the ground. Soil is not a true elastic substance in that stress and strain are not linearly related, strains are not fully recoverable after reduction of stress, and strains are not independent of time. Nevertheless, strains in soil generally increase as stresses increase. Formulas developed for pile loads involve many simplifying assumptions; if they took into account all possible vagaries of soil properties and behavior they would be too complicated for general engineering use.

Perhaps the oldest and most widely used method of estimating the load capacity of driven piles is to use a driving formula or dynamic formula. All such formulas relate ultimate load capacity to pile set (the vertical movement per blow of the driving hammer). There are over 450 pile-driving formulas with varying degrees of reliability on file with the editors of the *Engineering News Record* (Poulos and Davis, 1980). One of the most commonly used dynamic pile formulas, known as the Engineering News Record formula, is shown below (modified to metric units from Lambe and Whitman, 1969; Bowles, 1982):

$$P_u = \frac{W_r h}{s + 0.025 \text{ m}}, \tag{5.12}$$

where:

$P_u$ = ultimate pile capacity (N),

$W_r$ = weight of the ram (N),

$n$ = height of fall of the ram (m),

$s$ = amount of pile penetration per blow (m) during final 15 cm of driving.

For example, suppose a ram is used which has a mass of 5 metric tons. (The force of this mass due to gravity is 49,000 N.) Assume the free-fall distance is 1 m, and the final penetration rate is 2 cm/blow. Calculate the pile bearing capacity using an F.S. = 6. From Eq. (5.9),

$$P_u = \frac{(49{,}000 \text{ N})\,(1 \text{ m})}{0.02 \text{ m} + 0.025 \text{ m}},$$

$$= 1{,}900{,}000 \text{ N}.$$

If F.S. = 6, then the pile load should be no more than:

$$(1/6)\,(1{,}900{,}000) = 320{,}000 \text{ N}.$$

There are objections to using simple formulas such as Eq. (5.12) because they neglect time-dependent aspects of the transmission of the shock wave from the hammer through the pile tip into the soil (Peck et al., 1974). While there are many pile equations which predict the capacity of a pile to support a vertical load, experience and economic practicality are used by practicing engineers. Often the pile is simply hammered until refusal, or until a certain minimum number of blows is required to drive the pile a given distance, at which point the pile is considered adequately founded. Pile rafts consisting of groups of piles below a single concrete pad have been successfully used in cities with poor foundations such as London and Mexico City.

## 5.8 Foundation Case Histories

Throughout history there have been many foundational failures. Some failures occur during an earthquake, and others are due to a variety of geologic, hydrologic, and loading factors. The Leaning Tower of Pisa in Italy is perhaps the best example of a foundation failure. In addition to the Leaning Tower of Pisa, two catastrophic dam failures are discussed below—the St. Francis Dam in California (1928) and the Teton Dam in Idaho (1976). Both caused great loss of life and property, and both were the result of geologic foundation conditions which were ignored, improperly interpreted, and incorrectly treated. By way of contrast, the Brooklyn Bridge has a good foundation.

### *5.8.1 The Leaning Tower of Pisa, Italy*

The Leaning Tower of Pisa, completed in 1350 A.D., rests on a circular slab foundation 19.2 m in diameter. It exerts a unit pressure of 88 tons/m$^2$ on surficial deposits consisting of 3.1 m of clayey sand underlain by 6.4 m of sand (Legget, 1973). Beneath the sand is brackish clay, which is believed to be causing the differential settlement. In hindsight, it is unfortunate that no recognition was given to the engineering geology of the tower (Fig. 5.16). The following description of the foundation problems of the tower is given by Bolt et al. (1977):

> Construction of the tower was begun in 1173 at a site about 50 m distant from the cathedral of Pisa for which it was intended to be the bell tower. The separation distance probably indicates a recognition on the part of the builders of the problems involved with the subsoil at that area, as the cathedral itself, built 100 years earlier, shows unmistakable evidence of substantial differential settlement and subsequent repairs.
>
> Some time after construction of the tower commenced, it was observed to be leaning and construction was stopped just above the third gallery or level in 1185. The tower remained in this condition for almost 100 years, until 1274 when work recommenced. Another 10 years of effort saw the structure up to the seventh level and almost finished. Only the highest level, the chamber for the bells, still remained to be added, but this period of construction had caused the tilt of the tower to accelerate greatly, so work was stopped again. By this time, it is estimated that the cornice at the seventh story was out of plumb with respect to the base by almost a meter. This had increased to over 1.5 m by 1350 at which time a final 5 years' effort saw the construction of the bell chamber or gallery and completion of the entire structure.
>
> The tilting has continued, and now amounts to slightly over 5 m at the seventh floor cornice, while still increasing at a rate of about 0.02 mm a year. The angle of inclination of the 59 m high tower (measured from the foundation base) is nearly 5°. Each time that construction was recommenced an attempt was made to straighten the tower. Thus, in its final form, the tower is slightly banana shaped, with the form of a tree situated on a hillside which is creeping downslope. A fact which is less often observed about the Tower of Pisa is that it has also settled on the average more than 2 m since it was built, and the entrance to the structure is now this distance below ground surface.
>
> The tilting of the tower is attributed to the presence in the substrate of a layer 2 m thick of compressible clay whose upper surface is only half a meter below the massive foundation. The clay overlies a 4 m sand layer, which, in turn, rests on a

**"...and we can save 700 lira by not taking soil tests."**

Figure 5.16 Cartoon (from Robinson and Spieker, 1978).

thicker clay stratum. For such a massive structure, the original foundation was placed at a very shallow depth of only 2 m below ground surface. The consequence of the tilting is that the pressures on the soil on the downhill side are increased and on the other side decreased. The maximum and minimum pressures are now 9.8 and 2.7 kg/cm$^2$ respectively. Greater settlement accompanies the higher stress and the lower movement is therefore inherently unstable.

It seems likely that the foundation suffered a partial bearing capacity failure in the early stages of construction, perhaps because of shearing in the thin clay layer or in the clay and the underlying sand, which would have caused a sudden tilting immediately apparent to the builders and may have been the cause of the first halt in the work. With the passage of time, the excess pore pressures dissipated, the clay consolidated and became stronger under the load, so that it was able to take the increased stresses of the second period of construction. This must have been the most exciting time, with the increase in the rate of tilting and settlement. At about the time of construction of the 5th story in 1278, the structure must have been very close to complete collapse due to foundation failure. It was saved only by the extreme slowness of the progress, which permitted the clay to pick up strength as it drained and consolidated under the increasing load. The continued movements now are probably the consequence of viscous flow or creep of the underlying soil material at a relatively constant volume under the very high stresses to which it is subjected.

The more recent attempts to rescue the Leaning Tower of Pisa, including the injection of concrete into the foundation in 1935, have only made matters worse. Over the past series of years, a series of committees have been set up to find a solution. Their findings have reinforced the conviction that something must be done, but they have produced no agreement as to what needs to be done. The last committee disbanded in 1980.

### *5.8.2 The Brooklyn Bridge, New York*

The Brooklyn Bridge, built during 1870–1883, is one of the most successful of early American engineering projects (Steinman, 1945; McCullough, 1972). The suspension bridge connects Brooklyn to New York City, and has a span of nearly 500 m across the East River. The construction of the foundations for the two massive towers presented considerable difficulty. The engineer in charge, Washington A. Roebling, designed a large pneumatic caisson to allow for excavation of mud and unconsolidated sand and gravel below the water level.

A pneumatic caisson is essentially a large diving bell. It is a large, airtight loop, having a roof and sides but no bottom. Roebling's caisson was built of solid timber. The walls were tapered to a cutting edge at the bottom, shod with iron casting. After it was floated into position, masonry was added to the top, and compressed air was injected to keep water from entering the chamber. Men, working in the 3 m high chamber, dug out the mud, sand, and gravel so as to permit an even sinking of the caisson under the superimposed weight of the rock masonry that was simultaneously being added upon its roof. Large boulders of glacial drift (basalt, quartz, gneiss, and sandstone) were encountered under the cutting shoes of the caisson as it settled. These boulders had to be broken apart by hand, and caused considerable delay in the foundation work.

The caissons were approximately 30 m $\times$ 50 m in plan, and about 7 m thick. The caisson for the New York tower eventually came to bear on firmly cemented gravel about 0.3 m above crystalline bedrock, at a depth of 24 m below the river bottom. At the Brooklyn tower, bedrock occurs at 28 m depth, but the caisson excavation terminated at a resistant gravel layer at 14 m. After excavation ceased, the entire working chamber was filled with concrete. The Brooklyn Bridge towers were then built with about 105 tons of granite and limestone to a height of about 100 m. The towers have not deflected or settled since they were built.

The successful engineering of the Brooklyn Bridge was countered by the medical problems for the workers. Little was known in the last century about nitrogen introduced into the bloodstream when a worker enters normal atmospheric pressure after having worked in the higher atmospheric pressure of the pneumatic caisson. In 1872, Colonel Roebling was carried from the caisson, insensible due to an attack of the bends. He remained painfully paralyzed the rest of his life.

By way of contrast, the Golden Gate Bridge, built in San Francisco in 1933, had less severe foundation problems. The Golden Gate Bridge is the tallest and largest suspension bridge ever built. The massive concrete piers and anchorage are founded on firm bedrock, withstanding the scouring produced by tidal currents through the Golden Gate channel.

### *5.8.3 The St. Francis Dam, California*

The St. Francis Dam was located in San Francisquito Canyon, about 73 km northwest of Los Angeles. It was completed in 1928, and was an immense concrete gravity arch dam, the largest built to date anywhere in the world. It was 63 m high and 214 m long. The bedrock at the site consisted of the Precambrian Poloma Schist on one side, and an Oligocene fanglomerate (Sespe Formation) on the other side. A fault zone, consisting of 2 m of sheared and brecciated rocks separate the sedimentary and metamorphic rocks. On March 12, 1928, at 11:57 P.M., just as the reservoir was reaching complete capacity for the first time, the dam failed suddenly, and a peak discharge of about 22,000 $m^3/s$ was released down the valley (Outland, 1977). Over 500 lives were lost and property damage exceeded $10 million.

The foundational failure is believed to be due to the soluble nature of the sedimentary rocks and the fault breccia. Gypsum, a very soluble mineral (Table 3.1) was found throughout the fanglomerate. According to the April 5, 1928, *Engineering News-Record* (Anon., 1928):

> Above the fault plane the conglomerate is traversed in various directions by intersecting fractures, some of which contain small seams of clay gouge and others are filled with gypsum. . . . When dry the rock is moderately hard and fragments of considerable size can be broken out and trimmed down with a hammer to specimen size. When, however, a piece of the rock is placed in water a startling change takes place. Absorption proceeds rapidly, air bubbles are given off, flakes and particles begin to fall from the sides of the immersed piece and the water becomes turbid with suspended clay, and usually in from fifteen minutes to an hour a piece the size of an orange has disintegrated into a loose sand and small fragments covered by muddy water.

It is clear that percolating water from the reservoir entered the foundation near the fault contact, resulting in the softening of the rock until a blowout occurred, quickly followed by collapse of the dam.

The failure of the St. Francis Dam, probably more than any other factor, precipitated serious interest in engineering geology in the United States. Clements (1966) describes the geology of the site, and offers the following commentary to the sad state of affairs on that fateful evening:

> The writer was a graduate student at the California Institute of Technology at the time of the failure of the dam, and for some time had been mapping the southeast portion of the Old Tejon Quadrangle, in which the (dam) site is located. He had mapped a fault across the ridge and up San Francisquito Canyon, but when he found that it passed under the St. Francis dam he began to doubt his own competence, for he thought that surely no one would build a dam across so large and obvious a fault. He was seeking further evidence in the canyon below the dam the day before it failed. It was raining and he decided not to camp that night at his favorite spot under a large cottonwood tree. Instead he drove around and went up Charley Canyon, the next

canyon to the west, where the small model T got stuck in the mud. After finally extricating it, he gave up in disgust and returned to Los Angeles. That night the dam gave way.

Failure of the dam could have occurred from any one of three causes: (1) slipping of the schists on the easterly side of the canyon along the planes of schistosity; (2) slumping of the fanglomerate and associated rocks on the westerly side of the canyon as the result of their becoming soaked with water, or (3) seeping of water under pressure along the fault and the washing out of the gouge. When failure occurred, chunks of the westerly section were carried more than a mile down-stream, most of the easterly section collapsed almost in place, and the center section remained standing.

The evidence suggests that failure occurred as the result of seepage along the fault plane until finally the soft gouge was washed out. The stream of water pouring through quickly enlarged the opening by attack on the weak, water-soaked fanglomerate. The entire westerly abutment then gave way causing collapse of the westerly section of the dam, pieces of which were carried far down stream by the sudden rush of water.

As the water gushed from the now enlarged opening it swirled across the canyon and undercut the schist of the easterly abutment. With the removal of the support of the base the schist slid down into the canyon carrying the easterly section of the dam with it. Although later surveys seem to indicate that the central section of the dam was moved slightly, it remained standing.

The many inquiries that followed the disaster brought out no evidence that either the construction or the design of the dam was faulty. Although suggestions were made that the structure had been dynamited or that movement on the fault had caused the failure, these were discarded for lack of evidence. The consensus of opinion of all competent engineers and geologists was that the dam failed because of adverse geological conditions at the site which either were unrecognized or ignored. As a result, more than 500 persons lost their lives and more than $10,000,000 worth of damage was done to property.

### *5.8.4 The Teton Dam, Idaho*

The Teton Dam was built in 1976 by the U.S. Bureau of Reclamation. It was an earth fill dam constructed across a deep canyon on the Teton River, a tributary to the Snake River in southeastern Idaho. The dam was 94 m high and 960 m long. The Teton Dam failed just as the reservoir was filled for the first time. Figure 5.17 shows a series of photographs by Mrs. Eunice Olson who happened to be visiting the site on the morning of June 5, 1976. A wet spot developed in the fill very close to the foundation at the right abutment at 9 A.M. At 10:30 A.M. two bulldozers began attempting to push material into the seep, which began sloughing material and eroding back into the fill. Both dozers were soon lost. At 11 A.M. a whirlpool developed in the reservoir near the right embankment shoreline. At 11:55 A.M. the dam completely collapsed. It took only 5 hr for the major portion of the 1-$km^3$ reservoir volume to empty (Farina, 1977). An enormous wall of water swept over the flood plain below, killing 14 persons and causing an estimated property damage at $1 billion. The USGS estimated the peak discharge of the flood at 57,000 $m^3/s$, which equals the nation's highest peak discharge ever observed (Mississippi River flood of February 17, 1937).

The bedrock foundation at the Teton Dam site consists of rhyolite, a welded ashflow tuff, of Cenozoic age. High angle joint systems were recognized, and drill

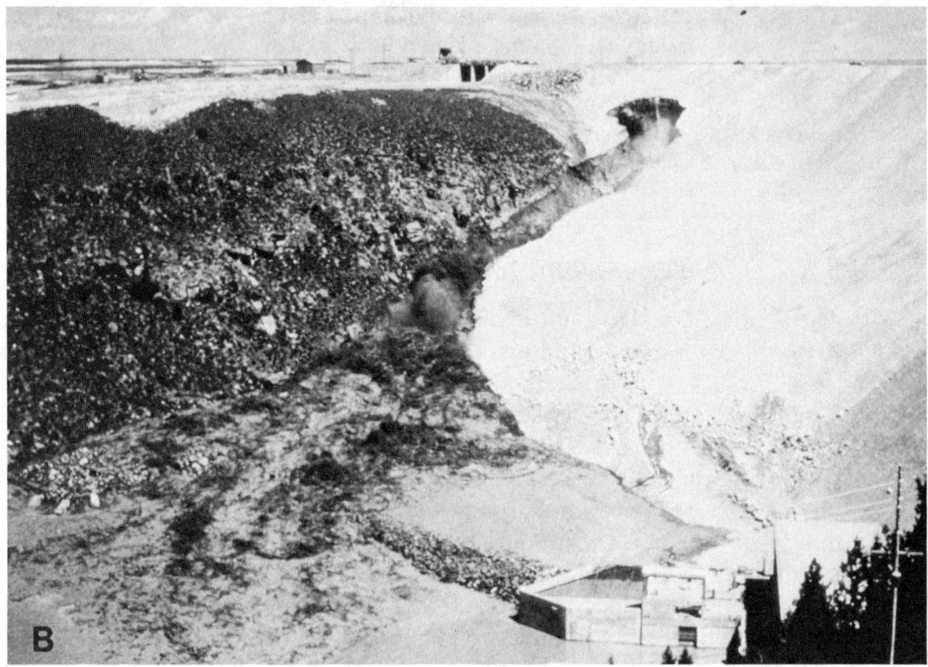


Figure 5.17 Teton Dam, ID. This sequence of photographs was taken by Mrs. Eunice Olson, who just happened to witness the failure. **A**. 11:00 A.M. Initial leak (*A*) is visible on the right abutment. A D-8 caterpillar (*B*) is moving down to try and stop the leak. **B**. Approximately 11:30 A.M. The leak has enlarged to a large crevasse. Two caterpillars were lost into the widening hole. **C**. Approximately 11:40 A.M. The crevasse has eroded almost to the top of the dam. The power plant below the dam is almost completely inundated. **D**. 11:57 A.M. Complete collapse of dam. One 18-ton caterpillar was subsequently found 11 km downstream.

C

D

tests showed they were quite permeable. Test holes consumed up to 30 l/s of water. The bedrock was so intensely jointed at places that there was concern as to whether a grout curtain could actually make the foundation watertight. On the right abutment, 20 m of jointed rock was removed. The total grouting program required 36,000 lineal meters of drilled hole, 14,000 $m^3$ of cement plus sand and other additives. Grouting of the dam was conducted in an attempt to seal the fractures; however, it is now believed that the grouting was insufficient or may have actually opened more fractures by hydraulic fracturing because of high injection pressures. Further, the rapid erosion of the earthen dam itself was attributed to the use of loess for the embankment material for the impervious core material. Loess, a highly erodable wind-blown silt, was apparently selected at the last minute to make the benefit/cost ratio of the entire dam project more attractive.

Despite the U.S. Government's experience in dam building, complex geologic conditions at the Teton site coupled with poor engineering decisions caused the dam failure. According to an independent review committee (Chadwick et al., 1976), " . . . the design of the dam did not adequately take into account the foundation condition and the characteristics of the soil used for filling the key trench."

Following the dam failure, criticism of the U.S. Bureau of Reclamation was made for unheeding local geologists, Robert C. Curry of the University of Montana and David L. Schleicher of the USGS, who warned of the poor foundation and danger of flooding if the dam collapsed (Boffey, 1976). More severe criticism was made to the effect that the USBR risked construction short cuts in order to keep construction costs low: "BuRec's design professionals . . . gambled and lost when extra-heavy runoff filled the dam's reservoir faster than problems could be detected and repairs made. But it is important for the public to know that this kind of pork barrel monstrosity, built on jerry-rigged economic 'analyses' to benefit local real-estate and farming interests with good political connections gets forced upon us just about every day" (Ross, 1977).

## Problems

1. (From Bouwer, 1978) For the soil sample described below, calculate the following parameters: *a.* water content by weight, *b.* porosity, *c.* void ratio, *d.* saturation percentage (ratio of water volume to total void volume), and *e.* bulk density. Data is for an undisturbed core sample of sandy soil taken above the water table. The net weight of the sample is 419 g before drying and 371 g after drying. The core sample is 10.19 cm high and has a 5 cm diameter. Assume the solids are quartz grains ($\rho = 2.65$ g/$cm^3$).

2. Organic clay from New London, CT, and gumbo clay from Arkansas both have a liquid limit of about 65%, but the plasticity index of the Connecticut clay is about 25% whereas the Arkansas clay is about 50%. Plot data on Figure 5.6 and discuss significance.

3. (From Terzaghi and Peck, 1948.) A sample of dense, dry sand is subject to a triaxial test. The angle of internal friction is about 37°. If the minor principal stress (the confining pressure) is $2 \times 10^3$ N/$m^2$, at what value of the major principal stress (the compressive stress) is the sample likely to fail?

4. Solve Problem 3 on the assumption that the sand has a slight cohesion, equal to $1 \times 10^2$ N/m$^2$.

5. Refer to Fig. 5.11. Below what approximate depth should foundations be anchored so that the highest seasonal value of water content ($W_{max}$) does not exceed the liquid limit?

6. (From Lambe and Whitman, 1969.) A sieve analysis on a soil yields the following results:

| Sieve | 3 in. (7.6 cm) | 2 in. (5.1 cm) | 1 in. (2.5 cm) | 0.5 in. (1.2 cm) | #4 | #10 | #20 | #40 | #60 | #100 | #200 |
|---|---|---|---|---|---|---|---|---|---|---|---|
| Cum. % passing by wt | 100 | 95 | 84 | 74 | 62 | 55 | 44 | 32 | 24 | 16 | 9 |

   **a.** Plot the particle size distribution as a curve on the graph below.
   **b.** Classify the soil according to the scale shown on the graph.
   **c.** Determine the uniformity coefficient.
   **d.** Make a general comment on the suitability of this soil as a drainage material behind a concrete retaining wall. (Hint: to drain properly, the soil must be very permeable. To mitigate frost heave, less than 3% of the soil should be finer than 0.02 mm in size.)

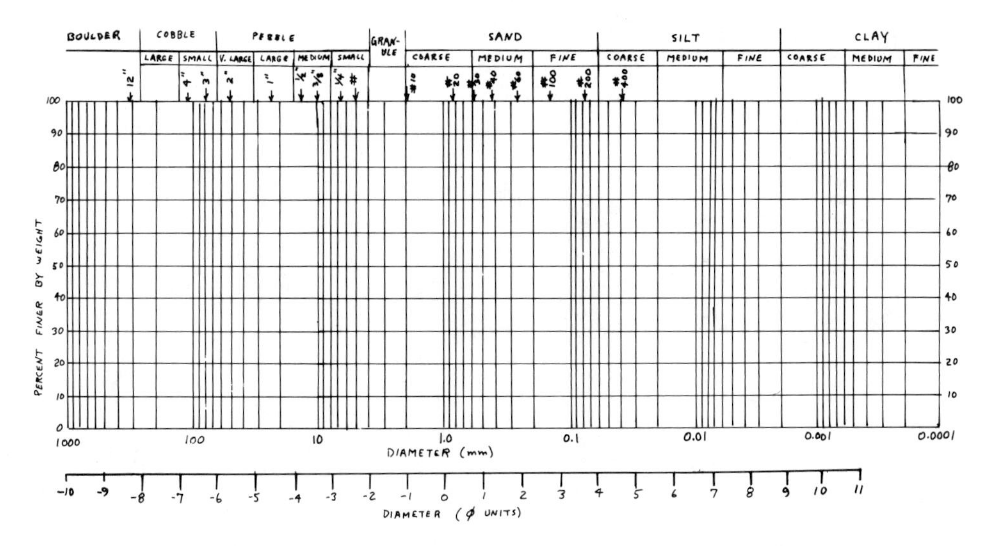


7. (From Lambe and Whitman, 1969.) Given the following principal stresses (N/m$^2$) from standard triaxial tests upon a dense, well-graded coarse quartz sand. Solve for $c$ and $\phi$ using a Mohr Circle.

| Confining pressure ($\sigma_3$) | Peak axial stress ($\sigma_1$) |
|---|---|
| 718 | 3640 |
| 1440 | 7090 |
| 2870 | 14,800 |
| 5740 | 29,000 |

**8.** The concrete retaining wall below is acted on by sand having a unit weight of 14,000 $N/m^3$ and an angle of internal friction ($\phi$) of 35°. The friction angle ($\phi'$) between this sand and the face of the wall is 30°. Determine the magnitude, direction, and point of application of the resulting force, in newtons per linear meter of wall.

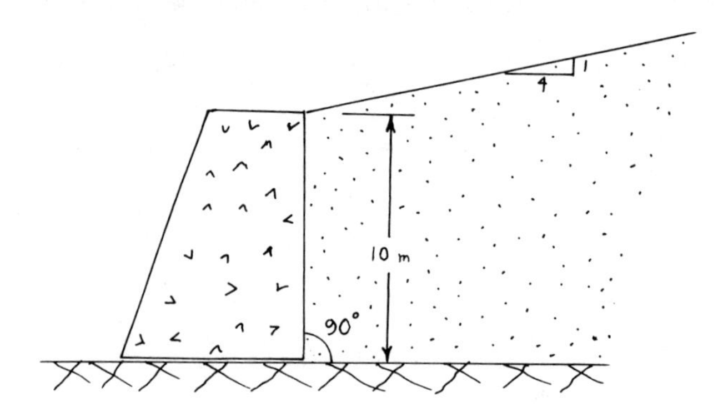

# 6

# Mass Wasting

> Few phenomena gave me more delight than to observe the forms which thawing sand and clay assume in flowing down the sides of a deep cut on the railroad through which I passed on my way to the village. . . . When the frost comes out in the spring, and even in a thawing day in the winter, the sand begins to flow down the slopes like lava, sometimes bursting out through the snow and overflowing it where no sand was to be seen before. Innumerable little streams overlap and interlace with another. . . . It is wonderful how rapidly yet perfectly the sand organizes itself as it slows, using the best material its mass affords to form the sharp edges of its channel.
>
> There is nothing inorganic. . . . The earth is not a mere fragment of dead history, stratum upon stratum like the leaves of a book, to be studied by geologists and antiquaries chiefly, but living poetry like the leaves of a tree, which precedes flowers and fruit—not a fossil earth, but a living earth. . . .
>
> Henry David Thoreau

## 6.1 Mass-Wasting Processes

Flowing sand, landslides, and rock falls are all examples of mass wasting, the general term for gravitationally induced downslope movement of debris. Mass-wasting processes operate on sloping lands; eventually the debris reaches a stream or gully where fluvial processes continue the erosional process.

Mass-wasting processes are more important than fluvial processes in terms of the total geomorphological "work" performed in eroding the land. Consider the Grand Canyon: The volume of the canyon eroded by mass wasting on the canyon slopes far exceeds the volume of rock actually scoured by the downward corrosion of the Colorado River and its few tributaries in the canyon. The Colorado River does, nevertheless, remove all the products of erosion, including debris scoured in the stream bed as well as debris delivered to the stream by mass-wasting processes.

Many schemes have been devised to classify mass-wasting processes. Nature abhors classification; no single classification scheme is perfect. Figure 6.1 is a classification of mass-wasting processes used by the Highway Research Board. This scheme considers falls, slides, and flows in bedrock, soils, and unconsolidated deposits. For example, the process described above by Thoreau would be a sand flow. For the purpose of this book, and as a descriptive convenience, mass wasting is simply divided into four processes: rockfall, creep, landslide, and debris flow. The types of processes are gradational in this classification. The moisture

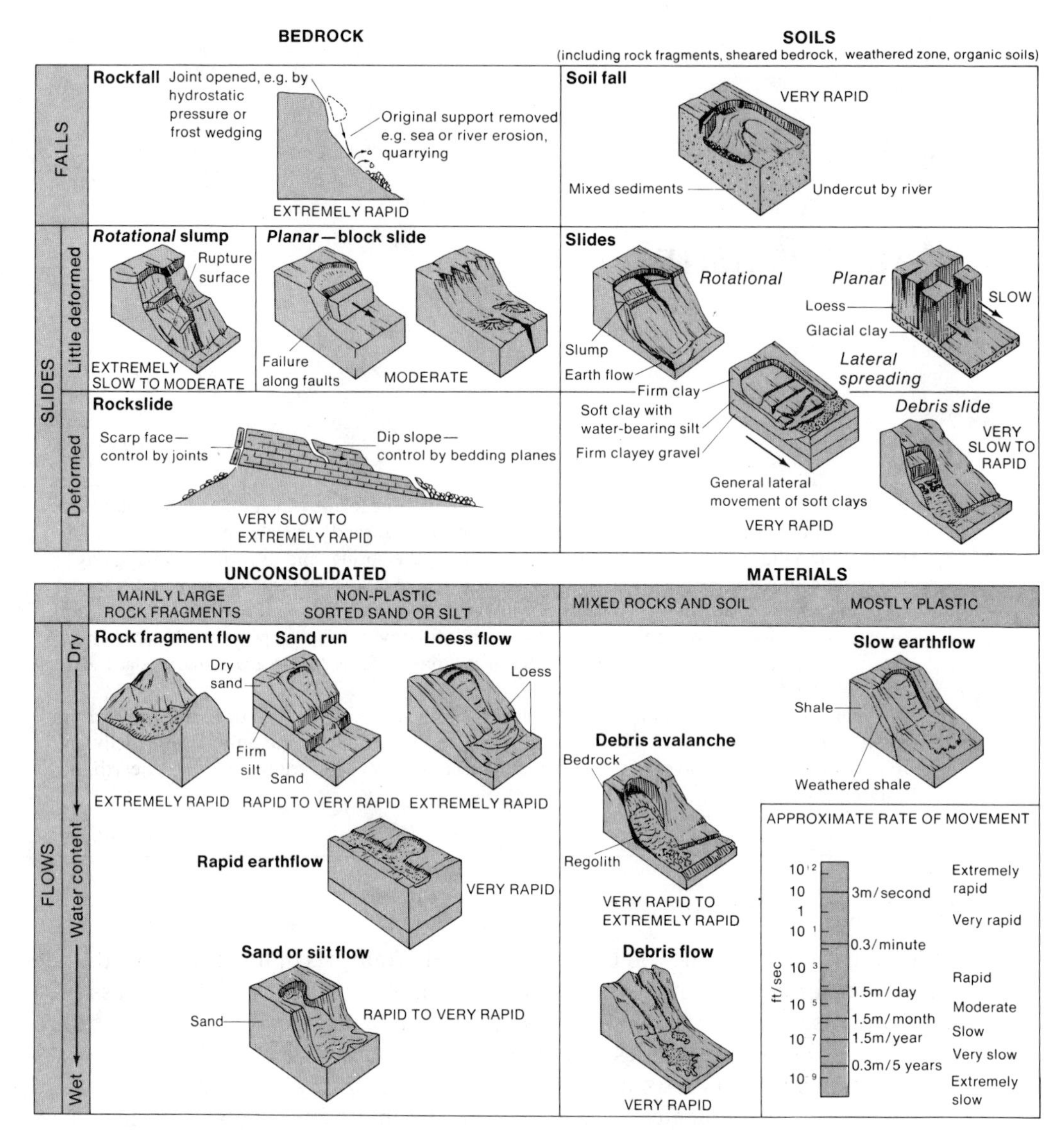


Figure 6.1 Classification of landslides (from Varnes, 1958).

content increases from rockfall to debris flow, and, ultimately, a very wet debris flow grades into a very turbid stream.

## 6.1.1 *Rockfall*

Rockfall occurs where a cliff of bedrock is exposed to weathering and pieces of the bedrock are pried loose and fall, bounce, or roll to rest at the base of the cliff. The resulting deposit is called talus (Behre, 1933; Schumm and Chorley, 1966; Rapp, 1976). In an irregularly eroded cliff, talus cones may develop in the gully head areas (Fig. 6.2). In more extensive cliffs, coalescing talus cones form a more continuous blanket of debris referred to as scree (Jeffreys, 1932; Andrews, 1961).

Figure 6.2 Extensive talus cones on the sides of Bear Butte, SD. The bedrock is an exhumed Eocene andesite laccolith. Note light-colored lobate-formed tongues of cobble-sized talus (*A*) formed as a result of landslide-like failure.

Talus will not rest at an angle greater than the angle of repose. The angle of repose is measured similarly to dip of sedimentary rocks, i.e., from the horizontal, and is typically about 33–35° (Van Burkelow, 1945; Morisawa, 1966; Chandler, 1973). Bedrock composition, as well as talus particle size, angularity, and sorting, has some effect on the angle of repose; large angular boulders may rest at up to 40° (Rahn, 1966; Frankfort, 1968). The angle of repose of open-pit mine waste rock is approximately 38° (Hoek and Bray, 1974).

One of the world's most famous talus slopes is the one which formed an impediment to prospectors during the Klondike gold rush of 1898. The snow-covered talus slope, called the Scales, was the only access to Chilkoot Pass and the Yukon River. The pass was guarded by the Canadian Northwest Police, who required every prospector to carry 1 ton of supplies up the slope and into the interior.

Talus is limited in areal extent, and hence is rarely encountered in most engineering works except in mountainous areas. Movement of talus can be initiated either by overloading the top slope or undercutting the base. The resulting failure resembles a landslide, with fresh surfaces exposed on the talus surface (Fig. 6.2). Sand accumulating in an hour-glass sand-clock illustrates this form of landslide failure. In terms of engineering geology, any disturbance of talus is extremely risky, and may result in failure. Some talus cones may have an internal zone of weathered boulders and soil and/or colluvium. This matrix may add to the short-term stability of road cuts. In time, talus may weather and allow for vegetation to take root, thus helping stabilize the slope (Rahn, 1969).

The stability of slopes susceptible to toppling of an individual rock block can be mathematically analyzed using a free-body diagram of a unit thickness (Fig.

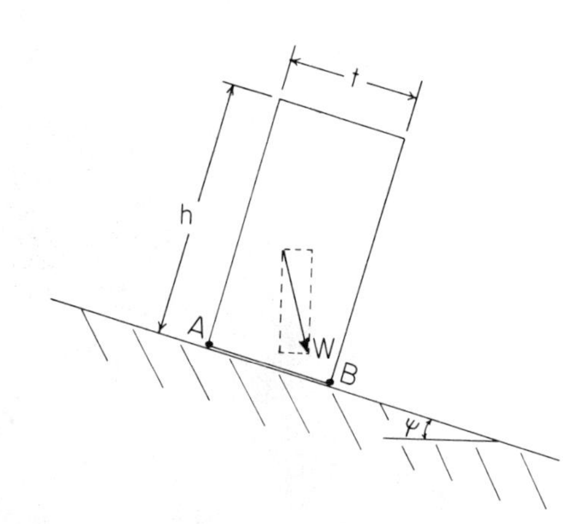


Figure 6.3 Stability conditions for a rock parallelepiped on an inclined slope (after Zanbak, 1983). The block will rotate if $t/h < \tan \Psi$.

6.3). Assuming that the block is stable against sliding, then the block will rotate (topple) when the vertical component of the weight vector lies within the supporting basal plane *AB*. A computer program for toppling analysis using varying column dimensions, slope geometry, and discontinuity orientation is given by Zanbak (1983).

Places in the Alps are subject to devastating rockfalls (Heim, 1932). In 1717, a massive fall of ice-clad rocks estimated at about 18 million $m^3$ traveled at velocities of 125–160 km/hr, plunging down from the crest of Mont Blanc in northwestern Italy. Prehistoric and historic landslides indicate the hazards of living in this recently glaciated mountainous area. Little is known about the frequency or cause of the rockfalls. Porter and Orombelli (1981) describe rockfall hazards in the Alps. The rockfalls are triggered by snow avalanches in many places, and have been known to change into rock avalanches or debris flows as they hurtle into the habitated valleys below.

Rockfall grades into other forms of mass wasting processes such as landslides, rock avalanches, and debris flow. The flow of high-speed rock avalanches apparently involves a mechanical fluidization process, whereby the debris moves as a single mass at first, but upon gaining kinetic energy becomes less viscous and flows as a fluid. The rock avalanche then thins and spreads at a rate which is dependent upon the debris viscosity (McSaveney, 1978).

Melosh (1983) believes that fluidization can be facilitated by acoustical (sound) waves generated within the moving debris itself. Fluidization by entrapped dust particles has also been theorized to account for large rock avalanches such as those found on the moon (Howard, 1973).

The May 31, 1970 earthquake triggered a $5 \times 10^7$-$m^3$ rock fall which turned into a very rapidly moving (~300 km/hr) landslide (debris avalanche) near Mt. Huascaran, Peru. Most of the city of Yungay was buried, and 18,000 people were killed (Browning, 1973; Hays, 1981).

## *6.1.2 Creep*

Creep is the slow, downslope movement of soil, weathered rock, or other surficial debris. Movement is by quasiviscous flow, occurring under shear stresses sufficient to produce permanent deformation, but too small to result in a discrete failure surface such as a landslide.

Figure 6.4 Surface expression of creep. **A**. Steamboat rock, near Dayton, WY. Large boulders of Ordovician Bighorn Dolomite are creeping down this steep slope over shales of the Cambrian Gros Ventre Formation (visible in road cut). **B**. Cemetery near Mt. Hope, CT. The cemetery is on glacial till which is creeping down the slope.

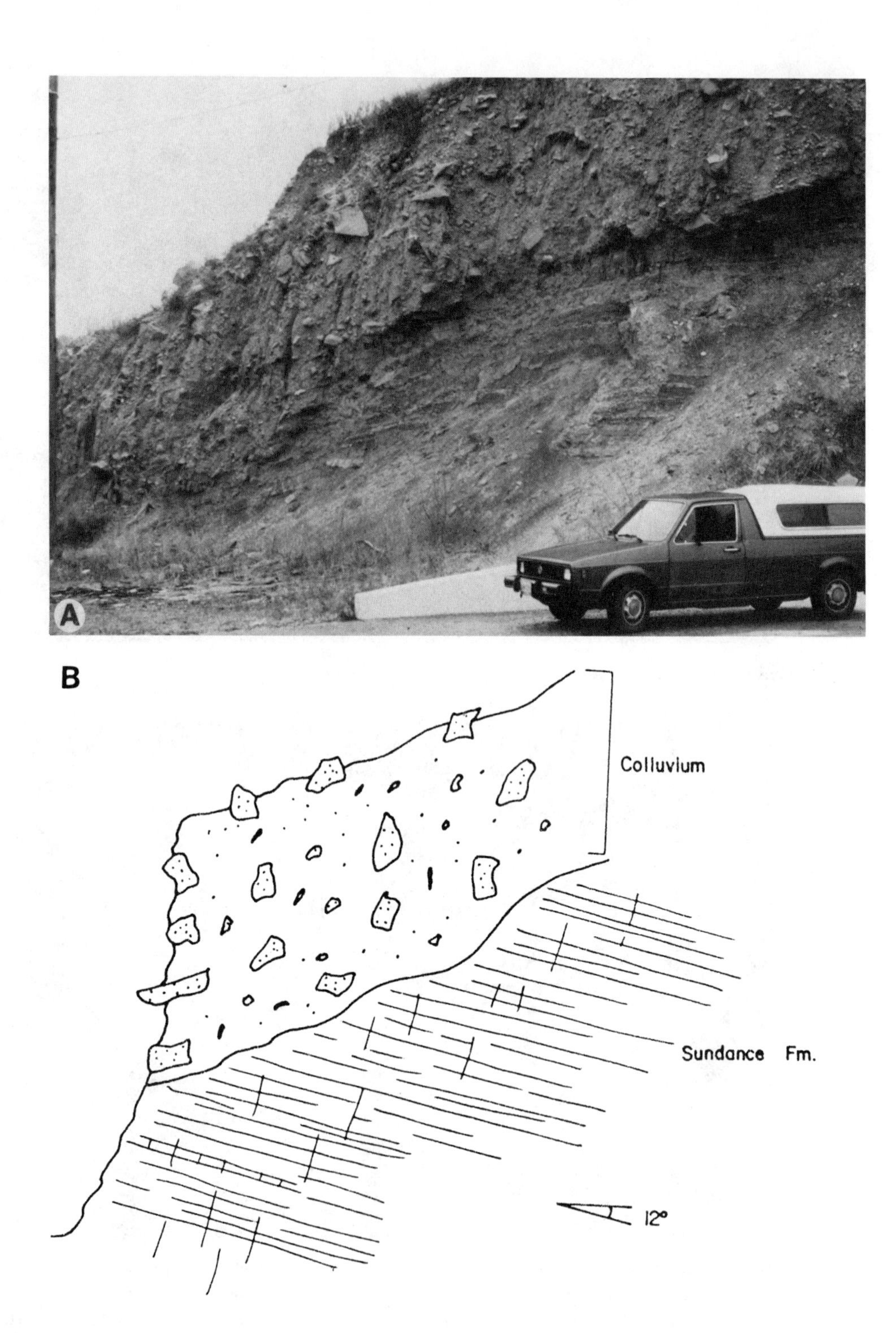


Figure 6.5 Instability of colluvium as illustrated by an exposure made at a parking lot in Rapid City, SD. **A**. Photo taken February, 1983. Note vertical bluff of colluvium. **B**. Sketch showing geology. **C**. Photo taken June, 1983. The colluvial slope failure broke the small reinforced concrete retaining wall.

Creep may be caused by numerous phenomena. For example, in terrains subject to frost, rock or soil may get heaved up perpendicular to the slope upon freezing, and returned vertically downward during a thaw. Thus, over time, material on a slope is likely to move downhill. Micro-landslides and random movements such as overturned trees, etc., all contribute to creep (Campbell, 1941; Capp, 1941; Parizek, 1957; Yen, 1969; Potter and Moss, 1968; Radbruch-Hall, 1978). Evidence of creep is manifest in observations such as erratic boulders in a clayey matrix juxtaposed on foreign bedrock on a hillslope (Fig. 6.4A). In an unkept cemetery (Fig. 6.4B), the tops of tombstones tend to rotate downhill due to the differential movement of surficial materials, with more displacement near the top and less displacement at the bottom. Trees, posts, and telephone poles are similarly displaced. This type of soil displacement is analogous to the shearing of a deck of cards on an inclined surface.

Creep differs from the process of landslide in that 1) a landslide has a definite detachment plane and 2) landslides usually move at faster rates, up to several hundred km/hr. Swanston and Swanson (1976), for example, found creep rates of only about 1 cm/yr in forest regions of the Pacific Northwest.

Deposits of creep are called colluvium (Sharpe and Desch, 1942; Hartshorn, 1958; Lattman, 1960). At the base of some talus or scree slopes, talus may grade into colluvium (the gradational material is called taluvium). Colluvium resembles clayey glacial till in that it is generally unsorted and unstratified. Colluvium typically contains angular pieces of rock in a clayey matrix (Fig. 6.5). The novelist Jack London (1945) describes the labors of a gold prospector locating a bedrock vein on a colluvial hillslope; the vein was located by identifying gold-bearing float exposed in pits which were dug progressively up the slope.

Colluvium is a media with which engineering geologists are frequently involved. As a rough generalization, the whole earth's surface is covered by a blanket of colluvium, with only bedrock occasionally cropping out. (Of course, other forms of surficial deposits such as alluvium, till, sand dunes, etc., also occur.)

In general, houses and small structures can be built on gently-sloping colluvial areas with no major adverse effects during the lifetime of the structure. This is because the natural rate of movement of creep is very slow. However, unsupported vertical cuts in colluvium at temporary excavations such as tunnel portals, basements, or along road cuts may result in cave-ins (Fig. 6.5). Another problem with colluvium is that the effects of frost heave or seasonal moisture changes in expandable clays are compounded in colluvium.

Colluvium itself may be subject to landslides. For example, in the Cincinnati, OH area, Fleming et al. (1981) report that landslide problems rarely involve bedrock (Ordovician shale and limestone), but typically occur only in colluvium. Inclinometer measurements show that two types of landslides may be generated. Where the colluvium is thin (less than approxiamtely 5 m) and occurs on steep slopes, the entire thickness of colluvium fails and moves downslope by a combination of sliding and flowing. Where colluvium is thick, landslides typically are discrete slumps. Rates of movement of these colluvial landslides were found to be fairly slow, ranging from 0.3 to 50 cm/day. Savage and Chleborad (1982) present mathematical models for creeping landslide deposits. In these models, the mass is assumed to possess Bingham properties. (A Bingham substance has the property that below a certain stress, called the yield stress, it behaves elastically. Above the yield stress it behaves viscoplastically.)

Figure 6.6 Engineering geology associated with colluvium in alluvial valleys. **A.** Cross section of Beaverdam Run, PA (after Lattman, 1960). **B.** Cross section of hillslope, near Weirton, WV (after D'Appolonia et al., 1967).

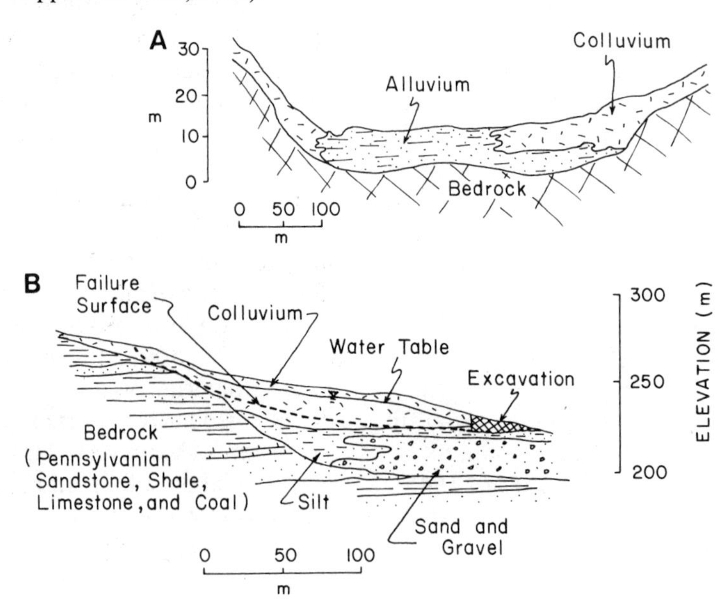

Figure 6.7 Nappelike structure in Cretaceous Fall River sandstone caused by creep. Excavation for hillside home at Rapid City, SD. Downslope is to left.

Colluvium may interfinger with or overlie alluvium (Fig. 6.6A). A large excavation near Weirton, WV, revealed about 20 m of colluvium on top of alluvium (silt, sand, and gravel) deposited by an abandoned loop of the Ohio River (Fig. 6.6B). An ancient landslide was observed in the colluvium and special design features were necessary to ensure safe construction (D'Appolonia et al., 1967).

Bedrock may, in fact, be drawn into the creep process. Figure 6.7 shows sandstone bedrock deformed along shale layers by creep. The presence of this nappelike structure in a hillslope should alert the engineering geologist to potential foundation problems.

## *6.1.3 Landslides*

*Characteristics.* Landslide is the process whereby a distinct mass of rock or soil moves downslope due to gravity. Landslides, unlike creep, have one or more distinct failure surfaces. Landslide velocities (Fig. 6.1) are typically on the order of 1 m/day to perhaps 300 km/hr as in the special case of air-entrapped landslides generated by earthquakes (Sharpe, 1938; Eckel, 1958; Hubbert and Rubey, 1959; Kern, 1963). Shreve (1968) believes that the prehistoric Blackhawk Landslide in California slid at immense speeds on a layer of compressed air that was trapped beneath the debris. The landslide, consisting of 320 million $m^3$, traveled 8 km across a desert floor at an average slope of 2.5°. [Hsu (1975) challenges the air-lubrication hypothesis in favor of a debris flow mobilized by collision of particles within a dust-laden cloud.]

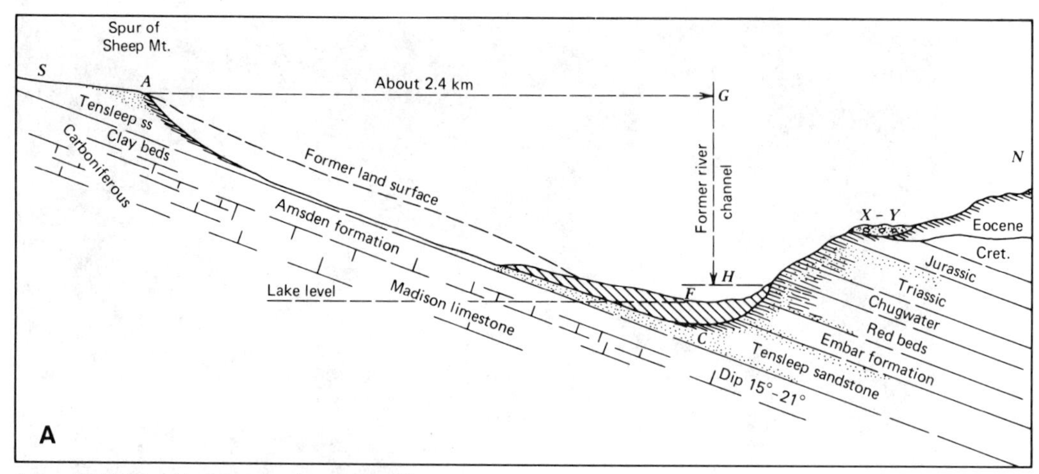


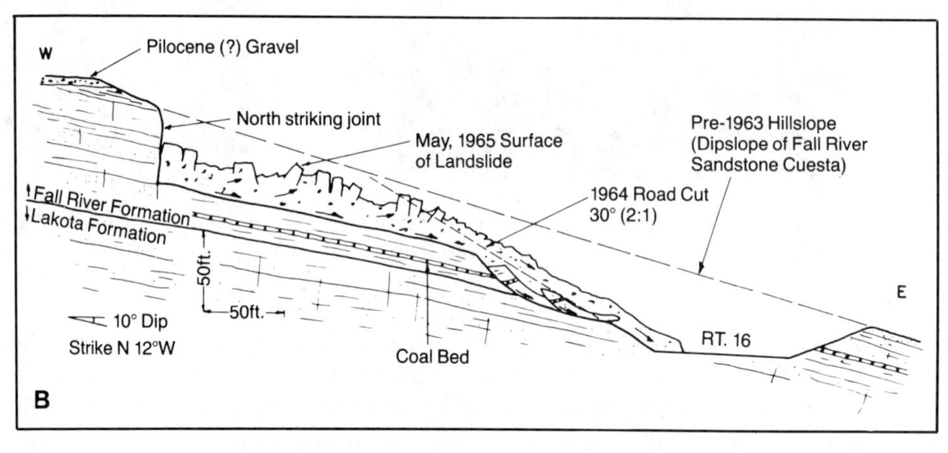


Figure 6.8 Translation landslides. **A.** Diagrammatic cross section (north-south) illustrating the damming of the Gros Ventre River Valley, WY, by the landslide (Alden, 1928). **B.** Landslide in Cretaceous rocks near Rapid City, SD (from Rahn et al., 1981).

Landslides can be devastating to life and property. For example, in 1903 an enormous section of limestone at Turtle Mountain suddenly crashed down onto the coal mining town of Frank, Alberta, killing 70 persons (Daly et al., 1912).

Some landslides move slowly but may be very costly. For example, at the Palos Verdes Hills area of Los Angeles, along the Pacific coast, an ancient landslide complex affects several square kilometers of seaward-dipping tuffaceous Miocene shales. According to Bolt et al. (1977): "The appearance of the whole area is that of a rumpled carpet thrown on the hills. In spite of the obvious evidence of instability, no measures were taken to prevent the establishment of housing subdivisions and the construction of houses in the area in the 1950's." Movement since 1956 is continuing at a rate which increases in the winter rainy season, and averages from 1 to 3 cm/day in an average year (Merriam, 1960). Three ages of landsliding have been described by Jahns and Vander Linden (1973), the youngest

Figure 6.9 Madison Canyon landslide, MT.

initiated by wave erosion by the Pacific Ocean at the base, and human activities including loading of ancient landslides by highway fill, and the introduction of excessive water from cesspools, swimming pools, and lawn watering. Urban development of the Palos Verdes Hills increased dramatically in the early 1950s (Jahns, 1958); many expensive homes were built on the old slides, which were temporarily stable, but wave erosion reactivated them. The largest slide, the Portugese Bend landslide, is 1100 m wide, 1280 m long, and 80 m thick. Easton (1973) reports that the rate of movement accelerates following a soaking rain. The landslides are also affected by earthquakes and possibly by the buoyant effect of high tides on the submerged toe of the slide.

Many landslides are completely natural in origin, such as the Gros Ventre landslide (Fig. 6.8A) in Wyoming (Alden, 1928). Many are triggered by earthquakes (Keefer, 1984). The 1959 Hegben Lake, Montana, earthquake caused an enormous slide (Fig. 6.9) which killed 28 campers and blocked the Madison River (Hadley, 1964). The 1964 Alaskan earthquake triggered devastating landslides in the Bootlegger Cove Clay in Anchorage, as well as an underwater landslide which caused a tsunami at Valdez (Hansen, 1965; Coulter and Magliaccio, 1966).

Some Laramide orogeny thrust sheets may be giant gravitational-induced landslides (Pierce, 1963; Wise, 1963; Scholten, 1974). An ancient landslide deposit at Saidmarreh, Iran, is 20 km$^3$ in volume (Watson and Wright, 1969; Bolt et al., 1977). The largest known Quaternary landslide has recently been identified at the base of Mt. Shasta, California (Crandell et al., 1984). The debris-avalanche deposit has an estimated volume of 26 km$^3$. On the moon there are large landslides in Craters Tycho, Aristarchus, Copernicus, and other recent craters. One landslide on Mars has a volume of 100 km$^3$ (Lucchitta, 1978).

Landslide activity typically increases after prolonged rains, or after the frost leaves the ground in the spring. The novelist Fritsch (1980) describes the ominous hazard of landslides which weighs heavily on the minds of people who live in the steep slopes of the Swiss Alps:

> More serious than the collapse of a dry-stone wall would be a crack across the grounds, narrow at first, no broader than a hand, but a crack—That is the way landslides begin, cracks appearing noiselessly, not widening, or hardly at all, for weeks on end, until suddenly, when one is least expecting it, the whole slope below the crack begins to slide, carrying even forests and all else that is not firm rock down with it.

Man may trigger a landslide in a number of ways. Figure 6.10A shows rotational landslide scarps along the shore of the Bighorn Reservoir in Montana. Prior to the construction of the Yellowtail Dam in 1965, colluvium and old landslides covered much of the bedrock (horizontal Cambrian shales). Renewed movement occurred in these mass-wasting deposits due to buoyancy associated with the rising reservoir (Dupree et al., 1979).

Man may cause landslides by providing the triggering mechanism for a slope that was nearly ready to fail. The most familiar case is where a cut is made at the base of an unstable slope (Fig. 6.8B). Huge artificial fills are subject to landslides (Fig. 6.10B). Land use practice affects slope stability. In steep slopes having colluvial soils in Ohio, Riestenberg and Sovonick-Dunford (1983) found that removing forest vegetation facilitated landsliding because of the diminished effectiveness of tree roots which hold soil onto the bedrock. Man may trigger landslides by various other mechanisms such as increasing the pore pressure and/or buoyancy of unstable slopes adjacent to man-made reservoirs (Figs. 6.10A and 4.20).

*Classification of landslides.* For simplicity, landslides can be classified into two groups: 1) translational, and 2) rotational.

Translational landslides (Fig. 6.8) slide along a bedding plane or other plane of weakness in the rock. The entire mass moves parallel to this plane, as illustrated in the rock slide in Fig. 6.1. Probably the largest translational landslide to occur in historic times happened in 1925 in the Gros Ventre River Valley in Wyoming (Fig. 6.8A). About 40 million m$^3$ of Pennsylvania Tensleep Sandstone slid on a bedding plane inclined about 20°. Heavy rain and stream undercutting at the base of the mountain slope contributed to the failure (Voight, 1978). A 45 km$^2$ lake formed behind the debris; two years later the dam was overtopped by heavy spring runoff, and six people drowned in the ensuing flood. The 1903 landslide disaster at the mining town of Frank, Alberta, was a translational slide which developed along a prominent set of joints which dip into the valley.

Immense landslides occurred during the construction of the Panama Canal. The giant excavation, completed in 1914, caused translational sliding along tuffaceous beds in gently dipping, faulted early Cenozoic shales and sandstones (Lutton et al., 1979). The giant Culebra Cut across the continental divide (Fig. 6.11) was begun by the French government, but finally completed by the United States. The French excavated 15 million m$^3$, and the United States excavated 73 million

Figure 6.10 Man-induced landslides. **A**. Colluvium and landslide deposits overlie Cambrian shale along the edge of the Bighorn Reservoir near Lovell, WY. The recent landslide scarps (see numerous scarps at *A* and large scarp at *B*) were caused when the reservoir was created. **B**. Landslide in newly placed highway fill near Platte, SD (photo by Vernon Bump, South Dakota Dept. of Transportation).

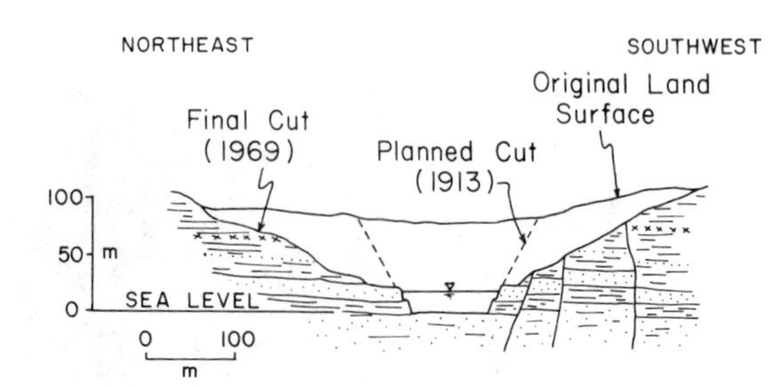


Figure 6.11 Cross section of the Culebra Cut for the Pan-American Canal (after Lutton et al., 1979).

$m^3$. The worst slide was the "Cucaracha" slide located on the east bank of the Culebra Cut. On October 4, 1907, after days of unusually heavy rain, it slid to the bottom of the cut, burying two steam shovels. For days afterward, it moved about 4 m/day. Major Gaillard, the engineer in charge, noted: "It was, in fact, a tropical glacier—of mud instead of ice" (McCullough, 1977).

Rotational landslides ("slumps" or "slips") develop in surficial deposits or weathered rock (Figs. 6.1 and 6.12). They may be natural in origin or induced by man. The failure surface is roughly arcuate in cross section (spoon-shaped in three dimensions).

At some places the occurrence of a landslide at the base of a hill may leave the hillslope above unsupported, and so another landslide occurs (Crandell, 1951). The chain-reaction type of stair-stepping slope failure up a hill is called progressive failure.

*Mathematical analysis.* Mathematical models of landslide stability are presented in Terzaghi (1950b; 1962), Scheidegger (1961), Stout (1969), Jaeger (1972), Schuster and Krizek (1978), and Huang (1983). Most problems involving the stability of slopes are associated with the design and construction of unbraced cuts for highways. A very generalized rule-of-thumb is given by Terzaghi and Peck (1948):

Figure 6.12 Small slump ("slip") near Elkins, WV. The landslide is spoon-shaped in three dimensions. *A* is the scarp and *B* is the toe of the landslide.

> Experience has shown that slopes of 1.5 (horizontal) to 1 (vertical) are commonly stable. . . . Therefore, a slope of 1.5 to 1 can be considered the standard for highway or railroad construction. The standard slopes for flooded cuts such as those for canals range between 2:1 and 3:1. Steeper than standard slopes should be established only on rock, on dense sandy soil interspersed with boulders, and on true loess.

Landslides can occur in bedrock as well as surficial deposits. The compressive strength of rock is high, but Terzaghi (1950b) points out that it is not the compressive strength of rock that governs the landslide susceptibility of rock. If unweathered rock has an unconfined compressive strength ($S_u$) of $35 \times 10^6$ N/m$^2$, and its density ($\rho$) of 2700 kg/m$^3$ giving a unit weight on earth of 26,700 N/m$^2$ for every meter height ($w$), then the critical height ($H_c$) of a vertical slope would be, roughly:

$$\begin{aligned} H_c &= \frac{S_u}{w}, \\ &= \frac{35 \times 10^6 \text{ N/m}^2}{26.7 \times 10^3 \text{ N/m}^2 \text{ per m}}, \qquad (6.1) \\ &= 1310 \text{ m}. \end{aligned}$$

In reference to this calculation, Terzaghi (1962) points out that:

> For intact rocks such as granite, $H_c$ is several times greater. Yet no vertical slopes with such height exist, and many gentler slopes with a much smaller height than $H_c$ have failed. This fact indicates that the critical height of slopes on unweathered rock is determined by the mechanical defects of the rock such as joints and faults, and not by the strength of the rock itself.

Slope stability problems in anisotropic rocks present unique solutions. For example, sedimentary rocks along the Pennsylvania Turnpike pose severe stability problems where the road alignment is parallel to the strike of the rocks in areas where deep cuts are made. Kilburg et al. (1980) found that the design of stable slopes was possible by use of anchor bolts, terracing and the designing of cuts which correspond to the dip of the strata, which averages 17°.

Road cuts in slate, phyllite, and schist in the central Black Hills, SD, offer practical guidelines for small excavation in rocks containing a major planar anisotropy (cleavage). Where the strike of the cleavage is parallel to the highway, vertical cuts can be made if the dip is away from the highway, but cuts no greater than the dip of the cleavage can be made if cleavage dips into highway. Vertical cuts can be made if cleavage is perpendicular to the highway (see Fig. 6.13). If a second planar anisotropy exists (joints) the design of stable road cut slopes must be guided by joint orientation (see Jaeger, 1972, for example).

Obviously, Eq. (6.1) is inadequate to design vertical cuts in man-made excavations. Hoek and Bray (1974) collected empirical data on the slope height for excavations in various types of rocks, and found that heights greater than 100 m can rarely be achieved in vertical cuts, although heights approaching 300 m can be made with 45° cuts.

The design of slopes for open-pit mines is very important because the economics of an ore deposit is determined to a large degree by how much overburden

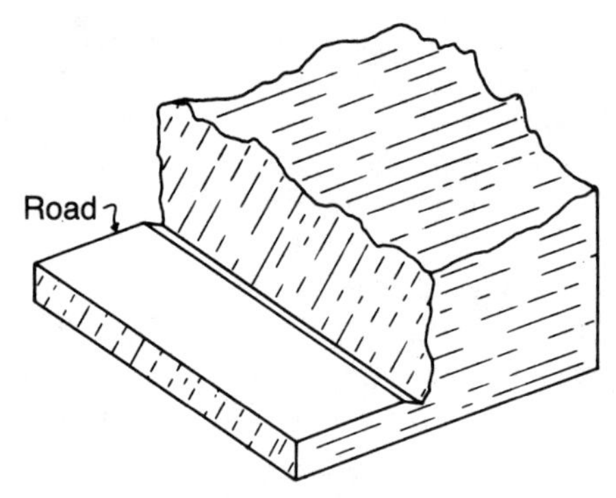


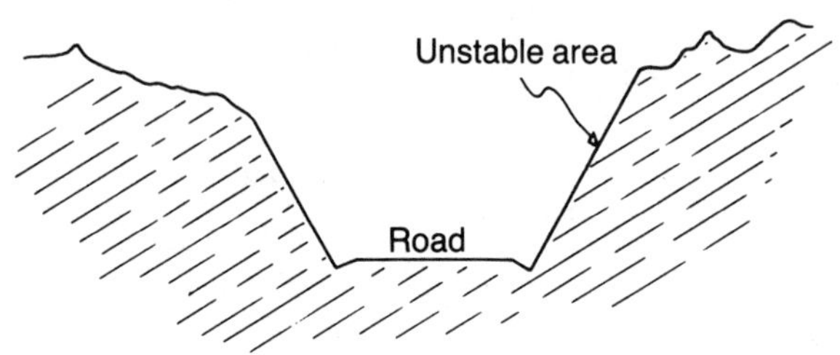


Figure 6.13 Sketch illustrating relationship between road cut stability and orientation of prominent planar anisotropy in rock (from Rahn et al., 1981).

must be removed so as to expose the ore. The slope angles of natural hills can be used as a guide to design man-made cuts, but since mines are worked for usually only a few decades, overconservative design of slopes will result if recourse is made to design of slopes equal to natural slopes. It is possible in many places to design huge open-pit mines at angles greater than 38°, the average angle of repose of talus. Terracing, with ~50 m elevation increments at an angle of say 50°, may be successful because, even if an individual terrace bluff collapses (to an angle of repose), the flat arm below the bluff catches the debris and the overall mine slope still remains in place.

The role of rock anisotropies is very important in the design of stable slopes. Figure 4.24 shows an example of a stereographic solution of a possible wedge failure due to two prominent planar anisotropies on a deep cut. The use of stereoplots in analysis of slope stability is explained in Chapter 4.

The following example of a free-body diagram solution for the effect of planar anisotropies on the stability of a slope is given by Hoek and Bray (1974). Consider a slope cut at some angle $\psi_f$ where the plane on which sliding can occur strikes parallel to the cut (example: strike is parallel to the highway) and dips at an angle $\psi_p$ less than the cut face. Assume in third dimension that a slide is not restrained. Figure 6.14 shows a unit thickness of this slide, so that the sliding area can be reduced to a length ($A$) as shown.

$$A = \frac{H - Z}{\sin \psi_p}. \tag{6.2}$$

For this method of analysis it will further be assumed that a vertical tension crack exists, whose strike is also parallel to the cut (Fig. 6.14). It will be assumed

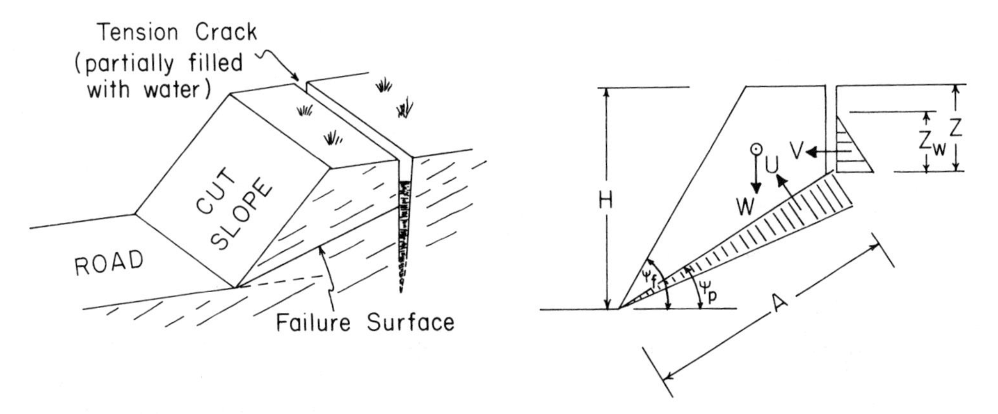


Figure 6.14 Geometry of a slope with a tension crack in the upper slope surface (after Hoek and Bray, 1977).

that ground water fills the crack to a depth $Z_w$, so that hydrostatic pressure (force $V$) acts to push the slide down the hill. Because the water pressure drops to zero where the sliding surface daylights in the slope face, the (uplift) pressure distribution induced by water in the sliding surface is shown by $U$ in Fig. 6.14.

It will be assumed that no moments exist; i.e., the force $U$, $V$, and $W$ (weight of the sliding block) all act through the centroid of the sliding mass. The unit weight of water is $\gamma_w$ and the rock is $\gamma$.

The shear strength of the sliding surface is defined by Coulomb's law [Eq. (4.17)]: $\tau = c + (\sigma - \mu) \tan \phi$. The factor of safety (F.S.) is given by the ratio of the total resisting force to the total force tending to induce sliding. The forces are analyzed in the direction of the plane of sliding:

$$\text{F.S.} = \frac{\text{resisting force}}{\text{driving force}}, \tag{6.3}$$

$$= \frac{cA + (W \cos \psi_p - U - V \sin \psi_p) \tan \phi}{W \sin \psi_p + V \cos \psi_p}. \tag{6.4}$$

The three forces involved ($V$, $U$, and $W$) can be solved using trigonometry from Fig. 6.14:

$$V = \frac{1}{2} \gamma_w Z_w^2, \tag{6.5}$$

$$U = \frac{1}{2} \gamma_w Z_w A, \tag{6.6}$$

$$= \frac{1}{2} \gamma_w Z_w \frac{H - Z}{\sin \psi_p},$$

$$W = \left[ \frac{1}{2} \gamma H^2 \right] \cdot 1 - \left( \frac{Z}{H} \right)^2 \cdot [\cot \psi_p - \cot \psi_f]. \tag{6.7}$$

Solution to these equations can be tedious, and may involve trial and error analysis. For example, it may be desired to determine the cut slope angle $\psi_f$ given a

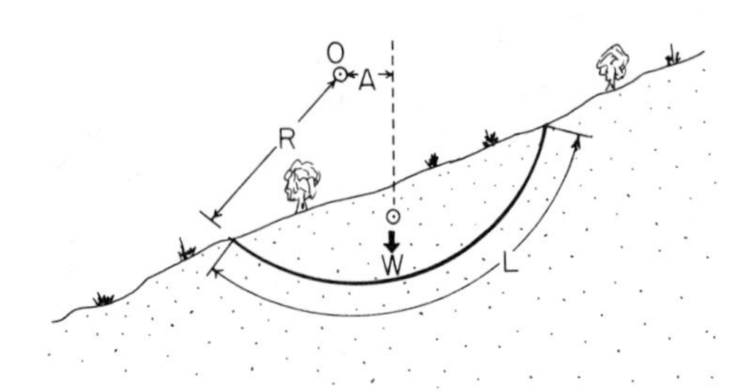


Figure 6.15 Idealized cross-sectional sketch of a rotational landslide.

range of possible heights ($H$) or assumptions of the rock properties tan $\phi$ and $\gamma$. A useful aspect of this approach is that a parameter variability analysis (sensitivity analysis) will show which parameters are most critical for the design of safe slopes.

Another useful result of the free-body diagram approach to landslide stability is that it allows for the quantitative determination of rock bolts or buttresses necessary to increase the stability of a rock slope in order to achieve a desired safety factor. A rock bolt would be anchored at some angle into the hillslope, and the resolution of the force of the bolt holding the rock in place would be added to the resisting force. (Note that an anchor bolt not only exerts an uphill force on the landslide, but also adds a normal load so that frictional resistance along the failure plane is increased.) Earthquake loads can also be analyzed by a free-body diagram model where the force of the earthquake equals the mass of the landslide times the acceleration of the earthquake. Lateral components of earthquake acceleration as well as vertical can be addressed.

A rotational landslide, such as may occur in surficial deposits, can be very simply analyzed in terms of the balance of moments (Peck et al., 1974). Consider the sketch of a rotational landslide of unit width (Fig. 6.15) where the landslide of circular cross section has a force due to its weight ($W$) and a length of arc ($L$) whose circle center is at 0. The driving moment ($M_d$) created by the weight is ($W$) times the moment arm ($A$). The resisting moment ($M_r$) is the shear force [$L$ times the shear strength of the soil ($S$)] times the radius of the circle ($R$). At the moment of failure,

$$M_d = M_r, \tag{6.8}$$

$$W \cdot A = L \cdot S \cdot R. \tag{6.9}$$

From (6.9) it can be seen that landslide instability would be created by 1) a reduction of $L$ due to, say, an excavation at the base of the hill, 2) an increase in $W$ due to loading the top of the hillslope with an embankment, or 3) a decrease in $S$ as may occur due to increased moisture.

Changes leading to instability are precisely what man has done at many places. Figure 2.6 is an air photo stereo pair showing an area where there has been a reduction of $L$, in this case an excavation at the base of an ancient landslide. Parts of the urbanized hillslopes at Pacific Palisades near Los Angeles are referred to as "Heartbreak Hills" (Jahns, 1958), where slides occur in steep Tertiary shale hillslopes which have been bulldozed into roads and landscaped lots, supplemented by leaking swimming pools and watered lawns. Earthquakes also contribute to the overall instability.

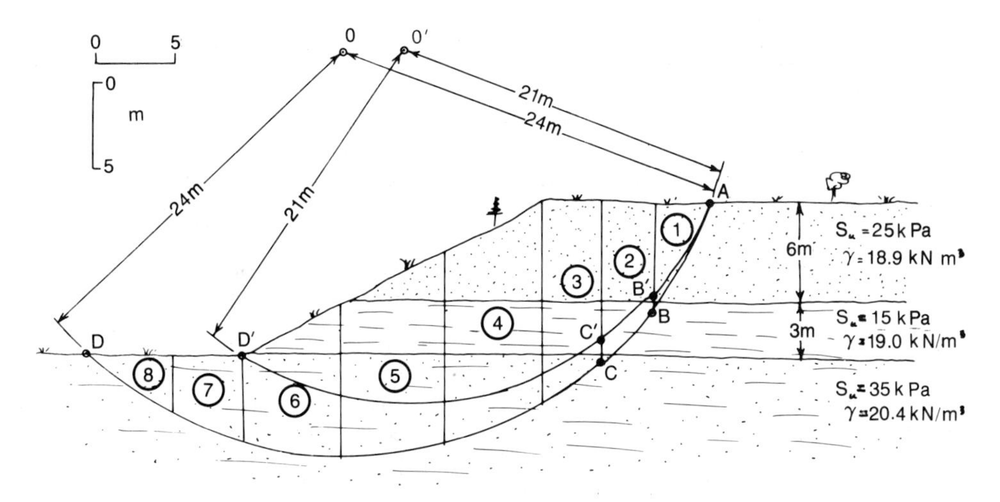


Figure 6.16 Landslide cross-sectional sketch showing two trial arcs used for method of slices (modified from Schuster and Krizek, 1978).

A widely accepted analytical solution of the above balance of moment concept is the method of slices. This is a method of landslide analysis in which the moment of each of several components of a landslide is calculated. The driving and resisting moments are totalled, and the factor of safety is computed. Several trial runs using different failure surfaces can be analyzed. The F.S. calculation which yields the lowest value is the surface on which a slide is most likely to occur.

Consider the landslide cross section shown in Fig. 6.16 having a unit thickness of 1 m. Assume the debris consists of three layers of known density and shearing strength as shown. First assume a trial failure plane #1 *(ABCD)*. The slide can be arbitrarily divided into a convenient number of slices, and the moment of each computed (Table 6.1). Accurate drafting and measurement of moment arms is necessary. The sum of the eight driving moments is 29,507 kN-m. The resisting moment is then determined, and is equal to the shearing strength (in this case the so-called undrained shear strength $S_u$ is used) times the length of arc in each slice times the moment arm. The sum of the resisting moments is calculated (Table 6.1) to be 34,500 kN-m. The factor of safety, F.S., is equal to the ratio of resisting moment to driving moment:

$$\text{F.S.} = \frac{M_r}{M_d}. \tag{6.10}$$

In this case:

$$\text{F.S.} = \frac{34{,}156}{29{,}507} = 1.16.$$

Next, assume a trial failure arc #2 *(AB′C′D′)*. Six slices are selected and a new factor of safety of 1.27 is determined (Table 6.1). From this it can be shown that a landslide is more likely to occur in arc *ABCD* than *AB′C′D′*.

Table 6.1 Table for Calculation of Moments in Method of Slices Shown in Figure 6.16

Trial 1 (slip surface *ABCD)*

DRIVING MOMENT

| Slice | Unit thickness (m) | Width (m) | Height (m) | Density (kN/m³) | Force (kN) | Lever arm (m) | Driving moment (kN · m) |
|---|---|---|---|---|---|---|---|
| 1 | 1 | 3 | 3 | 18.9 | 170 | 20 | 3400 |
| 2 | 1 | 3 | 7.5 | 18.9 | 426 | 17.5 | 7455 |
| 3 | 1 | 3.5 | 10.5 | 19.1 | 703 | 14 | 9842 |
| 4 | 1 | 6 | 12 | 19.4 | 1400 | 9 | 12,600 |
| 5 | 1 | 6 | 10 | 19.7 | 1182 | 3 | 3546 |
| 6 | 1 | 6 | 7 | 20.1 | 844 | −3 | −2532 |
| 7 | 1 | 4 | 4 | 20.4 | 326 | −8 | −2608 |
| 8 | 1 | 6 | 1.5 | 20.4 | 183 | −12 | −2196 |
| | | | | | | Total | 29,507 |

RESISTING MOMENT

| Unit thickness (m) | Su (kN/m²) | Arc distance (m) | Lever arm (m) | Resisting moment (kN·m) |
|---|---|---|---|---|
| 1 | 25 | *AB* = 7.22 | 24 | 4332 |
| 1 | 15 | *BC* = 4.44 | 24 | 1600 |
| 1 | 35 | *CD* = 33.6 | 24 | 28,224 |
| | | | Total | 34,156 |

$$\text{F.S.} = \frac{34{,}156}{29{,}507} = 1.16$$

Trial 2 (slip surface AB′ C′ D′)

DRIVING MOMENT

| Slice | Unit thickness (m) | Width (m) | Height (m) | Density (kN/m³) | Force (kN) | Lever arm (m) | Driving moment (kN·m) |
|---|---|---|---|---|---|---|---|
| 1 | 1 | 3 | 2.75 | 18.9 | 156 | 16 | 2495 |
| 2 | 1 | 3 | 7.5 | 18.9 | 425.25 | 15 | 6378 |
| 3 | 1 | 3.5 | 8.5 | 18.9 | 562.28 | 9 | 5060 |
| 4 | 1 | 6 | 8.5 | 19 | 969 | 5.5 | 5330 |
| 5 | 1 | 6 | 7 | 19 | 821 | 0 | 0 |
| 6 | 1 | 6 | 2.5 | 19.7 | 295.5 | −5.5 | −1625 |
| | | | | | | Total | 17,638 |

RESISTING MOMENT

| Unit Thickness (m) | Su (kN/m²) | Arc distance (m) | Lever arm (m) | Resisting moment (kN·m) |
|---|---|---|---|---|
| 1 | 25 | *AB′* = 6.67 | 21 | 3500 |
| 1 | 15 | B′C′ = 4.44 | 21 | 1400 |
| 1 | 35 | C′D′ = 23.89 | 21 | 17,560 |
| | | | Total | 22,460 |

$$\text{F.S.} = \frac{22{,}460}{17{,}638} = 1.27$$

The process of trial arcs continues until the lowest value for Eq. (6.10) gives the most likely landslide failure arc. From a theoretical point of view, any time Eq. (6.10) is less than unity the slope is unstable, and movement is imminent.

An alternative method of calculation of the resisting force is to substitute a value of shear stress ($\tau$) from Coulomb's equation [Eq. (4.18)] for the value of undrained shear strength ($S_u$) as shown in this example. A recent modification of the method of slices uses computer programs which repeatedly solve for the factor of safety by moving the center or radius for a circular failure surface. The sequence is repeated until the failure surface for the minimum factor of safety is determined (Cross, 1982).

The method of slices can be used to calculate the factor of safety for any arc through the hillslope in question. Changes in hillslope geometry due to roadcuts or embankments can also be determined. Design slopes for man-made embankments can be made by this method. The method is particularly effective in the design of earth-fill dams because the engineering properties (density, shear strength) and geometry of the fill are accurately known. Coates (1967) illustrates how pore water pressure changes the values of each slice in this analysis. The added effects of earthquake forces can be analyzed by this method (Attewell and Farmer, 1976). Where all or part of the landslide is below the water table, uplift hydrostatic forces must also be considered. In the case of partial saturation, each slice has an additional force, directed upward and normal to the failure plane (Lambe and Whitman, 1969; Attewell and Farmer, 1976). More sophisticated methods of landslide calculations for rock slopes can be found in Morgenstern and Price (1965), Hamel (1971), Hoek and Bray (1977). For surficial deposits, see Schuster and Krizek (1978), and references contained therein. Yu and Coates (1979) give examples of the finite-element method for analysis of the stress and deformation of rock slopes.

*Examples of landslides.* One of the world's most catastrophic landslides happened on October 9, 1963, at Vaiont Dam in Italy. About 3000 people lost their lives. The catastrophe was caused by a landslide which fell into the reservoir created by the world's second highest dam, 266 m high. The dam was constructed by SADE (Societa Adriatica di Elettricita, Venezia) as part of its extensive hydroelectric system in northeastern Italy. The dam did not fail, but the rising water in the reservoir triggered an immense translational landslide (240 million $m^3$) that fell into the reservoir (Mencl, 1966). A giant wave (70 m high) traveled across the reservoir, wiping out the town of Casso on the opposite side of the reservoir (Gaskill, 1965). The wave also overtopped the dam and a wall of water hit the town of Longarone about 0.5 km downstream.

Figure 6.17 is a map and cross section of the Vaiont landslide area. A strong set of rebound (unloading) joints roughly parallel the bedding of the Cretaceous limestone on the failure slope; both joints and beds dip into the valley. Alternating limestone and seams of montmorillonite clay partings are included in the bedrock. The hillslope was known to have had landslides in the past, and the catastrophic landslide of October 9 may have been a slide reactivation rather than a first-time event.

The sequence of events leading up to the Vaiont Dam failure began when water first started filling the reservoir in 1960, and alarming cracks high on the slopes

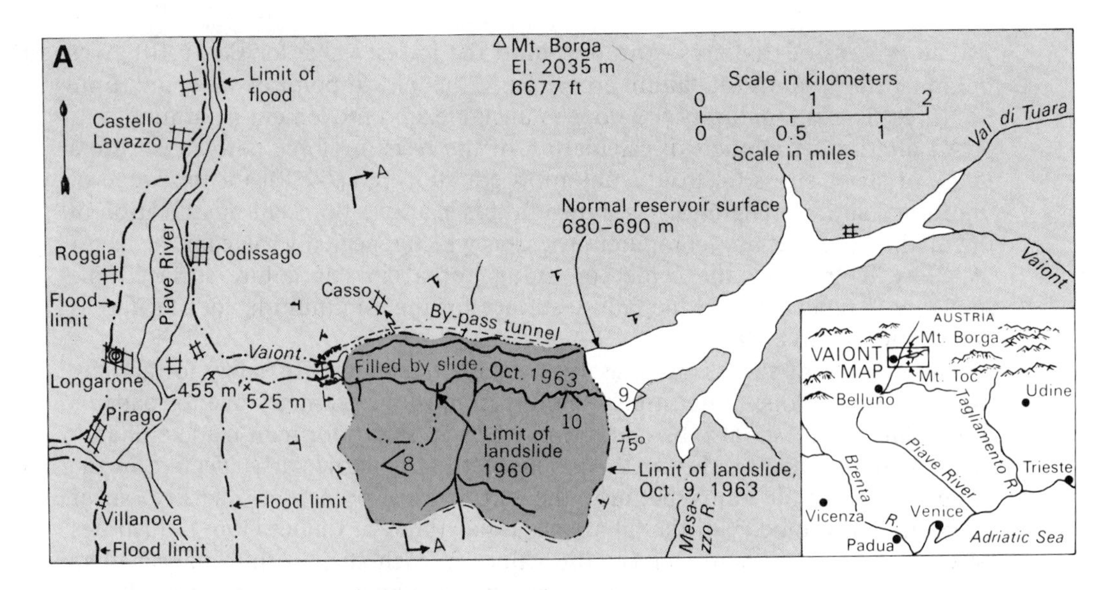


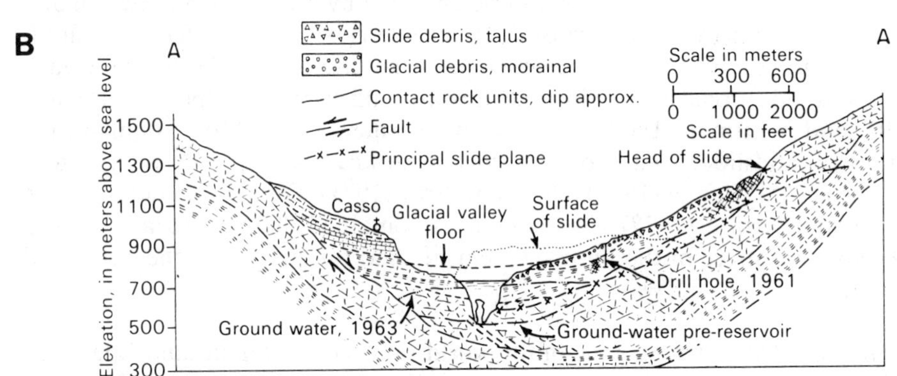


Figure 6.17 Vaiont Dam, Italy (from Kiersch, 1964. (Reprinted with permission of the American Society of Civil Engineers.) **A.** Map. **B.** Cross section (vertical exaggeration = 3.3X).

of Monte Toc were noticed. People living nearby distrusted the stability of Monte Toc, which anchored the dam's left side and rose over 1000 m above the new reservoir. Monte Toc was given a new name by local inhabitants: "la montagna che cammina" (the mountain that walks). Engineers from SADE were somewhat concerned about small landslides for fear they might clog the reservoir basin and reduce its storage capacity. No one envisioned a danger to human life. In 1963 the reservoir was filling for the first time, causing a rise in the water table (Fig. 6.17B), allowing for hydrostatic (buoyancy) uplift in the hillslope. Heavy rains during early October added weight to the hillslope. Slow creep was observed (Table 6.2), but engineers did little to alert the public to any danger. In early October, grazing animals became uneasy and moved off the mountain. On October 8, due to an increasing creep rate, worried engineers attempted to quickly lower the reservoir by letting water out through outlet tunnels. In the evening darkness of October 9, the slide occurred. The following description is from Kiersch (1964):

Table 6.2 Creep Data for Vaiont Landslide Area

| Date | Rate of creep (approximate) | Type of creep |
|---|---|---|
| September 18–24 | 1 cm per day | Transient creep |
| September 25 to October 1 | 10–20 cm per day | Quasi-viscous creep |
| October 2–7 | 20–40 cm per day | Quasi-viscous creep |
| October 8 | 40 cm | Creep to failure |
| October 9 | 80 cm (before collapse) | Creep to failure |

Modified from Kiersch, 1964.

The reservoir contained about 135 million cu m of water at the time of the disaster.

On October 9, the accelerated rate of movement was reported by the engineer in charge. A five-member board of advisers were evaluating conditions, and authorities were assessing the situation on an around-the-clock basis. Although the bypass outlet gates were open, verbal reports describe a rise in the reservoir level on October 9. This is logical if lateral movement of the left bank had progressed to a point where it was reducing the reservoir capacity.

Those who witnessed the collapse included 20 SADE technical personnel stationed in the control building on the left abutment and some 40 people in the office and hotel building on the right abutment. But no one who witnessed the collapse survived the destructive flood wave that accompanied the sudden slide at 22 hours 41 min 40 sec (Central European Time). However, a resident of Casso living over 260 m above the reservoir, and on the opposite side from the slide, reported the following sequence of events:

- About 10:15 P.M. he was awakened by a very loud and continuous sound of rolling rocks. He suspected nothing unusual as talus slides are very common.
- The rolling of rocks continued and steadily grew louder. It was raining hard.
- About 10:40 P.M. a very strong wind struck the house, breaking the window panes. Then the house shook violently; there was a very loud rumbling noise. Soon afterward the roof of the house was lifted up so that rain and rocks came hurtling into the room (on the second floor) for what seemed like half a minute.
- He had jumped out of bed to open the door and leave when the roof collapsed onto the bed. The wind suddenly died down and everything in the valley was quiet.

Observers in Longarone reported that a wall of water came down the canyon about 10:43 P.M. and at the same time a strong wind broke windows, and houses shook from strong earth tremors. The flood wave was over 70 m high at the mouth of Vaiont canyon and hit Longarone head on. Everything in its path was destroyed.

As the water shot down the short gorge, it picked up rocks and debris. Ahead of it raced a strange icy wind. In 6 min the flood thundered across the Piave Valley, and Longarone vanished from the earth.

The Vaiont dam disaster was caused by adverse geologic features which should have been recognized prior to construction, and was triggered by a rising water table affecting the delicately balanced stability of a steep rock slope. The slide reached a very high velocity, perhaps 90 km/hr (Jaeger, 1972). It has been proposed that the high velocity of the slide required a low coefficient of friction, which may have been caused by increased pore pressure due to a frictional heat-induced fluid pressure enhancement (Voight and Faust, 1982). Following the dis-

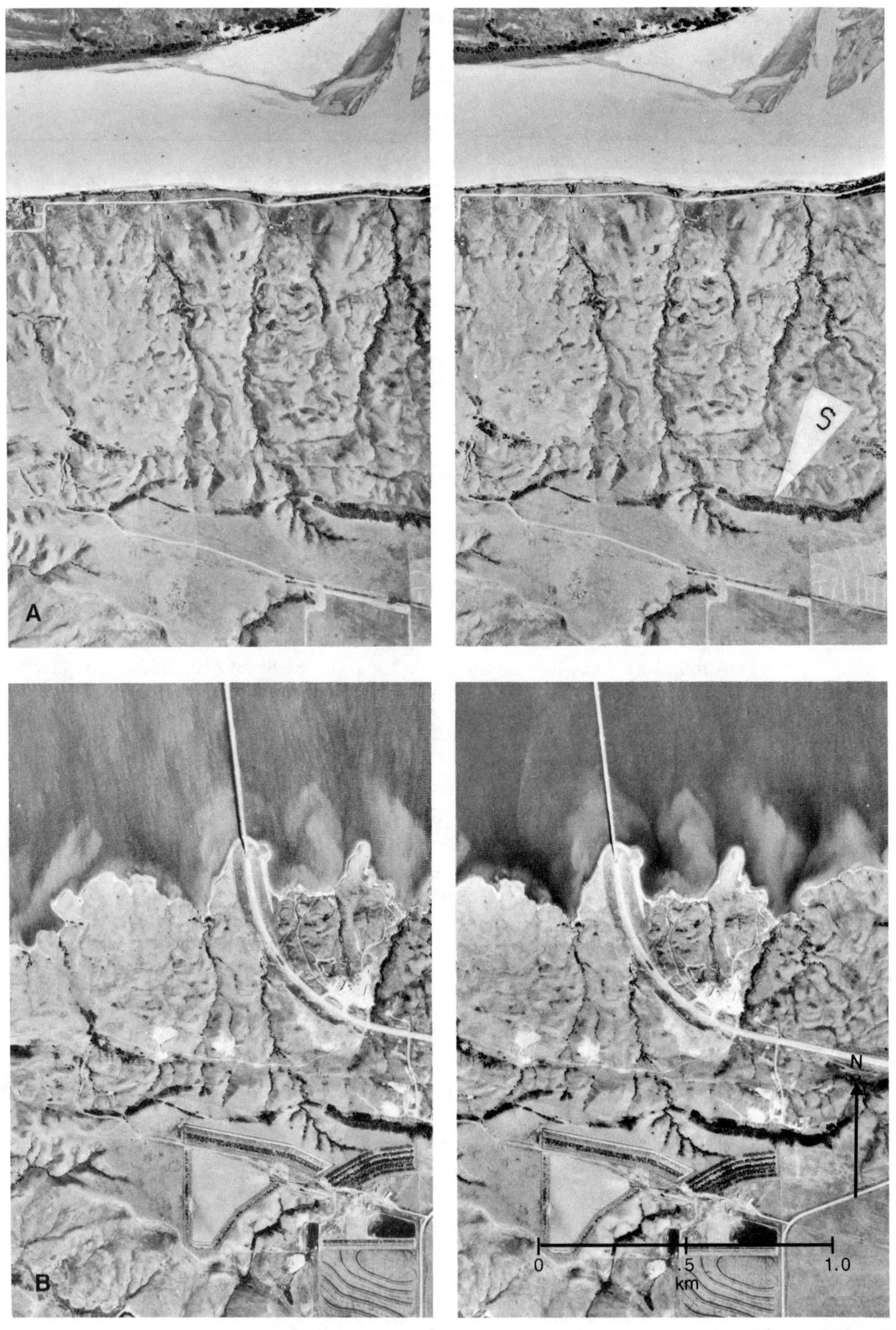


Figure 6.18 A,B

aster, geologists and engineers were blamed not only for the disaster, but also for not issuing an alert to the public.

Hillslopes adjacent to reservoirs having water level fluctuations are commonly subject to landslides. For example, following the 1976 breach and collapse of Teton Dam, Wyoming (see Chapter 5), many landslides occurred on the reservoir slopes. According to USGS engineering geologist Robert Schuster, 34% of the valley slopes submerged by the reservoir failed after breaching and subsequent drainage of Teton Reservoir. Drawdown of reservoir levels has been known to cause damage to earth-filled dams (Sherard et al., 1963). Most landslides develop when the reservoir is lowered for the first time. Others develop after years of operation, when an unprecedented drawdown occurs. At the Belle Fourche dam in South Dakota, for example, landslides occurred on the upstream dam face in 1931, 22 years after the dam had been in operation. The landslides developed in the embankment fill, a highly plastic clay with a 2:1 slope, following an unprecedented rate of reservoir drawdown. Analytical techniques for analyzing dam stability under rapid drawdown are presented in Terzaghi and Peck (1948).

Fluid pressure is very important in the stability of earth or submarine slopes (Postma, 1984). Landslides associated with a heavy rainfall on August 18, 1972, in West Virginia were found to be related to geologic strata and to pore water pressure (Everett, 1979). The rain infiltrated secondary joints in horizontally bedded shale until the relatively impermeable Sandstone Tuscarora was reached. The water then moved horizontally, issuing at hollows under a mantling regolith. The role of pore pressure in landslide movement is discussed in Chapter 4, and is succinctly stated by Terzaghi and Peck (1967):

> The slides are preceded by a sudden, but temporary or local, increase in pore-water pressure in the zone of sliding. . . . The weight of the overburden is temporarily transferred to the water, whereupon the effective pressure and the corresponding shearing resistance along a potential surface of sliding are reduced and a slide occurs.

Figure 6.18 illustrates an interesting landslide problem at Oahe Reservoir in South Dakota. Landslides are abundant in the Pierre Shale and glacial drift in the breaks along the Missouri River (Crandell, 1958; Erskine, 1973). After Oahe Reservoir was formed in 1963, the Rt. 14 bridge across the reservoir showed signs of distress due to a landslide on the east abutment. Inclinometer measurements indicate the bluff slope is moving at 10–15 cm/yr (Gardner and Tice, 1977). Movement of the slide seems to be correlated to the level of the Oahe Reservoir.

*Landslide correction.* It is easier to properly design and engage in good construction in landslide-prone areas than to correct a landslide problem. Hoexter et al. (1978), for example, show how engineering geologists can present maps to

Figure 6.18 Aerial photographs (stereo pair) of Missouri River area, South Dakota, before and after construction of Oahe Reservoir. **A.** Photos taken July 21, 1950. The letter *S* indicates the scarp of one of many landslides in the Pierre Shale which here is thinly veneered with till. The Missouri River flows westerly at this location. **B.** Photos taken July 8, 1968. The landslide identified above has become activated by the reservoir and associated rise of water table, causing continued maintenance problems for the Rt. 14 bridge abutment. In this photo, large waves are eroding the shoreline, causing sediment plumes.

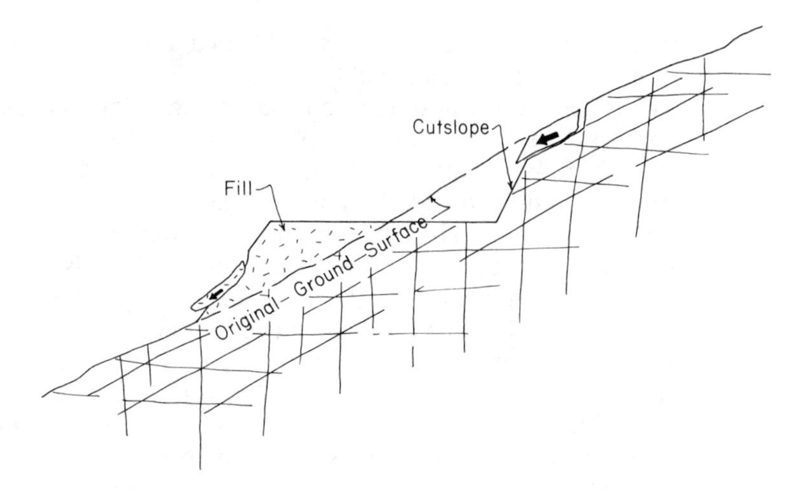


Figure 6.19 Sketch showing landslide potential created by cut and fill.

planners with three broad units (stable, unstable, and potentially unstable areas) that can help prevent landslides from occurring.

If it is necessary to construct a highway or other structure in an area where landslides are suspect, a geotechnical engineer should determine the factor of safety of the slope involved. In highway engineering, slope designs generally require safety factors in the range of 1.25–1.50 (Gedney and Weber, 1978). To render a slope more stable, generally it is necessary to either reduce the forces tending to cause movement or increase the forces resisting the movement.

Consider a man-made excavation for an interstate highway which includes both a cut and a fill (Fig. 6.19). The potential of a landslide is created 1) in the natural soil or rock in the upslope area and 2) in the artificial fill and natural soil or rock in the downslope area. Experience indicates there are several ways to correct landslide hazards:

*a. Balance moments.* In this method the upslope area can be brought to precut stability by removing additional material on the top of the landslide. A volume removed should be roughly equal to the volume originally excavated in the cut. Figure 6.20 shows a cross section of an area where an interstate highway in South Dakota was constructed in 1963, cutting off the toe of an ancient landslide. Massive failure ensued. To correct the problem of a slowly moving unstable mass, a second cut was made on the upper part of the landslide to balance rotational movements. The slope has been stable ever since.

A similar type of balance of moments was achieved in the construction of an Italian roadway overlying a soft clay base (Terzaghi and Peck, 1948). The weight of the highway fill on the soft clay caused a landslide in the underlying portions of the clay. To avoid this problem a gravel counterweight was added extending out on the clay adjacent to the roadway so as to adequately compensate for the rotational landslide moments.

*b. Drains.* Landslide potential can be reduced by keeping moisture out of the upslope or downslope areas. Near-horizontal wells may be installed, allowing for drainage. Royster (1973) shows how the installation of approximately 15,000 m of horizontal drain pipes, along with twelve 30-m deep vertical wells, was effec-

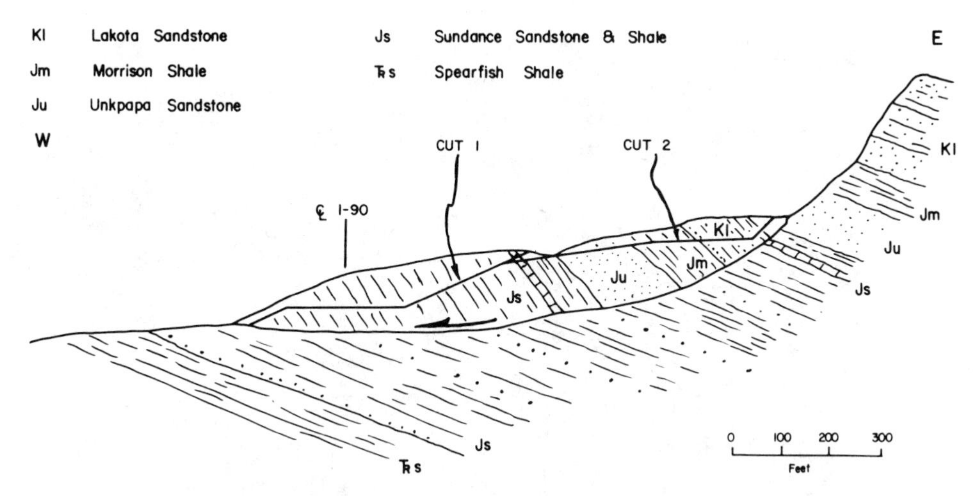


Figure 6.20 Landslide along I-90, 3 km south of Sturgis, SD (from Rahn et al., 1981).

tive in lowering the water table and stabilizing colluvium and weathered shale landslides along Rt. I-75 in Tennessee. Horizontal drains have been used successfully to prevent or correct landslides in Tennessee (Royster, 1977) and North Carolina (Glass, 1977). The theoretical increase in the factor of safety for cut slopes using horizontal drains and pumped wells is described by Paige-Green (1981).

*c. Impervious layer.* Asphalt paving or other impervious covers over the upslope or downslope areas helps stabilize landslides by keeping moisture out, thus reducing the weight and preventing the shear strength from diminishing.

*d. Buttresses.* To prevent landslide movement, large walls can be built upslope, just below the cut, or below, at the base of the fill on the downslope areas. The walls can be constructed of concrete, reinforced earth, or gabions (wire baskets filled with rocks). The mass of the walls help resist landslide forces. Figure 6.21 illustrates a landslide area on Rt. I-40 near Rockwood, TN. The installation of a reinforced earth buttress wall with a rock mat underdrain successfully stabilized the landslide. Reinforced earth is a construction material that involves the use of soil backfill with cement additive, tied together with thin metal strips or wire mesh to form a retaining wall that is capable of restraining large imposed loads (Vidal, 1979; Godfrey, 1984). For small roads in mountainous terrain, wire meshes are available which can be incorporated into the fill and interlocked at the exposed fill face, allowing for the emplacement of fill with a very steep slope.

*e. Rock bolts.* In bedrock, incipient translational slides can be secured by rock bolts. Spalling rock may be secured or mitigated by buttresses, wire mesh, or concrete canopies (Hoek and Bray, 1977).

*f. Miscellaneous methods.* In some places it is more expedient to allow the landslide to move, and then simply excavate and remove it. This is particularly useful during the construction of a roadcut when no hazards are presented, and earth-moving machines are available.

Various schemes have been attempted in order to treat unstable slopes with injected chemicals. The ion-exchange technique consists of treating clay minerals along the plane of potential movement. Cation replacement can increase the soil

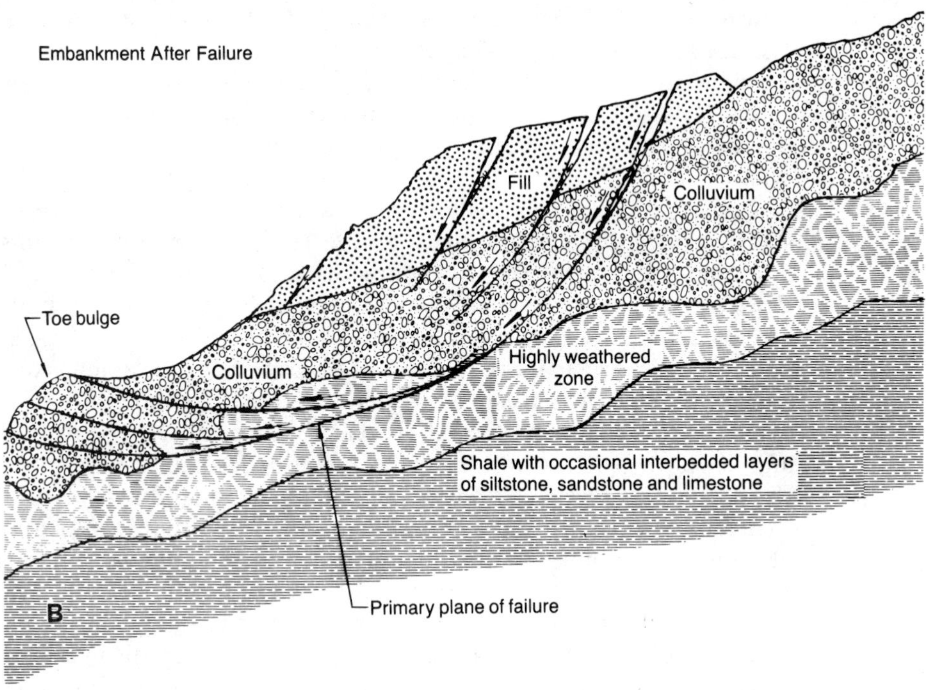


Figure 6.21 Landslide correction on I-40 in Tennessee (from Royster, 1974). **A**. Ground view of road failure. **B**. Geologic cross section illustrating shear failure in colluvium and fill.

shear strength by 200% or 300% (Gedney and Weber, 1978). Electro-osmosis causes the migration of pore water between direct current electrodes. In response to a potential difference, soil water tends to flow away from the anode and towards the cathode, thereby helping to reduce pore water and consolidate the soil.

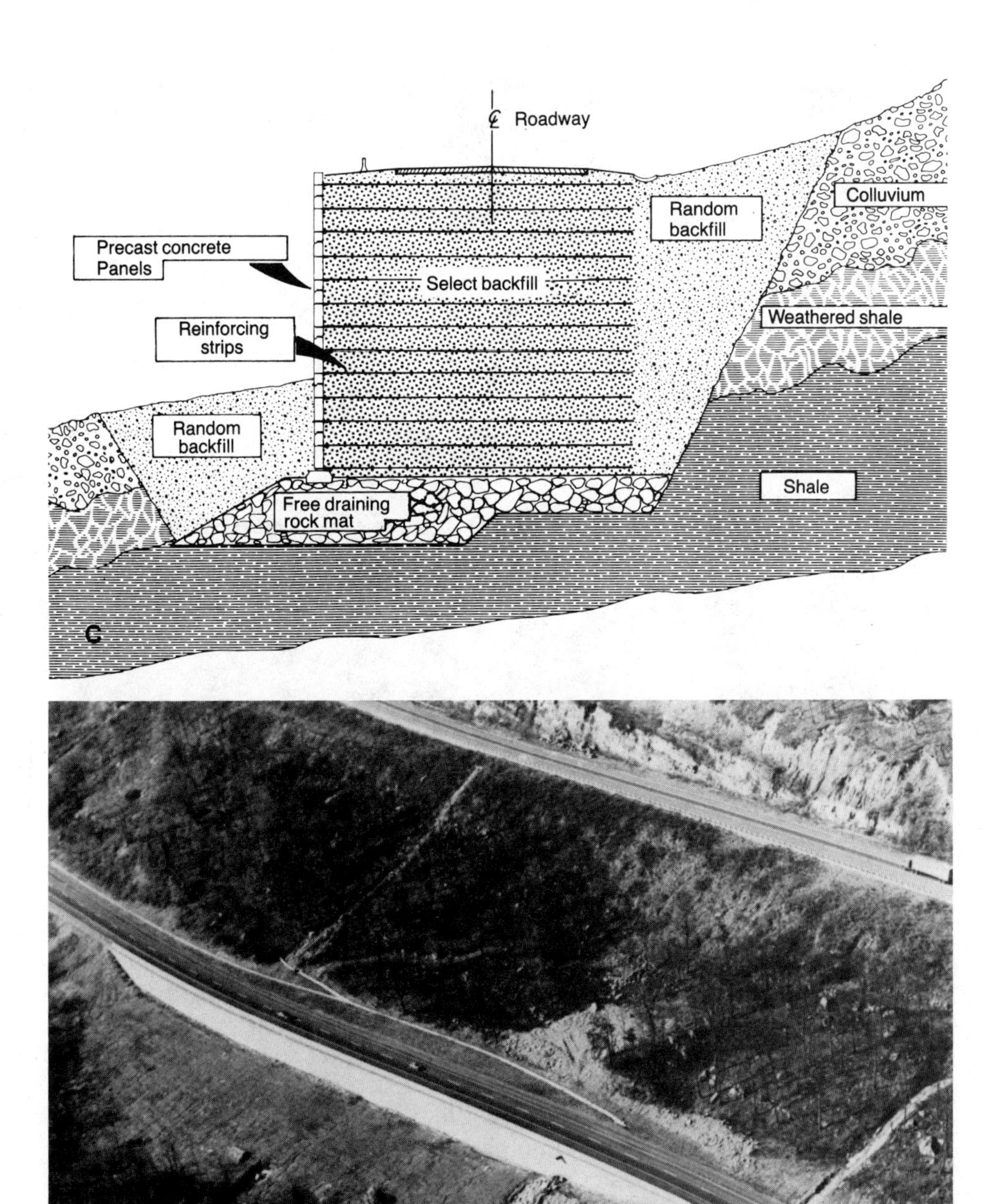


C. Cross section of reinforced earth wall used to correct fill failure. D. Aerial view of final correction to fill failure using reinforced earth.

---

Ground freezing can provide temporary ground support and ground water control during excavation of difficult soil conditions (Braun and Nash, 1985). In the excavation for the power plant at the Grand Coulee Dam, WA, earth movements plagued the construction effort (Bailey, 1980). A temporary solution to a slow moving landslide was found by freezing it until construction was finished.

Figure 6.22 Debris-flow deposits on surface of alluvial fan in Death Valley, California.

## *6.1.4 Debris Flow*

Debris flow, also generally known as mud flow or earth flow, is a form of mass wasting whereby wet debris moves downslope as a viscous fluid. According to some classifications, earth flow is the general name for this type of mass wasting; debris flow is a coarse-grained earth flow, and mud flow is a fine-grained earth flow. Debris-flow deposits typically have 50% of the solids larger than sand, whereas mud-flow deposits contain mostly sand and finer particles. Debris-flow deposits may contain 90% by weight solids, and have densities of 2–2.5 g/cm$^3$. Boulders up to 30 tons are known to have traveled 2 km during one debris flow event. A debris flow in motion has the consistency of wet concrete. Wet debris flows may reach velocities of 10 m/s. A very slow 27 hectare earth flow in Oregon has a velocity of about 1 m/month (Hicks and Lienkaemper, 1983).

A debris-flow deposit typically is unsorted and resembles glacial till. The deposit consists of a clayey matrix containing large angular boulders and rubble (Fig. 6.22). Debris flows are a common form of mass wasting in mountainous regions (Bryan, 1923; Blackwelder, 1928; Jahns, 1949; Blackwelder, 1954; Sharp and Nobles, 1953; Denny, 1961; Kennedy, 1969; Costa and Jarrett, 1981; Jones et al., 1984).

*Examples of debris flows.* Debris flows present a greater form of risk of death and injury in southern California than all other kinds of slope failure combined (Campbell, 1975). In southern California (Fig. 6.23) debris flows commonly occur after a period of several wet winters which have loosened landslides in the headward part of canyons. Development in the hillside areas of southern California

Figure 6.23 Debris-flow deposit covering car in 1978, La Crescenta, CA (USGS photo). The mudflow deposited over 15,300 $m^3$ of debris from a drainage area of only 0.6 $km^2$.

accelerated following World War II, and will continue due to the decrease in available flatland space. As a result of hillside construction, many structures are located on or below hazardous slopes. Campbell (1975) presents a discussion of the causes and mechanics of soil slips and related debris flows during the heavy rains of January 18–26, 1969. Heavy rains in the San Francisco Bay area on January 4, 1982, resulted in 27 fatalities and $280 million damage. Geologists have made it clear that enough knowledge already existed to prevent much of the misery and property loss. However, unless those who live on the land are required to utilize the information available (such as USGS slope stability maps), or do so by their own initiative, natural occurrences will continue to take an unnecessary and tragic toll.

Communities located near the apex of alluvial fans in the state of Utah have a history of debris flow disasters (Wooley, 1946), but the 1983 season was one of the worst. A mudslide on April 14 in Spanish Fork Canyon formed a 6-km lake, submerging the town of Thistle where people were forced out. One night in June 1983, a U.S. Forest Service employee stationed in Coldwater Canyon, high above the town of North Ogden, heard trees snapping and boulders rolling. One hundred residents fled for shelter in a Mormon center moments before a river of mud slithered into three houses. Only a week previously mud and rocks moved down Rudd Creek Canyon, burying five houses and damaging 100 more in Farmington, UT. In May 1983, a landslide in Fairview Canyon blocked a stream, forcing the evacuation of 1100 residents of Fairview, UT. Just after Memorial Day, when a dam of debris in Ward Canyon broke and set water and mud pouring into Bountiful, UT, 1000 Mormon volunteers showed up to place sandbags in critical locations. The 1983 debris flows in Utah were triggered by torrential rains that fell on soils already saturated by melting snow from a record winter snowfall. Landslides cascaded into the canyons, creating earth dams. Water ponded behind the dams and eventually the unstable materials gave way, forming a debris flow. Lamented Davis County, Utah, Deputy Sheriff Harry Jones: "We can control the water, but the mud just goes where it wants to. All we can do is try to anticipate where it is going and then try to get out of the way." The floods, landslides, and debris flows were so extensive that 22 of Utah's 28 counties were declared national disaster areas.

The failure of a coal spoil pile (a tip) at Aberfan, Wales, Great Britain, on October 21, 1966, resulted from a landslide in a 60-m high pile of mine waste materials following heavy rains. The landslide turned into a mud flow which swept 1 km downhill at a velocity of about 25 km/hr into a school and several houses. In the disaster 144 people lost their lives, including 116 children (Bishop et al., 1969; Perloff and Baron, 1976). A tribunal was appointed to inquire into the disaster and revealed that no soil mechanics investigations had been carried out prior to construction of the tips, and no safety inspections were ever made. The tribunal found that the tips were dumped onto gently dipping sandstone and shale beds, where springs occurred. A combination of high rainfall several days before the slide, interruption of local drainage, and occurrence of a high water table as evidenced by springs beneath the spoils combined to develop high pore water pressure, which changed the nature of the lower spoils from a solid to that of a heavy liquid of density approximately twice that of water. The debris moved downward, carrying the upper part of the spoils that were not saturated, and the whole mass moved rapidly down the hillside, transforming into a debris flow, spreading out sideways into a layer approximately 7 m thick.

On February 26, 1972, a dam made of coal mine refuse failed in the Middle Fork of Buffalo Creek (alias Buffalo Run), West Virginia. The failure caused a debris flow up to 7 m high which traveled down the narrow Buffalo Creek valley at about 8 km/hr. The disaster killed 118 people, and property damage was $50 million (Wahler, 1973). The dam itself was made of coarse coal refuse (largely sandstone and shale fragments). Coal processing wastes (sludge) collected behind the dam. The dam was not constructed on the basis of engineering plans, and before the disaster there had been numerous small landslides and evidence of piping. Following the failure, it was found that the dam failure may have been caused by 1) piping, 2) landslide of the dam itself, or 3) overtopping due to inadequate spillway. In all probability the failure was by landsliding due to high pore pressures developed as water rose behind the dam following heavy rains. Thus the mechanism of failure had similarities to Aberfan. However, because the coal refuse dumped into Buffalo Creek had a reservoir of water backed up behind it, the moving debris resembled a debris-laden flood rather than a debris flow such as Aberfan.

Probably the most tragic debris flow (variety ice and rock avalanche) occurred at Yungay, Peru, on May 31, 1970. Triggered by an earthquake, at least $5 \times 10^7$ $m^3$ of ice and rock traveled rapidly down a glacial valley from Mt. Huascaran, the highest peak in the Peruvian Andes. Part of the debris became air-launched at an abrupt topographic break-in-slope, and debris rained down on the valley below. Some boulders were hurled 4 km through the air. Preceding the debris were air

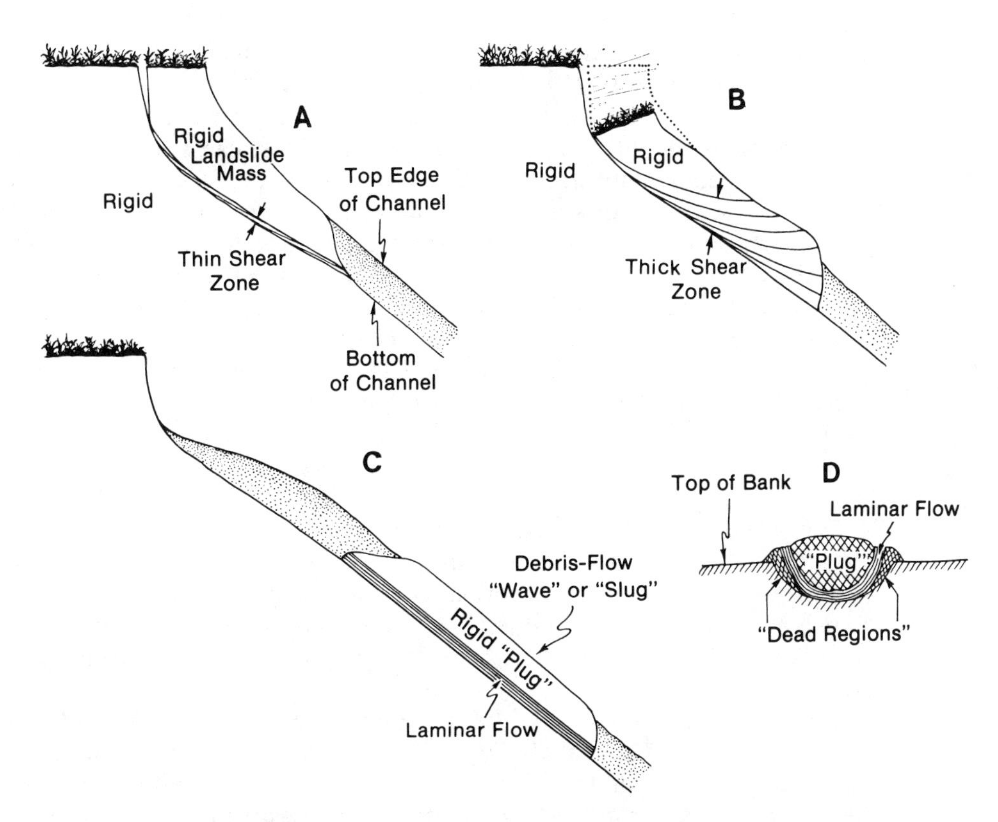


Figure 6.24 Idealized representation of transition from landsliding to debris flow (from Johnson and Rahn, 1970). **A**. Longitudinal section of landslide block as it begins to move into channel by shear along a thin zone. **B**. Longitudinal section of transition stage between landsliding and debris flow. Rigid material is behind front that is being raised by shear throughout a thick zone. **C**. Longitudinal section of debris flow. **D**. Transverse section of center of debris-flow "slug." Materials in center (the "plug") and at sides (the "dead regions") are rigid, although the former is moving. Zone between "dead regions" and "plug" is deforming, and its width depends upon properties of debris and size of channel.

blasts capable of felling trees. The debris avalanches killed about 18,000 people at Yungay, 14 km away from (and 3.7 km lower than) the source, reaching the town in only 3 min (Browning, 1973; Plafker and Ericksen, 1978; Varnes, 1978; Hays, 1981).

*Debris flow characteristics.* Debris flows may be very dense, approximately that of concrete ($\rho = 2.6$ g/cm$^3$), and hence they are capable of floating huge boulders as well as any man-made obstacles such as a car, house, or barn (Wooley, 1946). Typically debris flows originate in mountainous gulleys during a heavy rain. As a landslide occurs and moves into the gulley, it becomes more fluid (less viscous) by the increase in water from runoff in the adjoining slopes and gulleys. It also loses strength because of the thixotropic effect, where motion (vibrations, etc.) causes a firm material to dramatically lose shear strength. (The thixotropic effect can be observed by making disturbances in fine-grained, wet beach sand.)

As the landslide moves down into the gulley, it may retain a central rigid plug (Fig. 6.24) but ultimately transforms to a viscous fluid that may spread out onto an alluvial fan at the base of the mountains. For discussions of debris-flow viscosity, see Leopold et al. (1964) and Johnson (1970).

Debris-flow deposits make up a sizeable portion of many alluvial fans in semiarid regions. Frequently they become reworked by streams on the alluvial fan. Debris-flow deposits are similar to lava flows in that they have a lobate terminus and have levees and distributary channels. Prehistoric debris flows associated with volcanic eruptions (lahars), having volumes exceeding 2 $km^3$, reached far out onto the lowland near Tacoma, WA (Crandell and Waldron, 1969). During the May 18, 1980, eruption of Mt. Saint Helens, Washington, the mixing of vented steam, melted glaciers, and volcanic ash formed lahars that raced at 50 km/hr to the town of Toutle, about 25 km away. The lahars covered houses and cars with over 10 m of hot mud (Findley, 1981). Only four out of a total of 42 bridges spanning the Toutle River were not destroyed or badly damaged by the lahars (Schuster, 1983).

*Debris flow hazards.* Temporal prediction of debris flows can be made on a statistical basis. For example, based on radiocarbon dating of wood buried by rock avalanches, and from measurements of weathering-rind thickness of surficial clasts, Whitehouse and Griffiths (1983) show that large (1 $km^2$ area) rock avalanches in the Southern Alps of New Zealand can be expected about every 100 yr.

Submarine mud flows have been documented. In 1968, the world's tallest structure, Shell Oil Company's Cognac drilling platform, was completed in more than 100 m of water off the Mississippi Delta. (As an indication of the low shear strength of sediments at the Cognac site, it is interesting to note that each of the 24 2.1-m-diameter piles was driven to a design depth of 150 m in only about 1 hr.) In 1969, the platform collapsed during Hurricane Camille. It is all but certain that a submarine mud flow triggered by the storm waves was largely responsible for the loss (Bea, 1971).

Recognition of areas subject to debris flows can be based on topography, climate, and studies of surficial deposits. In the semiarid southwest United States, alluvial fans adjacent to mountain ranges which contain clayey deposits and subject to heavy seasonal rains should be suspect. Land-use planning should discourage urbanization of active debris flow areas. All structures should be built with precaution in debris flow areas: an example is the 1970 California Department of Water Resources canal on the western side of the San Joaquin Valley where mud-flow deposits on alluvial fans were encountered. To mitigate collapsing soils, the alignment was adjusted so that mud-flow deposits were bypassed in some areas and precompacted by water flooding in areas where canal gradient required crossing of hazardous areas (Curtin, 1973). Another consideration in design of the aqueduct was to provide structures that would prevent debris flows from entering the aqueduct (Conwell and Fuqua, 1960). To prevent catastrophic flooding in the event that the canal is overrun by a debris flow, water level recorders tied into a central computer can automatically shut off canal water discharge.

The densely populated European Alps offer a number of alternatives to the problem of reducing hazards due to debris flows. These include active (engineering) and passive (zoning) alternatives. Protective structures (walls built to confine or deflect debris flows, sediment catchment dams, and retention basins) have been developed in the Alpine countries during the past 300 years. In the French Alps, special hazards maps are available at 1:20,000 scale showing three colored zones: red (to be avoided), orange (geotechnical study required), and green (no obvious instabilities). Urbanization of the Alps is proceeding at a rapid rate, and because of the abundance of mass-wasting deposits, it is impossible to find completely safe places. For example, the modern ski village of Clusaz, France, is built on an ancient rock avalanche deposit. New apartment buildings are located on historic debris-flow deposit at Briancon, France. The mining town of Radmer a.d. Hasel, Austria, was buried by a rock avalanche in 1540; new houses are now built on the same deposit. In the nineteenth century, rock falls prompted a relocation of Alt-Felsberg, Switzerland, to Neu-Felsberg, although today both Alt-Felsberg and Neu-Felsberg continue to expand below a dangerous cliff subject to mass wasting. Eisbacher (1982) points out that absolute safety cannot be guaranteed in the Alps:

> Confronted with the question of what is being done in the Alpine countries to circumvent the possibility of catastrophic debris flows with a very low rate of recurrence, say one event in 1,000 years, dozens of workers answered with the same words: " . . . if we considered this possiblity there would be no more subdivisons anywhere. . . ."

## 6.2 Permafrost

Many engineering problems associated with permafrost involve accelerated mass-wasting processes. The term permafrost is a contraction of the words permanently frozen ground. It is defined as the thickness of soil or rock which has been frozen for 2 or more years. The permafrost table is the upper surface of the permafrost. The active layer is the zone above the permafrost table which thaws in the summer and freezes in the winter. Unfrozen areas in permafrost are called talik.

The thawed active layer (called a molisol by geomorphologists) is capable of movement of debris (gelifluction or solifluction) on very gently dipping slopes. The molisol essentially slides on the permafrost table. Washburn (1967) showed rates of frost creep and gelifluction in Greenland ranged from 1 to 4 cm/yr, with highest values on wetter soils. In favorable sites, the rate of movement can reach 0.3 m/yr (Corte, 1969). The term periglacial refers to cold areas near glaciers or in high latitudes or elevations where permafrost is apt to occur, although not all periglacial areas are necessarily underlain by permafrost (Embleton and King, 1968; King, 1976). Unusual landforms are found in periglacial environments, such as stone polygons and stripes, pingos, rock glaciers, and other forms of solifluction (rapid soil-flow features). Some landforms in the continental United States are relic periglacial features from the Pleistocene epoch. Rock glaciers (Fig. 6.25) and boulder fields (Fig. 6.26) are special forms of mass wasting in permafrost areas (Ives, 1940; Smith, 1953; Wahrhaftig and Cox, 1959; Rapp, 1962; Washburn, 1967; Curry, 1966; Potter, 1969; Vick, 1981; Giardino and Vick, 1985). Some rock glaciers may be rock-covered stagnant glaciers (Potter, 1974).

Permafrost forms when the mean annual air temperature is low enough to maintain a mean annual ground-surface temperature below freezing. Figure 6.27 shows permafrost areas in the northern hemisphere. Approximately one-fifth of the world's land area is underlain by permafrost. About 50% of Russia and Can-

Figure 6.25 Aerial photograph of a rock glacier in the Sunlight Basin, Wyoming.

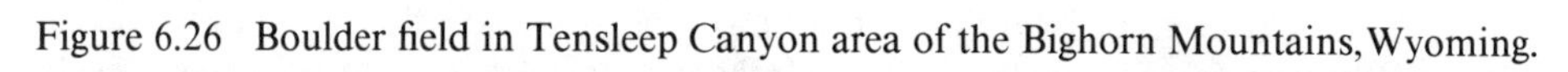

Figure 6.26 Boulder field in Tensleep Canyon area of the Bighorn Mountains, Wyoming.

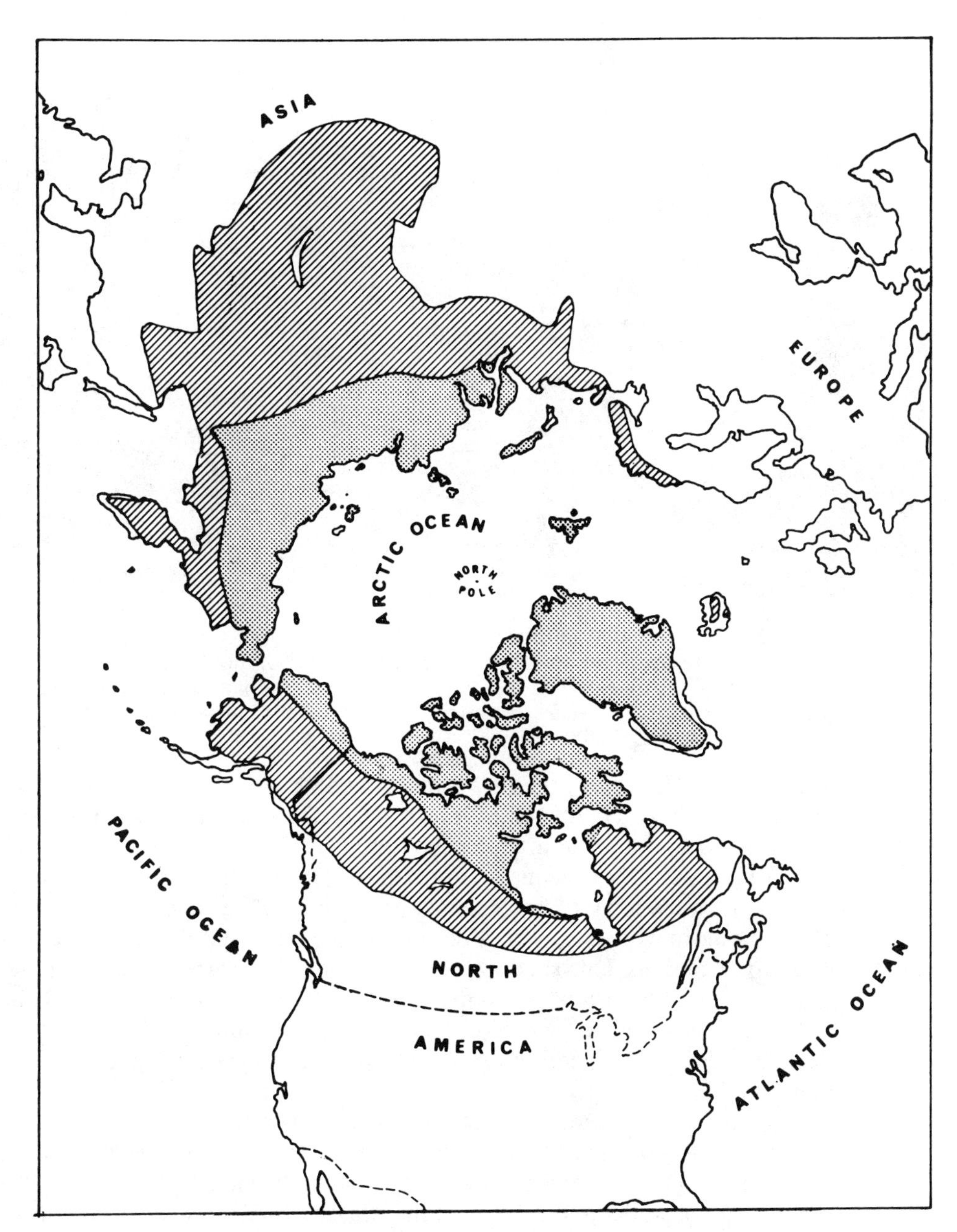


Figure 6.27 Extent of permafrost in the northern hemisphere (from USGS). ▩ Area underlain by continuous permafrost. ▨ Area underlain by discontinuous permafrost.

ada and 90% of Alaska are underlain by permafrost. The thickness of permafrost varies. In Siberia it approaches a thickness of 1.7 km. In Alaska permafrost varies from 0.6 km at the North Slope to zero near Anchorage (Ferrians et al., 1969).

Most permafrost developed during the Pleistocene epoch, and may be thousands of years old. In Siberia an extinct bison was uncovered in permafrost which

Figure 6.28 Differential subsidence of roadbed to Copper River and Northwestern Railway near Strelna, 120 km northeast of Valdex, AK (USGS photograph). The thermal equilibrium of the fine-grained sediments underlying the roadbed was disrupted during construction and the permafrost started to thaw. Maintenance and use of the railroad was discontinued in 1938.

was dated at 31,000 BP; the meat was still edible. Most permafrost is degrading (i.e., getting thinner) and the extent of permafrost is decreasing in conjunction with worldwide warming in Holocene time.

Frozen ground is a virtually impermeable layer that restricts recharge, discharge, and movement of water. It acts as a confining layer, and limits the volume of sediments in which water may be stored. The water supply and sewage disposal in a town such as Fairbanks, AK, is complex. Ordinary water wells must be located in talik or below permafrost. Water distribution and sewage collection lines can be located in permafrost, but, in order to prevent freezing, the water must be kept moving continuously through main pipe lines and even through individual house pipes.

Engineering problems are wide ranging in permafrost areas (Péwé, 1966). Changes in the surface environment such as the clearing of vegetation, the building of roads and other construction may lead to thawing of permafrost. It is now illegal to drive across public lands in the Alaskan tundra (except when the ground is solidly frozen) because dead vegetation can cause increased melting and changes in the permafrost environment. The thawing of ice wedges and subsequent subsidence followed by heaving associated with later freezing and frost action are responsible for major engineering problems in the Arctic region (Fig. 6.28). Undulations in the Alaskan Highway are the result of permafrost melting in areas where the permafrost includes discontinuous lenses of ice. The blacktop road absorbs solar energy and causes melting of the permafrost. Large trucks using the Alaskan Highway have extra wheels in the center of the trailer to help maintain the structural integrity of the trailer, due to the bumps in the treacherous highway.

When permafrost thaws in surficial deposits such as glacial clay or silt, what was once solid land becomes a veritable quagmire incapable of supporting any structure. Permafrost that is ice-rich can cause extremely serious problems such as liquification of soils, differential settlement of ground surface, instability of slopes, accelerated erosion, and disruption of drainage. A heated building may literally almost sink out of sight! In the past, inadequate knowledge about permafrost has resulted in large maintenance costs and, in some cases, abandonment of highways, railroads, and other structures.

The futility of exploiting the Arctic lands by what engineers call brute force was obvious very early in oil industries' experiences on Alaska's North Slopes, as broken chunks of very expensive but carelessly placed pavement now attest. Even the floor of the Arctic Sea can be icy, subject to freeze-thaw cycles. Warmth can be an erosive agent. If uninsulated, a heat source or heat conductor near the permafrost melts its own foundation. Designing offshore drilling platforms presents an engineering challenge.

Of recent interest is the Trans-Alaska pipeline, built in 1975–1979, from Point Barrow to Valdez, AK. The 1273-km-long line is 1.22 m in diameter, and carries $3 \times 10^5$ $m^3$ per day of oil at a temperature of 70–80°C. (The oil comes out of the ground from a depth of 2440 m and has a temperature of 82°C due to the earth's heat at the great depth from which the oil is obtained. The oil remains hot in the pipeline due to frictional turbulence.) Of concern is the problem that could arise if the hot pipe melts permafrost. A pipe buried 2 m in permafrost and heated to 80°C will thaw a cylindrical region up to 10 m in diameter in a few years (Lachenbruch, 1970).

About 50% of the Trans-Alaska pipeline is underlain by competent bedrock or by surficial deposits that contain little or no ice. In the hazardous permafrost areas, construction was done in a fashion so that the hot pipe would not melt the permafrost. Of particular concern was differential settlement that could occur where the pipeline passes across ice wedges which thaw more quickly than surrounding sediments (Isaacs and Codel, 1973). It is believed that melted permafrost could result in mud flows, which could allow for settlement of the pipe, possibly causing rupture. The spilled oil would freeze, and any oil in the pipeline which was not flowing would freeze, too.

Three construction modes were used for the pipeline, 1) above ground pile supports where the bottom of the piles become frozen into the permafrost, 2) above ground on gravel embankments, and 3) special burial using mechanical refrigeration and insulation to prevent or limit permafrost thawing. Because of permafrost, the length of the pipeline simply emplaced in a trench was drastically reduced to about 50% of the total distance (Moening, 1973). As of 1984, the pipeline has retained its integrity.

In general, there are two methods of construction for a structure in permafrost regions:

1. Active method. In areas of thin permafrost, the vegetation and permafrost are cleared, the ground is allowed to thaw, and spread footings are used. Slip casing can be used in drill rigs so drill stems telescope together as settlement of the platform occurs.
2. Passive method. In most places, especially in thick permafrost areas, an attempt is made to preserve the permafrost by insulation. For drilling plat-

forms in the north slope of Alaska, 15 m of gravel is sufficient insulation. The ground can even be refrigerated by mechanical means (Phukan and Andersland, 1978).

In nonpermafrost areas, freezing ground (ordinary frost) can also cause engineering problems (Bryan, 1934). Frost heave is a common problem in the northern continental United States. In the midwest, frost in north-facing slopes (particularly those lacking insulating snow accumulation) may reach 1.5 m depth, and south-facing slopes may have 1 m of frost. Installation of water and sewer pipes should take frost into account. Ice expands as it is frozen, and, because unconsolidated debris is inhomogeneous and has varying permeability, frost tends to form lenses and causes uneven heaving. Frost heaving causes improperly drained retaining walls to tip over, as well as causing improperly subdrained roads to break up in the spring.

## Problems

1. Lay a protractor on the outline of the scree shown in Fig. 6.2. What is the angle of repose?

2. Plot prefailure creep rate data for the Vaiont Dam landslide area (Table 6.2) on semilog paper using time (days) as arithmetic scale ($x$ axis) and rate (cm/day) as logarithmic scale ($y$ axis).
   - **a.** Derive an equation of best fit for a straight line through this plot using the form of $y = be^{dx}$.
   - **b.** Predict when failure should occur (when rate $\simeq$1 m/day). (Hint: Use September 20 as day 0. Plot five points, using simply the average value of the creep rate for the middle day in each time interval given in Table 6.2.)
   - **c.** Does the above analysis suggest this catastrophe could have been predicted?

3. Use the method of slices to calculate the factor of safety for the same slope shown in Figure 6.16 if all material has density of $1.89 \times 10^4$ N/m$^3$ and $S_u = 2.5 \times 10^4$ Pa. Use two arcs: *ABCD* and *AB′C′D′* as shown.

4. Beaty (1956) collected data which seems to indicate that the number of observed landslides on the hills near San Francisco is dependent upon exposure direction:

| Exposure | No. of slides |
|---|---|
| NW | 30 |
| NE | 24 |
| SW | 15 |
| N | 14 |
| E | 10 |
| SE | 10 |
| S | 5 |
| W | 4 |
| | N = 112 |

   - **a.** Use a "Chi-Square Test" to determine if, in fact, there is a significant difference between the above observed data and data expected if the distribution of landslides were equally distributed among exposure direction. (See Eisenhart (1977) or an applied statistics book for an explanation of this statistical test. Use $\alpha = 0.05$.)

**b.** If exposure direction is related to number of landslides, explain why this happens. (Hint: Assume homogeneous and isotropic rocks. Remember also this is only a limited set of data.)

**5.** Consider the freebody diagram for an incipient translational rock slide shown below. Use Eq. (6.3) to determine the F.S. of forces in the plane of sliding. Assume a proposed 10 m deep roadcut is proposed in hard sedimentary rocks where the bedding planes strike parallel to the highway and dip at 46° toward the highway. Laboratory tests on cores containing bedding plane discontinuities show that the friction angle ($\phi$) is 41° and cohesion is 34,500 N/m$^2$. The density of the rock 2800 is kg/m$^3$.

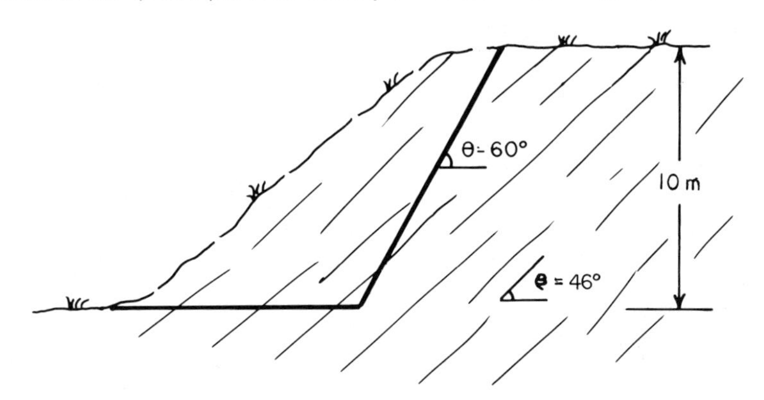


**a.** Assume no tension crack exists, and the slide is all above the water table. Determine the factor of safety for a 60° road cut.

**b.** Suppose the contractor uses explosives to excavate the roadcut, and in so doing destroys any cohesion in the bedding planes. Determine the factor of safety.

**c.** An alternative roadcut design is to increase the resisting forces by using rock bolts. Determine the rock bolt shear strength required per m of highway for a F.S. = 1.5. (Assume no cohesion. Assume the rock bolts are installed perpendicular to bedding, consisting of loosely installed bolts having 6.5 cm$^2$ cross-sectional area of steel whose shear strength is $3.5 \times 10^8$ N/m$^2$. Determine how many bolts are needed per 1 m length of roadcut.)

**d.** Rock bolts not only provide stability by providing shearing resistance directly across the bolt itself, as described in *c* above, but also by providing tension in the bolt, and thus squeezing the two sides of a failure plane together. In this way the frictional resistance to sliding is increased. If the bolts designed in *c* above develop $70 \times 10^6$ N/m$^2$ tension, what is the actual F.S. for the slope?

# 7

# Ground Water

All the rivers run into the sea, yet the sea is not full; unto the place from whence the rivers come thither they return again.

King Solomon, 960 B.C.
(Ecclesiastes 1:7)

The water of rivers comes not from the sea, but from the clouds.

Leonardo da Vinci, 1510 A.D.

## 7.1 Introduction

Hydrology, the study of water, includes the subjects of surface water, ground water, meteorology, oceanography, glaciology, and related fields (Meinzer, 1923, 1942). Hydrology is an interdisciplinary engineering science: agricultural engineering, civil engineering, forestry, limnology, meteorology, glaciology, geology, marine science, and other disciplines have contributed to the study of water. This chapter briefly describes the hydrologic cycle, emphasizing the role of ground water as part of that cycle. Aspects of surface water are discussed in Chapter 8.

### *7.1.1 The Hydrologic Cycle*

The concept of the hydrologic cycle is that precipitation returns again to the atmosphere by evaporation and transpiration (Fig. 7.1). This fundamental concept was recognized by King Solomon (see quote above). It is unfortunate that during the Dark Ages the concept of the hydrologic cycle was distorted to the idea that water somehow left the oceans, not by evaporation, but a subterranean connection, through Hades, to hot springs where all rivers supposedly had their source (Krynine, 1960).

Scientific truth was stifled until the Renaissance. Leonardo da Vinci and the French scientist Bernard Palissy (1580) understood and promoted the concept of the hydrologic cycle. But it was not until Pierre Perrault (1674) measured the annual precipitation ($\simeq$50 cm) of the drainage basin of a tributary to the Seine River near Aignay le Duc (drainage area 22 km$^2$), as well as the discharge of the stream ($\simeq 10^7$ m$^3$/yr), that it was demonstrated that precipitation (PPT) to the

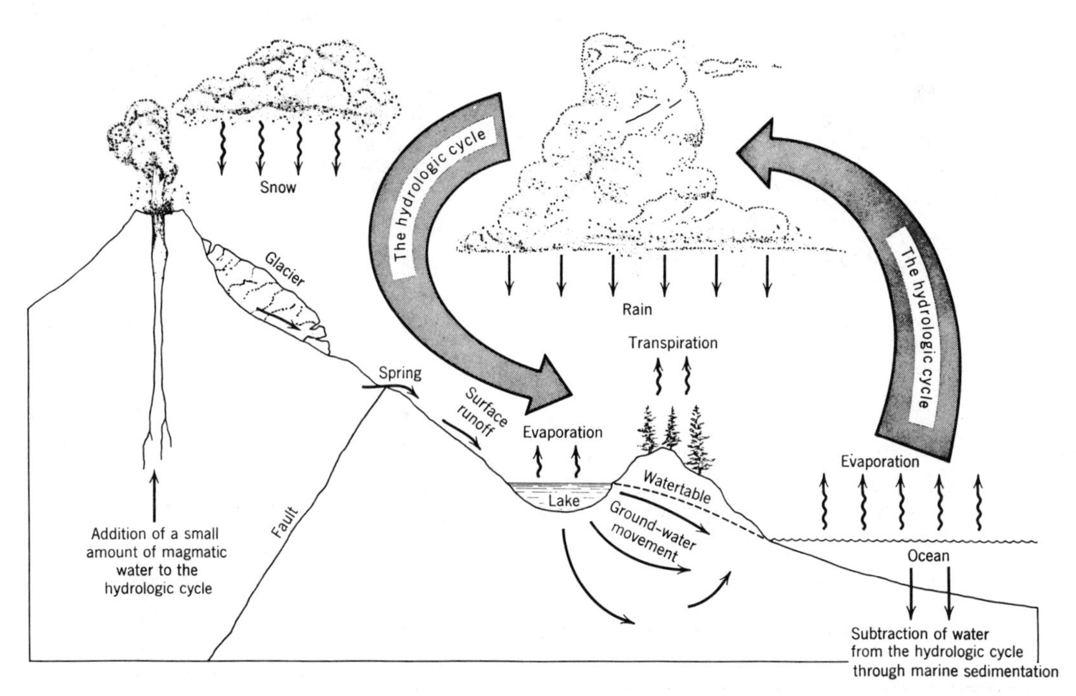


Figure 7.1 The hydrologic cycle (from Davis and DeWiest, 1966).

basin was more than enough to supply the annual surface water runoff (SWRO) of this stream basin:

$$\text{inflow} = (\text{drainage area}) \cdot (\text{PPT}) = (1.22 \times 10^8\ \text{m}^2) \cdot (0.5\ \text{m/yr}) \cong 6 \times 10^7\ \text{m}^3/\text{yr}.$$

Perrault showed that yearly precipitation to the basin was approximately six times the yearly runoff of the stream. In 1686, Mariotte repeated the experiment for the Seine River at Paris (Rodda, 1976).

As a rough estimate, it can be assumed that the remaining five-sixths of the precipitation to the Seine River watershed was consumed in the watershed by transpiration and evaporation, collectively called evapotranspiration (ET). Thus, the hydrologic budget for the Seine River watershed can be simply expressed as:

$$\text{inflow} = \text{outflow},$$

$$\text{PPT} = \text{SWRO} + \text{ET}, \tag{7.1}$$

$$6 \times 10^7\ \text{m}^3/\text{yr} = 1 \times 10^7\ \text{m}^3/\text{yr} + 5 \times 10^7\ \text{m}^3/\text{yr}.$$

In some stream basins, water also leaves by ground water underflow (GWUF), moving below the surface of the ground through permeable beds. For example, in southeastern New Jersey, where seaward dipping permeable sandstones occur, streams are virtually absent, and a large amount of ground water discharges into the Atlantic Ocean as submarine springs. These springs may be responsible for the erosion of submarine valleys with steep-headed basins, a process called spring sapping (Robb, 1984). Fresh water springs occur offshore of karst areas of Florida and Greece (Ghikas et al., 1983). Kohout et al. (1977) describe a flowing artesian well 40 km from the Florida coast which discharges fresh water into the Atlantic Ocean.

The entire United States has an average annual PPT of 76 cm and SWRO of 23 cm (Langbein, 1949; Leopold and Langbein, 1960). About 40% of this is SWRO from the Mississippi River (Leopold, 1974). The total GWUF to the oceans surrounding the United States is probably less than 1 cm/yr. Therefore, the U.S. water budget, expressed as depths of water per year, is:

$$\begin{aligned} \text{PPT} &= \text{SWRO} + \text{GWUF} + \text{ET}, \\ 76\text{ cm} &= 23\text{ cm} + (\sim 1)\text{ cm} + 52\text{ cm}. \end{aligned} \tag{7.2}$$

### *7.1.2 Water: A Peculiar Mineral*

Water, in its solid form, is a mineral. It is a naturally occurring, inorganic chemical compound with a definite crystal lattice.

Water is a peculiar mineral because of the following characteristics:

1. Water exists in three natural states: solid (ice), liquid (water), and vapor (water vapor).
2. Water is more dense as a liquid than as a solid. Water expands 9% by volume when it freezes. (Were it not for this phenomena, ice freezing in a lake would sink, eventually allowing for a lake to become frozen solid instead of only freezing to form an insulating cover.) The maximum density of water, 1 g/cm$^3$, occurs at 4°C.
3. Water has unusual thermal characteristics. The specific heat is the heat required to raise the temperature 1°C for 1 g mass of a substance. The specific heat of water is high: 1 cal/g/°C. This compares to ~0.2 cal/g/°C for sand or rock. The contrast between the specific heat of water and land masses has a marked affect on the climate. Near sea coasts or large water bodies, temperatures are somewhat constant in contrast to inland areas (especially deserts) where temperatures fluctuate markedly.

   The heat of vaporization (the heat required to make a given mass of liquid change to vapor) is also unusually high for water: 540 cal/g. This is why perspiration is such an effective cooling mechanism and why evaporative coolers work so well.

   The heat of fusion (the heat required to make a given mass of solid change to liquid) is comparatively high for water: 80 cal/g.
4. Properties such as high surface tension, high viscosity (which decreases with temperature), and low electrical conductivity are other unusual characteristics of water (Davis and Day, 1961).

### *7.1.3 The World's Water Supply*

Table 7.1 lists the volume of water occurring on earth (the hydrosphere). The world's oceans, which cover about 67% of the earth's surface, contain 97.2% of the earth's water. About 2.15% consists of glacial ice, most of which is in Antarctica and, to a lesser degree, Greenland. Ground water is about 0.62% of the world's water. In the United States the volume of ground water is estimated at $1.26 \times 10^{14}$ m$^3$ (Todd, 1980).

Table 7.1 World's Water Distribution

| Water category | Volume ($10^3$ km) | Percentage of total water |
|---|---|---|
| Atmospheric | 13 | 0.001 |
| Surface | | |
| Salt water in oceans | 1,320,000 | 97.2 |
| Salt water in lakes and inland seas | 104 | 0.008 |
| Fresh water in lakes | 125 | 0.009 |
| Fresh water in stream channels (average) | 1.25 | 0.0001 |
| Fresh water in glaciers and icecaps | 29,000 | 2.15 |
| Water in the biomass | 50 | 0.004 |
| Subsurface | | |
| Vadose water | 67 | 0.005 |
| Ground water within depth of 0.8 km | 4200 | 0.31 |
| Ground water between 0.8 and 4 km depth | 4200 | 0.31 |
| Total (rounded) | 1,360,000 | 100 |

From Bouwer, 1978.

## 7.2 Precipitation, Infiltration and Evapotranspiration

### *7.2.1 Precipitation*

The average temperature decreases from about +10°C at the earth's surface to −60°C with increasing elevation in the troposphere (the lower 10 km of the atmosphere where most weather occurs). At the dew point, condensation of saturated water vapor occurs. Because decreasing temperature forms dew (or a cloud droplet), all that is needed to form clouds, and ultimately rain, is for air to rise. Ideally, this is accomplished by three mechanisms, giving rise to three types of precipitation (Bataan, 1962).

*Orographic precipitation.* When moist air moves laterally across the earth's surface and encounters a mountain range, it flows up over the mountain and down the other side. Rain occurs in the zone of cooling, rising air. Air descending from the mountains warms up, and its moisture-holding capacity increases; as a result, deserts form on the lee (downwind) sides of mountain ranges. A comparison of average precipitation in the United States (Fig. 7.2) with a topographic map of the United States (particularly the Pacific coastal or Sierra Mountain region of California) offers convincing documentation of orographic precipitation.

*Convective precipitation.* On a hot summer day, the sun heats the surface of the earth differentially, triggering local rising air over hot areas such as plowed ground, sandy beaches, etc. By late afternoon these thermals can give rise to towering cumulus clouds, often reaching 8 km high (well below freezing temperatures), allowing intense rain and hail to form. In desert areas, such as southern Arizona, nearly all the ground water recharge is accomplished by these rare but very intense thunderstorms.

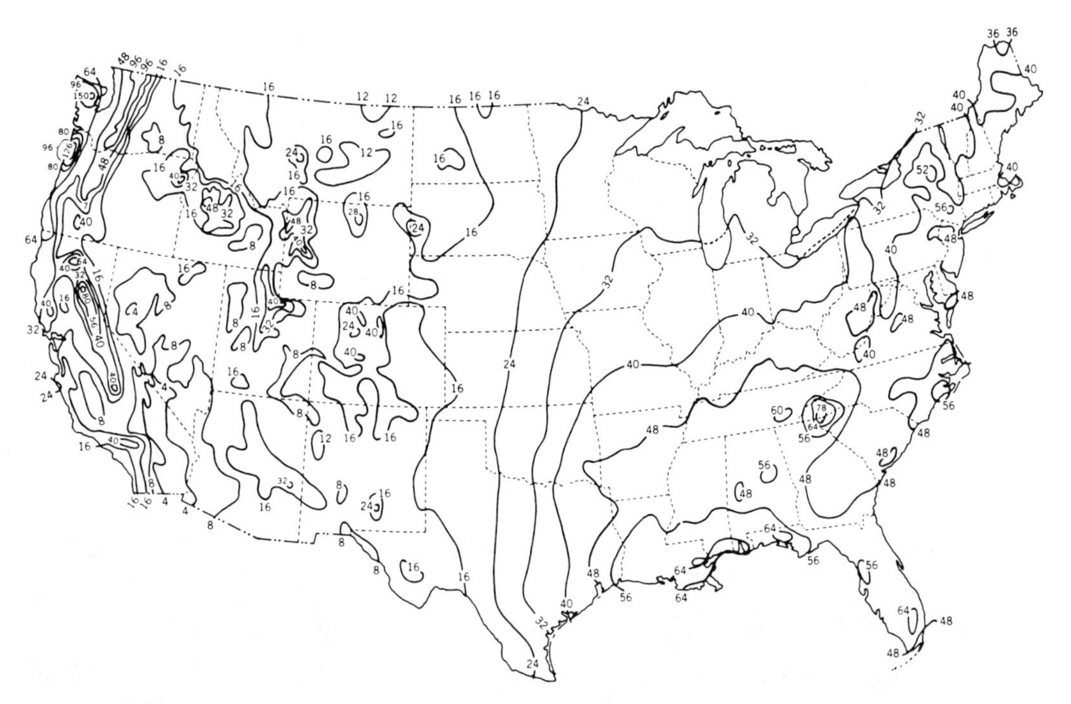

Figure 7.2 Isohyetal map of United States. Lines show average precipitation, in inches (from Linsley and Franzini, 1979). (1 in = 2.54 cm)

*Cyclonic precipitation.* Cold and warm fronts, as well as low or high atmospheric pressure fronts move across the earth, triggering weather changes. For instance, a cold front entering an area will slide beneath the warmer, less dense air, and push it up. Stratus rain clouds develop along the front, causing precipitation over a wide belt. When warm, moist air is forced violently upward just ahead of a strong cold front, a line of thunderstorms (a "squall line") may occur.

## *7.2.2 Infiltration*

Water that falls as precipitation either runs off the land surface as surface water runoff or infiltrates into the soil or rock. Much of the infiltrated water returns to the atmosphere shortly thereafter by evaporation or transpiration. Some moisture continues downward to the water table (Fig. 7.3). Water above the water table, in the zone of aeration, is called vadose water; below the water table, in the zone of saturation, the water is called ground water (or phreatic water).

Figure 7.4 shows photographs of the zone of aeration and saturation. Figure 7.4A shows the Redwall Limestone in the Grand Canyon. The rock is high and dry. (Note solution cavities formed along joints. These joints show why drilling productive water wells in limestone can be a hit or miss proposition. At some places aerial photographs may reveal fracture traces (linear surface expressions of jointed or faulted rocks at depth) and may be used to optimize water well yield (Lattman and Parizek, 1964).) Figure 7.4B shows ice formed on road cuts in schist along the Merritt Parkway in southwestern Connecticut. Water seeping along

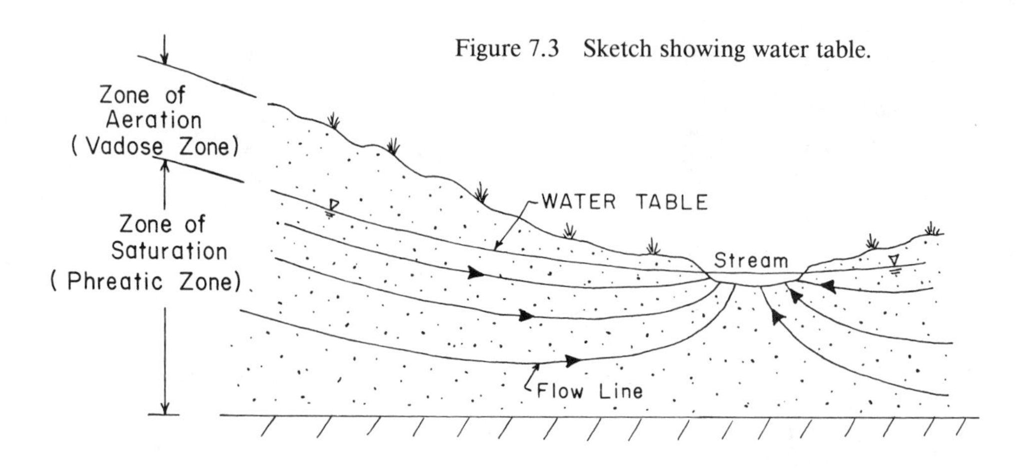


Figure 7.3 Sketch showing water table.

Figure 7.4

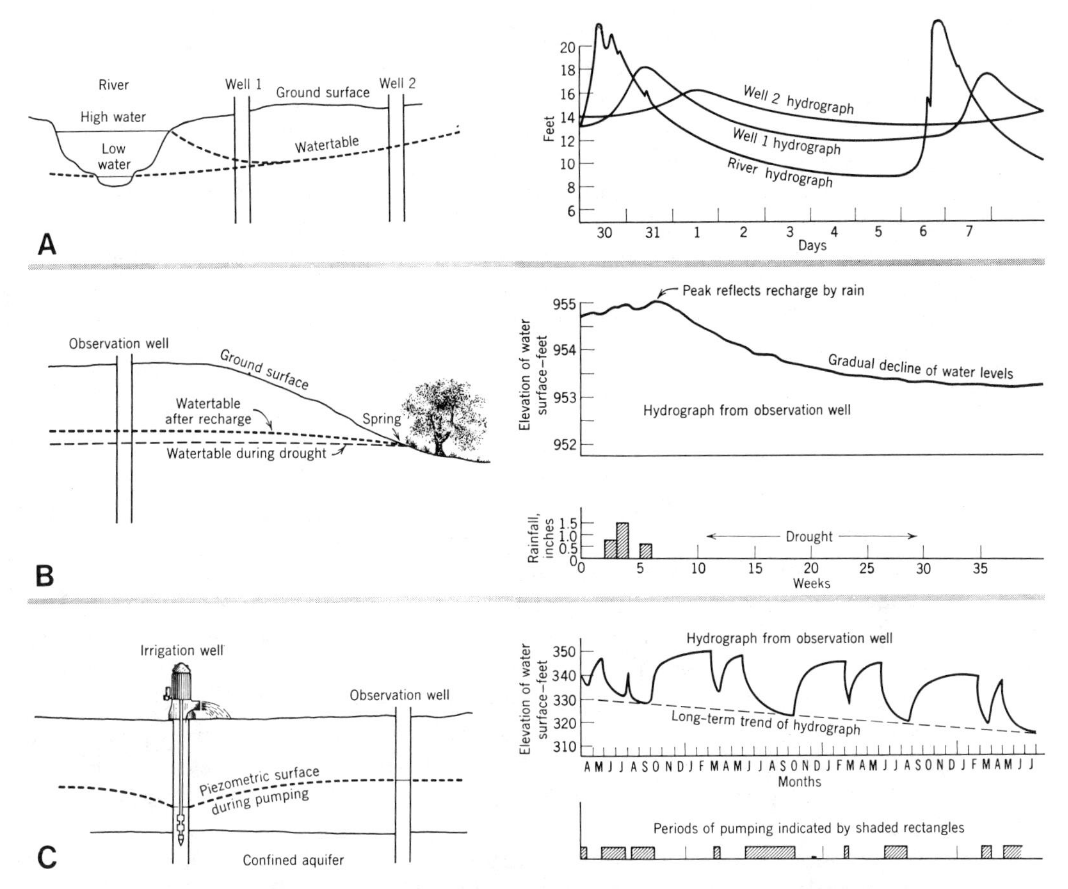


Figure 7.5 Water-level fluctuations caused by (**A**) changes in the water surface in a nearby river, (**B**) gradual discharge of an aquifer during a drought, and (**C**) discharge of an irrigation well (from Davis and DeWeist, 1966).

small cracks in the zone of saturation seeps out in this deep road cut. At depths over 100 m below the land surface, cracks such as these are generally sealed shut due to lithostatic stress and hence are unproductive to wells.

A saturated zone, called the capillary fringe, occurs directly above the water table. To be technically accurate, the water table is defined as the level below which there is a hydrostatic increase in water pressure. (This is analogous to the water pressure increase while descending below a lake surface.) Practically speaking, the water table is simply the level at which water stands in a shallow well under nonpumping (static) conditions. This level correlates to the level of springs, streams, swamps, and lakes.

The water table fluctuates due to several reasons. Hydrographs showing water table fluctuations due to floods, pumping, and droughts are shown in Fig. 7.5. In

---

Figure 7.4 Photographs illustrating zone of aeration and saturation. **A**. Redwall Limestone in the Grand Canyon, Arizona. Note solution-enlarged joints in the zone of aeration. **B**. Ice on Connecticut road cut in schist. In the winter, ground water seeps into the road cut from the zone of saturation.

the United States water tables tend to be high in the spring due to recharge from winter precipitation following thawing of the ground.

Infiltration rates are influenced by numerous factors such as soil or rock permeability, slope of the land surface, vegetation, and man's use of the land. Much of the United States, which originally had a forest or prairie grass cover, has been converted to crop land or urban areas in the past 200 years. As a consequence infiltration is reduced, leading to lower ground-water recharge as well as increased flooding. In the urban areas surrounding Washington, DC, for example, there has been a pronounced decline in the flow of springs and a consequent diminution of fish in the streams (Williams, 1977). Chapter 8 treats the subject of the effects of urbanization on stream runoff more extensively.

### *7.2.3 Evapotranspiration*

Evapotranspiration is difficult to measure directly, and is usually quantified by assuming it is the one unknown in the hydrologic budget [Eq. (7.2)].

For a given locality, evaporation can be determined by placing a large ($\simeq$1.2 m diameter) water-filled can in an open area. The can is open at the top, and the amount of water which evaporates over a year's time is measured. For example, if the yearly precipitation is 60 cm, yet the water level in the can is 20 cm lower at the end of the year, then the evaporation must have been 80 cm. Because a can may heat up in the sun, evaporation-can data must be multiplied by a factor of about 0.7 in order to convert to equivalent loss over a pond or reservoir. Figure 7.6 shows yearly potential evaporation rates (i.e., potential for evaporation if a lake exists) for the United States.

A comparison of Fig. 7.2 (precipitation) and Fig. 7.6 (evaporation) will allow for a determination of the net water loss or gain from a lake or reservoir. In the southwestern United States, evaporation is much greater than precipitation (Meyers, 1962). In the northeastern United States, precipitation is greater than evaporation. It follows that in the northeastern United States it will be ludicrous to call an industrial liquid waste pond an evaporation pond. After a year, more water would be added due to precipitation than would be diminished due to evaporation. (In these areas unlined ponds which consume industrial liquid waste water would be more appropriately called seepage ponds because of the water which seeps out the bottom.)

Evaporation consumes great quantities of fresh water from reservoirs in desert areas. For example, Lake Powell on the Colorado River and Lake Nasser on the Nile River evaporate about 3 m of water annually. On the positive side, however, evaporation can be put to man's use. For example, Egyptian engineers plan to construct a canal from the Mediterranean Sea 56 km to the Qattara Depression in the Sahara Desert, a large natural depression 134 m below sea level at its lowest point. A large hydroelectric plant will be built, taking advantage of the difference in elevation and the fact that the Qattara Depression will fill up only to a level 60 m below sea level, at which point evaporation from the lake will be equal to canal discharge.

Evapotranspiration in desert areas may consume vast quantities of water. Plants whose roots extend into the water table (phreatophytes) act as ground water pumps during the transpiration process. For example, cottonwood and

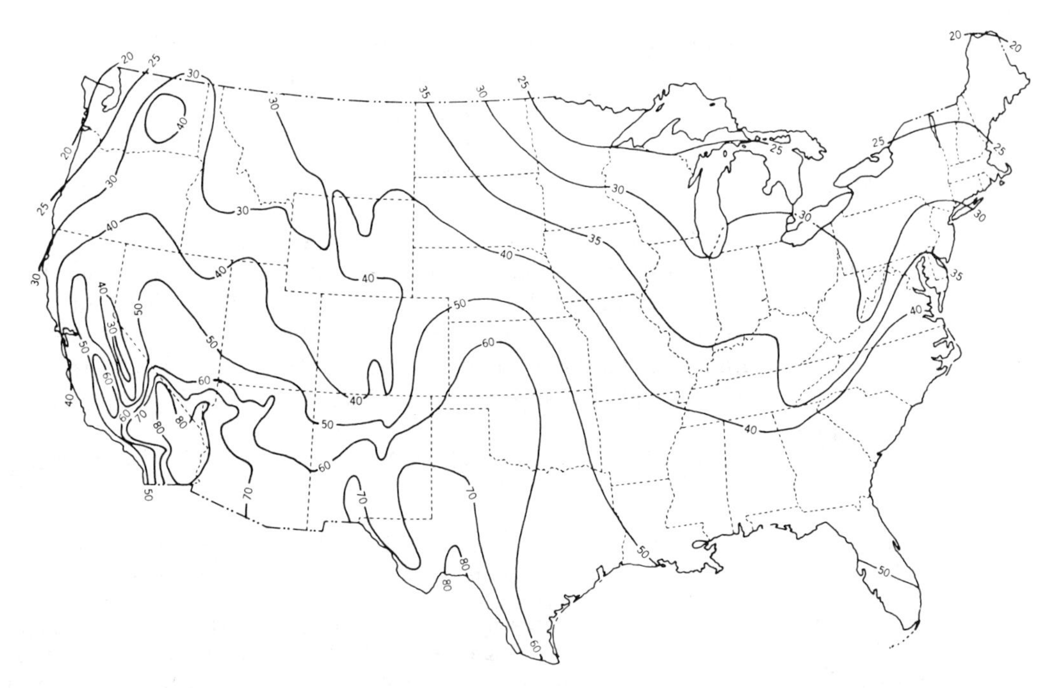

Figure 7.6 Mean annual reservoir evaporation rates (in inches) for the United States (from Linsley and Franzini, 1965). (1 in = 2.54 cm) (For detailed map of western states, see Meyers, 1962.)

mesquite tree roots in the southwestern United States can extend over 20 m depth. Salt cedar, a plant inadvertently introduced into the United States in the nineteenth century, has developed extensive thickets on many southwestern flood plains. Robinson (1958) estimates that phreatophyte water consumption in the western United States is about $3 \times 10^{10}$ $m^3$/yr, which is more than the discharge of the Colorado or Missouri River.

# 7.3 Ground Water Concepts

## *7.3.1 Origin of Ground Water*

Precipitation which infiltrates below the root zone and reaches the water table then moves down the gradient of the water table (Fig. 7.3). Ground water, which has its origin as precipitation, is called meteoric water. Almost all ground water is meteoric in origin.

Some ground water found in sedimentary rocks may be the same water which was present during the time when the sediments accumulated. This is called connate water. Water in some buried Paleozoic and Cretaceous aquifers in the midwestern United States, for example, has high chloride content, which represents, in part, water from the sea which covered the area hundreds of millions of years ago.

Juvenile water, also known as magmatic water, is water given off during volcanic eruptions or during the cooling of the underground magma. In Yellowstone Park, WY, it was originally believed that perhaps 12% of the water emanating from hot springs and geysers was magmatic water, although more recent studies indicate that nearly all the water originates as meteoric water, and that very little originates as magmatic water (Keefer, 1971).

## *7.3.2 Springs*

Springs or seeps form where the water table intersects the land surface. Springs can be formed by a number of geologic conditions such as those illustrated in Fig. 7.7. Cavernous limestone may have numerous sinkholes and springs; aquifers such as this can become contaminated by bacteria because there is little opportunity for filtration. The discharge of the spring can be used to determine the drainage area contributing to the spring, if ground water recharge rates are known (see Chapter 3).

In semiarid or arid areas, large springs have become the basis for the founding

Figure 7.7 Origin of springs (from Davis and DeWiest, 1966). Springs localized by (**A**) a surface depression that intersects the water table, (**B**) infiltration of rain water into coarse and permeable landslide rubble, (**C**) permeable sandstone overlying impermeable shale, (**D**) a fault that offsets impermeable beds against permeable beds in alluvium, (**E**) a fault that forms an open fractured zone in brittle rock, (**F**) sheet structure in granitic rock, (**G**) outcrop of an artesian aquifer, (**H**) dominant jointing in one direction, and (**I**) outcrop of permeable gravel and basalt overlying impermeable granitic rock.

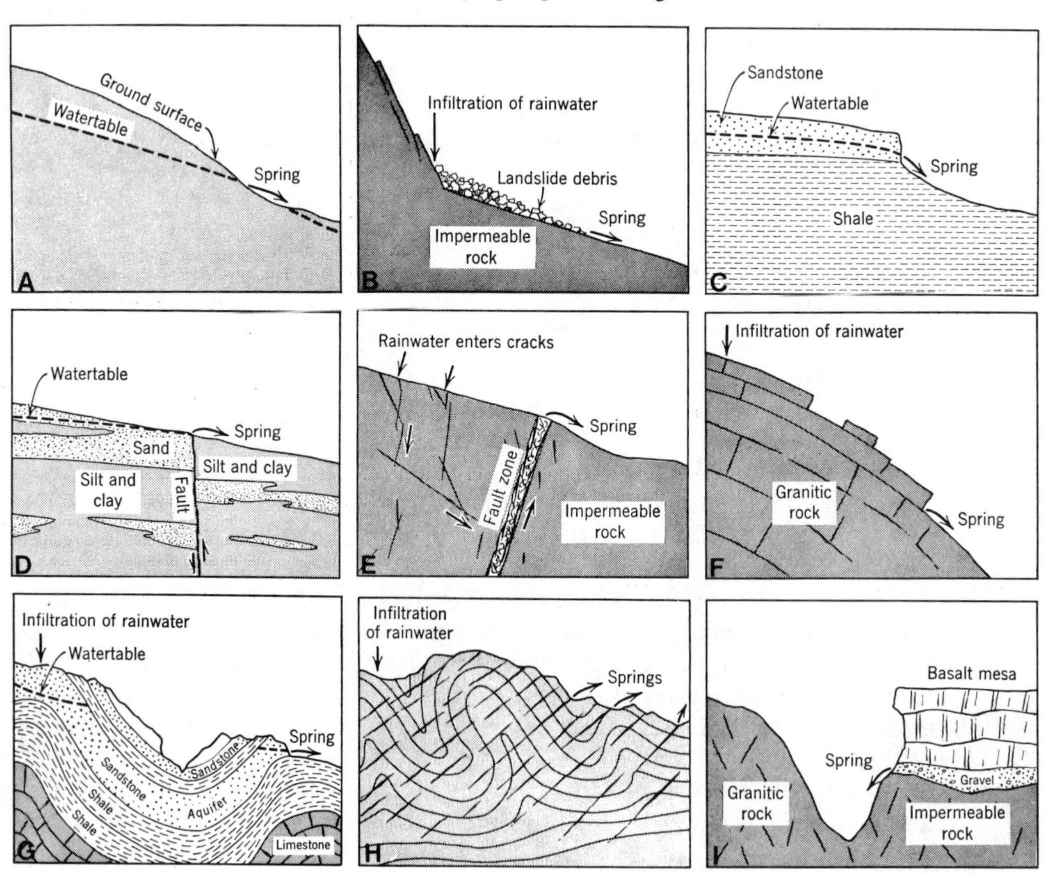

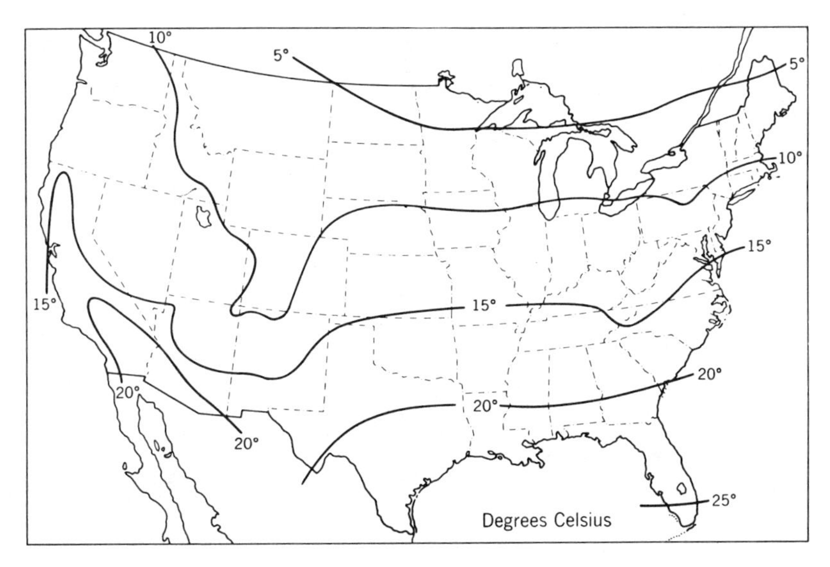


Figure 7.8 Approximate temperature of ground water in the United States at depths of 10–20 meters (from Todd, 1980).

of settlements. For example, the origin and growth of Jerusalem was intimately tied to the spring of Gihon. Many ranches and homesteads in the western United States were located near springs.

Figure 7.8 shows the temperature of shallow ground water in the United States. This is roughly the same as the average annual air temperature, and also the same as the temperature of the soil at about 2 m depth. Most springs would be at this temperature or slightly warmer.

At greater depths, the temperature of ground water increases according to the geothermal gradient, which is about 1°C/50 m. Hot springs typically originate where circulating meteoric water becomes warmed at depth, rises, and discharges. Geysers, such as those which occur in Yellowstone Park, are due to ground water becoming heated in open cracks or chambers near volcanic rocks. When the boiling temperature is reached, water in the upper part of the crack or chamber boils and is expelled from the chamber. The temperature at which water boils decreases with decreasing pressure; therefore when the hydrostatic pressure is decreased in the plumbing system below, a chain reaction of boiling in the entire chamber results, and there is an explosive release of all the water. The chamber then slowly refills with cool ground water and the process repeats, sometimes with clocklike regularity as in the case of Old Faithful Geyser in Yellowstone Park.

## *7.3.3 Water Wells and Artesian Aquifers*

There are numerous references concerning the technology and construction of various types of water wells (Johnson, 1966; Davis and DeWeist, 1966; Gibson and Singer, 1971; McCray, 1982). In general there are two popular techniques of drilling water wells: the cable tool and the rotary method. The cable tool method utilizes a large chisel-like drilling tool which is rhythmically lifted and dropped, whereas the rotary uses a power driven rotary table, and drilling mud is pumped

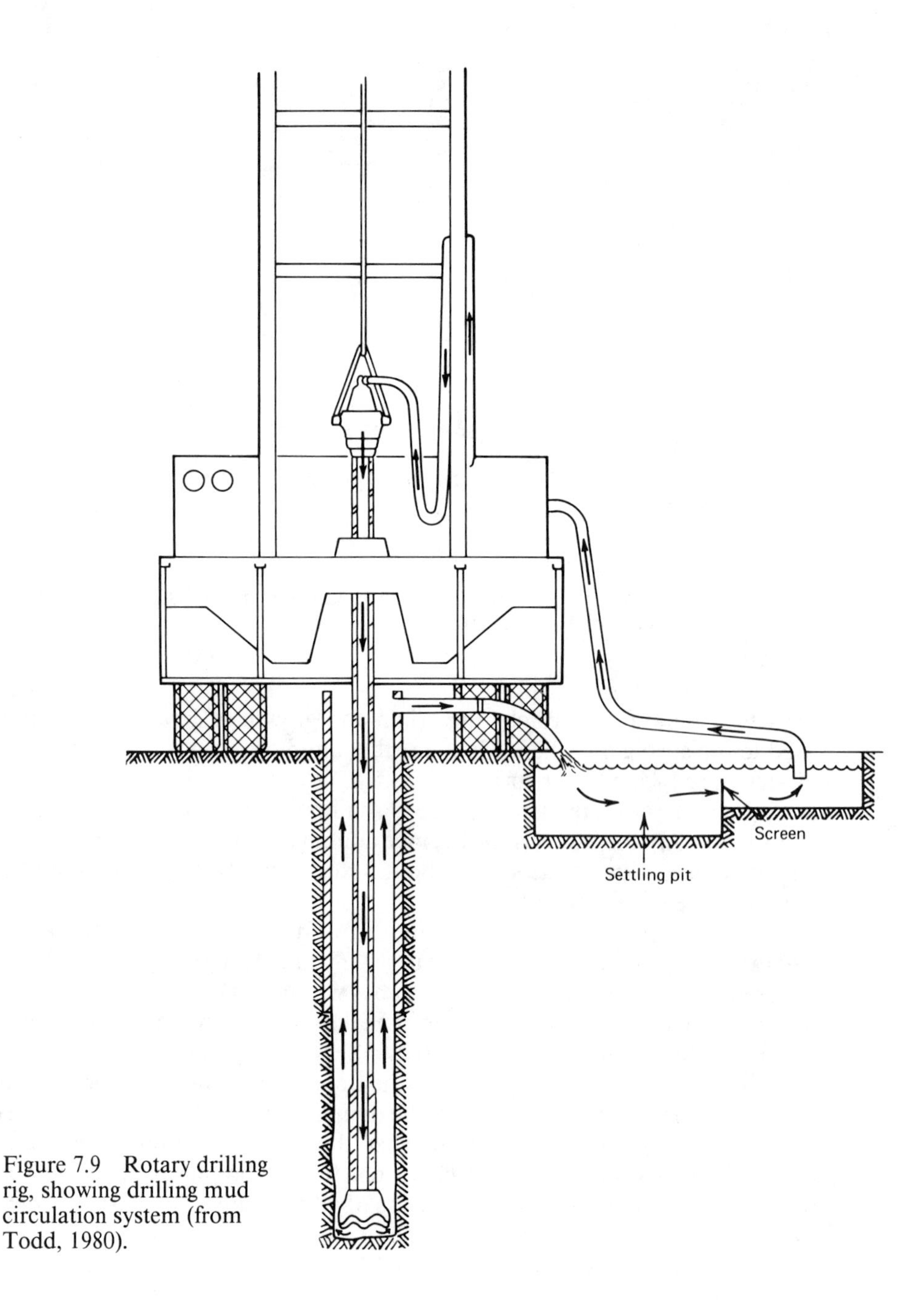


Figure 7.9 Rotary drilling rig, showing drilling mud circulation system (from Todd, 1980).

through the rotating drill pipe so that broken rock and drilling mud flows upward out of the hole (Fig. 7.9).

Shallow ground water is typically under unconfined conditions, i.e., there is no geologic strata which prohibits ground water movement upward. Rather, there is a simple hydrostatic increase in water pressure with depth.

Artesian conditions generally require the presence of a "confining" layer. Sedimentary rocks, for example, cover most of the earth's surface; they consist of

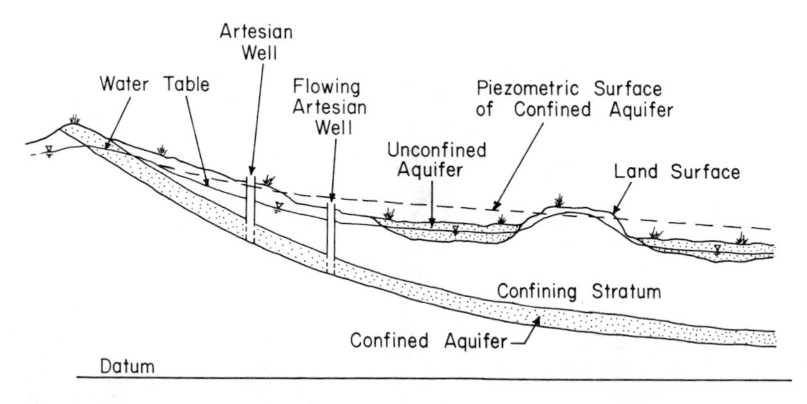


Figure 7.10 Schematic cross section illustrating unconfined and confined aquifers.

layered sequences of rocks of widely varying permeability. At many places a permeable unit (aquifer) may be capped by a rock having low permeability (aquitard). If a water well is drilled into the aquifer, and the static (nonpumping) water level rises above the top of the aquifer, the well is called an artesian well (Fig. 7.10). The artesian water level is also called the potentiometric surface or piezometric surface. If the water level rises above the land surface, it is called a flowing artesian well. Examples of geologic causes of artesian conditions are shown in Fig. 7.11.

Figure 7.11 Geologic controls for artesian wells. **A.** stabilized sand dunes. **B.** Crystalline rock. **C.** Complexly folded and fractured sedimentary rocks. **D.** Horizontal sedimentary rocks. **E.** Glacial deposits (from Davis & DeWeist, 1966).

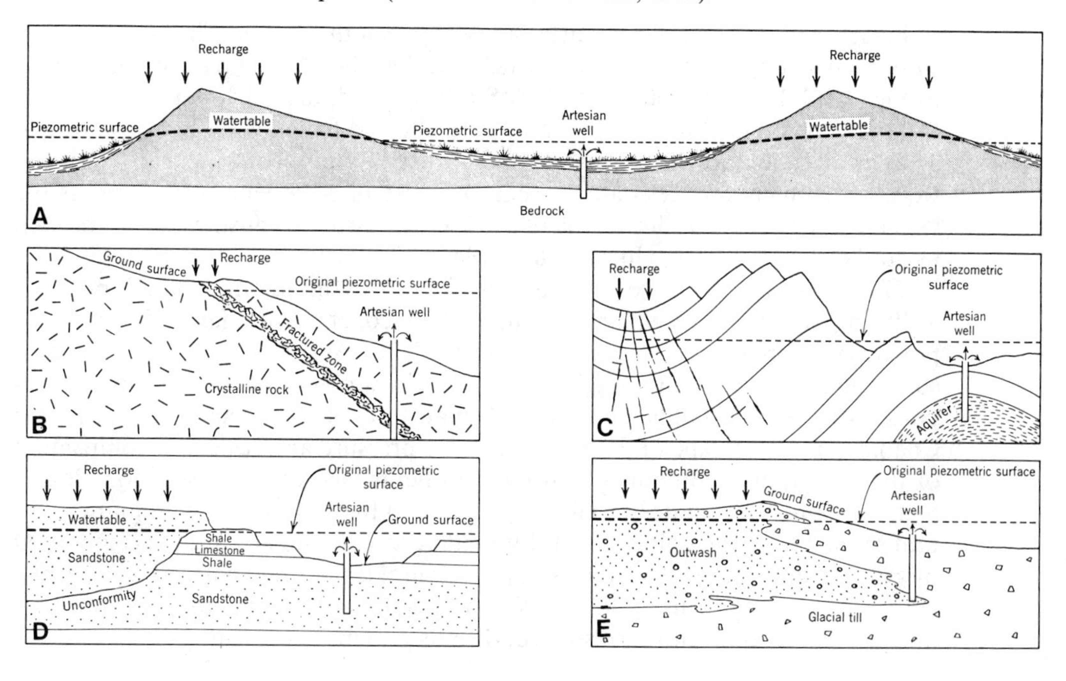

Figure 7.12 Flowing artesian well near Mitchell, SD (from Darton, 1909).

Figure 7.12 shows a flowing artesian well in South Dakota. The artesian pressure is caused by the Pierre Shale which overlies the Dakota Sandstone (similar to Fig. 7.11D). Many wells in eastern South Dakota, drilled about 400 m to the Dakota Sandstone, flowed at first and were allowed to run wild. But now, due to a declining potentiometric surface, they must be pumped. Practically all deep wells in sedimentary rocks are artesian, because in a sequence of sediments the water at the bottom of the well is of sufficient pressure to push water in the well up to the elevation of the local water table.

The hydrostatic increase of fresh water is 9800 $N/m^2$ per m depth. Some deep wells, in particular oil wells in the Gulf of Mexico, encounter fantastic pore fluid pressures, approaching 23,000 $N/m^2$ per m depth. Where this occurs in unconsolidated rocks, the entire weight of the superincumbent rocks is essentially supported by the pore fluids, and the rock matrix loses all resistance to shear (see Chapter 4). Any escape of fluid reduces pore pressure and results in compaction of the fine-grained components of the sediments (see Chapter 9). Anomalously high fluid pressures in deep wells may be due to the lithostatic loading from natural sediments over geologic time. Lithostatic stress onto any strata is supported by grain-to-grain contact as well as fluid pressure. If the strata has very low permeability, but is subject to rapid deposition of sediments above, the fluid pressure may build at depth. Thus artesian conditions may be generated.

The ancient Persians dug gently sloping tunnels (qanats) into alluvial fans. The fans consist of alternating aquitards (mud-flow deposits) and aquifers (alluvium). The qanats intersect water-saturated permeable layers near the mountains. The qanats are about 1 m wide and 2 m high, and some extend for more than 40 km underground. The qanats of Iran total 280,000 km of underground channels (Leveson, 1980). The hand-dug tunnels are connected by a series of vertical well-shafts which serve to provide for ventilation and removal of mined debris. In 1968, qanats supplied 75% of all the water used in Iran.

## 7.4 Ground Water Flow

### *7.4.1 Darcy's Law*

In 1856, the French hydrologist Henry Darcy measured the discharge of water through inclined cylindrical tubes containing sand (Fig. 7.13). For a given sand specimen with a cross-sectional area ($A$), he found that the velocity of water in the tube is directly proportional to the head loss ($H$) and inversely proportional to the length ($L$) of the cylinder. The conversion coefficient for this relationship is called the hydraulic conductivity ($K$) of the permeable media. Hydraulic conductivity is defined as the discharge in $m^3$/day through 1 $m^2$ cross section of material having $H/L$ of unity. The velocity thus measured is called the seepage velocity or Darcy velocity ($V_d$):

$$V_d = K\frac{H}{L}. \tag{7.3}$$

The hydraulic gradient $H/L$ (referred to as $i$ by some authors) is dimensionless; thus $K$ has the same unit as velocity, such as meters/day in the metric system. In the English system $K$ has units of gallons/day/$ft^2$ (a unit also known as a "Meinzer"). Typical values of $K$ are given in Table 7.2.

From fluid mechanics, the "law of continuity" states that discharge ($Q$) equals the velocity times cross-sectional area:

$$Q = VA. \tag{7.4}$$

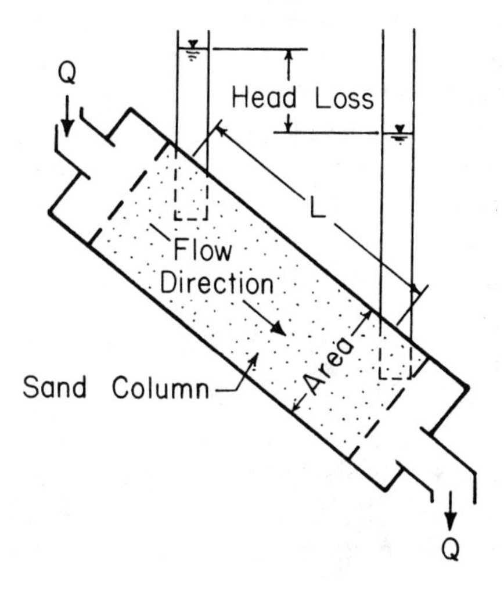


Figure 7.13 Darcy Law apparatus.

Table 7.2 Representative Values of Hydraulic Conductivity

| Material | Hydraulic conductivity (m/day) | Type of measurement[a] |
|---|---|---|
| Gravel, coarse | 150 | R |
| Gravel, medium | 270 | R |
| Gravel, fine | 450 | R |
| Sand, coarse | 45 | R |
| Sand, medium | 12 | R |
| Sand, fine | 2.5 | R |
| Silt | 0.08 | H |
| Clay | 0.0002 | H |
| Sandstone, fine-grained | 0.2 | V |
| Sandstone, medium-grained | 3.1 | V |
| Limestone | 0.94 | V |
| Dolomite | 0.001 | V |
| Dune sand | 20 | V |
| Loess | 0.08 | V |
| Peat | 5.7 | V |
| Schist | 0.2 | V |
| Slate | 0.00008 | V |
| Till, predominantly sand | 0.49 | R |
| Till, predominantly gravel | 30 | R |
| Tuff | 0.2 | V |
| Basalt | 0.01 | V |
| Gabbro, weathered | 0.2 | V |
| Granite, weathered | 1.4 | V |

After Todd, 1980.
[a]H is horizontal hydraulic conductivity, R is a repacked sample, and V is vertical hydraulic conductivity.

Thus, multiplying both sides of Eq. (7.3) by A:

$$Q = KA\frac{H}{L}. \tag{7.5}$$

Equation (7.5) is useful for determining the discharge in an aquifer. For laterally moving ground water, the ratio $H/L$ is the slope of the water table. Thus it may be possible to determine the discharge through an aquifer if hydraulic conductivity, cross-sectional area, and slope of the water table are known. Consider the cross-sectional sketch shown in Fig. 7.14. Assume $K = 25$ m/day and the head loss ($H$) is 20 m in a 1 km horizontal distance (L) between the two wells. (Because the horizontal distance is very close to the inclined distance between wells in natural situations, the horizontal distance can be used for $L$.) Assuming the aquifer has a thickness ($b$) of 30 m and a width of 5 km, then the discharge is:

$$\begin{aligned} Q &= KA\frac{H}{L}, \\ &= (25 \text{ m/day})\,((30 \text{ m})\,(5{,}000 \text{ m}))\,\frac{20 \text{ m}}{1000 \text{ m}}, \\ &= 75{,}000 \text{ m}^3/\text{d}. \end{aligned}$$

It should be emphasized that the true velocity ($V_t$) of ground water is not the same as the Darcy velocity ($V_d$). The true velocity cannot be obtained by dividing discharge by cross-sectional area because the aquifer is not an open tube, such as

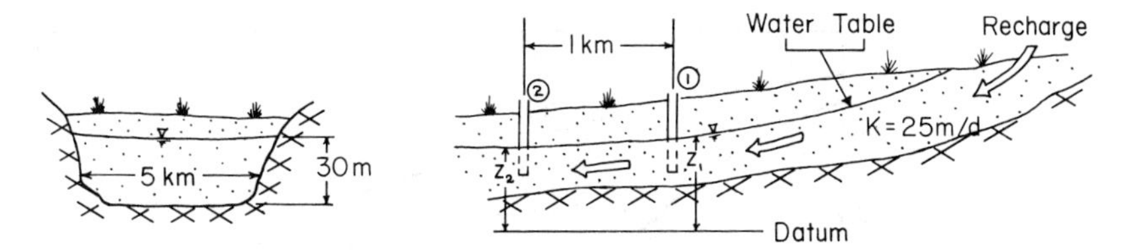


Figure 7.14 Transverse *(left)* and longitudinal *(right)* cross section of unconfined aquifer to illustrate discharge determination (after Todd, 1980).

a stream or empty pipe, but is largely soil or rock. Ground water moves only through the interstices of the aquifer. The true velocity equals the Darcy velocity divided by the porosity ($n$) of the medium:

$$V_t = \frac{V_d}{n}. \tag{7.6}$$

In this case, if $n = 0.2$:

$$V_t = \frac{Q/A}{n} = \frac{\dfrac{75{,}000\ \text{m}^3\text{day}}{(30\ \text{m})(5000\ \text{m})}}{.2} = \frac{0.5\ \text{m/day}}{.2} = 2.5\ \text{m/day}.$$

[Note: In Eq. (7.6), $n$ should technically be the "effective" porosity of interconnected openings, which is less than the overall porosity especially for fine-grained materials. See discussion of porosity below.]

There is a very large range in the velocity of movement of ground water. In an extensively, deeply buried artesian carbonate aquifer in the Great Plains, Back et al. (1983) report that water has moved 260 km during the past 20,100 yr based on isotope analysis. This indicates the ground water moved at an average velocity of 13 m/yr. In gravel aquifers, velocities may exceed 70 m/day (Davis, 1979). Based on a study of radioactive isotopes disposed in gravel beds within basalt in southern Idaho, tritium has moved about 13 km in 30 yr, and Iodine-129 about 3 km. Based on dye tests, ground water in limestone caves in Kentucky may move 8 km in 1 day (Parfit, 1983).

## 7.4.2 Transmissivity

Transmissivity ($T$) is a useful parameter which defines the hydraulic conductivity of a vertical strip of aquifer that is 1 m wide. Transmissivity is equal to $K$ times the saturated thickness of the aquifer (b):

$$T = Kb. \tag{7.7}$$

Common units of $T$ are $\text{m}^2$/day in the metric system and gpd/ft in the English system.

From Darcy's law, since cross-sectional area of an aquifer equals width ($W$) times thickness ($b$), it can be seen that Eq. (7.5) becomes:

$$Q = TW\frac{H}{L}. \tag{7.8}$$

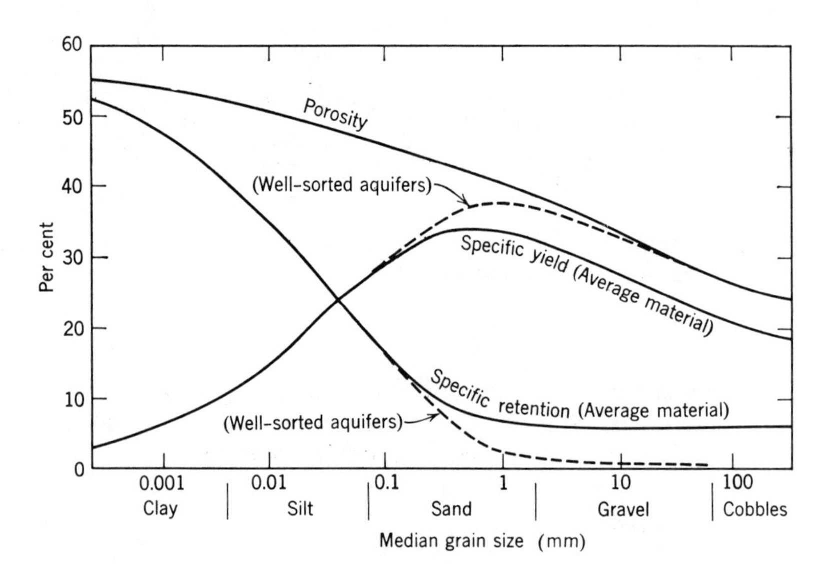


Figure 7.15 Relationship between median grain size and water-storage properties of alluvium from large valleys (from Davis and DeWiest, 1966).

## *7.4.3 Porosity*

Porosity ($n$) is the percent of voids in a soil or rock. Below the water table, all voids are filled with water. Not all of this ground water is available to wells, however. Water which is free to drain out by gravity, and hence available to wells, is called gravitational water. The remaining water clings to soil or rock particles and is either held by capillary forces (capillary water) or by minute electrostatic or molecular forces (hygroscopic water).

Effective porosity, which supplies gravitational water, is very nearly the same as total porosity for coarse-grained aquifers such as sand and gravel, but may be only a small percent of the total porosity for fine-grained rocks such as shale. Specific yield ($S_y$) is a term which is the same as effective porosity, but is more commonly used when dealing with an aquifer. Ground water not available to wells (i.e., the water retained by capillary and hygroscopic forces) is defined by the specific retention ($S_r$). Thus:

$$n = S_y + S_r. \tag{7.9}$$

Typical values of $S_y$, $S_r$, and $n$ are shown in Fig. 7.15.

The coefficient of storage, or storativity ($S$), for an aquifer is the same as the specific yield for an unconfined aquifer. The storage coefficient also takes into account the volume of water which can be released from a confined aquifer (see Fig. 7.16). The storage coefficient is defined as the ratio of the volume of water released from storage per unit decline of potentiometric surface per square unit of area of aquifer. For highly artesian aquifers, $S$ may range from 0.001 to 0.00001; for unconfined aquifers $S$ may range from 0.3 to 0.01. The very low values of $S$ for confined aquifers explain why these aquifers are subject to large declines in the potentiometric surface from only modest water withdrawals.

In summary, transmissivity ($T$) and storage ($S$) are considered the two basic

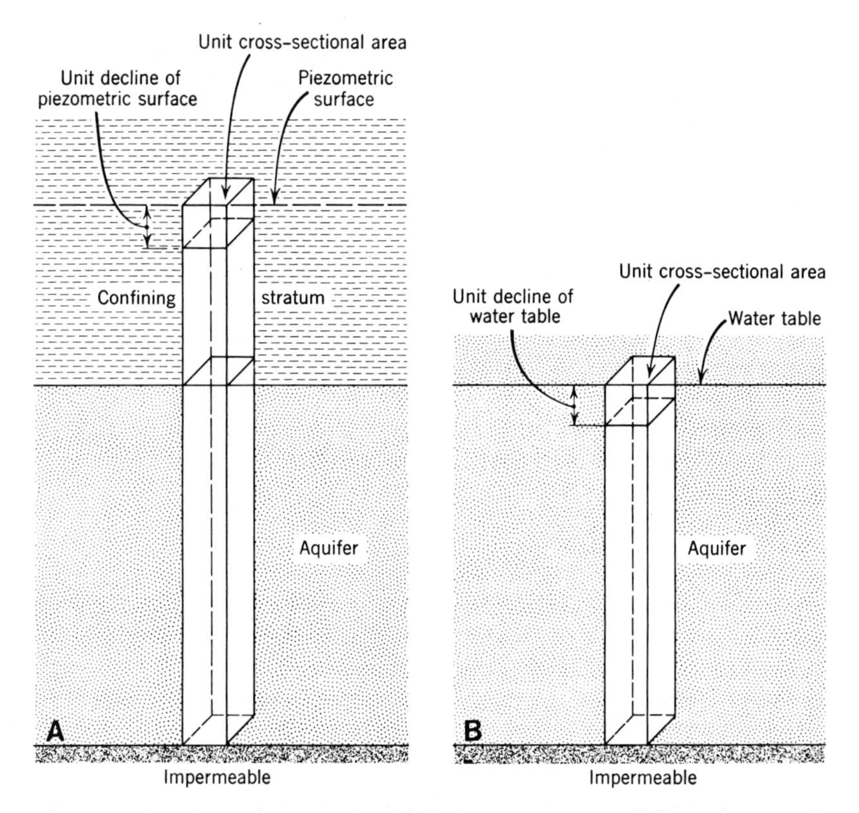


Figure 7.16 Illustrative sketches for defining storage coefficient of (**A**) confined and (**B**) unconfined aquifers (from Todd, 1980).

aquifer constants. They are analogous to permeability and porosity, respectively, but are more specifically defined.

## *7.4.4 Flow Nets*

Consider Fig. 7.17 which is a map showing the static level of unconfined water in three shallow wells: 50, 54, and 66 m above sea level. Lines connecting the three wells are drawn and divided into increments according to the differences in elevation between the two wells at the end of the lines. Equipotential lines are then

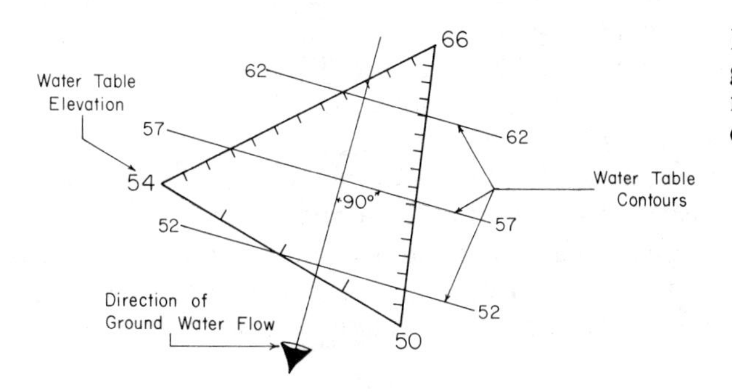


Figure 7.17 Estimate of ground water contours and flow direction from water table elevations in three wells.

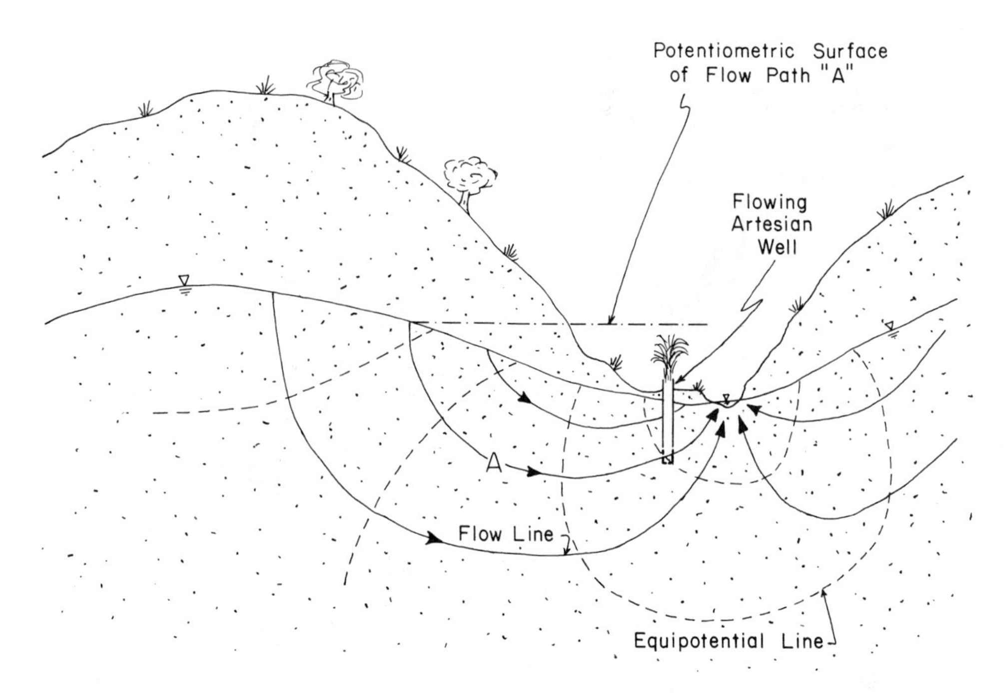


Figure 7.18 Cross section showing relationship of artesian water well level to the flow net in an unconfined aquifer.

drawn, representing contour lines of the water table. These lines should be roughly parallel to each other. Because ground water flows downhill, flow lines can then be constructed perpendicular to the equipotential lines. The boxlike grid thus constructed is called a flow net. For anisotropic aquifers, the direction of ground water flow is not parallel to the hydraulic gradient, but the direction of flow can be determined graphically if the ratio of hydraulic conductivities in $x$ and $y$ coordinates are known (Fetter, 1981).

Figure 7.18 illustrates a flow net construction in a cross-sectional view of an aquifer which is homogeneous (same material) and isotropic (same directional properties). (Sand dunes are the best example; see Fig. 7.11A.) Ground water moves more or less vertically downward in the recharge area as it sinks to lower potentials. The pressure of each potential line is defined by the level where the potential line intersects the water table. Thus the static water level in a well, open to the aquifer only at the end of the well, would *drop* in the recharge area (Saines, 1981). In the area where lateral movement is indicated no equipotential lines would be intersected. In the discharge area (valley bottom, swamp, or lake), a well encounters higher potentials as it is drilled. In Fig. 7.18, a well drilled in the bottom land has a high (artesian) pressure as indicated. Thus flowing artesian wells would be expected even in this unconfined situation.

This theory has been put to practical use. Water levels in observation wells at different depths were measured at the Nevada atomic bomb test site to illustrate areas of ground-water recharge, discharge, and lateral transport (Blankennagel and Weir, 1973; Winograd and Thordarson, 1975). (Contamination of ground water in recharge areas is particularly undesirable because the water moves to other areas.)

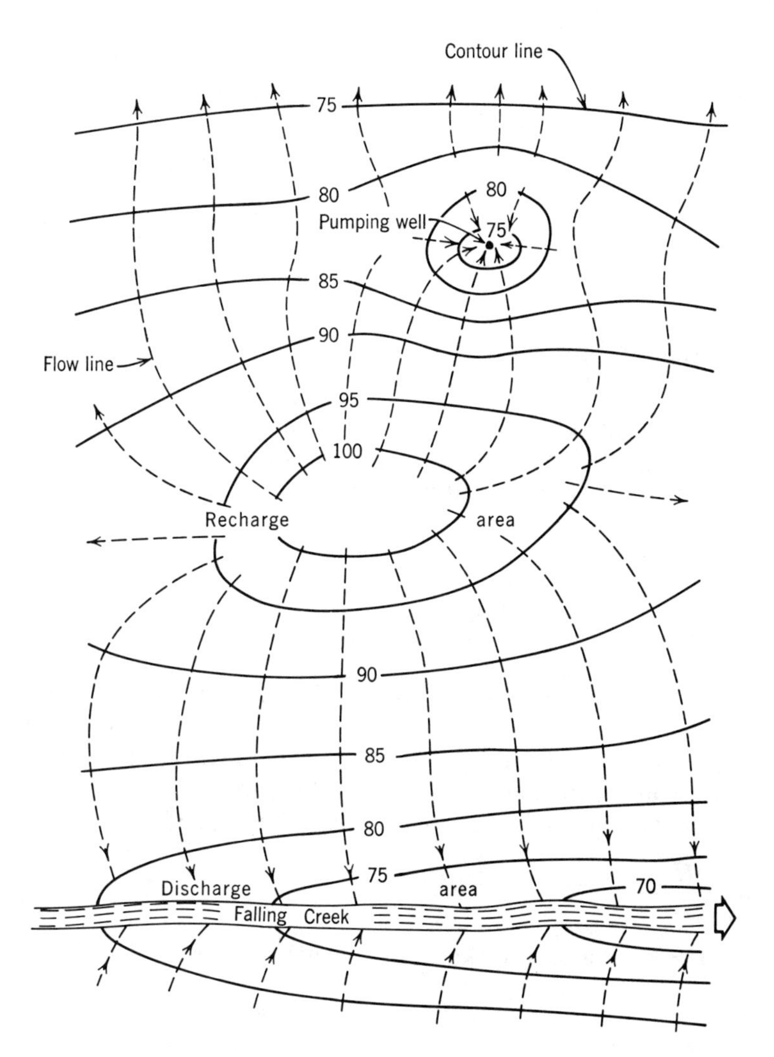


Figure 7.19 Identification of recharge and discharge areas on a map showing water level contours and flow lines (from Heath and Trainer, 1981).

Figure 7.19 is a plan view of water table contours near a river and pumping well. The movement of ground water away from an area indicates that recharge occurs there. The convergence of flow lines indicates the presence of a discharge area; this could be due to effluent (gaining) streams, springs, or pumping wells. In Fig. 7.19, the equipotential lines near the stream are U-shaped, opening downstream. This indicates an effluent stream. If the equipotential lines open upstream, it is an influent (losing) stream.

A flow net can be used to quantitatively determine ground-water discharge. By using Darcy's law, and making the reasonable assumption that flow is laminar, this approach can be used to predict the flow conditions for a given two-dimensional situation, using the graphical method of Forchheimer (Muskat, 1947). For example, if a cross-sectional flow net is constructed (Fig. 7.20) from equipotential lines having an equal pore pressure difference or equipotential drop ($dH$), and

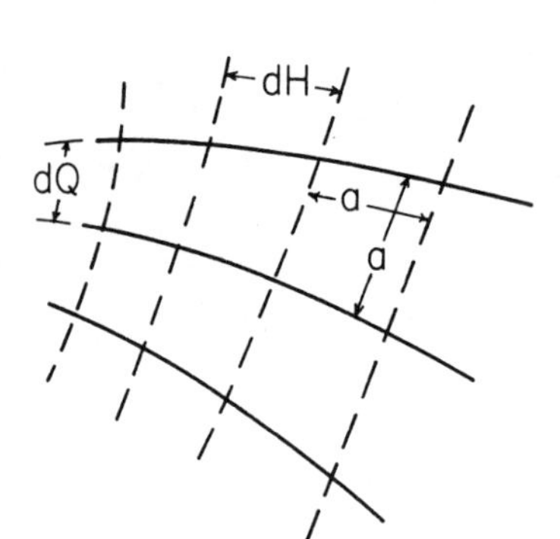


Figure 7.20 Portion of a flow net.

from flow lines separated so that each intermediate flow tube has an equal discharge flow rate ($dQ$). If $N_f$ represents the number of flow tubes and $N_d$ the number of equipotential tubes in the entire net, then:

$$dH = \frac{H}{N_d}, \tag{7.10}$$

$$dQ = \frac{Q}{N_f}, \tag{7.11}$$

where $H$ is the total hydraulic head loss and $Q$ is the total discharge for a unit thickness of flow net.

$N_f$ and $N_d$ are measured so that each element contained between equipotential and flow lines is approximately equal-sided. Then, for a given element of side length "a", each discharge tube has a flow rate as determined from Darcy's law:

$$dQ = KA\frac{dH}{dL}. \tag{7.12}$$

In this case, $A = a$ for a unit thickness of flow net. Hence,

$$dQ = Ka\frac{dH}{dL}.$$

Because the flow net has an incremental $dL = a$,

$$dQ = KdH.$$

From Eq. (7.10),

$$dQ = \frac{KH}{N_d}.$$

From Eq. (7.11),

$$\begin{aligned} Q &= dQN_f, \\ &= KH\frac{N_f}{N_d}. \end{aligned} \tag{7.13}$$

The selection of the number of $N_f$ and $N_d$ is somewhat arbitrary and depends on the way in which the sketched lines on the scale drawing work out from a trial and error process (see Cedergren, 1977, for a detailed review).

In summary, if permeability and head loss are known, then the discharge through a permeable deposit can be estimated by constructing a flow net and

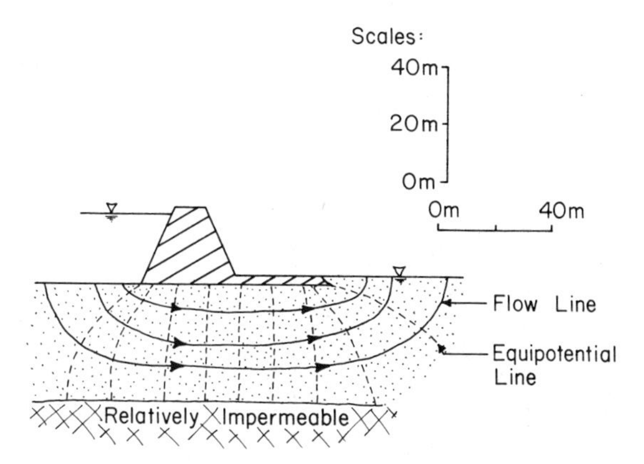


Figure 7.21 Flow net for seepage under a concrete dam.

determining the ratio of flow lines to equipotential lines. For example, suppose that seepage under a concrete dam is to be determined (Fig. 7.21). Assume the hydraulic conductivity ($K$) of the substrate is 30 m/day and the head loss ($H$) is 20 m from the reservoir level to the tail water. The dam is 40 m in length. From Fig. 7.21, there are four flow tubes and 10 equipotential tubes. From Eq. (7.13) we have:

$$Q = KH\frac{N_f}{N_d},$$

$$= (30 \text{ m/day})(20 \text{ m})\frac{4}{10},$$

$$= 240 \text{ m}^2\text{/day per unit thickness.}$$

For a 40-m-long dam:

$$Q_{total} = (40 \text{ m})(240 \text{ m}^2\text{/day}),$$

$$= 9600 \text{ m}^3\text{/d.}$$

For anisotropic media, for instance where the vertical permeability ($K_v$) is much less than the horizontal permeability ($K_h$), then a flow net is drawn so that the vertical exaggeration is equal to $(K_h/K_v)^{0.5}$. (This creates a transformed section where isotropic medium has an equivalent hydraulic conductivity $K' = (K_hK_v)^{0.5}$.) (For additional explanation, see Hubbert, 1940; Davis and DeWeist, 1966; Attewell and Farmer, 1976; Cedergren, 1977; Freeze and Cherry, 1979; Todd, 1980; or Das, 1983.) Hálek and Svek (1978) present mathematical analyses of various types of ground water flow situations including analysis of three dimensional flow nets.

# 7.5 Pump Tests

## 7.5.1 *Methods of Determining Hydraulic Conductivity*

In order to determine aquifer discharge rates and ground water velocities, it is necessary to know the hydraulic conductivity ($K$) of the medium. If the aquifer thickness ($b$) is known, then the hydraulic conductivity can be determined from transmissivity ($T$) by Eq. (7.7).

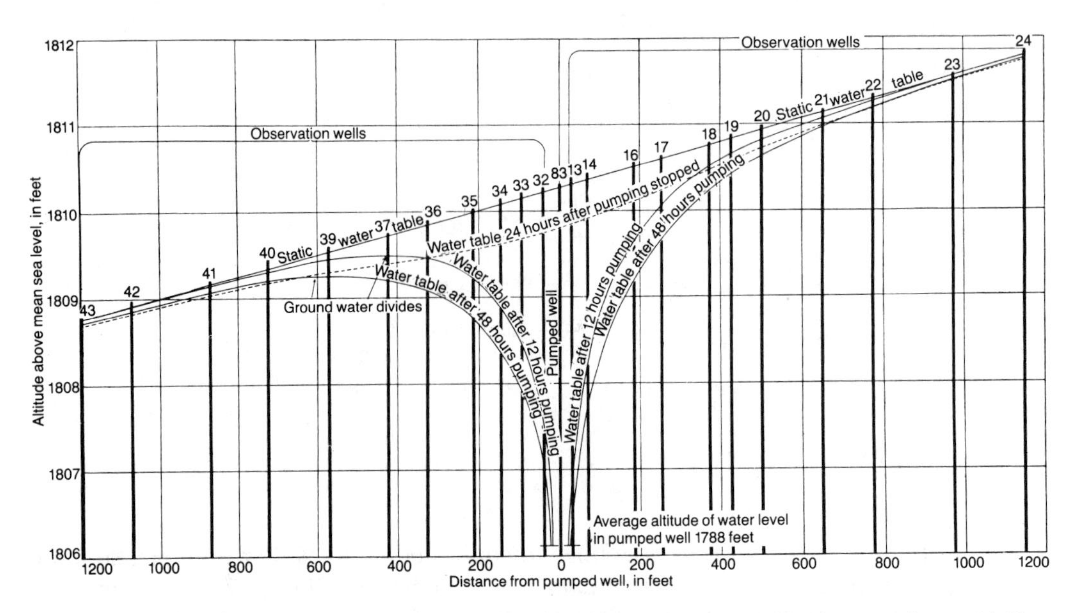


Figure 7.22 Cone of depression measured in Nebraska alluvium (from Meinzer, 1942).

Hydraulic conductivity can be determined in the laboratory by taking a sample of the aquifer and repacking it in a cylinder and essentially repeating Darcy's experiment. The apparatus, called a constant head permeameter, requires no special technical skills to assemble or to obtain results. The main problem is in packing the sample in the cylinder (commonly $\simeq$7.5-cm radius, $\simeq$15-cm height) so as to duplicate field density. I have found that dropping the cylinder gently several times for each 2.5 cm layer placed into the cylinder seems to work better than hand tamping with a blunt tool. The laboratory value of hydraulic conductivity ($K_l$) is liable to be in error by an order of magnitude from the in situ (in place) hydraulic conductivity. $K_l$ is generally lower than the in situ $K$ because the sample's integrity (e.g., sorting) is destroyed, whereas in situ deposits may have well-sorted and stratified layers. It is much better, therefore, to determine hydraulic conductivity in situ. The underlying principle for determining $T$ and $S$ in situ is a pump test, where the pumped well creates drawdown (a cone of depression) in the water table or potentiometric surface (Fig. 7.22). The rate of drawdown is mathematically analyzed to determine $T$ and $S$ as described below. If $T$ and $S$ are known, the mathematics can also be used to predict the drawdown at any place due to different pumping situations. The methods are presented below in a practical form. For more details, the reader can consult a number of excellent books on ground water hydraulics (Ferris et al., 1962; Davis and DeWeist, 1966; Bouwer, 1978; Walton, 1970; Freeze and Cherry, 1979; Fetter, 1980; Todd, 1980, Heath and Trainer, 1981).

## *7.5.2 Thiem Method*

As a well is pumped, the water table (or piezometric surface) is lowered. The cone of depression produced is parabolic in shape (Fig. 7.22). The outer limit of the cone defines the area of influence of the well at any given time.

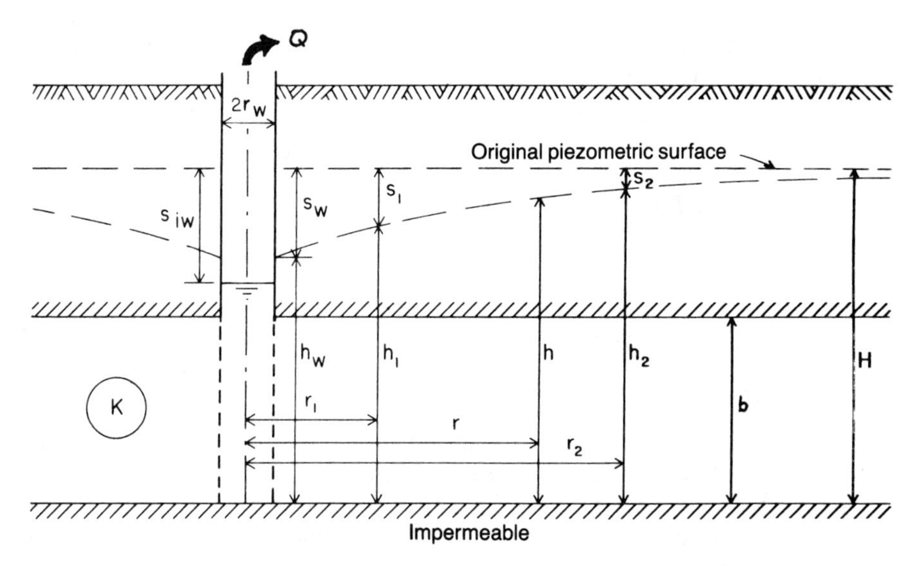


$Q = K2\pi b\,(dh/dr)$,

where

$Q$ = flow from well (volume/time)
$K$ = hydraulic conductivity of aquifer
$r$ = radial distance from well center
$b$ = height of aquifer
$dh/dr$ = hydraulic gradient (slope of piezometric head $h$ at distance $r$ from pumped well)

Figure 7.23 Geometry and symbols for pumped well in confined aquifer (modified from Bouwer, 1978).

In order to mathematically analyze the cone of depression it must be assumed that 1) water flow is horizontal into the well, and 2) the velocity of water moving to the well is proportional to the slope of the water table immediately above the point in question. These assumptions are called the Dupuit assumptions.

At equilibrium conditions (after the cone of depression has stabilized), the well discharge ($Q$) is the same as the quantity of water discharging through the area ($A$) of the cylindrical shell of the well screen, and also through any cylindrical shell of the aquifer at a distance ($r$) from the pumped well. The velocity is governed by Darcy's law where $V = Kdh/dr$ (see Fig. 7.23 for explanation of symbols). Because $Q = VA$, the discharge through the cylindrical surface at any distance is:

$$Q = K\frac{dh}{dr}(2\pi rb). \tag{7.14}$$

Rearranging terms, we obtain:

$$Q\frac{dr}{r} = 2Kb\,dh.$$

Because $T = Kb$,

$$Q\frac{dr}{r} = 2\pi T\,dh. \tag{7.15}$$

Integrating between two points at different distances from the pumping well ($r_2$, $h_2$ and $r_1$, $h_1$), we obtain:

$$Q(\ln r_2 - \ln r_1) = 2\pi T(h_2 - h_1),$$

$$Q \ln \frac{r_2}{r_1} = 2\pi T(h_2 - h_1),$$

or

$$T = \frac{Q \ln \dfrac{r_2}{r_1}}{2\pi(h_2 - h_1)}. \tag{7.16}$$

Equation (7.16) is the Thiem or equilibrium equation named after the father-son team of Adolph and Gunther Thiem, who developed the equation in the latter part of the nineteenth century. From Fig. 7.23, the difference between the heads ($h_2 - h_1$) of the two observation wells is the same as the difference between the drawdown ($s_2 - s_1$) of the two observations wells, so that, to solve for $T$, the difference in drawdowns is simply plugged into Eq. (7.16) instead of $h_2 - h_1$.

Jacob (1946) introduced a simplified graphical solution as shown below. (See texts listed above for theoretical analysis.) This solution, involving a semilog plot of drawdown vs. distance or time, is also referred to by many as Jacob's method instead of the Thiem method. The method, as outlined below, can be used to solve for $T$ and $S$ using either a distance-drawdown plot or a time-drawdown plot:

Figure 7.24 Thiem method, distance-drawdown solution (data from Fetter, 1980).

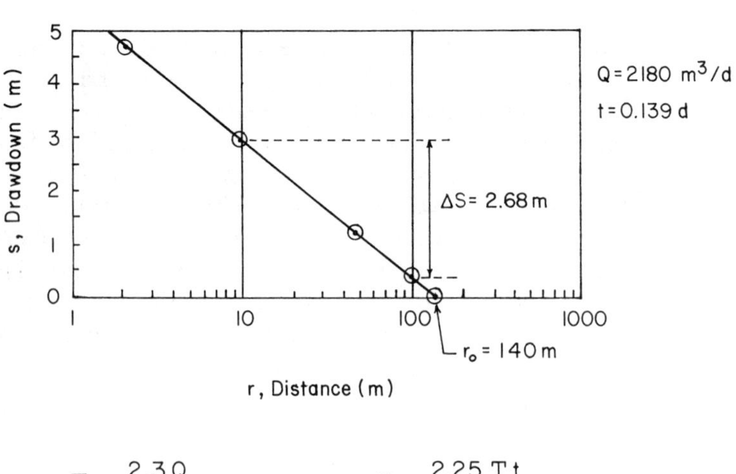


$$T = \frac{2.3Q}{2\pi \Delta S} = \frac{2.3(2180\ \mathrm{m^3/d})}{2\pi(2.68\ \mathrm{m})} = 297\ \mathrm{m^2/d}$$

$$S = \frac{2.25\,Tt}{r_0^2} = \frac{2.25(297\ \mathrm{m^2/d})(0.139\ \mathrm{d})}{(140\ \mathrm{m})^2} = 0.0047$$

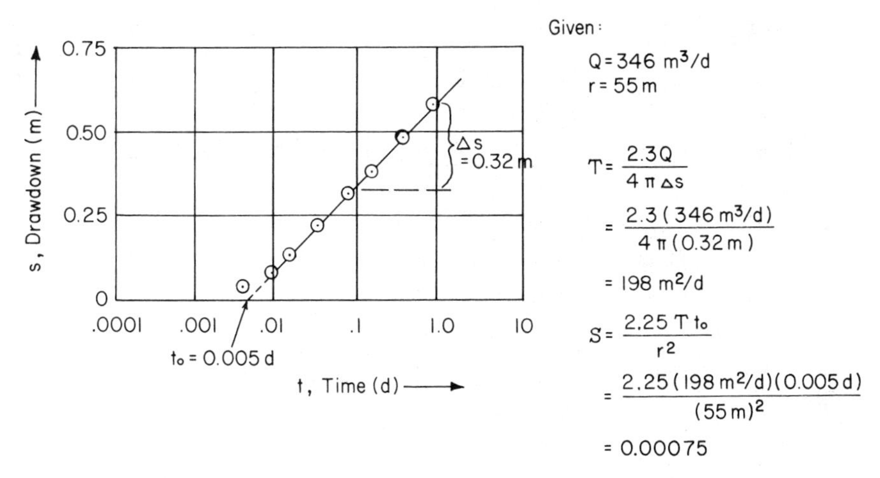


Figure 7.25 Thiem method, time-drawdown plot (data from Freeze and Cherry, 1979).

*Distance-drawdown method.* The drawdown in several observation wells is plotted as the ordinate on the arithmetic scale, and the distance to the pumped well plotted as the abscissa on the logarithmic scale (Fig. 7.24). A line is drawn through the points, and extended to zero drawdown at distance $r_0$. The drawdown per log cycle is determined ($\Delta s$), and Eq. (7.16) reduces to the following:

$$T = \frac{2.3Q}{2\pi(\Delta s)}. \tag{7.17}$$

The coefficient of storage ($S$) is determined as:

$$S = \frac{2.25Tt}{r_0{}^2}, \tag{7.18}$$

where $t$ is the time since pumping began.

*Time-drawdown method.* The Jacob semilog solution can also be applied to time-drawdown data for a single observation well. Figure 7.25 shows the drawdown of a well located 55 m from a well being pumped at 346 m$^3$/day. Similar to the distance-drawdown method, a log cycle drawdown ($\Delta s$) is determined from the plot and the line extended to a time of zero drawdown ($t_0$). $T$ and $S$ can be computed (Fig. 7.25) from the following two equations:

$$T = \frac{2.3Q}{4\pi\ \Delta s}, \tag{7.19}$$

$$S = \frac{2.25Tt_0}{r^2}. \tag{7.20}$$

## 7.5.3 Theis Method

The Thiem method described above assumes equilibrium has been reached, and the cone of depression is not being lowered any longer. In 1935, C. V. Theis, an American geologist, studied the Ogallala aquifer in the Texas-New Mexico area.

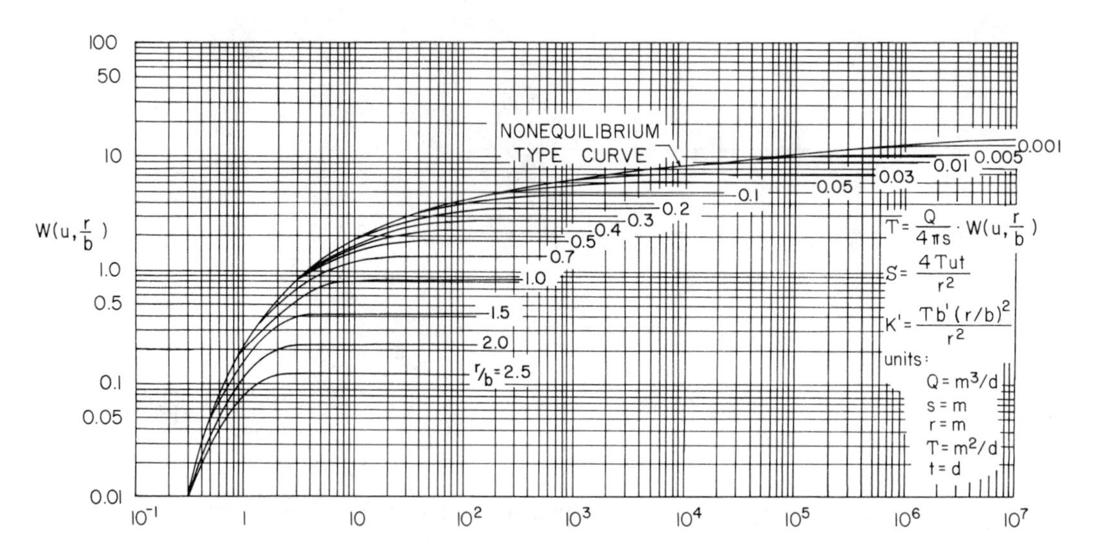


Figure 7.26 Type curve and leaky artesian curves (after Walton, 1960).

He recognized the need for a transient model to predict drawdown. Theis reasoned that equilibrium conditions may never be achieved because small drawdowns and release of some ground water from storage seem to continue even after long periods of pumping. A physicist, Clarence Lubin, gave Theis the idea of using a heat-flow model which analyzes the rate of temperature change in a metal slab. The method, which bears Theis's name, is now widely accepted. The mathematical derivations for the Theis method are very complex, and the reader can refer to books by Hantush (1964), Davis and DeWeist (1966), Walton (1970), Todd (1980), and other texts on ground water for a more detailed explanation. The resulting equations developed by Theis (1935) for drawdown ($s$) at distance ($r$) from a pumped well in a confined aquifer are:

$$s = \frac{Q}{4\pi T} \int_{\mu}^{\infty} \frac{e^{-\mu}}{\mu} \, d\mu, \tag{7.21}$$

where $s$ is the drawdown of the piezometric surface, $Q$ is the well discharge, and the term $\mu$ is a dimensionless mathematical term defined as:

$$\mu = \frac{r^2 S}{4Tt}. \tag{7.22}$$

For the solution of these equations, $W(\mu)$ is defined as the integral in Eq. (7.21), and a "type curve" (Fig. 7.26) is used to relate $\mu$ to $W(\mu)$.

To solve for $T$ and $S$, it is necessary to plot time ($t$) as the abscissa and drawdown as the ordinate on log-log paper of exactly the same log interval scale as the type curve. Keeping the axes parallel, the data points are superimposed onto the type curve (using a light table or window as an aid). When a match is achieved, an arbitrary point is picked (a match point) from which four parameters are derived: $W(\mu)$, $1/\mu$, $t$, and $s$. (To simplify calculations, it is suggested to choose $W(\mu)$ and $1/\mu$ both equal to unity.) Transmissivity can be calculated as:

$$T = \frac{Q}{4\pi s} W(\mu), \tag{7.23}$$

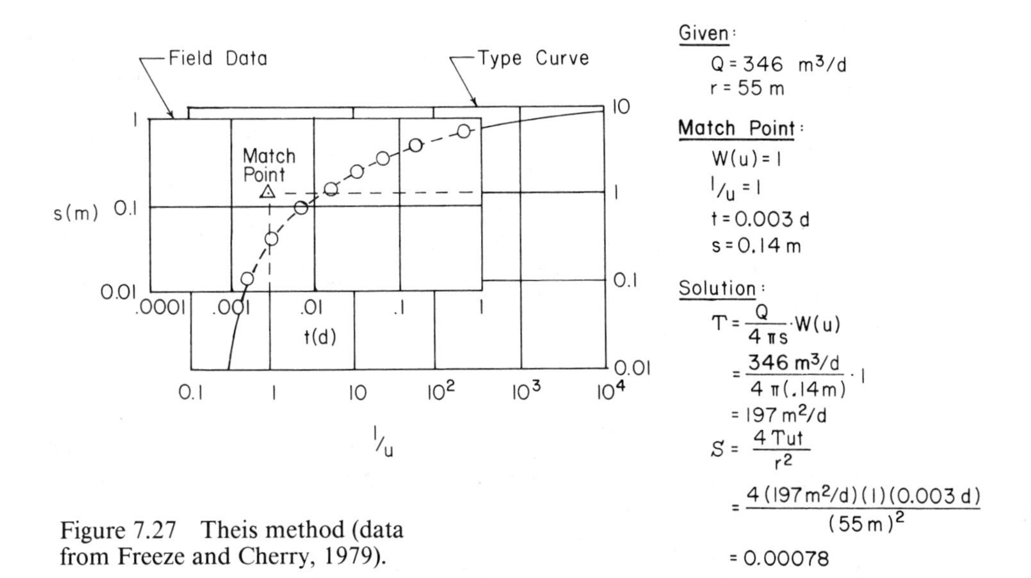


Figure 7.27 Theis method (data from Freeze and Cherry, 1979).

and the value of $T$ is then used to calculate storage:

$$S = \frac{4T\mu t}{r^2}. \tag{7.24}$$

In Fig. 7.27, $T$ and $S$ are determined from a pump test where $Q = 346$ m$^3$/day and the observation well is 55 m from the pumped well.

The Theis method is superior to the Thiem method where time-drawdown data are available because the calculation of $S$ by this method has been found to be more accurate than by the Thiem method (Walton, 1970).

If $T$ and $S$ are known, the Theis method or the Thiem method can be worked in reverse order to predict drawdowns at any place and time with any pumping rate (Rahn and Paul, 1975). In the Theis method, $\mu$ can be solved by using Eq. (7.24). The corresponding $W(\mu)$ can then be determined from the type curve, and the drawdown determined from Eq. (7.23).

The time-drawdown plot may deviate from the type curve if the aquifer being pumped: 1) is near a recharge or barrier breakdown, 2) has "delayed" yield from a water-table aquifer or 3) has leakage of water derived from an adjacent aquitard (Fig. 7.28). Walton (1960, 1970) shows that the "leakance" ($K'/b'$, where $K'$ is the vertical hydraulic conductivity in an aquitard of thickness $b'$) may be calculated by matching the time-drawdown plot to the $r/B$ leaky curves as shown in Fig. 7.26.

The use of hand-held calculators or microcomputers allows for rapid analysis of pumping test data. Time-drawdown field data can be displayed graphically on a microcomputer screen simultaneously with computer curves, eliminating the necessity for manual curve-matching techniques (Dumble and Cullen, 1983).

## *7.5.4 Limitations of Pump Tests*

In situ pumping tests are the accepted ways to determine $T$ and $S$. However, there are a number of assumptions inherent in the Thiem and Theis methods. They are as follows:

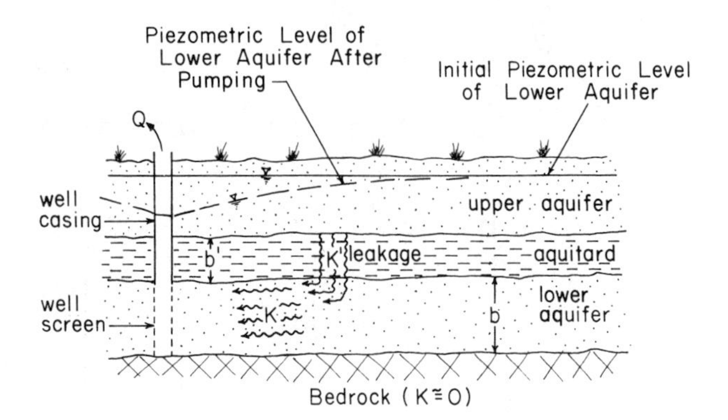


Figure 7.28 Diagrammatic sketch illustrating leakage from an adjacent aquitard into an aquifer.

1. The aquifer is homogeneous, isotropic, and is infinite in aerial extent.
2. The aquifer and water table are horizontal, and bounded by a horizontal impermeable base.
3. The well is pumped at a constant rate.
4. Water is released instantaneously from storage.
5. The well completely penetrates the aquifer.

Obviously these requirements are never achieved in the real world, but that fact does not obviate the usefulness of the methods. In the first place, due to plotting of data and graphical manipulations, the degree of accuracy is rarely greater than 10%. Secondly, there are some mathematical methods to circumvent shortcomings in the assumptions. For instance, if the aquifer has a barrier boundary on one side (example: an alluvial aquifer bounded by a granitic mountain), then an "image" well can be placed an equal distance on the other side of the boundary and opposite to the pumped well, pumping at the same rate. Then the drawdown at any point in the aquifer area is simply the algebraic sum of the drawdown due to the real plus imaginary well. A recharge boundary (lake or large stream) can be simulated by a recharge well (Ferris et al., 1962).

## *7.6 Dewatering*

Pump test methods described above can be used in engineering geology applications such as studies to determine well yields. The discharge, spacing, and depth of dewatering wells (Powers, 1981) at a construction site where high water table conditions may exist can be determined by analytical methods (see Problem 7.10) or by specially prepared nomograms (Hunt, 1983).

Dewatering is a process of removing water from a foundation pit by draining it to a sump or by pumping ground water from wells. The purpose of dewatering is to keep the excavation dry by lowering the water table. Dewatering can be one of the most difficult operations in deep foundations. The choice of dewatering techniques depends to a large degree on the permeability of the foundation subsoil (Fig. 7.29). Open pumping is the method whereby ground water is allowed to flow into trenches along the perimeter of the pit. For less permeable subsoils, drilled wells can be installed outside the pit perimeter, creating overlapping cones of depression. Other methods of dewatering are used only rarely. Silts, for example, have low permeability, and seepage, if any, can be controlled by sheet pile walls. Clays are practically impermeable and dewatering usually serves merely to remove precipitation that falls in the pit.

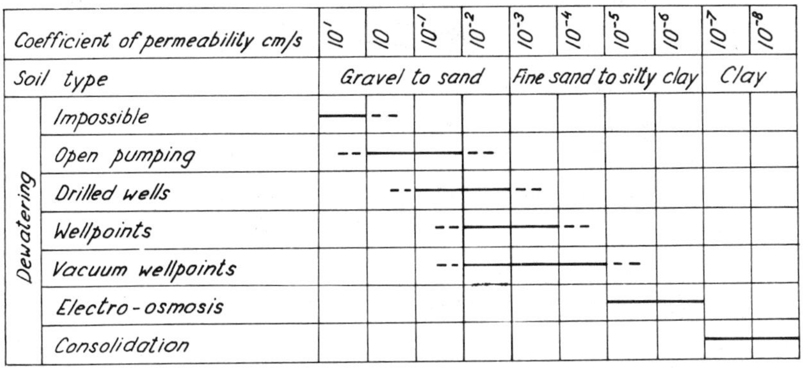

| Coefficient of permeability cm/s | | $10^{1}$ | $10$ | $10^{-1}$ | $10^{-2}$ | $10^{-3}$ | $10^{-4}$ | $10^{-5}$ | $10^{-6}$ | $10^{-7}$ | $10^{-8}$ |
|---|---|---|---|---|---|---|---|---|---|---|---|
| Soil type | | Gravel to sand | | | | Fine sand to silty clay | | | | Clay | |
| Dewatering | Impossible | — | -- | | | | | | | | |
| | Open pumping | -- | — | — | -- | | | | | | |
| | Drilled wells | | -- | — | — | -- | | | | | |
| | Wellpoints | | | -- | — | — | -- | | | | |
| | Vacuum wellpoints | | | -- | — | — | — | -- | | | |
| | Electro-osmosis | | | | | | | — | — | | |
| | Consolidation | | | | | | | | | — | — |

Figure 7.29 Suggested methods to be used for dewatering sediments of varying permeability (from Bazant, 1979).

## 7.7 Drain Spacing

Where the water table is very close to the land surface or where irrigation in low-permeability soils is to be practiced, it is often necessary to install underground drains. These typically consist of plastic pipes or concrete tile installed at depths of 1–3 m. Ground water may also be collected by pumps or open ditches (Fig. 7.30).

High water tables are incompatible with most croplands not only because tractors get stuck in swampy areas, but because evaporation in the capillary fringe can lead to the precipitation of salts, ultimately ending in soil salinization. In the Indus Valley in West Pakistan, for example, 9 million hectares of land are irrigated by canals in the largest irrigated region in the world. In the nineteenth century the British began this immense irrigation program, but did not line the

Figure 7.30 Agricultural drains installed in a glacial kettle in Wisconsin. The former peat bog is used for growing grass for lawns.

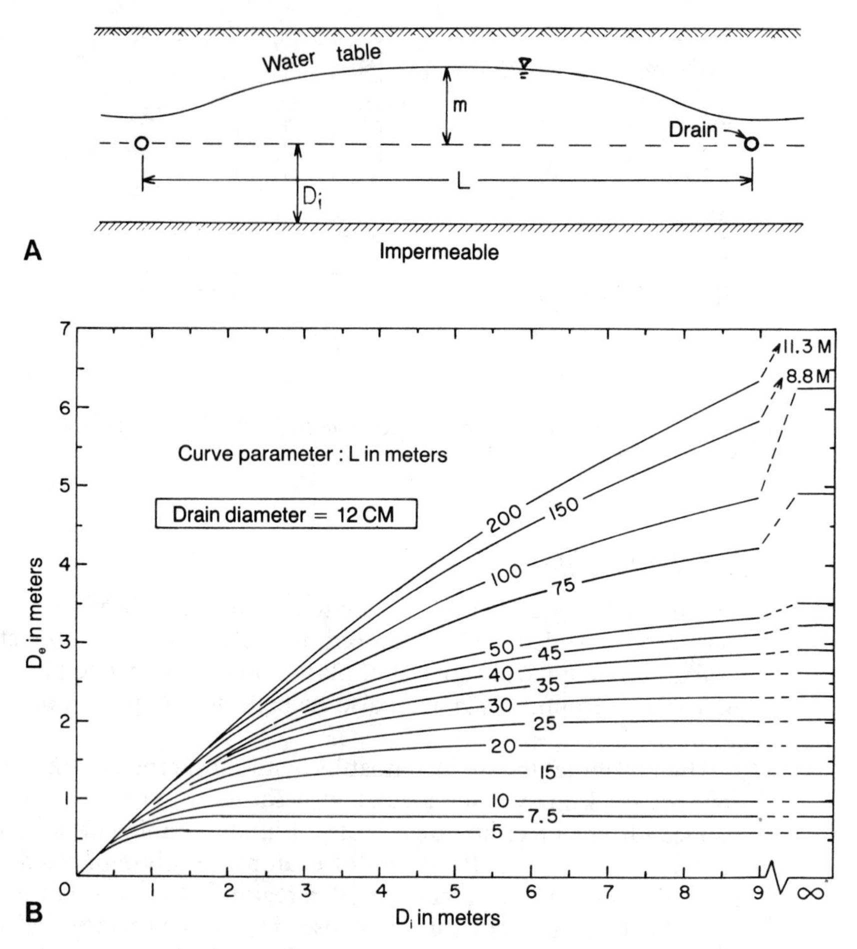


Figure 7.31 Drain spacing (from Bouwer, 1978). **A**. Schematic cross section showing symbols. **B**. Effective depth $D_e$ of impermeable layer below drains as a function of its actual depth $D_i$ for different drain spacings $L$.

65,000 km of canals, or account for drainage from the low, nearly flat irrigated land. Today, waterlogging and soil salinization has ruined much of the land. According to Calder (1970), "In flying over large tracts of this area one would imagine that it was an Arctic landscape because the white crust of salt glistens like snow."

The thickness of the zone of aeration under irrigated lands depends on the crop, soil type, source of water, and salinity of the water. In general, a depth of 1 m is satisfactory for most crops, but should be deeper in heavy soils or in areas where irrigation with water containing excessive salts occurs.

Numerous equations and analytical models have been developed for the purpose of designing drain spacing (Kirkham et al., 1974; U.S. Bureau of Reclamation, 1978; Luthin, 1978). Hooghoudt (1940), a Dutch hydraulic engineer, devised an equation relating the drainage rate to drain spacing (Fig. 7.31):

$$V_a = \frac{4\,Km(2D_e + m)}{L^2}, \tag{7.25}$$

where:

$V_a$ = drainage rate (rainfall rate, or rate of application of irrigation water, m/day),

$K$ = hydraulic conductivity of soil (m/day),

$D_e$ = effective depth of impermeable layer below drain center ($m$),

$m$ = height of water table above drain centers midway between drains ($m$),

$L$ = distance between drains ($m$).

The relationship between actual depth ($D_i$) and effective depth ($D_e$) to impermeable layer is shown in Fig. 7.31B.

The U.S. Bureau of Reclamation (1978) has developed methods of determining hydraulic conductivity of soils above the water table using shallow well pump-in tests or ring permeameters. These tests are necessary to predict subsurface drainage requirements for irrigated lands.

## 7.8 Ground Water Quality

The National Water Well Association (Campbell and Lehr, 1973) defines pollution as "an impairment of water quality by chemicals, heat, or bacteria to a degree that does not necessarily constitute an actual public health hazard, but that does adversely affect such waters for normal domestic, farm, municipal, or industrial use. The term contamination denotes impairment of water quality by chemical or bacterial pollution to a degree that creates an actual hazard to public health."

### *7.8.1 Bacterial Quality*

The subject of bacteria in ground water was reviewed by Romeo (1970) and recently updated by Gerba et al. (1975), Wood and Ehrlich (1978), and Bitton, et al. (1983). Traditionally, about 30 m is considered adequate separation of a domestic well from a septic tank leach field in fine sand or deposits of lower permeability. Because ground water generally moves slowly through rock or fine-grained soil interstices, bacteria are filtered out or die off due to the length of time required to travel between any two points. For common bacterial strains, the $T_{90}$, the time required for 90% of the bacteria to die off (i.e., so that there is a 1 order of magnitude reduction), ranges from 2 to 6 days (Sinton, 1980). Because of slow velocity, most ground water is pure from the point of view of bacterial contamination. Only isolated cases of virus or bacterial contamination have occurred in the United States; these cases are generally limited to rapid unfiltered flow in cavernous limestone or coarse gravel aquifers (Davis and DeWeist, 1966; Davis, 1979; Craun, 1979). Viruses have been identified as the cause of outbreaks of disease in some ground water systems. It is typical of those outbreaks that investigators have detected sources of sewage contamination in close proximity to wells (Lippy, 1981). Dye tests often show only a few hours time lapse between a septic tank and well in these cases. Viruses are extremely small (in the range of 5 $\mu$m), which probably enables them to penetrate aquifers whereas bacteria are trapped or die. The fact that ground water is generally bacterially pure even in urbanized areas is rather surprising in view of the number of septic tanks and polluted streams, and means that most subsoils are amazing bacterial purifiers.

In fact, Schmidt (1979) believes that "it is now becoming widely known that the adverse effects of septic tanks were greatly exaggerated in many areas in order to promote sewering."

In undeveloped countries, diseases transmitted by water-borne bacteria are common. In remote villages in Nicaragua, for example, I have seen people drinking from swamps and streams where pigs wallow. Figure 1.4 shows a properly constructed hand pump; but the lid was thoughtlessly pushed aside because the pump handle was broken. The villagers lower dirty buckets into the water, thus contaminating the well water. It is obvious that the villagers do not understand the concept of germs. A large percentage of Nicaraguan children die due to tuberculosis and other water-borne diseases. In 1984 the World Health Organization reported that diseases associated with contaminated water killed about 50,000 people per day.

### *7.8.2 Chemical Quality*

The chemical quality of ground water depends on the type of host rock, the length of time in residence, and solubility reactions. Chapter 3 contains weathering reactions whereby ions are introduced into ground water. An example illustrating the effect of time on ground water quality is in the Grand Canyon of Arizona. Water from Vasey's Paradise spring contains 163 ppm total dissolved (TDS); the spring receives its recharge from the Kaibab Plateau, which is only a short distance from Vasey's Paradise. In contrast, spring flow discharged into the canyon of the Little Colorado River has moved long distances in the subsurface and contains 2320 to 3970 ppm TDS (Cooley, 1976). Other examples showing the gradual solution of carbonate rocks, as indicated by increasing bicarbonate content, calcite saturation, and $^{14}C$ age downgradient are documented by Talma (1981), Toran (1982), and Back et al., (1983).

Natural ground water has a wide range and diversity of dissolved ions present. In areas of high precipitation where rocks of low solubility such as quartzite or granite occur, near-surface ground water is likely to contain very low concentrations of dissolved ions. Where ground water circulates slowly in soluble rocks such as carbonates or gypsiferous shales, possibly mixing with connate water, ground water may be rendered undrinkable due to dissolved minerals.

Table 7.3 lists the chemical composition of representative samples of ground water. Table 7.4 shows drinking water standards for dissolved substances. The maximum contaminate level for many substances (i.e., 0.01 mg/l for selenium) are subject to change as new evidence on toxicology is found (Hammer, 1981). Many substances are not harmful; indeed, high calcium levels in water are associated with low incidences of cardiovascular disease.

Water chemistry data is usually expressed as mg/l, which is nearly the same as ppm except the former is weight/volume and the latter is weight/weight. For samples of water near fresh water density, the two parameters are essentially the same.

There are three anions commonly present in ground water (sulfate, bicarbonate, and chloride) and three cations (calcium, magnesium, and sodium). When expressed in terms of milliequivalents per liter (meq/l), also referred to as equivalents per million (epm), the data can be drawn as bar graphs or map symbols called Stiff Diagrams (Fig. 7.32). The concentrations in meq/l are equal to the

Table 7.3 Chemical Analyses of Ground Water

| | Source | | | | | | | | |
|---|---|---|---|---|---|---|---|---|---|
| | 1 | 2 | 3 | 4 | 5 | 6 | 7 | 8 Acid hot | 9 Basic |
| Contents | Snow | Well | Well | Well | Well | Well | River | spring | spring |
| $SiO_2$ | 0.2 | 135 | 42 | 36 | 7.2 | 14 | 14 | 216 | 3970 |
| Al | 0.06 | — | — | 0.1 | — | 0.0 | 0.02 | 56 | 1.8 |
| Fe | 0.02 | 0.06 | 0.03 | 0.0 | 0.33 | 9.0 | 0.01 | 33 | 0.00 |
| Mn | — | — | 0.05 | 0.0 | — | 0.0 | — | 3.3 | 0.00 |
| Ca | 0.9 | 5.2 | 0.4 | 173 | 288 | 696 | 9.6 | 185 | 7.3 |
| Mg | 0.2 | 1.0 | 0.5 | 32 | 41 | 204 | 13 | 52 | 2.6 |
| Na | 0.42 | 28 | 51 | 40 | 37 | 462 | 3.2 | 6.7 | 10,900 |
| K | 0.1 | 3.6 | 3.0 | 2.6 | — | 9.1 | 0.5 | 24 | 135 |
| Li | — | — | — | — | — | — | — | — | 3.2 |
| $NH_4$ | — | — | — | — | — | — | — | — | 148 |
| $HCO_3$ | 4.0 | 50 | 120.0 | 218 | 267 | 72 | 89 | 0.0 | 0.0 |
| $CO_3$ | 0.0 | 0.0 | 0.0 | 0.0 | 0.0 | 0.0 | 0.0 | 0.0 | 5560 |
| OH | 0.0 | 0.0 | 0.0 | 0.0 | 0.0 | 0.0 | 0.0 | 0.0 | 1430 |
| $SO_4$ | 0.7 | 7.8 | 10 | 268 | 190 | 2130 | 5.0 | 1570 | 267 |
| Cl | 0.6 | 3.5 | 7.2 | 135 | 94 | 850 | 3.5 | 3.5 | 7180 |
| F | 0.0 | 12 | 0.0 | 0.1 | 0.3 | 1.5 | 0.1 | 1.1 | 3.0 |
| Br | — | — | — | — | — | — | — | — | 11 |
| I | — | — | — | — | — | — | — | — | 5.7 |
| $NO_3$ | 0.2 | 0.2 | 2.4 | 7.5 | 518 | 0.0 | 0.5 | 0.0 | 0.0 |
| $PO_4$ | — | — | — | 0.0 | — | 0.0 | 0.0 | — | 4.3 |
| B | — | 0.3 | — | — | — | — | — | — | 242 |
| Dissoved solids | 5.4 | 309 | 177 | 820 | 1306 | 4400 | 94 | — | 31,200 |
| Conductance ($\mu$mhos) | 10 | 164 | 214 | — | — | 5610 | 141 | 4570 | 36,800 |
| pH | 6.0 | 7.6 | 7.9 | 7.4 | — | 7.3 | 7.3 | 1.9 | 11.6 |
| Temperature (°F) | — | 107 | 78 | — | — | 73 | 63 | — | 54 |
| $H_2S$ | — | — | — | — | — | — | — | — | 1000 |

From Davis and DeWeist, 1966.
1. Snow from near Mestersvig, Greenland (From *Air Force Survey in Geophysics* 127, 1961.)
2. Well water, Yellowstone National Park (From *U.S. Geol. Survey Water Supply Paper* 1475-F, 1962.)
3. Well water, Gulfport, Mississippi (From *U.S. Geol Survey, Water Supply Paper* 1299, 1954.)
4. Well water, Santiago, Chile (From *Inst. Investigaciones Geol. Boletin* 1, 1958.)
5. Well water, Chase County, Kansas (From *State Geol. Survey Kansas,* **11,** 1951.)
6. Well water, near Carlsbad, New Mexico (From *U.S. Geol. Survey, T.E.I.*, 1962.)
7. Smith River, northern California (From *California Dept. Water Resources Bull.* 65–59, 1961.)
8. Hot spring, Sulfur Springs, New Mexico (From *U.S. Geol. Survey Water Supply Paper* 1473, 1959.)
9. Mineral Spring, near Mt. Shasta, California (From *Geochim, et Cosmochim, Acta,* **22,** 1961.)

concentrations in mg/l divided by the ratio of atomic (or molecular) weight to valence:

$$\text{meq/1} = \frac{\text{mg/l}}{\text{mol wt/valence}}. \tag{7.26}$$

For example, 40 mg/l of the anion sulfate ($SO_4^=$) equals

$$40 \div \frac{[32 + 4(16)]}{2} = \frac{40}{48} = 0.83 \text{ meq/l}.$$

Table 7.4 Drinking Water Standards

| Substance or property | Public Health Service, 1962 | | | World Health Org. 1963 | | |
|---|---|---|---|---|---|---|
| | Desirable maximum limit | Absolute maximum limit | EPA interim 1975 | Maximum acceptable | Maximum allowable | Natl. Acad. Sci. Natl. Acad. Eng. 1972 |
| Alkyl benzyl sulfonate (ABS, LAS, methylene—blue active substances | 0.5 | — | — | 0.5 | 1 | 0.5 |
| Ammonium nitrogen | | | | | | 0.5 |
| Arsenic | 0.01 | 0.05 | 0.05 | — | 0.05 | 0.1 |
| Barium | — | 1 | 1 | — | 1 | 1 |
| Cadmium | — | 0.01 | 0.01 | — | 0.01 | 0.01 |
| Calcium | — | — | — | 75 | 200 | — |
| Chloride | 250 | — | — | 200 | 600 | 250 |
| Chromium (hexavalent) | — | 0.05 | 0.05 | — | 0.05 | 0.05 |
| Color (Pt-Co units) | — | — | — | 5 | 50 | 75 |
| Copper | 1 | — | — | 1 | 1.5 | 1 |
| Cyanide | 0.01 | 0.2 | — | — | 0.2 | 0.2 |
| Fluoride[a] | 0.6–0.9 | 0.8–1.7 | 1.4–2.4 | — | — | 1.4–2.4 |
| Iron ($Fe^{2+}$) | 0.3 | — | — | 0.3 | 1 | 0.3 |
| Lead | — | 0.05 | 0.05 | — | 0.05 | 0.05 |
| Magnesium | — | — | — | 50 | 150 | — |
| Magnesium and sodium sulfates | — | — | — | 500 | 1000 | — |
| Manganese ($Mn^{2+}$) | 0.05 | — | — | 0.1 | 0.5 | 0.05 |
| Mercury | — | — | 0.002 | — | — | 0.002 |
| Nitrate nitrogen[b] | 10 | — | 10 | — | — | 10 |
| Nitrite nitrogen | — | — | — | — | — | 1 |
| Organics | | | | | | |
| Carbon chloroform extract | 0.2 | — | — | 0.2 | 0.5 | 0.3 |
| Carbon alcohol extract | — | — | — | — | — | 1.5 |
| Pesticides | | | | | | |
| Aldrin | — | — | — | — | — | 0.001 |
| Chlordane | — | — | — | — | — | 0.003 |
| DDT | — | — | — | — | — | 0.05 |
| Dieldrin | — | — | — | — | — | 0.001 |
| Endrin | — | — | 0.0002 | — | — | 0.0005 |
| Heptachlor | — | — | — | — | — | 0.0001 |
| Heptachlor epoxide | — | — | — | — | — | 0.0001 |
| Lindane | — | — | 0.004 | — | — | 0.005 |
| Methoxychlor | — | — | 0.1 | — | — | 1 |
| Toxaphene | — | — | 0.005 | — | — | 0.005 |
| Organo phosphorus and carbamate insecticides | — | — | — | — | — | 0.1 |
| Herbicides | | | | | | |
| 24-D | — | — | 0.1 | — | — | 0.02 |
| 2,4,5-TP (Silvex) | — | — | 0.01 | — | — | 0.03 |
| 2,4,5-T | — | — | — | — | — | 0.002 |
| pH (units) | — | — | — | 7–8.5 | — | 5–9 |
| Phenolic compounds (as phenol) | 0.001 | — | — | 0.001 | 0.002 | 0.000001 |
| Selenium | — | 0.01 | 0.01 | — | 0.01 | 0.01 |
| Silver | — | 0.05 | 0.05 | — | — | — |
| Sulfate | 250 | — | — | 200 | 400 | 250 |

Table 7.4 Drinking Water Standards *(continued)*

| Substance or property | Public Health Service, 1962 Desirable maximum limit | Public Health Service, 1962 Absolute maximum limit | EPA interim 1975 | World Health Org. 1963 Maximum acceptable | World Health Org. 1963 Maximum allowable | Natl. Acad. Sci. Natl. Acad. Eng. 1972 |
|---|---|---|---|---|---|---|
| Total dissolved solids | 500 | — | — | 500 | 1,500 | — |
| Zinc | 5 | — | — | 5 | 15 | 5 |

From Bouwer, 1978.

[a]Maximum fluoride levels are given in relation to annual average daily maximum air temperature because ingestion of water increases with temperature. The following range is recommended by the National Academy of Sciences and National Academy of Engineering (1972):

| | |
|---|---|
| 26–32°C | 14 mg/l F |
| 22–26°C | 1.6 |
| 18–22°C | 1.8 |
| 15–18°C | 2.0 |
| 12–15°C | 2.2 |
| 10–12°C | 2.4 |

[b]Nitrate-nitrogen limits are also expressed in terms of nitrate (10 mg/l $NO_3$-N corresponds to 45 mg/l $NO_3$).

Because a water solution has a neutral valence, the sum of the cation milliequivalence should equal the sum of the anion milliequivalence.

Computerized models may be constructed for water analyses (Faust and Aly, 1981). These models are based on 1) the ionic concentration, 2) the molarity charge and solubility products of the ions, and 3) the mass balance of the water

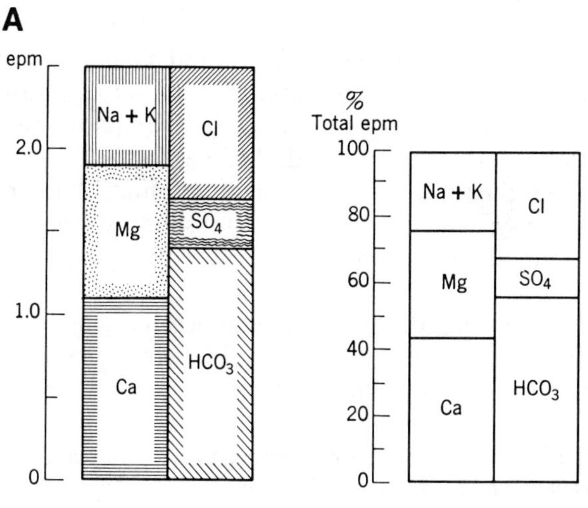


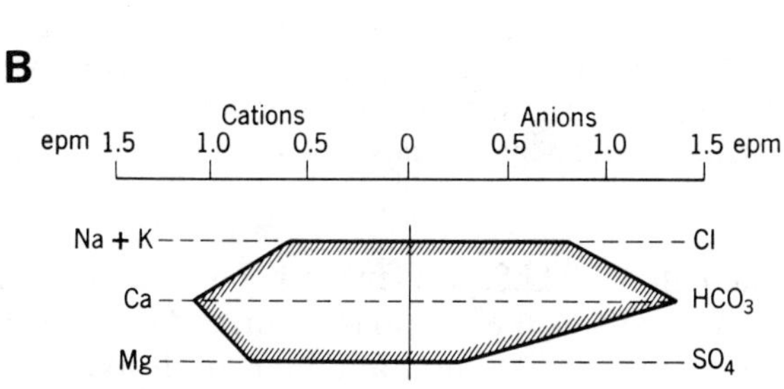


Figure 7.32 Water quality diagrams (from Davis and DeWeist 1966). **A**. Bar graphs expressed as epm and percentage of epm. **B**. Stiff diagrams for same analysis as **A**.

sample, which requires the sum of the free and complex species to be equal to the total concentration. The computer program, WATEQ, uses analytical expressions for equilibrium constants over the 0–100°C range; the program rewritten into FORTRAN is called WATEQF. The program SOLMNEQ has a data base of tabulated equilibrium constants over a range of 0–300°C.

### *7.8.3 Contamination and Dispersion*

Unlike bacteria which are filtered or die in most ground water systems, chemicals may remain dissolved in water. There are thousands of examples of man-made chemical contamination of ground water, ranging from nontoxic but widespread buildup of chlorides near highways salted for winter ice, to local farm wells contaminated by barnyard bacteria or nitrate, to sanitary landfills often containing illegally dumped toxic chemicals, to horrid examples of organic chemical compounds such as Love Canal near Niagara Falls, NY.

The tragedy of Love Canal unfolded in the 1970s. For 40 years, Hooker Chemical and Plastics Co. (acquired in 1968 by Occidental Petroleum Corp.) and other chemical companies dumped chemical wastes into the abandoned Love Canal near Niagara Falls, NY. About 20,000 tons of chemical wastes were dumped into the canal before it was covered in 1953 and sold to the Niagara Falls Board of Education. In 1953 the open trench was filled in, the land sold to developers, homes built around the dump, and an elementary school built directly over the site where the poisons were buried. Residences built adjacent to the former dump frequently had basements flooded by contaminated seepage, including several toxic chemicals, such as the banned pesticides Mirex and Lindane. (Hooker knew of the problem, which included chemical burns to children playing near the dump in 1958, but did almost nothing to warn the public.) Birth defects, miscarriages, liver damage, cancer, and a host of health problems were rampant among the 97 families living immediately adjacent to the landfill (Fine, 1980). New York State officials are concerned about the presence of the deadly chemical Dioxin, (a gram of which in a drinking water system can kill tens of thousands of people), because Love Canal drains into Lake Ontario. Since 1978, more than 700 families have been evacuated.

A problem with many toxic hazards, as well as engineering geology hazards in general, is that people tend to minimize them for fear their property values will go down. Regarding Love Canal, Ross (1981) states:

> . . . the landfill was either improperly built, or its safeguards were breached by construction of a school, homes, and sewers on and around the site. Then, a succession of rainy years pushed increasing amounts of toxic materials to the surface. . . . In 1978, the New York State Department of Health released a health survey of the area, suggesting some possible effects. . . . The state never released the medical data upon which its study was based, however, on the grounds that this would violate resident's privacy.

In 1980, noting that one-half of all U.S. residents rely on ground water—which is often not treated or disinfected—as their primary source of drinking water, the U.S. Council on Environmental Quality cited hundreds of drinking-water wells which have been recently closed because of contamination by toxic organics. Data collected by the U.S. Environmental Protection Agency shows serious contamination of drinking-water wells in almost every state. New Jersey has its share of problems; a 1981 study of 670 wells representative of ground water in New Jersey found that 111 of them contained elevated levels of toxic organic chemicals. In 1980, Rockaway Township, NJ, discovered all three of its public wells were contaminated with tetrachlorethylene (TCE) and other chemicals. George Pinder, of Princeton University, testified in 1981 that several Atlantic City, NJ, drinking water supply wells could become contaminated by toxic chemicals from Price's Pit, an abandoned waste dump in Pleasantville. Price's landfill in New Jersey, a sand and gravel quarry from 1960 to 1968, was used in following years as a dump. In 1972, about $3.4 \times 10^4$ $m^3$ of toxic and flammable chemical and liquid wastes were dumped at the landfill both in drums and by direct discharge to the ground (Gray and Hoffman, 1983). The landfill is 1 km upgradient of a public water supply well for Atlantic City.

There are numerous site-specific studies of ground water contamination. For instance, ground water contamination by industrial organics was described by Roux and Althoff (1980) and Althoff et al., (1981). Seepages from sanitary landfills are described by Back and Baedecker (1979) and Johnson et al. (1981). Cherkauer (1980) found contaminated ground water below a fly-ash disposal landfill in Wisconsin. Pettyjohn (1982) describes ground water contamination by oil field brines in Ohio. The ineffective role of regulatory agencies is well-documented in the excellent report on ground water pollution by hazardous wastes (Geraghty and Miller, 1977).

Drinking water supplies have been endangered in many states. For example, in 1982 one of Tucson's water wells was tagged unfit to drink because of high TCE. The solvent TCE has caused the closing of over 12 wells near Tucson. The U.S. Air Force owns much of the affected area south of the Tucson airport (formerly operated by Hughes Aircraft Co.). Increasing concentrations of nitrate in Long Island's ground water (Porter, 1980) is of concern because excessive concentrations may be a health hazard to infants. The predominant sources of nitrogen are septic tanks, lawn and agricultural fertilizers, animal wastes, and landfills. In Missouri, leaking drums of dioxin were found in 1982 on a site believed to be the disposal ground for the now defunct Northeast Pharmaceutical and Chemical Co. Ground water contamination in Mojave River alluvium near Barstow, CA, is the result of percolation of industrial wastes and municipal sewage for over 60 years. Wells have been abandoned due to high concentrations of detergents, nitrogen, chromium, phenols, and other chemicals. The degraded ground water is moving at about 0.4 m/day in very permeable river-channel deposits, and now extends 6.4 km downgradient (Hughes, 1975).

Another horrendous example of ground water contamination is near Denver, CO, where, in 1943, the Rocky Mountain Arsenal of the U.S. Army Chemical Corps, began to manufacture war materials including nerve gas. In 1951, the facilities were leased to the Shell Oil Co. for the production of pesticides. Unlined holding ponds on permeable eolian deposits were built to contain toxic solutions of chlorides, chlorates, salts of phosphoric acid, fluorides, and arsenic. The "evaporation ponds" resulted in wide spread contamination of ground water (National Research Council, 1984). After nearby farmers reported livestock sickness and crop damage, a governmental study showed that shallow wells had become contaminated with an assortment of chemicals (Walton, 1959). It had taken 7 years for contaminated ground water to travel 5 km from the holding ponds to the nearest farm. The most mysterious feature of the episode was the discovery of the weed-killer 2,4-D in some of the wells and holding ponds. None of this weed killer had been manufactured at the plant. Apparently it formed spontaneously in the holding ponds from the other substances discharged at the arsenal (Carson, 1962). Konikow (1977) presents an example of a solute-transport computer model applied to the chemical pollution of ground water at the Rocky Mountain Arsenal. In 1983 the U.S. Justice Dept. sued Shell Oil Co., seeking $1.85 billion to cover costs of cleaning up contamination on the army property. (Under the 1980 "Superfund" law, the federal government can clean up pollution caused by abandoned chemical dumps, and then can seek to recover these funds from the responsible party.)

Dye tests in a sinkhole area of the Ozarks has shown that ground water may travel as fast as 25 km in 3 months in karst terrain. Small communities in Missouri and Pennsylvania have used sinkholes as convenient dumps, unconcerned that sinkholes serve as important ground water recharge points.

A complex ground water contamination problem occurs when numerous small (nonpoint) sources of contamination occur. For example, the buildup of nitrate in ground water in heavily fertilized farmland areas has been recognized in many areas. In the rapidly expanding irrigated farmlands in central Nebraska, for example, nitrates used in fertilizers may move downward through sandy soils at velocities of about 2 m/yr, reaching ground water in a few years. The drinking water limit for nitrate-nitrogen is 10 mg/l. Most of the wells in Merrick, Buffalo, and Hall Counties, Nebraska, exceed this limit. Average concentrations in Holt County are increasing at a rate of more than 1 mg/l per year.

In the United States, there are presently enough seeping landfills, industrial storage facilities, leaking gasoline tanks, and nonpoint sources of pollution such as fertilizer and pesticide compounds already loose in the environment (Geraghty and Miller, 1975) to keep engineering geologists and hydrogeologists employed for decades. Of the 400,000 gasoline tanks buried underground in the U.S. in the last 25 years, 25% are believed to be leaking. In Connecticut, the USGS recently found 3600 sources of possible contamination, including more than 300 road salt and 3200 petroleum storage facilities, fly-ash and sewage disposal facilities, and 185 active landfills. According to the Environmental Protection Agency, over 90% of the hazardous waste generated in the United States today is handled improperly. There are an estimated 20,000 sanitary landfills in the United States which have been established with no regard to the future contamination of ground water. In the United States, as federal and state regulations stiffen with respect to anti-pollution laws for surface water, more and more wastes will undoubtedly be disposed of by dumping into the ground. Because of the slow movement of ground water, contamination is slow to be discovered. But once chemically contaminated, ground water is extremely difficult to purify.

A point source of ground water contamination will disperse downgradient. Dispersion is the process whereby a flowing fluid spreads substances into the environment. Atmospheric dispersion is easy to see if one watches a plume from a smokestack spread as it migrates in a gentle breeze. Because of dispersion, some contaminants in ground water can be expected to move downgradient faster or more slowly than the average velocity determined by Darcy law equations (Fig. 7.33). The reader may consult Fried (1975), Freeze and Cherry (1979), or Todd

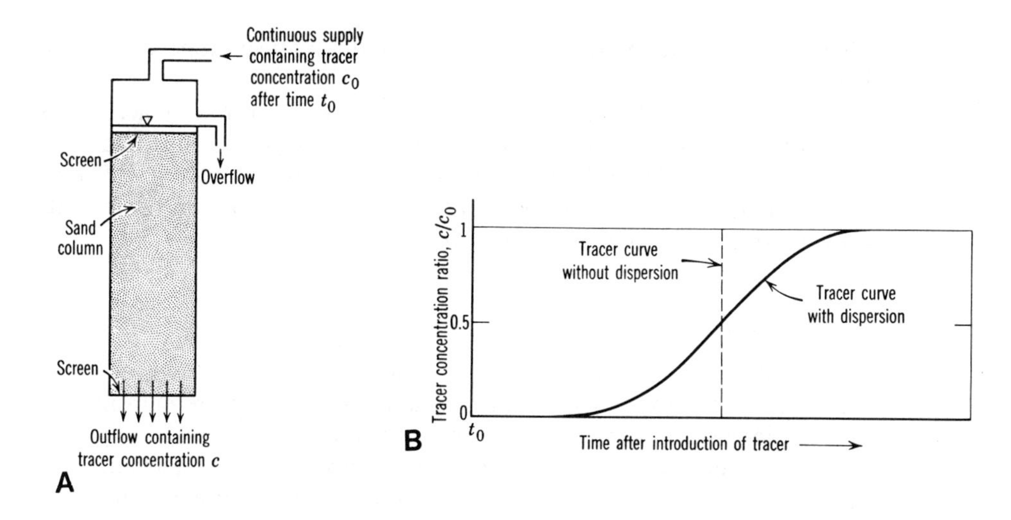


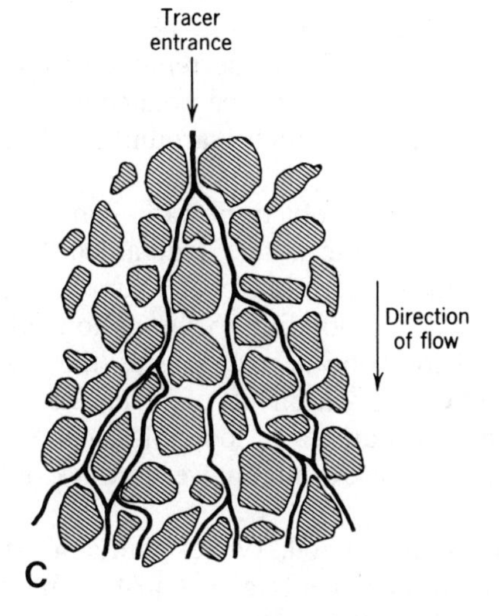


Figure 7.33 Sketches illustrating dispersion (from Todd, 1980). **A**. Laboratory apparatus to illustrate longitudinal dispersion of a tracer passing through a sand column. **B**. Dispersion curve from sand column. **C**. Sketch illustrating lateral dispersion. **D**. Map illustrating tracer distribution resulting from dispersion due to a continuous injection of tracer. **E**. Map illustrating tracer distribution resulting from dispersion due to injection of a single slug of tracer.

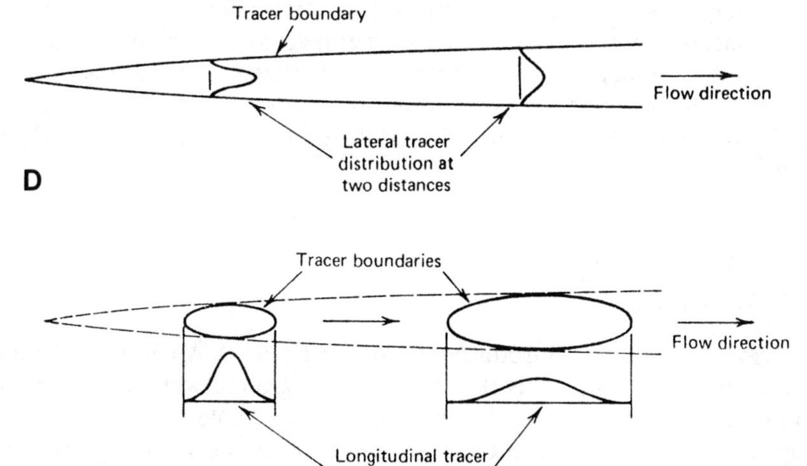

(1980), or Gillham and Cherry (1982) for mathematical models of ground water dispersion.

In addition to waste dilution by dispersion, dilution may be accomplished by "sorption" of dissolved ions onto soil particles. A general formula relating the velocity of an ion ($V_i$) relative to the velocity of ground water ($V_w$) involves the use of a laboratory-derived parameter called the distribution coefficient ($K_d$) as follows (from Davis and DeWeist, 1966):

$$V_i = \frac{V_w}{1 + K_d \rho_b / n} \tag{7.27}$$

where $\rho_b$ is the bulk (dry) density and $n$ is the porosity of the soil. Values of $K_d$ range from 0 for ions such as chloride and sulfate which remain in solution to 1000 for ions such as thorium that are readily sorbed on soil.

Current methods of remedial treatment of point sources of ground water contamination include the use of interceptor wells to pump contaminated ground water to the surface where it is circulated through filters such as activated charcoal. Readily-absorbed contaminants such as gasoline may be extracted in this manner. Water containing volatiles such as TCE can be removed by aeration devices. In some places, contaminated ground water can be isolated by bentonite-impregnated soil barriers which are emplaced in ditches around shallow waste seepage sites.

The U.S. Government has responded to the problem of water contamination. To prevent discharge of pollutants into navigable waters, federal legislation initiated the Clean Water Act of 1972. To protect drinking water supplies, the Safe Drinking Water Act of 1974 was enacted. Responding to the nationwide crisis caused by uncontrolled and improper disposition of toxic wastes, the U.S. Congress passed the Resource Recovery and Conservation Act of 1976. The programs are a step in the right direction, but have a number of loopholes such as failure to deal with abandoned toxic waste sites or lack of manpower to track industrial chemicals and prosecute offenders.

M. Gordon Wolman (National Research Council, 1982c) points out that few individuals have specialized in the combined fields of chemistry and ground water, disciplines essential to understand the dispersion of toxic wastes: "As in so many facets of environmental management, geologists will be called on to deal with engineers, biologists, public officials, and citizens as the siting of waste-disposal facilities moves to the point where it counts, that is, where disposed sites must be in someone's backyard. The task is immense, the job clearly for the geologist."

## 7.9 Basin Evaluation

### *7.9.1 Safe Yield*

The term safe yield was introduced by Meinzer (1923). A common definition of the safe yield of an aquifer is the rate at which ground water can be withdrawn without causing a long-term decline of the water table or piezometric surface (Bouwer, 1978). As a general concept, safe yield is the pumping rate which an aquifer can sustain indefinitely. In its broadest sense, the concept of safe yield can be considered as the rate at which ground water can be withdrawn without producing undesirable effects (such as seawater intrusion) or causing economic or legal problems (Gass, 1981b).

The hydrologic cycle can be considered to be in a state of dynamic equilibrium (Fig. 7.1). The safe yield could be conceived as the rate of replenishment of the aquifer if it were all intercepted and used. If, however, this water was withdrawn and consumed (for example, by irrigation or exportation from the area), ground-water discharge to the stream would totally cease. Drying up a stream would be viewed as undesirable, and it could be argued that the safe yield had been exceeded. Viewed in the context that any interruption of ground water would ultimately affect a downgradient situation, any withdrawal of water can be said to be in excess of the safe yield because it is impossible to consume any water without affecting something else. In this sense, the term safe yield is a myth, and some hydrogeologists suggest that use of the term safe yield be discontinued (Anderson and Berkebile, 1977).

Because of vagaries in the definition of safe yield, the term has been dropped by the USGS. Nevertheless, the term still has merit as a general concept, or as a site-specific term which can be defined by the researcher. For instance, in Rhode Island (average annual PPT $\simeq$115 cm) the sustained withdrawal of more than 2.7 $m^3/s$ of water could theoretically be obtained from glacial deposits of ice-contact stratified drift and alluvium. However, this withdrawal would cause a serious reduction of the flow of perennial streams, so that the safe yield of Rhode Island's ground water would have to be less than 2.7 $m^3/s$ (Rahn, 1968).

Ground water supplies are being seriously depleted in many places. Ground water withdrawals from the Cambrian-Ordovician aquifer in the Chicago area exceed the safe yield ("practical sustained yield") of the aquifer, equal to $2.5 \times 10^6$ $m^3$/day (Visocky, 1982). Even with the proposed import of Lake Michigan water to Chicago suburbs, computer projections show there will still be future ground water withdrawals far in excess of the practical sustained yield.

For an oceanic island such as Bermuda, where only a few hundred meters of fresh ground water "float" on top of porous coral saturated by salt water, the concept of safe yield is particularly relevant (Vacher, 1974). Here only a certain amount of fresh ground water may be withdrawn annually or else salt water intrusion will occur. In the Miami, FL, area, seawater intrusion due to ground water withdrawal also occurs.

Much attention has centered on ground water withdrawals from the Ogallala aquifer, Pliocene sandy alluvium found in the western Great Plains. The description of how farmers became reliant upon the Ogallala for irrigation, its legal battles, and potentially dark future is described by Bittinger and Green (1980). Located in the semiarid High Plains of the United States, the Ogallala aquifer contains enormous quantities of fresh ground water. Some $6.5 \times 10^4$ $km^2$ of land are irrigated with Ogallala ground water. About 36% of the ground water used in the United States is pumped from the Ogallala aquifer. Water table declines have increased in the past 30 yr. The Ogallala is being overdrawn at the rate of $1.7 \times 10^{10}$ $m^3$/yr, an amount equal to the annual flow of the Colorado River (McCray, 1982). According to the USGS, the saturated thickness of the aquifer has declined more than 10% under ¼ of the total aquifer area. Nebraska, which has the largest percent of the aquifer ground water volume, has over 65,000 irrigation wells. In Oklahoma and Texas, much of the water is used to grow cotton and corn. Economics may eventually solve the problem of ground water decline because, although irrigated cotton may return 20 times the income of dry-land cattle grazing, the electricity costs from increasingly greater pumping lifts associated with the dropping water table may eventually make it uneconomic to irrigate. The U.S. Army Corps of Engineers has prepared four separate plans for canals to bring water from the Missouri or Mississippi River to the High Plains. The 1982 costs of these projects range from $13.4 to $40 billion each. A major obstacle to the canal projects is that using water from these canals would be at least 10 times more expensive than what farmers currently pay for water. Nevertheless, in 1984 the prospect of drying up irrigated agriculture with attendant job and population losses are putting pressure on elected officials to build one of these canals.

Southern Arizona has severe ground water problems. Withdrawals for municipal and agricultural use in the Tucson area amount to $4.7 \times 10^8$ m$^3$/yr, but ground water recharge is only $1.0 \times 10^8$ m$^3$/yr; thus the aquifers are overdrafted by about $3.7 \times 10^8$ m$^3$/yr (Foster and Wright, 1980). In the desert basins of southwestern Arizona (average annual PPT $\simeq$ 10 cm) large withdrawals of ground water for irrigation have exceeded the safe yield (Miller, 1958; Harshbarger et al., 1966), and are causing widespread lowering of the water table which in turn causes the land to subside (see Chapter 9). The concept of safe yield is useful in Arizona because it conveys man's abuse of a slow, renewable resource—even to the point of jeopardizing his own environment. The Central Arizona Project, under construction in 1984, is a gigantic project to pump $1.5 \times 10^9$ m$^3$ of water annually from the Colorado River to the Phoenix and Tucson area (Morrison, 1983b). As of 1983, the 500-km aqueduct was about half-finished. The water to Arizona will be gained at the expense of California's use of the river. Even the gigantic Central Arizona Project will not be enough to stop the decline of ground water in southern Arizona because the overdraft of ground water is about twice the amount to be supplied by the canal. Eventually, agriculture (which uses 89% of the water consumed in Arizona) will have to be curtailed. Editorializing about the use of ground water to grow subsidized crops, and the construciton of the gigantic federally sponsored Central Arizona Project, Jay Lehr, editor of *Water Well Journal* said in 1982:

> And don't cry for Arizona, the most wasteful state of them all. The state is a desert where agriculture should never have been king. The crops were often unnecessary and more often unprofitable, but for federal subsidies which allowed a truly finite water supply to be squandered. This poor planning brought about the most absurd boondoggle this nation will ever know, "The Central Arizona Project." When it's over and the dust settles in Arizona, you will have paid in excess of $4 billion to bring water down from the Colorado River into Tucson. Never mind that south doesn't always mean down (Tucson is 2500 feet above the Colorado intake), and never mind that there isn't enough water in the Colorado to satisfy current claimants in its flow (ask the Californians). Drowned scenic Indian land and hunting grounds will yield obscene quantities of water to an unbelievable desert evaporation rate. . . . No, don't cry for Arizona. The state is laughing all the way to the bank with your bucks.

In the future, the utilization of ground water in conjunction with surface water will become important. Conjunctive use is the coordinated development of ground and surface water in order to minimize the overdevelopment of either resource. In Orange County, California, for example, about one-half of the imported Colorado River and Feather River aqueduct water is applied to infiltration basins and ponds for ground water recharge (Hammer and Elser, 1980). The recharged water maintains good quality water for wells, and helps prevent seawater intrusion.

### *7.9.2 Ground Water Models*

A model is anything which simulates the real world. The Theis and Thiem analytical methods discussed above are mathematical models. Physical, electrical, and computer models are also used in ground water studies. Prickett (1975) describes the use of many kinds of ground water modeling techniques, including an example of numerical analysis using the finite difference method for determining declines in the piezometric surface of the Cambrian-Ordovician aquifer in northeastern Illinois.

Physical models (Peterson et al., 1978) include devices such as sand-filled plexiglass and wood watertight boxes where dye can be used to qualitatively trace movement from contamination source to a pumping well. Physical models showing the movement of saline water relative to drain spacing in an irrigated area are described by Maasland and Bittinger (1963). The flow lines aid in drawing flow nets, from which a quantitative evaluation can be made. Due to the time required to construct physical models and the problem of storage of the devices, physical models are rarely used except for educational purposes. Another form of physical model involves the flow of viscous fluids through the interface between the two closely spaced parallel plates. This model, called a Hele-Shaw model, is particularly useful to depict the two-dimensional investigation of problems of multiple-fluid flow such as arise in the case of simultaneous flow of fresh and salt water or oil and water.

Electrical models were in great use during the 1950s and 1960s. The USGS mathematician Skibitzke (1961) pioneered the development of electrical models used by the USGS. For these models, electrical current is analogous to ground water discharge, voltage analogous to head, conductance analogous to permeability, and capacitance analogous to storage. In its simplest form, a conductive paper (sold commercially under the name Teledeltos paper) is used to simulate the aquifer. The paper is cut to the geographical limits of the aquifer (plan or cross-sectional view), and current is sent through the paper with electrodes simulating the boundaries of constant head. Equipotential lines can be determined on the paper by using a probe. In the 1960s, giant resistance/capacitance networks were made by soldering together thousands of resistors and capacitors on large poster boards. The array could then be excited by electrical current, elaborate boundary conditions simulated, and drawdowns from various pumping schemes predicted (Fig. 7.34). Electrical models have been largely replaced in the past decade by computer models because computers are more versatile and less time is required to construct computer models (Prickett, 1979). Electrical models, once built, are inflexible in terms of changing $T$ and $S$. There is also the problem of finding room to store electrical models.

Space does not permit a detailed explanation of computer models in this book. However, the following example of a finite difference model is given to illustrate the kind of techniques which can be utilized by a computer. Consider a well pumping near a river (elevation 100 m) and bounded by a mountain as shown in Fig. 7.35. It is assumed that zero drawdown occurs at the river (a recharge bound-

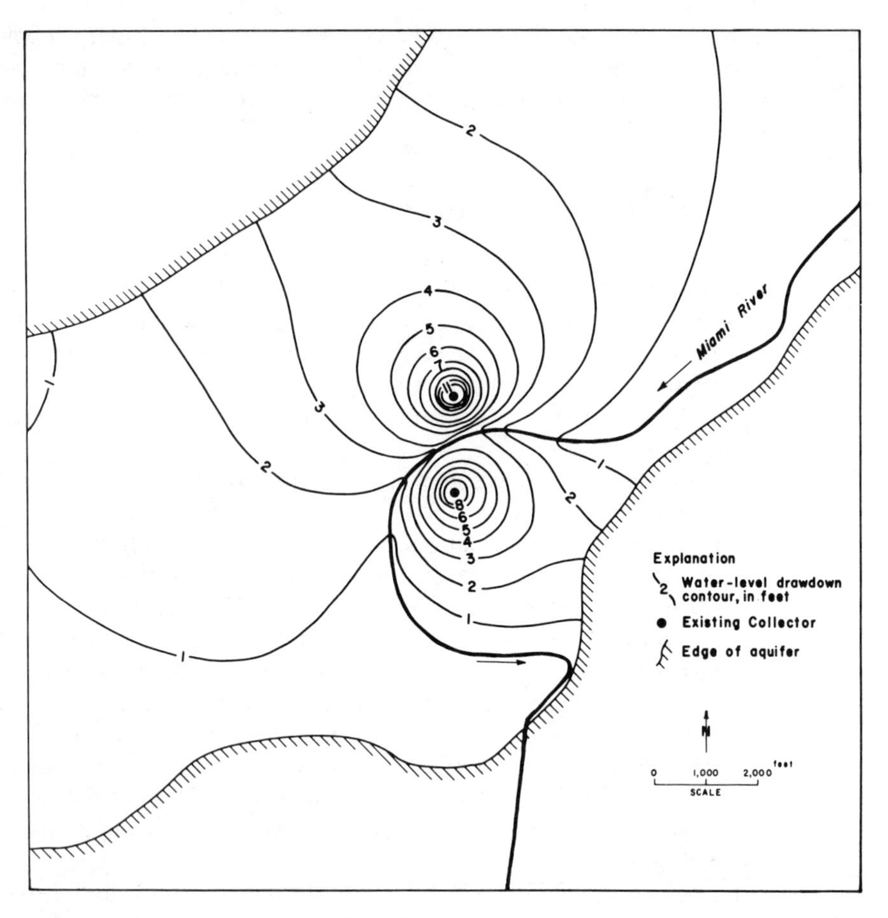


Figure 7.34 Electrical model results show drawdown contours for well system based on analog model. (From Walton, et al., 1967.)

ary) and a pumping level of 0 m elevation occurs at the well. An array of nodes can then be constructed to form a grid, where the elevation of the water table at each node is defined as the average of each surrounding node. This can be done manually, starting with assumed initial values at the nodes, and repeating the averaging over the array several times until the values reach some unchanging level of accuracy (say two significant figures). Figure 7.36 shows a computer program written to solve for the head at each node in this example. Wang and Anderson (1982) give additional examples of the application of (*a*) finite-difference computer models used to predict ground water heads by iterative solution of Laplace equations and (*b*) other techniques such as the finite element method. Bair and O'Donnell (1983) give a computer program that predicts the well spacing for a series of dewatering wells used to lower the water table for foundation excavation.

The main problem with ground water models, including mathematical models described in this chapter, is not the math but the geology. Many mathematical models are available in the literature, which in almost all cases vastly exceed requisite geologic data. Of particular importance are the basic hydrologic properties

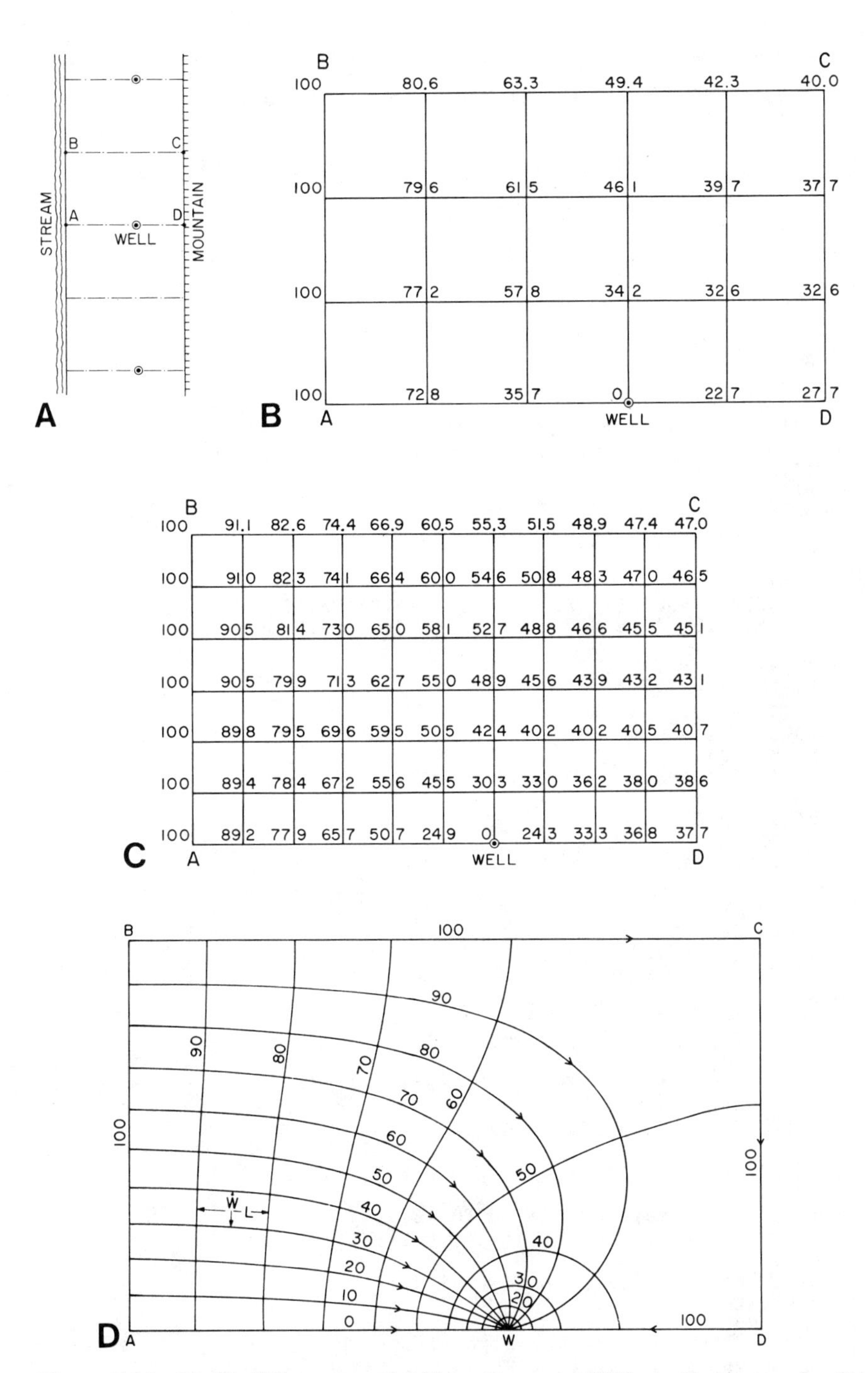


Figure 7.35 Finite difference model (from Bouwer, 1978). **A**. Geometry of well in valley between stream and mountain range. **B**. Solution of H values in percentage of total potential difference between stream and well for coarse network and fine network. **C, D**. Equipotentials and streamlines for flow system in the rectangle *ABCD* of 7 × 10 grid *E*.

```
      PROGRAM FDM (DATA,OUTPUT,TAPE5=DATA,TAPE6=OUTPUT)
      DIMENSION H(100,100)

      WRITTEN BY ZUHAIR HAFI.  MODIFIED BY A. D. DAVIS.

      READ THE INPUT PARAMETERS
    1 READ(5,90,END=600) NI,NJ,NIMAX,DEMAX
      IF(NI.EQ.4.AND.NJ.EQ.6) GO TO 17
      IF(NI.EQ.7.AND.NJ.EQ.11) GO TO 18
      N=3
      GO TO 19
   17 N=1
      GO TO 19
   18 N=2
   19 JA=NJ-(2**N)
      WRITE(6,100) NI,NJ,NIMAX,DEMAX
      ESTABLISH INITIAL GUESSES FOR HEAD
      DO10 I=1,NI
      H(I,1)=100.0
      DO10 J=2,NJ
      IF(I.EQ.1.AND.J.EQ.JA) H(I,J)=0.0
      H(I,J)=20.0
   10 CONTINUE
      CALCULATE SUCCESSIVELY BETTER APPROXIMATIONS FOR HEAD AT ALL
      NODES, ITERATING UNTIL SATISFACTORY CONVERGENCE IS ACHIEVED.
      NITER=0
    8 DEL=0.0
      NITER=NITER+1
      DO20 I=1,NI
      DO20 J=2,NJ
      HH=H(I,J)
      IF(I.EQ.1.AND.J.EQ.JA) GO TO 2
      GO TO 3
    2 H(I,J)=0.0
      GO TO 40
    3 IF(I.EQ.1.AND.J.NE.NJ) GO TO 4
      GO TO 5
    4 H(I,J)=(H(I,J-1)+H(I,J+1)+2.0*H(I+1,J))/4.0
      GO TO 40
    5 IF(I.EQ.1.AND.J.EQ.NJ) GO TO 6
      GO TO 11
    6 H(I,J)=(2.0*H(I,J-1)+2.0*H(I+1,J))/4.0
      GO TO 40
   11 IF(I.NE.NI.AND.J.EQ.NJ) GO TO 12
      GO TO 13
   12 H(I,J)=(H(I-1,J)+H(I+1,J)+2.0*H(I,J-1))/4.0
      GO TO 40
   13 IF(I.EQ.NI.AND.J.NE.NJ) GO TO 14
      GO TO 15
   14 H(I,J)=(H(I,J-1)+H(I,J+1)+2.0*H(I-1,J))/4.0
      GO TO 40
   15 IF(I.EQ.NI.AND.J.EQ.NJ) GO TO 16
      H(I,J)=(H(I-1,J)+H(I+1,J)+H(I,J-1)+H(I,J+1))/4.0
      GO TO 40
   16 H(I,J)=(2.0*H(I,J-1)+2.0*H(I-1,J))/4.0
   40 DEL=DEL+ABS(H(I,J)-HH)
   20 CONTINUE
      STORE ITERATIONS IF COMPUTED VALUES SHOW LITTLE FURTHER CHANGE
      OR IF NUMBER OF ITERATIONS IS TOO LARGE.
      IF(DEL.LE.DEMAX) GO TO 7
      IF(NITER.LE.NIMAX) GO TO 8
      GO TO 9
      PRINT VALUES OF THE ITERATION COUNTER,
      NUMBER OF ITERATIONS, AND THE FINAL HEAD
    7 WRITE(6,200) NITER
      DO30 I=1,NI
      WRITE(6,300) (H(I,J),J=1,NJ)
```

Figure 7.36

```
30 CONTINUE
   GO TO 1
   COMMENT IN CASE NUMBER OF ITERATIONS EXCEEDS MAXIMUM
 9 WRITE(6,400)
   DO50 I=1,NI
   WRITE(6,700) (H(I,J),J=1,NJ)
50 CONTINUE
   GO TO 1
   FORMATS FOR INPUT AND OUTPUT STATEMENTS.
90 FORMAT(I2,I3,I4,F6.4)
100 FORMAT(10X,//36HHEAD DISTRIBUTION IN THE REGION WITH/10X,5H NI =,
  1I4/10X,5H NJ =,I4/10X,19HMAXIMUM ITERATION =,I5/10X,30HTOLERANCE T
  20 STOP ITERATIONS =,E12.2//)
200 FORMAT(9X,44HCONVERGENCE CONDITION HAS BEEN REACHED AFTER,I4,11H I
  1TERATIONS/10X,52HTHE VALUES OF HEAD AT DIFFERENT POINTS ARE GIVEN
  2BY //)
300 FORMAT(1H0,2X,22F6.1/)
400 FORMAT(10X,41HNO CONVERGENCE.  CURRENT VALUES OF H ARE://)
700 FORMAT(1H0,2X,22F5.1//)
600 STOP
    END
```

Figure 7.36 Computer program to solve example shown in Figure 7.35.

of rocks. For example, is a fault a ground water barrier boundary or recharge boundary?

In a sense elaborate mathematical models are a case of overkill, because in the real world the geology is very complex. Baski (1979) uses the following example: "... the data, aquifer and confining bed characteristics, and assumptions can easily vary by a factor of 2 to 10. This emphasizes the absurdity of many computer printouts which have answers with four or five significant figures and no reliability indicated. Does this mean an accuracy of plus or minus one part in ten thousand? No! It doesn't mean that. But it could be misleading." (The concept of engineering accuracy was perhaps better conveyed to engineering students in the first half of the nineteenth century, who solved problems with a slide rule.) Nevertheless, analytical models and computers are here to stay. This is a challenge to the engineering geologist. He must be armed with all the models that his discipline affords; but in the end, judgments must be made that require experience, a keen perception of geology, and plain common sense.

## Problems

**1.** Oahe Reservoir, located on the Missouri River in north-central South Dakota has an area of 1420 km$^2$.

**a.** What is the annual net loss (evaporation minus precipitation) from this reservoir (cm)?

**b.** The average discharge of the Missouri River just below the reservoir is 650 m$^3$/s. What percent of the Missouri River is lost due to net evaporation loss at Oahe Reservoir?

**2.** A plastic bubble tube is lowered to a depth of 30 m into an observation well. The bottom end of the tube is open and is in the water. Air is pumped into the tube, and

$5 \times 10^4$ N/m$^2$ air pressure is required to make air bubbles begin to come up in the well water. What is the depth to water?

**3.** A 5000-m-deep offshore oil well is drilled into a highly geopressured zone in the Gulf of Mexico. The well is capped, and develops $27 \times 10^6$ N/m$^2$ pressure at the well head. If the density of the brine fluids is 1.1 g/cm$^3$, to what elevation above the well head would the column of water rise if it were in an open-ended pipe?

**4.** Cascade Spring is in the southern Black Hills, SD, and has a constant discharge of 0.66 m$^3$/s. Recharge to the spring occurs in the limestone plateau near Jewel Cave, consisting of an area of 860 km$^2$ (Rahn and Gries, 1973). The average precipitation is 46 cm/yr. What is the average ground water recharge rate for the recharge area (cm/yr)?

**5.** See Fig. 7.14.
   **a.** Determine the discharge through the aquifer if the head loss between well 1 and 2 is 10 m.
   **b.** If $\alpha = 0.2$, how many days will it take for water to move from well 1 to well 2?

**6.** The map below (from Wilson and Gerhart, 1979) shows the potentiometric surface on the Florida aquifer in west-central Florida. If $T = 8000$ m$^2$/day, and $b = 400$ m, compute the rate of ground water discharge (m$^3$/day) into the Gulf of Mexico between Sarasota and a point 50 km to the south-southeast. (Hint: use either a flow net construction or an analytical analysis using Eq. (7.5). Consider only the area downgradient from the 16 m equipotential line.)

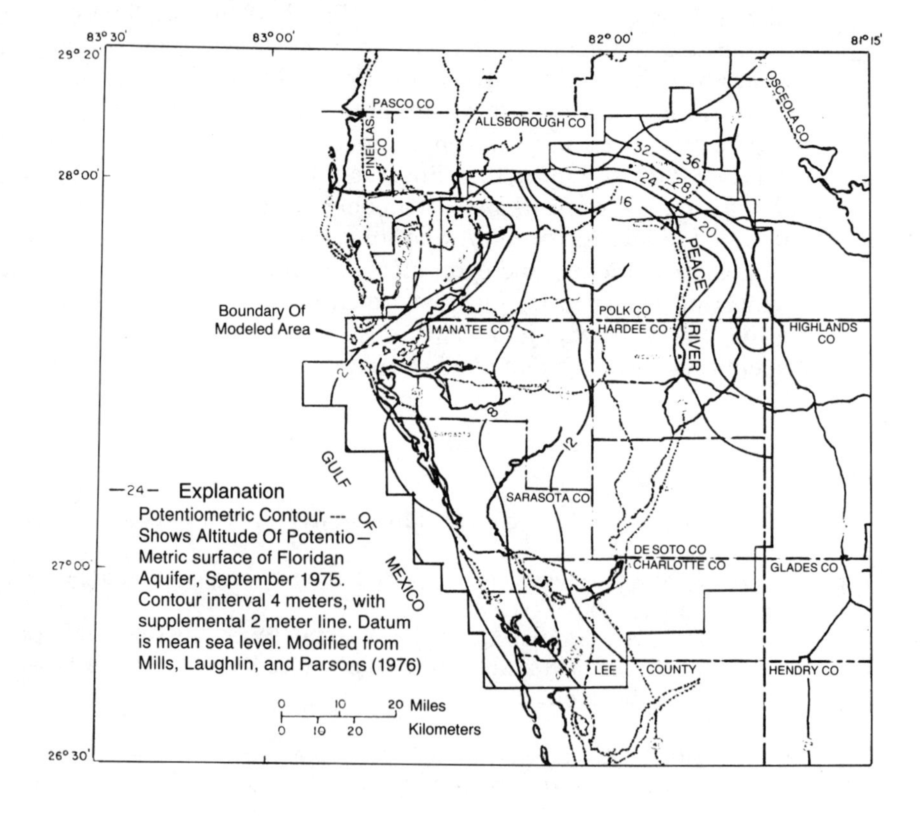

7. The hydraulic conductivity ($K$) of the 100 m long earth fill dam shown below is 0.1 m/day. Seepage lines and equipotential lines through the dam for two construction techniques (with and without a drain) are shown (after Attewell and Farmer, 1976).

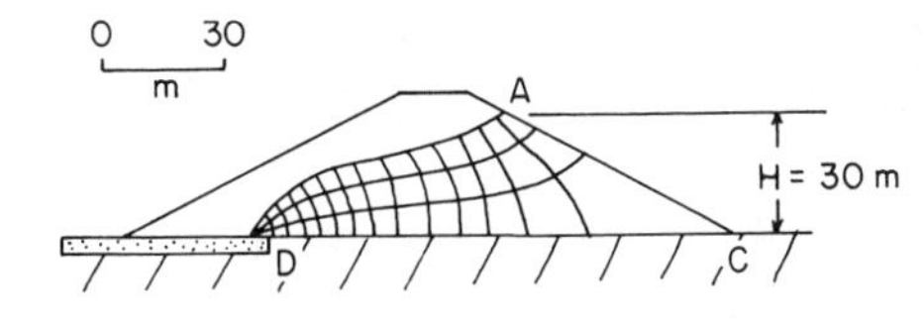

a) Earth dam with under-drainage

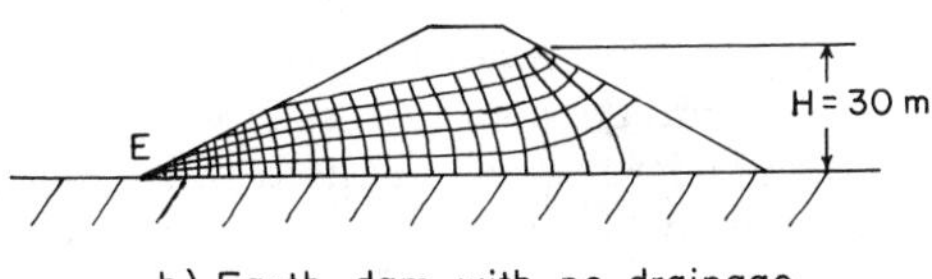

b) Earth dam with no drainage

**a.** Determine the seepage rate through the dam under both situations. Assume isotropic and homogeneous media. (Note: $AC$, the upstream face of the dam having an underdrain, is an equipotential line. Flow lines are drawn to intersect it at right angles. The base of the dam $CD$ is assumed to be relatively impermeable and therefore it constitutes a flow line.)

**b.** What conditions exists in the area $E$ for the dam having no drainage?

8. Use Fig. 7.21 to determine the seepage rate under the 40-m-long dam. Assume a uniform cross section head loss of 20 m, and anisotropic conditions exist where:

$$K_h = 30 \text{ m/day and } K_v = 3 \text{ m/day.}$$

9. Assume that an aquifer has $T = 200 \text{ m}^2\text{/day}$ and $S = 7.5 \times 10^{-4}$. Determine the drawdown 100 m away from a pumping well whose discharge is 1728 $\text{m}^3$/day. Solve for drawdown at 0.3, 3.0, and 30 days. (Hint: plot time-drawdown curve on semilog paper using Jacob equations (7.19) and (7.20). Then read drawdown off the curve for the three times given.)

10. Assume it is desirable to dewater a large sandy area for the foundation excavation of a large building. Assume that $T = 30 \text{ m}^2\text{/day}$ and $S = 0.01$. Dewatering wells are to be installed on a square grid. Each well is to pump at 100 $\text{m}^3$/day. Determine the proper well spacing so that the drawdown everywhere after 20 days is at least 10 m. (Hint: plot distance-drawdown curve on a semilog paper using Jacob equations (7.17) and (7.18). Assume that the drawdown in the center of a square array of wells is the algebraic sum of the drawdown from all four neighboring wells. Use trig identity for solution of drawdown in center of a square where the side of the square is the well spacing.)

11. The following observation well data is from a pump test in the Dakota Sandstone at Wall, SD, (from Gries et al., 1976). The aquifer is at 1 km depth. The observation well is 747 m from the pumped well, which is discharging at a constant 657 $\text{m}^3$/day.

**a.** Solve for $T$ and $S$.

**b.** Using $T$ and $S$, predict the drawdown 1 km away after 1 yr at a pumping rate of 5 l/s.

| Time (days) | Drawdown (m) |
| --- | --- |
| 0.052 | 0.058 |
| 0.063 | 0.12 |
| 0.073 | 0.19 |
| 0.090 | 0.28 |
| 0.139 | 0.55 |
| 0.215 | 1.01 |
| 0.347 | 1.59 |
| 0.623 | 2.56 |
| 0.972 | 3.66 |
| 1.805 | 4.58 |

**12.** Assume the saturated thickness of a leaky artesian aquifer is 18 m. A fully penetrating well is installed into the aquifer and pumped at 5500 $m^3$/d. During pumping, water leaks into the aquifer from an aquitard which has a saturated thickness of 9 m. The drawdowns given below were observed at a distance of 640 m from the production well (modified from Walton, 1970).
Compute the transmissivity ($T$), the hydraulic conductivity ($K$), and the storage coefficient ($S$) of the aquifer, and the coefficient of vertical conductivity ($K'$) for the aquitard.

| t<br>Time after pumping started (days) | s<br>Drawdown (m) |
| --- | --- |
| 0.035 | 0.006 |
| 0.042 | 0.009 |
| 0.049 | 0.012 |
| 0.056 | 0.018 |
| 0.063 | 0.021 |
| 0.069 | 0.027 |
| 0.139 | 0.070 |
| 0.208 | 0.110 |
| 0.278 | 0.131 |
| 0.347 | 0.152 |
| 0.417 | 0.171 |
| 0.486 | 0.183 |
| 0.556 | 0.198 |
| 0.625 | 0.207 |
| 0.694 | 0.213 |
| 1.000 | 0.229 |
| 2.000 | 0.230 |
| 3.000 | 0.250 |

**13.** (From Bouwer, 1978). A tile-drainage system is to be installed to reduce high water table conditions. The tile diameter is 12 cm, and the drains will be installed at a depth of 1.5 m. The hydraulic conductivity of the soil is 0.7 m/day with an impermeable layer at 4.5 depth. Irrigation is 0.7 cm/day. What is the spacing of the drains so that the water table is no less than 0.5 m below the land surface?

# 8

# Fluvial Processes

---

These dry details . . . give me an opportunity of introducing one of the Mississippi's oldest peculiarities—that of shortening its length from time to time. If you will throw a long, pliant apple paring over your shoulder, it will pretty fairly shape itself into an average section of the Mississippi River, that is the nine or ten hundred miles stretching from Cairo, Ill., southward to New Orleans, the same being wonderfully crooked, with a brief straight bit here and there at wide intervals.

At some forgotten time in the past, cutoffs were made. . . . These shortened the river, in the aggregate, seventy-seven miles.

Since my own day on the Mississippi, cutoffs have been at Walnut Bend, and at Council Bend. These shortened the river, in the aggregate, sixty-seven miles. . . .

Therefore, the Mississippi between Cairo and new Orleans was twelve hundred and fifteen miles long one hundred and seventy-six years ago. It was eleven hundred and eighty after the cutoff of 1722. . . . Consequently its length is only nine hundred and seventy-three miles at present.

Now, if I wanted to be one of these ponderous scientific people and "let on" to prove what had occurred in the remote past by what has occurred in a given time in the recent past, or what will occur in the far future by what has occurred in late years, what an opportunity is here! Geology never had such a chance, nor such exact data to argue from! Please observe:

In the space of one hundred and seventy-six years the Lower Mississippi has shortened itself two hundred and forty-two miles. That is an average of a trifle over one mile and a third per year. Therefore, any calm person, who is not blind, or idiotic, can see that in the Old Oolitic-Silurian Period, just a million years ago next November, the Lower Mississippi River was upward of one million three hundred thousand miles long and stuck out over the Gulf of Mexico like a fishing rod. And by the same token any person can see that seven hundred and forty-two years from now the Lower Mississippi will be only a mile and three-quarters long, and Cairo and New Orleans will have joined their streets together and be plodding comfortably along under a single mayor and a mutual board of aldermen. There is something fascinating about science. One gets such wholesale returns of conjecture out of such a trifling investment of fact.

Mark Twain

---

## 8.1 Fluvial Geomorphology

The science of fluvial geomorphology contains many relevant applications to engineering, Mark Twain's spoof notwithstanding.

### *8.1.1 Ephemeral and Perennial Streams*

Stream channels in arid regions rarely contain water flowing in them all the time. Streams which only flow during floods or wet seasons are called ephemeral streams. The Gila River below Phoenix, AZ, for example, has a drainage area of

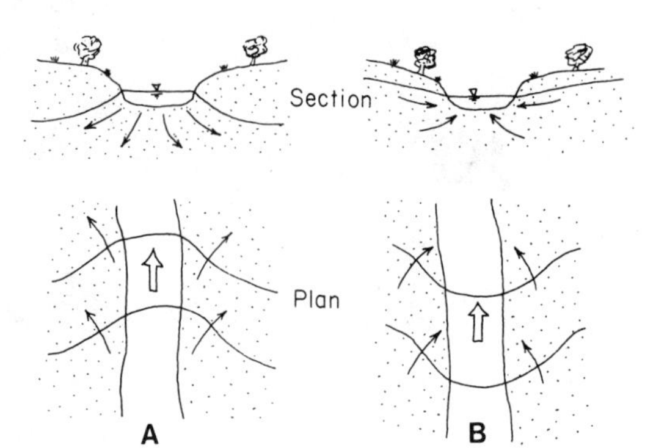


Figure 8.1 Water table contours as related to ground water flow directions (after Todd, 1980). **A**. Influent (losing) stream. **B**. Effluent (gaining) stream.

150,000 km$^2$, and yet flows so infrequently that major highway crossings do not have bridges, but rather are paved right across the bed of the channel. In humid regions most streams flow all the time. They are called perennial or live streams. Some streams have perennial and ephemeral reaches.

Where a stream gains water from ground water, it is called a gaining or effluent stream; where it loses water to ground water, it is called a losing or influent stream. Figure 8.1 illustrates the cross-sectional and plan view of the relationship between ground water, and influent or effluent stream conditions. Bedrock constrictions may cause a decrease in the cross-sectional area of alluvium and hence decrease in ground-water discharge beneath a stream; in this way an ephemeral stream may become perennial (Fig. 8.2). Water may sink into the ground where more permeable rocks are encountered, such as cavernous limestone or wide alluvially filled valleys. In semiarid regions, some streams may be perennial in the mountains but become ephemeral where they debouch onto alluvial fans in the arid valleys.

A stream hydrograph (Fig. 8.3C) is a plot of discharge versus time. For a perennial stream, there are two components of discharge, 1) that part which is due to overland runoff immediately following rain or melting snow and 2) base flow, that

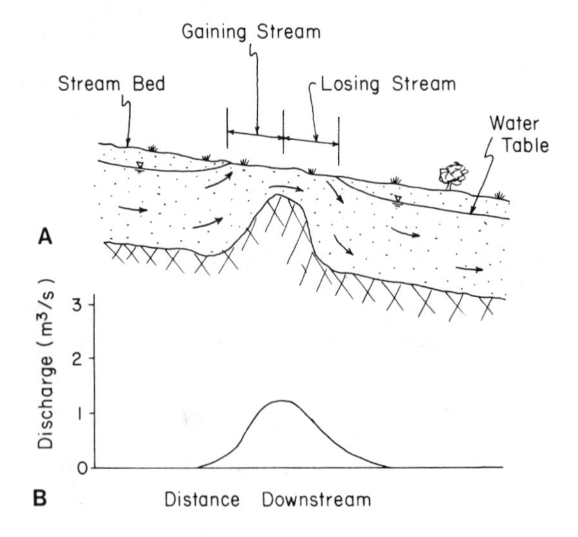


Figure 8.2 Illustration showing ground water/surface water relationships during low stream discharge (after Todd, 1980). **A**. Cross section along a stream channel in an alluvial valley. **B**. Discharge plotted as a function of distance along the stream channel.

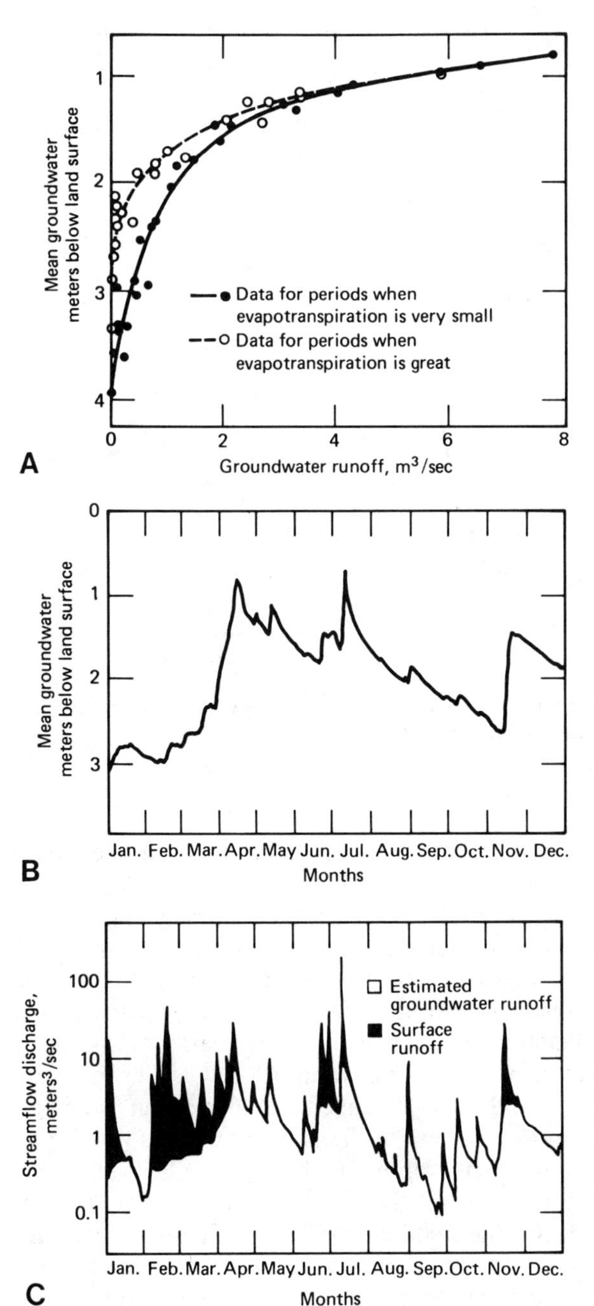


Figure 8.3 Base flow for Panther Creek near Peoria, IL (drainage area: 246 km$^2$). **A.** Rating curves on mean ground water stage versus ground water runoff (base flow). **B.** Mean ground water stage for 1951. **C.** Stream-flow hydrograph for 1951 showing surface runoff and base flow components. (From Todd, 1980.)

part which is due to the contribution from ground water. Streams with a large base flow typically occur in humid areas and are underlain by permeable rocks. Figure 8.3 illustrates how the overland runoff and base flow of a stream in Illinois varies during different months of the year. During the spring, the water table is higher and the base flow is greater.

Alluvial valleys are capable of temporarily absorbing flood water. This water is slowly released after a flood subsides. The resulting bank storage causes a

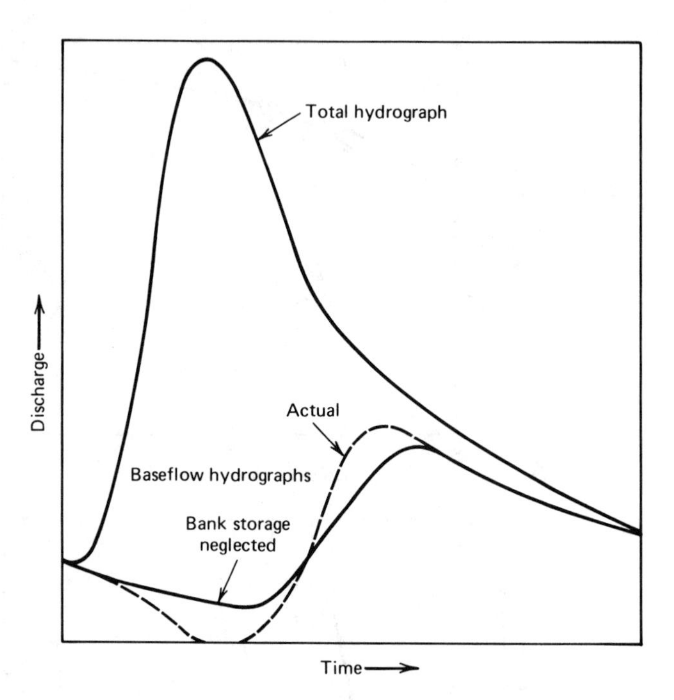


Figure 8.4 Schematic diagram of the variation of base flow during a flood hydrograph with and without effects of bank storage (from Todd, 1980).

skewed hydrograph (Fig. 8.4), in that water slowly drains from the alluvium, prolonging the high discharge.

Stream hydrographs are affected by the size of the drainage basin, amount and intensity of precipitation, topography, geology, and land use. Figure 8.5 illustrates how the peak discharge of streams is affected by the size of a drainage basin. The three hydrographs show the progression of the flood crest recorded at several gaging stations in the Susquehanna River basin, Pennsylvania, following the heavy rain accompanying Hurricane Agnes on June 19–21, 1972. In the tributary stream, Bald Eagle Creek, the discharge peaked at 143 $m^3/s$ on the night of June 22, indicating relatively quick response to the precipitation. In the Juniata River, a major tributary, the flood crested about 12 hr later at 3540 $m^3/s$. Far downstream on the Susquehanna River, the flood peaked at 30,500 $m^3/s$ another 12 hr after the Juniata River peak. The flood caused over a $1 billion damage at Harrisburg and other Pennsylvania communities. Because of the lag time at Harrisburg, there was adequate warning time, and there were few deaths.

## *8.1.2 Stream Dynamics*

Streamflow can be divided into regimes based on velocity. For slow-moving streams, the Reynold's number ($N_R$) is an appropriate velocity parameter, and is defined as:

$$N_R = \frac{VR}{\nu}, \tag{8.1}$$

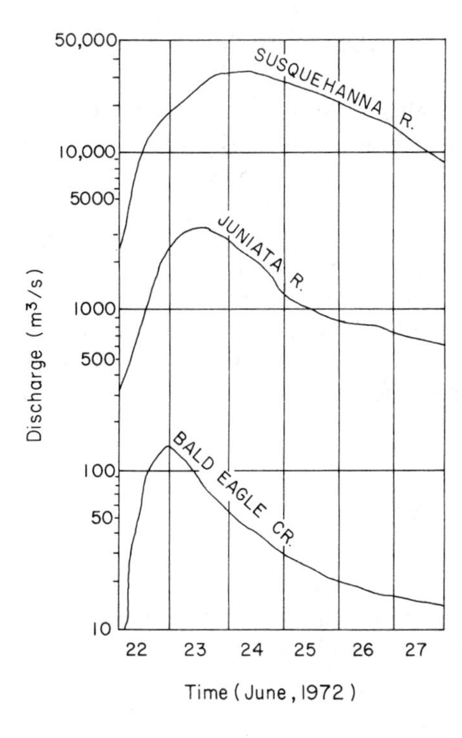


Figure 8.5 Hydrographs of streams in the Susquehanna River basin, PA, as affected by the June 1972 floods (after Ritter, 1978). Bald Eagle Creek at Tyrone has a 115 $km^2$ drainage basin area and had a peak discharge of 143 $m^3/s$. The Juniata river at Mapleton has a 5280 $km^2$ drainage basin area and had a peak discharge of 3540 $m^3/s$. The Susquehanna River at Mariatta has a 67,600 $km^2$ drainage basin area and had a peak discharge of 30,500 $m^3/s$.

where:

$V$ = average velocity (m/s),

$R$ = hydraulic radius (= cross-sectional area divided by the distance along the wetted perimeter), nearly equal to average depth ($D$),

$\nu$ = kinematic viscosity ($1.4 \times 10^{-6}$ $m^2/s$).

If $N_R$ is greater than approximately 2000, turbulent flow exists. In this case, water flow paths include some upward moving vectors, thus allowing for transport of suspended load. If $N_R$ is less than approximately 2000, laminar flow exists. In this case water moves in parallel layers and is incapable of transporting suspended load. Practically all rivers exhibit turbulent flow.

For higher flow regimes, the Froude number ($N_F$) is defined as:

$$N_F = V(gD)^{-0.5} \tag{8.2}$$

where:

$D$ = average depth,

$g$ = gravitation acceleration (9.8 $m/s^2$).

If Froude's number is less than 1, the regime is called subcritical or (tranquil) flow, and streamflow greater than a Froude's number of 1 is called supercritical (or rapid or jet) flow. Where a stream at supercritical flow suddenly becomes subcritical, a standing wave, called a hydraulic jump, is formed. Stream bed patterns in sandy channels change in accordance with the Froude number. Normal dunes, which migrate downgradient, occur for subcritical flow, but violent antidunes whose forms move upgradient, occur in supercritical flow (Simons and Richardson, 1963).

In the past several decades American geomorphologists have studied the variables relating stream geometry relative to discharge and velocity. The basic relationships between the hydraulic geometry of streams were first documented by Leopold and Maddock (1953) and Leopold and Miller (1956). Later work, primarily by U.S. Geological Survey geomorphologists, is summarized by Leopold et al., (1964). For the purpose of comparing flows at one station or comparing one stream to another, discharge ($Q$) serves as the independent variable at any station, and changes in average width ($w$), depth ($d$), and velocity ($v$) are the dependent variables. At any given location along a stream, the width, depth, and velocity were observed by numerous field measurements to increase as a power function so that:

$$w = aQ^b, \tag{8.3}$$

$$d = cQ^f, \tag{8.4}$$

$$v = kQ^m, \tag{8.5}$$

where $a$, $c$, $k$, $b$, $f$, and $m$ are constants for the particular stream. Because $Q$ equals the product of velocity times width times depth, then:

$$\begin{aligned} Q &= (aQ^b)(cQ^f)(kQ^m), \\ &= ackQ^{b+f+m}. \end{aligned}$$

It follows that both $(a \cdot c \cdot k)$ and $b + f + m)$ must be equal to 1.

Leopold and Maddock (1953) found that the average at-a-station values of $b$, $f$, and $m$ for western U.S. streams were 0.26, 0.40, and 0.34, respectively. The values of $b$, $f$, and $m$ are 0.5, 0.4, and 0.1, respectively, for downstream changes along any stream.

The value of $Q$ used in these relationships is the average discharge, which is derived from stream gaging data collected over a long period of time. From inspection of Eq. (8.5), stream velocity increases downgradient. This is contrary to casual observations where one is used to thinking of a rushing mountain stream which seems to slow down as its gradient flattens as it approaches the sea. Actually the average velocity of the Mississippi River at New Orleans is faster than mountain streams in the headwaters (Leopold, 1953).

### *8.1.3 Scour*

A stream transports clastic debris as suspended load and bed load. Leopold and Maddock (1953) found that the stream's suspended load-carrying capacity ($L$) is exponentially related to the discharge ($Q$), as follows:

$$L = jQ^n, \tag{8.6}$$

where:

$L$ = suspended load transport rate at a given stream station,

$Q$ = discharge,

$j$ = a constant,

$n$ = exponential constant, about 2.2–2.5.

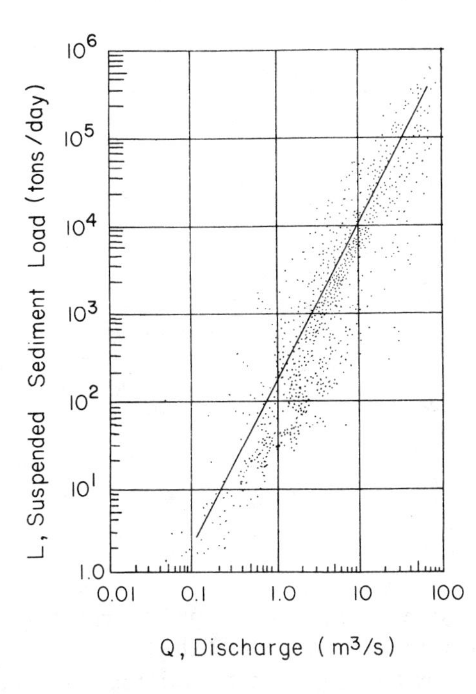


Figure 8.6 Relationship between suspended load to discharge in Powder River at Arvada, WY (from Leopold and Maddock, 1953).

Figure 8.6 illustrates the relationship between suspended load and discharge for the Powder River, WY. The wide scatter in the plot is due to several factors. Among the more important is that when a river stage is rising (waxing) during a flood, it carries more sediment due to rainwash and gulley erosion by small tributaries than it does during the waning period of a flood. It is during the waning period of flooding that much scouring at bridges or other structures along the main channel occurs because the sediment/discharge ratio is relatively low.

Most of the geomorphological work performed by a stream takes place during floods. During a flood the stream channel is eroded (or scoured) downward as well as laterally. Scour is rarely directly observed because an observer sees only the muddy flood water carrying trees and other floating debris. During a flood, clastic debris such as sand, gravel, and boulders are moved along the channel bottom. As the flood subsides, the stream bed aggrades back to its original elevation. Figure 8.7 shows the profile of the San Juan River before, during, and after a flood.

Scour is increased by an obstruction. After a flood has subsided, inspection of the channel shows that almost any obstruction such as a large boulder or bridge pier typically produces a scour hole around the obstruction. Local scour around an obstruction, such as a pier or abutment, is caused by vortex flow system such as a hydraulic jump initiated by the contraction in flow in that region. Theoretical analysis of water movement is unable to predict the magnitude of this scour (Henderson, 1966; Jansen et al., 1979). Posey (1974) established a rule of thumb for fluvial scour near an obstruction: for every meter of river rise, there are 3 m of river scour. Practically speaking, if exploratory test drilling for a bridge foundation reveals a thin veneer of alluvium over bedrock, it should be assumed that the alluvium is periodically transported; hence the bridge foundation should be firmly founded into the bedrock below the alluvium (Fig. 8.8.).

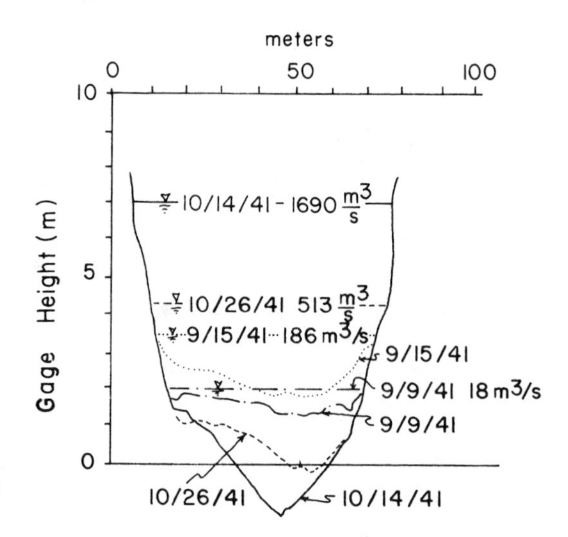


Figure 8.7 Changes in the cross section of the San Juan River during a flood (after Leopold et al., 1964).

Terzaghi and Peck (1948) describe the scour of the Colorado River near Yuma, AZ, where scour of about 11 m has been known to accompany a river rise of only 4 m. The base of an old bridge pier at Yuma was established at a depth of 3 m below the bottom of the river channel because at that depth the river bed contained large boulders. The boulders were so tightly wedged that further excavation was impossible without blasting, thus the base of the pier was established at that depth. Yet, the first high water after construction caused the pier to fail.

Rahn et al. (1981) reported that 4 m of scour occurred beneath a bridge abutment across a small ephemeral stream in Rapid City, SD. This was evidenced by a pickup truck buried in gravel at that depth below a concrete bridge abutment during a 1972 flood.

Figure 8.8 Scour on Boxelder Creek, South Dakota, caused the failure of the westbound Interstate-90 bridge structure during the Rapid City 1972 Flood. A waterfall created by debris blocking the upstream bridge (not shown) scoured alluvium below the concrete abutments of this downstream bridge.

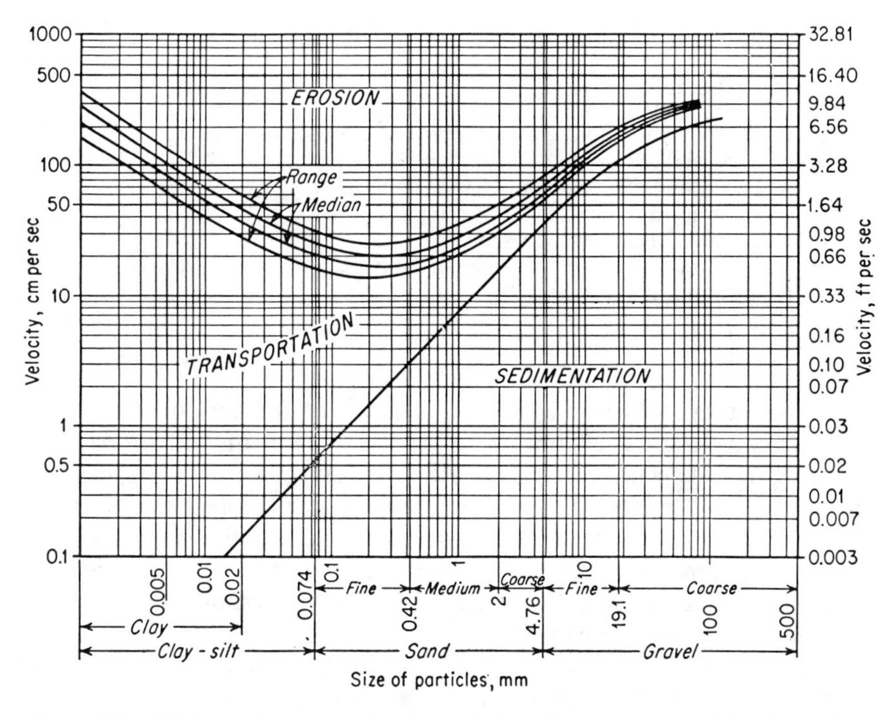


Figure 8.9 Hjulstrom's curves showing relationship between particle size and water velocity required for erosion, transportation, and deposition (from Krynine and Judd, 1957).

## *8.1.4 Sediment Transport*

The faster a stream flows, the more capable it is to carry large-sized debris. Thus, stream competency is related to velocity. In the early part of this century it was believed that the weight of the largest particle moved by a stream varied as the sixth power of the stream velocity (Rubey, 1938). Hjulstrom (1939) developed empirical curves relating these variables (Fig. 8.9), which is a logical extension of Stoke's Law and the Impact Law which govern the settling velocity of particles. (See Gibbs et al. (1971), Jansen et al. (1979), or Komar (1981) and references contained therein for empirical formulas governing settling velocities.) From Fig. 8.9, for example, a stream velocity of 50 cm/s would erode a 0.02-mm particle found in its bed. If the velocity dropped to 20 cm/s, it would not erode particles of this size, but would continue to transport them if they were already in motion. If the velocity dropped below 0.15 cm/s, the particles would settle out. As shown in Fig. 8.9, there is an exception from the general trend (large velocities transport large particles) with respect to the competence of very fine-grained soils having coherence such as clay.

The physics governing the movement of debris by a stream is quite complex. The rate of change of velocity (shear) is a very important parameter. Consider two streams: the Mississippi River at New Orleans and Rapid Creek at Rapid City. Assume both have the same average velocity, say 10 km/hr. The Mississippi River, with an enormous discharge, is not competent to transport anything larger than sand, whereas Rapid Creek has a much lower discharge, and yet is capable

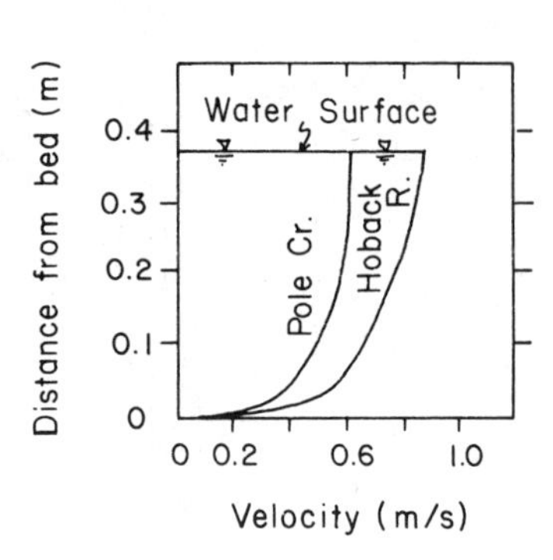


Figure 8.10 Current-meter measurements of velocity in two streams, plotted as functions of depth (after Leopold et al., 1964). The two streams have the same depth but differ in mean size of bed material ($D_{50}$) and in slope ($S$). Pole Creek, near Pinedale, WY, has $D_{50}$ = 0.13 m and $S$ = 0.003, whereas the Hoback River, near Bondurant, WY, has $D_{50}$ = 0.13 m and S = 0.005.

of transporting cobbles and boulders. The reason for this apparent contradiction is that the velocity a few millimeters above the stream bed is much greater for Rapid Creek than the Mississippi River. In other words, the shearing stress of Rapid Creek is greater, and allows for its greater competency. Recognizing the importance of rate of change of velocity to sediment transport led to the development of numerous stream competency formulas including the idea of critical tractive force (Bruun and Lackey, 1962). Figure 8.10 shows contrasting shear situations in Wyoming streams.

There is much uncertainty governing the hydraulic conditions necessary to entrain a given particle size. There are many complex analytical techniques used to predict the mobility of a stream bed in terms of hydraulic variables (Chow, 1959; Bridge, 1981; Strand and Pemberton, 1982; Tharp, 1983). Interrelated parameters such as "sorting" have been found to affect stream competency (Andrews, 1983). One simple model for sediments greater than 1 cm diameter assumes that the shear stress ($\tau$) exerted by moving water on a stream bed can be expressed as the tractive force per unit area, or:

$$\tau = \gamma g R S, \tag{8.7}$$

where $\gamma$ is the unit weight ($10^3$ kg/m$^3$ for water), $g$ is gravitational acceleration (9.8 m/s$^2$), $R$ is the hydraulic radius (cross-sectional area divided by wetted perimeter), and $S$ is the energy slope of the river. For coarse sediments (greater than 1 cm), Costa and Baker (1981) show that the critical shear stress necessary to entrain sediments ($\tau_{cr}$) is related to competency as:

$$D = 65\left(\frac{\tau_{cr}}{g}\right)^{0.54}, \tag{8.8}$$

where $D$ is the particle size (mm) and $\tau_{cr}$ has units of N/m$^2$.

As an example, suppose the cross-sectional area and the wetted perimeter of a stream are 200 m$^2$ and 100 m, respectively, yielding a hydraulic radius of 2. The slope is 0.002. From Eq. (8.7), the shear stress is 39 N/m$^2$. Using Eq. (8.8), the maximum particle capable of being moved is 137 mm.

Stream competency studies have practical value. Costa (1983) determined the theoretical paleohydraulic flash-flood velocity and discharge for Colorado Mountain streams. For example, he found ~2 m diameter boulders in Holocene alluvium excavated for the Justice Center in Boulder, CO. This size corresponds to a velocity of 7.6 m/s, which in turn corresponds to a discharge of 1,500 m$^3$/s at this site, equivalent to about five times the historic peak discharge of Boulder Creek.

Sediment transport in alluvial channels is important to engineers engaged in problems dealing with river control for navigation. In the case of canal construction for irrigation, it is important that the canal bed be neither eroded nor subject to sediment deposition by sediment-laden water (Lane, 1955).

For more information on the complex subject of fluvial geomorphology and sediment transport, see Gilbert (1914), Leliavsky (1955), Leopold et al. (1964), Graf (1971), Schumm (1961, 1977), Hooke (1979), Jansen et al. (1979), and Smith and Smith (1984).

## *8.1.5 Meanders*

The Mississippi is one of the world's largest meandering rivers. While it is possible to qualitatively describe the physiography of a meandering stream, data and theories relating to the physics of fluid flow (quantitative geomorphology) are not universally accepted (Leopold, 1960). In many cases, hydraulic engineers have been fooled completely. A case in point is the Rhine River in Germany and its main distributary branch, the Waal River in the Netherlands. Complexes of weirs, groins, revetments, and other channelization measures have been built in the past century to provide for navigation in selected areas, only to have the river deposit sediments in unexpected areas and thus provide continual problems (Verstappen, 1983). Engineers who tamper with a river often have a tiger by the tail.

Some hazards associated with a naturally meandering stream are predictable from basic geologic principles. In a small meandering stream (Fig. 8.11), the outer portion (undercut bank) of a meander is subject to continual erosion whereas the

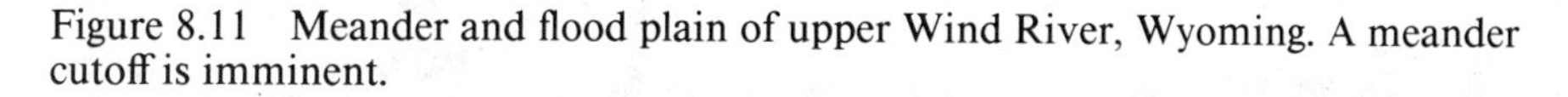

Figure 8.11 Meander and flood plain of upper Wind River, Wyoming. A meander cutoff is imminent.

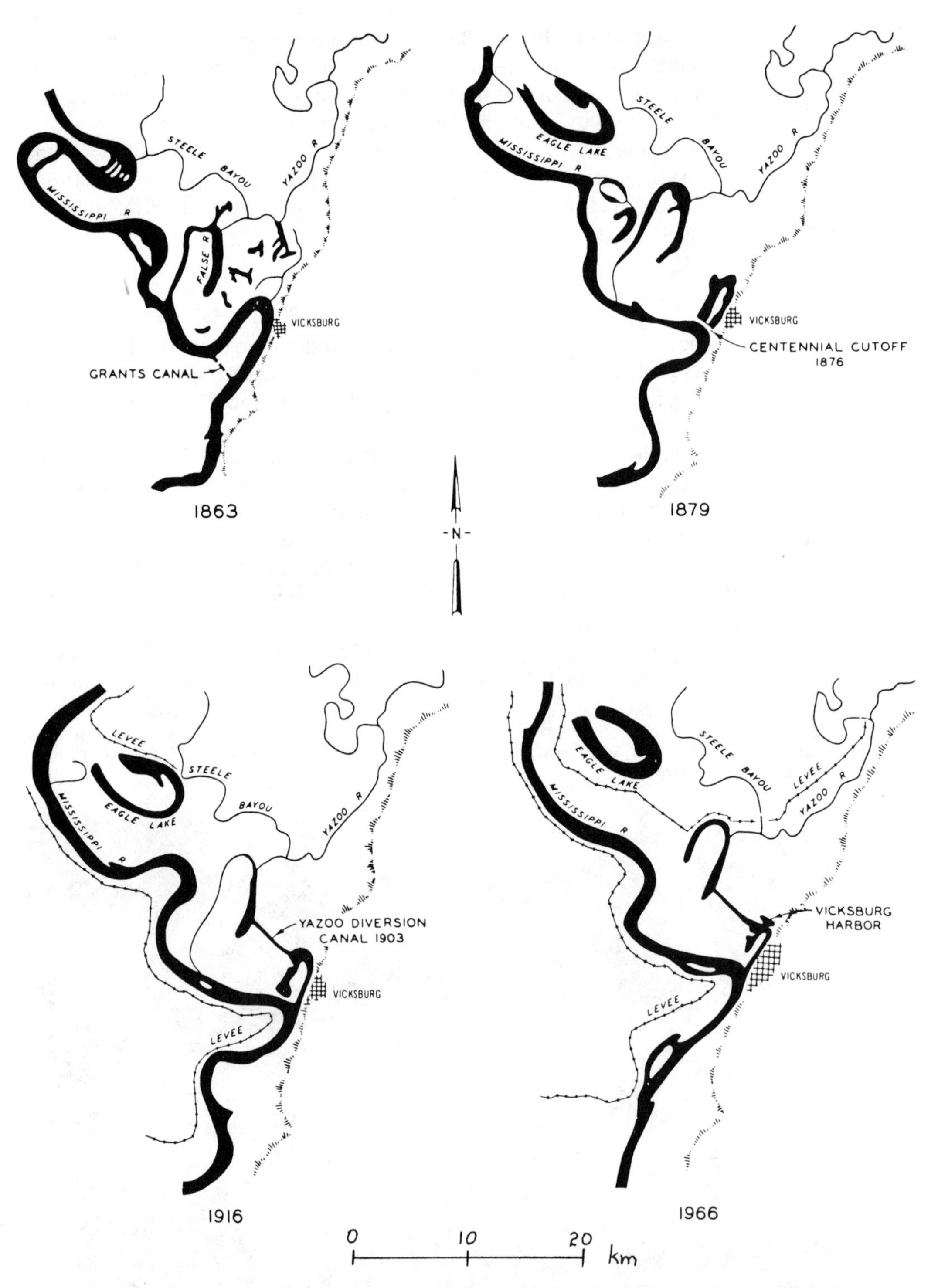


Figure 8.12 Major changes in the Mississippi River in the Lower Yazoo Basin between 1863 and 1966 (from Kolb and Steinreide, 1967).

inner portion (point bar) is gradually expanded by accumulation of bedload debris. It is amazing how often man foolishly builds houses or other structures on the edge of an undercut meander bluff.

The same principle of meander erosion and deposition also holds true with larger meandering rivers. Consider the Mississippi River, the largest river in North America. In the United States, the Corps of Engineers is responsible for the construction and operation of dams, levees, and locks on navigable rivers such as the Mississippi. In the 1940s, large scale physical models of the Mississippi River were constructed at Vicksburg, MS, to predict the kinds of artificial meander cutoffs, walls, etc., necessary to maintain navigable conditions on the River. Figure 8.12 shows the natural and man-made changes in the Mississippi River between the years 1863 and 1966. Managing this river presents a real challenge and financial investment. After the construction of riparian industrial facilities, locks, and canals, there is considerable incentive to keep the river in the same place, and the Corps of Engineers is engaged in this effort.

As a stream's morphology is constantly changing, viewed in the context of geologic time, the challenge to man is essentially a no-win situation because change is part of a river's behavior. However, the expenditure of large sums of money for dams and fluvial structures in many places on the Mississippi River has, to date, produced satisfactory results for man's immediate interests.

### *8.1.6 A Graded Stream*

One obstacle to modeling fluvial systems is that many fluvial parameters are interdependent. This means there is no clear-cut dependent or independent relationship. In general, precipitation and size of the drainage basin are independent stream variables, whereas channel width, depth, and meander characteristics are dependent stream variables. Longitudinal profile (slope) of the channel, velocity, and size of debris carried appear to be interdependent. An example of this interdependency is slope and debris size. Coarse debris (e.g., cobbles and boulders) in a stream bed seem to cause a steepening of the stream gradient, be it in the longitudinal profile of a stream (Hack, 1957), or the local water-surface topography in riffle and pool sequences (Prestegaard, 1983). Where coarse debris enters a stream bed (example: the Platte River emptying sand into the Missouri River) a steepening of the gradient results, along with increase in the velocity of the Missouri River. Rapids in the Colorado River in the Grand Canyon occur where tributary canyons have deposited boulders in the main channel. Thus, sediment size appears to control slope. Yet it can also be observed that where the gradient of a stream is low (example: the Mississippi River in Louisiana), the stream is not competent to carry sand or larger-sized debris. In this latter case the slope appears to control the sediment size, not vice versa.

In some respects a river is analogous to a living thing. It reacts to pressures or changes in conditions imposed upon it. Mackin (1948) provides a succinct statement about the dynamic balance of stream variables:

> ... A graded stream is one in which, over a period of years, slope is delicately adjusted to provide, with available discharge and with prevailing channel characteristics, just the velocity required for transportation of the load supplied from the drainage basin. The graded stream is a system in equilibrium; its diagnostic charac-

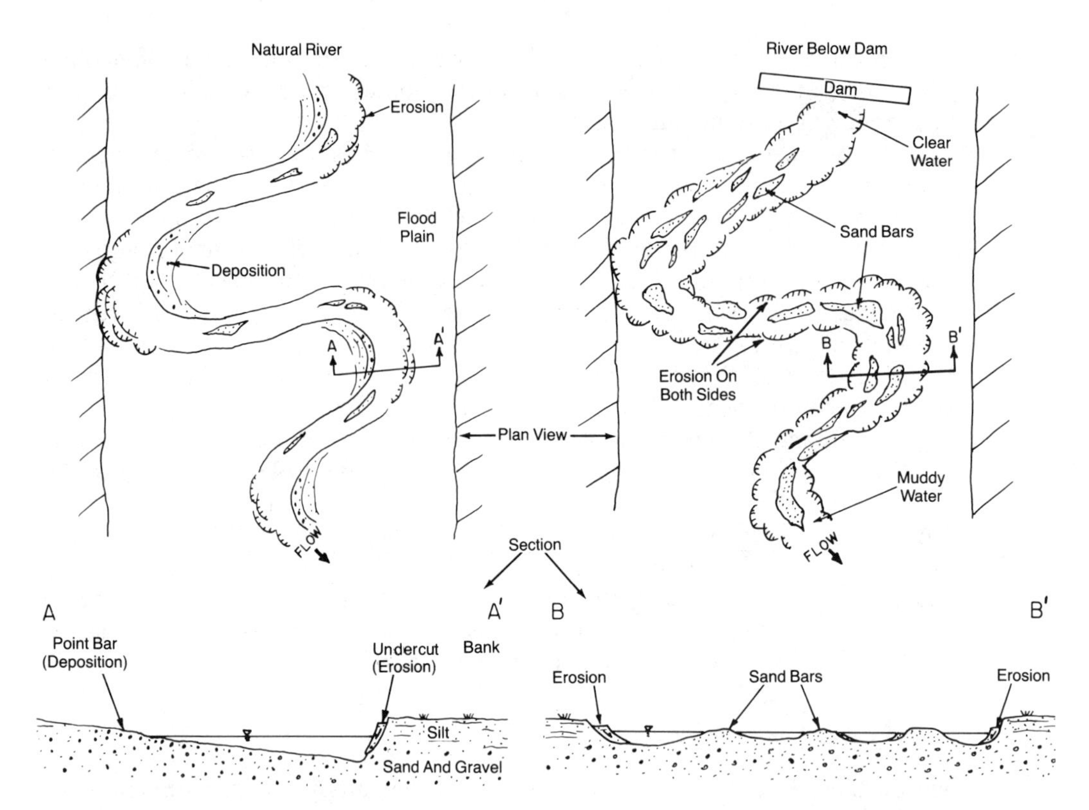


Figure 8.13 Idealized sketch map and cross section showing changes in fluvial geometry caused by a dam (from Rahn, 1977).

teristic is that any change in any of the controlling factors will cause a displacement of the equilibrium in a direction that will tend to absorb the effect of the change.

Although Mackin's qualitative model does not take into account all stream parameters such as changes in width, depth, roughness, and channel and meander morphology, it still provides the framework from which some useful general predictions can be made. For instance, consider the effect of construction of a dam on the capacity of a stream which has attained a graded profile. Sediment is deposited in a delta in the upstream reservoir. The water flowing in the reach below the dam has very approximately the same velocity as before, but because the water below the dam lacks sediment, the river has an increased capacity to carry debris. Therefore, it begins to erode the channel below the dam in an effort to achieve its former equilibrium. This may cause excessive erosion downstream or scouring around downstream bridges (Lane, 1955, 1956; Schumm, 1971; Williams and Wolman, 1984). Along the Missouri River, for example, floods no longer occur immediately below the six large main stem dams completed in the 1960s. This allows for urbanization of the flood plain downstream, but at the expense of land lost by reservoir inundation, and, more recently, land lost by erosion downstream. About 2000 hectares (ha) of valuable cropland have been eroded in the 88-km reach below Gavin's Point Dam on the Missouri River, South Dakota, between 1973 and 1979. (Some new land in the form of worthless sand bars has accumulated in the channel because high velocity floods no longer occur, and coarse debris such as sand and gravel is left behind, resulting in the formation of sandy islands and a shallow, braided stream (Fig. 8.13). Large discharges, absent since dam construction, are no longer present to provide adequate stream competency to remove the coarse debris.) The Missouri River, once capable of navigation by large steam-powered paddle-boats, has been changed into an ever-widening, shallow channel full of sand bars so that navigation by motorboats is tenuous.

Another example of erosion below a dam is the Nile River in Egypt. Since the Aswan Dam has been built, sediment no longer discharges through the lower reach of the Nile and into the Mediterranean Sea. As a result there has been increased erosion around bridges and small barrier dams downstream, and erosion of coastal areas which are now deficient of sand. Similar effects have been observed on the Gulf coast due to dams on the Brazos River, TX (Mathewson and Minter, 1981).

## 8.2 Stream Discharge Characteristics

### *8.2.1 Stream Gaging Techniques*

From fluid mechanics, the law of continuity states that fluid discharge ($Q$) in a pipe or open channel is equal to the average velocity ($V$) times the cross-sectional area ($A$):

$$Q = VA, \tag{8.9}$$

where:

$Q = \mathrm{m^3/s}$,

$V = \mathrm{m/s}$,

$A = \mathrm{m^2}$.

A crude way to measure stream velocity is by the float method. The velocity is obtained by measuring the velocity of tracers or partially submerged styrofoam balls. This method usually yields values of discharge in excess of the true discharge because surface velocity is slightly greater than the average velocity (Fig. 8.14). For this reason some floats have chains, rocks, or submerged devices attached to allow for a better estimate of velocity (Jansen et al., 1979).

Stream discharge can be accurately measured by a weir, a man-made constriction which creates a small waterfall in a stream channel. The depth of flow (stage) through the weir is related to discharge by an empirical formula. Weirs may be V-shaped, rectangular, or a combination of shapes. Published tables showing stage versus discharge exist for conventionally shaped weirs (Streeter and Wylie, 1979).

Discharge may be measured by a Parshall flume, a device placed in a channel which, because of a wasp-shaped design, causes supercritical flow to occur (Leopold et al., 1964). The water depth in the constricted area is related to discharge as published in empirical tables.

Current meters may be used to obtain stream discharge without the aid of constricting structures. The conventional USGS method is to select a stream length

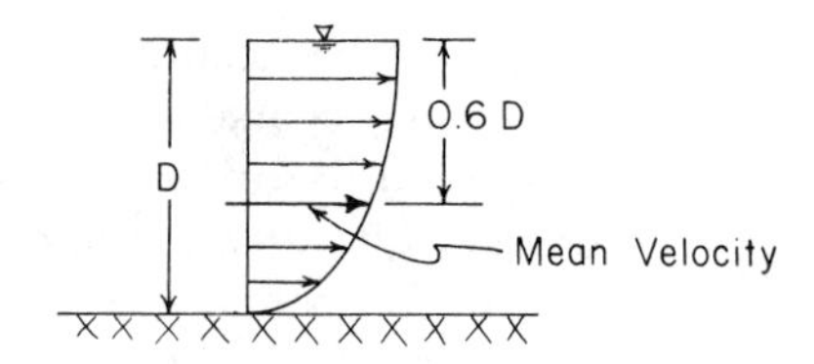


Figure 8.14 Relationship of position of mean velocity to the distribution of velocity with depth in a river.

(reach) where uniform currents occur, such as at the lower end of a large pool. The cross section of the stream is divided into 20 equal increments and a current meter (typically a small anometer or propeller device) is used to determine the velocity. The velocity is measured at 0.6 times the depth below the stream surface, where the mean velocity occurs (Fig. 8.14). In large streams the average of 0.2 D and 0.8 D is used instead of 0.6 D. The velocity at each of the 20 stations is multiplied by the depth at that station and the incremental width; the sum of the 20 values determines the total stream discharge.

For permanent recording of stream discharge, it is necessary to build a structure to record the stage of the water, typically by using a floating bulb connected mechanically to a clock-driven pen-and-ink recorder, or a pressure-sensitive manometer device. A formula is established relating stage to discharge by periodic measurement with a current meter as described above. The U.S. Geological Survey is the agency responsible for stream gaging in the United States, and maintains about 10,000 stream gaging stations which operate continuously. Stream discharge for a typical USGS stream-gaging station is calculated by a digital computer system where the stage is measured approximately every 15 min; the data are published in USGS Water-Supply Papers as average daily or average monthly discharge. Table 8.1 shows average monthly discharge data for two streams in the Black Hills, SD.

Velocity can be calculated indirectly. In order to determine the peak discharge of a flood after the high water crest has passed, for example, the slope-area technique is used by the USGS, where the velocity is established by the Chezy-Manning equation:

$$V = \frac{1}{n} R^{2/3} S^{1/2}, \tag{8.10}$$

where:

$n$ = roughness of stream bed (ranges from about 0.013 for concrete to 0.02 for corregated metal pipe or smooth channels to 0.06 for rough channels),

$R$ = hydraulic radius (m) equal to the cross-sectional area ($A$) divided by the wetted perimeter ($P$) of the cross-sectional area,

$S$ = slope of channel.

After a flood, slope and cross-section characteristics can be measured in the field by observing high water levels such as debris lodged in trees. Roughness can be estimated from manuals (Barnes, 1967). For example, assume that high-level water marks left from a flood in a 1.8-m-wide rectangular channel indicate that the water had been 1.1 m deep. With $S$ = 0.002 and $n$ = 0.012, determine $V$, $Q$, and $N_F$.

The hydraulic radius can be calculated as:

$$R = \frac{A}{P} = \frac{(1.8)(1.1)}{1.8 + 1.1 + 1.1} = 0.5.$$

From Eq. (8.10):

$$V = \frac{(0.5)^{2/3}(0.002)^{1/2}}{0.012} = 2.3 \text{ m/s}.$$

From Eq. (8.9):

$$Q = (2.3 \text{ m/s})(1.98 \text{ m}^2) = 4.5 \text{ m}^3\text{/s}.$$

From Eq. (8.2):

$$N_F = (2.3 \text{ m/s})(9.8 \text{ m/s}^2) \cdot (1.1 \text{ m}))^{-0.5} = 0.7.$$

## *8.2.2 Runoff*

In order to compare the discharge of one stream basin to another, it is convenient to multiply the average discharge ($m^3$/s) times $3.1536 \times 10^7$ s/yr divided by the drainage basin size ($m^2$) in order to arrive at annual runoff in meters. This is in keeping with the conventional description of annual precipitation in units of length such as meters.

Figure 8.15 Graph illustrating effects of clear-cutting a forest watershed in South Carolina on the annual water budget (from Dils, 1957).

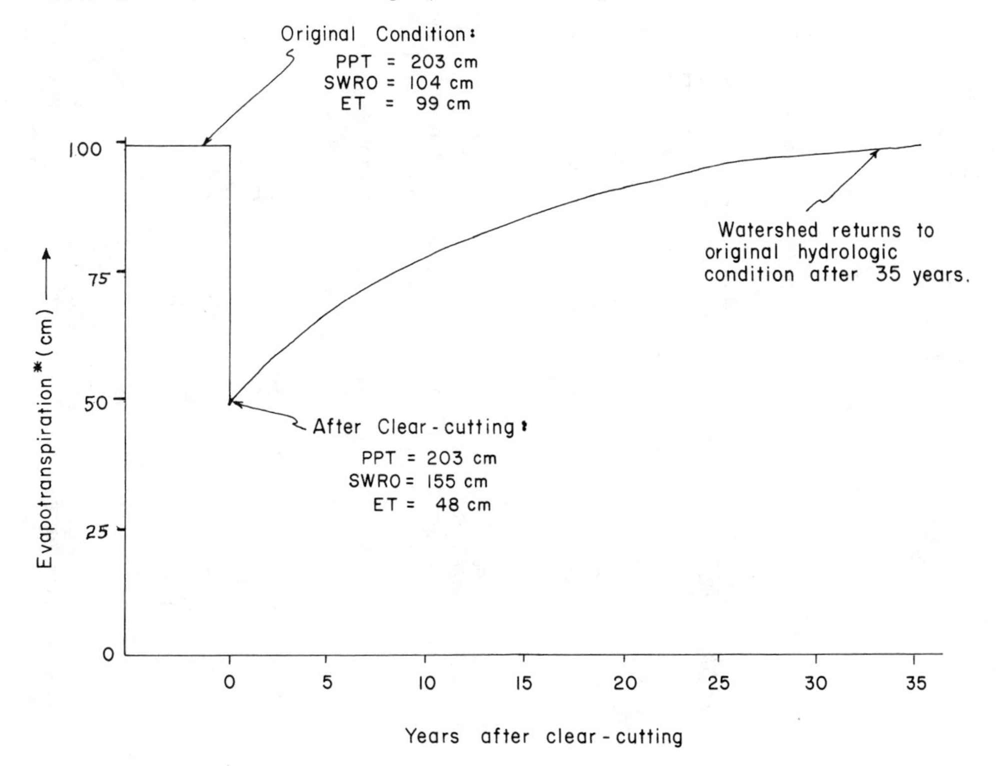

Table 8.1 Mean Monthly Stream Discharge for (A) Battle Creek and (B) Castle Creek in the Black Hills Region, South Dakota, for Years 1962–1982

A. Battle Creek at Keystone, SD (USGS Station No. 06404000); drainage area = 171 sq km
Mean monthly discharge ($m^3/s$)

| | Year | | | | | | | | | | | | | | | | | | | | |
|---|---|---|---|---|---|---|---|---|---|---|---|---|---|---|---|---|---|---|---|---|---|
| Month | 1962 | 1963 | 1964 | 1965 | 1966 | 1967 | 1968 | 1969 | 1970 | 1971 | 1972 | 1973 | 1974 | 1975 | 1976 | 1977 | 1978 | 1979 | 1980 | 1981 | 1982 |
| January | 0.000 | 0.035 | 0.054 | 0.036 | 0.053 | 0.049 | 0.058 | 0.029 | 0.004 | 0.023 | 0.018 | 0.046 | 0.024 | 0.000 | 0.038 | 0.020 | 0.019 | 0.015 | 0.012 | 0.027 | 0.014 |
| February | 0.001 | 0.059 | 0.085 | 0.028 | 0.039 | 0.048 | 0.068 | 0.021 | 0.009 | 0.033 | 0.018 | 0.034 | 0.034 | 0.010 | 0.052 | 0.040 | 0.027 | 0.022 | 0.027 | 0.018 | 0.039 |
| March | 0.013 | 0.199 | 0.072 | 0.032 | 0.123 | 0.071 | 0.095 | 0.038 | 0.029 | 0.097 | 0.099 | 0.135 | 0.073 | 0.040 | 0.050 | 0.100 | 0.119 | 0.070 | 0.092 | 0.041 | 0.084 |
| April | 0.061 | 0.543 | 0.163 | 0.130 | 0.316 | 0.118 | 0.088 | 0.123 | 0.274 | 1.09 | 0.138 | 1.05 | 0.101 | 0.364 | 0.147 | 0.325 | 0.146 | 0.141 | 0.210 | 0.042 | 0.104 |
| May | 1.03 | 0.641 | 0.350 | 1.89 | 0.174 | 0.262 | 0.082 | 0.086 | 0.498 | 1.99 | 0.375 | 0.686 | 0.085 | 0.294 | 1.39 | 0.109 | 2.51 | 0.078 | 0.145 | 1.27 | 0.569 |
| June | 1.82 | 2.39 | 0.815 | 2.55 | 0.054 | 3.67 | 0.462 | 0.219 | 1.15 | 1.67 | 5.57 | 0.375 | 0.049 | 0.591 | 2.02 | 0.045 | 0.711 | 0.132 | 0.108 | 0.174 | 0.903 |
| July | 1.30 | 0.596 | 0.720 | 1.24 | 0.038 | 0.535 | 0.164 | 1.17 | 0.094 | 0.190 | 0.442 | 0.165 | 0.030 | 0.146 | 0.389 | 0.040 | 0.153 | 0.924 | 0.036 | 0.114 | 0.241 |
| August | 0.412 | 0.109 | 0.116 | 0.364 | 0.081 | 0.113 | 0.091 | 0.119 | 0.094 | 0.046 | 0.213 | 0.134 | 0.003 | 0.060 | 0.223 | 0.080 | 0.086 | 0.437 | 0.027 | 0.097 | 0.070 |
| September | 0.100 | 0.171 | 0.051 | 0.136 | 0.074 | 0.077 | 0.063 | 0.016 | 0.038 | 0.041 | 0.106 | 0.060 | 0.003 | 0.000 | 0.054 | 0.039 | 0.024 | 0.086 | 0.004 | 0.007 | 0.052 |
| October | 0.146 | 0.082 | 0.035 | 0.132 | 0.116 | 0.060 | 0.031 | 0.041 | 0.061 | 0.063 | 0.076 | 0.053 | 0.009 | 0.021 | 0.052 | 0.053 | 0.027 | 0.039 | 0.052 | 0.039 | 0.137 |
| November | 0.086 | 0.084 | 0.037 | 0.100 | 0.071 | 0.070 | 0.038 | 0.048 | 0.034 | 0.061 | 0.081 | 0.069 | 0.024 | 0.013 | 0.043 | 0.050 | 0.034 | 0.078 | 0.037 | 0.030 | 0.075 |
| December | 0.059 | 0.063 | 0.037 | 0.088 | 0.046 | 0.052 | 0.038 | 0.026 | 0.018 | 0.047 | 0.046 | 0.029 | 0.004 | 0.041 | 0.043 | 0.041 | 0.029 | 0.056 | 0.065 | 0.022 | 0.052 |

B. Castle Creek above Deerfield Reservoir (USGS Station No. 06409000); drainage area = 215 sq km
Mean monthly discharge ($m^3/s$)

| | Year | | | | | | | | | | | | | | | | | | | | |
|---|---|---|---|---|---|---|---|---|---|---|---|---|---|---|---|---|---|---|---|---|---|
| Month | 1962 | 1963 | 1964 | 1965 | 1966 | 1967 | 1968 | 1969 | 1970 | 1971 | 1972 | 1973 | 1974 | 1975 | 1976 | 1977 | 1978 | 1979 | 1980 | 1981 | 1982 |
| January | 0.111 | 0.122 | 0.209 | 0.237 | 0.313 | 0.283 | 0.322 | 0.269 | 0.243 | 0.265 | 0.302 | 0.267 | 0.245 | 0.238 | 0.235 | 0.220 | 0.255 | 0.283 | 0.244 | 0.268 | 0.180 |
| February | 0.142 | 0.148 | 0.191 | 0.227 | 0.322 | 0.288 | 0.283 | 0.257 | 0.237 | 0.283 | 0.266 | 0.258 | 0.243 | 0.227 | 0.239 | 0.283 | 0.272 | 0.339 | 0.269 | 0.246 | 0.189 |
| March | 0.169 | 0.245 | 0.181 | 0.223 | 0.440 | 0.417 | 0.372 | 0.278 | 0.254 | 0.294 | 0.409 | 0.277 | 0.361 | 0.221 | 0.269 | 0.333 | 0.328 | 0.398 | 0.302 | 0.257 | 0.228 |
| April | 0.333 | 0.431 | 0.428 | 0.479 | 0.594 | 0.473 | 0.378 | 0.406 | 0.454 | 0.664 | 0.451 | 0.456 | 0.384 | 0.554 | 0.350 | 0.213 | 0.386 | 0.468 | 0.468 | 0.297 | 0.346 |
| May | 0.330 | 0.400 | 0.470 | 0.893 | 0.504 | 0.532 | 0.350 | 0.451 | 0.787 | 0.546 | 0.456 | 0.515 | 0.322 | 0.557 | 0.291 | 0.442 | 0.862 | 0.414 | 0.165 | 0.297 | 0.716 |
| June | 0.342 | 0.756 | 0.588 | 0.974 | 0.330 | 0.546 | 0.342 | 0.330 | 0.437 | 0.403 | 0.356 | 0.448 | 0.241 | 0.384 | 0.372 | 0.308 | 0.630 | 0.364 | 0.319 | 0.231 | 0.418 |
| July | 0.212 | 0.347 | 0.392 | 0.711 | 0.361 | 0.412 | 0.294 | 0.409 | 0.311 | 0.325 | 0.319 | 0.347 | 0.249 | 0.274 | 0.291 | 0.267 | 0.459 | 0.381 | 0.239 | 0.229 | 0.309 |
| August | 0.167 | 0.236 | 0.308 | 0.574 | 0.392 | 0.325 | 0.246 | 0.302 | 0.314 | 0.314 | 0.330 | 0.314 | 0.259 | 0.236 | 0.254 | 0.266 | 0.361 | 0.350 | 0.260 | 0.230 | 0.326 |
| September | 0.144 | 0.278 | 0.279 | 0.482 | 0.350 | 0.378 | 0.280 | 0.240 | 0.308 | 0.325 | 0.316 | 0.330 | 0.234 | 0.253 | 0.248 | 0.259 | 0.353 | 0.311 | 0.264 | 0.215 | 0.286 |
| October | 0.168 | 0.245 | 0.294 | 0.465 | 0.378 | 0.353 | 0.262 | 0.308 | 0.308 | 0.344 | 0.325 | 0.325 | 0.253 | 0.266 | 0.234 | 0.302 | 0.372 | 0.278 | 0.262 | 0.230 | 0.379 |
| November | 0.157 | 0.236 | 0.236 | 0.417 | 0.370 | 0.339 | 0.263 | 0.280 | 0.283 | 0.322 | 0.260 | 0.288 | 0.286 | 0.240 | 0.238 | 0.240 | 0.325 | 0.300 | 0.276 | 0.204 | 0.246 |
| December | 0.149 | 0.189 | 0.243 | 0.347 | 0.288 | 0.308 | 0.260 | 0.278 | 0.288 | 0.336 | 0.227 | 0.241 | 0.302 | 0.256 | 0.233 | 0.229 | 0.271 | 0.193 | 0.250 | 0.194 | 0.323 |

From USGS Water-Supply Papers 1917 and 2117 and yearly USGS "Water Resources Data for South Dakota."

Runoff increases with increasing precipitation, steepness of topography, imperviousness of the land, and increased use of the land by man. Agricultural scientists have demonstrated the effects of land management practice on runoff. For example, Fig. 8.15 shows the change in runoff by clear-cutting a small forest watershed. The graph shows that clear-cutting has caused an increase of runoff from 1.04 m/yr to 1.55 m/yr. The progressive conversion of a watershed from forest to rangeland, plowed fields, or urban areas results in significant increases in surface water runoff. This can be advantageous in allowing for more available water to fill a downstream reservoir, or can be disadvantageous due to changes in channel erosion, or the decreased infiltration into the soil and resulting decrease in ground-water recharge.

Geologic control of runoff is illustrated by the data for the two Black Hills streams shown in Table 8.1. By inspection, the discharges for the entire 21 years of record for Battle Creek show a greater seasonal variation than for Castle Creek. Both stream basins have nearly the same precipitation (51 cm/yr), topography (hilly), and land use (pine forest with some open prairie). The reason for the difference in discharge pattern is that Battle Creek drains Precambrian granite and schist where infiltration and ground water recharge is very low; as a result, Battle Creek is a "flashy" stream, having large floods and very low base flow. Castle Creek, on the other hand, drains Paleozoic limestone; precipitation can easily infiltrate into the karst terrain and recharges the limestone aquifer that supplies large springs at the headwaters of Castle Creek and helps maintain a large base flow.

Flow-duration curves reduce a stream's discharge characteristics to a single curve, which can be easily used for comparison with other streams. To construct a flow-duration curve, discharge data is first ranked in an array. The cumulative percentages are then plotted on probability paper with discharge as the ordinate. For example, Fig. 8.16 shows flow-duration curves for Connecticut streams. Streams draining areas with 100% stratified drift have nearly flat flow-duration curves due to the stready discharge because of the permeable nature of stratified drift. Streams draining areas of low-permeability glacial till have great variability in discharges. The slope of the flow-duration curve generally is an indication of the ground water potential of the basin. Flat-sloping curves have high infiltration and high base flow, indicating that permeable deposits are present.

Most streams in the United States have high discharge in the spring and low discharge in the fall. Figure 8.17 shows high and low discharge for Spring Creek, central Pennsylvania. In the northern United States, melting snow and lack of active vegetation cause most large discharges in the spring of the year. In the fall, when vegetation has depleted soil moisture, stream discharges are typically low.

### *8.2.3 Floods*

Most alluvial valleys contain a distinct stream channel as well as an adjoining flood plain. Following heavy precipitation, the discharge of a stream may increase to the point where the channel becomes filled. This level is called bankfull stage (Fig. 8.17). A higher discharge would cause water to flow out onto the flood plain, resulting in flooding. Studies of many natural rivers show bankfull stage can be expected about every 1.5 yr on the average (Leopold et al., 1964; Costa, 1978).

The principle characteristics of flood inundation in a stream valley are shown

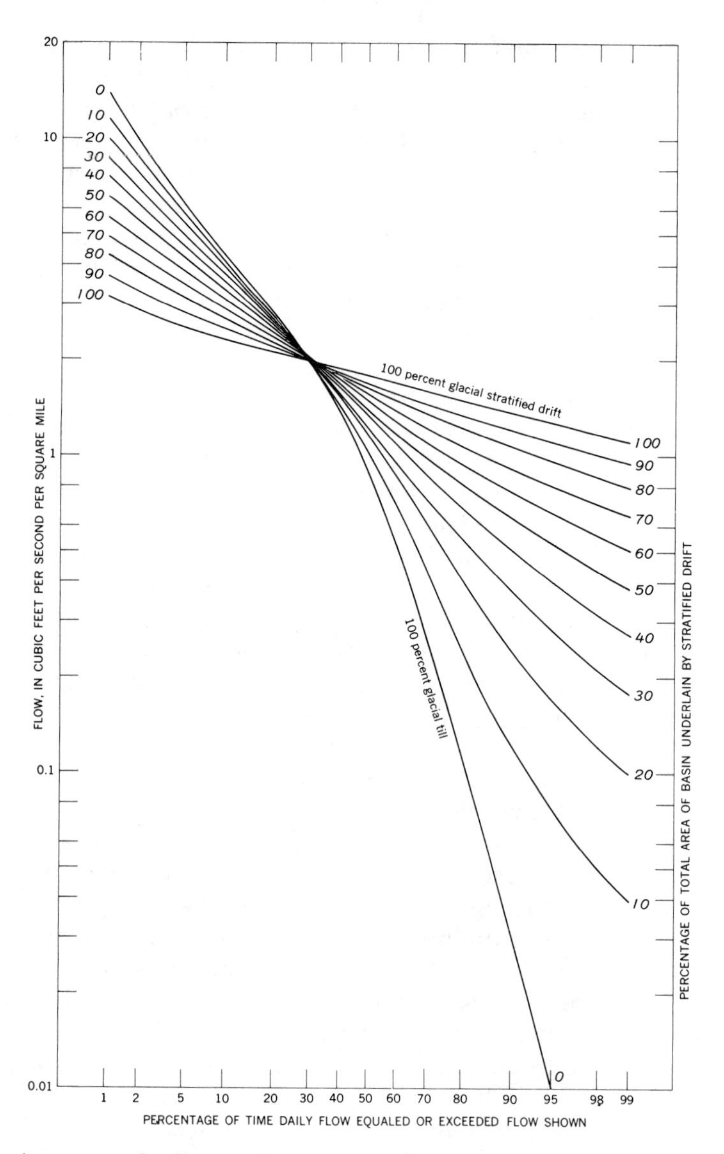


Figure 8.16 Flow-duration curves for Connecticut streams (from Thomas, 1966). The curves represent composite data for streams ranging from 30 to 4000 km$^2$ in drainage area.

in Fig. 8.18. Discharges that overtop the natural channel banks and occupy low-lying flood plains adjacent to the stream may occur at intervals of approximately 2–3 years in a natural stream system. Greater floods, which occur at less frequent intervals, extend over the remainder of the flood plain. A basic feature of new federal flood plain management policies in the United States is the area inundated by the 100-yr flood. Figure 8.19 is a map clearly identifying the 100-yr flood-prone area for part of the Napa, CA area. Flood plains constitute about 5% of the land area in the United States.

Throughout historic time, floods have wreaked havoc upon farms and cities. One of the most tragic floods included the destruction of priceless works of art at

Figure 8.17 Spring Creek, central Pennsylvania. **A**. High discharge, March 1964. The stream is approaching "bankfull stage" since the channel is nearly filled to the level of the flood plain. **B**. Low discharge (base flow), September 1964.

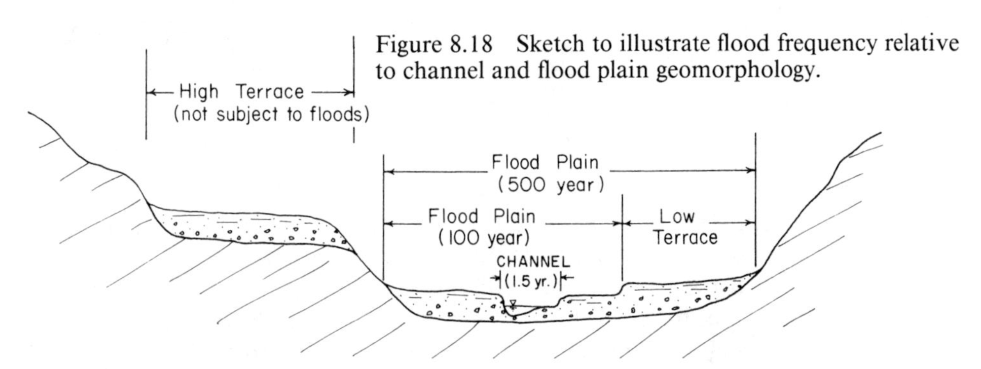


Figure 8.18 Sketch to illustrate flood frequency relative to channel and flood plain geomorphology.

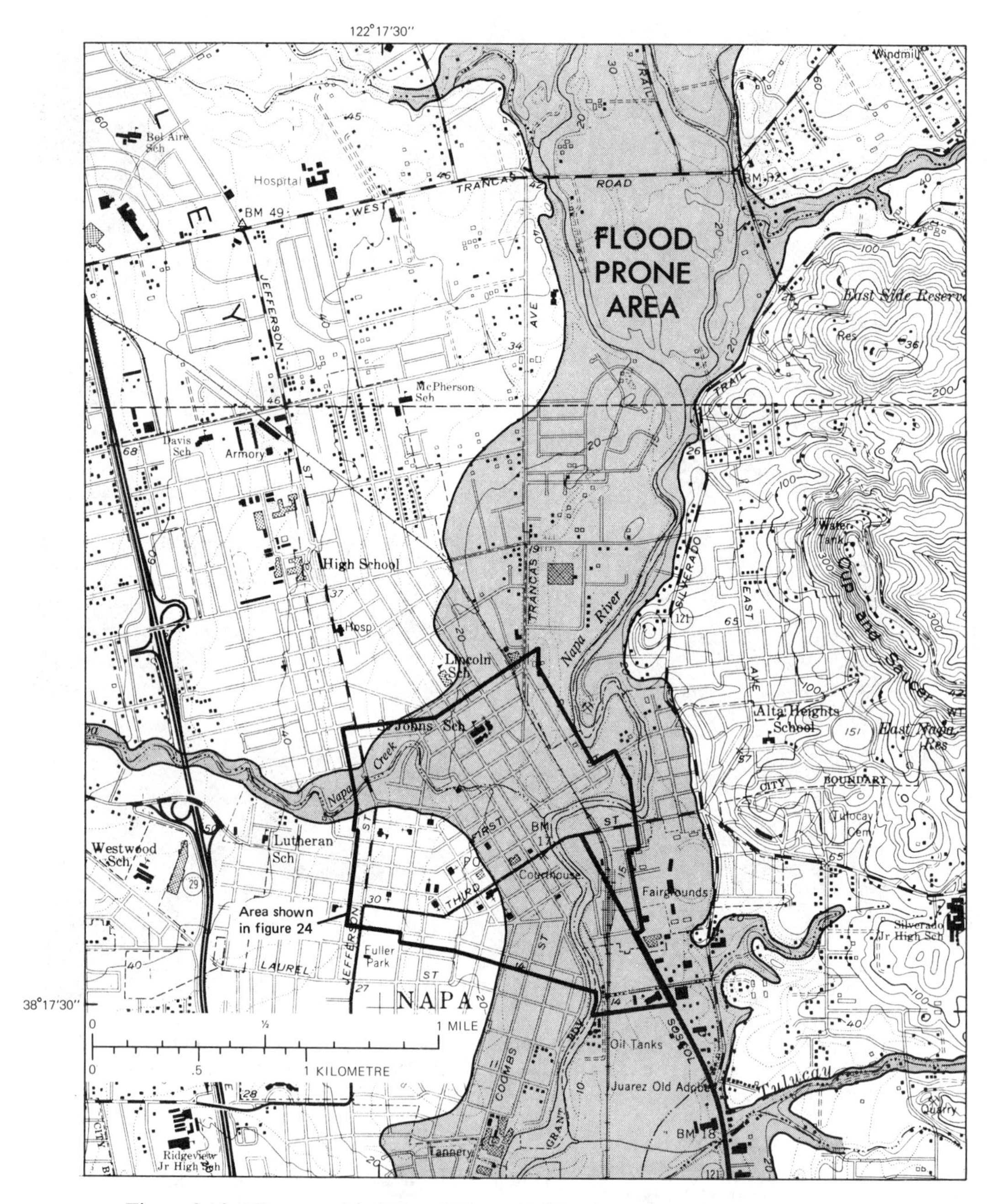


Figure 8.19 Topographic Map of Napa, California area, showing flood-prone areas (from Waananen et al., 1977).

Florence, Italy, in 1966. The city, founded on the flood plain of the Arno River, had a flood of nearly the same magnitude in the year 1333.

Floods can occur any time, although in some rivers there is a greater likelihood during certain times of the year. The Nile River in Egypt, for example, regularly floods during the summer months. Streams in the Rocky Mountains flood in the spring, and coastal California streams flood in the winter months.

Table 8.2 Peak Discharge for Each Year of Record, 1928–1958, Seneca Creek at Dawsonville, Maryland

| Year | Discharge ($m^3/s$) | Rank order ($N$) | Recurrence interval (years) $(N+1)/M$ |
|---|---|---|---|
| 1928 | 107.6 | 4 | 8.00 |
| 1929 | 45.3 | 24 | 1.33 |
| 1930 | 41.5 | 26 | 1.23 |
| 1931 | 49.0 | 23 | 1.39 |
| 1932 | 39.1 | 28 | 1.14 |
| 1933 | 263.3 | 2 | 16.00 |
| 1934 | 68.2 | 13 | 2.46 |
| 1935 | 40.2 | 27 | 1.19 |
| 1936 | 57.2 | 19 | 1.68 |
| 1937 | 79.9 | 11 | 2.91 |
| 1938 | 64.6 | 14 | 2.29 |
| 1939 | 60.9 | 17 | 1.88 |
| 1940 | 49.3 | 22 | 1.45 |
| 1941 | 36.8 | 29 | 1.10 |
| 1942 | 41.3 | 25 | 1.28 |
| 1943 | 102.5 | 6 | 5.33 |
| 1944 | 75.3 | 9 | 3.56 |
| 1945 | 59.7 | 18 | 1.78 |
| 1946 | 83.2 | 7 | 4.57 |
| 1947 | 56.3 | 20 | 1.60 |
| 1948 | 56.3 | 21 | 1.52 |
| 1949 | 63.4 | 16 | 2.00 |
| 1950 | 64.6 | 15 | 2.13 |
| 1951 | 68.5 | 12 | 2.67 |
| 1952 | 79.6 | 8 | 4.00 |
| 1953 | 207.6 | 3 | 10.70 |
| 1954 | 35.1 | 30 | 1.07 |
| 1955 | 74.2 | 10 | 3.20 |
| 1956 | 424.7 | 1 | 32.00 |
| 1957 | 27.2 | 31 | 1.03 |
| 1958 | 103.1 | 5 | 6.40 |

From Leopold, 1974. The drainage area is 262 $km^2$.

Flood frequency can be evaluated statistically, using the concept of recurrence interval. The method outlined below assumes that stream discharge data is available; if not, there are techniques for estimating flood flows on ungaged streams (Dalrymple, 1960; Leopold and Skibitzke, 1967; Thomas and Benson, 1969; Baker, 1977; Baker et al., 1979; Costa, 1983; Tharp, 1983). Table 8.2 shows the peak discharges which occurred every year on Seneca Creek, MD, between the years 1928 and 1958. (Note: these are instantaneous peaks, not average discharge for the day.) The data is ranked in an array as shown in Table 8.2 and the recurrence interval ($T$) is defined as:

$$T = \frac{N + 1}{M}, \tag{8.11}$$

where:

$N$ = the number of events (31 in this case),

$M$ = position in array, where one equals the largest flood.

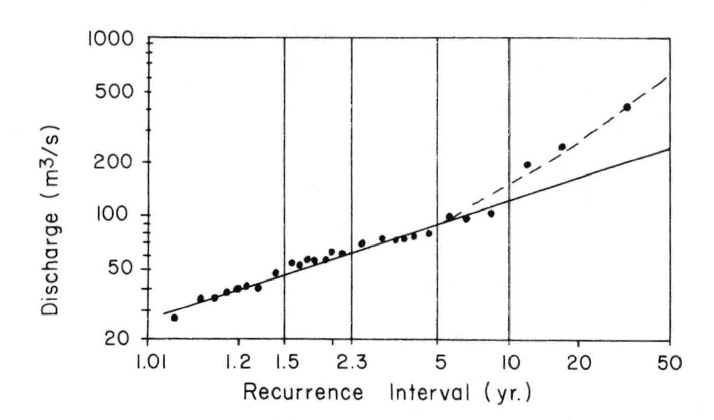


Figure 8.20 Flood frequency plot of data from Table 8.2.

The 31 values of $T$ are then plotted on graph paper with discharge as the ordinate (Fig. 8.20). The data may be plotted on log-log paper or semi-log paper (discharge on arithmetic scale), whichever best allows the points to fall closer to a straight line. (A special type of paper (Gumbel Paper), which stretches the abscissa log interval between 1 and 10 yr, is also commonly used.)

Once plotted, the recurrence interval curve can be extrapolated beyond the plotted points. There is, however, no completely accurate way to make these predictions. As shown in Fig. 8.20, the data could be fitted by a straight line, or a curved line. The curved line better fits all the data; accordingly a 50-yr flood would have an estimated discharge of approximately 630 $m^3/s$. Extrapolating gaging station records to long intervals may lead to conflicting determination of flood risk and design criteria for dam spillways. The evaluation of a stream discharge record containing a single extraordinary flood is difficult, because the large flood may not line up with the remaining data points; this could lead to a divergence of the estimation of the 100-year flood (Costa and Baker, 1981). Extrapolation of limited flood data containing one very large flood discharge presents a dilemma. In many cases it simply is not possible to accurately estimate the recurrence interval for a large flood based on only limited years of record (Baker et al., 1979).

[Note: because the technique outlined above uses only the single largest flood every year, it should be borne in mind that the recurrence interval represents the interval during which a flood of a given magnitude will occur on an annual maximum. This would be most appropriate for rivers such as the Mississippi, Nile, or Mekong, which have only one seasonal peak. For a small stream subject to more than one peak per year, it would be more appropriate in terms of flood probability to rank all the large discharge events within a given interval regardless of the water year (Réméniéras, 1967).]

One problem with recurrence interval flood determination is that the general public is liable to think that a 100-yr flood means it will be exactly 100 years until the next 100-yr flood! The fallacy of this reasoning was emphasized by the Connecticut River at Hartford: in 1936 it had a 400-yr flood, and only 3 years later a 200-yr flood.

## *8.2.4 Effects of Urbanization on Runoff*

Man can affect the hydraulic regime of a river by changing the discharge characteristics as well as the sediment load. A vivid example was presented by Gilbert (1917), who described the effect of gold placer operations on the Yuba River in

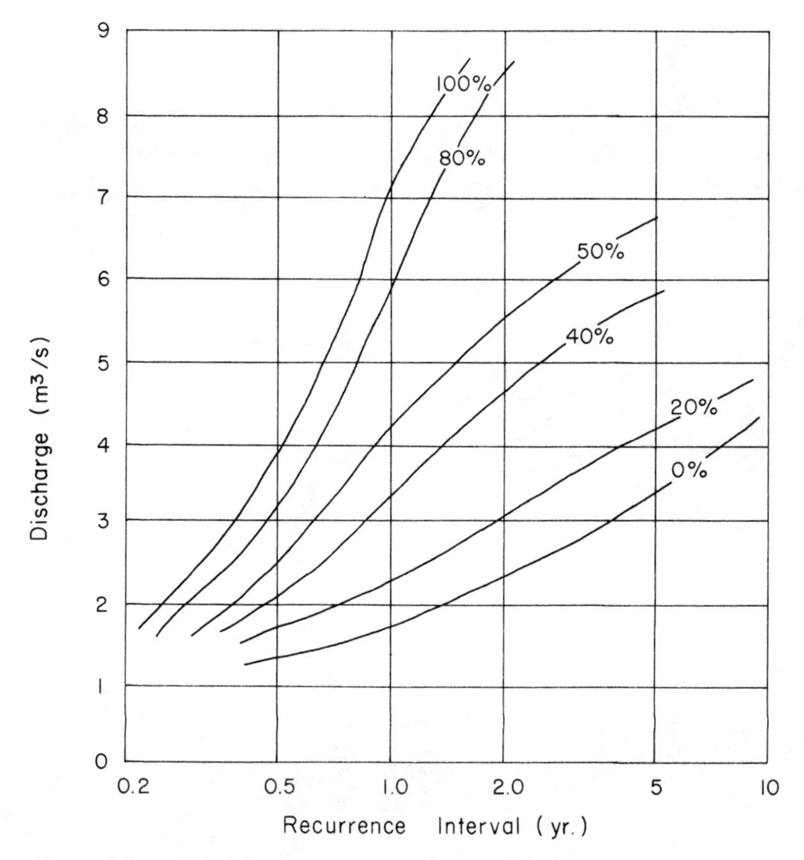


Figure 8.21 Flood-frequency curves for a 2.6 km$^2$ basin in various states of urbanization. The hydrologic influences of urbanization are approximated by the percentage of area served by storm sewers (after Leopold, 1972).

California. In the latter part of the nineteenth century, hoses were used to direct streams of water and erode gravel deposits. The gravel was washed into sluice boxes, a process called hydraulicking. A vastly increased sediment load was brought into the Yuba River, causing aggrading conditions that lead to the inundation of farm land and to an increase of flood hazards at Marysville. Dikes and levees at Marysville were insufficient to keep up with the rate of aggradation, reaching 0.3 m/yr. After several floods, endangered downstream landowners were able to legally stop the hydraulicking.

Rivers in developed countries have suffered environmental degradation not only because of municipal sewage and industrial waste, but also because of other effects of urbanization (e.g., the replacement of natural fields and forests by man-made rural and urban area). Figure 8.21 shows that urbanization caused discharge to increase several fold. Floods become more frequent and sediment loads increase.

Urbanization also affects runoff by causing the peak discharges to occur in a shorter period of time after a storm. For example, Fig. 8.22 shows a street in Rapid City, SD, that was poorly planned and constructed in an ephemeral gully. A thunderstorm dropped 6.6 cm of rain in 1 hr on June 15, 1972, and two people drowned when the water rushed down the street at a velocity of about 50 km/hr.

Figure 8.22 Meade Street, Rapid City, SD (from Rahn, 1973). **A**. Inverted crown street looking upgrade. **B**. Storm of June 15, 1972, looking downgrade. Note standing waves, indicative of supercritical flow.

Because of urbanization, less water is able to infiltrate into the soil, which may cause a decline in the water table. On Long Island, New York, the effects of urbanization and pumpage of ground water on lowered water tables has led to local regulations requiring developers to build artificial recharge areas. In Washington, DC, urbanization had led to vanishing springs and streams (Williams, 1977), as the natural forested ground-water recharge areas have been paved over.

## 8.3 Dams

The purpose of constructing a dam is flood-control, irrigation, electric power generation, recreation, or a combination of the above. In the eastern United States, the U.S. Corps of Engineers is responsible for the construction of most large dams. These are generally constructed for downstream flood control. In the western United States, the U.S. Bureau of Reclamation is responsible for the construction of most large dams. The water is used largely for irrigation and electric power generation. Other governmental agencies such as the Bonneville Power Administration (Columbia River) and the Tennessee Valley Authority (Tennessee River) as well as state governments (e.g., California Department of Water Resources: Feather River Project) and private groups are also responsible for dam construction.

The large dams in the western United States provide water for the conversion of arid areas into productive areas that today supply much of the nation's agricultural produce. These dams and irrigation projects were the dream of the first director of the U.S. Geological Survey, John Wesley Powell, who first rode the untamed Colorado River rapids of the Grand Canyon in 1869.

### *8.3.1 Benefits and Costs*

In recent decades, dam construction has received vigorous opposition from environmentalists and from engineers. Lehr (1980) pointed out that ground water can serve as a source of water supply without the environmental impacts associated with dam construction:

> ... I am more convinced than ever that this nation doesn't need any more dams. The 50,000 dams built in the U.S. during the past 100 years have inundated scenic valleys, flooded valuable forests, farmland and wildlife habitats, displaced rural residents from their ancestral land, and caused social hardship as the rural taxbase is reduced. Even more important is this fact: dams are no longer a cost-efficient method of producing water supply.

In terms of topography and economics, the best dam and irrigation sites in the United States have been utilized. The Colorado River, for example, the largest river in the southwestern United States, has so many dams and irrigation withdrawals that today practically none of its water reaches the sea. Benefits of the construction of new dams (e.g., irrigation, recreation, power, etc.) are decreasing, but the costs of dam construction (finding ideal topographic sites with small land acquisition costs) is increasing.

Generally speaking, if a governmental dam-building agency finds a benefit/cost ratio (B/C) greater than unity, the agency feels it has a mandate to ask Congress for money to construct the dam. Today B/C ratios for potential dam sites have decreased to the point where few dam projects are cost effective. Past experiences show that final completion costs usually far exceed predicted costs (Table 8.3). Regarding a study of 147 Corps of Engineers dams, Morgan (1971), a retired Corps of Engineers employee, states: "There was a consistent and persistent overstatement of benefits and understatement of costs. On pure economic grounds—leaving the environmental aspects aside—about half the projects should never have been built."

The big dam-building era in the United States may be reaching an end, but planning for future dams and the repair of existing dams continue to consume the energies of engineering geologists. The remainder of this chapter explores some of the more salient points of dam construction and engineering geology applications to dams and reservoirs.

Table 8.3 Cost Increases of U.S. Army Corps of Engineers Projects

| Name of project | Cost estimate at time project was authorized (dollars) | Amount spent through fiscal year 1966 (dollars) | Percent overrun |
|---|---|---|---|
| Whitney (Texas) | 8,350,000 | 41,000,000 | 391 |
| John H. Kerr (North Carolina and Virginia) | 30,900,000 | 87,733,000 | 185 |
| Blakely Mountain (Arkansas) | 11,080,000 | 31,500,000 | 184 |
| Oahe Reservoir (South Dakota) | 72,800,000 | 334,000,000 | 359 |
| Jim Woodruff (Florida) | 24,139,000 | 46,400,000 | 92 |
| Chief Joseph (Washington) | 104,050,000 | 144,734,000 | 39 |
| Fort Peck (Montana) | 86,000,000 | 156,859,000 | 82 |
| Clark Hill (Georgia and South Carolina) | 28,000,000 | 79,695,000 | 185 |
| Bull Shoals (Arkansas) | 40,000,000 | 88,824,000 | 122 |

From Morgan, 1971.

## *8.3.2 Design of Dams*

There are two general types of dams: concrete and earth-fill. Concrete dams are built in narrow canyons where a relatively small volume of concrete is necessary. Concrete dams are either 1) gravity-type, where the mass of the concrete is sufficient to support lateral hydrostatic forces of the reservoir, or 2) arch type. The Grand Coulee Dam on the Columbia River in Washington and the Hells Canyon Dam on the Snake River in Idaho are gravity dams (Fig. 8.23). Concrete arch

Figure 8.23 Hells Canyon Dam, Snake River, Weiser, ID. This is a concrete gravity type dam. It is 100 m high and was built by Idaho Power Co. in 1968.

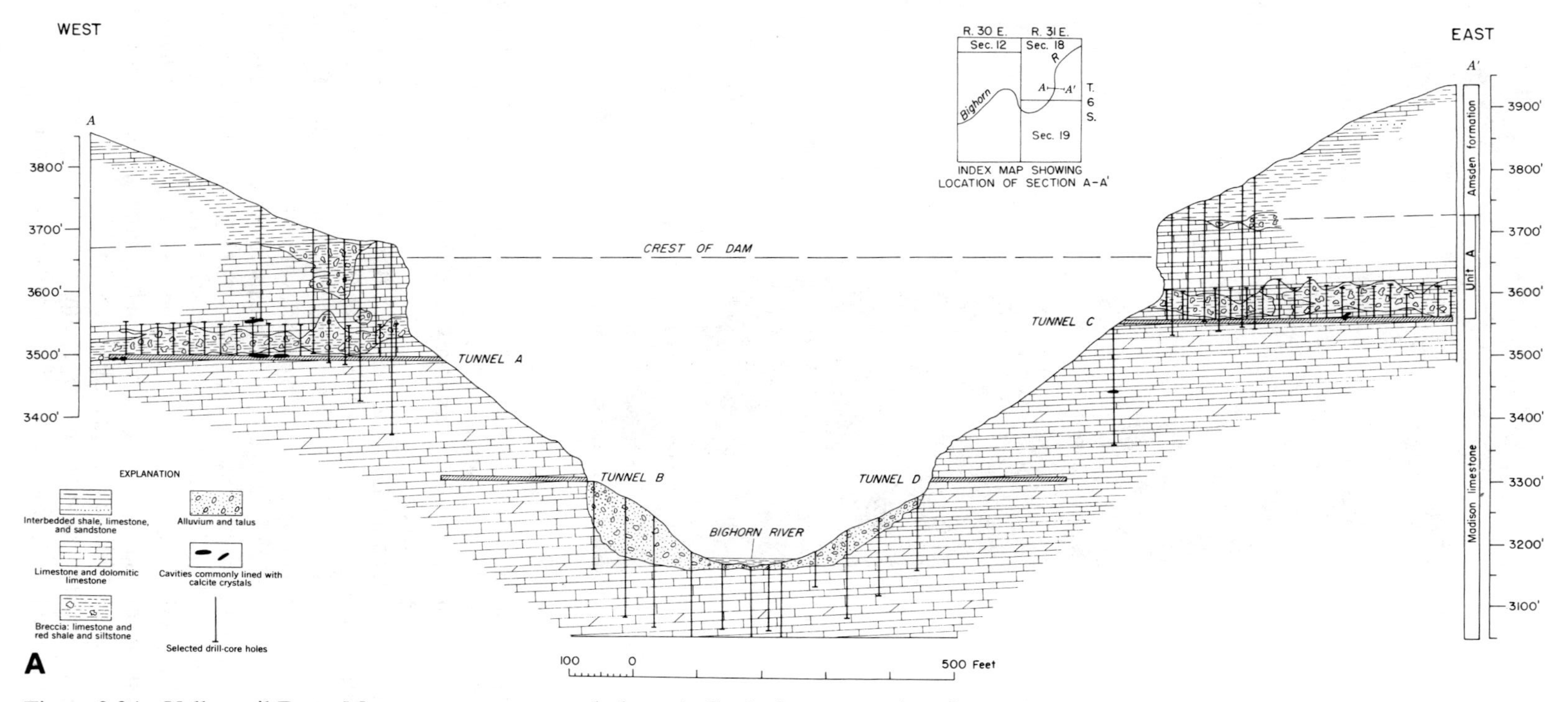


Figure 8.24 Yellowtail Dam, Montana, a concrete arch dam. **A**. Geologic cross section (from Richards, 1955). The brecciated and cavernous zones in the Madison Limestone were difficult to grout. **B**. Concrete arch dam. The Bighorn Reservoir is barely visible in the background; the Bighorn River visible in canyon below the dam.

Figure 8.24.
(*continued*)

dams utilize a relatively thin arch which can be keyed into a rock valley wall so that compression in the arch supports the lateral hydrostatic load (Fig. 8.24). Famous examples are the Vaiont Dam (Italy) and the Hoover and Glen Canyon Dams on the Colorado River.

Earth-fill dams such as the Oroville Dam, California (Fig. 8.25) require a suitable source of earth for the dam. Broad valleys having an absence of firm bedrock are better suited for earth fill dams than concrete dams. Figures 8.26 and 8.27 show cross sections of earthen dams. Three zones are typically used, 1) the central core consists of clayey soil having low permeability to prevent leakage, 2) the downstream zone is a coarse material (sand or gravel) to help keep the water table (once established within the dam) as low as possible in the back face, and 3) the remainder of the dam is composed of readily accessible unconsolidated deposits. If the foundation consists of permeable material, a core may be extended into the foundation to reduce leakage (Figs. 8.26B and 8.26C).

Earth fill is typically emplaced by large dump trucks (Fig. 8.28). Sheeps-foot rollers compact the fill to achieve a suitable density determined by soil tests. The

Figure 8.25 Oroville Dam, California, constructed in 1963. The dam impounds 4.3 × $10^9$ $m^3$ of water for irrigation, flood control, and recreation, and provides 800 megawatts of hydropower. Oroville is the highest earth-fill dam in the world (236 m high) and was completed in 1968. It is believed to have triggered a magnitude 5.7 earthquake in 1975. (Photograph by California Department of Water Resources.)

upstream dam face may be protected from wave erosion by emplacement boulders (rip-rap).

The downstream zone, or drain, helps prevent seepage and possible landslides in the back face, as well as reduce the hydrostatic uplift within the dam. Dams are subject to hydrostatic pressure due to the impounded reservoir. Figure 8.29A shows a sketch of hydrostatic conditions for a retaining wall impounding a volume of water. A translational force is imparted to the structure. A dam must resist this force by shearing forces within the dam itself, and by frictional forces at the base of the dam or by a keyway. In the case of concrete arch dam, hydrostatic pressure is resisted by distributing the stresses into the adjacent bedrock hillslopes. In earthquake-prone areas, special design criteria for earth-fill and concrete dams must be utilized (Clough and Chopra, 1977; Attewell and Farmer, 1976).

Figures 8.29B and 8.29C illustrate how a drain adds to the stability of an earthen dam. A large volume of the dam is saturated if there is no drain (Fig. 8.29B). Archimedes' principle indicates that the resultant weight of a mass sub-

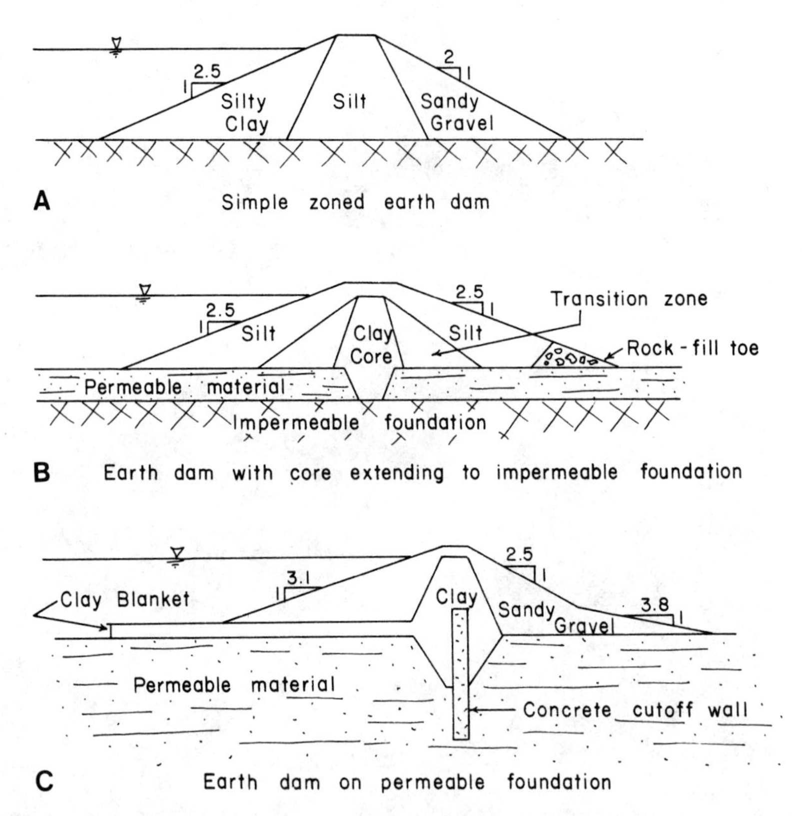


Figure 8.26 Cross sections of typical earth dams (after Linsley and Franzini, 1979). **A**. Simple zoned earth dam. **B**. Earth dam with core extending to impermeable foundation. **C**. Earth dam on permeable foundation.

Figure 8.27 Cross section of Pipestem Dam on the James River, 5 km west of Jamestown, ND, (after U.S. Army Corps Engineers, Omaha District, Design Memorandum JP-3, 1969).

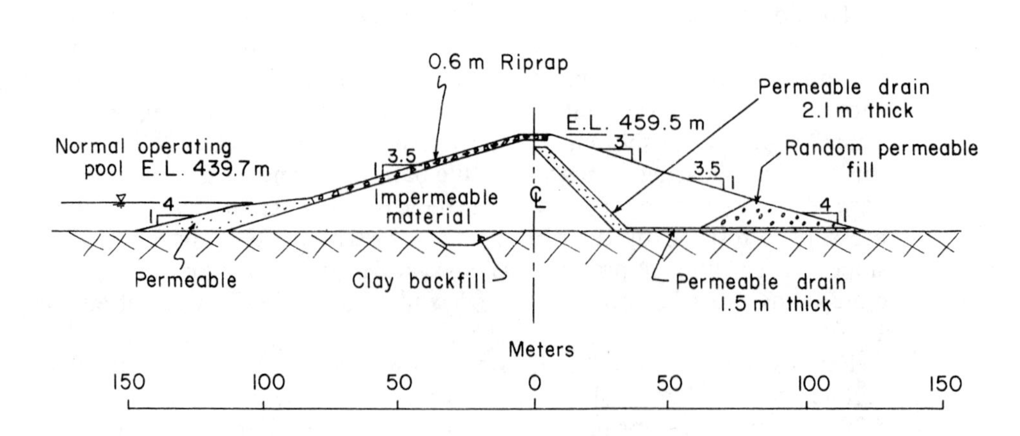

Figure 8.28 Construction of Frenchman Dam, a small earth dam in northeastern California. Note coarse drain material on right (downstream side of dam). A "belly dump" brings earth to the proper location where a bulldozer and sheeps-foot roller spreads it into 20 cm deep layers.

Figure 8.29 Hydraulic considerations for a dam. **A**. Pressure on a retaining wall. **B**. Water table in an earth dam. **C**. Water table in a dam having a permeable drain.

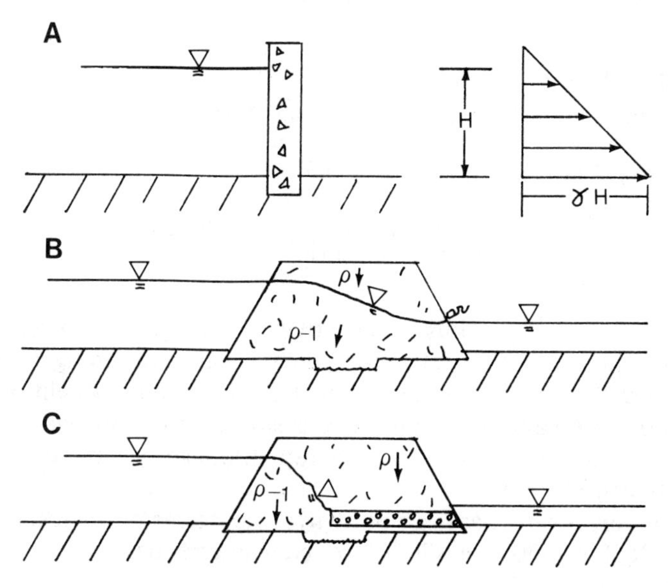

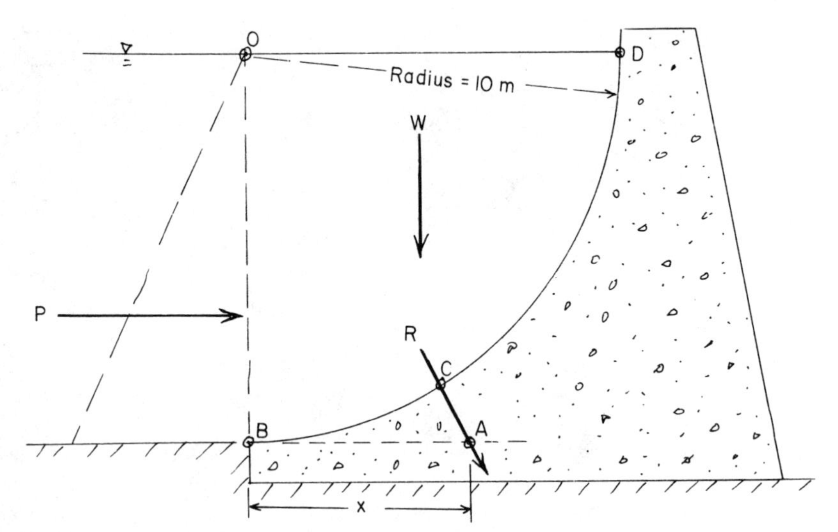


Figure 8.30 Determination of hydrostatic forces on a concrete dam.
Problem: What is the resultant hydrostatic force ($R$) on a unit section of a concrete dam having a circular upstream face as shown?
Solution:
(1) Determine the vertical weight ($W$) of the quarter-cylindrical block of water $BCDO$:

$$W = (10^4 \text{ N/m}^3)\left(\frac{\pi\,(10\text{m})^2}{4}\right) = 7.85 \times 10^5 \text{ N per m.}$$

(2) Determine the horizontal force ($P$) of the rest of the reservoir:

$$P = (10^4 \text{ N/m}^3)\left(\frac{10\text{m}}{2}\right)(10\text{m}) = 5 \times 10^5 \text{ N per m.}$$

(3) Equilibrium requires that:

$$\Sigma F_x = 0\,;\therefore\, R_x = P = 5 \times 10^5 \text{ N per m;}$$
$$\Sigma F_y = 0\,;\therefore\, R_y = W = 7.85 \times 10^5 \text{ N per m.}$$

(4) From Pythagoras:

$$R = (R_x^2 + R_y^2)^{1/2} = 9.31 \times 10^5 \text{ N per m.}$$

(5) The distance x can be located by summing moments about $B$:

$$P\left(\frac{10\text{m}}{3}\right) + \text{W}\,(4/3\pi)\,(10\text{m}) - R_y\,(x) = 0;\; x = 6.37 \text{ m}$$

(6) The point $C$ on the upstream dam face can be located graphically or by combining the moment relationship with the equation of a circle.

merged in a fluid is buoyed up. Thus the effective dam weight is reduced by $\rho - 1/\rho$ over an unsaturated condition. Therefore, the resistance to sliding is less with no drain than where there is a drain (Fig. 8.29C). An additional benefit of drains is the prevention of a seep or a spring on the back slope, which can lead to sloughing and landslides. Figure 8.27 shows an elaborate drainage system built into the Pipestem Dam, North Dakota.

The hydrostatic forces on a dam can be analyzed in terms of the resolution of forces. For example, Fig. 8.30 shows how a concrete gravity dam can be analyzed in terms of hydrostatic forces.

Much information is available in the literature on dam geology; for example, see Walters (1962), Chow (1964), U.S. Bureau of Reclamation (1973), and Linsley and Franzini (1979). Sherard et al. (1963) illustrates many kinds of earth-fill dams built throughout the world; lessons learned from dam failure such as foundation settlement, landslides, and solution of foundation rock are illustrated. Design criteria for dams in earthquake-prone areas include the recommendation that an earth dam may survive a displacement along a fault whereas a rigid concrete dam may not. Mathematical analysis of stresses of dam foundations is described by Jaeger (1972). Zienkiewicz (1968) used the finite element method to compute stresses of dam plus rock foundations. These methods show that an area of tension may exist in the region upgradient for a gravity dam. Since rock masses may be incapable of withstanding tension, the dam can be redesigned using a computer program that assumes "no tension" solution, and stresses in the foundation are redistributed accordingly.

### *8.3.3 Spillways*

A dam should be capable of withstanding any conceivable flood. This is usually accomplished by the construction of a spillway, but may also include, to some degree, storage of the flood within the reservoir between normal pool and dam crest elevation.

For earthen dams, it is desirable to have the spillway built on bedrock on the side of the valley (Fig. 8.25), and not on the earthen dam itself. This is because the water descending the spillway during a flood is capable of vigorous erosion. If the spillway is located on the earthen dam, adequate protection of the dam should be made by using concrete or other erosion-resistant material.

Spillways should have adequate freeboard to pass the maximum project flood, which is normally about the 1000-yr flood. Freeboard area is the average width of the spillway times the freeboard (the distance between normal pool elevation and the top of the dam). Spillway freeboard may change with time. For instance, in older, poorly compacted earth dams, settlement of the earth fill may take place, reducing the elevation of the top of the dam, thus the freeboard area of a spillway (located on adjacent bedrock) will be correspondingly reduced.

The spillways of most large dams in the United States are designed with respect to the maximum probable flood. This discharge is determined by a meteorological estimate of the physical limit of rainfall over the drainage basin. The U.S. Army Corps of Engineers uses a standard project flood as a basis of its studies, which is usually about 50% of the probable maximum flood (Linsley and Franzini, 1979). The discharge values obtained by these techniques may seem large, but a dam spillway should always be safe, especially if the consequences of dam failure would be catastrophic. For example, two older dams (Pactola Dam and Deerfield Dam) on Rapid Creek upstream from Rapid City, SD, were studied by the U.S. Bureau of Reclamation following record-breaking rainfall in other stream basins of the Black Hills on June 9, 1972. When Deerfield Dam was built in 1941, the "inflow design flood" (IDF) was calculated at 700 $m^3/s$, based on a rainfall accumulation of 5 cm over a 2-hour duration. In 1981 the USBR used a revised rainfall accumulation of 69.9 cm over a 72-hour duration, which resulted in a IDF of 4421.5 $m^3/s$. As a result of these studies, and because of the urban area downstream, it was decided in 1983 to widen both spillways.

Care must be taken to insure that spillways remain unobstructed during a flood. The dam at Canyon Lake in Rapid City, SD, failed on June 9, 1972, because the spillway, although 30 m wide, was obstructed by a pedestrian walkway supported by three columns. Floating debris (trees, cars, etc.) lodged in these columns and rendered the spillway useless. As a consequence, the earthen dam was overtopped and washed out. Scores of people lost their lives as a wall of water swept through Rapid City.

The greatest dam disaster in the United States, the Johnstown Flood, occurred on the South Fork of the Conemaugh River, PA, on May 31, 1889. The following account from the *Engineering News Record* (Anon., 1889) summarizes the tragic events:

> The engineer in charge of the reconstruction of the dam is stated to have been the late Gen. James N. Morehead, in regard to whose professional standing and record we have no information. He was never a member of the American Society of Civil Engineers nor a subscriber to this journal, and is not known by reputation to us or to a number of well-informed engineers whom we have consulted. Nevertheless, he may of course have had an excellent professional record, and moreover, it seems tolerably clear that the primary cause of failure lay in no part of the work which he reconstructed, but in lack of sufficient spillway. The considerable leakage reported from the dam, however, would indicate that the reconstruction was none too secure, as would also its low cost, and it may well be that its lack of the substantial solidity of the old structure aggravated the disaster by aiding the dam to go out all at once instead of gradually. It is now forever too late to determine these facts with exactness, although further light may be shed on them by our pending investigation on the ground, or testimony at the official investigation.
>
> The excessive and phenomenal rainfall, which is the second contributing cause of the disaster, and is more fully described below, began Thursday night, May 30, after several days of moderate prior rains, and continued almost till the dam gave way. The lake, although discharging its best through the contracted spillway, gradually rose, at the rate of a foot an hour, it is stated, for some hours before the break, implying that the crest of the dam was some 7 ft or so above the normal level, as the old records indicate, until, at 2:30 P.M. on Friday, May 31, water began to run over the crest. The end then was certain, and could not be long delayed, no earth dam being capable of sustaining such discharge over it. It took half an hour for the increasing current to gradually cut away the earth support from the lower side of the rubble heart wall, and then, at 3:00 P.M., the dam gave way. The breach once made was doubtless instantly enlarged to nearly its final dimensions, 250 ft across. In fact, some of the eye-witnesses state that it "burst with a report like thunder," which it is quite possible would have been the effect upon the eye and ear, even if the real course of events was gradual but very rapid disintegration in the torrent.
>
> That the chances for life of any one in the way of the flood after the dam had once given way were very poor, is evident when we remember that, as in all such cases, the flood advances, not with a comparatively shallow advance guard, but in a solid wall in front, which strikes a house or human being with terrible velocity and directly downward rather than from the side. All accounts agree that the water did in fact advance "like a wall 30 or 40 ft high", and it is wholly in accordance with physical laws that it should do so, in fact it cannot well do otherwise. The water which first flows out, being retarded by the rough surface, trees, rocks, houses, animals, and other obstacles, speedily loses its own velocity, but furnishes an almost frictionless surface over which other water can slide, like ice down a plank, with almost the full theoretical velocity due to the fall. Thus the top of the flood is continually moving

> much faster than the bottom, and falls over the end of it when it reaches it, to be itself retarded and to surrender a large part of the energy due to its velocity in tearing up the ground and beating down, not driving forward, any unfortunate creature or structure which stands in its path. The bottom of the flood is relatively stationary, the top has its full theoretical velocity due to the fall, and the average velocity of advance hence becomes slightly more than half the top velocity; the water, in fact, rolling over itself at the front very much as a wheel rolls, except that the lower part of the flood-wheel never rises again when it once strikes the ground. We have seen precisely this mode of advance on a small scale (say a "wall" 10 ft high) in mountain, floods which were not caused by the bursting of reservoirs, but merely from cloud-bursts in the mountain tops miles away. The fall from the reservoir to Johnstown having been about 250 ft, the actual time taken by the flood to reach Johnstown (15 to 17 min) corresponds very closely with that which this theory requires, as we may show more in detail later; and it requires that the top of the water should have very nearly the theoretical velocity due to a fall of 250 ft, which is about 87 miles per hour; that due to the 160 ft fall of Niagara Falls being about 70 miles per hour. Allowing for all the frictional losses, a man (or house) which sustained the direct impact of this torrent had about the same chance of resisting it and escaping alive as he would in sustaining the impact of Niagara Falls itself.

The fact that inadequate spillways pose the most serious threat to the integrity of dams in the United States was emphasized by a 1981 inspection of dams by the Army Corps of Engineers. Of the 8639 dams in a high-risk category (where high loss of life or extreme economic damage could occur if the dam failed), 2884 dams were found to be unsafe. Inadequate spillways accounted for 2351 of the unsafe dams.

### *8.3.4 Foundations*

Suitable bedrock for a dam site is of the utmost importance. Bedrock below the dam should be able to support the dam and reservoir load, and should have low permeability. After the proper bedrock is located, unconsolidated debris should be removed along the dam center line before the dam is built. This will allow for the seating of a core of low permeability into a keyway. Loose debris (colluvium, alluvium, etc.) may be removed by scrapers, hoses, etc.

An example of poor foundation is afforded by the St. Francis Dam, California, which was built astride a fault and the left abutment was founded on soluble gypsiferous fanglomerate. (See Chapter 5 for description.) Another example of a poor geologic foundation is the dam at Malpasset, France, which burst in 1959, killing about 450 people. The geologist in charge of investigating the site apparently had not been informed that the proposal for a gravity dam had been dropped in favor of a bold, thin-arch dam. Consequently, he failed to comment on finding fissured gneiss masses at a higher level on the dam foundation. The dam collapsed when the gneiss gave way under the thrust from the filled reservoir (Jaeger, 1972).

At dam sites underlain by permeable bedrock, grouting may be required to prevent leakage. For a grouting operation, open holes are drilled along the dam centerline, and grout, a mixture of cement and water, is pumped into the holes (Fig. 8.31). The cement sets up and seals the cracks. Care must be taken not to exceed rock uplift pressures which would heave up the foundation. The design of

Figure 8.31 Grouting operation. Cement and water are mixed at tank (**A**). Pump (**B**) sends grout to injection wells (**C**) along the dam centerline. Note dark area (**D**) where grout has burst through to surface through the rhyolitic bedrock. (Location: Frenchman Dam, Portola, CA.)

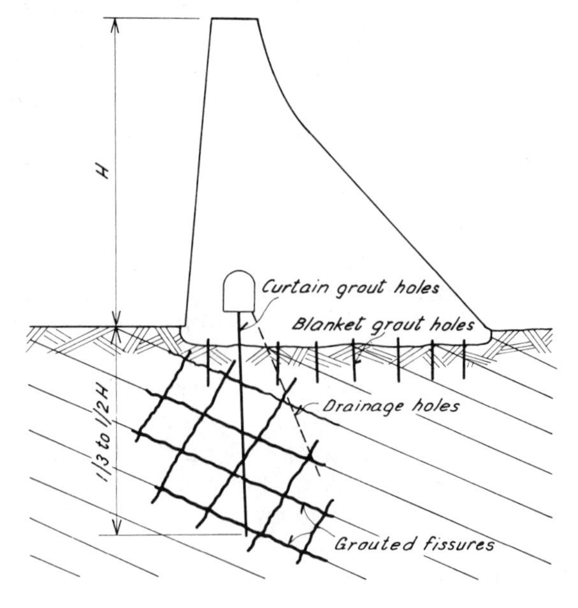


Figure 8.32 Cross section showing a grout curtain into bedrock below a dam (from Bazant, 1979).

a grout curtain is adapted to the conditions encountered during the work. Figure 8.32 shows two types of grout curtains. According to Bazant (1979):

> A procedure suitable for blanket grouting, i.e. grouting of relatively shallow holes drilled in a grid pattern in the footing bottoms of dams, is the single-stage grouting carried out along the whole depth of the holes. Each hole is drilled, cleaned, pressure washed, water pressure tested, and grouted in one stage. At the beginning of the operation, the holes are drilled far apart (5 to 25 m); in the next step, at about half the initial spacing. This split-spacing drilling is continued as necessary until the grout take is small. Usually the primary and the intermediate secondary systems are adequate for the purpose; sometimes, however, a third system is needed with the spacing reduced to about 1.5 m.
>
> Curtain grouting is achieved as stage grouting that is carried downwards. The operation begins by drilling a hole to a depth dictated by the geological conditions, i.e. about 3 to 5m, until the first fissure is encountered; this is followed by cleaning of the hole, pressure washing of fissures, water pressure testing, and grouting at a low pressure. The grout is allowed to set, the hardened grout is drilled out, the hole deepened for the next stage, washed out and grouted again, this time under a higher pressure. The cycle is repeated until the grout take decreases to the prescribed amount. To prevent grouting of the previous stage, a packer is inserted to plug the lower end of the grouted zone, thus allowing each seam to be grouted separately. Stage grouting, which is carried out upwards, is cheaper than that which proceeds downwards, but can only be used if the grout does not flow past the packer and leak to the overburden.

Most open bedrock joints are near the surface. Figure 8.33 illustrates how the specific capacity (water well discharge divided by drawdown) diminishes with deeper wells in different terrains. This is because deeper fractures are sealed shut by virtue of lithostatic pressure. Figure 8.33 (curve *C*) illustrates pregrouting water injection tests on a dam foundation, and indicates that deeper holes will require less grout. It can be seen from these data that most fracture permeability in rocks is confined to the upper 100 m of rock.

Attewell and Farmer (1976) supply data on different types of grouting compounds and additives, with examples of grouting operations in different geologic and engineering applications. Klosterman et al. (1982) showed that grouting of dolomite at the Meramec Dam site, Missouri, caused a 97% reduction in the specific capacity of a test well installed in the grouted area, offering convincing documentation of the reduction of ground water movement by grouting.

In some cases it is almost impossible to grout successfully due to the abundance of solution cavities such as those found in a limestone. For example, the Tennessee Valley Authority dams built in the 1930s required enormous volumes of grout. Hand-dug excavations were required, and open caves were filled with concrete. During the reservoir filling, mattresses were reportedly used to stuff into the open cracks as water was sucked underground. Another example is Anchor Dam, near Thermopolis, WY, built in 1961. The reservoir is located on the Tensleep Sandstone overlying Mississippian Limestone. According to Voight (1974), it has never held any water, and underground cavities are large enough to cause gravity anomolies in parts of the reservoir area.

Lugeon (1933) mentions several cases where geologists were either not consulted or overlooked the problems relative to European dams built on permeable

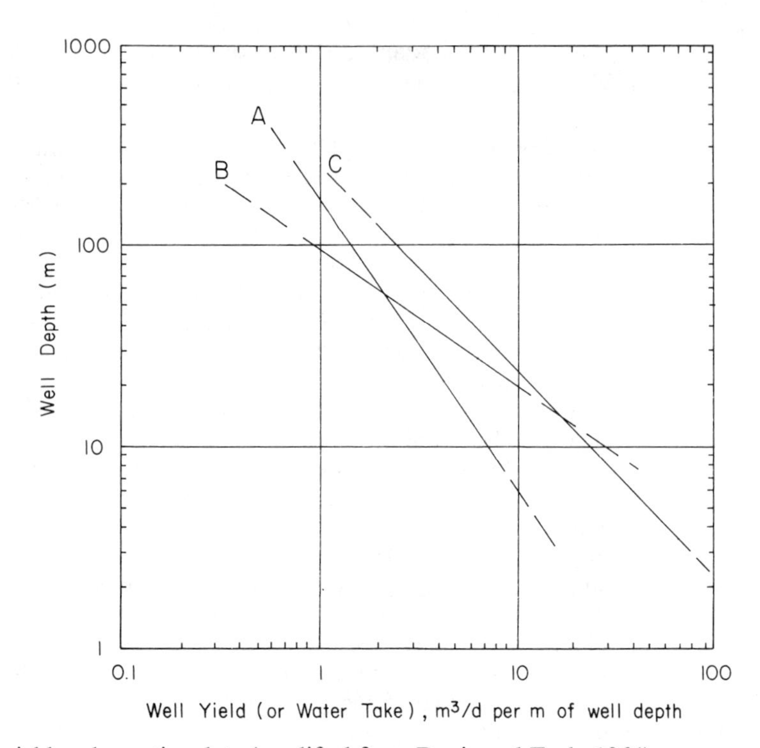


Figure 8.33 Well yield and grouting data (modified from Davis and Turk, 1964). Curve *A* shows the yields of 2336 wells in crystalline rock (granite and schist) in the eastern U.S. Curve *B* shows the yields of 239 wells in granodiorite in the Sierra Nevada, California. Curve *C* is water injection data for 385 test holes in amphibolite at the Oroville Dam, California.

rock. The Camarasa Dam in Spain (1920) is founded almost entirely on fissured dolomite. Lugeon measured the water loss at 11 $m^3$/s. A grout curtain was recommended and consumed 40,000 tons of cement, 20,000 tons of ashes, 130,000 tons of sand and gravel (Jaeger, 1972). Another example is Cold Brook Dam, South Dakota, built by the U.S. Army Corps of Engineers in 1960 on the Madison Limestone about 7 km from Wind Cave National Park. Although records of the volume of cement pumped into the limestone are not available, several water wells were rendered useless immediately downstream in the outskirts of Hot Springs as grouting was in progress. To date no reservoir of any consequence has formed behind the dam due to continued leakage. There are three other Corps of Engineers dams in the Black Hills that are also built on the Madison Limestone; none of them hold water either.

Yellowtail Dam, near Hardin, MT, was built by the USBR and completed in 1965. The dam rises 151 m above the Bighorn River. That dam, built on the cavernous Madison Limestone (Fig. 8.24), has required continuous grouting to the present day. In spite of this grouting, water still seeps out of the Madison Limestone into the canyon below the dam. At the Soap Creek oil field, about 15 km downvalley, piezometric levels in oil wells in the Madison Limestone have increased; it is not clear whether this increase is due to the hydraulic pressure from 1) grouting, or 2) the reservoir water itself.

The improper installation of a grout curtain can have disastrous consequences. In 1979, the Teton Dam in Idaho failed due to an inadequate grouting operation. A description of this dam failure is given in Chapter 5.

### *8.3.5 Sedimentation*

Rates of erosion in the world vary widely due to natural and man-made factors. Prior to human modifications, the world's land masses were probably eroding at approximately 2.5 cm/1000 yr. Presently, the rates have nearly doubled. Streams in the United States have a present average annual dissolved load of 35 tons/km$^2$ of drainage area, and an average annual suspended load of 86 tons/km$^2$ (Curtis et al., 1973; Leifeste, 1974). Due to lack of vegetation, semiarid regions generally have greater erosion rates than humid or arid regions (Langbein and Schumm, 1958). See Figure 8.34.

Sediment rates on rivers vary greatly. The Hwang Ho River in China, with a drainage area of $0.77 \times 10^6$ km$^2$, averages 4000 m$^3$/s in discharge, and $1{,}900 \times 10^6$ tons/yr of sediment; this is equivalent to 15,000 mg/l of sediment as a proportion of discharge. The Amazon River in Brazil, the world's largest river in terms of discharge, with a drainage area of $7.0 \times 10^6$ km$^2$, averages 100,000 m$^3$/s in discharge, and $900 \times 10^6$ tons/yr of sediment; this is equivalent to only 290 mg/l of sediment as ppm of discharge. One-fifth of the total discharge of all the world's rivers runs down the Amazon. Yet the Hwang Ho carries more sediment, testimony to the erosivity of the semiarid loess-covered drainage of its basin compared to the jungles of Brazil.

Agricultural scientists have developed the Universal Soil Loss Equation, which is a crude empirical relationship between the longtime average annual soil losses causes by sheet and rill erosion (Olson, 1981; Mildner, 1982). The equation is expressed as:

$$A = R \cdot K \cdot L \cdot S \cdot C \cdot P, \tag{8.12}$$

where:

$A$ = average annual soil loss per unit land area,

$R$ = measure of erosional forces of rainfall and runoff,

$K$ = soil erodibility,

$L$ = slope-length factor,

$S$ = slope-steepness factor,

$C$ = vegetation cover,

$P$ = crop support practice factor (reflects benefits of contour or strip cropping).

Numerical values for each of the six independent factors have been derived from analysis of data from test plots (for example, see Wischmeier and Smith, 1978).

Reservoir survey data provide an excellent source for determining sediment yield rates for the United States. Strand and Pemberton (1982) found that the size of the drainage base is of prime importance to determine the sediment yield:

$$Q_s = \frac{1098}{A^{0.24}} \tag{8.13}$$

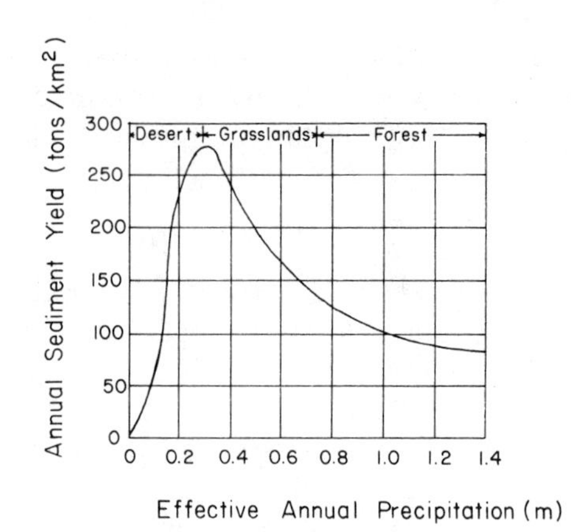


Figure 8.34 Average annual sediment yield as it varies with effective precipitation (after Langbein and Schumm, 1958).

where:

$Q_s$ = annual sediment yield ($m^3/km^2$),

$A$ = drainage area ($km^2$).

Man can increase erosion rates by several orders of magnitude due to urbanization (Fig. 8.35). The increased sediment load in streams draining urban areas causes downstream reservoirs to fill with sediment, limiting their usefulness for recreation and water supply (Kautzman and Cavaroc, 1973). Another example of sedimentation problem is the modern city of Rome, which was once a seaport, but silting of the Tiber River has raised the level of its bed so that today the city is far from the coast and the water level of the river is now at house top level. The $600 million Mangla reservoir in West Pakistan had an original life expectancy of 100 yr, but is now expected to be completely filled with silt within 50 yr

Figure 8.35 Effect of construction intensity and drainage area on sediment yield (after Wolman and Schick, 1967). Most of the data are from the Baltimore and Washington, DC, metropolitan areas.

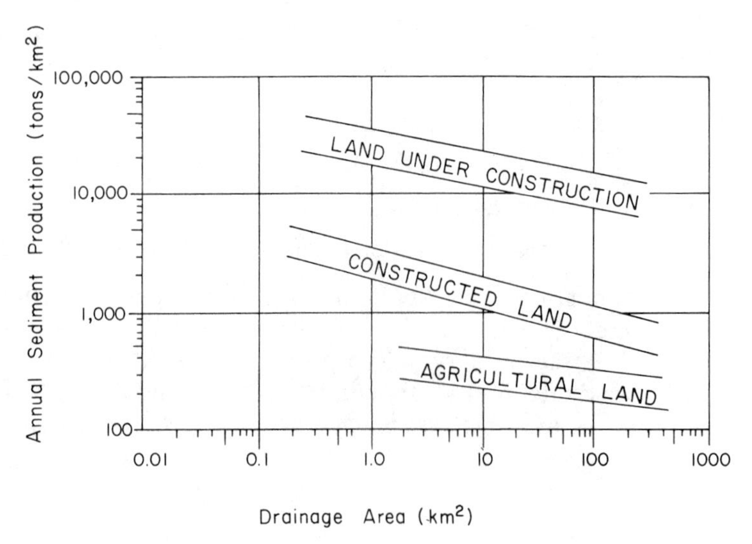

Table 8.4 Rates of Sediment Accumulation in Selected Reservoirs in the United States

| Name and location | $W$, net drainage area, ($km^2$) | C, original capacity ($10^6$ $m^3$) | Sediment production rate ($t/km^2/yr$) | Annual loss of storage (%) |
|---|---|---|---|---|
| Schoarie (Prattville, NY) | 800 | 78.5 | 77 | 0.07 |
| Roxboro (Roxboro, NC) | 19 | 0.6 | 159 | 0.69 |
| Norris (Norris, TN) | 7238 | 2520 | 160 | 0.05 |
| Bloomington (Bloomington, IL) | 155 | 8.2 | 182 | 0.50 |
| Crab Orchard (Carbondale, IL) | 410 | 82.7 | 701 | 0.45 |
| Abilene (Abilene, TX) | 250 | 12.7 | 97 | 0.19 |
| Dallas (Denton, TX) | 2967 | 222 | 463 | 0.72 |
| Mission (Horton, KN) | 20 | 2.3 | 1380 | 1.20 |
| Morena (San Diego, CA) | 279 | 82.1 | 868 | 0.31 |
| Roosevelt (Globe, AR) | 14,770 | 1870 | 395 | 0.25 |
| Mead (Boulder City, NV) | 404,100 | 38,440 | 311 | 0.33 |
| Arrowrock (Boise, ID) | 5560 | 343 | 61 | 0.09 |

After Linsley and Franzini, 1979.

as a result of accelerated soil erosion. The effects of land conservation practices on sediment yield of watersheds has been documented by the U.S. Soil Conservation Service. For example, the rate of sediment yield to Waco Reservoir in Texas was reduced by 38% as a result of land-use adjustments and conservation practices on less than half of its 4300 $km^2$ drainage area (Bruun and Lackey, 1962).

The effects of sedimentation are evident at deltaic areas. For example, the Tigris and Euphrates Rivers join at a location which today is more than 160 km from the Persian Gulf. The water continues to the Gulf as the Shatt-al-Arab River. In the Seventh Century B.C., however, the shoreline of the Persian Gulf was about 300 km northwest of its present shoreline, and the Tigris and Euphrates Rivers entered the Gulf as separate entities (Mirsky, 1983).

The stream's sediment load consists of dissolved minerals and clastic sediment (suspended plus bedload debris). Almost all clastic sediment carried by a river drops out in the delta at the upper end of a reservoir. The trap efficiency (percent of clastic sediment trapped) of most reservoirs is about 90–100% (Heinemann, 1981; Trujillo, 1982). As a reservoir becomes filled, some turbidity currents will allow suspended debris to pass beyond the dam and, as a result, the trap efficiency decreases. Trap efficiency is greatest for large reservoirs on small streams, and smallest for small reservoirs on large streams.

Table 8.4 shows rates of sediment accumulation in U.S. reservoirs. Figure 8.36 shows typical patterns of sediment accumulation behind a dam. Trap efficiency decreases with age as the reservoir capacity is reduced by sedimentation. The construction of upstream dams, on the other hand, extends the life of a downstream reservoir. Boulder Dam on the Colorado River, for example, accumulated 1.5 billion tons of sediment during the 10 years following its construction in 1935. Its rate of sedimentation has been reduced recently because of the construction of Glen Canyon Dam upstream.

The life expectancy of a dam (yr) can be roughly determined by dividing the volume of the reservoir ($m^3$) by the yearly volume of debris brought in by the

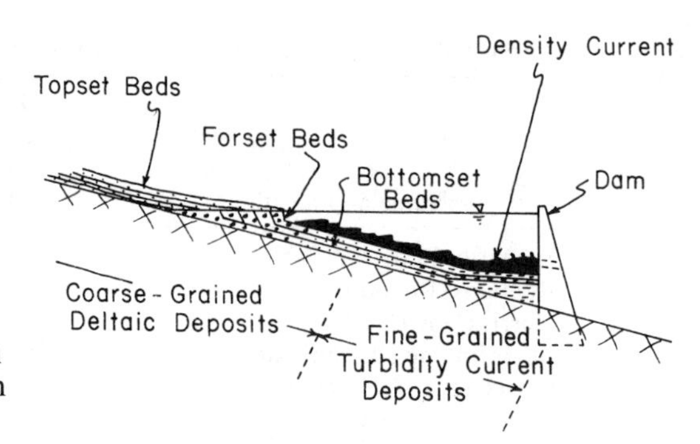


Figure 8.36 Schematic cross section illustrating sediment accumulation in a typical reservoir.

river ($m^3$/yr). Deltaic material has surprisingly low density, with an average of about 800 kg/$m^3$ for fresh sediments and 1300 kg/$m^3$ for older sediments (Linsley and Franzini, 1979). The life expectancies of some California reservoirs have already been reached; debris now fills the entire reservoir. (Locally these places are known as debris dams.)

The following example (modified from Krynine and Judd, 1957) shows how the trap efficiency of a reservoir (the percentage of sediment volume retained) affects the total sediment accumulation in a reservoir.

> *Trap efficiency.* The percentage of the total inflow of sediment retained in a reservoir usually is termed its trap efficiency. This ability of the reservoir to trap silt depends on the ratio between storage capacity and inflow, the age of the reservoir, the shape of the reservoir basin, the type and method operation of the outlets, the grain size of the sediment, and its behavior under various conditions.
>
> The trap efficiency, or TE, of an existing reservoir can be determined in terms of the ratio of sediment inflow to outflow as follows:
>
> $$\mathrm{TE} = \frac{S_i - S_o}{S_i}, \tag{8.14}$$
>
> wherein $S_i$ is the volume of sediment that enters the reservoir every year and $S_o$ is the volume of sediment that leaves the reservoir every year past the dam.
>
> If the reservoir was designed merely to pass the major part of an annual flood with little or no storage, the trap efficiency would be very low, as there would be little sediment trapped behind the dam. Conversely, a desilting basin would have a trap efficiency of almost 100%. (Theoretically 100% trap efficiency is not possible; however, actual measurements on some dams have shown 100% trap efficiency.)
>
> The trap efficiency of a proposed reservoir can be roughly estimated by dividing the proposed reservoir capacity $C$ ($m^3$) by the expected annual inflow ($I$) of water ($m^3$). This ratio $C/I$ can be compared with special curves (such as shown in Fig. 8.37) to obtain an estimated trap efficiency. In most cases, the trap efficiency is assumed at about 95% or more for proposed reservoirs. With these data available, an approximation of the useful life of a reservoir ($R_L$) can be computed:
>
> $$R_L = \frac{C}{S_i \times \mathrm{TE}}. \tag{8.15}$$

Krynine and Judd (1957) give the following example of reservoir life computation based on surveys of Lake Corpus Christi, Texas:

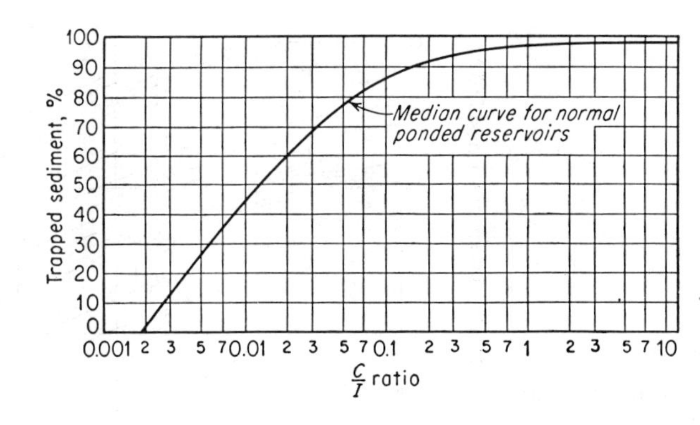


Figure 8.37 Curve illustrating how trap efficiency is related to C/I ratio (from Krynine and Judd, 1957).

| | |
|---|---|
| Reservoir capacity in 1942 | 54,034,000 $m^3$ |
| Reservoir capacity in 1946 | 48,604,000 $m^3$ |
| Therefore the average annual sediment deposit ($S_i - S_o$) is | 1,358,000 $m^3$ |
| $S_o$ = Measurement of annual average sediment carried past dam | 481,000 $m^3$ |
| Average annual water inflow | 911,049,000 $m^3$ |

Theoretically, $\dfrac{C}{I} = \dfrac{54{,}034{,}000}{911{,}049{,}000} = 0.059.$

According to Fig. 8.37, the trap efficiency would be theoretically about 79%. Actual computations, since the data were available, from above, shows that:

$$\mathrm{TE} = \frac{S_i - S_o}{S_i} = \frac{1{,}358{,}000}{1{,}846{,}000} = 74\%.$$

By using the real data, the reservoir life remaining in 1942 was:

$$R_L = \frac{54{,}034{,}000\ \mathrm{m}^3}{(0.74)(1{,}846{,}000\ \mathrm{m}^3/\mathrm{yr})} = 40\ \mathrm{yr}.$$

This period of life would be applicable only if the trap efficiency remained the same. Usually the trap efficiency decreases during the life of the reservoir. In order to utilize the TE versus $C/I$ ratio curve for a reservoir filling with sediment (decreasing $C$), it is necessary to break the calculations into increments, evaluating TE successively as $C$ decreases.

Figure 8.38 shows a series of empirical curves developed by Stall and Lee (1980) for sedimentation rates into Illinois reservoirs. The factor $C/W$ is the ratio of capacity/watershed, where units of reservior capacity are $m^3$ and watershed is $m^2$. For example, if the reservoir has a capacity of $75 \times 10^6$ $m^3$ and the watershed area drainage area is 686 $km^2$, then

$$C/W = \frac{75 \times 10^6\ \mathrm{m}^3}{686 \times 10^6\ \mathrm{m}^2} = 0.11\ \mathrm{m}\ (\text{or } 110 \text{ expressed as mm}).$$

From Fig. 8.38, a $C/W$ = 110 mm with watershed area of 686 $km^2$ indicates an annual capacity loss of about 0.15% by volume.

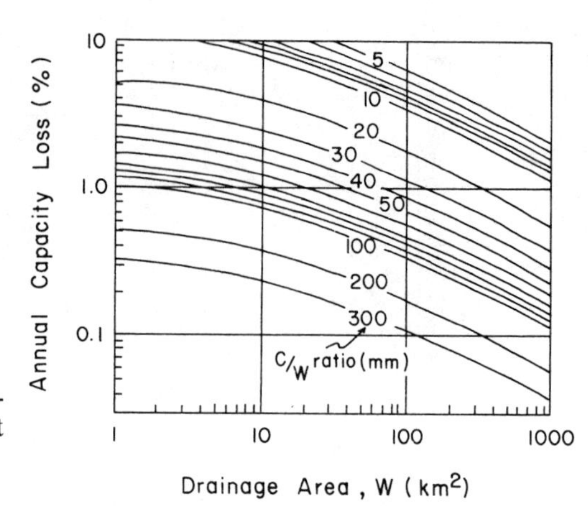


Figure 8.38 Generalized graph for estimating annual capacity loss to sediment for Illinois reservoirs (from Stall and Lee, 1980).

Ultimately any reservoir will completely fill with debris. The process may be slowed by construction of another reservoir just upstream, as is the case with six mainstream dams on the Missouri River in Montana, North Dakota, and South Dakota. However, even the Missouri River reservoirs are subject to filling by sediment because of tributary stream feeding directly into the reservoir and because of wave erosion on the hillslopes composed of Cretaceous shales and glacial drift. Koopersmith (1980) has shown that for the years 1966–1976 the average rate of shoreline erosion of Lake Sharpe in South Dakota, a mainstem reservoir built in 1966, is 3.4 m/yr. The eroded debris is transported by longshore drift and accumulates in the bays along the shore.

### *8.3.6 Environmental Impacts*

The history of bureaucracy in the United States shows that when a federal agency has been created to accomplish some intangible goal, be it NASA's exploring space or EPA's protection of the environment, there is an incentive from within the agency to keep growing. Most water projects are sponsored by the federal government. Many federal water projects, even though unsound economically or environmentally, are built anyway because Congressmen, fueled by pressure from agency personnel and special interest groups, tend to support each other's favorite pork barrel projects. (See Chapter 14.)

The benefits of dam construction are well documented and distributed by the dam-building agencies. These include hydroelectric power generation, flood control, irrigation, and recreation. The negative impacts are not so well known and include the following:

1. Dams inundate fertile land. (Flood plain alluvial soil is the most productive soil on earth.)
2. Dams eventually fill up with alluvium. As a related issue, erosion below dams occurs (see discussion in Sect. 8.1.6 above).
3. Dams do not eliminate all floods, and distant downstream protection may be exaggerated. Flood protection is afforded only within a short proximity below a dam. Despite flood-protection facilities, damage from floods in the United States is actually increasing (Table 8.5).

Table 8.5 Great Floods in the United States from 1889 to 1979

| Date | Location | Lives lost | Estimated damages[a] |
|---|---|---|---|
| May 1889 | Johnstown, Pennsylvania, dam failure | 3000 | — |
| September 8, 1900 | Hurricane—Galveston, Texas | 6000 | 30 |
| May–June 1903 | Kansas, Lower Missouri, and Upper Mississippi River | 100 | 40 |
| March 1913 | Ohio River and Tributaries | 467 | 147 |
| September 14, 1919 | Hurricane—south of Corpus Christi, Texas | 600–900 | 22 |
| June 1921 | Arkansas River, Colorado | 120 | 25 |
| September 1921 | Texas rivers | 215 | 19 |
| Spring of 1927 | Mississippi River valley | 313 | 284 |
| November 1927 | New England rivers | 88 | 46 |
| March 12–13, 1928 | St. Francis Dam failure, southern California | 450 | 14 |
| September 13, 1928 | Lake Okeechobee, Florida | 1836 | 26 |
| May–June 1935 | Rupublican and Kansas Rivers | 110 | 18 |
| March–April 1936 | Rivers in Eastern United States | 107 | 270 |
| January–February 1937 | Ohio and Lower Mississippi River basins | 137 | 418 |
| March 1938 | Streams in southern California | 79 | 25 |
| September 21, 1938 | New England | 600 | 306 |
| July 1939 | Licking and Kentucky Rivers | 78 | 2 |
| May–July 1947 | Lower Missouri and Middle Mississippi River basins | 29 | 235 |
| June–July 1951 | Kansas and Missouri | 28 | 923 |
| August 1955 | Hurricane Diane floods—Northeastern United States | 187 | 714 |
| December 1955 | West Coast rivers | 61 | 155 |
| June 27–30, 1957 | Hurricane Audrey—Texas and Louisiana | 390 | 150 |
| December 1964 | California and Oregon | 40 | 416 |
| June 1965 | South Platte River basin, Colorado | 16 | 415 |
| September 10, 1965 | Hurricane Betsy—Florida and Louisiana | 75 | 1420 |
| January–February 1969 | Floods in California | 60 | 399 |
| August 17–18, 1969 | Hurricane Camille—Mississippi, Louisiana, and Alabama | 256 | 1421 |
| July 30–August 5, 1970 | Hurricane Celia—Texas | 11 | 453 |
| February 1972 | Buffalo Creek, West Virginia | 125 | 10 |
| June 1972 | Black Hills, South Dakota | 237 | 165 |
| June 1972 | Hurricane Agnes floods—Eastern United States | 105 | 4020 |
| Spring 1973 | Mississippi River basin | 33 | 1155 |
| June–July 1975 | Red River of the North basin | <10 | 273 |
| September 1975 | Hurricane Eloise floods—Puerto Rico and Northeastern United States | 50 | 470 |
| June 1976 | Teton Dam failure, southeast Idaho | 11 | 1000 |
| July 1976 | Big Thompson River, Colorado | 139 | 30 |
| April 1977 | Southern Appalachian Mountains area | 22 | 424 |
| July 1977 | Johnstown—western Pennsylvania | 78 | 330 |
| April 1979 | Mississippi and Alabama | <10 | 500 |
| September 12–13, 1979 | Hurricane Frederic floods—Mississippi, Alabama, and Florida | 13 | 2000 |

[a]In millions of dollars.
From Hays, 1981.

4. In arid areas dams evaporate a lot of water. For example, the Powell Reservoir formed by Glen Canyon Dam on the Colorado River evaporates 9% of the Colorado River, while the residual water is proportionally enriched in dissolved salts. The average salinity of the Colorado River, for example, increases from less than 50 mg/l in the headwater to about 900 mg/l at Imperial Dam, the last point of major diversion in the United States. Coupled with the return irrigation water from the Gila Valley and the closure

Figure 8.39 Fort Meade dam in Sturgis, SD. The dam was ruined on June 9, 1972, due to an inadequate spillway that forced flood water to cascade over the rock-faced earth dam. A thin concrete upstream wall over the dam miraculously prevented complete failure. The Corps of Engineers is pumping water out of the reservoir prior to demolishing it.

of Glen Canyon Dam in 1962, the salinity of Colorado River water entering Mexico increased from 800 mg/l to 1500 mg/l (El-Ashry, 1980). This water is practically unfit for irrigation. Throughout the world, soil salinity problems associated with saline irrigation water, poor drainage, and mismanaged irrigation is seriously affecting about 40,000 km$^2$ or one-third of the world's irrigated land.

5. Dams can fail. Figure 8.39 is a picture of a failed dam in South Dakota. (Also see the example of the Teton Dam, Idaho (Chapter 5).) Another example is the November 6, 1977, dam failure at Toccoa, GA. The 8-m-high dam, built in 1937, failed following heavy rains, and killed 39 people at Toccoa Falls Bible College. The dam had never been inspected by state or federal officials. Jansen (1980) describes the lessons learned fron an analysis of 41 significant dam failures or accidents worldwide. Most dams fail due to an inadequate spillway.
6. Dams can leak. (Examples are given in Section 8.3.4 above.) Lake Nasser, built in 1964 in Egypt, loses an estimated $15 \times 10^6$ m$^3$/yr of water into the Nubian Sandstone. The dam's planners assumed that the Nile's incoming sediment would eventually seal the bedrock. But the sediment has been settling mostly in the lake's deep center, along the old riverbed, and the Nubian Sandstone lining the reservoir's entire 500 km-long western bank is capable of absorbing enormous quantities of water.

7. Dams can cause unexpected physical changes. An example is at the town of Niobrara, NE. The construction of Gavins Point Dam on the Missouri River in 1955 formed Lewis and Clark Reservoir, which led to the creation of a large delta where the Niobrara River empties into the reservoir. Consequently, the water table rose adjacent to the deltaic area and forced the abandonment of the town of Niobrara in 1977. The new townsite was built on a bluff south of the original townsite. (The cost of this relocation was not part of the original B/C ratio! According to Griggs and Gilchrist (1983), Gavin's Point dam was originally supposed " ... to be built *above* the mouth of the Niobrara River, but politicians downstream persuaded the Army Corps of Engineers to build it near their town about 65 km east of the original site.")
8. Dams cause many kinds of ecological changes. These include displacement of wild life from the reservoir area, disruption of fish spawning areas, and unusual, unexpected ecological changes such as have occurred at the Aswan High Dam, Egypt. Malaria-carrying mosquitoes on Lake Nassar are increasing. Scorpions and crocodiles swarm the lake's islands and shores in their flight from encroaching waters. Parasite-carrying snails infest canals (causing 10% of the fatalities in Egypt). The sardine and shrimp fishing on the Mediterranean coast has decreased, and the rodent population in Cairo has increased due to the lack of flushing of sewer systems from periodic floods. A more serious problem with the Aswan High Dam is that drainage of the more than 5000 $km^2$ acres of the nation's rich farm land is below the dam. Much of that land has become increasingly saline. Further, the natural silt which formerly enriched the soil on farms of the Nile delta is now trapped in Lake Nassar; thus, it has been necessary to increase the use of imported fertilizers. Because of the sediment discharge decline into the Mediterranean Sea, the delta is being eroded and the coastline is advancing inland, in some sectors as much as 2 m/yr. The great reservoir was supposed to be filled by 1970, but by 1980 was only half full. The Nubian Sandstone, exposed along the new lake's 500-km west bank, is absorbing much of the water. Further, high winds cause tremendous evaporation, about 15 billion $m^3$/yr, which is 50% greater than originally estimated (Griggs and Gilchrist, 1983). The great dam has extracted an environmental price, despite the flood control and drought-prevention benefits extolled by William Shenouda, Egyptian Undersecretary of State, responsible for the dam: "We are forever saved from the cycle of 7 lean years and 7 fat years that Joseph encountered in Biblical times."

Some dams, built as a source of water supply, are not economically justified. For example, the total estimated cost of dams, collectors, and filtration plant for the Warm Springs Dam, nearly completed in Sonoma County, California, is $442 million. Ground water reportedly could have provided the same amount of water from this area for only $10 million, but geologists were not consulted. The blame for this mistake (Boudreau, 1982) is " ... the result of having lawyers ... in the decision-making positions, and the fact that all the advice they were receiving came from the top layer of bureaucrats in the (California) Department of Water Resources, who are all engineers and remarkably ignorant of groundwater matters." Boudreau (1983) further criticizes the large dams and canals and related projects built by the California Department of Water Resources (DWR) as part of the State Water Plan (SWP): "Although the SWP was supposed to be for moving water from Northern California to the Los Angeles area for urban use, the high cost of pumping the water 3000 feet over the Tehachipi Mountains will make it uneconomic for many uses. In the overview, the SWP is a device for supplying low-cost irrigation water to corporate farming interests in the southern part of the San Joaquin Valley. . . . The beaver mentality of many engineers often leads them to building something that is totally uncalled for."

TVA's Tellico Dam on the Little Tennessee River offers a sad commentary on dam building in the United States today. The dam was completed despite the fact that few people wanted it. In 1979 the dam was closed and the inundation of one of the richest valleys and most highly used trout streams in the southeast began. TVA's own studies showed that more jobs could have been created by leaving the valley for agriculture and forestry, rather than by flooding it. The $138 million project has essentially no electric power-producing generation, and yet the reason for its existence was allegedly for power production. It is unfortunate that the rare snail darter, an insignificant fish, became involved during the much-publicized environmental impact hearings. The issue was much greater than that. John Gibbons, director of the U.S. Congress Office of Technology Assessment, said, "I think Tennessee's Congressional delegation misled people. The record is very clear that the amount of power produced by that dam will be negligible. We had a dam with a lot of money invested in it, and in order to keep the momentum going, the delegation tried to focus on something else. And the something else was the snail darter." The bureaucratic push was echoed by newsmen such as Paul Harvey who indicated that environmentalists were not acting in the nation's best interest: "Some of those scaredy-cats are helping a 3-inch fish hold up a $100 million power project."

## 8.4 Other Flood Control Structures

### *8.4.1 Levees*

As a flood-swollen stream rises and begins to overflow its banks, stream velocity decreases, largely because of riparian vegetation along the channel banks. As a result, some of the clastic load drops out, forming a natural levee. Natural levees are largely composed of sand and silt, built adjacent to the channel. Upon overtopping a natural levee, a flooding river will deposit fine silt and clay farther onto the flood plain, where they settle out due to lessening velocity caused by vegetation or oxbow lakes. Fine-grained flood plain deposits formed by overbank accretion in this manner make wonderful soils for agricultural production, but they often have undesirable engineering properties. They tend to be mixed with organic matter from swamps, have high natural water contents, and are much more plastic and compressible than other types of alluvium such as levee deposits or gravel areas formed at meander point bars.

A common method of flood control is to build a dike on the natural levee so as to help contain the stream flow in its channel. Along the Yellow River in China, dikes have been built by farmers since 603 B.C. People living in flood plains in some places in the United States have artificially increased levee heights so that today many streams have earth or concrete levees more than 10 m high.

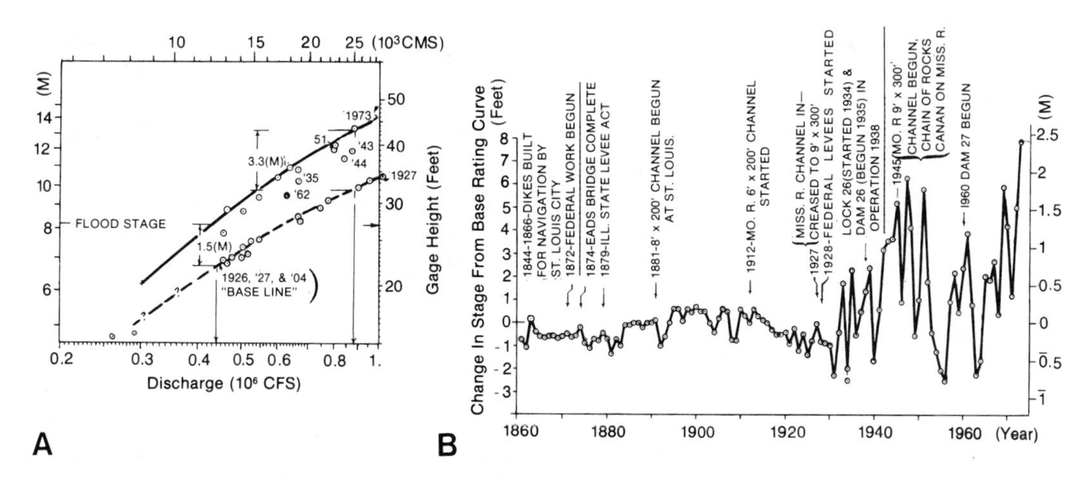


Figure 8.40 Stage-discharge relationships for the Mississippi River (from Belt, 1975). **A**. Maximum annual stage and discharge relations including base (1861–1927) and 1973 preliminary rating curves, Saint Louis. **B**. Relation of change in stage from base rating curve to time at Saint Louis.

For example, Three River, TX, a town of 2500 people, is a "walled city," completely encircled by a 7-km-long levee 6 m high built by the U.S. Army Corps of Engineers.

Artificial levees have been built by many private and quasigovernment agencies along the Mississippi River. Some 3200 km of artificial levees exist in the lower Mississippi River Valley (Linsley and Franzini, 1979). While these structures serve in a limited way to protect part of a flood plain from small floods, they promote a false sense of security in that they get overtopped by larger floods. North of St. Louis, 39 levees were breached or overtopped by floods in 1973 (Bolt et al., 1977). Once an artificial levee is overtopped, a harmful deposit of sand may be spread over an arable field. Bank failures were noted in 1973 not only from levees being overtopped, but by piping under the levees. A high stream levee promotes ground-water movement under the levee, producing sandboils on the lower parts of the adjacent flood plain, eventually leading to failure (Noble, 1976).

Any encroachment on the flood plain interferes with the natural stream flow pattern during high discharges. Construction of a levee on one side of a river adds to the magnitude of flooding on the other side of the river or upstream (Waananen et al., 1977). There is no uniform levee code for the Mississippi River. Rationale for levee construction in the St. Louis, MO, area was promoted by the U.S. Army Corps of Engineers with the idea that flood water would be quickly carried out of the area. This only compounded the problem for residents of Charles County downstream, as illustrated by Figure 8.40. The 1973 Mississippi River flood broke the stage (river level) records between Burlington, IA, and Cape Girardeau, MO. Yet the discharge was not as great as several other floods including the 1927 flood. The curves show that post-levee floods have much greater stages than pre-levee floods. With no levees, the 1973 flood would probably have crested 3.3 m below the stage that was observed. If the 1927 flood were to happen today, the flood would be deeper than it was in 1927.

A large weir may be installed in a levee so as to spill water into the flood plain when the river is near the top of the levee. Many countries provide this form of bypass so that the flood plain can be farmed during the summer months, but may receive flood waters during the rainy season. An example is the Colusa bypass weir along the Sacramento River, California. The water is passed by wiers in the levee onto the flood plain, allowing for silt-laden water to saturate the alluvial soil, providing ground-water recharge. This scheme is a form of flood-plain management, where specific areas are inundated with the help of man.

## *8.4.2 Channel "Improvements"*

An engineering solution to floods includes channelization, i.e., digging a deep, straight ditch through the flood plain in place of the natural stream (Fig. 8.41). This may locally reduce the amount and duration of flooding. Channelization allows for the speedy passage of flood water because, as the stream gradient and hydraulic radius is increased and roughness decreased, the velocity is increased according to the Chezy-Manning equation [Eq. (8.11)]. This increased velocity may cause physical problems elsewhere, however. Emerson (1971) shows that scour leading to bridge failure, and increased flooding occurred below a channelized reach of the Blackstone River, Missouri.

Channelization is an ecological disaster (Ruhe, 1970; Keller, 1976; Reuss, 1973). A meandering stream with alternating pools is transformed into a bulldozed or concrete ditch with trapezoidal or rectangular cross section. Aesthetically, they can hardly be called an improvement. Deep, cool pools of water at meanders or behind a large rock, capable of supporting aquatic life, are transformed into a continuous shallow gravel reach, where the shallow water is warmed by the sun and hence is unable to maintain cool-water aquatic biota such as trout. There are many ugly examples of channelized streams in the United States. For example, concrete-lined channels move the San Antonio River through San Antonio, TX. Hot Springs, SD, once a health spa and resort town, has its once-beautiful Fall River captured in an ugly concrete ditch through the entire downtown area. Costa and Jarrett (1981) show that stream channelization efforts in mountainous areas of Colorado have been ineffective because the debris-flow deposits fill the channels with mud and rock debris, blocking the channel and causing subsequent flow in new directions.

## *8.4.3 Flood Plain Management*

Floods are one of man's greatest natural hazards (Table 8.4). Flood damage is not only a problem for the United Stated. Ancient civilizations struggled to earn a living from the fertile land adjacent to the Nile, Tigris, Euphrates, Indus, and Yellow Rivers. About 900,000 Chinese lost their lives when the Yellow River flooded in 1887; another 100,000 died when the Yangtzee River flooded in 1911 (Tank, 1983a).

In 1966, the Arno River ripped through Florence, Italy, one of the foremost art centers of the world. "Firenze bella" (beautiful Florence) is situated on the flood plain, with commercial stores crowded along the river banks and even in bridges across the river. The 1966 flood washed away the Niccolo Bridge and deposited mud in many cathedrals and museums, ruining ancient works of art.

Figure 8.41 Stream channelization projects. **A**. Deadman Creek at Strugis, SD—a rectangular concrete channel. **B**. St. Joseph's Creek at Downer's Grove, IL—a trapezoidal ditch.

In 1974, Brisbane, Australia, was inundated by the Brisbane River. Noting the recurrence of these floods, the Australian Minister for Environment and Conservation, Dr. Moss Cass stated: "Flood plains are for floods. . . . Except in relation to small areas, there is simply nothing man can do to prevent widespread flooding and drought. It is simply impractical to build dams to prevent inundation of all flood plain areas. . . .

Legget (1973) adds this historical note concerning European floods:

> It is interesting to note that in Central Europe the builders of medieval towns generally avoided building in the flood plains. They were rightly afraid of floods, and they knew well the difficulty of construction on the wet ground found adjacent to rivers. The old parts of the older cities of Europe will usually be found to be located on the higher ground provided by river terraces, well above the flood-water levels. This wise practice, not based consciously on geology but utilizing the same kind of "horse sense" that the application of geology properly involves, has been widely forgotten in more recent years. Study of the growth of European cities will show that from the recent years of the second half of the last century, intensive building on flood plains has proceeded under the pressures of imprudent development of urban living.

Historically, flood problems have resulted in the demand for dam construction. For example, when a 1965 flood hit Denver, the Corps of Engineers responded with the $85 million Chatfield Dam on the South Platte River. According to the Corps of Engineers, "When the flood of June 16, 1965, swept through the area, it left in its wake millions of dollars in damages and 13 lost lives. Homes were tossed around like toothpicks. The flood triggered intense activity on the part of citizens and state and congressional officials to expedite construction of the Chatfield Dam."

As flood and subsequent dam construction activities throughout the United States are repeated year after year, it is important to examine another alternative, and that is simply not to build on the flood plains in the first place. Flood plain management is an alternative to hard engineering structures such as dams, levees, and channel improvements. The basic idea of flood plain management is for a governing agency to zone a flood plain in such a way as to prohibit residences and/or other structures. The land can be used as agricultural areas, parks, railways, and highways, and limited commercial buildings which would be little affected by floods.

There is no clear-cut way to exactly identify a flood plain. The 100-yr flood inundation is determined by calculating the 100-yr flood discharge, and dividing this discharge by an assumed average velocity to arrive at the cross-sectional area of flow, from which a stage can be ascertained. The amount of land inundated by this river stage is determined by cross-sectional topographic profiles of the valley. In general, the 100-yr flood inundation area coincides fairly well with what geologists map as Quaternary alluvium. Historical records of inundation offer additional evidence of flood hazard areas, as do flood plain vegetation studies (Costa and Baker, 1981).

Land-use planning for flood-prone areas in the United States is slowly receiving long-overdue acceptance. Comprehensive planning involves complex multijurisdictional involvements (Waananen et al., 1977). Local flood plain management schemes have recently been put into use along the South Platte River, at Littleton, CO, the Kickapoo River at Soldier's Grove, WI, Indian Bend Wash, Scottsdale, AZ, and the upper Charles River, MA.

It is instructional to follow the development of the flood plain management program in Rapid City, SD, following the devastating June 9, 1972, flood. Prior to the flood, there was complete lack of concern about flood hazards. The flood caused the loss of 238 lives and great economic hardship (Rahn, 1975). In the following years the city planning department, through a $48 million federal grant,

Figure 8.42 Flood plain of Rapid Creek, Rapid City, SD. **A**. June 10, 1972. Destroyed houses after a catastrophic flood. **B**. Floodway in 1980. The area was converted to a park.

purchased all homes and most commercial establishments on the flood plain. All remaining homes were removed, and the land was converted into recreational areas including a golf course, soccer and baseball fields, tennis courts, a bike path, and a low maintenance parkway (Fig. 8.42). Thus, future floods will pass relatively unobstructed through the town and will present little hazard to people or residences (Rahn, 1984).

Local governing bodies usually do not have sufficiently restrictive measures to prevent encroachment of flood plains. Boulder, CO, for example, exemplifies a community built to a large degree on the flood plain of a major river. Despite dozens of flood plain surveys and numerous floods, skepticism of the flood hazard remains, and efforts to create a serious flood plain management has received little success. The flood plain is still occupied, and serious potential for a catastrophe remains (Leveson, 1980).

To help reduce the hazards of floods and other natural phenomena in the United States, the Federal Emergency Management Agency (FEMA) was created in 1979. Its goals are to 1) encourage local governing bodies to develop flood plain management programs and 2) consolidate federal agencies involved in disaster relief. As a method of encouraging zoning of flood-prone areas, the federal government can withhold issuance of flood insurance and other government favors until the local government adopts a flood zone policy. Flood insurance is generally not available from private companies. In 1968, the U.S. Congress passed the National Flood Insurance Act (P.L. 448), which provides insurance if the community involved sets up and enforces standards for land-use control in flood-prone areas. The voluntary program was not very effective, as attested by the 1972 flood in Rapid City, SD, where only 29 houses out of 1200 lost had flood insurance policies. In 1973, Congress passed the Flood Disaster Protection (P.L. 93–234) which requires that communities must participate in the National Flood Insurance Program or else federal financial assistance and loans would not be available for structures in flood-prone areas. By and large, the program has promise of coercing local governments to look at the flood problem. There are still drawbacks; for example, participation is wholly at the discretion of local governments. Another drawback is that the basis of proving that a structure is safe is that it must be shown to lie above the 100-yr flood elevation. This may only set the scene for very great damage in the event of a flood having a greater magnitude than the 100-yr flood. The basis of FEMA's flood program is the "100-yr" flood plain, a term which unfortunately is not always understood or accepted by local officials who are unfamiliar with the sophisticated methods involved in defining this area. Another drawback of the national flood insurance program is that policyholders do not cover the cost of large deficits produced by increasing costs of payments for flood losses. For example, in 1983 a deficit of over $100 million was accrued, which was underwritten by the U.S. Treasury with money from general taxpayers.

Whether flood plain management works or not ultimately depends on local government. Don Barnett, mayor of Rapid City during 1972, stated it clearly:

> I think the first thing that we have to do is have the courage to say "no" to a developer. Say: "No, it's too damn close to the flood way. You can't build there," and write a flood plain ordinance that has some teeth in it where you can stop the guy. Some developers want to go in, build, make their dough, cut and run. You have to have the courage to say no to that potential investment in your flood way. That takes courage because the schools need the property tax base. Your county needs it. Your city needs it. Your municipality or special district get their money from property taxes. They need to strengthen their tax base, but strength in the wrong place is wrong because it endangers life and property in the future.

Bue (1967) succinctly points out that "floods are natural and normal phenomena. They are catastrophic simply because man occupies the flood plain, the high-water channel of a river."

### 8.4.4 Flood Plains and Land-Use Planning

Dam construction has traditionally been the method of reducing floods in the United States. An environmental and more acceptable alternative is flood-plain management, the restriction of the use of flood-prone lands in order to minimize flood damage. As discussed by Hoyt and Langbein (1955), "The primary goal of flood zoning must be to prevent unwarranted constrictions that reduce the ability of the channel to carry water and thereby add to the height of floods. . . . Most agricultural, grazing particularly, represents justifiable use of flood lands. . . . Home building, on the other hand, is an example of the most unwise use of flood land."

In addition to flood hazards, flood plains present other engineering problems. Due to high water tables, foundation excavations may require pumping, and foundation walls must be designed to be watertight. In some cases buildings must be designed to withstand hydrostatic uplift. Water and sewer line trenches are excavated with difficulty. Septic tanks and drainage fields may be below the water table and operate ineffectively. Alluvium may contain irregular deposits of clay, sand, or gravel, making variable foundation conditions. Because the alluvium is largely saturated, liquid limits may be reached and the bearing capacity may be limited and variable.

A related issue is the continued destruction of arable land by urbanization or reservoir inundation. The world's food supply is limited, and flood plains are the most fertile areas in the world. According to the "Global 2000" report, due to a rising world population, increased urbanization, and destruction of the world's farmlands, the amount of arable land will decrease in the world from 0.4 to 0.25 hectares/person between 1974 and the year 2000. Increased crop yields cannot be expected to keep pace with increased population solely because of the development of improved crops and better fertilizers (Evans, 1980). Viewed in the context of world food supply, it is urgent that a new land ethic be developed. In this sense, flood-plain management is a better alternative than dam construction.

Streams and flooding need to be recognized as a part of the physical environment. Alternatives to traditional dam construction must be evaluated. Is a flowing river more valuable aesthetically and/or recreationally than the kilowatts produced by hydropower (Fig. 8.43)? Do people need to be forcibly regulated to avoid places where geological hazards such as floods exist? Or is construction to be allowed anywhere, even on the apex of alluvial fans (Fig. 8.44)? Once built, should the residents of these endangered houses expect the government to construct dams for their protection? Would a sound management program eliminate the need for expensive, hard engineering structures, and at the same time provide for flood protection and the best use of the land?

Dams have their place in a civilized world. They serve numerous useful purposes, and engineering geologists will continue to be involved in the design and construction of more dams. Nevertheless, flood-plain management is a more environmentally acceptable alternative to flood control than dam construction.

Figure 8.43 Hell's Canyon, Snake River, ID. One of the last reaches of swiftly flowing rivers in the western United States.

Figure 8.44 Urbanization of alluvial fan at Georgetown, CO.

## Problems

1. See Table 8.5, Great Floods in the United States.
   - **a.** On arithmetic graph paper, draw a histogram showing dollar flood loss per decade. What is the general trend of flood loss in the United States?
   - **b.** In the generalized answer to a above, what factors are not considered?
   - **c.** Since the number of flood-control dams in the United States has increased dramatically during this period, have dams been the solution to flood damage? Why or why not?

2. Derive an equation similar to Eq. (8.6) for the Powder River, Wyoming (Fig. 8.6).

3. See Table 8.1, stream discharge data for Battle Creek and Castle Creek, SD.
   - **a.** Construct flow-duration curves for these two streams using Fig. 8.16 as an example. Use three-cycle log-probability paper. Plot ordinate as $m^3/s$ per $km^2$ of drainage area. (Hint: it is necessary to rank all the data initially, but about 10 points will suffice to plot each curve.) What is the median discharge ($m^3/s$ per $km^2$) for each stream?
   - **b.** Considering low flow characteristics, what stream would be the better trout fishing? Why?
   - **c.** Calculate the average (mean) discharge ($m^3/s$ per $km^2$) for each stream.
   - **d.** In these two instances, why are the median values different from the mean values?
   - **e.** What geologic differences exist on these two watersheds that cause the difference in the discharge characteristics between these two streams? (Hint: see text for discussion of hydrogeologic controls for these two streams.)

4. One of the unresolved problems of flood forecasting is the precise determination of how often a flood (having a given discharge) can be expected to occur. Because only a small percentage of all the streams in United States have been gaged for more than a few decades, it is difficult to determine, for example, the 100-yr or 500-yr flood. Consider the data below which is the instantaneous peak discharge for Rapid Creek in Rapid City, SD, in each of 32 yr of record:

| Water year | Discharge ($m^3/s$) |
|---|---|
| 1951 | 4.05 |
| 1952 | 73.6 |
| 1953 | 4.30 |
| 1954 | 3.90 |
| 1955 | 9.23 |
| 1956 | 3.68 |
| 1957 | 12.3 |
| 1958 | 2.29 |
| 1959 | 2.32 |
| 1960 | 2.32 |
| 1961 | 2.82 |
| 1962 | 37.1 |
| 1963 | 5.41 |
| 1964 | 7.59 |
| 1965 | 17.4 |
| 1966 | 3.96 |
| 1967 | 12.4 |
| 1968 | 5.61 |
| 1969 | 3.94 |
| 1970 | 6.94 |

| | |
|---|---|
| 1971 | 10.9 |
| 1972 | 885.0 |
| 1973 | 4.9 |
| 1974 | 14.6 |
| 1975 | 2.61 |
| 1976 | 18.0 |
| 1977 | 5.49 |
| 1978 | 12.3 |
| 1979 | 4.75 |
| 1980 | 3.37 |
| 1981 | 3.74 |
| 1982 | 7.48 |

**a.** Included in the data is the exceptionally large flood of 1972, which killed 238 people in the Rapid City area. Based simply on the number of years of record and the fact that the 1972 flood occurred one time during this interval, how often would a flood of this magnitude be expected?

**b.** The simplistic approach described above can be better treated by a mathematical treatment such as shown in Table 8.2 and Fig. 8.20. Using the graph paper below, plot a recurrent curve for the Rapid Creek flood data. (Hint: suggest that an eye-ball best fit straight line be plotted through the points except ignore the 1972 flood. Assume that, because of its magnitude, the 1972 flood does not nicely conform to the rest of the data, and, as such, can be ignored.) What is the 100-yr flood discharge?

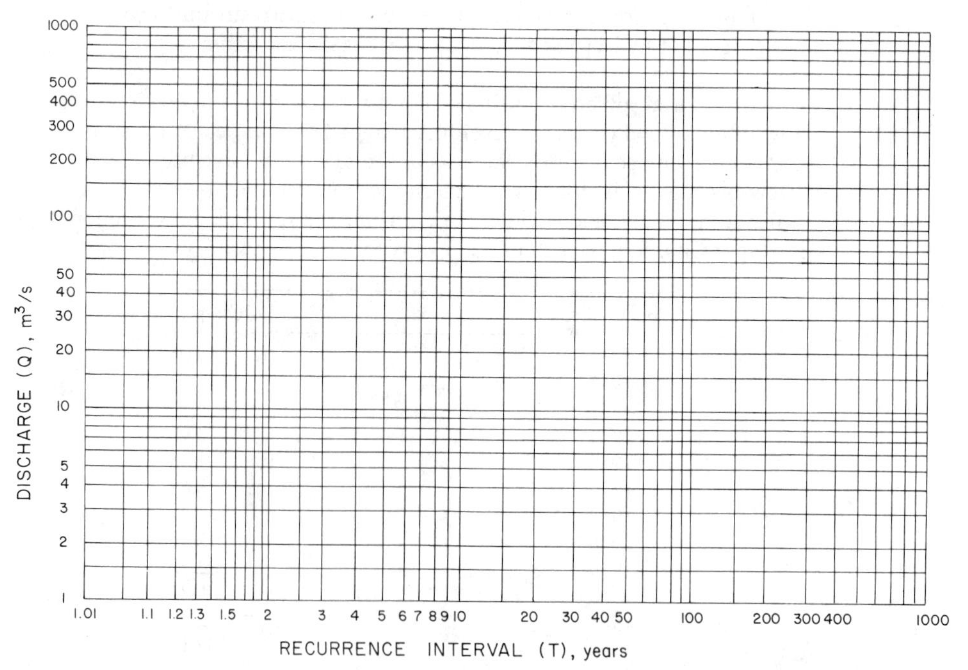


**c.** What is the recurrence interval of the 1972 flood, using the curve thus drawn?

**d.** Given the range of values from a and c above, what can be stated relative to the frequency of floods as large as the one which occurred in 1972?

**5.** See the graph showing the relationship between stage and discharge for the Mississippi River at St. Louis, MO (Fig. 8.40). Assume flood stage is defined as 8.1 m depth.

**a.** If a flood similar in discharge to the 1927 flood were to occur today, how many meters would this rise above flood stage?
**b.** How many meters above flood stage actually occurred in 1927?
**c.** Explain the discrepancy between a and b.

**6.** A study of the Highland Creek Reservoir, California, by the USGS (Trujillo, 1982) shows that the reservoir volume diminished from $3.956 \times 10^6$ m$^3$ to $3.912 \times 10^6$ m$^3$ due to sedimentation. The reservoir surveys, made in December 1965 and April 1972, indicate that a storage capacity loss of $0.44 \times 10^5$ m$^3$ occurred during the 6.3-yr period. The delta sediments were found to have a density of 1.12 g/cm$^3$. Measurements by the USGS revealed that a total of 8890 metric tons of sediment was discharged downstream from the reservoir during the 6.3 year period.
**a.** What is the trap efficiency?
**b.** What year will the reservoir be filled with sediment assuming the trap efficiency does not change?

**7.** Angostura Dam was built on the Cheyenne River, South Dakota, in 1956. The Cheyenne River carries into the reservoir an average suspended load of $1.45 \times 10^6$ metric tons per year. The average discharge of the Cheyenne River is 4.8 m$^3$/s. Clear water flows over the spillway and into the irrigation canal downstream. The volume of Angostura Reservoir when first formed was $2 \times 10^8$ m$^3$.
**a.** Determine the life expectancy of Angostura Reservoir. Assume trap efficiency is a constant 100% and the density of the deltaic sediment is 0.88 g/cm$^3$.
**b.** Use the theoretical trap efficiency curve (Fig. 8.37) to determine the trap efficiency of the Angostura Reservoir for initial conditions. If TE does not change with time, compute the life expectancy.
**c.** Erodible Cretaceous shale and Quaternary eolian sand and terrace alluvium constitute much of the shoreline on the downwind area of Angostura Reservoir. Since wave erosion is severe, should the life expectancy be increased or decreased from that calculated in b above?
**d.** If TE decreases with time, as the reservoir loses water storage due to deltaic sediments, will the life expectancy be increased or decreased?

**8.** See data for the Bloomington Reservoir, Illinois (Table 8.4).
**a.** Determine the $C/W$ ratio.
**b.** Use Fig. 8.38 to determine the annual reservoir capacity loss.
**c.** What is the observed annual capacity loss as published in Table 8.4?

**9.** Water flows down an open canal that has a slope of 0.015. The canal has a trapezoidal cross section with a bottom width of 3 m and side slopes of 2:1 (2 m horizontal for every 1 m vertical). The water is 2 m deep. The roughness coefficiency is 0.042. Calculate the velocity and discharge using the Chezy-Manning equation.

**10.** Determine the depth ($D$) and velocity ($V$) of water flowing through a trapezoidal concrete ditch where the bottom is 4.5 m wide and the side slopes are 1 (vertical) to 2 (horizontal). The discharge ($Q$) is 8.5 m$^3$/s, the roughness ($n$) is 0.02, and the slope ($S$) is 0.0009. (Hint: this problem can be solved by trial. Assume a $D$ (suggest starting at 1 m) and solve for $A$, $P$, $R$, $V$, and $Q$. If $Q$ thus calculated is $< 8.5$ m$^3$/s, repeat the process with a larger $D$, etc. For an engineering calculation of the final answer, values of $Q$ vs. $D$ can be plotted as a straight line on arithmetic paper, and a value of $D$ can be determined from the intercept where $Q = 8.5$ m$^3$/s. For greater precision, a series of $Q$ vs. $D$ calculations can be made, developing a synthetic rating curve which is not a perfect straight line (Tharp, 1983).)

**11.** Refer to the hydrographs of three streams in Pennsylvania (Fig. 8.5).

**a.** Determine the ratio of *peak* discharge ($m^3/s$) to drainage area ($km^2$) for each stream.

**b.** How are the three flood peaks related temporally?

**c.** Why are the three ratios different?

**12.** Consider a 1-m unit length of concrete dam as shown below: Draw a free-body diagram for the dam, assuming water is 30 m deep as shown.

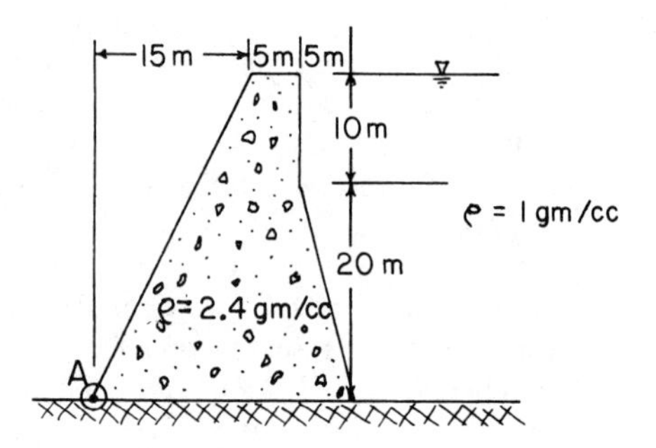


**a.** Is the weight of the dam sufficient to prevent the dam from being rotated around point $A$ by the water? What is the F.S.?

**b.** If the coefficient of friction between the concrete dam and the bedrock foundation is 0.3, will the frictional force be sufficient to prevent sliding? What is the F.S.? Is a key-way necessary?

**13.** Refer to Figure 8.33, data on grouting and well yield in crystalline rocks. (Assume water *injection* would be the same as well *yield* data for a well developed in the same place.)

**a.** What is the average water yield ($m^3$/day per m depth) for all 10 m deep wells in these locales?

**b.** What is the average water yield ($m^3$/day per m depth) for 100 m deep wells?

**c.** Explain why *a* is different from *b*.

# 9

# Land Subsidence

> Ode on Venice: . . . When thy marble walls are level with the waters . . .
>
> Lord Byron

## 9.1 Introduction

The surface of the land is subsiding over wide areas near Phoenix, AZ, Houston, TX, Fresno, CA, as well as at other parts of the world such as Tokyo, Mexico City, London, and Venice. This subsidence is caused by the withdrawal of large volumes of ground water from unconsolidated sediments.

Land subsidence also occurs due to the solution of soluble rocks; in some cases, this subsidence may be triggered by man. Land subsidence may also occur over mined land.

This chapter summarizes engineering geology considerations for these destructive processes.

## 9.2 Land Subsidence Due to Withdrawal of Fluids

Subsidence due to ground-water withdrawal develops under two contrasting environments and mechanics. The most extensive occurrence is in unconsolidated to semiconsolidated sediments containing confined or semi-confined sand and gravel aquifers interbedded with clayey a quitards. Another occurrence is in soluble rocks containing caves or sinkhole deposits; a lowered water table may cause erosion of cave fill or cave collapse due to loss of buoyant support from ground water.

### *9.2.1 Mechanism*

Extensive land subsidence is caused by the withdrawal of fluids (e.g., ground water or petroleum) from weakly consolidated artesian aquifers. The key to understanding this phenomena is the concept of aquifer elasticity. Figure 9.1 illustrates how the total stress (lithostatic overburden) on an aquifer ($S_t$) is supported by stress on the grain-to-grain skeleton of the aquifer (effective stress = $S_k$) as well as the water pressure in the aquifer ($S_w$):

$$S_t = S_k + S_w. \tag{9.1}$$

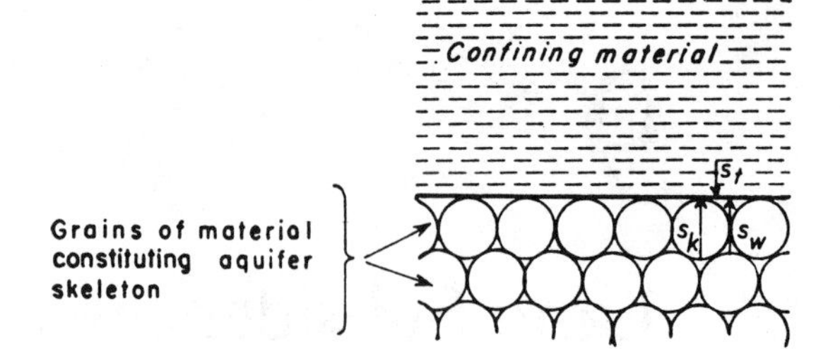


Figure 9.1 Microscopic view of forces acting at artesian aquifer interface (from Ferris et al., 1962).

If the aquifer skeleton is slightly elastic (visualized as a coiled spring), then any increase in overburden pressure will cause a corresponding additional load on both the water and aquifer skeleton. The water pressure is thus increased due to the extra load. (For instance, the passage of a large freight train above this aquifer may cause the water level in an artesian well to rise.)

An increase in atmospheric pressure will cause a slight lowering of the water level in an artesian well (Fig. 9.2) because the additional loading at the aquifer interface is borne by both the aquifer skeleton and the water. The change in water pressure at depth is not as great as the additional atmospheric loading the water receives at the open water surface at the well, so the water level in the well drops (Todd, 1980). Changes of atmospheric pressure produce proportional changes in artesian levels in these kinds of wells. The term barometric efficiency (B.E.) is the ratio of the change in potentiometric level to change in atmospheric level:

$$\text{B.E.} = \frac{s_w}{s_b}, \tag{9.2}$$

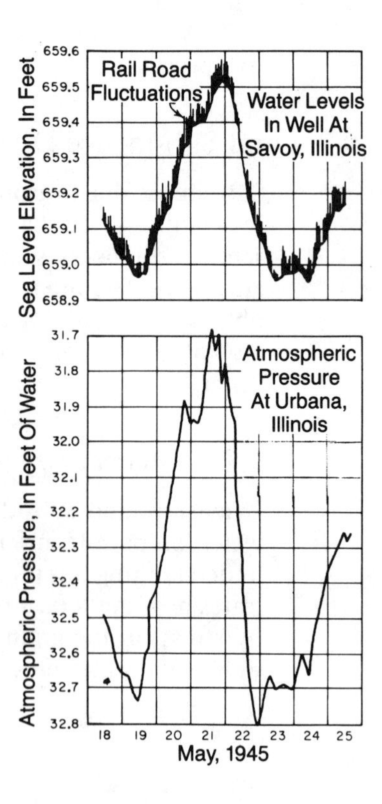


Figure 9.2 Barometric pressure changes in an artesian well (modified from Walton, 1970).

where $s_w$ is the change in water level observed in a well tapping the aquifer and $s_b$ is the corresponding change in atmospheric pressure, both expressed in meters of water. The B.E. for elastic artesian aquifers may reach 75%. An artesian aquifer which is not elastic (example: rigid limestone) has a B.E. = 0.

The B.E. is a general indication of how susceptible an area is to land subsidence. If the fluid pressure is greatly reduced due to large ground-water withdrawal, then the grain-to-grain contact may be increased beyond the aquifer's ability to spring back elastically, and hence permanent land subsidence will occur. The gradual loss of porosity in sediments is referred to as compaction by geologists and consolidation by engineers.

Within a series of compacting sediments there are fine-grained aquitards as well as coarse-grained aquifers. Recent work by the USGS (Poland, 1981) involves the installation of vertical extensometers, and shows that compaction within clayey aquitards rather than sandy aquifers is responsible for most of the land subsidence. The aquifer may remain saturated, but, as the artesian level drops, the decrease in head in the coarse-grained aquifers creates a hydraulic gradient from the clayey aquitards to the aquifers. The aquitards have low permeability and high compressibility. Therefore, the vertical escape of water and consequent decrease in pore pressure is slow and time-dependent. The ultimate compaction in the aquifer is large and, for all extensive purposes, permanent.

### *9.2.2 Examples of Land Subsidence*

*San Joaquin Valley, CA.* California has a greater area affected by land subsidence than any other state. The southern part of the Great Valley, California, is intensely irrigated by ground water. At least 11,000 km$^2$ of this important agricultural area has subsided more than 0.3 m since the 1920s, creating the world's largest area of intense land subsidence. According to Poland et al. (1975), an area 113 km long has subsided more than 3 m, with a maximum of 8.5 m. (See Figs. 9.3 and 9.4.) The volume of subsidence amounts to 20 km$^3$ or about half the volume of the Great Salt Lake.

The rate of sinking is slow, and is generally not observed until underground pipes crack, well casings fail by compression, irrigation canals fail to flow correctly (or flow the wrong way), or, in the case of coastal areas, shorelines are inundated.

The relationship between land subsidence and ground-water withdrawals is clear. Subsidence occurs where ground-water withdrawals exceed the safe yield; in other words, ground water is being mined. The potentiometric surface decline south of Bakersfield, for example, is more than 100 m. Figure 9.4A shows the potentiometric water level in wells near an area where accurate bench mark elevations were made over a period of several decades. The ratio of the artesian level change to land subsidence for a typical area may have a ratio of 20:1 for several years, but may increase to 10:1 at severely affected areas (Lofgren, 1963; Poland and Evenson, 1966; Poland, 1969). This increase in ratio with increasing drawdown in artesian head is characteristic of much of the subsiding area near Bakersfield and Fresno. It suggests a cumulative increase in delayed compaction of the strata as head pressures decline.

The obvious way to stop further land subsidence in California and elsewhere is to stop overpumpage of ground water. To date, this has not proven successful

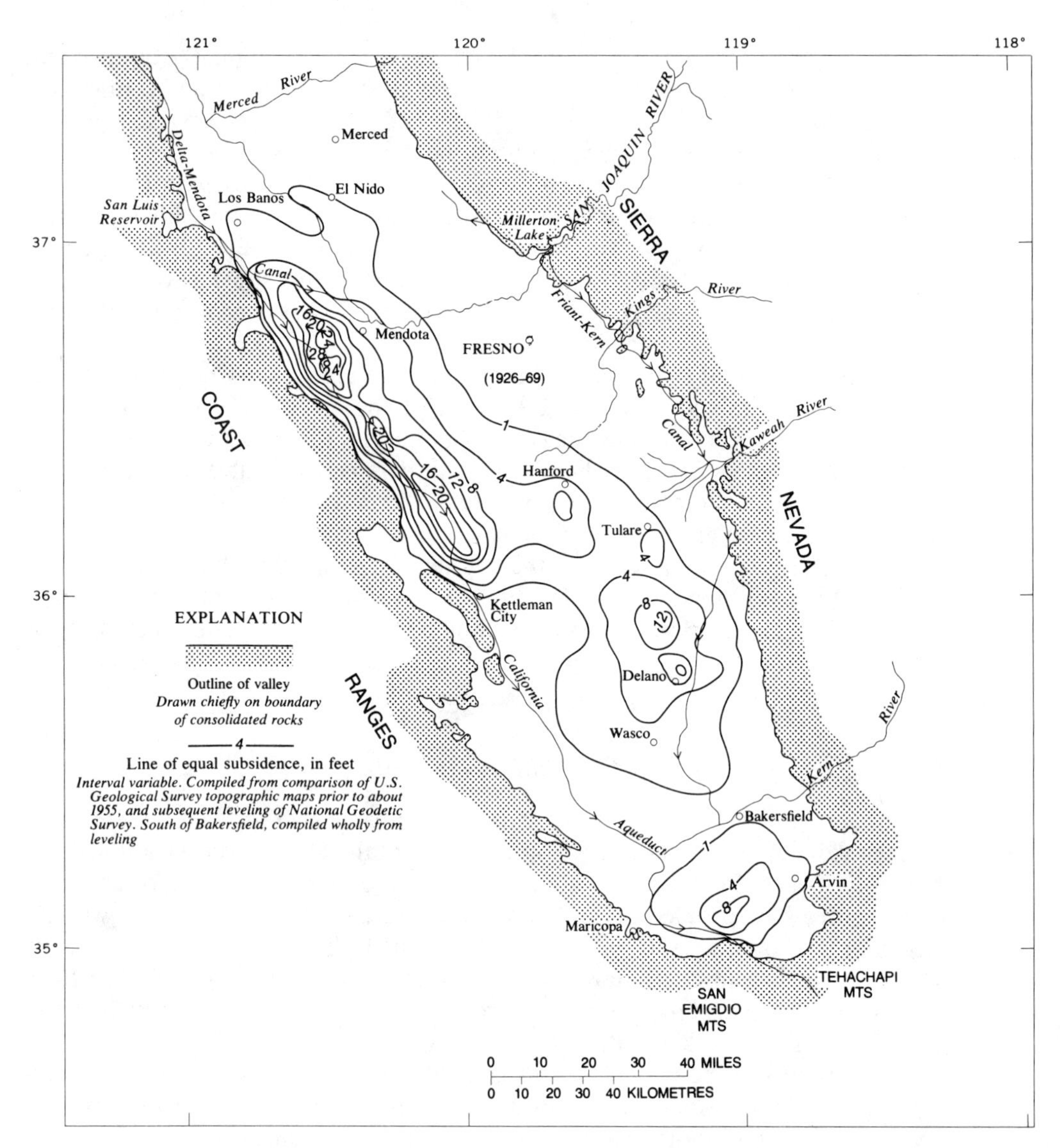


Figure 9.3 San Joaquin Valley land subsidence contours 1926–1972 (from Poland et al., 1975).

because regulations must be made and enforced, and, in general, Americans resist regulation. The San Joaquin Valley artesian water level decline is being slowed, however, by the import of surface water in large aqueducts.

*Houston, TX.* A large area southeast of Houston, near Baytown and Galveston, has subsided more than 2.4 m since 1970 (Gabrysch and Bonnet, 1975). The subsidence is particularly acute near Galveston Bay and along the Houston ship channel where more than 60 km$^2$ of low-lying land has been permanently inundated (U.S. Geological Survey, 1982).

Subsidence is extremely serious in the Houston area because this coastal area

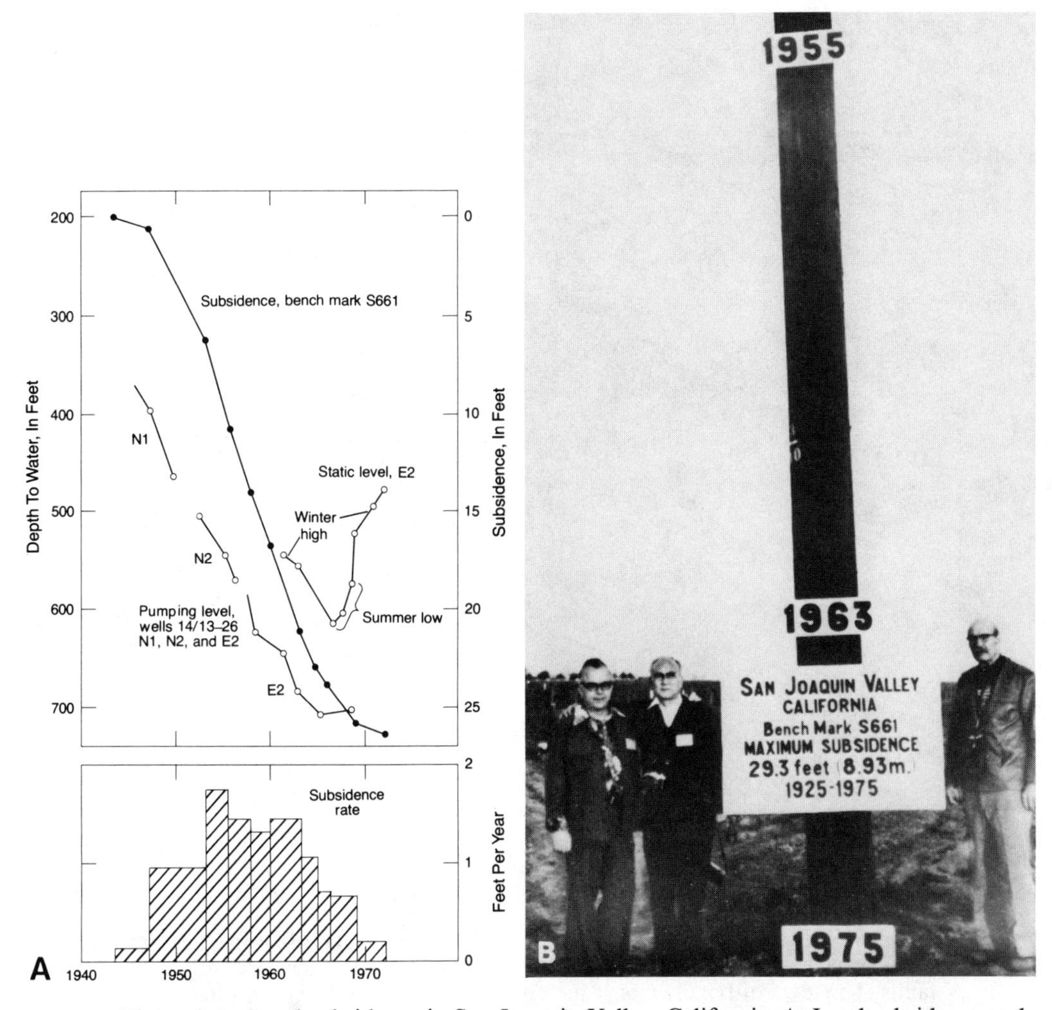


Figure 9.4 Land subsidence in San Joaquin Valley, California. **A**. Land subsidence and artesianhead decline 16 km southwest of Mendota, San Joaquin Valley, California (from Poland et al., 1975). **B**. Photograph in 1975 showing former elevation of land near Mendota in 1955 and 1963 (USGS photo).

is subject to very high tides, such as those produced by Hurricane Carla in 1961 and Hurricane Alicia in 1983. All coastal areas at elevations less than 4 m above mean sea level can be flooded by these tides. A 2-m tide in 1973 caused subsidence-related property damage of over $53 million; 2-m tides can be expected about every 5 yr. (See Chapter 10 for additional discussion of coastal hazards.)

Ground water is used extensively for municipal and industrial supply in Houston (Jorgenson, 1981). If surface water is not soon substituted for ground water as a source of water supply for the Baytown area, it is estimated that ground-water levels will continue to decline at a rate of 1.8 m/yr, and hence by the turn of the century the land would subside about 5 m below the original land surface of the Baytown area. Skeletons of the Brownwood subdivision in Baytown, a pennisula jutting into Galveston Bay, now stand door-hinge deep in the salt water. As much

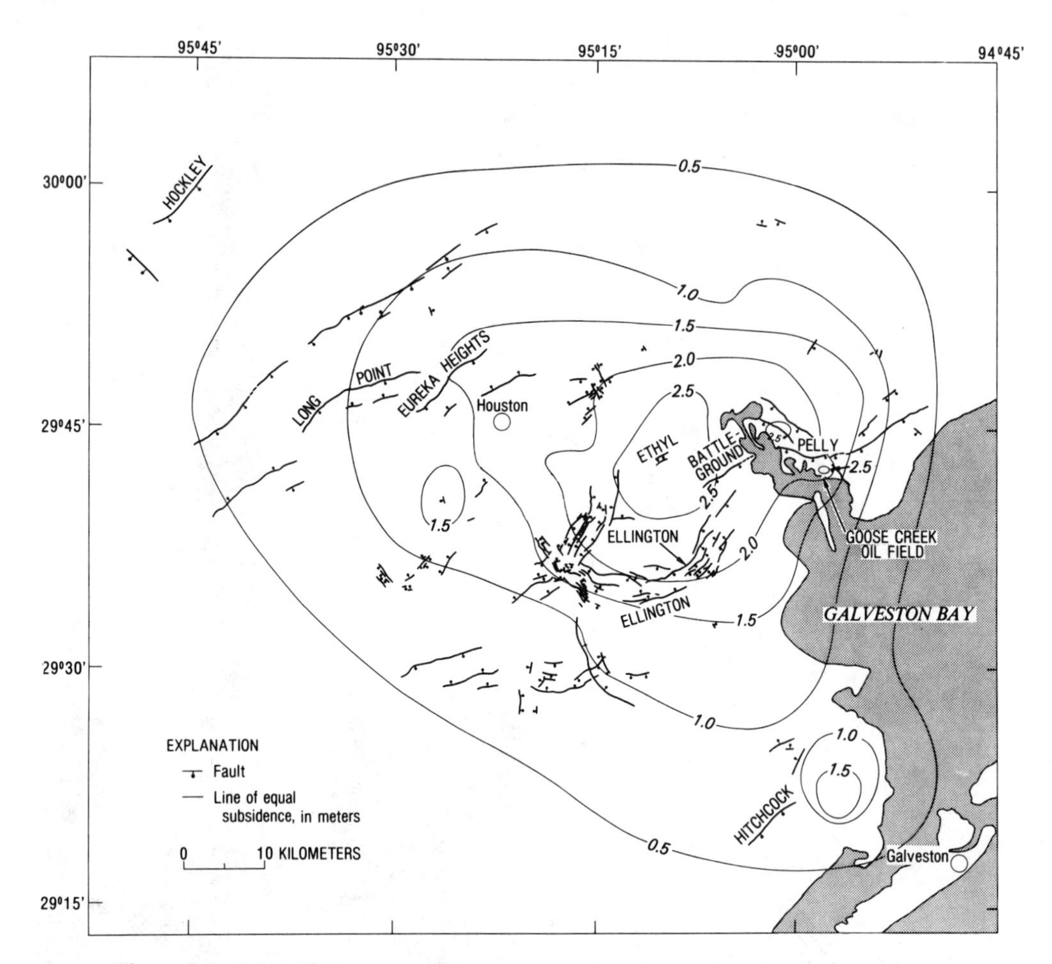


Figure 9.5 Map of Houston–Galveston area showing surface faulting and cumulative subsidence (from U.S. Geological Survey, 1982). The letters identify 14 major active faults. The town of Baytown is at the northwestern end of Galveston Bay.

as 4000 $m^3$/day of water is presently being pumped from the 7800 $km^2$ Houston area for municipal, industrial, and irrigation purposes. In 1975, the Harris-Galveston Coastal Subsidence District was created and charged with ending the land subsidence by regulating the amount of ground-water withdrawal and the spacing of new wells. As a result of the District, ground-water pumping has been reduced.

A number of active and potentially damaging faults are associated with land subsidence in the Houston area (Castle and Youd, 1972). Rigid structures, such as highways, runways, and building foundations, were built astride the faults and now commonly display the consequences of horizontal extension. More than 150 historically active faults with an aggregate length of 500 km are associated with land subsidence near Houston (Verbeek et al., 1979). Scarps locally exceed 1 m in height and can be traced across the land for a distance as long as 16 km. Movement has damaged roads, buildings, airport runways, sewers, and other utilities (Clanton and Amsbury, 1975). Sanitary engineers in the Houston area mitigate the surface faulting problem by placing an active sewage pipe within a large diameter pipe as a precautionary measure where crossing a fault. Figure 9.5 shows the location of major surface faulting in Houston. Some geologists believe the faults are simply natural growth faults in late Cenozoic deltaic strata, which are compacting naturally due to the weight of additional overlying sediments, as well as compacting because of ground-water withdrawals within the fault-controlled hydrological "compartments" (Kreitler, 1977). The coincidence of surface faulting scarps, ground-water level decline, and land subsidence suggests that the subsidence is largely a man-made hazard, however.

To reduce future land subsidence, ground-water supplies have been augmented by surface water. As a result ground-water levels have recovered beneath some of the coastal areas near Galveston Bay. Ground-water levels continue to decline in the western Houston area, however. In 1982 the U.S. Geological Survey studied movement of growth faults (Fig. 9.5) and found that seven monitored faults in the area of water level recovery had stopped moving. By contrast, the five monitored faults in the area where water level has continued to drop have displacements averaging 1 cm/yr.

*Phoenix, AZ.* Large areas of irrigated desert in southern Arizona are subsiding due to ground-water withdrawals. About 310 $km^2$ of land southeast of Phoenix has subsided more than 3 m between 1952 and 1979 (Fig. 9.6). The areas affected are the large intermountain basins, called bahadas. More than 133 $km^3$ of ground water has been withdrawn from southern Arizona since 1915, much more than could be replaced by natural recharge. As a result, ground-water levels have been declining as much as 120 m in places. The water level decline, in turn, has triggered land subsidence and earth fissures (Robinson and Peterson, 1962). As the land subsides, the porosity of the alluvium making up the basins is permanently reduced, diminishing the effectiveness of the aquifer. In 1979, USGS hydrologist Robert L. Laney discussed Arizona's problem:

> . . . as the land subsides the pore space in the alluvium where water is stored is reduced. In the area of greatest subsidence the underground storage capacity is estimated to have been reduced by more than 160 billion gallons.

Earth fissures are associated with the subsiding ground (Schumann and Poland, 1969; Holzer et al., 1979). Most of the cracks appear to be simple tensional breaks with little offset. They are several meters wide, up to 20 m deep, and many extend for 1 km in length (Bouwer, 1977). Holzer (1980) described an earth fissure 16 km long near Picacho, AZ. Earth fissures have caused damage to irrigation systems, railroads, and Interstate 10 near Picacho, AZ. Earth fissures and water table decline have contributed to the abandonment of irrigated farmlands in central Arizona. The lowering of water levels causes greater pumping expenses due to increases in lift and resulting higher energy costs. If the water table drops below the bottom of the well, the well must be deepened or replaced. Abandoned farmlands have led to wind deflation and dust-related accidents on Interstate 10 in central Arizona (Hyers and Marcus, 1981).

Earth fissures are particularly evident along the geomorphological boundary between the pediment and bahada (Figs. 9.7 and 9.8). Figure 9.7 is a 1963 aerial photograph of an area west of Casa Grande, AZ, where earth fissures occur downgradient from an isolated mountain. (Physiographically, an island-mountain such

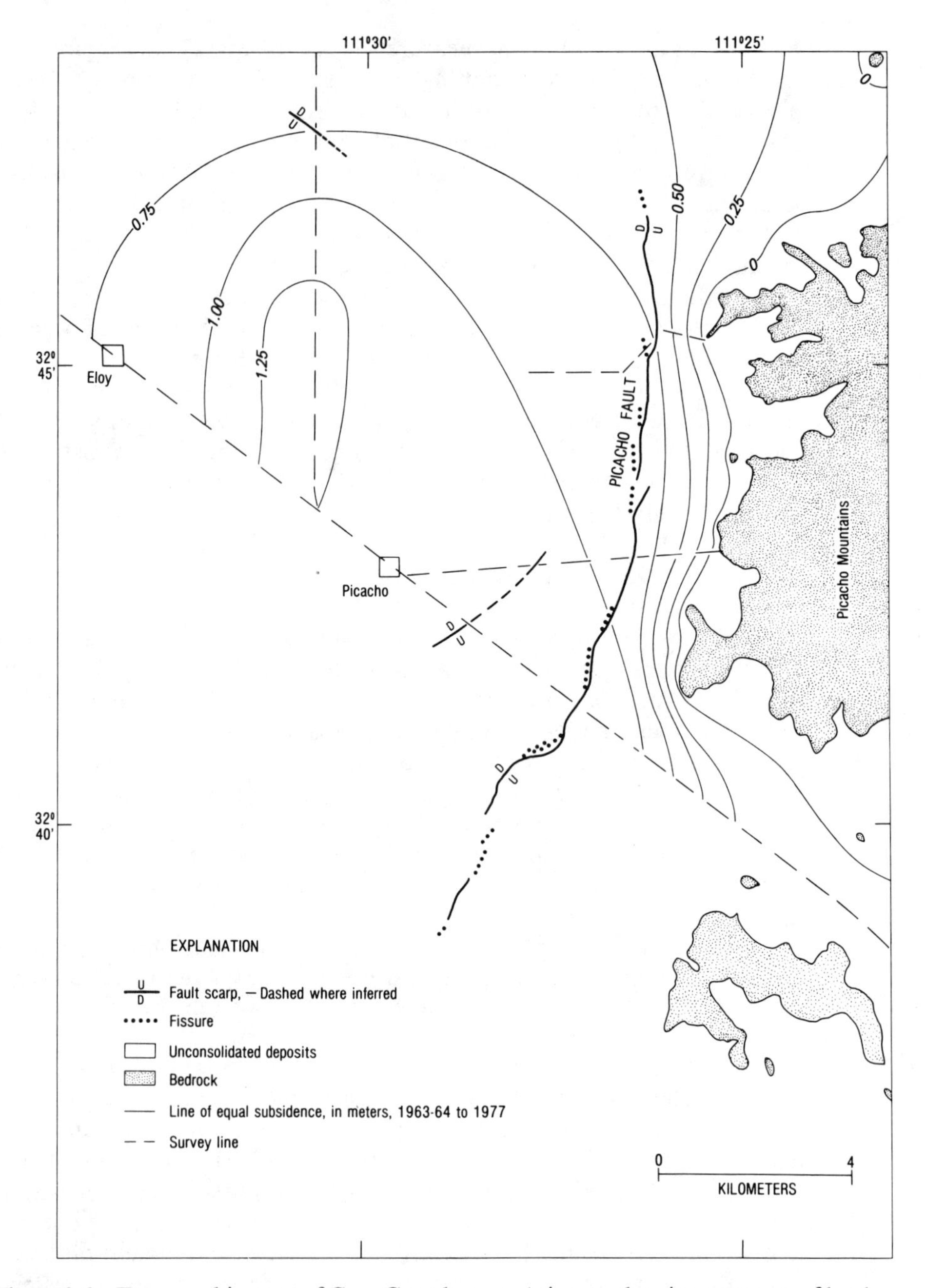


Figure 9.6 Topographic map of Casa Grande area, Arizona, showing contours of land subsidence (m). Heavy black lines are earth fissures. (From Holzer, 1984.)

as this is called an inselberg.) The fissure is about 10 m deep in places despite the fact that the fissure trends perpendicularly to the drainage of numerous ephemeral washes (arroyos), which deposit sand and gravel in the fissure during infrequent cloudburst floods.

Southern Arizona lies in the "Basin and Range" physiographic province. Figure 9.8 is an idealized cross-sectional sketch of the southern Arizona area showing the relationship of geomorphology to earth fissures. During the mid to late Cenozoic era, tension in the earth's crust produced grabens and horsts throughout much of the southwestern United States. The basins consist of 1) the alluvially filled graben (the "bahada") and 2) the gently sloping bedrock cut surface of the horst (thinly veneered with alluvium at most places), which is called the pediment (Rahn, 1967). The stratigraphy of the deposits in the bahada typically consists of middle Tertiary sand and gravel at depth overlain by fine-grained lacustrine and playa deposits and Holocene sand and gravel (Prokopovich, 1983). As pumping occurs, compaction is believed to occur in the fine-grained units separating confined aquifers. The old fault-line contact between the alluvially veneered pediment and bahada is normally not discernible in the field, but can be determined by drilling or geophysical methods (see Chapter 12). As the alluvial basins are dewatered by man, the bahada subsides, causing cracks, particularly near the pediment boundary. The idea that these earth fissures are tensional breaks resulting from differential settlement along the edge of a buried pediment was first advanced by Heindl and Feth (1955). It should be emphasized that the fault separating the pediment and bahada is not reactivated, but the earth fissures simply occur above the fault due to differential subsidence.

Earth fissures rarely occur in straight lines separating the bahada-pediment boundary, as shown in the sketch in Fig. 9.8B, because this is an idealized sketch. Cape et al. (1983) show that complex tectonic patterns of block faulting occur, including tilted-fault blocks, and listric fault blocks (mega-landslides from mountains which slid into adjacent troughs). Seismic reflection data in Nevada (Anderson et al., 1983) also show that basin fills are underlain by complex patterns of buried fault blocks. Jachens and Holzer (1983) studied the history of earth fissure development near Casa Grande, AZ. At places the earth fissures are located out in the bahada, and may be related to areas of intense pumping or to buried bedrock highs. The occurrence of the fissures at points of maximum convex-upward curvature in a subsiding alluvial sequence is believed to be analogous to the tensile cracks produced in the middle of a beam under a uniform load (Fig. 4.5A).

Hydrologists generally agree that southern Arizona is running out of ground water. Arizonans consume some 3.1 $km^3$ more ground water annually than is being replenished by nature. (See discussion of safe yield in Chapter 7.) In 1980, Arizona Governor Bruce Babbitt stated that "ground water is the single most important issue before this state." Water withdrawal for agricultural use accounts for 90% of the ground water being used today in the United States. Many of these crops are valuable, such as citrus fruit, melons, and grapes. However, many Arizona farms grow cotton, a government-subsidized crop not essential to the American economy.

Presently, the only control over Arizona ground water is a loose, ineffective law approved in 1948. The Colorado River aqueduct ("Central Arizona Project") presently being constructed in southern Arizona will divert 1.5 $km^3$ of water annually into this thirsty area. This will slow the rate of subsidence in some Arizona basins, but the imported water will only amount to less than one half of the estimated overdraft (Poland, 1981). As the population grows and water levels continue to drop, it will probably become necessary for the state of Arizona to enforce rigorous farming conservation practices, or to put agriculture out of business by purchasing farms.

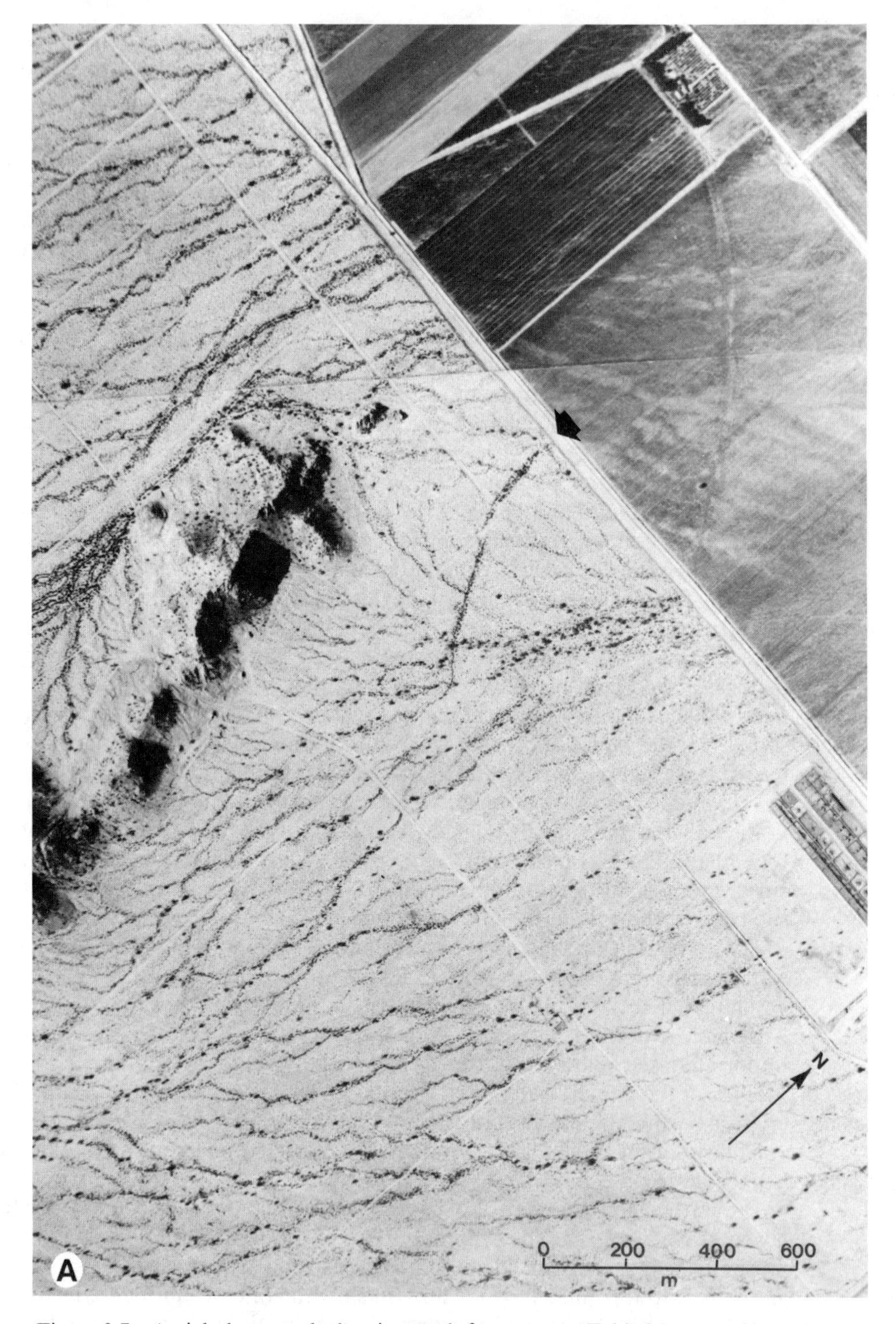


Figure 9.7 Aerial photograph showing earth fissures near Table Mt. area about 32 km southwest of Casa Grande, AZ. **A**. Small scale photo. Arrow locates earth fissure.

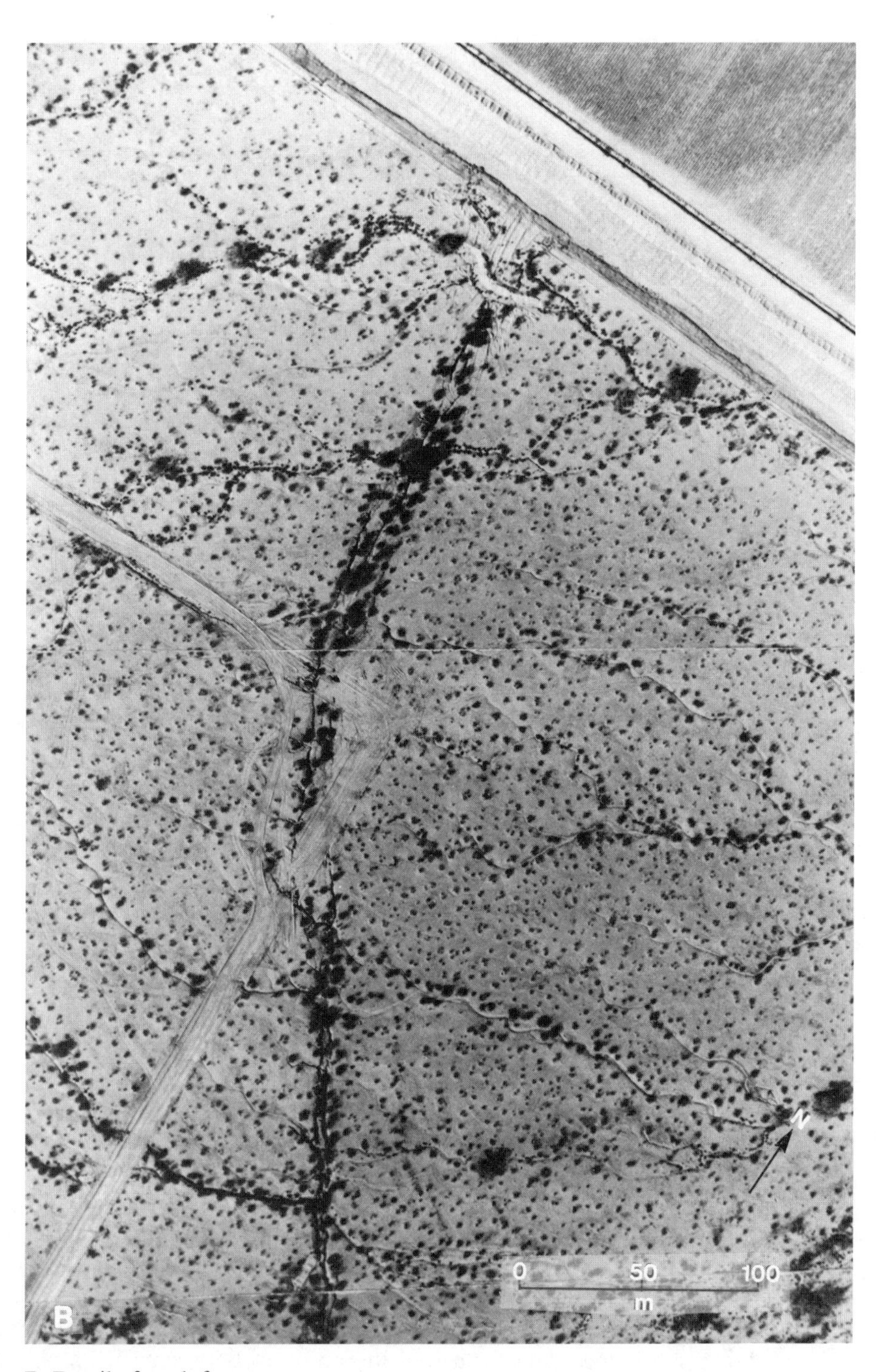


**B**. Detail of earth fissure.

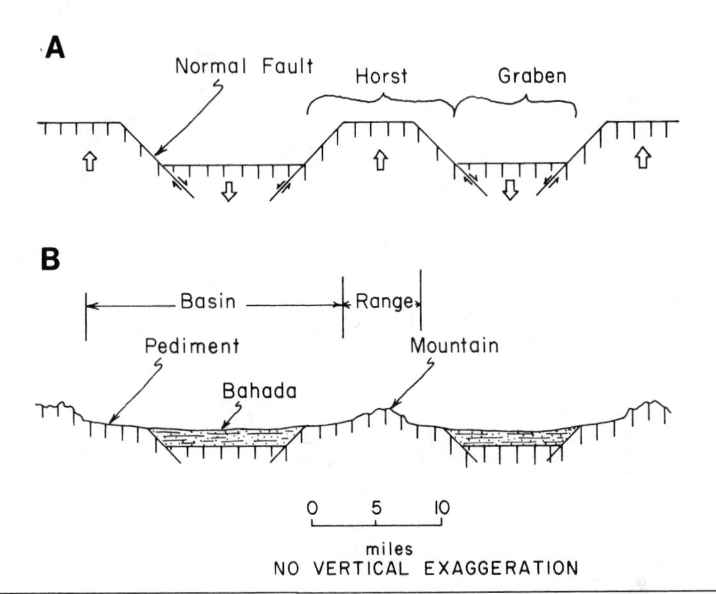


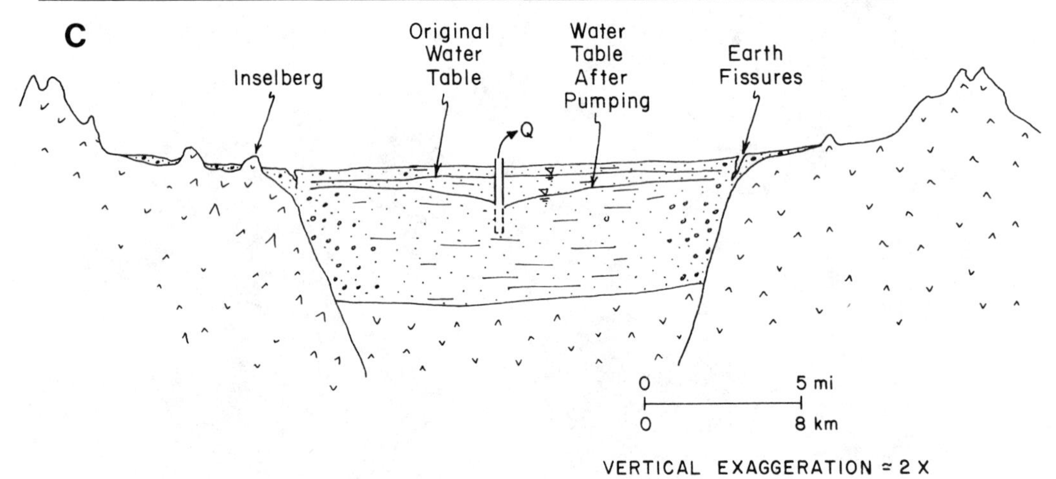


Figure 9.8 Sketch illustrating relationship of earth fissures to ground water decline and geomorphology of southern Arizona. **A** and **B**. Idealized evolution of basin and range (modified from Rahn, 1966). **C**. Earth fissure formed due to decline of ground water.

*Other places in the United States.* Subsidence in the Las Vegas, NV, valley (as much as 1.2 m between 1935 and 1970) has been linked to large-scale pumping of ground water. Ground-water levels have been lowered as much as 55 m in some areas (Holzer, 1978). Ground cracks have occurred in residential areas causing damage to roads, houses, and other structures.

In the Santa Clara Valley, the city of San Jose has subsided about 4 m. Figure 9.9 shows subsidence data for a well in San Jose. Between 1912 and 1967 the bench mark subsided 3.86 m. The USGS has installed compaction recorders which show that compaction of sediments between the bottom and top of a test hole was equal to the amount of lowering of the land surface, thus demonstrating that subsidence is actually due to compaction of the sediments (Poland and

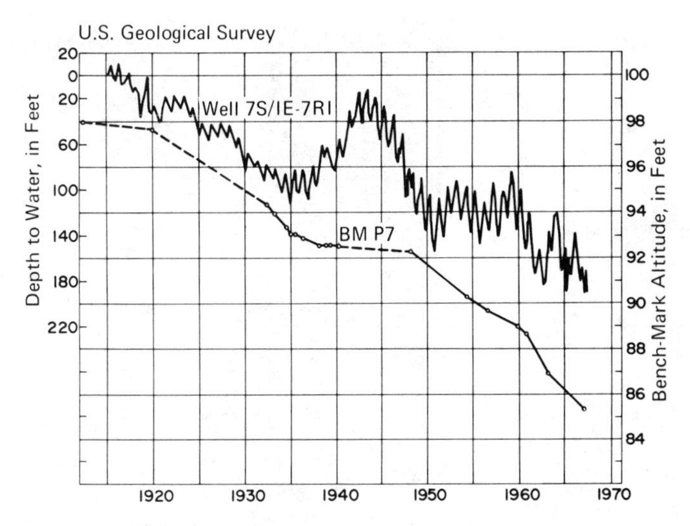


Figure 9.9 Land subsidence data for San Jose, CA (from Poland, 1981).

Davis, 1969). Fowler (1981) estimates damages in Santa Clara at $100 million. These damages include buckling of water-well casings, lost capacity of sanitary and storm drainage sewers as a result of change in slope, and flooding from the San Francisco Bay of roads, railroads, and developed areas, thus requiring construction of levees and other protective works.

In Santa Clara County, the land subsidence problem is critical because the area is near sea level. Millions of dollars of public funds have been spent in levee construction to prevent flooding. Local legislators passed special taxes on ground-water withdrawals. As a result, pumping has declined, and the rate of land subsidence slowed from about 0.3 m/yr in 1961 to 0.02 m/yr in 1970.

Although land subsidence may be slowed or even stopped by reducing ground-water pumping, the land does not rebound. Apparently the loss of porosity in the sediment is permanent.

In New Orleans, LA, subsidence due to ground-water withdrawals and possibly natural regional subsidence from gently seaward dipping Cenozoic strata have caused decline of the city up to 0.55 m. This situation is particularly hazardous because the city is essentially at sea level. Parts of New Orleans are 2 m below sea level. New Orleans is protected from inundation by the Mississippi River by huge artificial levees. Rain which falls into downtown must be pumped out of the city at considerable expense. On several occasions during heavy rains in 1983, the extensive storm drainage pumping system was unable to keep pace with the rain. Waist-deep water caused electricity outages, and stranded motorists switched to boats.

Petroleum withdrawal at the Wilmington oil field has caused enormous land subsidence along the Pacific coast at Long Beach, CA. The U.S. Navy shipyards have subsided up to 9 m causing the shipyard to sink below sea level (Fig. 9.10). This required the $100 million construction of levees and sea walls to keep the Pacific from inundating the shipyard. Apparently, the subsidence has been stopped by injecting salt water into the oil-bearing formations to restore fluid pressure (Allen, 1973).

TOTAL SUBSIDENCE
1928 TO 1968
SIGNAL HILL
LONG BEACH
TERMINAL ISLAND
PIER A
PIER J
A
2
4
6
10
14
22
27
29

Figure 9.10A

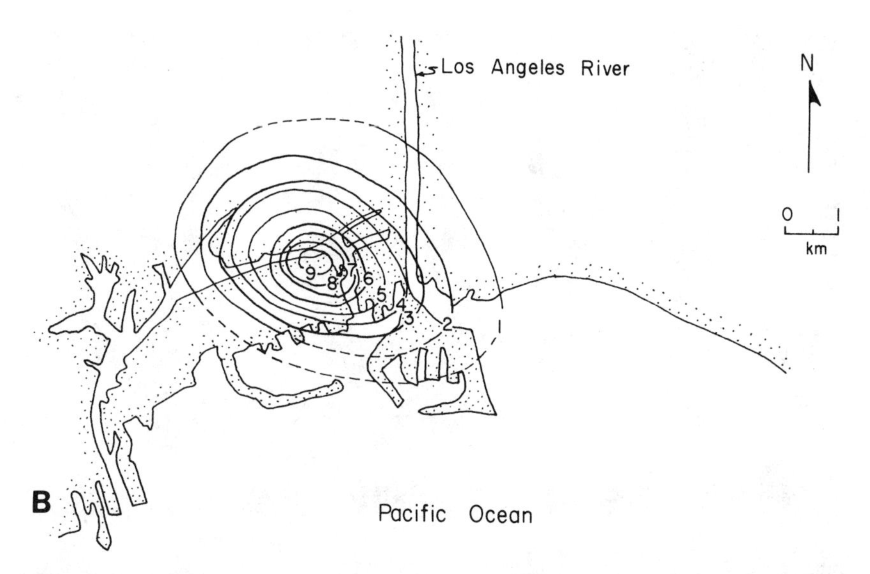


Figure 9.10 Subsidence data for Long Beach, CA. **A**. Aerial view of the city showing subsidence contours (ft) from 1928 to 1965 (from Mayuga and Allen, 1966). Photo compliments of Lawrence Olson, City of Long Beach. **B**. Map of the coastal area showing subsidence contours, in meters (from Randall et al., 1983).

Small, local subsidence problems associated with petroleum extraction has been known for some time (Thom, 1927; Holzer and Bluntzer, 1984). In some places, flowing sand moves up the well casing, resulting in local ground subsidence. Historical faulting at southern California oil fields may be related to oil-field development rather than deep tectonic forces (Yeats et al., 1981).

*Land subsidence elsewhere in the world.* Land subsidence due to withdrawal of underground fluids has been documented at Lake Maracaibo, Venezuela; London, England; Niigata, Japan; Venice, Italy; and other places throughout the world.

Venice, Italy, is a renaissance city built on sand bar islands in the delta area of the Po River. Venetians built the city about 3 km off-shore as protection from the Huns. Today the city is linked to the mainland by causeways. A unique and attractive feature of Venice is the absence of automobiles; the mode of conveyance is walking or by boats around the tidal canals.

Unfortunately, all is not well in this beautiful city. Venice is sinking into the Adriatic Sea. Quaternary coastal plain sediments from the Po River delta are over 1000 m thick beneath Venice, consisting of unconsolidated sand, silt, and clay. The upper 300 m of sediments are tapped by artesian wells in the Marghera and Mestre region (Gambolati et al., 1974). To slake the thirst of new industries on the mainland, some 20,000 water wells have been drilled, tapping the seaward dipping artesian deltaic sediments under Venice. As a result, the fabled city of palaces and churches is sinking at a frightening rate, about 0.5 cm/yr. Massive sea walls proved fruitless as the storm of November 1966 broke through and submerged Venice under 2 m of water.

Figure 9.11 Canals and apartment houses in Venice, Italy. Note deterioration of steps and door due to rising sea level and land subsidence.

In addition to land subsidence per se, there has been a worldwide rising of sea level in Holocene time. Since the Pleistocene glaciers reached their last great advance about 14,000 yr BP, sea level has risen about 100 m. In this century the rate of worldwide (or eustatic) sea level rise is about 1 mm/yr (Flint, 1957). So the problem with Venice's battle with the sea is partially natural (eustatic sea level rise) and partially man-made (subsidence due to ground-water withdrawal).

Presently, very high tides inundate the Piazza del San Marco in Venice. In the beautiful Ca'd'Oro, constructed in 1421 A.D., high tides now cover the ground floor (Bolt et al., 1977). The lower stories of many buildings are abandoned (Fig. 9.11), and inhabited by rats or cats. (This health situation is worsened by the fact that Venice lacks any form of sewage treatment. Raw sewage is merely dumped into the canals with the hope that tidal currents will carry it away.) In 1966, an *aqua alta* storm inundated 80% of Venice and caused irreparable damage to architecture and works of art.

Efforts to control ground-water withdrawals near Venice were initiated in 1969, and this has reportedly slowed the rate of land subsidence. In 1984, the Italian government proposed the construction of dikes with movable barriers which could be raised in the event of an aqua alta. These structures would reduce the tidal currents which presently flush sewage and chemical wastes from the lagoons, however. Presently, it is not known if Venice can be saved. It seems tragic that in a few decades this city may have to be abandoned.

The population of Mexico City is growing astronomically; it was 1 million in 1922, and by 1980 was nearly 10 million. Water has been withdrawn for urban use at a rate far exceeding natural recharge, and extensive compaction of alluvial and lacustrine deposits under the city has occurred. Subsidence of up to 8 m has taken place. In downtown Mexico City, some buildings on shallow foundations have settled nearly one story, so that the tops of casings for deep wells, formerly at street level, now protrude up to the second story windows of adjacent houses (Poland and Davis, 1969). Land subsidence will continue in Mexico City because in 1983 officials granted permission to drill 24 new wells to provide ground water for the city.

Due to land subsidence and poor soils, the design of structures in Mexico City is challenging. In downtown Mexico City, the old Palace of Fine Arts has settled 3 m. Mexico City's first skyscraper, the Tower Latino Americana, is a 43-story building located across the street. Completed in 1951, the Tower had extensive test borings to a depth of 70 m, and detailed soil tests were made. A foundation structure consists of a concrete slab supported on end-bearing piles driven through volcanic clay to a good bearing on a stratum of sand 34 m below the surface. The structure has been a complete success, with only small settlements as expected from soil mechanics calculations (Zeevaert, 1957).

Since the early 1970s, over 11,000 wells have pumped ground water from beneath Bangkok, Thailand. Currently about $1.3 \times 10^6$ $m^3$/day is being withdrawn. The city of 5.5 million is sinking at rates up to 10 to 30 cm/yr. Bangkok is built on a flat deltaic plain at an elevation of 0.5 to 1.5 m above sea level. Prinya Nutalaya, professor at geotechnical engineering at the Asian Institute of Technology, warns: "If nothing is done, all of Bangkok will be under water by the turn of the century." The problem is compounded by monsoon rains which cause raw sewage to flood city streets. A 1983 plan was suggested that would pipe water some 100 km to the city, and the private water wells would be shut down. The plan is reportedly too costly for the city's budget.

Land subsidence also occurs in other parts of the world (International Association on Land Subsidence, 1977). In Tokyo, for example, the land on which 2 million people live has subsided below mean high tide level.

## 9.3 Land Subsidence in Terrains Underlain by Soluble Rocks

Land subsidence in carbonate terrain is due to the chemical weathering of soluble rocks. Limestone, dolomite, and gypsum are the most common soluble sedimentary rocks (see Chapter 3). Solution is a natural process, but man-induced changes in the hydrologic regime may accelerate solution and land subsidence.

The term karst has long been used to define the sum of phenomena characterizing regions where soluble rocks are exposed (Stringfield et al., 1977; Boegli, 1980; Jennings, 1983). The word karst comes from the Slovenian area of Yugoslavia, which has distinctive limestone landforms and hydrogeologic features. Features such as sinkholes, long dry valleys, caves, deep water table, large springs, and disappearing streams are typical of karst terrains. Karst areas form by the circulation of vadose and phreatic waters. As solution occurs [Eq. (3.3)], a cavity is gradually created. The term sinkhole (or sink) in a karst terrain is used to designate a depression formed by sinking or collapse of the land surface where solution of the underlying rocks has formed cavities or other openings (Stringfield et al., 1979). The Serbian word doline, meaning a little dole or valley, is commonly used as the equivalent of sinkhole.

Solution of large volumes of limestone requires geologic time. Thus structural

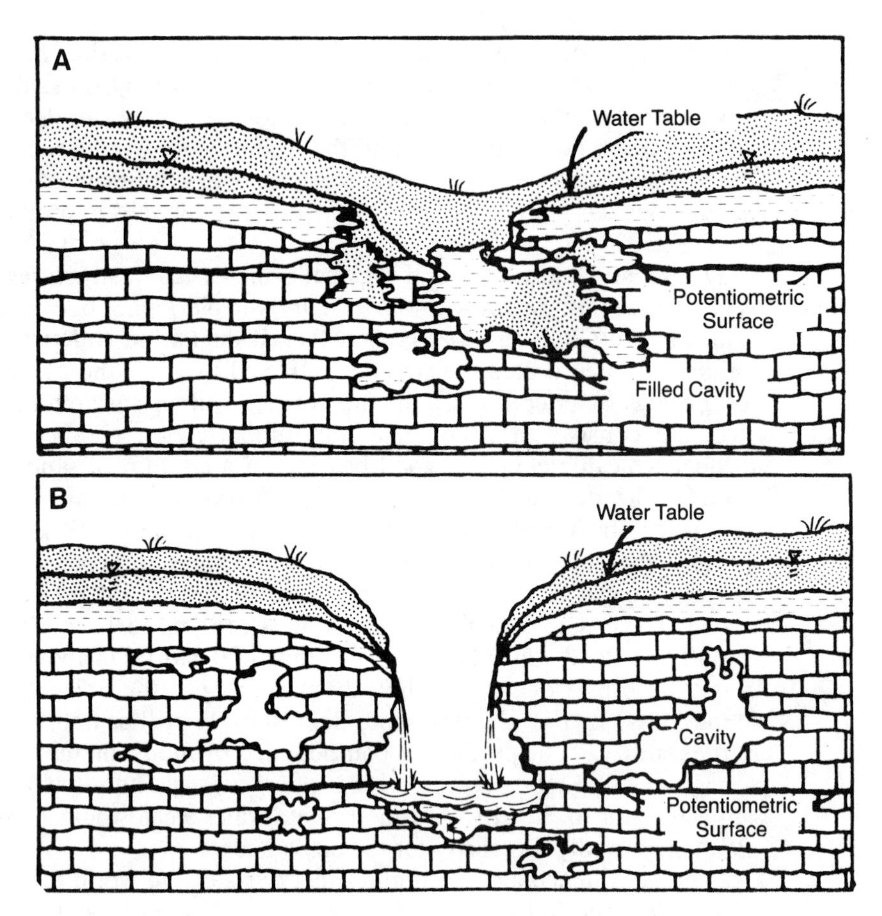


Figure 9.12 Sketch showing origin of sinkholes (from Gass, 1981a). **A**. Solution sink. **B**. Collapse sink.

failures are generally related to existing voids or to terra rossa soil, rather than to solution of sound rock per se. Foundation problems in carbonate terrain (Lifrieri and Raghu, 1982) commonly result from 1) differential settlement due to an irregular terra rossa soil contact (Fig. 3.8), 2) piping of soil (often associated with a change in hydrologic regime) into an existing cavity, or 3) collapse of a bedrock bridge over a cave.

Two types of sinkholes may develop over a solution cavity. In areas of thick regolith, the land surface may gradually subside, forming a solution sink, a gentle swale in the land surface (Fig. 9.12A). In areas of thin regolith, especially in areas of very active solution, a collapse sink may suddenly occur (Fig. 9.12B). These sinks are often instantaneous or unpredictable, thus constituting a geologic hazard. The sudden formation of a sinkhole in the wilderness is of little concern, " . . . but the problems come where man's development is intensive. . . . " says Barry Beck, director of the newly formed Florida Sinkhole Research Institute.

Regions in the United States noted for karst features include:

1. steeply dipping Paleozoic carbonate rocks in the Appalachian Ridge and Valley province, particularly in Pennsylvania,

Figure 9.13 A large collapse sink formed on May 9, 1981, in Winter Park, FL (from Windham and Campbell, 1981; photo by Rich Deuerling). This large sinkhole measures about 100 m in diameter. In this photo, taken May 13, the water level was about 13 m below the land surface. Note destroyed municipal swimming pool above the sinkhole.

2. gently dipping to horizontal Paleozoic carbonate rocks in the midwest plateau and lowland areas in Indiana, Kentucky, and Tennessee,
3. gently dipping to horizontal Paleozoic and Cenozoic limestones in Alabama, Florida, and Georgia, and
4. ancient paleokarst regions, now exhumed and containing some modern karst development in the western United States.

Figure 9.13 is an aerial photograph of a large collapse sink in Winter Park, FL. In 1981, over the period of a few hours, a house, several cars, and half of a municipal swimming pool fell into this sinkhole. After several months, water rose to approximately the same level as lakes in the area. Geotechnical investigations, including test borings, indicated that the sinkhole was caused by ravelling of terra rossa soil into a cave, and was not due to rock collapse of a cavity roof. The formation of sinkholes such as this should not come as a surprise to a geologist because 95% of the lakes of central Florida are former sinkholes. The Winter Park, FL, sinkhole collapsed following a dry decade, and the year 1981 was 38 cm short of rainfall when the sinkhole collapsed. The lowered water table probably triggered the collapse.

Some sinkholes are triggered by man's activities. If the water table is lowered

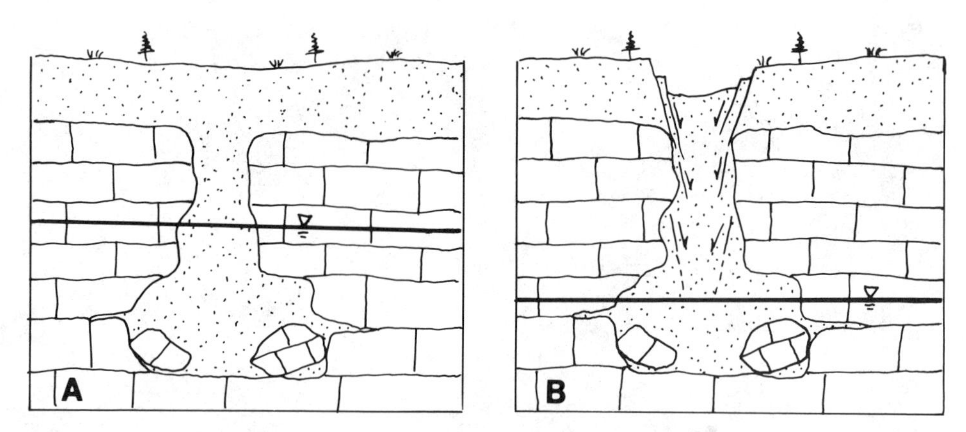


Figure 9.14 Sketch showing origin of land subsidence due to compaction of a mud-filled cavern. **A**. Original water table. **B**. Lowered water table results in compaction of terra rossa and cave fill material.

in a region where a natural covern exists, the overlying hydrostatic uplift afforded to the cavern roof (either bedrock or cave-fill regolith) is lost, and collapse may follow. (See Chapter 4 for related discussion.) At Saucon Valley, PA, for example, damaging sinkholes are triggered by large water-table declines associated with a nearby underground zinc mine (Fig. 3.9). If the cavern is completely filled with mud, a sinkhole may be generated by compaction (Fig. 9.14) or subterranean erosion of the cave-fill (Fig. 9.15). Newton (1984) estimates that 4000 sinkholes formed in Alabama since 1960; many were triggered by a declining water table.

A famous sinkhole area is the Far West Rand area, about 65 km west of Johannesburg, Republic of South Africa, where gently dipping Precambrian dolomites overlie gold-bearing conglomerates. The dolomites have a deep residual soil (Fig. 9.15). Because of pumpage of water from deep gold mines, the soil choking the vertical slot above a large cavern starts desiccating and eroding from the bottom upward (Foose, 1967). A temporary arch forms above the pipe until a catastrophic collapse and sinkhole forms. Note that in the Fig. 9.15, there is no solution of bedrock or change in the bedrock topography; only regolith piping causes land collapse. Where small gradual subsidence occurs in Johannesburg, South Africa, highway piers have adjustable jacks to provide future adjustment in case of small subsidence over old gold mine workings.

Parizek (1971a, 1971b) studied environmental problems in a carbonate terrain in central Pennsylvania. Due to urban sprawl, storm-water runoff may be collected and discharged into carbonate rocks. The residual soils erode out, forming conduits in the rock. Subsurface erosion may proceed until land subsidence or sinkhole development takes place. Knight (1971) describes problems of industrial plant foundations in Pennsylvania where sinkholes have developed adjacent to clusters of recharge wells installed to dispose of storm water runoff. The development of sinkholes was caused by erosion of soil into the cavernous bedrock.

The failure of numerous recreational ponds and industrial lagoons has been shown to be caused by downward-moving water leading to piping failures. The slow, grain-by-grain erosion of residual soil proceeds until a piping hole may break into the pond or lagoon floor, and the water suddenly drains in a matter of

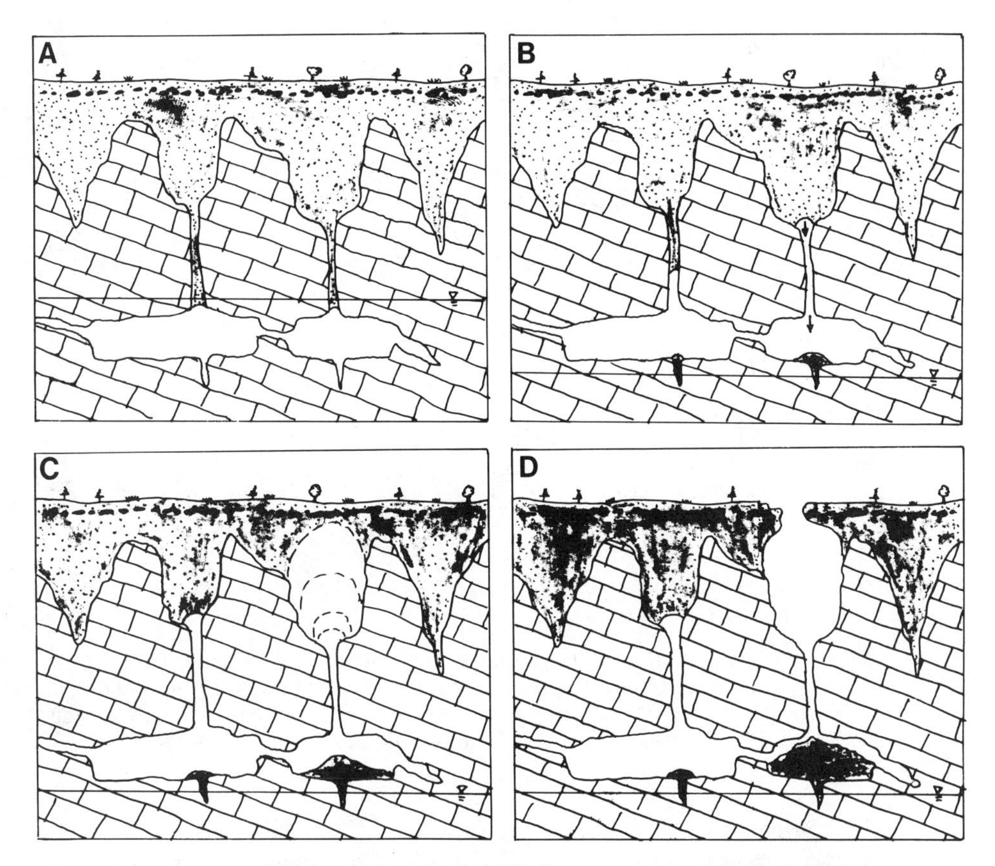


Figure 9.15 Progressive development of sinkhole over a cavernous limestone due to ground water withdrawal in South Africa (after Brink, 1979). **A**. The equilibrium situation before the lowering of the water table. **B**. The lowering of the water table. There is active subsurface erosion (piping into the open cave) as the slot is flushed out by a process of headward erosion. **C**. The progressive collapse of the roof of the vault, possibly temporarily arrested by a ferruginised pebble marker. **D**. The collapse of the last arch to produce a sinkhole surrounded by concentric tension cracks.

hours. Because of differential settlement in the reservoir floor, rigid liners such as asphalt or concrete are ineffective. Where the lagoons carry sewage effluent, or chemical and organic wastes, serious ground-water contamination occurs.

Figure 9.16 illustrates how a house foundation may be damaged in a carbonate terrain by a variety of subsidence-related processes. Foundation damage may be all but eliminated by compaction around a house foundation (Fig. 9.16A), by controlling storm runoff (Fig. 9.16B), or by preventing piping leaks (Fig. 9.16C). Catastrophic sinkhole collapse (Fig. 9.16E) is difficult to design against, although large voids with potentially unstable roofs can be avoided by geophysical surveys and test borings.

Sinkholes have been created by leaky water impoundments in Missouri (Aley et al., 1972), and Pennsylvania (Parizek, 1969) and elsewhere. Sewage and industrial waste lagoons are known to have drained suddenly, rendering them useless and causing ground water contamination. Many reservoirs leak because they are built on karst areas. Examples are the Euphrates Dam in Syria (Dzemeshko and

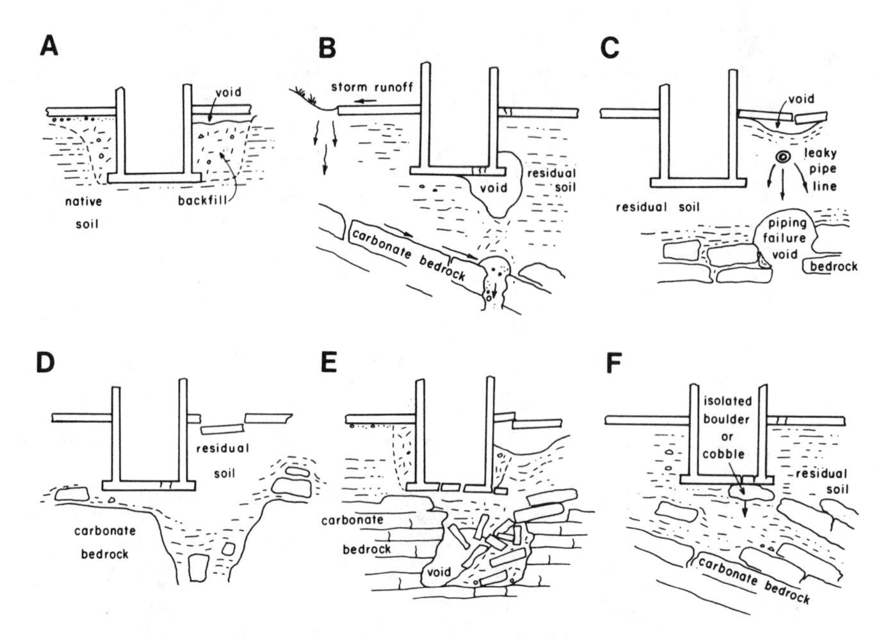


Figure 9.16 Foundation damage due to settlement in a carbonate terrain (from Parizek, 1969). **A**. Improper backfilling and compacting. **B**. Subsurface erosion by uncontrolled storm runoff. **C**. Leaky utility pipes. **D**. Differential settlement of soil of uneven thickness. **E**. Cave roof collapse. **F**. Differential settlement over a "floating boulder."

Parabutchev, 1979) and dams in Tennessee (Soderberg, 1979), Kentucky, China, Turkey, and many other places. Some small reservoirs in Missouri were rendered useless due to sinkholes that formed in the reservoir.

In order to allow for the quarrying of limestone at Hershey, PA, large quantities of ground water pumpage were required. By 1949, ground-water levels over 26 km$^2$ were reduced, and over 100 sinkholes appeared, some causing serious problems. Legal actions prompted a ground-water recharge program, and the quarry company initiated a program of grouting around the quarry to allow the surrounding water table to be restored (Foose, 1953).

Solution of halite can cause land subsidence (Ege, 1979). For example, numerous sinkholes, on the order of 1.5 km$^2$ in area and 15 m deep, have formed by collapse of the overlying sedimentary rock into cavities in bedded Permian salts in Texas. On June 3, 1980, a 110-m diameter sinkhole suddenly opened up in the Hendrick oil field near Kermit, TX (Baumgardner et al., 1980). Another example is in central Germany where a number of communities have had to be relocated due to subsidence caused by solution of underlying salt beds.

Solution features similar to those in limestone terrain have also been reported in areas underlain by gypsum or anhydrite. The existence of integrated subsurface solution channels in gypsum was demonstrated near Roswell, NM, when the Hondo Reservoir was built in 1907. Water in the reservoir area escaped rapidly through many sinkholes, and the reservoir was never able to function as a storage basin (Allen, 1969).

In western Rapid City, SD, sinkholes are associated with the solution of gypsum beds within Triassic shale. Ephemeral springs and drainages were carelessly bulldozed over during 1960 housing developments. Surface runoff began piping under streets and houses, and through the gypsum cave-fill labyrinths. Dozens of incidents have been created where sinkholes suddenly opened up or water mysteriously began appearing in people's basements. The same Triassic gypsiferous shale underlies nearby Spearfish, SD, where a sewage lagoon failed in 1976, 1 yr after construction. The raw sewage surfaced about 100 m away from the lagoon and flowed across neighboring pastures for 4 years until angry ranchers sued the U.S. EPA (because they approved the project).

Prediction of the precise location and time failure of a sinkhole is difficult. Windham and Campbell (1981) note that Florida sinkholes may be more common where the potentiometric surface of the cavernous Florida aquifer is below the local water table, so that near-surface ground water flow is downward into the Florida aquifer. Pumping down the potentiometric surface may provide the final triggering mechanism for a set of potentially unstable, preexisting physical circumstances.

Extensive research has been carried out to try to locate limestone cavities or cavities within overlying surficial deposits. Test drilling to locate cavities is expensive. Foose (1981) shows that a gravity survey may detect large (~10 m diameter) cavities if they are not too deep. Radar, developed to locate tunnels in Vietnam, has potential for cave location. Precisely evaluating the location of a future sinkhole is not possible, however. According to Windham and Campbell (1981), the results of geologic, hydrologic, and geophysical surveys " . . . can be meaningful when applied to the choice of two sites, qualitatively measuring the probability of sinkhole occurrence at one site against the other."

It is difficult to assess the monetary damages caused by land subsidence associated with solution. As in the case of swelling ground damage, home owners tend to quietly cover up the effects for fear that their property value would be diminished.

## 9.4 Land Subsidence Due to Underground Mining

### *9.4.1 Eastern U.S. Coal*

Where a large underground mine is made in horizontal strata, the amount of vertical land subsidence will ultimately be roughly equal to two-thirds of the thickness of the extracted layer. The surface area ultimately affected may extend laterally beyond the mined area by an amount equal to about one-half the depth of the extracted layer (Leveson, 1980).

The coal regions of Pennsylvania have severe land subsidence problems. Underground coal mining has been practiced in the Appalachian Mountains for over a century. Land subsidence is now occurring over numerous abandoned coal mines (Gray and Bruhn, 1984). The underground coal strata were never mined in their entirety; about 50% of the coal in place to allow for roof support. These remaining coal pillars eventually weather, lose their compressive strength, and slowly collapse. Collapse from coal mining has been documented in Pittsburgh, Scranton, and Wilkes-Barre, PA. For example, in 1982 a concrete seal over an 88 m deep coal mine suddenly gave way in downtown Scranton. A parking lot and crane disappeared into the hole. Partial collapse of roads is common in coal-mining districts of Pennsylvania as well as England and Scotland.

Fairmont, WV, like many Appalachian towns, sits atop a honeycomb of flat-lying bituminous coal mines. Every so often, one of the remaining coal pillars crumbles, allowing the earth above to shift and sink. Generally, subsidence above Fairmont mines is a slow process, and causes only inconveniences such as doors in houses that fail to open. Soon gas lines were pulled apart, electric lines snapped, and surface runoff discharges into old mined areas, creating sinkholes. Eventually many affected buildings in Fairmont were condemned. In 1983, over 1000 tons/day of grout was being injected into a 7-hectare abandoned mine area in downtown Fairmont.

(Another problem related to Appalachian coal mining is acid-mine drainage. Because of abundant pyrite in the coal, ground water becomes a weak sulfuric acid passing through abandoned coal mines and dumps. The pH of many streams is less than 3 (Biesecker and George, 1972). As a result, thousands of kilometers of streams in Appalachia are devoid of life.)

Land subsidence also occurs where old coal mines are on fire. For example, smoke and fumes belch forth continuously from cracks and old shafts in Centralia, an anthracite coal town of Pennsylvania. Here sedimentary rocks are intensely folded, and the coal locally dips about 40° beneath the town. Like many abandoned coal mines in Appalachia, the mines are on fire. The underground fires have plagued the town of 1100 since they started in 1962. Burning pillars and refuse hasten the collapse, and add to the hazard of carbon monoxide generation. Carbon monoxide, a poisonous gas, is particularly lethal because it is odorless. Carbon monoxide monitors are placed in the basements of many homes. The fume intensity may vary day by day as the land subsides and new cracks open up under different sections of town. In 1981, the main Centralia fire encompassed 56 hectares and was out of control despite repeated efforts by the U.S. Bureau of Mines to put it out. People who live near the coal fires suffer from various respiratory diseases and many have been hospitalized (Main, 1973). In 1982, the U.S. Department of Interior proposed to dig a 1-km-long trench, 150 m deep, through the heart of Centralia in an attempt to isolate the fire. In 1984, the federal government began to purchase 284 homes and businesses for Centralians who wanted to leave.

The U.S. Bureau of Mines has identified 26 coal mine fires at 17 abandoned sites in the Pennsylvania anthracite region. Many fires have been burning 50 years. Some of the underground fires can be identified in aerial thermal imagery due to a surface heat anomaly. In 1977, the U.S. Bureau of Mines identified 261 fires burning in abandoned coal mines in 16 states.

The Pennsylvania mine fires were started by spontaneous combustion, forest fires, lightning, burning trash, disgruntled union workers, and vandals. A fire burning at Shenendoah, PA, from 1956 to present, was started by bootleggers who dumped a burning stove down a coal mine shaft when their illegal operation was discovered.

There apparently is no way to put out the fires except by physical excavation which is financially impossible. There is no happy ending to this Pennsylvania story—an area of once beautiful woodland underlain by mineral wealth—now a landscape of waste dumps, where the air stinks, the water is contaminated, and the land is cracking apart.

Figure 9.17 Collapse features caused by undergound coal mining near Sheridan, WY (photo by U.S. Geological Survey). The pits result from collapse of the surface into voids left by underground mining at the Monarch mine. The mine was abandoned in 1914.

## *9.4.2 Western United States Coal*

Subsidence due to shallow underground coal mining near Sheridan, WY, has been documented (Dunrud and Osterwald, 1978). Figure 9.17 shows numerous sinkhole-like subsidence features over mined areas of flat-lying subbituminous coal beds. The mines have no more than 60 m of overburden. Interestingly, the depressions on the land surface outline the geometry of the post and pillar underground mining. The Monarch mine at Sheridan is burning underground due, possibly, to spontaneous combustion of the subbituminous coal. Remotely controlled seismometers have detected up to 2500 tremors per day caused by underground caving, surface collapse, or explosions in the mine. Seals and fire barriers installed by mine operators in the 1950s have been destroyed by explosions. A local contractor hired to fill the subsidence pits (in an effort to cut off the supply of oxygen) reported that his 90-ton bulldozer partially dropped into a cavity. The fire is threatening a major power line and Interstate 90.

Since the 1860s, approximately 130 underground mines have produced subbituminous coal from the Cretaceous Laramie Formation near Denver, CO. These operations have undermined over 60 km$^2$. The average thickness of the beds mined is about 1 m, and the workings are generally within 100 m of the surface. Some mines are on fire. The collapse associated with these mines occurs periodically, causing a characteristic typography called thermokarst.

Subsidence problems in Rock Springs, WY, date back to 1869, when the Union Pacific Railroad began mining coal to fire steam engines on the transcontinental route. Over 50 million tons of coal have been extracted from 150 ha beneath the growing town. In 1983, Mayor Keith West lamented "the whole core of the town is underlain with a network of old mine shafts and vaults. There are so many we don't even know where they are, until something up above gives way and falls in." The Union Pacific Railroad and other operators bear no actual liability for property damage resulting from subsidence because surface land was usually sold with specific provisions holding the mine operators not liable.

Land subsidence associated with underground coal mining is also noted in Beulah, ND, Decker, MT, Somercet, CO, and Raton, NM. Dunrud's (1976) studies in Colorado and Utah show that subsidence fractures can propagate through hundreds of meters of strata.

### *9.4.3 Other Mining Activity*

Butte, MT, and Lead, SD, suffer land subsidence from the effects of underground copper and gold mining, respectively. The city of Kutna Hora, Czechoslovakia, had vast underground silver mines, and the old mine workings have led to surface subsidence in many parts of the town. This subsidence has damaged many beautiful old buildings. Ground-water discharge through the mines has led to the solution of sulfate minerals, and the town lost its established water supply (Legget, 1973).

Lead and zinc mines in the Tri-State area of Missouri, Kansas, and Oklahoma hit their peak in World War II, but most mines are now abandoned. Tens of thousands of shafts and bore holes, as well as mines extending over about 1,500 km$^2$, have subsequently filled with ground water and are now overflowing, causing acid water to run down streams such as Tar Creek, OK. At Picher, OK, mine cave-ins have caused the abandonment of four blocks of businesses on both sides of Main Street.

An unusual beneficial effect of land subsidence was documented at Duisburg, West Germany. Channel straightening on the Rhine River caused channel degradation so that the immense harbor facilities were 2 m above river level. Mining of coal beneath the town was begun with the purpose of deliberately causing settlement of the whole area, which reached 1.75 m, compensating nicely for the river drop (Legget, 1973).

Underground mining of salt has contributed to sinkhole formation at Hutchinson, KS, and Grosse Ile, MO, and other places (Ege, 1984). A dramatic subsidence incident occurred in Louisiana, at Lake Piegneur, a 3-km$^2$ natural collapse-type lake that overlies a salt stock in Louisiana. Diamond Crystal Salt Co. had mined up to 1.5 million tons annually from salt beds below the lake. In 1980 a Texaco Oil Co. drilling rig was in the lake, and accidentally drilled into the working mine at 400 m depth. The lake drained within 3 hr into the mine in a powerful whirlpool (Martinez et al., 1981). The drill rig, 12 barges, one tug boat, and 20 ha of land (including residences and tourist attractions) were lost. Astonishingly, all 50 miners in the mine were evacuated in time, and there were no deaths. In 1984, Texaco agreed to compensate the salt company and damaged lake-front property owners in one of the most bizarre industrial accidents on record.

Underground mining can set in motion an interrelated chain of environmental and mining hazards such as land subsidence that not only affect the works of man and the environment, but can decrease the availability of the remaining mine reserves. In Colorado, coal mine subsidence is associated with environmental impacts such as vanishing streams and the production of methane gas (Dunrud, 1976).

Mine subsidence may be ameliorated by 1) leaving adequate supporting material in place, 2) filling the voids with rubble, concrete, or tailings, or 3) grouting (Gray and Meyers, 1970). In the future, mining companies may be required to balance production with subsidence reduction, not only for their own protection from cave-ins (or lawsuits), but in the interest of environmental protection and national conservation of minerals.

## 9.5 Land Subsidence Due to Miscellaneous Factors

### *9.5.1 Tectonism*

Subsidence can occur naturally due to tectonism, such as in the gentle downwarping along faults or synclines. Most of these processes operate too slowly to affect the safety of man, although in the 1964 Alaskan earthquake land subsidence as much as 2 m occurred suddenly over a broad coastal zone south of the Kodiak and Kenai Mountains. The land may noticeably shift in active volcanic areas such as Naples, Italy.

### *9.5.2 External Loads*

During the Pleistocene, portions of the earth's crust were subjected to loads from thicknesses of ice up to a few kilometers. Present for tens of thousands of years, the ice depressed the crust by the principle of isostacy. During the Holocene, this land is rebounding. The rate of uplift is as much as 1 m per century along the Baltic Sea. Uplift following the evaporation of Pleistocene Lake Bonneville in Nevada and Utah is evidenced by warped shorelines above the Great Salt Lake (Gilbert, 1890).

The effects of man-induced loads such as that caused by the weight of Lake Mead, Nevada, is also an isostatic adjustment. Following are two well-documented engineering geology situations relating to land subsidence believed due to loading.

*Baldwin Hills Reservoir, California.* On December 14, 1963, a large water supply reservoir for the City of Los Angeles suddenly failed. Five people were drowned, and 277 houses were destroyed. The Baldwin Hills Reservoir was built on the top of a small hill of Tertiary strata. Active faults are common in the area. Examination of the emptied reservoir revealed that 15 cm of vertical displacement had occurred (Fig. 9.18). The displacement apparently allowed water to work its way under the dam, eventually eroding a hole. The weight of the reser-

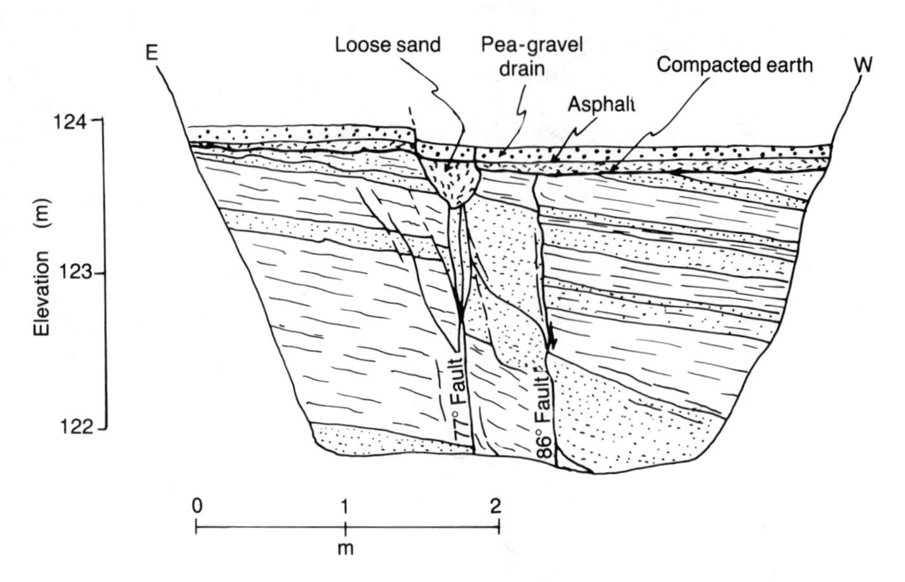


Figure 9.18 Baldwin Hills Reservoir, CA (from Kresse, 1966). Geologic cross section of failure area. Two steeply dipping faults, cutting through Pleistocene sand and clay, were found to have been displaced, offsetting the earth liner. This displacement resulted in sudden failure of the reservoir.

voir may have allowed for displacement along the fault. It is possible that water (pore pressure) also facilitated release of stored tectonic stress at Baldwin Hills. However, the sense of displacement along the faults (downward below the mass of the reservoir) suggests that simple isostatic adjustment may have caused movement along an old fault, and tectonic stresses were not involved.

Another possibility (Jansen et al., 1969) is that the fault movements were related to a widespread pattern of land subsidence due to nearby oil fields which have been occurring throughout the area for many years. Petroleum withdrawals from the nearby Inglewood oil field may have helped to cause subsidence and subsidence-related stress on the existing tectonic faults. Coupled with this was

Figure 9.19 Sketch showing London flood barrier. The insert sketch is a longitudinal profile showing the normal position of a barrier *(left)* being rotated 90° to its flood defense position *(right)*.

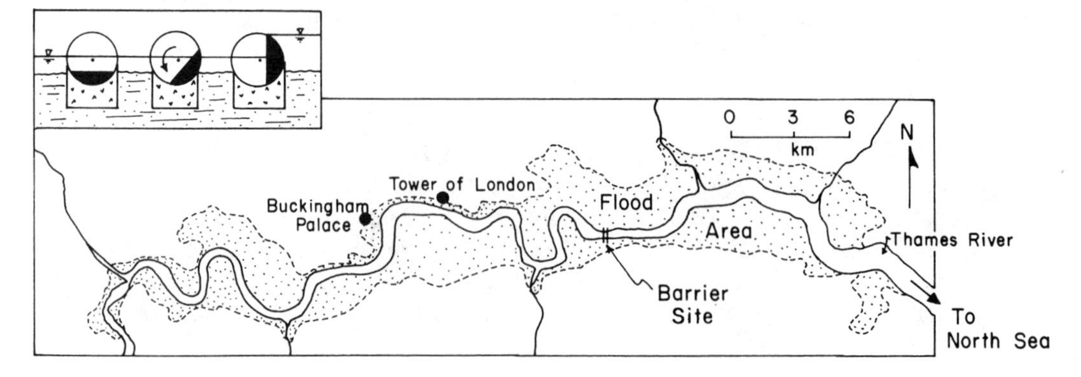

large-scale injection of water into the oil field. In the ensuing legal action, the City of Los Angeles brought suit against oil companies for extraction of petroleum, gas, and water at the nearby Inglewood oil field, contending that the collapse of the Baldwin Hills Dam was caused by settlement associated with fluid extraction. The matter was settled out of court.

*London.* Another form of subsidence is that caused by the weight of man-made structures. In London, heavy buildings were constructed along the weak, clayey banks of the Thames River. The land is sinking, due in part to the weight of the buildings, although the land is probably also subsiding due to ground-water withdrawals.

The problem goes back to early habitation of the area. Tradition held that anyone who reclaimed land from deltas or tidal areas should have rent-free use of that land. Thus, for centuries, people from London and other communities along the Thames have raised levee embankments. The height is now about 6 m above the mean water level of the Thames, which is nearly sea level.

Another factor affecting London's elevation relative to the sea is that eustatic sea level is rising, so that tidal flooding is more common. To make matters worse, Britain is slowly tilting, as an isostatic rebound associated with melting glaciers from the last great Ice Age. Northwest Britain is rising and southeast Britain is sinking at about 0.3 m per century (Leveson, 1980).

High tide levels of the Thames River are increasing at the rate of 0.8 m every 100 years. The metropolitan region of London, with a population of 1.2 million, is below the maximum elevation reached by a 1953 storm surge. London is in a precarious position relative to sea storms. According to the London Chairman of the Public Services and Safety Committee: "The flood now facing London would be the greatest natural disaster this country has ever faced. It would achieve in one night what months of Nazi bombing failed to do in World War II." Officials envision billions of tons of muddy water from the River Thames inundating 120 $km^2$ of London, threatening over 2 million lives.

To combat the flood hazard due to a combination of wind and tides, $1.5 billion was spent to construct flood control gates (Fig. 9.19). The project includes the construction of 10 giant D-shaped (half-cylindrical) flood gates across the 520-m width of the Thames River, 14 km downstream from the London Bridge. The floodgates can be rotated by steel discs pivoted between concrete piers. Under ordinary circumstances, the gates lie flat-face up in rounded troughs planted in the riverbed. If there is a threat of flooding, in 15 min the gates can be rotated 90°, turning them into five-story barricades against rising tides. The project, begun in 1975, was completed in 1983.

### *9.5.3 Deflation and Mud-flow Deposit Compaction*

Land subsidence of one kind or another has been observed in peculiar settings. In the delta region of west-central California, near Concord, sea-level deltaic islands in the Sacramento River are described as subsiding. This may be due in part to compaction, wind deflation, and by burning or dehydration of the peat-rich alluvial soil (Green, 1973). Stephens (1956) describes similar subsidence problems of drained peat soils in Florida.

Lofgren (1969) shows that many dry areas underlain by loess or loose deposits may compact when they become wetted (a process called hydrocompaction). For example, about 2 m of subsidence occurred when irrigation was begun in the alluvial terraces near Riverton and Cody, WY.

In the western side of the San Joaquin Valley, CA, mud-flow deposits form broad alluvial fans which are transversed by the California aqueduct. Mud flows differ from alluvium in that mud flows are not deposited in a subaqueous environment. Many mud flows have not been saturated since they formed; thus there was concern that the canal water may completely saturate the deposits for the first time, and settlement could occur. To prevent this, prior to construction of the canal, the mud-flow deposits were intensely watered to allow for all subsidence to occur (Poland, 1969). This precompaction of the aqueduct foundation cost the California Department of Water Resources an additional $20 million for the California aqueduct in the San Joaquin Valley (Curtin, 1973).

A number of older U.S. Bureau of Reclamation canals in the San Joaquin Valley have been severely damaged by subsidence caused by hydrocompaction or by irrigation overdraft of ground water (Prokopovich and Marriott, 1983). For example, the lower 50 km of the Delta-Mendota Canal has been notably affected. The existence of subsidence was recognized only after the canal construction had been completed. As a result, after a few decades of operation, bridges, check valves, and concrete canal linings became cracked and submerged.

## Problems

**1.** Calculate the barometric efficiency of the well shown in Fig. 9.2.

**2.** See Fig. 9.9.
 **a.** What is the ratio of land subsidence to artesian head decline?
 **b.** If the water table decline is halted, does the land elevation data indicate that the land rebounds?

**3.** The Salt River Valley in central Arizona is an alluvially filled bahada consisting of unconfined sand and gravel near the surface and confined sand and gravel at depth. In an area of $10^5$ ha, about $5 \times 10^8$ $m^3$ ground water is withdrawn for irrigation every year, and the potentiometric surface level drop is 3 m/yr (Bouwer, 1978).
 **a.** Assuming no water replenishment, and that all sediments being dewatered are unconfined, what is the specific yield? (Hint: see Sect. 7.4.3 for discussion of specific yield.)
 **b.** Assuming ground-water recharge from irrigation and periodic Salt River floods equals $4 \times 10^8$ $m^3$ per year and assuming water is taken from confined beds, what is the coefficient of storage?
 **c.** Assuming the confining beds being pumped behave elastically, what rate of land subsidence might be expected? (Hint: use Fig. 9.9 data for ratio of land/water table decline.)

# 10

# Engineering Geology of Coastal Regions

Every one then who hears these words of mine and does them will be like a wise man who built his house upon the rock; and the rain fell, and the floods came, and the winds blew, and beat upon that house, but it did not fall, because it had been founded on the rock. And every one who hears these words of mine and does not do them will be like a foolish man who built his house upon the sand; and the rain fell, and the floods came, and the winds blew, and beat upon that house and it fell; and great was the fall of it.

Jesus Christ
Matthew 7:24–27

As little-noticed as the Watergate burglary, 10 years ago, was a "tropical depression" in Yucatan. It blew to hurricane fury crossing the Caribbean, killed 12 people in Cuba and Florida, and . . . moved sluggishly north. Twenty-one persons died in Maryland.

Tropical Storm Agnes was such a freak that Baltimore never really faced up to the flood problem. Nor did much happen after the milder Tropical Storm David of 1979 did worse property damage to Mount Washington and drowned a tourist. . . .

Today, city storm drains fill with dirt, the budget not allowing for routine maintenance. Flood plain apartments on Frederick Avenue are being made ready for tenants. It would be nice, 10 years later, to say what this region learned from Agnes. But it appears we have learned very little.

*The Sunday Sun*
Baltimore, MD 1982

Vanity of vanities, says the Preacher; all is vanity.

King Solomon
Ecclesiastes 12:8

## 10.1 Basic Concepts

### *10.1.1 Shoreline Classification*

Coastal geomorphology classification traditionally includes two kinds of shorelines based on the relative position of the sea with respect to the land. Where the land is subsiding, and/or the sea is rising and advancing over the land, a shoreline of submergence is created; where the land is emerging from the sea, and/or the sea is withdrawing, a shoreline of emergence is created. Figure 10.1 illustrates examples of landforms associated with these two shorelines.

The scientific validity of this twofold coastal landform classification is doubtful because of the diverse rock types and geologic history of coastal areas (Shepard, 1963). For instance, the eastern seaboard of the United States appears to have a

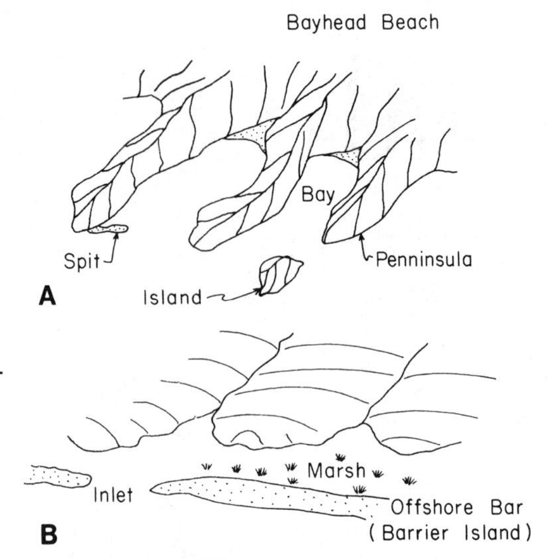


Figure 10.1 Idealized sketches illustrating shoreline geomorphology. **A.** Shoreline developed on mountainous terrain (the so-called "shoreline of submergence"). **B.** Shoreline developed on a wide coastal plain (the so-called "shoreline of emergence").

shoreline of submergence in the New England states and a shoreline of emergence in the central and southern area (Fig. 10.2). This is probably not because North America is bending in a weird manner, but rather that the rocks and geologic history of the two areas are different. New England is a glacially eroded coast, underlain largely by hard metamorphic rocks. The many islands, peninsulas, etc., are the result of weathering and glacial and coastal erosion of these rocks. On the other hand, a regressive sequence of nearly flat, readily erodible sediments dominates the coastal plain from New Jersey southward, and the coastline reflects the nature of coastal processes acting on these sediments. The barrier island seems to be largely the result of littoral transport of sediment (Kraft, 1980). The Pacific coastline of the United States exhibits mostly submergent features, but its physiography is complex. A myriad of rock types and complex geologic history dominates the coastal geomorphology. Especially important is the role of vertical movements from active faults which give rise to mountains, islands, or bay areas.

In spite of the lack of theoretical backing for emerging and submerging coastlines, this classification still serves a useful purpose in that it conveys a general physiography to the reader. The geologic processes and hazards are quite different in the two places. For example, hurricane and wave erosion are particularly hazardous on the low-lying barrier islands along the Atlantic and Gulf coasts of the United States.

For a more detailed examination of coastal landforms and processes the reader may consult Gilbert (1885), Shepard (1963), Van Dorn (1965), Ippen (1966), Shepard and Wanless (1970), Fisher and Dolan (1977), Swift and Palmer (1978), and Ritter (1978).

### *10.1.2 Beaches*

Waves are powerful agents of erosion along coasts, the strip of land adjacent to the sea (Fig. 10.3). Armed with sand, cobbles, and boulders, waves cut into the coast, eventually forming a sea cliff. Waves may be expected to reach the base of the sea cliff during exceptional storm events. Winter storms can cause extensive

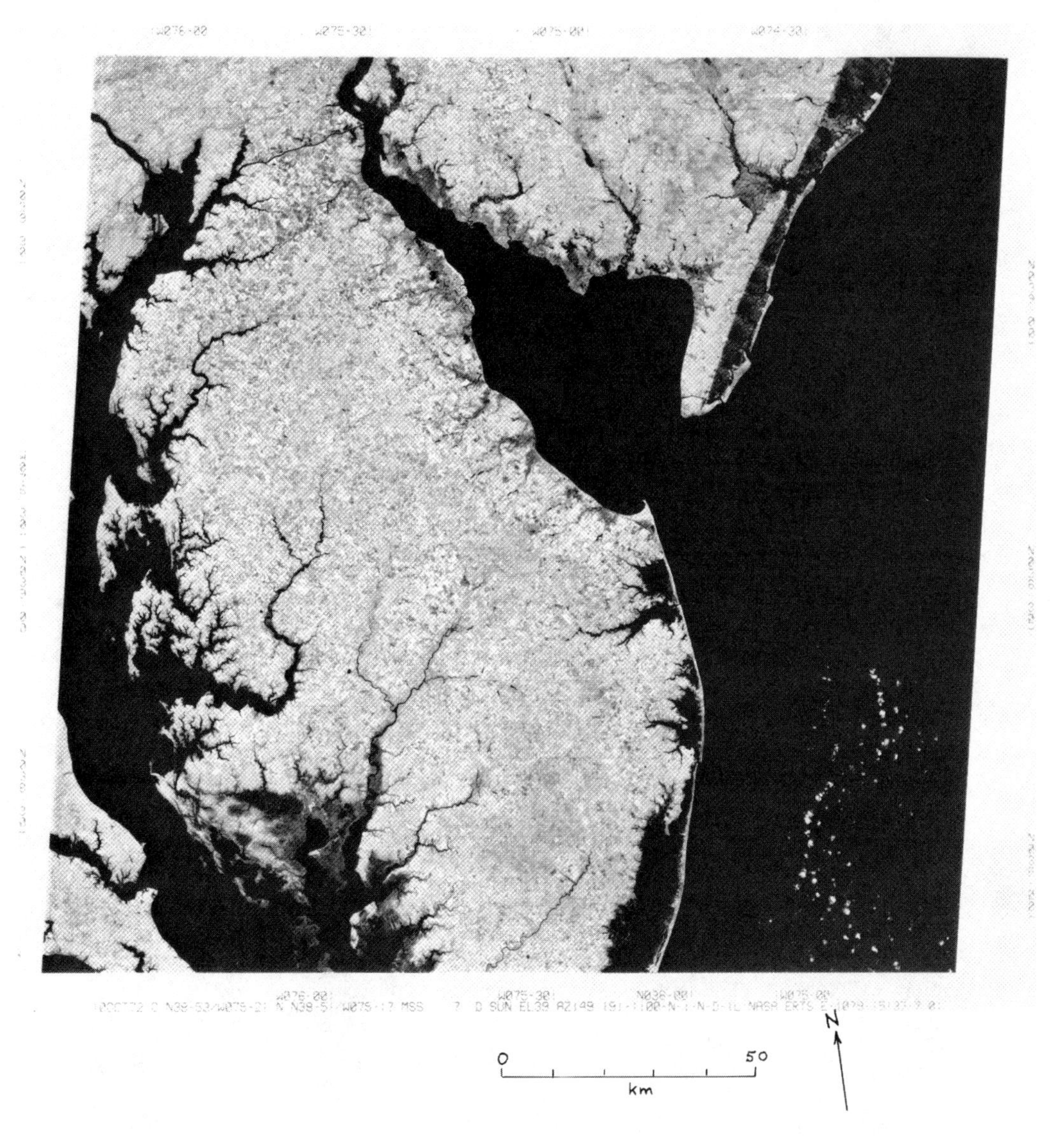


Figure 10.2 Landsat image of the Chesapeake and Delaware Bay areas. The narrow barrier island along the New Jersey, Delaware, and Maryland coast is clearly visible.

erosion and coastal damage. For instance, the great Ash Wednesday storm of March 7, 1962 produced waves over 10 m in height and caused millions of dollars of property damage along the mid-Atlantic coast (Dolan et al., 1980a).

The beach is a relatively gentle sloping surface that is washed by waves; along the ocean coast it includes the area between low tide and the sea cliff; it includes the shore (area between high and low tides) and berms (elevated parts of the beach formed during storm tides). (Note: some authors refer to the beach as the foreshore in contrast to the offshore which includes the seaward area from the foreshore to the depth where waves first "feel" the bottom; the backshore includes the area inland from the beach to the sea cliff.)

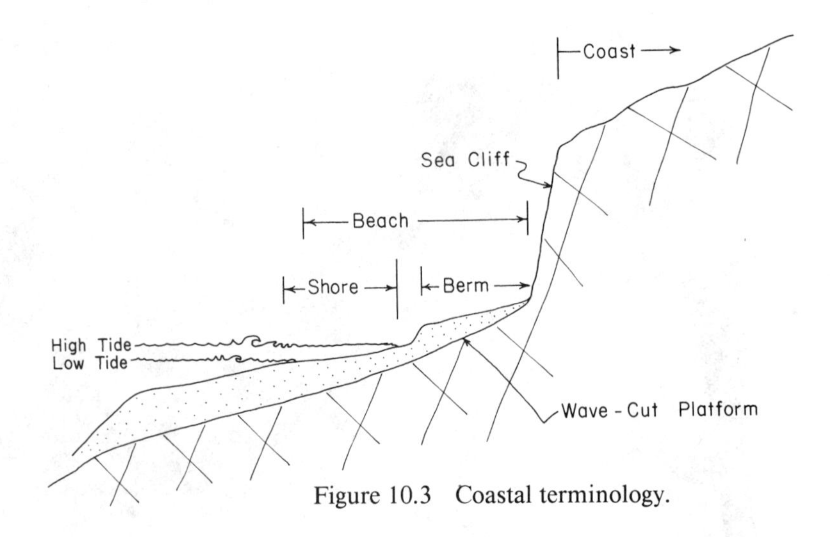


Figure 10.3 Coastal terminology.

About 90% of the world's oceanic beaches are underlain by sand. Sand or other beach material accumulates on a bedrock-cut terrace called the wave-cut platform (Fig. 10.3). The slope of a beach consisting of fine debris is less than the slope of a beach consisting of coarse debris. Beaches rarely form in high-energy environments; rather the sand is moved by waves and currents to low-energy environments such as a bay. In some places (example: La Jolla, CA), winter storms produce waves and currents that are competent to remove all but cobble to boulder-sized debris, whereas sand accumulates on these beaches in the summer due to gentler waves (Shelton, 1966).

The relative amount of wave energy impacting on any place along a shoreline depends on the coastline configuration. Figure 10.4 shows advancing wave fronts aligned roughly parallel to the coast. Orthogonals to the waves show the direction

Figure 10.4 Plan view sketch of wave attack on a coast.

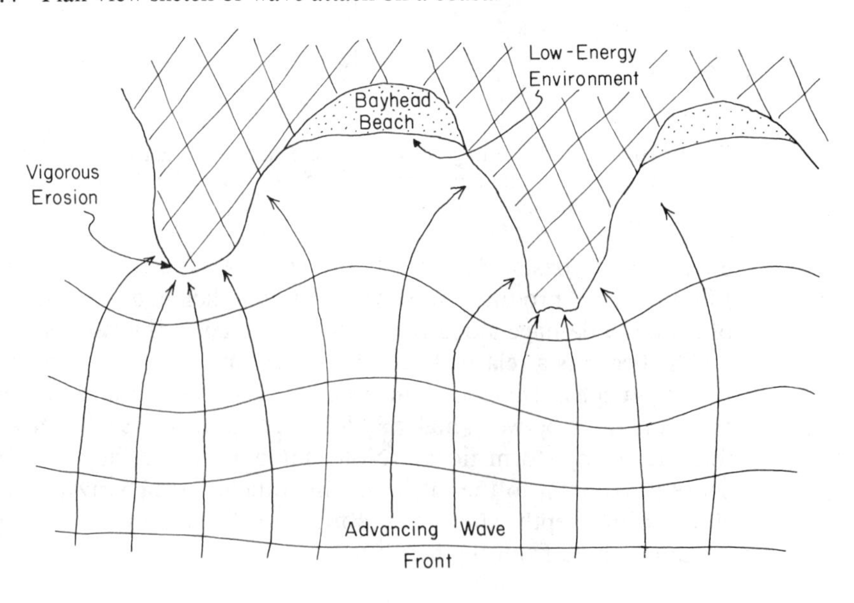

of movement of the waves. If each unit length of wave contains a certain amount of energy, it can be seen that headlands of the peninsulas receive more energy than the bays. Waves bend because as an advancing wave comes to shore it feels the bottom and begins to slow down. This happens at a depth of approximately half the deep-water wave length. The waves slow down first nearer the protruding peninsulas, and the wave fronts bend. Thus, the peninsula heads are areas of vigorous attack by waves, whereas the bays are low-energy environments where deposition predominates, forming a bay-head beach. Waves may refract in this fashion even along a straight coastline if a subterranean peninsula protrudes.

### *10.1.3 Currents and Tides*

A longshore current is produced where waves strike the coast obliquely (Fig. 10.5). As the wave breaks, water and sediment are washed obliquely up the beach in the direction of the advancing wave, but as water returns to the sea it tends to flow back perpendicularly to the coast line. So the resultant transport direction after a time is parallel to the coastline. This current, called longshore current, is a powerful transportation agent. The sediment moved by the longshore current is called longshore drift (or littoral drift). (How many unwary swimmers in the surf have inadvertantly participated in this process and have later found themselves a long distance from where they started!)

Longshore drift helps distribute sediments eroded by waves and sediment brought to the sea by rivers (Dyhr-Nielson and Sorenson, 1970). Deposits at improved inlets on the New York and New Jersey coast indicate rates on the order of 380,000 $m^3$ of longshore drift a year passing a given point (Bruun and Lackey, 1962). On the California coast, figures as high as 760,000 $m^3$/yr are probably representative. Large migrating bars may occur, such as Presque Isle, near Erie, PA, which has been moving northeasterly along Lake Erie since Pleistocene time. Toronto Island, Canada, is a 10-km-long spit that has been stabilized in its east to west littoral drift movements by the dredging of sand (used for local construction) and the construction of a large breakwater. The breakwater, created with suitable waste material, helps provide a safe harbor as well as a low-energy environment where sediment collects.

Sand for construction aggregate is a rapidly diminishing resource in many metropolitan areas. Sand has been mined in some beaches for construction projects.

Figure 10.5 Plan sketch of coast showing origin of longshore current.

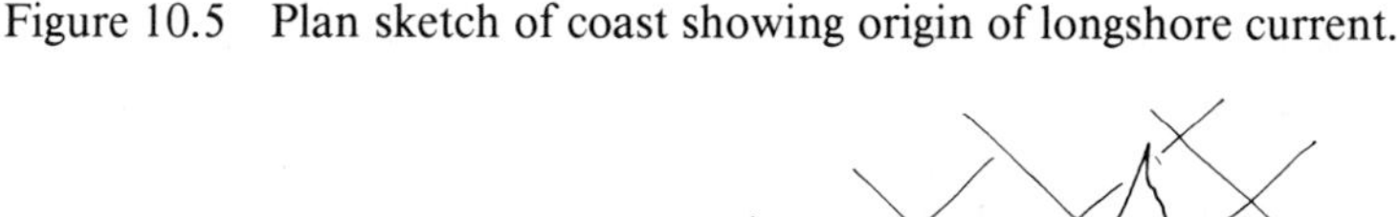

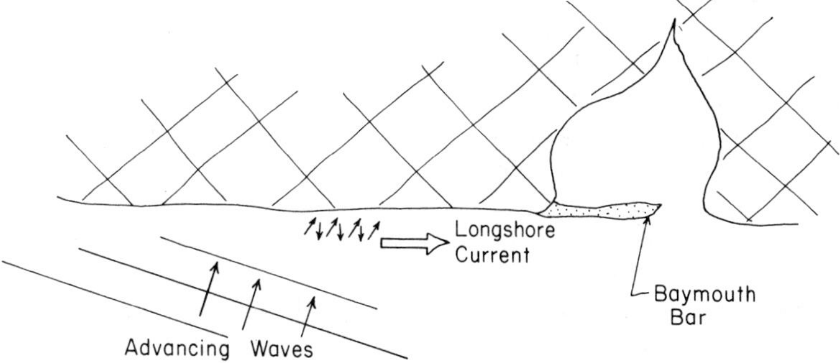

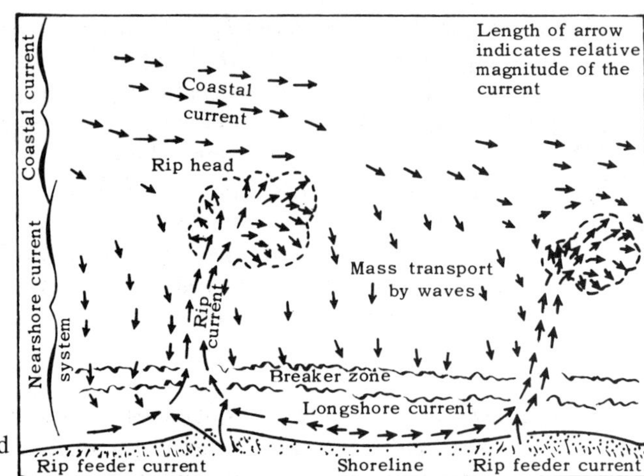


Figure 10.6 Plan view of shore currents (from Shepard and Inman, 1951).

Man has diminished the supply of sand to coastal areas by the construction of dams on rivers. Artificial beach-nourishment projects, such as offshore dredging projects by the Corps of Engineers, are used to supply sand to Connecticut shorelines (Coates, 1976).

In addition to longshore current, there may be other currents along a coast. Shepard and Inman (1951) divided marine currents into 1) coastal currents and 2) nearshore currents (Fig. 10.6). A coastal current flows roughly parallel to the shore, and constitutes a relatively uniform velocity in deeper water. A coastal current may be tidal or wind-driven in origin. Nearshore currents exist independently of the coastal current, and are associated with waves. Nearshore currents generated include a) the shoreward movement of water associated with waves, b) the longshore current, c) rip currents, and d) the longshore movement of the expanding head of the rip current. Rip currents occur where two longshore currents meet, for example, at a promontory; these currents are particularly dangerous to swimmers as they may carry them offshore. Rip currents are seldom greater than 10 m wide; if a swimmer is caught in one, he should swim parallel to the shore until out of the rip current. In cross section, currents in the breaker zone also include a strong seaward movement, the undertow, which compensates for the wave surge which advances shoreward.

Ocean tides, due to the gravitational attraction of the moon, can be unusually high when the sun is aligned with the moon (a spring tide). Atmospheric phenomena such as pressure changes or severe winds also create exceptionally high tides (known as Aqua Alta in Europe). A several-hour wind on Lake Michigan can raise the level of the lake 3 m. Along the Gulf Coast of the United States, large tides are associated with hurricanes.

Coastal physiography may cause unusual tidal currents. For instance, at the Bay of Fundy, Nova Scotia, it is believed that the 20-m high tides and associated tidal currents can be harnessed to generate electricity. In some deltas where river water enters the ocean, a tidal bore may occur (Fig. 10.7). This is a moving hydraulic jump, a wall of cascading water perhaps 1 m high that can surge forward at 15–25 km/hr and can upset small boats. Tidal bores are well-developed in a number of rivers, including the Amazon, the Yangtze, and the Elbe.

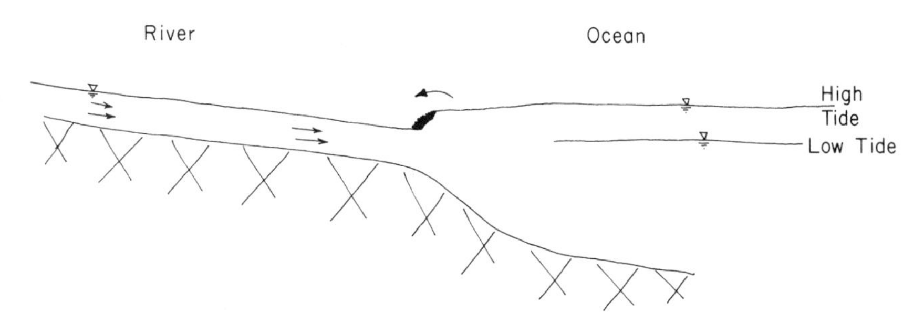


Figure 10.7 Cross-sectional sketch illustrating origin of tidal bore.

Major ocean currents are caused by coriolis force as well as temperature and salinity differences in the ocean. For example, cold temperature produces a cold, dense current which moves from Antarctica along the floor of the Atlantic Ocean as far north as Bermuda. Evaporation of the Mediterranean Sea causes the water to become more saline than ordinary oceanic water. The dense saline water forms a density current which travels through from the Straights of Gilbralter into the depths of the Atlantic Ocean. Fresh water floats on top of seawater; where a large river enters salt water (example: Susquehanna River at Chesapeake Bay) the river water may extend as a thin wedge over the salt water. The fresh water lens from the Amazon River extends over 100 km into the South Atlantic Ocean.

Density currents due to turbidity are common in oceans and are also known to occur in man-made reservoirs. A turbidity current may form when a landslide occurs in a muddy area on a subterranean slope, causing a large volume of the water to become sediment-laden. Being denser than the surrounding clear water, a density current is formed as the turbid material seeks a lower level. Turbidity currents are known to have travelled at 22 m/s and to have broken trans-Atlantic telephone cables following the November 18, 1929, earthquake near the Grand Banks (Longwell et al., 1969). Turbidity currents are capable of distributing sediment over a reservoir bottom. At Lake Mead, Arizona, muddy water that occasionally issues from the dam hydroelectric gates are evidence of turbidity currents that have transgressed the reservoir floor. The flows are several meters thick, and take about 1 week to traverse approximately 110 km from the steep front of the delta to the dam, a route along which the average slope is less than 0.001 (Shelton, 1966).

### 10.1.4 *Wave Celerity*

Consider the simple harmonic progression of waves as shown in Fig. 10.8. Wave celerity (velocity) is defined as the time required for successive waves to pass a fixed point. Thus:

$$C = L/T. \tag{10.1}$$

From mathematical theory, it can be shown (Ippen, 1966) that wave celerity is governed by the depth of water:

$$C^2 = \frac{gL}{2\pi} \tanh \frac{2\pi D}{L}, \tag{10.2}$$

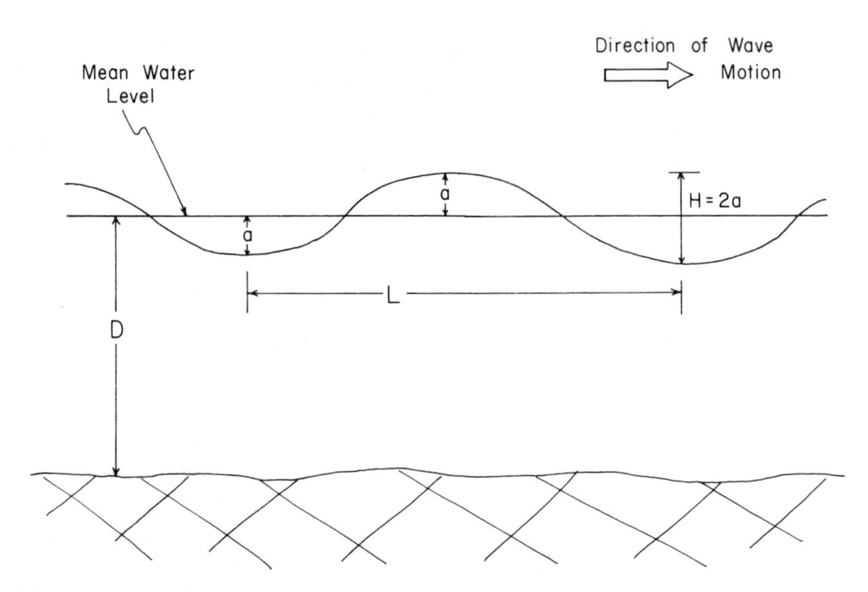


Figure 10.8 Wave terminology.

where:

$g$ = gravitational constant and

tanh = the hyperbolic trig function.

If the depth is known, the wave celerity and wave length can be solved by trial and error solution of Eq. (10.2) or by special tables (Ippen, 1966).

For shallow water conditions ($D/L < 0.04$), $\tanh 2\pi D/L$ is nearly the same as $2\pi D/L$, so Eq. (10.2) reduces to:

$$C^2 = \frac{gL}{2\pi}\frac{2\pi D}{L},$$

or,

$$C^2 = gD. \tag{10.3}$$

Therefore, in this case, wave celerity is independent of wavelength and depends only on depth. For instance, if the water is 5 m deep, the velocity of a medium to large wave will be:

$$C = [(9.8 \text{ m/s}^2)(5 \text{ m})]^{0.5} = 7 \text{ m/s}.$$

For deep water conditions ($D/L > 0.5$) $\tanh 2\pi D/L \simeq 1$, and Eq. (10.2) becomes:

$$C^2 = \frac{gL}{2\pi}. \tag{10.4}$$

In this case, wave celerity is independent of depth. Combining Eq. (10.1) and (10.4) yields:

$$C^2 = \frac{gL}{2\pi} = \frac{g(CT)}{2\pi}.$$

Therefore,

$$C = \frac{gT}{2\pi}. \tag{10.5}$$

From this equation it can be seen that in deep water long-period waves with great wavelengths and velocity will separate from short-period waves. For example, a wave with a period of 6 s will have a wavelength of 56 m and a velocity of 9 m/s, where a longer period wave such as 14 s will have a longer wavelength (303 m) and a faster velocity (22 m/s).

Practical applications of the above equations include consideration of amplitude changes as a wave comes to shore. From Eq. (10.3), because a wave velocity depends on depth, an advancing wave essentially "feels" the bottom and begins to slow down. Because wave momentum is conserved, the diminishing velocity is transferred into the creation of a higher wave. This continues until the wave is unstable, and it "breaks," forming a breaker.

Another facet of this theory is the explanation of tsunami dynamics. A tsunami may travel at hundreds of kilometers per hour on the open sea, where its amplitude is less than 1 m, but upon reaching a coast the *giant* wave is formed.

### *10.1.5 Wave Fetch*

Practically all waves are generated by wind. The distance of open water over which the wind blows is called the fetch. The transfer of energy from wind to waves, and the resulting wave growth, is not completely understood, and wave forecasting relationships are largely empirical rather than theoretical.

When the wind has been blowing for many hours over a large expanse of water, the maximum height of the waves is roughly equal to the square of the wind velocity, and is given by the formula (Howell, 1972):

$$H = 0.0023\ V^2, \tag{10.6}$$

where $H$ is height (m) and $V$ is wind velocity (km/hr). Thus, for example, a 100 km/hr wind will produce waves 23 m high.

The size of waves generated in an open body of water is also dependent upon the fetch. The Stevenson formula used in design of British sea defenses (Attewell and Farmer, 1976) is

$$H = 0.34\ F^{0.5}, \tag{10.7}$$

where $H$ is the wave height (m) and $F$ is fetch (km).

For short fetches and high windspeeds, Bretschneider (1957) shows that:

$$H = 0.0067\ VF^{0.5}, \tag{10.8}$$

where:

$H$ = significant wave height (m) (Note: significant height is the average height of the highest third of all waves present in a wave train),

$V$ = windspeed (km/hr),

$F$ = fetch length (km).

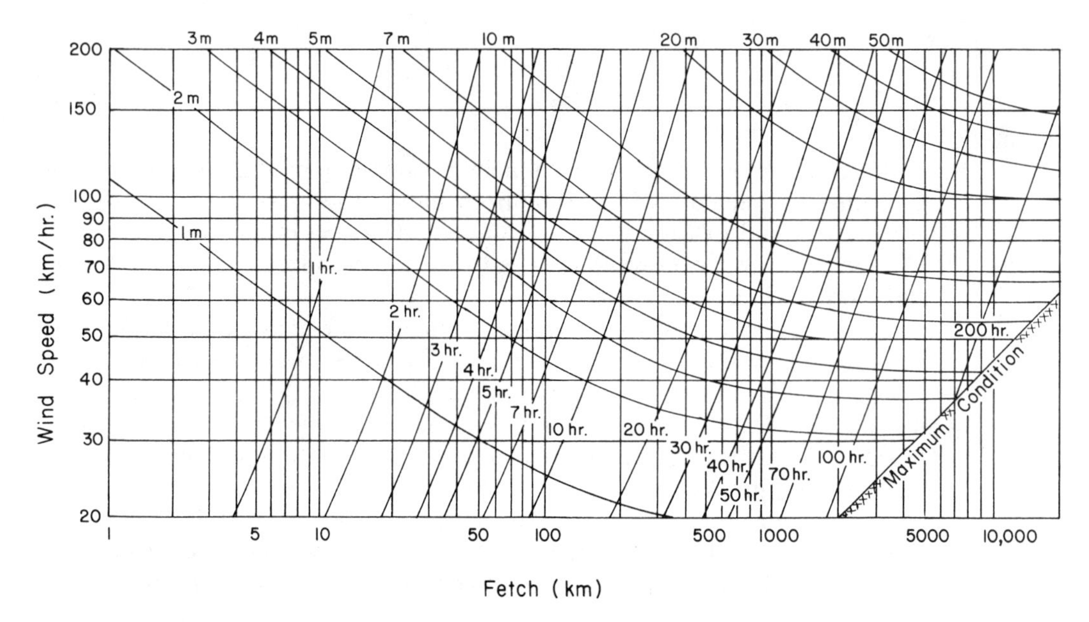


Figure 10.9 Relationship between wind velocity, fetch distance, and significant wave height (after Ippen, 1966; Linsley and Franzini, 1979). To find wave height, find the wind speed on the vertical axis, and move horizontally until either the vertical fetch line or diagonal wind duration line is intersected. For example, a 100 km/hr wind blowing for 2 hr across a 100-km-wide bay would produce waves 3.5 m high. The same wind blowing for 10 hr would produce waves 5.8 m high.

Wave-height data obtained from major reservoirs (Linsley and Franzini, 1979) show a modified formula for the significant wave height:

$$H = 0.005\ V_w 1.06 F^{0.47}, \tag{10.9}$$

where:

$H$ = significant wave height (m),

$V_w$ = wind velocity (km/hr) about 8 m above the water surface,

$F$ = fetch (km).

Wave heights can be determined graphically in Fig. 10.9. This figure is useful, for example, as a guide to determine the approximate height of wave erosion along shorelines, or the minimum elevation of rip-rap to be placed on an earth-fill dam. (Rip-rap should extend above the highest wave height because, upon impact, a wave will actually "run up" the slope to a height approximately twice its open-water height.) In addition to fetch-induced waves, wind setup (the slight tilting of the entire surface of a reservoir) may occur, causing even higher wave erosion to occur than predicted. Wind setup equations developed by Dutch engineers for the Zueder Zee are shown in Linsley and Franzini (1979).

## 10.2 Engineering Structures in Coastal Areas

Engineering structures along coasts are generally built to provide a safe passage or a harbor for boats, and to control littoral drift for purposes of beach stabili-

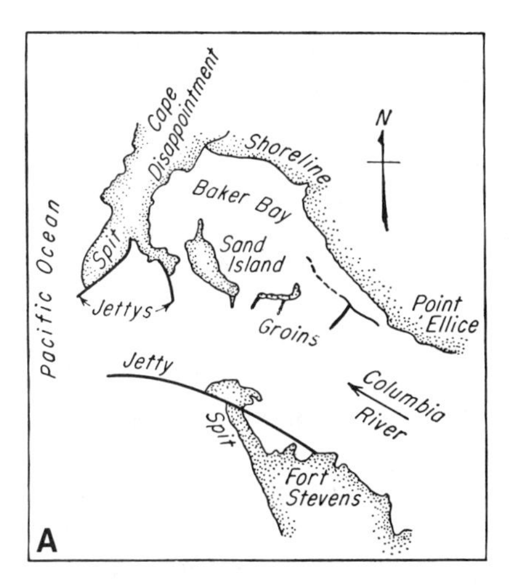


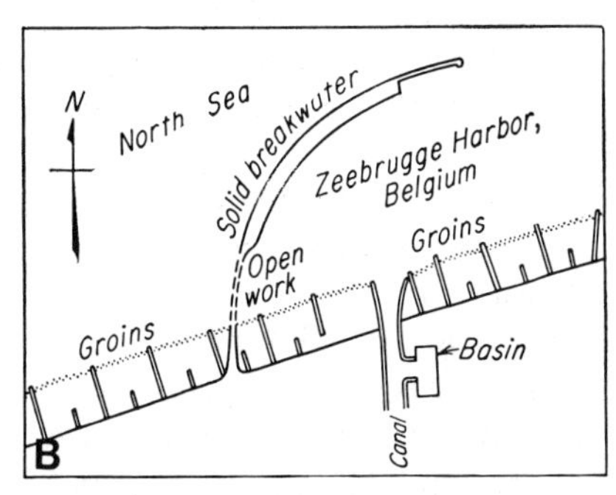


Figure 10.10 Coastal engineering structures (from Krynine and Judd, 1957). **A**. Jetties and groins at the inlet of a river. **B**. Breakwater (to protect the harbor) and groins (for stability of the shore line).

zation (Watts et al., 1977). Some structures are built to prevent erosion or enhance recreational opportunities.

Figure 10.10 illustrates the origin of common coastal engineering structures. Jetties, breakwaters, and groins are commonly made with boulders or large masses of concrete.

A jetty is usually built near the mouth of a river. Jetties concentrate river discharge so that stream velocity is maintained and sediment does not impede navigation at the river mouth.

Breakwaters provide a man-made harbor. At Chicago, for example, the natural shoreline of Lake Michigan is fairly straight and nature provides no safe mooring for boats. An enormous breakwater was constructed that eliminates the storm waves of Lake Michigan, providing safe harbor for thousands of pleasure crafts.

A groin is a structure oriented perpendicular to the coast which serves primarily to catch sediment being moved as longshore drift. The sediment accumulates mainly updrift of the groin. Some groins are built to enhance fishing opportunity or to provide a discharge point for city storm runoff or sewage.

Revetments are erosion barriers erected along the shore. Revetments commonly consist of concrete masses, large rocks, steel plates, old car bodies, etc.

Other forms of engineering structures have been used for unusual circumstances. For example, flood control gates have been constructed in the tidal area of the Thames River at London to protect subsiding areas from inundation (see Chapter 9).

A gigantic coastal project is taking place in the Netherlands, in the deltaic region along the North Sea formed by the confluence of the Rhine and Meuse Rivers. The North Sea area was inundated as the Ice Age ended, and has been eroding the sides of the English Channel between Britain and the Continent. When the Romans occupied Holland (Netherlands, meaning bottom lands) in 34 B.C., the present north coast of Holland was not islands but part of the mainland. In a dramatic reshaping of the land, the North Sea advanced inland via a series of storms between 1200 and 1300 A.D. to take possession of the area until the industrious Dutch began the slow process of repossession in the 1930s. The Dutch have now reclaimed about 69 $km^2$ from the sea. A polder is land that is maintained in a dry state by surrounding it with a broad, high dike, which keeps the sea out. The Schiphol Airport, where visitors to Holland typically unload, is 5 m below sea level. Because Holland has walled itself from the sea, it is particularly vulnerable to that sea. On February 1, 1953, a storm surge reached 3.5 m above high tide level. Dikes were overtopped in 89 places, 15 $km^2$ were flooded, and 2,000 people drowned (Danilevsky, 1983). The Delta Plan, now under construction, calls for the strengthening of all coastal dikes, and the closing off of the Delta estuaries from the sea. The construction of the Eastern Scheldt Storm Surge Barrier began in 1978. Only the estuaries leading to the parts of Antwerp and Rotterdam are to remain open. Due to opposition of fisherman and environmentalists concerned with estuary ecology, giant gates in the storm-surge barriers will be left open during normal tides but will close only when a storm in the North Sea threatens.

## 10.3 Coastal Erosion

### *10.3.1 Natural Coastal Erosion*

Coastal erosion leading to the formation of a sea cliff is a natural process. Typically, waves cut the base until the cliff becomes unstable and falls onto the beach where it is washed away by the waves. In some places the process of cliff retreat is very slow, particularly where hard rocks such as granite stand for centuries with little evidence of erosion. In some places in California, however, Cenozoic marine sandstone has been elevated above sea level; erosion can be vigorous on these unconsolidated sediments. A steep sea-cliff with absence of talus at the base is indicative of active marine erosion (Emery and Kuhn, 1982).

Along much of the shoreline of Lake Michigan and other places on the Great Lakes, erosion is severe due to the fact that 1) the cliff is composed of glacial till, and 2) waves have only recently ($\sim$10,000 yr) been brought to bear on the slopes, and little protective armor (boulders, resistant layers) has developed. Wave erosion at the base of the cliff initiates erosion, and is compounded by ground-water discharge and slumping (Sterrett and Edil, 1982). Rates of bluff-top retreat of 10 m/yr have been observed.

Since World War II, dozens of homes and hundreds of structures and public facilities have been damaged or destroyed as a result of sea cliff retreat along southern California's coastline. For example, at the Sunset Cliffs area of San Diego (Fig. 10.11), the bedrock is a Cretaceous sandstone and siltstone (Point Loma Formation) overlain by about 5 m of Quaternary marine terrace sand and gravel (Bay Point Formation). Erosion by wave action is concentrated on the cliff base. Cliff retreat is a result of 1) roof collapse of sea caves or surge channels or 2) mass wasting. The channels, which follow the joint planes, gradually evolve into caves by progressive basal undercutting. Excavation is aided by abrasion from sand and cobbles. Caves are widened at joint intersections. Erosion continues until the roof collapses.

According to Gerald Kuhn, a geologist from Scripps Institute of Oceanography, people in California "have been living in a fantasyland," referring to those

Figure 10.11 Aerial photograph taken in 1968 at Sunset Cliffs, San Diego, CA (from Kennedy, 1979). Erosion is accomplished by waves which form sea caves and tunnels which follow the strike of vertical joints.

who had built homes on cliffs overlooking the Pacific Ocean. "Many of the homes are built on ancient landslides," where the cliff is crumbling into the ocean.

Erosion rates along the Pacific vary with location and are episodic. Shepard and Wanless (1970), Kennedy (1979), and Kuhn and Shepard (1979) document the rate of erosion near San Diego, and attempts by city engineers to dump boulders and concrete blocks at cave portals to stop erosion. Photographic analysis shows that approximately 75% of the Sunset Cliffs sea cliff area in San Diego had undergone no appreciable change in 75 yr, but 5% had undergone rapid retreat (up to 3 m). The overall average rate of erosion is estimated at over 1.3 cm/yr. Studies of rates of sea cliff retreat along the southern California coast show that sea cliff retreat may average less than 0.5 cm/yr in dense igneous rock (andesite) while averaging as much as 50 cm/yr in poorly consolidated Pleistocene terrace deposits (Ploessel, 1973). During 1983, wave erosion from severe winter storms caused cliff undercutting and damage to thousands of houses along the southern California coast.

Short-term cliff retreat in southern California is episodic, and is related to meteorological conditions and bedrock composition. Analysis of old maps of Encinitas suggest that entire city blocks have disappeared into the Pacific (Kuhn and Shepard, 1979). Sante Fe railroad maps show that tracks have been repaired periodically due to bluff collapse. In 1940, a freight train went over the cliff after the tracks were undermined at Del Mar following heavy rains and large waves during a high tide.

In 1972, California passed a state-wide Coastal Protection Initiative. As a general rule, 7.6 m is now required for bluff-top setback for new structures. Much of the development of southern California's coast took place prior to 1972, however, and is not affected by this regulation.

Rapid rates of erosion often occur along the shores of man-made reservoirs. For example, Lake Sharpe, South Dakota, is the reservoir formed behind Big Bend Dam on the Missouri River. It was built in 1963. The bedrock is weak Cretaceous shale (Pierre Shale) overlain by glacial till and terrace alluvium. Koopersmith (1980) determined an average rate of coastal erosion for a 10-yr period was an astonishing 3.4 m/yr. The eroded debris contributes to the siltation of bays and causes a reduction of the life expectancy of the reservoir.

Ice Bay, AK, a potential staging area for offshore oil and gas development in the Gulf of Alaska, may experience a short life for that purpose because onshore facilities are vulnerable to the area's naturally rapid coastal erosion and siltation. A glacier which filled Ice Bay until 1904 has receded 40 km, and a large hooked spit formed in the Bay. The spit grew 6.6 km long between 1904 and 1980. Longshore currents have eroded adjacent shorelines at a rate of 37 m/yr, so that 8.2 $km^2$ of land has disappeared since 1941.

Since 1824, the Corps of Engineers has been trying to stop the natural migration of Presque Isle Peninsula, PA, a large sandbar on Lake Erie. Moving at a rate of about 8 m/yr since Pleistocene time, the Corps has spent millions of dollars building groins and replenishing sand to the eroding beaches.

Coastal erosion plagues many countries. In Bulgaria, for example, government resort complexes were built along sandy beaches that form part of the western edge of the Black Sea. Wave erosion has caused slope instability and landsliding, thwarting plans to establish a vacation area for eastern European countries which would bring hard currency to Bulgaria (Williams and Caldwell, 1981).

Records of the changes in the east coast of England are available for about 2000 yrs, and show that 215 $km^2$ of land has been lost since Roman times. At least 28 towns have slowly disappeared into the sea (Sheppard, 1912). Rates of erosion along selected coasts of Japan range from 0.1 to 5.5 m/yr (Horikawa and Sunamura, 1967).

Local geomorphologic history plays an important role on coastline erosion. In some places, land once covered by glaciers is rebounding due to isostacy. In the Hudson Bay area, for example, the land is rising at a rate of 2.5 cm/yr, and, if the trend continues, the bed of Hudson Bay will eventually become dry land. The northern side of the Great Lakes area is rebounding more than the south side; the north shore of Lake Superior, for example, is rebounding at 0.5 cm/yr, decreasing to 0 at Lake Erie. The tilting of Lake Michigan would ultimately allow for its natural drainage into the Mississippi River in the near geologic future. This tilting action compounds shore erosion problems in the southern edge of Lake Michigan.

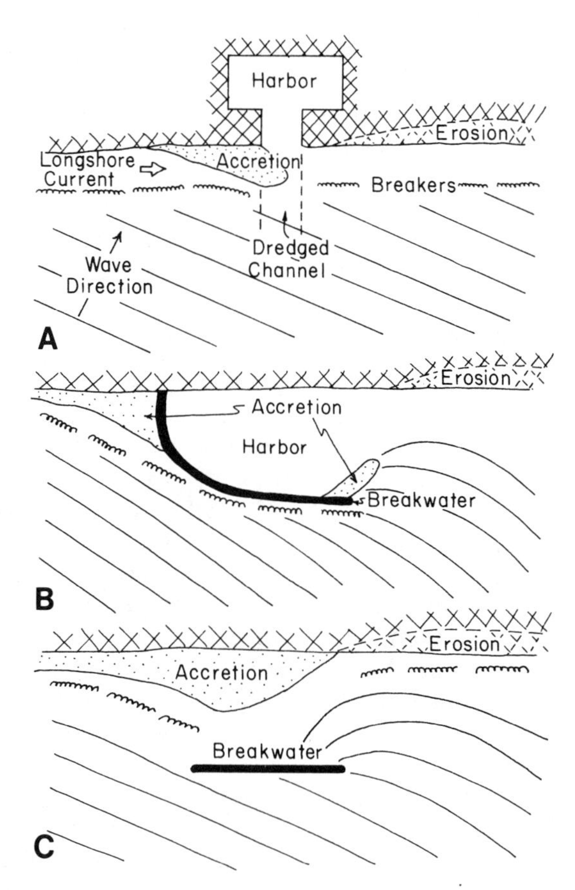


Figure 10.12 Sediment accumulation adjacent to engineering structures (from Ippen, 1966). **A**. Channel dredged through littoral zone. **B**. Shore-connected breakwater. **C**. Detached breakwater.

## *10.3.2 Coastal Erosion as Related to Man-made Structures*

When engineering structures such as jetties and groins are placed in a coastline area, the natural regime of waves, currents, and sediment transportation is changed.

The longshore drift of sediment is caused by the longshore current, and the direction towards which the sediment moves is called downdrift. The volume of debris involved demonstrates the tremendous problems engineers face when building structures in coastal areas.

Figure 10.12 shows typical patterns of sediment accumulation near man-made structures. If a harbor is constructed where a prominent longshore current occurs (Fig. 10.12A), a baymouth bar will begin to form at the updrift side of the harbor entrance. Thus, harbor dredging must be performed to keep the harbor open. If a breakwater is built (Fig. 10.12B) sediment will accumulate updrift. This sediment may have a beneficial effect of adding a beach; however, because the sediment is no longer passed downdrift, erosion of the coast can be expected there. If a detached breakwater is built (Fig. 10.12C), a beach will grow in the protected low-energy environment behind the breakwater. The detached breakwater at Santa Monica, CA, built in 1934, exemplifies the changes brought about because the breakwater disrupts the waves normally causing littoral drift (Johnson, 1957). At Santa Monica, a large detached breakwater was built of scrap automobile bodies.

Thus, the problem of disposal of these car bodies was found, at the same time providing a beach. The car bodies also provide a fish habitat.

Not all engineering structures work as perfectly as planned, however, The following examples of engineering geology coastal problems illustrate the range and variety of problems.

On a stable shoreline, sediment movement in the littoral zone can be thought of as being steady state (or dynamic equilibrium), in that over a long time the sediment brought into an area equals the amount taken away. Man-made structures upset this equilibrium. According to Inman and Brush (1973), two major types of man-made disruptions are the following.

1. Dams. The reservoirs behind dams on inland rivers act as settling basins for sediment which would ordinarily nourish beaches on the coastline. Mathewson and Minter (1981) show, for example, that dam and reservoir development on the Brazos River, TX, traps about 76% of all sand produced within the basin. This loss of sand, coupled with the decreased competency and capacity of the river due to reduction of large discharges, has resulted in increased erosion along the Gulf Coast of Texas. In 1983 property losses totalled almost 2 billion dollars due to severe storms and resulting coastal erosion in southern California. Douglas Inman of Scripps Institute of Oceanography ascribed much of the damage to the loss of beaches due to lack of sand replenishment because of dams constructed on every major coastal river. "If you had enough sand out there, you would not have a battered coast," he said.
2. Shoreline structures such as harbor breakwaters, jetties, groins, and landfills disrupt longshore drift. Figure 10.13 shows successive stages of shoreline conditions after emplacement of a groin. Immediately after construction (Fig. 10.13A), sediment will accumulate updrift of the groin. This sediment does not reach the shoreline downdrift, so the downdrift beaches are effectively starved. Erosion is initiated downdrift (Fig. 10.13B and 10.13C), and continues until the capacity of the groin to capture longshore drift is reached.

One of the best-documented examples of erosion downdrift of a man-made structure is the Palm Beach inlet in Florida. The inlet was dredged, and large jetties were built in the period 1918–1925. This blocked the southward littoral drift almost completely, and heavy erosion commenced on the south side of the inlet. Through a great many years, attempts were made to meet the erosion problem by construction of small groins along the downdrift reach, but, because there was no source of sediment coming into that area, the groins did not work. Artificial nourishment with sand from the bay was begun, and in 1958 a sand bypassing plant was built on the north side of the inlet for the purpose of pumping about 19,000 $m^3$/yr across the inlet. Bruun and Lackey (1962, p. 84) state: "It would probably have been better in the overall picture if groins had never been built. It is a well-known fact that a group of groins function as (sediment traps) and for this reason will always have adverse effects on erosion on the downdrift side. It may nevertheless not be fully recognized that groins will usually cause considerably more erosion than accretion!"

Rosenbaum (1976) reports that the length of coastline erosion downdrift of groins in Lake Michigan, near Milwaukee, WI, is about 3–5 times the distance by

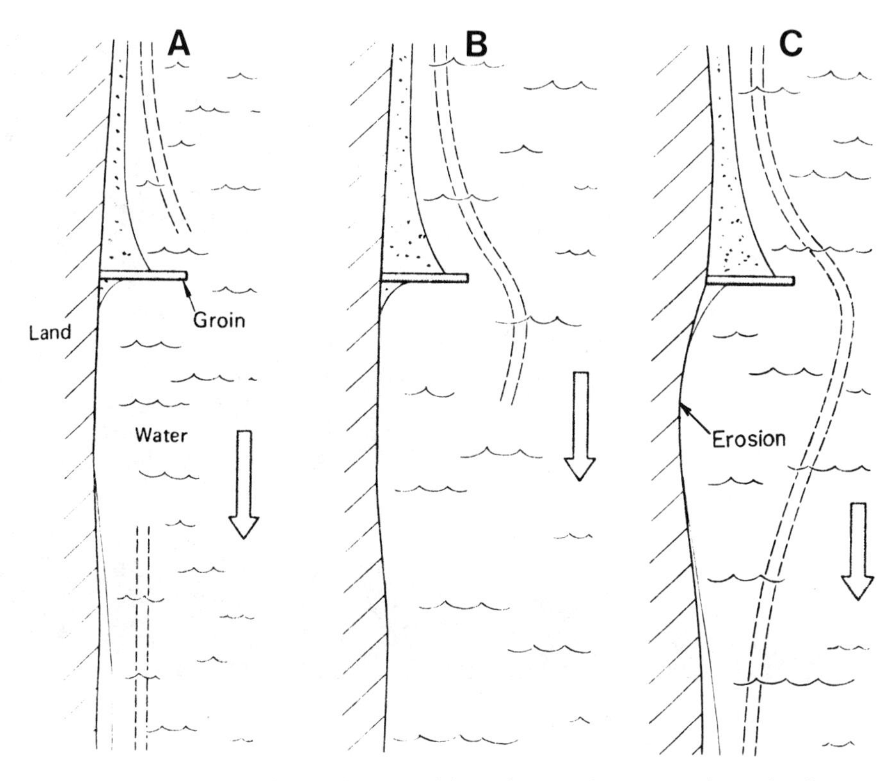


Figure 10.13 Stages of shoreline condition after emplacement of a groin (from Rosenbaum, 1976). Although not shown here, the land area downdrift of the groin may recede during the time of **A** and **B**. A limited region downdrift of the groin remains deprived of sediment nourishment, and may continue to erode indefinitely at time **C**.

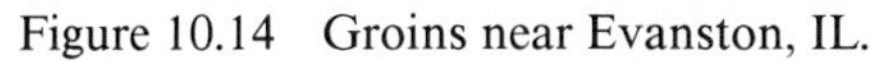
Figure 10.14 Groins near Evanston, IL.

Figure 10.15 Steel revetments downdrift from Figure 10.14, near Evanston, IL. Note eroded cliff down drift of revetment.

which the groin protrudes from the shore into the water. The groins along Lake Michigan (Fig. 10.14) are typically unsightly concrete blocks which collect sand on which accumulate thousands of dead alewives, a small fish inadvertantly introduced to the Great Lakes following the opening of the St. Lawrence canal. Severe erosion downdrift of the groins prompts construction of additional structures. Figure 10.15 shows steel revetments along Lake Michigan; due to severe erosion, there is no beach at all. Expensive houses are located above the sea cliff and are in jeopardy. Thus, man, in his ignorance, has transformed the beautiful Lake Michigan shoreline into an ugly one. The ugliness is compounded by lawsuits brought by parties suffering erosion (Lillevang, 1965).

The following account by Rosenbaum (1976) illustrates the complexity of the coastal erosion south of Milwaukee (see Fig. 10.16):

> The history of some major shoreline projects in Milwaukee, Wisconsin, typifies the way in which shoreline structures propagated. The federal government constructed the main breakwater in stages from 1881 to 1929. Construction progressed from north to south. By 1916, down drift erosion had become severe enough to cause the city of Milwaukee to construct a rubble mound breakwater approximately 1000 feet offshore, and extending southward parallel to the shore. It terminated at a position opposite the small breakwater of a local utility. The city's rubble breakwater was built from 1916 to 1931. It was observed that the shoreline opposite the end of the structure experienced severe erosion, and that "as the breakwater was extended from year to year, the point of greatest erosion on the shore kept pace."
>
> The utility's landfill and small breakwater, built in 1920, probably compounded erosion to the south until 1930–1931, when its effects would have been largely masked by the overlapping city breakwater.

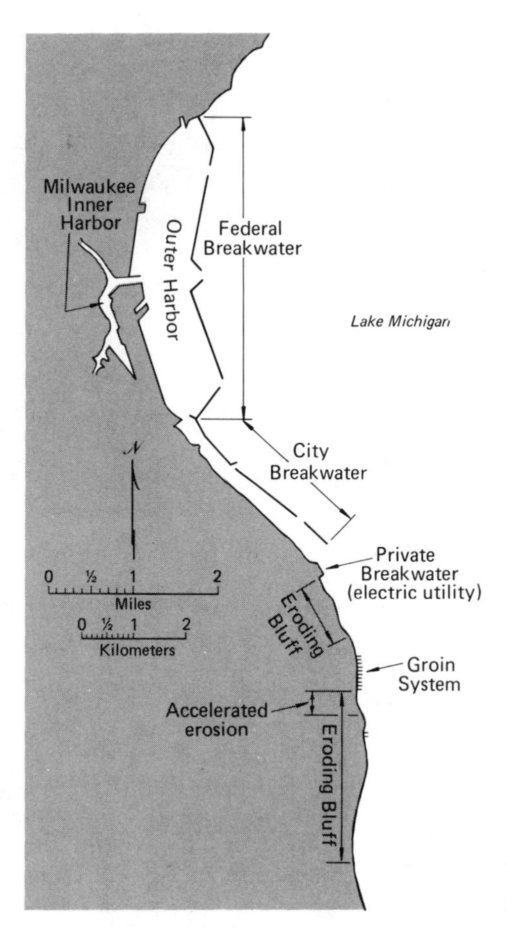


Figure 10.16 Lake Michigan shoreline structures at Milwaukee, WS (from Rosenbaum, 1976). The longshore drift is from north to south. The Federal breakwater was built in stages from 1881 to 1929. The city breakwater was constructed in stages from 1916 to 1930. The utility company's breakwater was built in 1920 and the groins in 1933–34.

> In 1933–1934 a system of eleven permeable groins was placed to protect the downdrift third of a 1½ mile section of eroding parkland that extended south of the termination of the new rubble breakwater. In turn, the groin system is currently responsible for accelerated erosion for at least ¼ mile to the south. To protect this and other eroding areas a citizen task force on lakefront planning has recently proposed construction of a series of offshore islands, each one being several miles long, up to a mile wide, and 3000–5000 feet offshore. If ever built, these islands, which function similar to a detached breakwater, would undoubtedly cause extremely severe erosion for many miles downdrift.

At Santa Barbara, CA, a large breakwater was constructed in 1930. The Miramar Beach Hotel sued the city of Santa Barbara because of the erosion of Miramar Beach 5 km downdrift (Tank, 1983b). Until a dredging program was initiated to move sediment past the harbor, beaches were affected for up to 38 km downdrift.

A large breakwater modified the wave-refraction pattern in Half-Moon Bay, California, focusing energy on the cliffs to the south. Mathieson and Lajoie (1983) report that rate of cliff retreat increased from about 0.05–0.3 m/yr between 1861 and 1960 to 1.0–2.0 m/yr between 1960 and 1980. According to this study, "the erodibility of these cliffs, which consists of Pleistocene and Holocene alluvial, dune, and beach deposits, was not considered during the breakwater planning

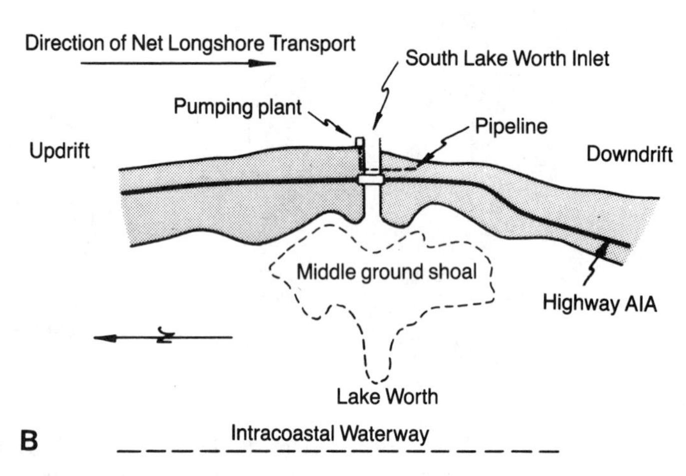


Figure 10.17 Fixed bypassing plant, Florida (from Rosenbaum, 1976). **A**. Pumping plant near outer end of North Jetty. **B**. Sketch map.

process." The severe storm during the winter of 1982–1983 added measurably to the problem.

Groins built on Westhampton Beach, NY, were constructed against the advice of engineers and geologists. The groins caused immediate severe erosion to the west. A $40 million federal beach-replenishment project was proposed to save threatened private homes (Pilkney, 1981).

It is ironic that engineering structures are actually responsible for accelerated erosion, because they are built to protect or to widen a beach. Studies by Johnson (1957), Hartley (1964), Bakker (1968), Dean and Jones (1974), and Rosenbaum (1976) show field, mathematical, and laboratory model confirmation of the problems of starved beaches.

Partial remedy to the problem of downdrift erosion can be made by the following:

1. Fixed hydraulic pumping stations: At South Lake Worth Inlet, Florida (Fig. 10.17), a fixed bypassing station was built on the updrift jetty. The process

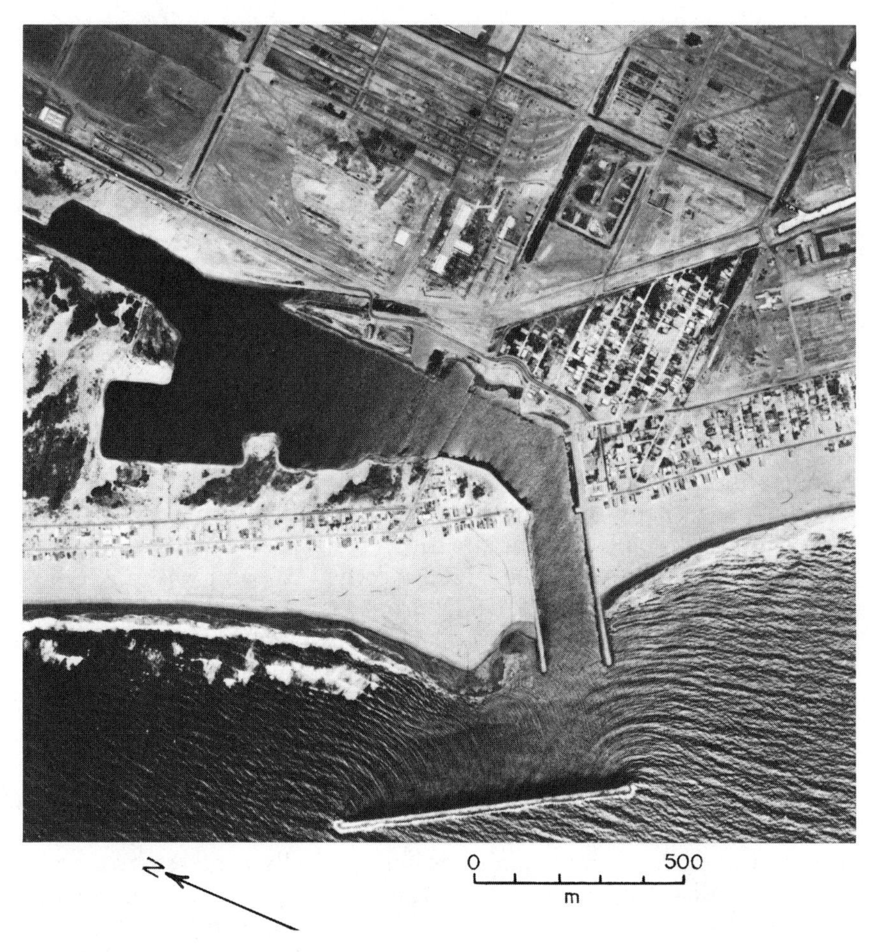


Figure 10.18 Sand bypassing at Channel Islands Harbor, Port Hueneme, CA. Note wave refraction around breakwater.

consists of hydraulically pumping sediment across the shoreline structure, rather than allowing the structure to collect sediment updrift or deflect the sediment away from the shoreline into deep water.

2. Detached breakwater: At Channel Islands Harbor, 1.6 km northwest of Port Hueneme, CA (Fig. 10.18), a detached breakwater was completed updrift of an older jetty system in 1962. The breakwater traps sediment which would otherwise be deflected offshore into a submarine canyon. The breakwater also provides shelter to the harbor from waves and reduces sedimentation in the harbor. Sheltered by the breakwater, a hydraulic dredge on a barge periodically pumps sediment downdrift of the Port Heuneme jetties. The floating dredge has moved about 10 million $m^3$ of sand between 1976 and 1977 (Mathewson, 1981).
3. Remove the existing groins: This has never been attempted, but this action would help to return the coastline to its original condition.
4. Zoning a natural shoreline to prohobit construction of updrift groins.

In summary, man has been slow to learn of tremendous destructive forces of coastal erosion. In ignorance of consequences, well-intentioned governmental agencies frequently recoomend engineering structures as a solution to erosion. These may work as expected; natural coastal erosion may be locally stopped, and wide beaches develop updrift of the structures. However, in most cases, rapid erosion will begin at a new location downdrift. If subsequent downdrift erosion were included in the initial benefit/cost analysis of these structures, it is doubtful they would be so attractive to public officials responsible for shoreline management.

## 10.4 Tsunami

### *10.4.1 Origin*

Tsunami is the Japanese word for large wave. Tsunamis are also called seismic sea waves (Wiegel, 1969). Commonly they are misnamed tidal waves; in fact, they have nothing to do with tides.

Tsunamis usually originate directly or indirectly from an earthquake (Fig. 10.19), although some tsunamis are caused by giant rockfalls or subterranean volcanic explosions. Probably the largest recorded tsunami was caused by the 1958 earthquake in Alaska, when a rockfall from a coastal mountain fell into a bay and displaced enough water to send a local wave about 500 m up onto an adjacent mountain (Miller, 1960). In Fig. 10.19A, a vertical displacement along a submarine fault causes a displacement of the ocean. Generally, dip-slip motion is required; strike-slip movement, such as occurred on the 1906 San Andreas fault, produced no tsunami. The waves radiate outward from the earthquake epicenter. In order to produce a tsunami, the earthquake epicenter must be shallow (less than 40-km depth) and of magnitude greater than 6. In Figure 10.19B, a submarine landslide occurs; these landslides are typically triggered by a nearby earthquake. The displacement of the landslide causes a tsunami to radiate outward from the landslide area. Failure of deltaic sediments such as at Valdez, AL, were responsible for the tsunami devastation of that city during the 1964 earthquake.

Fitz Roy, the captain of the Beagle, the famous ship which carried Charles Darwin to South America in 1835, describes a tsunami following an earthquake along the coast of Chile (Darwin, 1962).

> About half an hour after the shock . . . the sea having retired so much, that all the vessels at anchor, even those which had been lying in seven fathom water, were aground, and every rock and shoal in the bay was visible—an enormous wave was seen forcing its way through the western passage which separates Quiriquina Island from the mainland. This terrific swell passed rapidly along the western side of the Bay of Concepcion, sweeping the steep shores of every thing movable within thirty feet (vertically) from high water-mark. It broke over, dashed along, and whirled about the ships as if they had been light boats; overflowed the greater part of the town, and then rushed back with such a torrent that every moveable which the earthquake had not buried under heaps of ruins was carried out to sea.

A tsunami may be caused by a subterranean volcanic explosion, such as the August 27, 1883 explosion of Krakatoa. Krakatoa, a small active volcanic island, completely blew away in a gigantic volcanic explosion. The explosion was so loud the Vietnam Navy, hundreds of kilometers away, went out into the Pacific looking for mysterious cannons! A 42-m-high tsunami struck the coasts of Java and

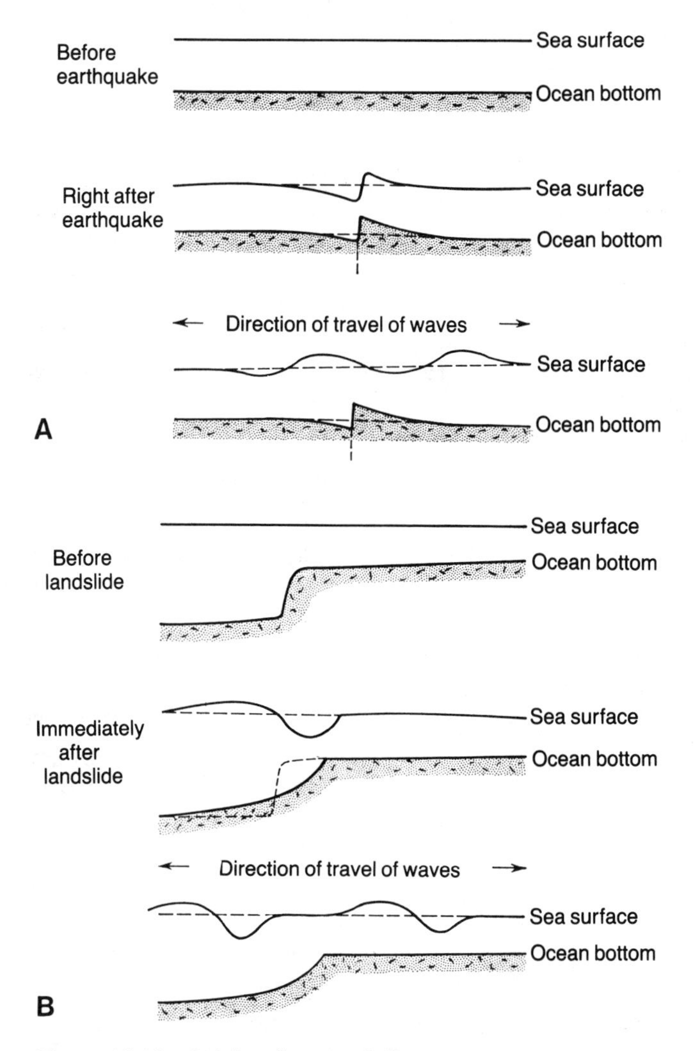


Figure 10.19 Origin of tsunami (from Howell, 1959). Vertical scale greatly exaggerated. **A**. Submarine fault displacement. **B**. Submarine landslide.

Sumatra. The land was low lying, and the wave funnelled through a straight at Merek, Java. Over 295 villages were destroyed and 36,000 people were killed (Furneaux, 1963).

Table 10.1 is a list of the world's largest tsunamis. The orogenic belt encircling the Pacific Ocean is active tectonically, and hence most of the world's tsunamis occur in the Pacific Ocean.

Some earthquakes cause a seiche, an oscillation of water in a lake or nearly closed body of water. As the earth moves slightly with the passage of earthquake waves, the oscillation may coincide with the natural resonance of a body of water. (The fundamental period of oscillation of Lake Geneva is 72 min. San Francisco Bay has a period of 39 min. The natural resonance of Loch Lomond is 10 min.) Some seiches are also caused by unusual tides, atmospheric pressure changes, or wind.

Table 10.1 Great Tsunamis of the World

| Date | Source region | Visual run-up[a] height (m) | Report from | Comments |
|---|---|---|---|---|
| 1500 B.C. | Santorin eruption | | Crete | Devastation of Mediterranean coast |
| November 1, 1755 | Eastern Atlantic | 5–10 | Lisbon, Portugal | 25,000–30,000 deaths |
| December 21, 1812 | Santa Barbara Channel, California | Several meters | Santa Barbara, California | Early reports probably exaggerated |
| November 7, 1837 | Chile | 5 | Hilo, Hawaii | |
| May 17, 1841 | Kamchatka | <5 | Hilo, Hawaii | |
| April 2, 1868 | Hawaiian Islands | <3 | Hilo, Hawaii | |
| August 13, 1868 | Peru–Chile | >10 | Africa, Peru | Observed in New Zealand, damaging Hawaii |
| May 10, 1877 | Peru–Chile | 2–6 | Japan | Destructive to Iquique, Peru |
| August 27, 1883 | Krakatoa eruption | | Java | 36,000 drowned |
| June 15, 1896 | Honshu | 24 | Sanriku, Japan | 27,122 people drowned 35-m high wave |
| February 3, 1923 | Kamchatka | About 5 | Waiakea, Hawaii | |
| March 2, 1933 | Honshu | >20 | Sanriku, Japan | 3022 deaths from waves |
| April 1, 1946 | Aleutians | 10 | Wainaku, Hawaii | 173 deaths |
| November 4, 1952 | Kamchatka | <5 | Hilo, Hawaii | |
| March 9, 1975 | Aleutians | <5 | Hilo, Hawaii | Associated earthquake magnitude 8.3 |
| May 23, 1960 | Chile | >10 | Waiakea, Hawaii | 250 deaths |
| March 28, 1964 | Alaska | 6 | Crescent City, California | 122 deaths overall $104,000,000 damage from tsunami |
| August 17, 1976 | Mindanao | > 5 | Phillipines | 917 dead |

[a] Run-up is the maximum inundation above sea level.
Modified from Bolt et al., 1977; and Costa and Baker, 1981.

During the November 1, 1775, Lisbon earthquake, Raleigh earthquake waves caused seiches on N–S oriented lakes and canals in Holland and Scotland, 1600 km away. Ship's moorings were broken by seiches as far away as 3200 km in Scandinavia. The tsunami at Lisbon itself was variously reported at 5–15 m high, and, combined with the earthquake and fires, killed 60,000 people of the total 235,000 people living in Lisbon.

## *10.4.2 Hydrodynamics*

Once formed, tsunamis travel at high speeds over great distances. Wave celerity on the open sea is nearly the speed of sound, i.e., 700–800 km/hr. Figure 10.20 shows the times of arrival of the tsunami which originated in Alaska. The tsunami reached Hawaii, 3600 km away in a little over 4 hr.

Out at sea a tsunami is low and long. The maximum height is usually less than 1 m, its wavelength is of the order of 100 km, and its duration at any station is of the order of 15 min. Even though a 1-m elevation change in sea level is small, the energy carried by a tsunami is enormous. To lift hundreds of square kilometers of sea 1 m requires a lot of energy.

Where a tsunami is triggered by an earthquake, the tsunami conveys about 10–100% of the earthquake energy. At sea the slope of the wave is so small that the tsunami is usually undetected by ships. It is only when the wave approaches land and begins to slow down that the energy is transformed into a large wave. An example of this was afforded by the April 1, 1946, tsunami at Hilo, HI. A magnitude 7.2 earthquake near Unimak Island, AK, set in motion a tsunami great enough to wipe out not only Scotch Cap Lighthouse and five men inside—10 m above sea level—but also the radio antenna perched at 31 m. The wave then bounded across the Pacific. Passengers in ships standing off port watched Hilo destroyed by waves that had passed unnoticed under the ships. Three waves at 12-min intervals reached 18 m height. Damage exceeded $6 million, and 159 people were killed.

From Eq. (10.3), wave velocity is controlled by the water depth. In deep oceans, such as the mid-Pacific, tsunamis are extremely long (~240 km) and have periods ≅ 1000 s, and in open water may only be 1 m high. The corresponding celerity exceeds 700 km/hr. In shallow water (e.g., 100 m deep), the velocity of wave will be reduced to:

$$C = (gD)^{0.5} = [(9.8 \text{ m/s}^2)(100 \text{ m})]^{0.5} = 33 \text{ m/s} = 120 \text{ km/hr}.$$

From conservation of momentum, it can be seen that in slowing from 700 km/hr to 120 km/hr, a tsunami must gain considerable mass. The wave gets higher. Decreasing water depth and the undertow have a braking effect at the bottom of the wave column, but the top continues to push forward, bunching higher and higher until it topples with tremendous force on the shore. Hydrodynamic equations are not available to predict the exact height of a large wave, but many tsunamis rise to heights of 20 m, and the highest known tsunami, in 1737 in Kanchatko, Siberia, reached 70 m high (Eiby, 1980). The 1964 Alaskan earthquake caused a submarine landslide-induced tsunami which sheared trees off 30 m above sea level at Valdez. The same tsunami reached Hawaii (3540 km away) in 4.5 hr and was 16 m high. It reached Chile (12,870 km away) in 18 hr and was 1.5 m high and reached Antarctica in a little over 22 hr.

The height of a tsunami also depends on local coastal topography. A tsunami may be funnelled into a bay, reflecting off shorelines, or turn into a bore (a moving hydraulic jump), capable of causing great damage as at Hilo, HI, in 1957.

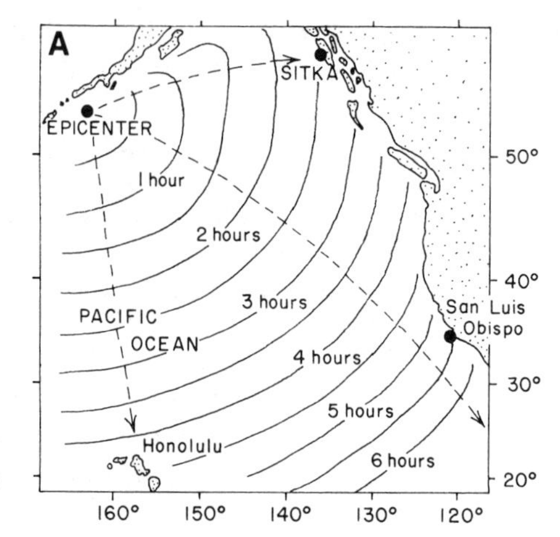


Figure 10.20 Record of the April 1, 1946, tsunami (after Shelton, 1966). **A**. Map of the North Pacific Ocean showing arrival times. **B**. Mareograms (water-level records) at four locations.

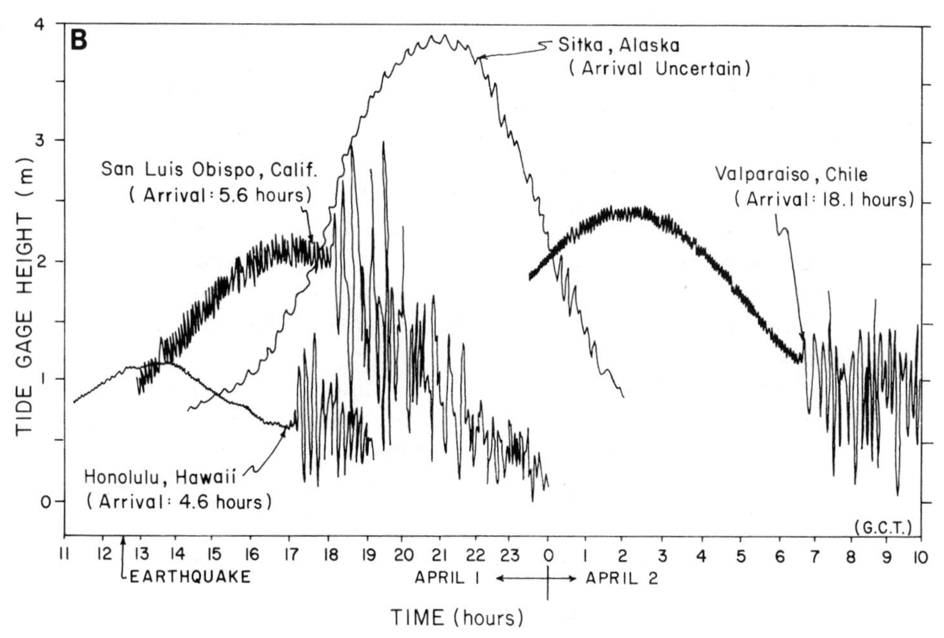


Reflected tsunamis can be deceptively dangerous. Eighteen hours after the 1946 tsunami hit Hawaii, waves hit again. Reflection of the tsunami off a submarine cliff near Japan coincided at Hawaii with arrival of reflected waves off another sea escarpment. The two sets of waves arrived in phase at Hawaii and caused a tsunami equivalent to, if not higher than, the parent tsunami (Bolt et al., 1977).

In addition to reflected tsunamis, it is possible for more than one parent tsunami to be initiated from a single earthquake because 1) earthquakes often occur as a series of fault displacements or 2) one tsunami may originate from a fault displacement and another from a nearby subterranean landslide. In 1964 the coincidental arrival of sets of waves *in phase* caused widespread destruction at Crescent City, CA, while cities closer to the tsunami origin suffered little damage (compare Fig. 10.20 mareogram at Sitka, AK, to San Luis Obispo Bay, CA).

The fact that tsunamis travel across the ocean at relatively uniform velocities is fortuitous in that it allows for the worldwide prediction of the time of the parent tsunami. An elaborate tsunami warning system is in operation in the Pacific. The Seismic Sea Wave Warning System operates by the cooperation of Japan, Russia, and the United States. The United States has a tsunami center at Honolulu, and 18 stations in the Pacific. The stations are alerted whenever an earthquake is detected. Water levels detect deviation from normal tides, and are used to give warning throughout Pacific communities.

One problem with this warning system is that people tend to be curious. For example, when it was announced in 1964 that a tsunami would hit San Francisco, thousands of people went to the beaches to watch! In Crescent City, 2-m-high waves occurred first, and people went down to the docks to inspect the damage. Then a 4-m-high wave came in and killed 10 people.

Due to a fairly uniform period with which tsunamis occur, common sense can be used to avoid disaster. On November 18, 1929, a postmaster in Newfoundland is credited with alerting people to danger when the sea withdrew, leaving the harbor exposed. Before the tsunami struck he had time to run the length of the village's single street, calling to people to flee to the hills, which they did, so that they were saved (Howell, 1959).

When the sea suddenly retreats, a prudent person should leave the beach. Many people on the Hawaiian Islands rushed out to see the bare reefs and collect fish and shells before a 1923 tsunami struck. Most of the 150 people killed were out on the beach.

Large tsunamis have contributed to the destruction of at least two civilizations. A combination of volcanic explosions, earthquakes, and tsunamis destroyed the Minoan civilization on Crete in 1400 B.C. Something similar caused the destruction of a civilization on Java in 1004 A.D.

In the future it is hoped that technological improvements in tsunami warning systems, cooperation among nations, and proper land-use planning in low-lying coastal areas may reduce tsunami hazards.

Planning for coastal hazards, as floods and flood-plain management, always contain the argument for construction of hard engineering structures to protect buildings already built in hazardous locations. It is hoped that in the future more recognition will be given to coastal hazards before construction of homes occurs. In Hawaii, for example, it was shown that the construction of ugly breakwaters of questionable durability in the northeast side of Hawaii would be better spent in properly planning for the location and design of houses to be safe from tsunamis (Adams, 1973).

## 10.5 Hurricanes and Land Use on Barrier Islands

A barrier island is a thin beach offshore from the mainland, typical of the shoreline of emergence (Fig. 10.1). In the United States 7 million km$^2$ of barrier islands stretch intermittently from Texas to New York City. Most of these are in Florida and Texas. Barrier islands are a very fragile environment, slender accumulations of sand built during the past few thousand years. The land is seldom more than 1 km wide and a few meters above sea level.

Barrier islands are subject to coastal flooding and wave erosion, which are most severe during storms associated with large waves and high tides. High spring tides

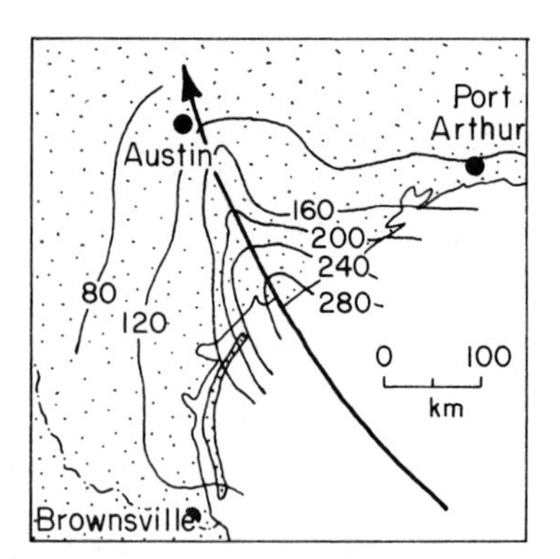


Figure 10.21 Contour map of wind velocities (km/hr) associated with the path of Hurricane Carla, September 10 to 12, 1961, along the Texas coast (after Mathewson, 1977).

are caused by the alignment of the earth, moon and sun (a position called syzygy). This happens twice each month. Because the lunar orbit is elliptical in shape, the moon attains its closest monthly approach to earth once each revolution. This is called the perigee. Ordinarily the passage of the moon through perigee does not coincide with syzygy. However, if it does, very high tides can be produced, called perigean spring tides. These tides can cause extensive flooding, such as the March 6, 1972, inundation of the entire mid-Atlantic coastline from the Carolinas to Cape Cod, resulting in the loss of 40 lives and over $500 million in property damage.

Since 1900, Atlantic and Gulf coast barrier islands have been crossed by more than 100 hurricanes. Three costly hurricanes were Frederick, which in September 1979 caused an estimated $700 million in damages along the Gulf coast near Mobile, AL; Agnes, which caused $2 billion in damages in 1972; and Camille, which destroyed $1.4 billion worth of property in 1969 (Dolan, et al., 1980a). Hurricanes are now rated on a scale of 1–5, based on a formula using wind speed and potential damage. Hurricane Camille in 1969 was the last category 5 storm to hit the United States; it claimed 255 lives when it assaulted the Mississippi coast.

Hurricanes generate tremendous winds and high tides. These storms occur in late summer and early fall on the eastern and Gulf coasts. Hurricane David, which struck the east coast in 1979, had winds of over 240 km/hr, the energy equivalent of several hundred 20-megaton hydrogen bombs every day. Figure 10.21 shows the wind velocities associated with Hurricane Carla which struck Texas on September 10, 1961.

Hurricane frequency can be predicted on a statistical basis. From Fig. 10.22, the probability of a large hurricane ($>$33 m/s winds) striking the Miami, FL, area, for example, is 16% in any year, and 7% in any year for a great hurricane (winds $>$56 m/s).

Continual erosion and deposition are part of the natural process in barrier islands formation (Machemehl and Forman, 1980). For instance, the Cape Hatteras, NC, Lighthouse, once over 1 km from the sea, now stands only 20 m from the shoreline. The lighthouse on Fire Island, near New York City, once on the island's western tip, is now 8 km from the island's western tip because of the migration of sand from one end of the island to another.

Sea level has been gradually rising in Holocene time (Fairbridge, 1960). The eustatic (worldwide) sea level change is associated with the melting glaciers. The

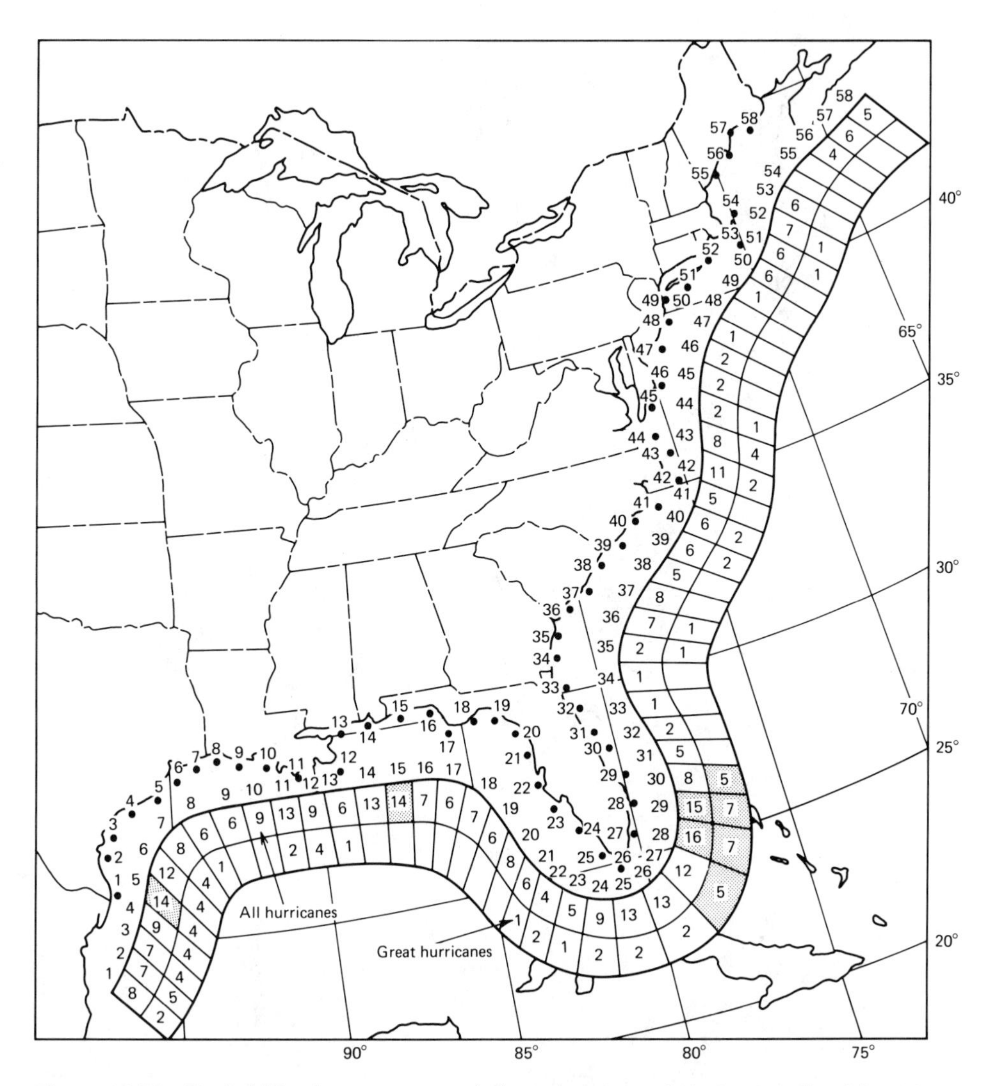


Figure 10.22 Probability (as a percentage) that a hurricane (winds exceeding 33 m/s) or great hurricane (winds exceeding 56 m/s) will occur in any one year in an 80 km segment of the U.S. Atlantic and Gulf coastlines (from Simpson and Laurence, 1971).

sea level was about 120 m lower 12,000 yr ago, at the end of the Wisconsin glaciation, and the shorelines of the Atlantic and Gulf coasts were 60–150 km seaward of their present positions (Dolan et al., 1980a). The eustatic sea level rise has been about 30 cm during the last 100 years. The highest sea level during the Sangomon interglacial was about 30 m above present sea level (Flint, 1957). (This corresponds to the "Surry" strandline 50 to 100 km inland along the south Atlantic states.) If all the world's ice were to melt, sea level would rise about 54 m. Because of eustatic sea level rise, beaches on U.S. barrier islands are gradually eroding, and the landward sides (salt marsh areas) are gradually migrating towards the mainland. In recent decades this movement has been landward, at a rate of about 1.4 m/yr for the Atlantic coast (Doland et al., 1980a). The addition

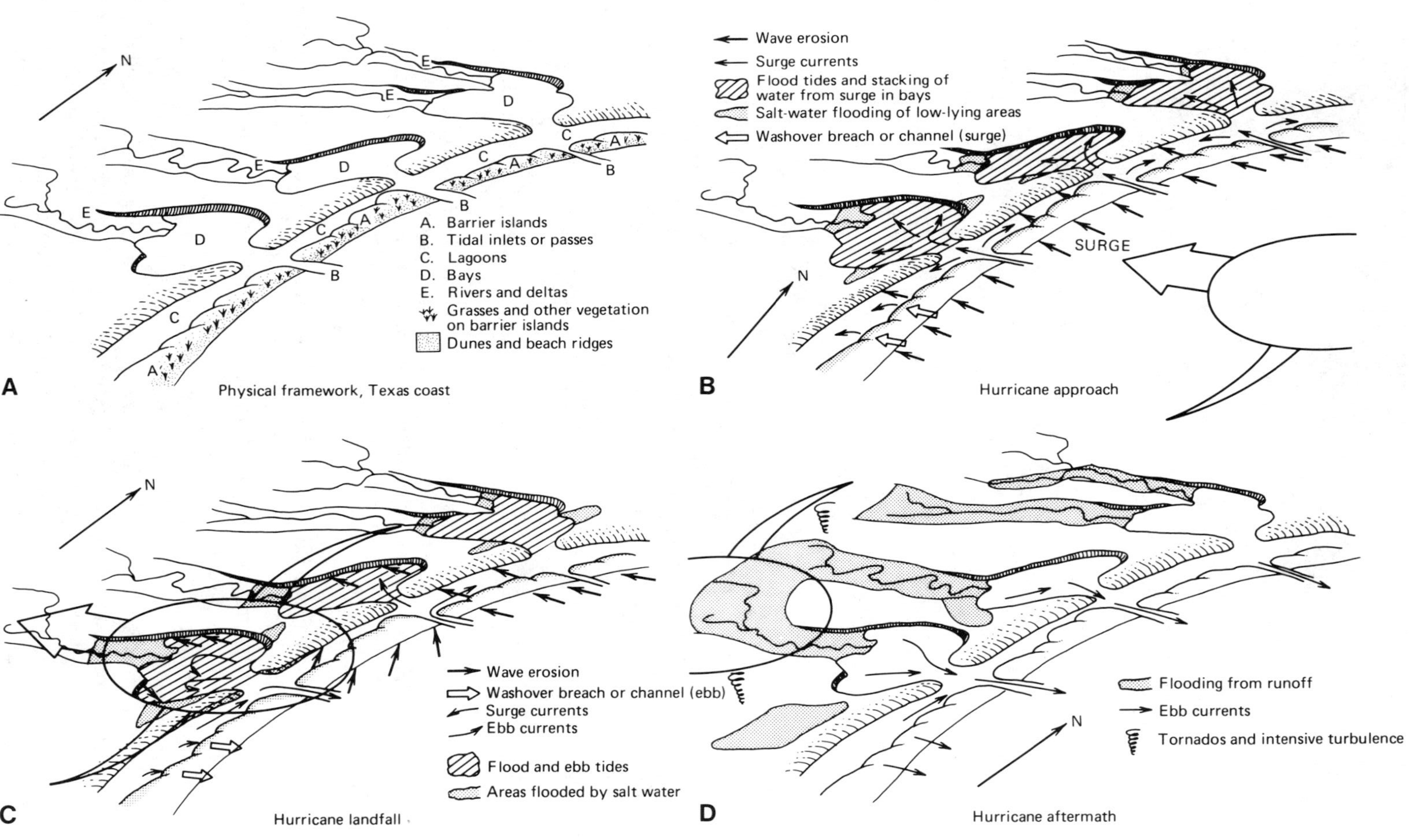


Figure 10.23 A highly generalized model of hurricane effects on the Texas coast (from Texas Bureau of Economic Geology).

of large volumes of carbon dioxide, smoke, and aerosols to the atmosphere could hasten worldwide warming (Sagan et al., 1980) which could cause dramatic melting of glaciers.

The physical and ecological environment of a barrier island is diagrammatically illustrated in Fig. 10.23. In a natural state, high waves occasionally erode the low dunes and carry completely across the island into the marsh. This overwash is an important part of the ecology because it replenishes sediments and creates new land on the lagoonal side of the barrier island.

Man's interference with barrier island processes has generally met with negative results. On Cape Hatteras, for instance, dunes were artificially stabilized between 1936 and 1940. Now the fencing and dune vegetation stabilization has caused problems in that the beach widths have narrowed from about 150 m to 75 m. The artificial stability has changed a dynamically stable ecosystem, capable of withstanding continual stress, to one that is destroyed when a major storm breaks the man-made dune. In recent years the barrier dune systems on Cape Hatteras have deteriorated, and large expenditures of resources are necessary to save highways and structures. The Ash Wednesday storm of 1962, for instance, opened a new inlet and damaged roads, costing about $2 million to replace. To put sandbags in this part of the entire Cape Hatteras dune system would cost many millions of dollars (Dolan et al., 1973). At the Cape Hatteras lighthouse, groins have been installed to prevent erosion, but with limited success. In 1980, Orrin Pilkney, geologist from Duke University said:

> There is no way to save the lighthouse except to move it. You could build the biggest revetment in the world around the lighthouse and come the next storm, it would probably disappear. The island will migrate right out from underneath it.

Man's interference with barrier islands was also documented by Mathewson (1977), who showed the effects of 1961 Hurricane Carla on Padre Island, TX. Because continuous artificially reinforced dune ridges were built with no gaps for wave overtopping, extensive damage was caused. The ineffectiveness of revetments to protect poorly located houses on the gulf coast of Florida is documented by Warnke (1973).

The Galveston, TX, "storm surge" of September 8, 1900, resulted in 6000 dead, the greatest loss of life of a natural disaster in the United States. Wind-driven waves, aided by a seiche, smashed 3000 homes. In the following years, a 6-m concrete sea wall was constructed. In August 1983, Galveston and Houston were struck by Hurricane Alicia, which left 18 dead and caused about $2 billion in damage. Following the 1983 disaster, Neil Frank, director of the National Hurricane Center, said that some condominiums have been built on the sea side of the wall and another one is being built there: "This has to be considered a very questionable development." But the construction is typical of what is occurring in many areas of the Gulf and Atlantic coasts. The developments are insured by federal flood insurance, so that inland taxpayers end up bailing out those whose property is damaged by coastal storms, because insurance payments do not cover claims paid to injured parties.

Storm surges (Jelenianski, 1978) are not unique to the United States. In 1970 a storm surge associated with a cyclone killed 100,000 people in the Bay of Bengal. In 1974, a hurricane ripped into Darwin, Australia, accompanied by winds up to 265 km/hr.

Despite the well established, long-term trend of landward migration of barrier islands and the effects of periodic storms, coastal-zone planning and development have been based largely on the concept that beaches and barrier islands are stable or that they can be engineered to remain stable. Urbanization of barrier islands is increasing at alarming rates. Man has attempted to draw a line and tried to prevent the sea from passing, instead of a strategy of submission and rebuilding. The results have been negative. Development of the American coastline has led to an endless program to protect investments. Property owners, who have built in a dynamic and ephemeral environment, demand governmental stabilization structures. The amount of urban and built-up land on barrier islands in the United States increased from about 36,000 ha in 1951 to 93,000 ha in 1981 (Lins, 1981). The federal government actually subsidizes barrier island development by providing low-cost flood insurance, disaster relief, and bridge and highway development to island communities. An example of the folly is Folly Island, SC. The island was opened in 1918, and has become a vacation spot for residents of Charleston, SC. Business is booming. In 1981 a developer obtained a permit for a 540-unit condominium and convention center only 30 m distant from the normal high tide. Residents of Folly Island (which has been hit by several storms this century including Hurricane David in 1979) are seeking a new million-dollar beach to replace their badly eroded one. Where there was formerly a wide beach, the pilings of some homes now are being splashed by high tide. The town wants the federal government to pay the entire bill for the new beach. The cost would be $1 million about every 5 yr.

During a northeastern storm in March, 1962, damage to property along the mid-Atlantic coast amounted to more than $500 million, and 32 lives were lost. This devastation was soon forgotten, however, and rapid development along the coast has continued since then. For example, Ocean City, MD, on a barrier island called Fenwick Island, had 150 businesses and 50 homes damaged or destroyed, yet rebuilding commenced almost immediately after the storm (Dolan et al., 1980b). The barrier island is now about completely urbanized. Meanwhile, the shore is rapidly dwindling as the sea slowly encroaches on high apartment houses lining the coast. In 1980, to help reduce coastal erosion at Ocean City, the U.S. Army Corps of Engineers developed a plan for the transportation of sand to Ocean City from an offshore shoal. The plan calls for an annual beach replenishment program of over $10^4$ $m^3$ of sand to replace anticipated seasonal deficiencies.

Concerning the present federal policies, in 1981 Laurence Rockerfeller stated

> It's folly to continue the current policies. I suppose you can say we had advance notice way back in the Sermon on the Mount. Jesus said, and there it is in the Bible, not to build on the sand. And here we are, not only building on the sand, but asking the general taxpayer to subsidize it.

The construction of dams on the Rio Grande River is one of the contributing causes of shoreline erosion of Padre Island, TX. Condominium development is now threatened in areas where, according to the Texas Bureau of Economic Geology, the land has disappeared at an average rate of 3 m/yr over the past century. In 1978 state officials restricted developers to building in back of the line where natural vegetation meets the beach.

Every hurricane costs the United States about $500 million in damages. Evac-

Figure 10.24 Barrier island at Gulf Shores, AL, following Hurricane Frederick on September 12, 1979 (U.S. Geological Survey photo). Note the foundations of homes destroyed by high winds and tidal surge on the ocean side of the island (background).

uation efforts are not very effective. Development in the Florida Keys is especially vulnerable. It would take 2 days to evacuate the Keys. The National Hurricane Center estimates that at least 6000 deaths would occur if a hurricane should ever strike there. In 1983, Neil Frank, director of the U.S. National Hurricane Center, warned that barrier islands were ill-prepared for evacuation. In New Jersey, he said " . . . there are not even comprehensive evacuation studies . . . On the New Jersey coast, it's just a string of barrier islands with hundreds of thousands of people living on them."

The Galveston-Houston area is very vulnerable (see Section 9.2.2). Land subsidence due to ground water withdrawals adds to the hurricane hazard, as well as allowing for salt water intrusion into streams (Jorgensen, 1980). According to a 1981 warning by Neil Frank of the National Hurricane Center, Galveston is growing in ways that ignore the hurricane risk. Present buildings extend beyond the protection of the city's man-made sea wall, and land subsidence on the Texas shore has made evacuation more difficult because many roads lie so low they would be covered with water from an ocean surge or hurricane.

In New Orleans, close to half a million people live on a flood plain that could be inundated up to 5 m deep during a hurricane; yet there is no comprehensive evacuation plan.

Evacuation of the 1 million people in the Tampa Bay area is one of the most difficult in the United States. Because of the high concentration of people living on the barrier island it would take between 14 and 17 hr to evacuate, whereas reliable hurricane forecasts can be made only 12 hr or less in advance. People who live on barrier islands generally are retired or older people. In Alabama, Dauphin Island's bridge to the mainland was destroyed by Hurricane Frederick in 1979 (Fig. 10.24).

According to a 1979 statement by William Wilcox of the U.S. Office of Disaster Response and Recovery:

It is part of the American philosophy that building and expansion is always good;. . . . It sort of goes back to the pioneer spirit that we are in a contest with nature.

Mathewson (1982) offers some sobering thoughts about man's pioneer spirit in hazardous areas:

The "Urban Pioneer" now faces an increasing number of losses that are related to natural processes; homes in flood plains are flooded, homes on hillsides are damaged in landslides, coastal homes erode away, stuctures in tectonically active regions are devastated during earthquakes, and so on. As the frequency and cost of these "Acts and God" increase, Man continues to apply large sums of money in disaster relief and engineering projects to provide protection against future disasters. Within the past 20 years, Man has begun to question whether or not this continual drain on the economy is justified . . . Land-use controls, based on geology as well as social factors, can be applied only when the public accepts the concept that "God is innocent" and that floods, landslides, earthquakes, and erosion are natural processes . . . Once a city government institutes policies that control development in geologically hazardous areas, then it is possible to develop performance standards that allow both the land developer and the public to attain their respective goals. A different conflict exists when Man has already urbanized geologically hazardous areas. Now the city must deal directly with the victims of the hazards, victims who have their life savings invested in their damaged property. How does a city government, responsible for the health and safety of its citizens, correct a problem? . . . The options are limited, expensive, and frequently difficult for the general population to accept . . . It is difficult to gain public support for a program that will retrofit the city to its geology, because many persons feel "they should not pay to protect fools who live in hazardous areas." Little do they realize that they are doing just that through increased maintenance costs incurred by the city, through higher insurance rates, and through disaster relief programs.

For better or worse, the U.S. government has initiated a flood insurance program (see Chapter 8). Although it could be argued that this program uses federal money in a way that actually encourages construction in hazardous areas, the program does have some merit in that in order for communities to qualify for federal flood insurance they are forced to exact codes that require making buildings more flood-proof. Currently, federal flood insurance standards require houses to be built above the highest water level expected from a once-in-a-century storm excluding wave height.

Care must be made to ensure proper design of coastal foundations (Morton, 1976). Current standards generally require only that pilings be securely anchored. Hurricane Frederick toppled 7-m pilings embedded 4 m in the ground. Mathewson (1981) reports that normal retaining walls behaved poorly (Fig. 10.25):

Hurricane Eloise hit the Florida Panhandle on September 23, 1975, and undermined and destroyed many "seawalls" that were intended to protect the high-value beachfront property. the loss of the "seawall" resulted in direct wave scour below foundations and corresponding failures. Why did these seawalls fail? They were not seawalls but retaining walls. Failure occurred because the structures were not designed for scour at their base or for wave-impact forces. Cross section A shows a typical coastal retaining wall that should be used along dredged channels and other environ-

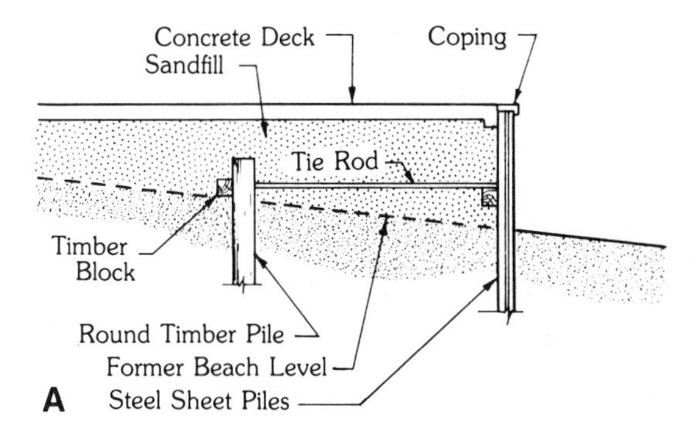


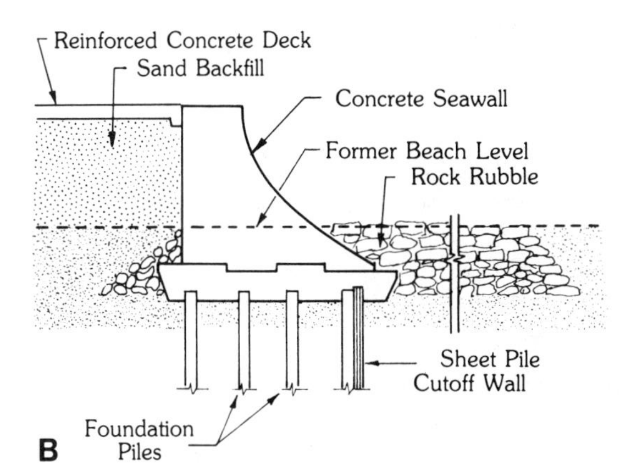


Figure 10.25 Retaining walls vs. sea walls (from Mathewson, 1981). **A**. Retaining wall. **B**. A sea wall is better able to withstand wave erosion.

ments of low wave energy but was used for beachfront protection. Notice that the only compressional member is the concrete deck. With flooding of the sand fill by storm surge and vibrationary loading of the wall face, the pore water pressure increases sharply with each wave impact. The combination of liquefaction of the confined saturated sand and horizontal forces from wave impact caused the only compressional member, the deck, to fail. Failure of the deck allowed excessive flexing of the face of the wall, which ultimately failed.

Compare the design of the retaining wall with a recurved seawall design (B) from the U.S. Army Corps of Engineers. The seawall is designed to redirect the impact of incoming waves and to absorb the wave forces. The rock rubble and sheet pile cutoff wall protect the foundation piles from scour. A seawall is designed to 1) absorb wave energy, and 2) resist foundation scour in order to protect the land behind it.

As discussed above, even the proper construction of sea walls may result in loss of beach, however. In Holocene time, there has been a general landward migration of barrier islands, so that sea walls will only lead to a narrower beach (Fig. 10.26). The Galveston, TX, seawall, America's mightiest, was built in response to the 1900 hurricane. As recently as 1965, a wide sand beach existed seaward from the wall. By 1984, the beach had essentially disappeared (Pilkney, 1981).

Discussing the effects of hurricanes on the Gulf Coast of Florida, Warnke (1973) states:

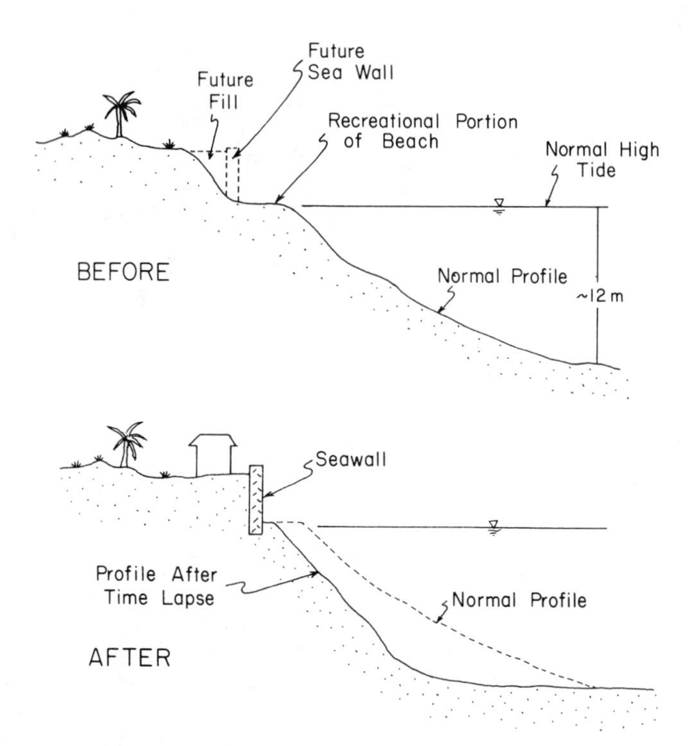


Figure 10.26 Two diagrammatic cross sections illustrating the effect of a sea wall on the profile of a beach (after Leveson, 1980).

> The public which is not familiar with these beaches is largely unaware of the dynamics involved. The main reason for this is the generally placid appearance of the sea (this is low-energy area) and the reluctance of developers to discuss the financial and proprietary consequences of beach erosion. . . . In any event, if the present beach-retrogression and population trends continue, severe material losses because of erosion of coastal property and the destruction of man-made structures by hurricane effects are unavoidable. Despite the usually placid and inviting appearance of the coast, hurricanes and hurricane surges are not "catastrophic" or unforeseeable events, but part of the physical processes which shaped and continue to shape the Gulf Coast.

Developments on barrier islands are disasters waiting to happen. Without federal subsidies to pave the way and absorb the risk of this development, the folly of private investment of money in these developments would become more apparent. The U.S. federal government often works in strange and contradictory ways. It pays subsidies to the tobacco industry and yet supports cancer research and discourages smoking. On barrier islands, the government helps preserve the environment through the U.S. National Park Service and yet encourages development by building roads and offering flood insurance. In 1981, Secretary of Interior James Watt said that "Taxpayers subsidize initial development, a hurricane sweeps the area, the taxpayers via government assistance encourage rebuilding, and the cycle begins again." As an example, in 1979 Hurricane Frederick swept across Dauphin Island in Alabama causing property losses totaling more than $2 billion. The federal government helped pay for the rebuilding, including 90% of the $34 million cost of replacing the bridge to the mainland. This alone amounted to $50,000 for every island resident.

In 1977, President Carter expressed his concern: "Coastal barrier islands are a fragile buffer between the wetlands and the sea. . . . Many of them are unstable and not suited for development, yet in the past the federal government has subsidized and insured new construction on them. Eventually, we can expect heavy economic losses from this short-sighted policy."

There are about 20 federal agencies which administer about 30 programs which affect barrier islands; half of these are pro-development, and one-fourth are protection-oriented. We spend millions to encourage development while we spend millions for protection. When nature proves us fools by washing away what we have built, we start all over again because we have federally insured what was bound to be destroyed.

## Problems

**1.** Consider a tsunami in the open sea which has a wave length of 50 km.
   **a.** What is its velocity?
   **b.** If this wave breaks in 10 m water, what is its velocity?

**2.** An earth dam, 100 m long, has an embankment slope of 1 vertical to 5 horizontal facing into the reservoir. The fetch is 10 km and the winds reach 80 km/hr. Assume that the reservoir level is constant and rip-rap is to be placed so that it extends from 1 m depth below minimum water level (see Fig. 10.8) to 1 m above maximum wave height. How many square meters of rip-rap are required to protect the dam from wave erosion? Explain why this is a minimum or maximum value.

**3.** On February 16, 1982, the world's largest offshore oil rig was destroyed by a North Atlantic storm. Eighty-four oil workers were drowned. Waves 15 m high were reported, as well as 73-km/hr winds which developed freezing spray.
   **a.** Use Eq. (10.6) to solve for the theoretical wind velocity required to produce 15-m waves.
   **b.** Use Eq. (10.7) to determine the theoretical fetch required to produce 15-m waves.
   **c.** Using a fetch determined from b above, and a wind speed of 73 km/hr blowing for a very long time, use Fig. 10.9 to determine the resulting wave height.

# 11

# Earthquakes

. . . and lo, there was a great earthquake . . . and every mountain and island were moved out of their place.

Revelation VI: 12,14

Oft the teeming earth
Is with a kind of colic pinch'd and vex'd
By the imprisoning of unruly wind
Within her womb; which for enlargement striving
Shakes the old bedlam earth and topples down
Steeples and moss-grown towers.

William Shakespeare

They walked toward Lisbon. . . . They scarcely set foot in the town when they felt the earth tremble under their feet; the sea rose in foaming masses in the port and smashed ships which rode at anchor. Whirlwinds of flame and ashes covered the streets and squares; the houses collapsed, the roofs were thrown upon the foundations, and the foundations scattered; thirty thousand inhabitants of every age and both sexes were crushed under the ruins. . . . Candide had been hurt by some falling stones; he lay in the street covered by debris. He said to Pangloss: "Alas, get me a little wine and oil; I am dying."

"This earthquake is not a new thing," replied Pangloss. "The town of Lima felt the same shocks in America last year; similar causes produce similar effects; there must certainly be a train of sulfur underground from Lima to Lisbon . . . all this is for the best."

Voltaire

Civilization exists by geological consent, subject to change without notice.

Will Durant

## 11.1 Earthquake and Seismic Waves

### *11.1.1 Epicenter and Focus*

An earthquake is the oscillatory movement of the earth's crust which follows a release of energy. This energy is usually generated by a sudden movement along a fault. When subject to tectonic stress, the rocks along a fault may first bend; then, when the stress exceeds the frictional resistance in the fault zone, the rocks suddenly move to a new position.

When an earthquake occurs, seismic waves travel in all directions from the focus, the region where the earthquake's energy originates. The foci of most earthquakes are concentrated in the earth's crust, which is about 50 km thick under continents. Earthquakes are classified according to focal depths a) 0–60 km are

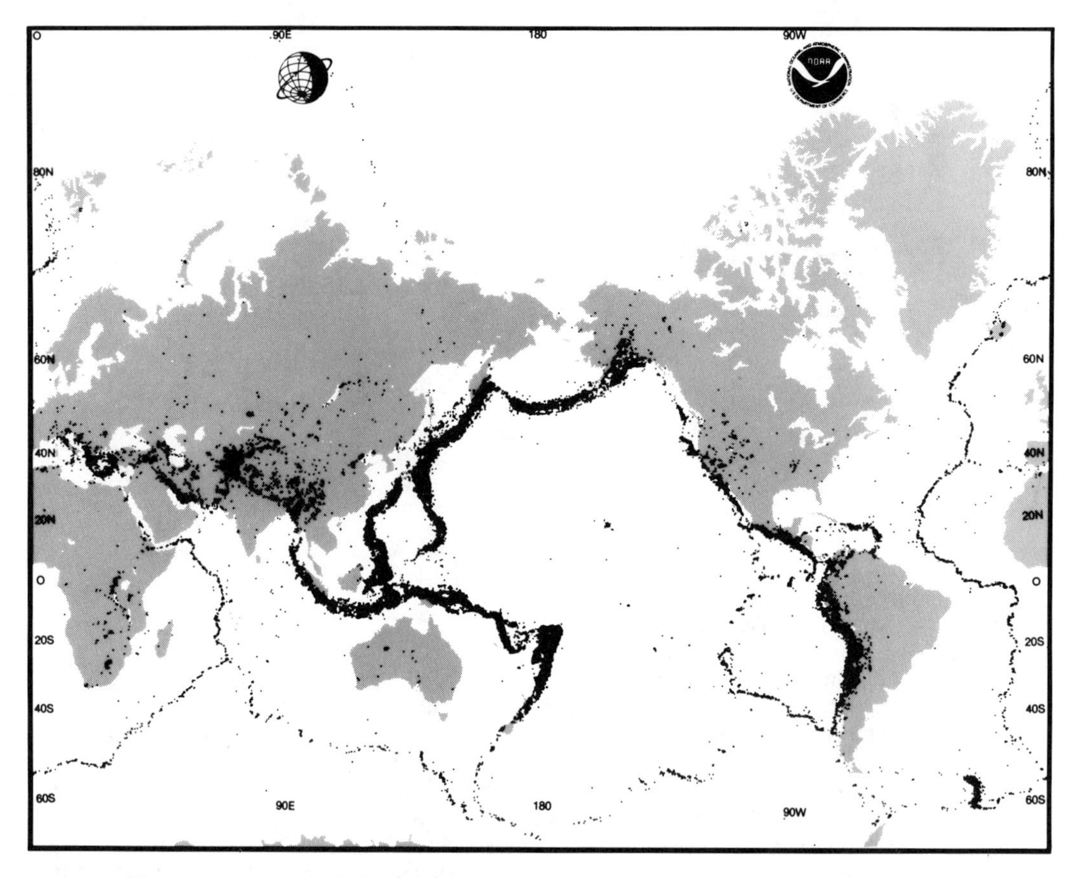


Figure 11.1 World Seismicity, 1984 (from NOAA).

shallow, b) 60–300 km are intermediate, and c) deep earthquakes may reach depths of 700 km. The epicenter of an earthquake is the point on the earth's surface directly above the focus.

The scientific study of earthquakes began with the Chinese as early as 132 A.D. Chang Heng, a Chinese astronomer and mathematician, invented an earthquake weathercock, an instrument consisting of an array of loosely held balls in the mouths of eight dragons facing at angles of 45° from one another. An internal mechanism, activated by even a slight tremor, opened the mouth of a dragon, releasing a ball which fell into the open mouth of a toad below, sounding an alarm. The direction of the earthquake could be determined by the orientation of the empty-mouthed dragon.

Until the 18th century, few factual descriptions of earthquakes were recorded, and the natural cause of earthquakes was unknown (see quotation from *Candide*). One popular theory in the Dark Ages was that earthquakes were caused by air rushing out of caverns deep in the interior.

Figure 11.1 shows the epicentral location of major historical earthquakes in the world. Most earthquakes occur along the edge of the Pacific Ocean and in other tectonically active areas. The epicenters outline the moving lithospheric plates (Cloud, 1980).

An earthquake is frightening as well as destructive. Most people who have lived in California a long time have felt an earthquake. I was in the Army in 1961, stationed at Fort Ord, CA, when at 11:23 P.M. on April 18, the earth moved. My three-story barracks was shaken in such a way that it caused all the beds to vibrate up and down. The lights were turned on and we could see the ceiling and floor moving laterally. The vertical columns were tilting back and forth, and it seemed there was a possibility that the entire barracks could collapse. The trembling lasted for a few seconds; then a minute later the process repeated. Many soldiers slept out on the lawn for the rest of the night! The earthquake was magnitude 5.6, and the epicenter was 30 km away.

Earthquakes in the world that have caused damage and loss of life are listed in Table 11.1. Enormous loss of life has occurred in some countries. For example, the death toll was over 600,000 in the earthquake of July 27, 1976, in China. Figure 11.2 shows the locations of damaging earthquakes in the United States since 1755. Total property damage in the United States due to earthquakes has been nearly $2 billion. The loss of life in the United States has been small considering the number of earthquakes and the continuously increasing population density in earthquake-prone areas. A repeat today of the 1906 San Francisco earthquake would probably cause losses in the order of tens of billions of dollars (Hays, 1979).

### *11.1.2 Intensity*

The intensity of an earthquake is based on observed effects of ground shaking on people, buildings, and natural features. Intensity varies from place to place within the affected region, depending on the distance from the epicenter and the type of bedrock.

Intensity is expressed by the Modified Mercalli Scale, a subjective measure which describes the severity of earthquake effects at a particular location. The Modified Mercalli Scale (Table 11.2) is composed of 12 increasing levels of severity that range from imperceptible shaking to catastrophic destruction (designated by Roman numerals I–XII). There is no real mathematical basis for earthquake intensity; instead, it is a ranking based on effects as observed by people.

After an earthquake occurs, intensity values can be plotted on a map. Lines connecting values of equal intensity are called isoseismal lines. Figure 11.3 shows isoseismal lines for two great earthquakes. Note that the New Madrid, MO, earthquake of 1811 produced an intensity VII for a larger area than the 1906 San Francisco earthquake. The New Madrid earthquake may have been a greater earthquake and probably occurred at a greater depth. The New Madrid earthquake actually came in three tremendous quakes—one on December 16, 1811, one on January 27, 1812, and one on February 7, 1812. Eyewitness accounts make it clear that cities would have been devastated if there had been any cities in the area (Penick, 1976). Chimneys toppled and thick stone walls cracked in St. Louis, Louisville, and Cincinnati—the latter city 640 km from the epicenter. The shaking awakend Thomas Jefferson in Monticello, VA, rang church bells in Washington, DC, and was felt as far away as Canada. The widespread high intensity makes this the Unites States' most violent earthquake. Since the New Madrid earthquake and the August 31, 1886, earthquake in Charleston, SC, both occurred in relatively stable parts of North America, their tectonic explanation is obscure.

Table 11.1 Earthquakes in the World which have Caused Damage or Loss of Life

| Year | Region | Deaths | Magnitude | Comments |
|---|---|---|---|---|
| 856 A.D. | Corinth, Greece | 45,000 | | |
| 1038 | Shensi, China | 23,000 | | |
| 1057 | Chihli, China | 25,000 | | |
| 1268 | Silicia, Asia Minor | 60,000 | | |
| September 27, 1290 | Chihli, China | 100,000 | | |
| May 20, 1293 | Kamakura, Japan | 30,000 | | |
| January 26, 1531 | Lisbon, Portugal | 30,000 | | |
| January 23, 1556 | Shensi, China | 830,000 | | |
| February 5, 1663 | St. Lawrence River | | | Maximum intensity X |
| November 1667 | Shemaka, Caucasia | 80,000 | | |
| January 11, 1693 | Catania, Italy | 60,000 | | |
| October 11, 1737 | Calcutta, India | 300,000 | | |
| June 7, 1755 | Northern Persia | 40,000 | | |
| November 1, 1755 | Lisbon, Portugal | 70,000 | | Great tsunami |
| February 4, 1783 | Calabria, Italy | 50,000 | | |
| February 4, 1797 | Quito, Ecuador | 40,000 | | |
| December 16,1811 | New Madrid, MO | Several | | Intensity XI |
| December 21, 1812 | Santa Barbara, CA | | | Maximum intensity X |
| June 16, 1819 | Cutch, India | 1543 | | |
| September 5, 1822 | Aleppo, Asia Minor | 22,000 | | |
| December 18, 1828 | Echigo, Japan | 30,000 | | |
| January 9, 1857 | Fort Teion, CA | | | Intensity X–XI |
| August 13, 1868 | Peru-Bolivia | 25,000 | | |
| August 16, 1868 | Ecuador-Columbia | 70,000 | | |
| March 26, 1872 | Owens Valley, CA | ~ 50 | | Large-scale faulting |
| August 31, 1886 | Charleston, SC | ~ 60 | | |
| October 28, 1891 | Mino-Owari, Japan | 7000 | | |
| June 15, 1896 | Japan | 22,000 | | Tsunami |
| June 12, 1897 | Assam, India | 1500 | 8.7 | |
| September 3 and 10, 1899 | Yakutat Bay, Alaska | | 7.8 and 8.6 | |
| April 18, 1906 | San Francisco, CA | 700 | 8.3 | |
| December 28, 1908 | Messina, Italy | 120,000 | 7.5 | |
| January 13, 1915 | Avezzano, Italy | 30,000 | 7.0 | |
| December 16, 1920 | Kansu, China | 180,000 | 8.5 | |
| September 1, 1923 | Kwanto, Japan | 143,000 | 8.2 | Great Tokyo Fire |
| December 26, 1932 | Kansu, China | 70,000 | 7.6 | |
| May 31, 1935 | Quetta, India | 60,000 | 7.5 | |
| January 24, 1939 | Chillan, Chile | 30,000 | 7.8 | |
| December 27, 1939 | Erzincan, Turkey | 23,000 | 8.0 | |
| June 23, 1948 | Fukui, Japan | 5131 | | |
| August 5, 1949 | Pelileo, Ecuador | 6000 | | |
| February 29, 1960 | Agadir, Morocco | 14,000 | 5.9 | |
| May 21–30, 1960 | Southern Chile | 5700 | 8.5 | |
| September 1, 1962 | Northwest Iran | 14,000 | 7.3 | |
| July 26, 1963 | Skopje, Yugoslavia | 1200 | 6.0 | |
| March 28, 1964 | Alaska | 131 | 8.6 | Tsunami |
| August 31, 1968 | Iran | 11,600 | 7.4 | Surface faulting |
| May 31, 1970 | Peru | 66,000 | 7.8 | $530,000,000 damage |
| February 9, 1971 | San Fernando, CA | 65 | 6.5 | $550,000,000 damage |
| December 22, 1972 | Managua, Nicaragua | 5000 | 6.2 | |
| February 4, 1975 | Liaoning, China | Some | 7.4 | Much damage |
| September 6, 1975 | Lice, Turkey | 2400 | 6.8 | 12,000 homes damaged |

Table 11.1 Earthquakes in the World which have Caused Damage of Loss of Life (Continued)

| Year | Region | Deaths | Magnitude | Comments |
|---|---|---|---|---|
| February 4, 1976 | Guatemala | 23,000 | 7.9 | Motagua fault break |
| May 6, 1976 | Friuli, Italy | 1000 | 6.5 | Extensive damage |
| July 27, 1976 | Tangshan, China | 600,000 | 7.6 | Great economic loss |
| March 4, 1977 | Roumania | 1500 | 7.2 | Felt Moscow to Rome |
| November 30, 1980 | Liona, Italy | 3000 | 6.8 | 700,000 homeless |

Modified from Bolt et al., 1977.

One reason that the ground shaking and associated ground cracking were so severe and widespread in both earthquakes is that unconsolidated Tertiary and Quaternary sediments underlie the two areas; unconsolidated sediments such as these seem to intensify destruction.

Studies of other earthquakes show that the isoseismal lines do not occur simply as bull's-eye circles with the epicenter at the center. Often, elongate patterns are oriented along the strike of the fault because the fault may be displaced along a considerable length of its total extent. Further, the type of bedrock affects the amount of damage. Figure 11.4 is an isoseismal map of the great 1906 California earthquake. Note the large intensities in the San Joaquin Valley, about 200 km southeast of the epicenter near San Francisco. The San Joaquin Valley is made up largely of alluvial deposits.

Figure 11.2 Location of damaging earthquakes in the United States (from Hays, 1979).

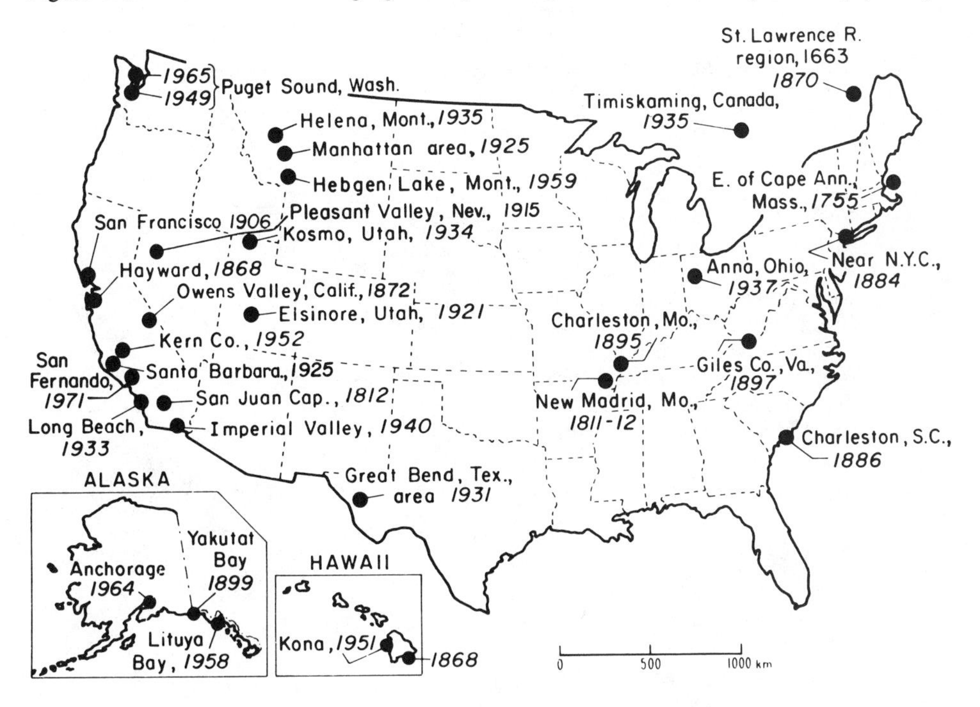

Table 11.2 Modified Mercalli Scale of Earthquake Intensity

| Scale | Description |
|---|---|
| I. | Not felt except by a very few under especially favorable conditions. |
| II. | Felt only by a few persons at rest, especially on upper floors of buildings. |
| III. | Felt quite noticeably by persons indoors, especially on upper floors of buildings. Many people do not recognize it as an earthquake. Standing motor cars may rock slightly. Vibration similar to the passing of a truck. Duration estimated. |
| IV. | Felt indoors by many, outdoors by few during the day. At night, some awakened. Dishes, windows, and doors disturbed; walls make cracking sounds. Sensation like heavy truck striking building. Standing motor cars rocked noticeably. |
| V. | Felt by nearly everyone; many awakened. Some dishes, windows broken. Unstable objects overturned. pendulum clocks may stop. |
| VI. | Felt by all, many frightened. Some heavy furniture moved; a few instances of fallen plaster. Damage slight. |
| VII. | Damage negligible in buildings of good design and construction; slight to moderate in well-built ordinary structures; considerable damage in poorly built or badly designed structures; some chimneys broken. |
| VIII. | Damage slight in specially designed structures. Considerable damage in ordinary substantial buildings with partial collapse. Damage great in poorly built structures. Fall of chimneys, factory stacks, columns, monuments, and walls. Heavy furniture overturned. |
| IX. | Damage considerable in specially designed structures; well-designed frame structures thrown out of plumb. Damage great in substantial buildings, with partial collapse. Buildings shifted off foundation. |
| X. | Some well-built wooden structures destroyed; most masonry and frame structures destroyed with foundations. Rails bent. |
| XI. | Few if any (masonry) structures remain standing. Bridges destroyed. Rails bent greatly. |
| XII. | Damage total. Lines of sight and level are distorted. Objects thrown into the air. |

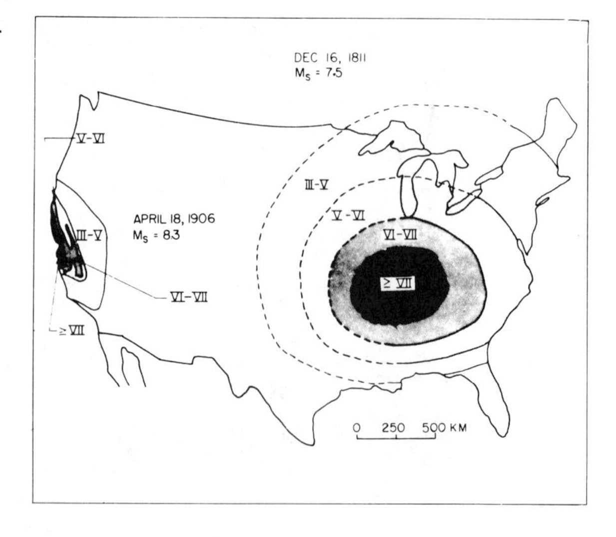


Figure 11.3 Comparison of isoseismal maps of 1811–1812 New Madrid, MO, and 1906 San Francisco, CA, earthquakes (from Hays, 1979). $M_s$ denotes surface wave magnitude, and roman numerals denote Modified Mercalli intensity. Dashed lines depict inferred isoseismal values.

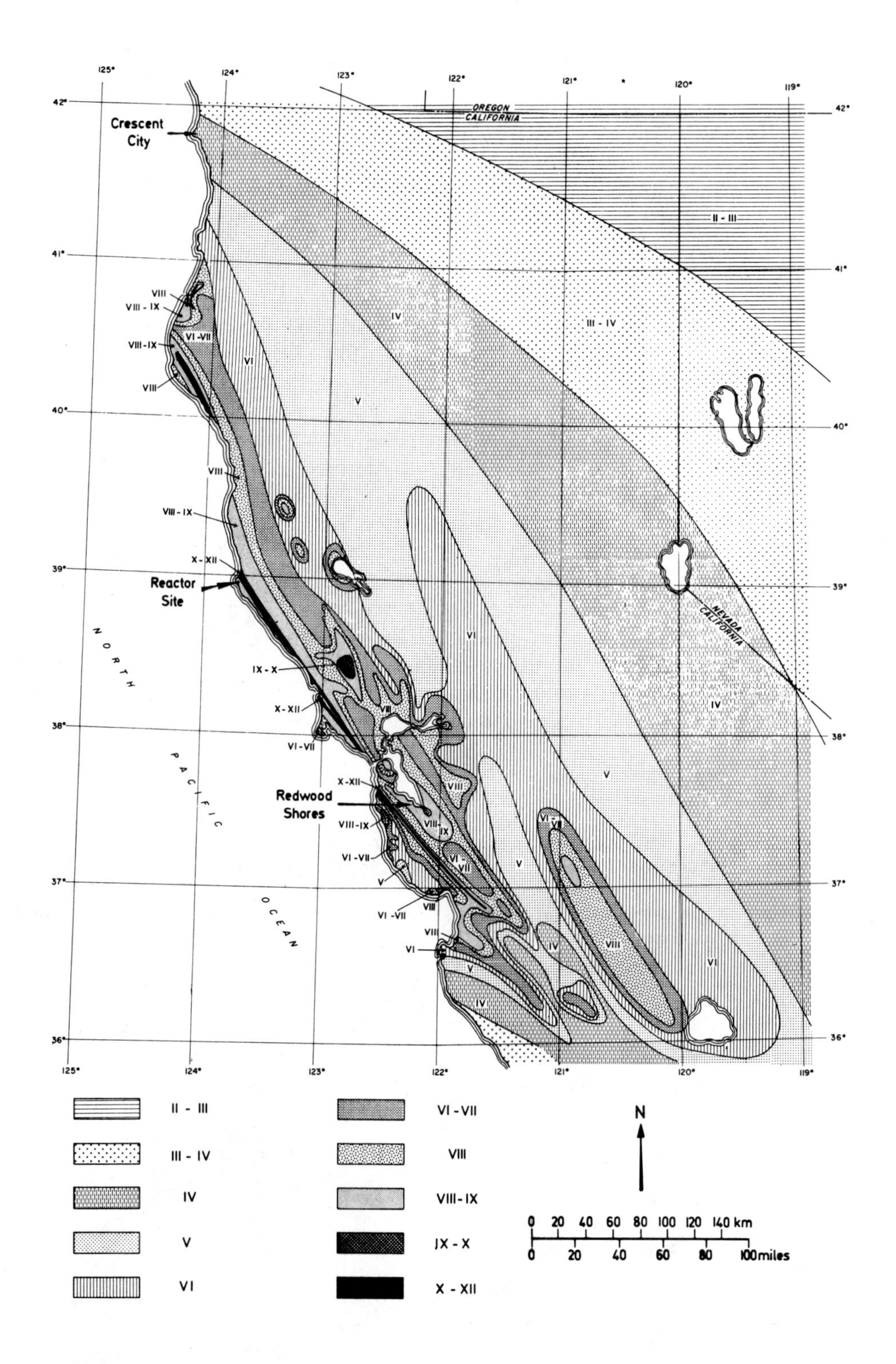


Figure 11.4 Detailed isoseismal map of 1906 California earthquake (from Bolt et al., 1977).

## *11.1.3 Seismic Waves*

*Seismograms.* The seismic waves radiating from the focus can be recorded by an instrument called a seismograph which produces a record known as a seismogram. Seismograms give data on the time, epicenter, and focal depth of an earthquake, and can provide estimates of the amount of energy that was released.

Nearly all seismographs use a heavy mass in suspension so that its inertia keeps it at rest while the earth is displaced (Fig. 11.5). Displacement of north–south, east–west, and vertical are usually made, using both short and long period displacement. Commonly the seismogram is made by a light beam projected onto photosensitive paper, so that the actual earth displacement is magnified by about 800×. Figure 11.6 is a seismogram of the 1939 Turkey earthquake.

*Seismic waves and the earth's interior.* There are two general types of vibrations produced by an earthquake: surface waves, which travel along the earth's surface, and body waves, which travel through the earth. Surface waves (L waves) usually have the strongest vibrations and probably cause most of the damage done by earthquakes. Surface waves travel to distant points more slowly than body waves, and consist of many complex types (Howell, 1959).

Body waves are of two types: compressional and shear. Because compressional waves travel at great speeds (~30,000 km/hr in the earth's crust), they reach seismograph stations first and are called primary waves, or simply P waves. Shear waves travel at about 14,000 km/hr, and hence reach the station later than P waves. These are called secondary, or S waves. P and S waves travel through the

Figure 11.5 Seismographs for measuring horizontal and vertical components of earthquake waves (from Longwell et al., 1969). **A.** Horizontal seismograph with recorder based on beam of reflected light. Upper left. Trace of reflected light on photographic paper wrapped around drum becomes visible only after photographic development. **B.** Vertical seismograph for recording long-period seismic waves.

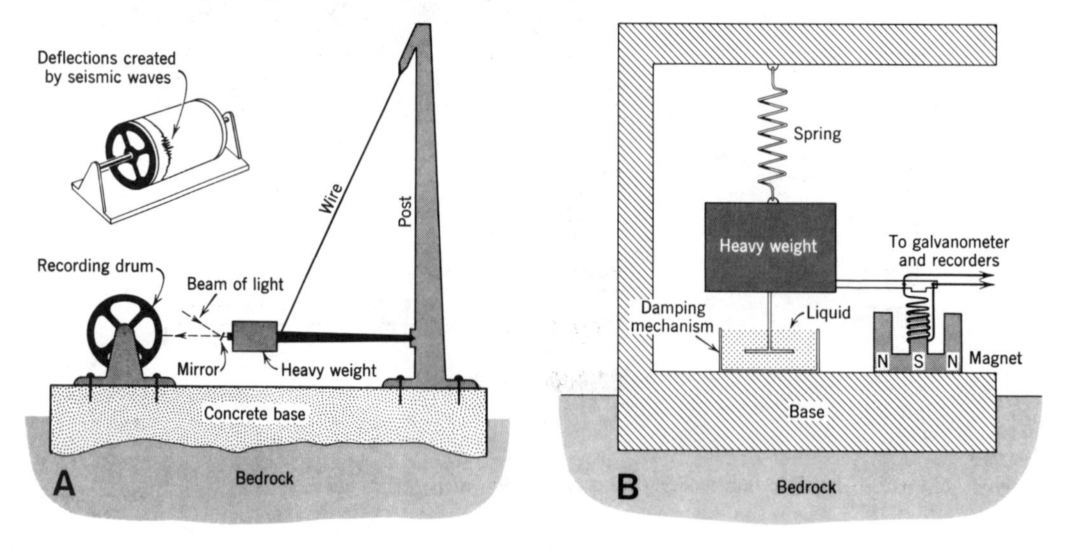

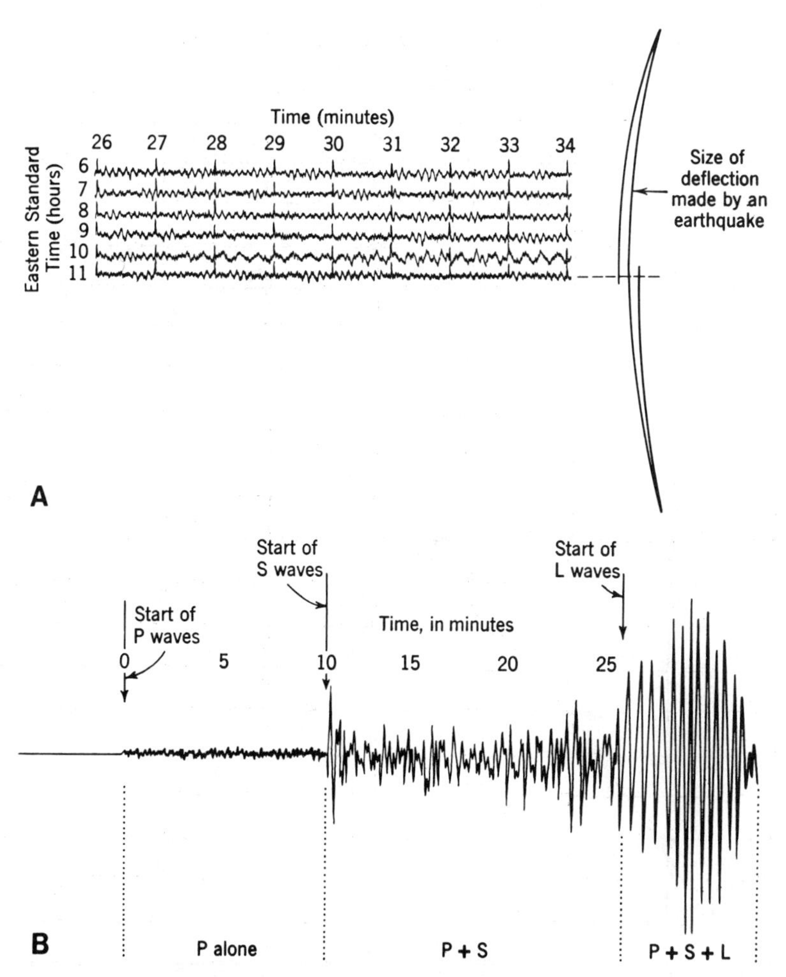


Figure 11.6 Examples of seismograms (from Longwell et al., 1979). **A.** Traces of microseisms recorded at Troy, NY, January 1, 1967, compared with deflection made by an earthquake on the same day. (Rensselaer Polytechnic Institute). **B.** Seismogram of earthquake recorded at Cambridge, MA, December 26, 1939, with epicenter at Erzincan, Turkey, shows small deflections made by P waves alone and large deflections made by P+S and P+S+L. Time elapsed between start of P and that of S, 10 minutes and 45 seconds, registers distance from epicenter to station as 88°30′, or about 9700 km.

entire earth, although shear waves do not travel through the earth's moltenlike core. The velocity of body waves increases with depth (see Chapter 12).

The difference in the times of arrival of the P and S waves at any place is a direct measure of the distance to the earthquake epicenter from the recording station. Figure 11.7 is a plot of P and S wave time versus distance to epicenter. These general curves were determined from numerous records all over the earth. For any given seismogram, the difference in time between the arrival of the P and S wave is determined. (This generally requires the skills of a seismologist.) A position on the ordinate scale is found on Fig. 11.7 which corresponds to this time

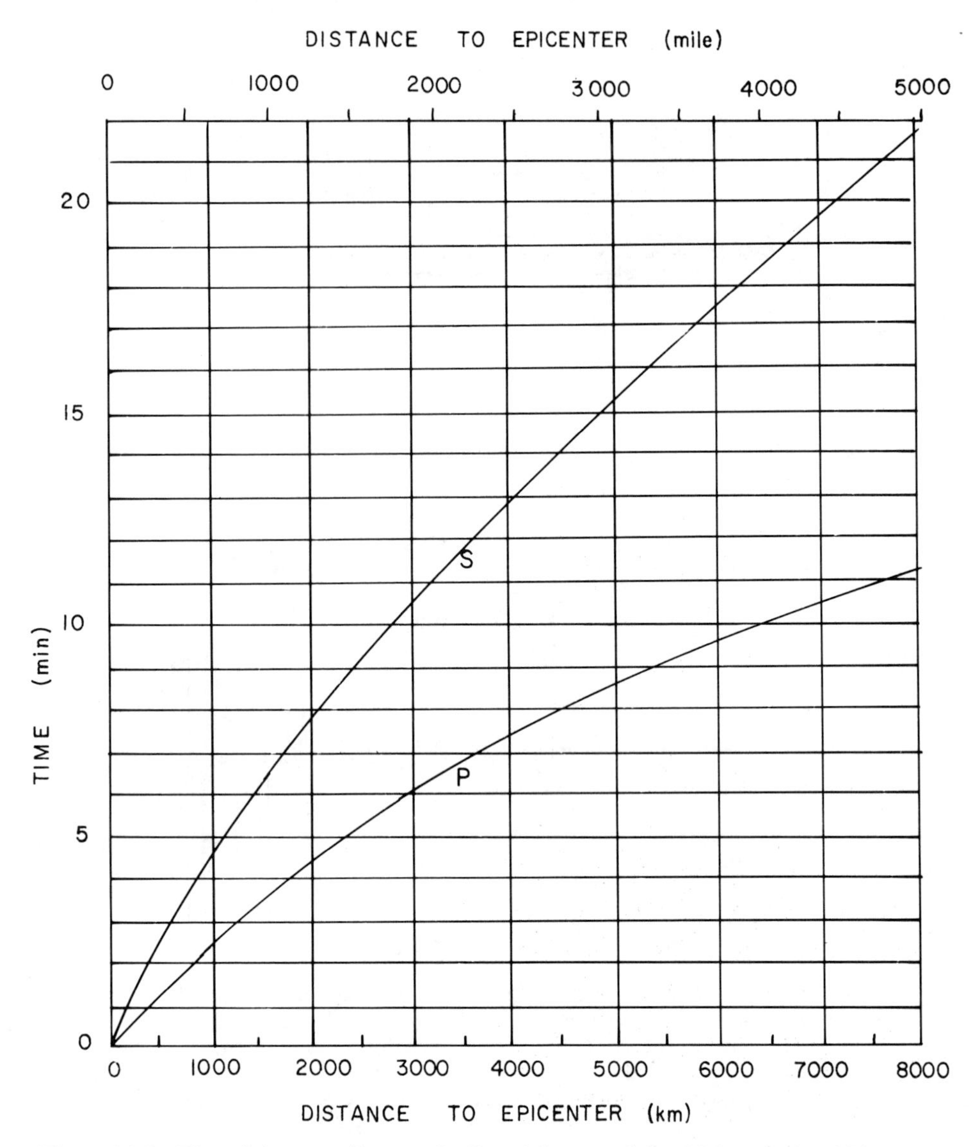


Figure 11.7 Plot of time vs. distance for P and S waves (after Nehru et al., 1972; Longwell et al., 1979).

difference between the P and S wave. The appropriate distance to the epicenter is then determined on the abscissa. Other waves such as the PcP (wave reflected off the earth's core) may also be used in this study. The precise epicentral location on the earth's surface is determined using epicentral distance calculations from three seismograph stations. The intersection of the three distances on a spherical global surface is the epicenter.

A Worldwide Standard Seismograph Network (WWSSN) maintains cooperation between many nations. The U.S. Geological Survey supports the operation of about 120 seismographic stations throughout the world. Data from these and other stations are analyzed at the Survey's National Earthquake Information Ser-

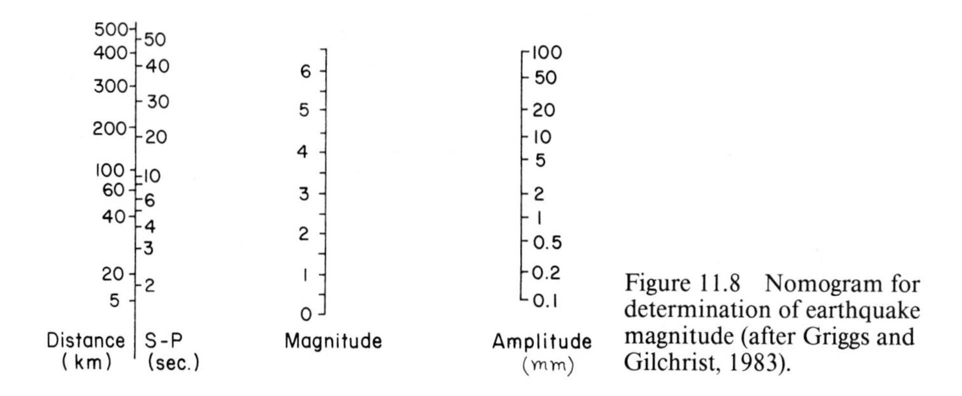


Figure 11.8 Nomogram for determination of earthquake magnitude (after Griggs and Gilchrist, 1983).

vice in Golden, CO, from which bulletins and lists of worldwide seismic activity are issued. The U.S. Department of Commerce also collects and disseminates geophysical data at the Geophysical Data Center in Boulder, CO.

*Earthquake magnitude and intensity.* The magnitude of an earthquake as devised by Richter (1935) is a parameter that is quantitatively determined from a seismogram. Specifically, magnitude ($M$) is defined as the $\log_{10}$ of the amplitude in microns ($10^{-3}$ mm) of the largest displacement on a seismogram, obtained by a standard Wood-Anderson torsion seismometer located 100 km from the earthquake epicenter. Because such a precise location is unlikely, interpretation is needed from stations whose epicentral distance is closer or further than 100 km. Figure 11.8 illustrates how a nomogram may be used to solve for magnitude given the epicentral distance. For example, consider a seismogram where the maximum amplitude is 23 mm, and the difference between the arrival time of the P and S wave is 24 seconds. Drawing a straight line between these points on Fig. 11.8 intersects the magnitude scale at 5.

In the past few decades seismologists have refined the measurement of earthquake magnitude by considering the type of wave used to determine the amplitude. When the largest amplitude of the body-wave pulse is used, the magnitude is called body-wave magnitude ($M_b$). When the largest amplitude of the surface-wave pulse is used, the magnitude is called surface-wave magnitude ($M_s$). The two magnitudes are related empirically (Howell, 1973) as:

$$M_b = 2.5 + 0.63\, M_s. \tag{11.1}$$

Magnitude is related to the amount of energy released by an earthquake. Several empirical relationships are discussed by Gutenberg and Richter (1956) and Richter (1958). The formula proposed by Bath (1966) is:

$$\mathrm{Log}_{10}\, E = 12.24 + 1.44\, \mathrm{M}, \tag{11.2}$$

where $E$ is the energy released in ergs.

Table 11.3 shows the average annual number of earthquakes of different magnitude on earth. It shows, for instance, that a "great" earthquake can be expected somewhere on earth about once a year. Some earthquakes come in swarms. Although there were no operational seismographs during the New Madrid earth-

Table 11.3 Worldwide Frequency of Earthquakes of Varying Magnitude

| Classification | Magnitude | Average number per year |
|---|---|---|
| Great earthquakes | 8 or more | 1.1 |
| Major earthquakes | 7–7.9 | 18 |
| Destructive shocks | 6–6.9 | 120 |
| Minor strong shocks | 5–5.9 | 800 |
| Damaging shocks | 4–4.9 | 6200 |
| Generally felt | 3–3.9 | 49,000 |
| Potentially perceptible | 2–2.9 | 300,000 |

From Howell, 1959.

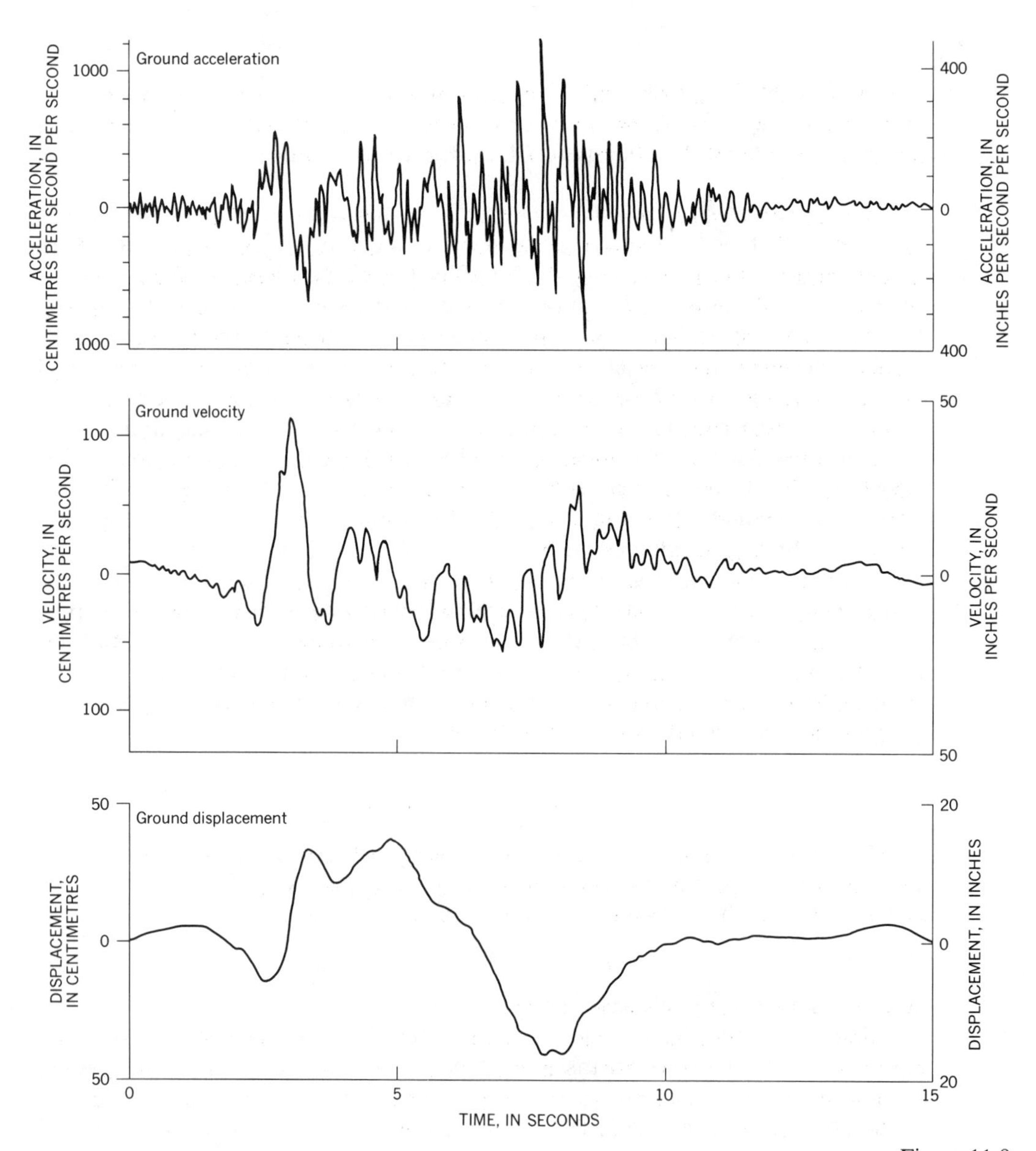


Figure 11.9

quakes of 1811–1812, Johnston (1982) estimates the magnitude of the three earthquakes at 8.6, 8.4, and 8.7.

For shallow focus earthquakes there is a rough relationship between magnitude and the maximum intensity ($I_{max}$) as judged from isoseismal line plots (Howell, 1970):

$$I_{max} = 2\,M_b - 4.6. \qquad (11.3)$$

*Earthquake acceleration.* In general, the first derivative of displacement with respect to time is velocity, and the second derivative is acceleration. Therefore, plots of velocity and acceleration can be derived from conventional seismograms (Fig. 11.9). An accelerogram can be obtained by employing an accelerograph, an instrument with a pendulum having a frequency higher than any of the vibrations they are expected to measure (Neumann, 1962). Accelerographs can be adjusted so that only strong motion will trigger a recording. Usually this triggering motion is caused by an acceleration greater than 0.01 g (1% of the acceleration of gravity). [As a very approximation, this acceleration is considered to be caused by a magnitude 4 or higher earthquake (intensity IV–V or higher). For earthquake in southern California, the acceleration at the epicenter is very roughly about 0.1 g for a magnitude 6.4 earthquake. Acceleration is diminished to about 10% at a distance of 50 km away from the epicenter (Green, 1973).]

Acceleration is of special importance in the design of structures. Buildings must, of course, always withstand a vertical 1 g, the natural pull of gravity on earth (9.8 $m/s^2$), but many buildings in the United States could not survive 0.1 g in a lateral direction.

Figure 11.9 shows the largest horizontal acceleration ever recorded, equal to 1.04 g (i.e., about 10.2 $m/s^2$). This occurred during the February 9, 1971, San Fernando earthquake.

## 11.2 Lessons Learned from Catastrophic Earthquakes

### *11.2.1 San Andreas Fault and the 1906 San Francisco Earthquake*

The San Andreas Fault is the longest and most active fault in the world. It stretches from near Yuma, AZ, to San Francisco, where it curves out into the Pacific Ocean. It is clearly seen in aerial photographs and satellite imagery. Tectonic models for the evolution of the earth's continents account for the development of the San Andreas and related faults as a complex transform boundary between the North American and Pacific plates (Carlson, 1982).

The San Andreas Fault is some 1000 km or more long, extending vertically into the earth to a depth of at least 30 km. In detail, the fault is a complex zone of breccia up to 1 km wide. Many smaller faults branch from and join the San Andreas Fault. Displacement along any fault occurs sporadically in time and space. Figure 11.10 shows where ground was displaced in the 1857 and 1906 earthquakes along the San Andreas Fault. As a result of the 1906 earthquake near

Figure 11.9 Records of N 14°E component of horizontal ground motion of Pacoima damsite for San Fernando, CA, earthquake of February 9, 1971 (from Page et al., 1975b). Velocity (center) and displacement (bottom) records are obtained by integrating acceleration record (top) once and twice, respectively.

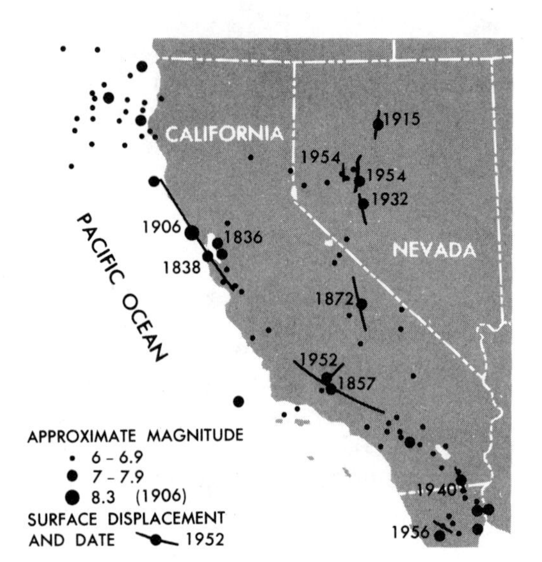


Figure 11.10 Map of California showing the epicenters of some large earthquakes in the California–Nevada region. Heavy lines show where surface of ground was broken (from U.S. Geological Survey).

San Francisco, roads, fences, and rows of trees that crossed the fault were offset by as much as 6.4 m. The total length of the fault where 1906 displacement or cracked ground occurred was about 300 km, starting at San Juan Bautista in the south and extending out to sea at Point Arena in the north.

The sense of displacement along the San Andreas Fault is right-lateral, based on the 1906 earthquake as well as observation of past displacement judged by offset streams, etc. Some sections of the fault have never been displaced in historic time. These sections would appear to be the most dangerous areas because, over geologic time, the displacement along the entire fault must be roughly equal from one area to the next.

Some sections of the fault move more or less continously, a phenomena called creep. At the Almaden Winery near Hollister, CA, for example, creep on the order of a fraction of a centimeter occurs every few weeks or months, averaging about 1 cm/yr. Other sections of the San Andreas Fault seem to store stress for a long time, releasing strain in a violent earthquake. Unfortunately for San Francisco, the fault seems to be "locked" in the San Francisco area. No movement has occurred since the 1906 earthquake.

The reason some sections of the San Andreas Fault glide easily past each other may be due to 1) the presence of serpentinite, a rock which has a great deal of plasticity under great pressure, and/or 2) the presence of metamorphic water, which facilitates movement in areas where the Franciscan rocks are confined by the Great Valley sequence.

Geologists generally agree that the total accumulated displacement along the fault may be about 500 km in the area near Los Angeles to San Francisco (Hill, 1981). If the fault has been in existence 100 million yr (based on geologic mapping), then a rough determination of its average rate of strain is:

$$\frac{5 \times 10^5\text{m}}{10^8\text{yr}} = 0.5 \text{ cm/yr.}$$

The offset of distinctive Cenozoic rock units can also be used to establish the rate of fault movement within fairly wide bounds. These offset rock units average

Figure 11.11 Effects of earthquake and fire from 1906 earthquake in San Francisco. Photograph by W. C. Mendenhall, showing Nob Hill, from corner of Van Ness and Washington Streets. Photograph compliments of Earthquake Information Bulletin.

the rate of movement over millions of years, and cannot distinguish between sudden slip (creating a large earthquake) and creep. Data for displaced Cenozoic strata near the San Andreas Fault suggest an average slip rate of 1–2 cm/yr over the past 20 million yr (Wesson et al., 1975). Estimates of slip rates for the southern San Andreas fault system (Keller et al., 1982) average 2–3 cm/yr for the past 12 million yr, although a slip rate of 6 cm/yr is suggested for the most recent 4 million yr. Based on an offset stream in central California, Sieh and Jahns (1984) calculated an average rate of offset of ~3.5 cm/yr during the Holocene. Precise surveying across the fault in areas where the fault is slipping freely shows the present rate of strain is about 5 cm/yr. The discrepancy between all of these published strain rates is not understood (Anderson, 1971).

Great earthquakes along the San Andreas Fault occurred in 1857 at Fort Tejon, in 1872 at Owens Valley, and 1906 in San Francisco. The San Francisco earthquake of April 18, 1906, is the most famous. It resulted in the loss of possibly 700 lives and damage estimated at $350 million to $1 billion (1906 dollars). Figure 11.4 is an isoseismal map of this earthquake. Although modern seismographs were not available in 1906, it is estimated that the earthquake magnitude was about 8.3 (Oakeshott, 1973).

The effects of the 1906 earthquake were thoroughly studied by a group of very competent geologists (Lawson et al., 1908). After the earthquake, tremendous fires raged for 3 days in the eastern part of San Francisco (Fig. 11.11). The fires were due to gas lanterns being knocked over in wooden buildings, broken water and gas mains due to landslides (Youd and Hoose, 1978), and gusty winds (Thomas and Witts, 1971). The most extensive ground breaking in the city was near the waterfront in areas underlain by natural bay mud and artificial fill. Following the earthquake, a report was published by the California Earthquake Investigation Commission containing information on the vibrational force of major soil and rock units, as evidenced by damaged buildings. The foundation coefficient for the least damage (solid rock) was defined as unity. Sand was 2.4, artificial fill 4.4–11.6, and marsh (12.0) was found to be the worst material for buildings.

Today most San Francisco residents refer to the 1906 earthquake disaster as a fire rather than an earthquake. (Subconsciously it is more calming to refer to it as a fire, because it implies the catastrophe would not be repeatable with today's better fire-fighting equipment.)

At the time of the 1906 earthquake, the tallest building in San Francisco was 19 stories, with a steel frame and brick walls. This building sustained some earthquake damage and was gutted by fire. Most large buildings in San Francisco survived due to their steel frames or massive nonstructural walls; only six of 52 major buildings in San Francisco were destroyed by the earthquake itself.

The population of the San Francisco Bay area has increased sixfold since 1906, and structures about tenfold. In spite of the fact that a large loss by fire is not likely to occur, Hays (1981) estimates that a repetition of the 1906 earthquake is expected to cause $24 billion in damage (1978 dollars). The loss of life, if the occurrence were during non-business hours, is estimated at about 3000–5000 (Page et al., 1975b). If the earthquake occurs during business hours, however, the loss of life could be about 10,000–20,000. Actually, San Francisco does not have to wait for another magnitude 8.3 earthquake to sustain damage. A shallow earthquake of magnitude 6–7 would cause considerable damage if close by. Additionally, tall buildings, whose natural periods of vibration are long, can be expected to respond considerably to great earthquakes some distance away.

A television documentary produced by the British Broadcasting Company is titled "San Francisco, City that Waits to Die." The title may be an overstatement of the case; nevertheless, San Francisco, more than any city in North America, is sitting on a time bomb. For fear of alarming the populace, the TV program was not shown on California television stations. As in the case of coastal erosion, Californians prefer to live in a fantasyland.

## *11.2.2 1964 Alaska Earthquake*

At 5:36 P.M., March 27, 1964, a catastrophic earthquake occurred approximately 130 km east of Anchorage, AK. The focus was 20–50 km deep. The magnitude is estimated between 8.4 and 8.75 (see Grants et al., 1964; Plafker, 1969; Eckel, 1970; and USGS Professional Papers 541–546 series.)

Land elevations were displaced over an area of 250,000 km$^2$. The sea floor subsided as much as 2.15 m, and a maximum uplift of 12 m occurred on Montague Island. Tsunamis were triggered (see Chapter 10).

The damage occurred in a sparsely inhabited area with only about 1 inhabitant/km$^2$. Nevertheless, the earthquake claimed 130 lives and caused about $300–$750 million in damage. These figures would have been larger in the height of tourist season or when schools and offices were occupied.

The worst destruction was in Anchorage, where giant landslides occurred (Fig. 11.12). The distribution of landslides at Anchorage was related to the Pleistocene-aged Bootlegger Cove Clay. Over 75 homes were destroyed in Anchorage's Tur-

Figure 11.12 Landslides at the Turnagain Heights area of Anchorage, AK. The March 27, 1964, earthquake caused numerous landslides which destroyed the houses nearest the sea cliff (lower left). (From Hansen et al., 1966.)

nagain Heights neighborhood. Before the earthquake, these homes overlooked the Knik Arm of the Pacific Ocean from a 21 m high bluff. Seismic waves caused the underlying silty clay to liquefy, transforming a block of land 2.4 km long and about 0.3 km wide into a jumble of landslides that slid out onto the tidal mud flats. The landslides at Turnagain Heights did not begin until about 2 min after the earthquake started. They did not develop during the period of maximum ground motions, but developed as a result of loss of strength (liquefaction) in the Bootlegger Cove Clay caused by a sequence of earthquake motions (Seed and Wilson, 1969). The physical properties of the Bootlegger Cove Clay is described by Miller and Dobrovolny (1959). A preliminary report and map were available to Anchorage officials in 1950, yet city planners disregarded the warnings contained in the report concerning the landslide possibilities of the clay. Before the earthquake occurred, Cederstrom et al. (1964) noted that this clay "commonly becomes quicksand when penetrated by the (well) drill." The Bootlegger Cove Clay was deposited as rock flour in a marine environment. Isostatic adjustment of the land, due to melting glaciers, resulted in uplift of the land only a few thousand years ago. Sodium chloride salts apparently were being flushed by fresh ground water, helping to destroy the clay's cohesiveness (Hanse, 1965), thus the clay liquefied during the earthquake.

The seismic waves caused worldwide fluctuations in the levels of artesian wells. At Belle Fourche, SD, the artesian level of a well fluctuated 7 m (Hansen and Eckel, 1977). The seismograph station at the South Dakota School of Mines and Technology was partially destroyed.

An important lesson learned from the Alaska earthquake is that damage is much more severe in unconsolidated sediments such as occur in Anchorage, but is less severe in hard rock areas such as near the epicenter. Whittier, for example, was closer to the epicenter but is on slate bedrock and had little damage. According to Eckel (1970): "In general, intensity was greatest in areas underlain by thick unsaturated, unconsolidated deposits, least on indurated bedrock, and intermediate on coarse gravel with low water table, on morainal deposits, or on moderately indurated sedimentary rocks of late Tertiary age."

Figure 11.13 Air photo of the 1971 San Fernando earthquake area (from U.S. Geological Survey, 1971). Heavy lines indicate cracked ground.

## *11.2.3 1971 San Fernando, California Earthquake*

Figure 11.13 is an aerial photograph of the San Fernando area. The intensity of the urbanization in the lowland in this northeast suburb of Los Angeles is visible. The epicenter of the earthquake was in a sparsely populated mountainous area about 14 km north of town. Generally, the lowlands in southern California are unconsolidated alluvial fan deposits, and the mountains are Tertiary sedimentary rock.

The fault which caused the earthquake was not the famous San Andreas Fault, which lies 30 km to the northeast. The earthquake was generated by the thrusting of a small (barely mappable) fault near the Garlock Fault. Displacement ranged from 2 to 3 m, and the focus was about 13 km deep.

The initial shock came at 6:01 A.M. lasted 60 seconds and took 64 lives. Total damage was about $1 billion (U.S. Geological Survey, 1971) with only approximately $46 million covered by insurance.

The earthquake triggered more than 1000 landslides and produced fractures and slumping across much of the ground surface underlying the cities of San Fernando and Sylmar. Transportation routes were blocked, buildings, bridges, and freeway overpasses collapsed (Fig. 11.14), and utilities were extensively damaged. The off-hour timing of the earthquake was fortuitous; loss of life would have been much greater had the earthquake occurred 1 or 2 hr later.

Figure 11.14 Collapsed VA hospital, San Fernando, CA (from Blair et al., 1979). The Medical Treatment Building of Olive View Hospital was completed in 1970. The south stair tower overturned and collapsed on the one-story portion of the hospital ambulance area. The black roof of the stair tower is visible.

Figure 11.15 Partial collapse of the lower Van Norman Dam following the 1971 San Fernando earthquake (U.S. Geological Survey photo).

The earthquake was classified as moderate in magnitude (6.6 on the Richter scale). There were many seismograph stations in the area. Of particular interest were the large accelerations produced by the earthquake. Figure 11.9 shows the recording of an accelerograph where over 1 g (horizontal) was recorded. This is the largest acceleration ever recorded by an earthquake, although acceleration greater than 1 g must have been produced by many other earthquakes such as the 1897 Assam, India, earthquake where large boulders were heaved up in the air (Howell, 1959).

The upper Van Norman Dam moved downstream about 2 m. The water level in the lower Van Norman Dam was about 7.6 m below the spillway when the earthquake struck; the earthern dam sustained landslides which damaged the dam to within 1.5 m of the water level (Fig. 11.15). The 40-m-high dam has a clay core surrounded by hydraulically filled sand. A large amount of the sand liquified after about 12 s of shaking, causing portions of the upstream face of the dam to break into blocks and slide into the reservoir. Had the dam failed, it could have made this earthquake the deadliest in U.S. history (U.S. Geological Survey, 1971).

The earthquake emphasized the hazards of non-earthquake-resistant construction. Older structures, particularly unreinforced masonry bearing wall-type with weak mortar, performed poorly. Collapse of the non-earthqauke-resistant skeleton concrete frame buildings with unreinforced hollow tile walls such as the Veterans Hospital in Sylmar resulted in the greatest loss of life. Systematic post-earthquake studies (Housner and Jennings, 1977) showed that many multistory buildings that had concrete shear walls were undamaged or had only small cracks. As a result of the San Fernando earthquake, there was evidence that Californian design criteria and methods needed to be reassessed for bridges, overpasses, buildings, and earth dams.

## *11.2.4 1972 Managua Earthquake*

At 12:30 A.M., December 23, 1972, a relatively small earthquake struck Managua, the capital of Nicaragua. It was assigned a Richter magnitude ($M_b$) of 5.6 and a surface-wave magnitude ($M_s$) of 6.2 (Brown et al., 1973). The epicenter was exactly in the center of the city and the focus was quite shallow, perhaps 2–5 km. Thus, despite its relatively small size, the fact that the earthquake was very shallow and exactly beneath the city caused severe destruction in downtown Managua. The most intense destruction occurred along several small fault displacements (Fig. 11.16).

Figure 11.16 Map of faulting observed after the 1972 Managua earthquake. The epicenters of the aftershocks (shown as dots) follow the general strike of the faults into Lake Managua. The small lakes are volcanic maars (after Bolt et al., 1977).

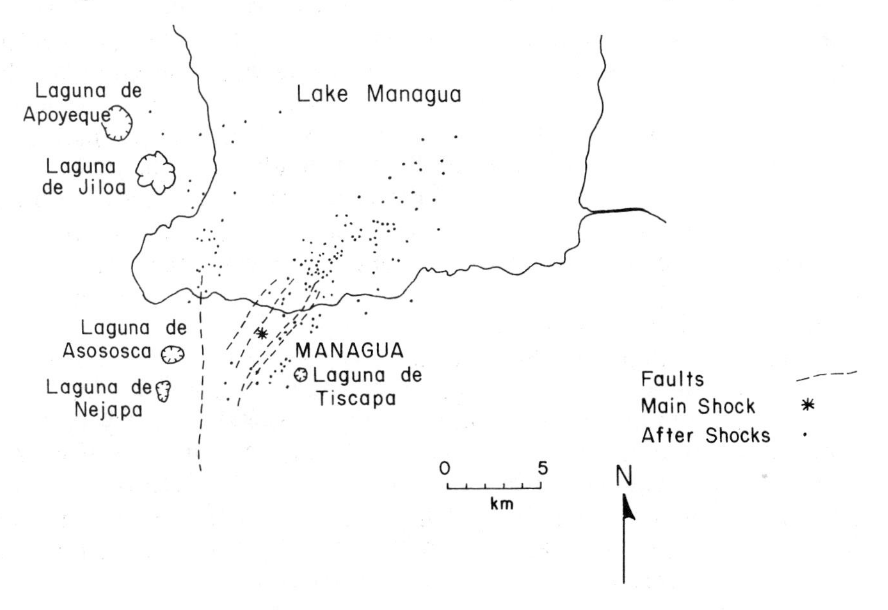

The population of Managua in 1972 was 420,000. The earthquake killed over 11,000 people, and 250,000 were left homeless. The total damage exceeded $500 million (Brown et al., 1973). This economic loss was severe for such a small and relatively poor country.

I visited Managua in February, 1973. The downtown area was devastated; from the air it looked as though a huge bomb had been detonated in the city center. Destruction in the approximately 300 blocks of downtown Managua was intense. Two months after the earthquake, the downtown area was still blocked off by armed guards to prevent looting. Everything inside the fence was in complete ruins, caused by both earthquake and fire. (Some property owners deliberately set fire to their buildings to collect fire insurance. Furthermore, shortly after the earthquake the downtown area was burned to help prevent the spread of disease due to the rotting bodies.) Almost all older buildings of masonry or rock wall construction had collapsed. Tarquezal (wood frame and adobe) structures behaved very poorly. Figure 11.17 shows damage to reinforced concrete buildings. Stress due to lateral displacement of the ground was manifested by chipped corners of the columns (Fig. 11.17A). It is not hard to see why buildings collapsed. Modern several-story reinforced concrete buildings "pancaked"; this included a three-story reinforced concrete Customs House office building astride one of the faults (Fig. 11.17B). (Similar pancaking-type failure of concrete apartment buildings occured in the July 20, 1967, earthquake in Venezuela.) Astonishingly, in the midst of the ruins, one building survived. Exactly at the epicenter, the new Bank of America, the tallest building in Managua, still stands, seemingly completely intact. The building was constructed of a steel skeleton using California earthquake standards, and was braced by concrete shear walls in a central core (Green, 1978).

The historic record of seismicity and geologic evidence of active Holocene faulting and volcanism shows that Managua is an unusually high risk area. In fact, a major factor in the decision to build the Panama Canal in Panama in 1906 was that Nicaragua was known to have great earthquake hazards. On March 31, 1931, a severe earthquake (magnitude 5.3–5.9) nearly destroyed the city and killed 1000 people. There are numerous active volcanoes in Nicaragua, El Salvador, and Guatemala. In Managua there are several maars, geomorphologic evidence of violent volcanic explosions. The source of the earthquake apparently was faulting in volcanic rocks beneath the city. Figure 11.16 shows the location of small faults observed after the earthquake. The epicenter of aftershocks are plotted on the map; they follow the general strike of the faults under Lake Managua.

Another reason for severe damage in Managua is the lack of firm bedrock. The city is built on unconsolidated alluvium and pyroclastic (volcanic) rubble. Although the earthquake happened at night, eyewitnesses reported seeing waves travel across the ground, and buildings and furniture heaved about in two seismic events separated by about 1 min.

Damaging earthquakes occur frequently in Nicaragua. The 1931 earthquake was very similar to the 1972 quake. According to Brown et al., (1973, p. 32): "Earthquakes comparable in magnitude to those of 1931 or 1972 can reasonably be expected within the next 50 years." Despite the fact that earthquakes seem to destroy Managua about every 50 years, the government proceeded to reconstruct the city at the same location. (Rationale for these types of decisions in a dictatorship such as that ruled by President Somoza in Nicaragua are often not subject to engineering analysis. In 1979 Somoza was overthrown.)

Figure 11.17 Reinforced concrete buildings damaged by Managua earthquake. **A**. Lateral movement damaged these columns, although complete failure of the buildings did not occur. **B**. Collapsed three-story reinforced concrete Custom's House office building (from Brown et al., 1973). The building "pancaked" when columns failed.

## 11.3 Seismic Risk and Earthquake Probability

It is one thing to discuss past catastrophic earthquakes, but another thing to try to predict where and when the next earthquake will happen. This section contains information pertaining to the a) spatial (geographic location) and b) temporal (time) prediction of future earthquakes.

### *11.3.1 Spatial Prediction*

Almost all earthquakes occur along active faults. The USGS defines an active fault as a "break in the earth's crust upon which movement has occurred in the recent geologic past and future movement is expected." Therefore, a key to identifying the location of possible future earthquakes is field mapping at sufficient scale (say 1:24,000 or larger), showing known or inferred active faults. In most of California, maps showing these faults are currently available, although there is a problem in ascertaining the question of activity. Certainly faults displacing glacial moraines, streams, etc., are active. Other faults may have no geomorphologic indication of activity, but have had earthquakes and/or displacements in historic time.

There is heated controversy among geologists concerning the activity of some faults. An example is the Rose Canyon Fault in San Diego (Gastil et al., 1979; Threet, 1979). The Rose Canyon Fault has had no seismic activity in historic time; however, Pleistocene ($\simeq$100,000 yr old) beach terraces are displaced about 50 m near Mt. Soledad, indicating fault activity in Pleistocene and probably Holocene time.

Recently the U.S. Geological Survey has published maps at 1:5,000,000 scale showing young faults known to have undergone movement in Quaternary time (the last 1.8 million yr) in the United States and Puerto Rico (Howard et al., 1978). Hatheway and McClure (1979) describe techniques for determining the date of the last movement along faults, as well as the general role of geology in the selection of sites for nuclear power plants.

The U.S. Nuclear Regulatory Commission (NCR) is vitally concerned with the safety of nuclear reactors (U.S. Nuclear Regulatory Commission, 1974). As an aid to safety design of nuclear power plants, the NRC has defined a "capable fault" as one which has exhibited "movement at or near the ground surface at least once within the past 35,000 years or movement of a recurring nature within the past 500,000 years." In 1982 the NRC suspended the license for the Diablo Canyon nuclear-power plant, along the Pacific Ocean near San Luis Obispo, CA. Structural design errors related to earthquake safety were found at the plant, which is near an offshore fault.

A widely used technique to predict future earthquake location is to plot epicenters of known earthquakes on a map and, by extension of the historic trends, make general inferences about future locations. In other words, if areas have been seismically active in the past, future earthquakes in these same areas can be expected. From inspection of Fig. 11.18, for example, one would conclude that San Diego is a lot safer than Los Angeles. One problem with this technique is that along a long fault such as the San Andreas Fault, segments with the most activity (many small earthquakes) may actually be a lot safer than an area where there

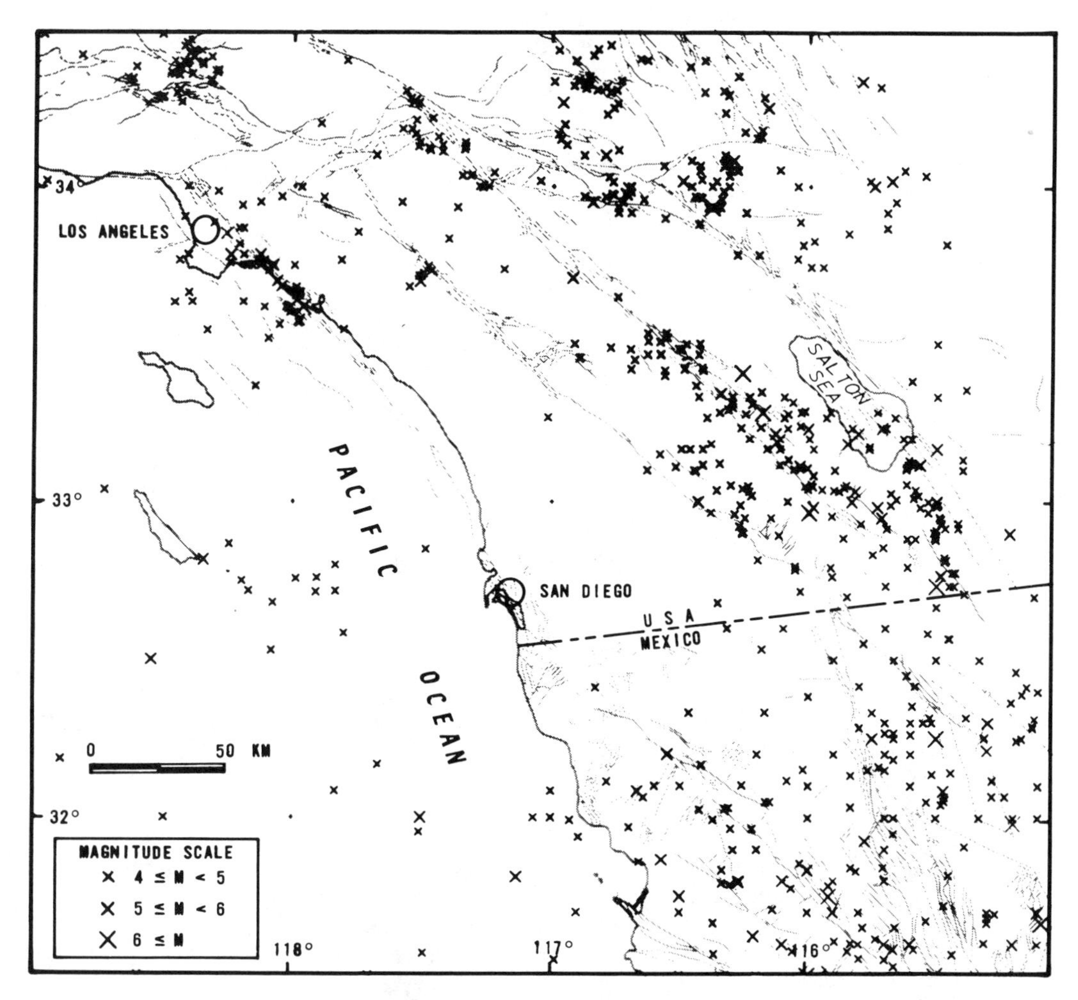


Figure 11.18 Seismicity of the San Diego region, CA, 1932–1977, for magnitude 4 earthquakes or greater (from Hileman, 1979).

has been historic seismicity, where the fault is locked and storing energy up for a catastrophic event.

Figure 11.19 shows how seismic data from geographical areas can be translated into probability curves. Crouse (1979) divided the southern California fault systems into five geographical provinces, and, based on past earthquake magnitude data, was able to assess the probability for future earthquakes of various magnitudes. The curves are compensated to eliminate the fact that some fault zones are bigger than others. It can be seen, for instance, that the San Andreas Fault is much more active and more likely to produce a large earthquake than the San Jacinto Fault.

From Newton's second law of physics, force and acceleration are directly related. During an earthquake, the acceleration of the ground beneath a structure is what causes forces to act on the structure. Hence, acceleration is an important parameter in earthquake engineering. Earthquake magnitude is related to acceleration in a general sense, in that bigger earthquakes usually produce bigger accel-

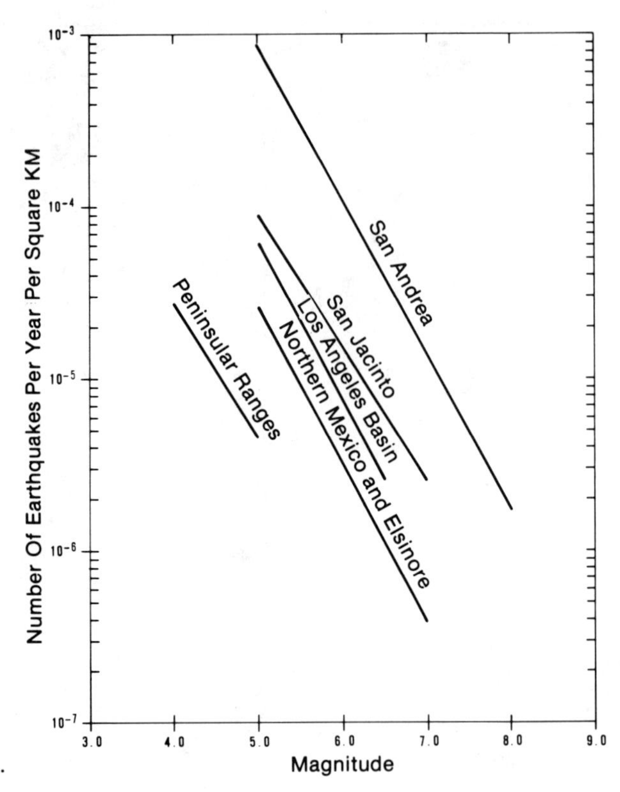


Figure 11.19 Earthquake recurrence curves for southern California (from Crouse, 1979).

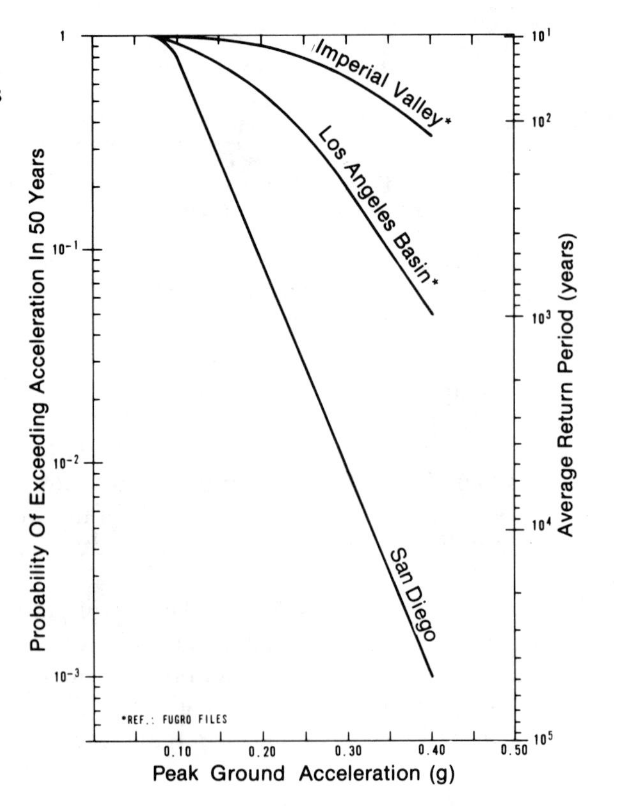


Figure 11.20 Average return periods and probabilities of exceeding ground accelerations in 50 years for southern California areas (from Crouse, 1979).

erations. In addition, acceleration is related to proximity to the epicenter. Donovan (1973) proposed the following empirical equation for acceleration on firm ground in southern California:

$$a = \frac{1080\ e^{0.5M}}{(R + 25)^{1.32}}, \tag{11.4}$$

where $a$ is the peak ground acceleration (cm/s²), $M$ is the earthquake magnitude, and $R$ is the hypocentral distance (km), the distance to the point in the earth where a fault begins to rupture.

From Eq. (11.4) and Fig. 11.19, Crouse (1979) was able to show the probability of various accelerations in three regions in southern California (Fig. 11.20). Clearly the Imperial Valley is the riskiest area, where ground acceleration of at least 0.29 g can be expected almost every other year.

Figure 11.21 Earthquake acceleration vs. intensity for different foundation conditions (from Leeds, 1973).

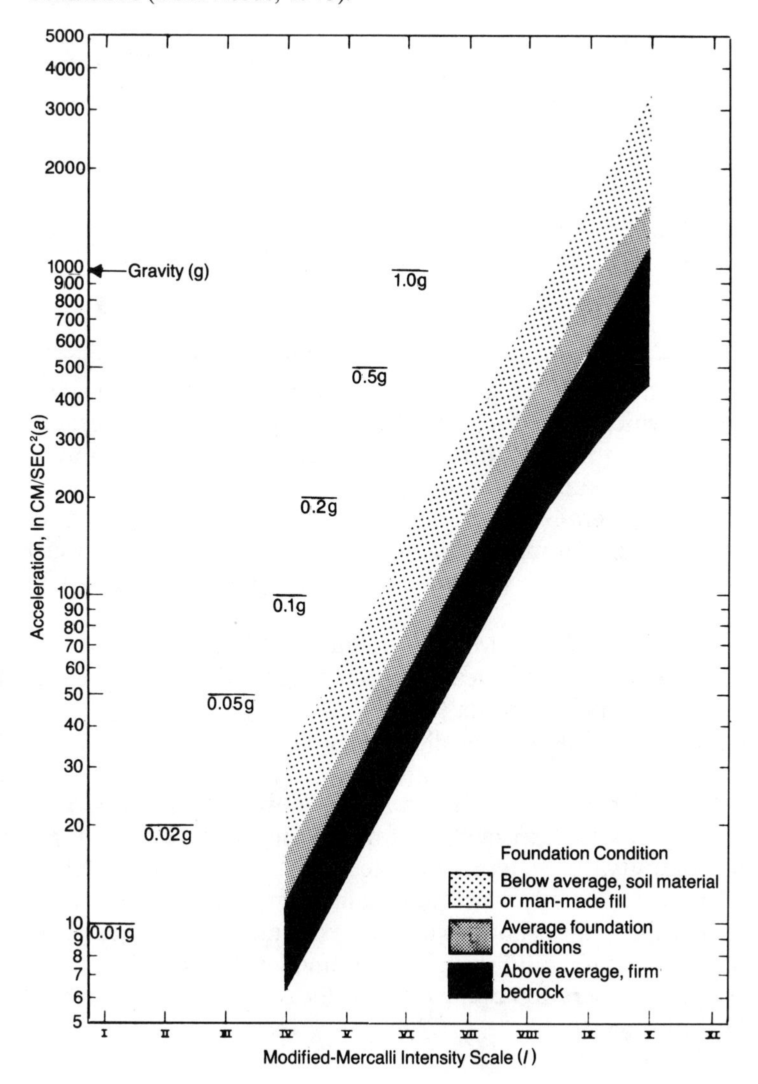

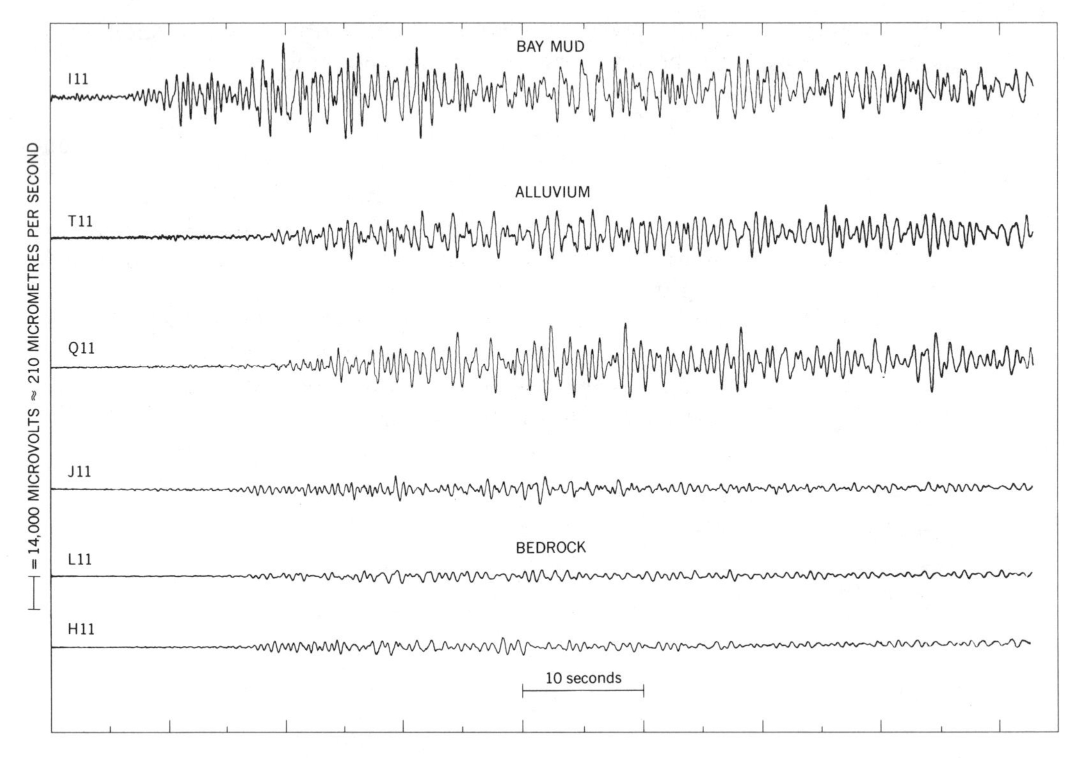


Figure 11.22 Recordings of horizontal ground motion in San Francisco produced by a nuclear explosion in Nevada (from Borcherdt et al., 1975).

The effect of bedrock on seismic waves is well known and described in the four case histories described above, as well as other earthquakes such as the 1775 Lisbon earthquake, which caused severe damage due to soft Tertiary strata under the city (Howell, 1959). It is difficult to quantify bedrock conditions; however, engineering geologists can supply qualitative data which can be useful. For example, Fig. 11.21 shows a generalized relationship between intensity and acceleration for three different foundation materials. Note that firm bedrock is expected to produce about one-third of the ground acceleration found in below average soil material.

The ground shaking of Bay Mud, alluvium, and bedrock near San Francisco was studied using comparative measurements of ground motion generated by distant nuclear explosions in Nevada (Fig. 11.22). Note that the Bay Mud had the largest displacements, followed by alluvium, and lastly bedrock (largely Tertiary sandstones).

Figure 11.23 is a map of the United States showing earthquake shaking hazards. The map shows the expected level of shaking, as expressed in percentages of the force of gravity, for shaking likely to occur at least once in 50 yr. All contours are expressed at the 90% probability level; in other words, there is only a 10% chance that the values shown would be exceeded in a 50-year period. Accelerations are estimated for solid rock; thus for unconsolidated deposits or artificial fill the maximum acceleration at a particular local area may be much larger than those predicted in Fig. 11.23.

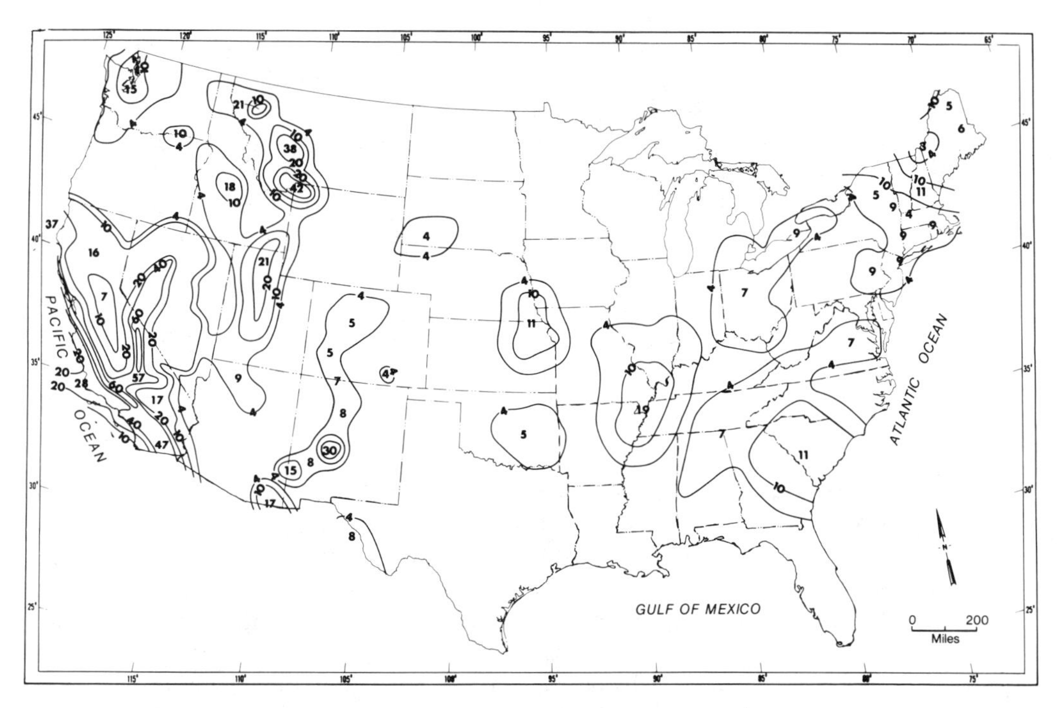


Figure 11.23 United States earthquake hazard map (from Algermissen and Perkins, 1976). Map shows expectable levels of earthquake shaking hazards. Levels of ground shaking for different regions are shown by contour lines which express (in percentages of the force of gravity) the maximum amount of shaking likely to occur at least once in a 50-year period.

## *11.3.2 Temporal Prediction*

Figures 11.19–11.21 illustrate general methods of statistically evaluating the time of occurrence of an earthquake. For example, from Fig. 11.23 every 50 yr a 0.19-*g* earthquake can be expected at St. Louis. Figure 11.23 is, of course, a very generalized map of a complex situation, and is not accepted by everyone. Costa and Baker (1981), for example, contend that a 100-yr historical record of earthquake activity for Colorado is unacceptable as a basis for seismic risk classification. Colorado appears on the risk map as a low seismicity area, and a nuclear power plant (St. Vrain) has been built under the assumption of only minor earthquake risk.

Although Fig. 11.23 portrays general (and contestable) earthquake statistical probability, of more immediate interest is *exactly* when will an earthquake occur? (Every once in a while a self-styled soothsayer will prophesy an exact date for the supposed destruction of California; there is little scientific value to these predictions.) Geophysicists have no crystal ball to predict the exact time of a future earthquake. The effectiveness of the following scientific methods for temporal earthquake predictions is being studied. These methods include:

a. Seismic wave velocity: Micro-earthquake swarms may be precursors of a large earthquake. It has also been observed that the difference in arrival time between P and S waves sometimes decreases from normal. It is believed that stress causes microfracturing about 1 year before an earthquake, so that the seismic velocity changes. Whitcomb et al. (1973) showed that in San Fernando the normal ratio $V_p/V_s$ is about 1.75, but decreases in the ratio of 6–15% are apparently caused by a drop in the P wave velocity because seismic waves (especially P waves) do not travel as fast through fissured rocks.

b. Radon gas: The fissure-fracturing dilatancy process theorized above may also explain phenomenon such as the reported release of radon gas just before some earthquakes. Scientists at the California Institute of Technology have experimented with the radon levels in about a dozen test wells scattered around southern California. Their studies indicate that the levels of radon concentration increased prior to several recent earthquakes in the area, including earthquakes of October 15, 1979 at Imperial Valley, January 1, 1979, at Malibu, and June 29, 1979, at Bear Lake. Presently there is not enough data to confirm this technique of earthquake forecasting.

c. Ground water levels: Ground water levels in wells have been known to drop prior to earthquakes in China, Japan, and California. It is theorized that prior to fault displacement, small cracks open up, allowing water to move into the new pores. Thus, ground water levels drop in the area. This phenomenon has been repeated in laboratory experiments (Sundaram et al., 1976). Kovach et al. (1975) strongly emphasizes the need for drilling additional observation wells on the San Andreas Fault zone to get information on static water levels and pressure gradients.

d. Crustal elevation changes: Dilatancy prior to an earthquake may cause widespread uplift in the area where an earthquake is about to happen. In the 1970s, widespread uplift occurred in the San Andreas Fault area about 65 km north of Los Angeles. The "Palmdale Bulge" included an area of 85,000 $km^2$. Precise surveys seemed to show that up to 0.45 m of uplift occurred between 1961 and 1977. The bulge seemed to stop growing between 1977 and 1980, and geophysicists are unsure of the significance.

e. Seismic activity: With an abundance of operating seismograph stations, it is possible to notice any unusual seismic activity. These anomolies may be precursors of an earthquake. Seismic gaps, i.e., places where anomolous quiescence occurs, can be harbingers of large earthquakes. It is interesting to note in this respect that China has 10,000 trained seismologists, over ten times the number of the United States. Presumably China's seismic field monitoring system is quite extensive.

f. Triangulation surveys: The accurate surveying of triangulation points in southern California shows that during 1979 Los Angeles got 5 cm closer to Navy Station, 240 km to the northeast. Although triangulation data presently is useful only for general tectonic information, some day the data may be used to pinpoint anomolies and predict earthquakes.

g. Electrical and magnetic field changes: To date, these phenomenon have not been effective precursors of earthquakes.

h. Animals: Canaries have been kept in cages deep in underground mines because miners believe they can sense when a cave-in may occur. Animals may be able to sense deep precursor breaking sounds or change in magnetic or electrical fields. There are any number of unconfirmed reports about strange animal behavior before earthquakes, although the data is inconclusive.

i. Strain-rate extrapolation: If sufficient reliable past rates of strain are known, a reasonable inference can be made as to future displacement. Consider the San Andreas fault, which we will assume has an average rate of displacement of 3 cm/yr (Lamar et al., 1973). In 1906, in the San Francisco area, the San Andreas fault moved about 6 m over a length of the fault that has not moved at all since. If the fault moves by lurching in 6-m increments, and no creep occurs along the fault, then a displacement should occur every 600 cm ÷ 3 cm/yr = 200 yr. Thus, the next earthquake in this location should occur in the year 2106. (Of course, this calculation is not very accurate because earthquakes are not known to be reliably rhythmic.)

Historic and prehistoric earthquake frequency data is not adequate to accurately predict the exact time of a future earthquake, however. Consider the 370-km-long Wasatch fault, which trends along the front of the Wasatch Mountains just east of Salt Lake City, UT. No large earthquakes have taken place along this fault since the settlement of the area in 1847. However, geologic evidence suggests that a moderate to large earthquake happens on the average about once every 50–430 years along the Wasatch fault zone (Hays, 1981). As early as 1884, USGS geologist G. K. Gilbert warned the residents of Salt Lake City of earthquake hazards (Wallace, 1980). Because the tectonic uplift of the Wasatch range occurs in small increments, characterized by alternate long periods of frictional resistance and then sudden sliding, Gilbert predicted that some day stress " . . . will overcome the friction, lift the mountains a few feet, and reenact on a more fearful scale the (1872 earthquake) catastrophe of Owens Valley." In 1933, structural geologist Bailey Willis was quoted in the Salt Lake Tribune (Cook, 1972) as warning:

> You live directly in an earthquake zone. The Wasatch Fault, which skirts the Wasatch Mountains, is a young, active fault. It has in the past given rise to many earthquakes and in the future will give rise to many more.

In conclusion, temporal earthquake forecasting at present is an uncertain science. As an example of this, consider the Chinese, who lead the world in earthquake forecasting studies. On February 4, 1975 an earthquake of magnitude 7.3 occurred near Haicheng in northeast China. Following a chain of precursors that had begun 5 yr earlier, the State Seismological Bureau was able to narrow down the probable time of occurrence of the future earthquake. Precursor observations included tilting of the land surface, changes in water levels in wells, strange animal behavior, and the migration of activity of small seismic disturbances (Hamilton, 1978). According to Tributsch (1982), before the earthquake hibernating snakes came out of their burrows, geese panicked, pigs butted walls, and " . . . there were many other anomolous phenomena. The great worker-peasant-soldier masses observed such conditions and played an important role in successfully predicting the earthquake." The final warning by Chinese seismologists came only a few hours before the quake destroyed or severely damaged 90% of the buildings in Haicheng. Thousands of lives were saved by the warning. This was the first successful prediction of a major earthquake, and a major feat in the annals of geophysics.

Over a year later, however, the Chinese were not as fortunate. On July 28, 1976, an earthquake of magnitude 7.8 occurred at Tangshan, a city also located in northern China. Tangshan was a large city built largely of reinforced brick. Precursors were not confirmed, and no warnings issued. More than 600,000 people were killed, and 75% of Tangshan's 916 multistory buildings, which were not built to withstand earthquakes, were flattened or severely damaged by the tremor; only four remained intact. This was possibly the greatest natural disaster in the history of mankind.

Kerry Sieh, a California geologist who studied displaced recent alluvial strata as a guide to the recurrence of earthquakes, said in 1981: "I think the chances of a great earthquake in southern California within the next 40 years are at least greater than 50%." That is not the kind of forecast in which civil defense officials can take much comfort. In fact, sociological research indicates that real estate values would plummet and political and socioeconomic problems would result following an official earthquake prediction in the United States. According to Costa and Baker (1981, p. 81): "It is not surprising then that the first successes in earthquake prediction should come from China. There, scientists and especially public officials are immune from lawsuits, insurance claims, and the need to appease the electorate."

In summary, the complexities of fault mechanics are too great to presently permit accurate temporal predictions about earthquakes. Gilbert (1909) said in reference to the time of the next 1906-type earthquake in San Francisco: "The hypothesis of rhythmic recurrence has no sure support from observation." His general assessment of the prospects for earthquake prediction apply as well today as it did in 1909. It is more likely that scientists would be able to pinpoint the location of an impending earthquake than to forecast the exact time of its occurrence: "The determination of times of danger belongs to the indefinite future. It still lingers in the domain of endeavor and hope."

## 11.4 Man-made Earthquakes

Man does not *make* earthquakes except by nuclear explosions. But in many cases man has changed conditions slightly at a place where tectonic stresses already existed, thereby supplying the trigger for the movement along an active fault (Rothe, 1973; Judd, 1974; Milne, 1976). The change in conditions provided by man can be brought about by the creation of a reservoir which causes isostatic loading. More commonly, the change in conditions is the increase in fluid pore pressure, which can be brought about in two ways, 1) artificial recharge by wells and 2) dam construction, and the resulting hydrostatic increase caused by the reservoir.

### *11.4.1 Artificial Recharge by Wells*

When it was found in 1961 that "evaporation ponds" were causing ground water to be contaminated by chemical warfare liquids at the U.S. Army Rocky Mountain Arsenal near Denver, CO, the decision was made to dispose of the liquids by drilling a deep disposal well. The well was drilled in 1961 to a total depth of 3674 m (Fig. 11.24A). The hole was drilled into largely Cretaceous and Paleozoic sedimentary rocks which were cased off. The lower 23 m was open to Precambrian granite (Evans, 1970).

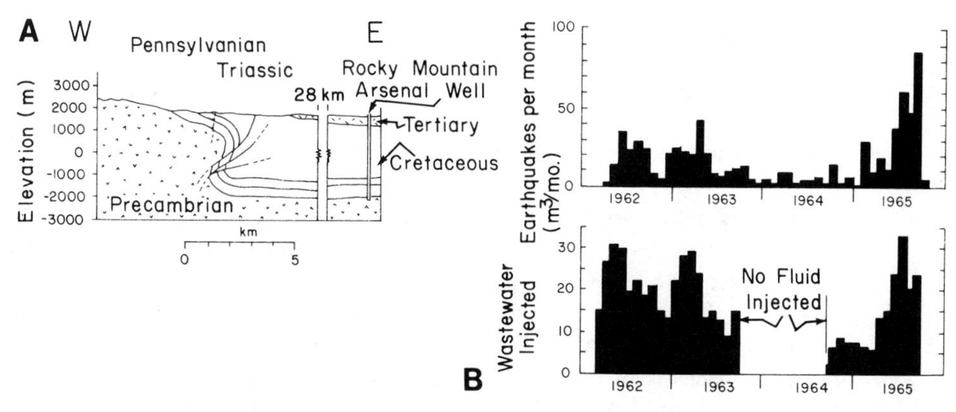


Figure 11.24 Induced seismicity at Rocky Mountain Arsenal, Denver, CO. **A**. Earthquake frequency and pumpage rates (after Healy, 1968). **B**. Geologic cross section (after Evans, 1970).

Injection of fluids (including nerve gas) began in 1962 at a rate of about 700 $m^3$/d until September 1963 when pumpage ceased. Then injection began a year later at about 250 $m^3$/d. Within a few weeks after injection started, Denver had its first earthquake. Since the start of the injection, 710 Denver area earthquakes were recorded after an elaborate seismic station was set up by the USGS. The majority of these earthquakes had epicenters within 8 km of the Rocky Mountain Arsenal well. The volume of fluid injected correlates to the frequency of earthquakes (Fig. 11.24B). The Rocky Mountain Arsenal well earthquake epicenters plot along a line extending northwesterly from the well. Apparently movement along a fault was propagating in that direction.

The fluid pore pressure in the Precambrian rocks was raised by the injection of liquids (Evans, 1970). Near the base of the hole (3356 m), the natural water pressure was measured to be $28.5 \times 10^6$ N/$m^2$; the brackish well water had a measured density of 1.06 g/$cm^3$. Using a hydrostatic gradient of 10,300 N/$m^2$ per m, the very act of filling the hole with water raised the pressure near the bottom of the hole to $34.6 \times 10^6$ N/$m^2$, or an increase of $6.1 \times 10^6$ N/$m^2$. In addition, pumping at the surface was used ($6.9 \times 10^6$ N/$m^2$ pressure) throughout much of the injection. Therefore, the pressure near the bottom of the hole was raised a total of $13.1 \times 10^6$ N/$m^2$. Apparently this rise in fluid pressure within the Precambrian rocks allowed movement to take place along a fault. Injection stopped after 1965. A few months later three sizeable earthquakes (magnitude 5.2, 5.1, and 5.0) occurred (Healy, 1968). Then all earthquakes in the Denver area ceased.

A water-injection program at the Rangely Oil Field in Colorado also triggered earthquakes (Munson, 1970; Raleigh et al., 1976). Waterflooding (secondary recovery) at the oil field started in 1957 in an effort to recover the maximum amount of oil from the reservoir. A seismic study showed epicenters of earthquakes centered in the reservoir. U.S. Geological Survey experiments demonstrated that pumping water underground until fluid pressures reached $28 \times 10^6$ N/$m^2$ would increase the number of earthquakes in the area from 1 or 2 per month to 30 or 40 per month. The study showed that earthquakes could be turned off and on.

### 11.4.2 Dam Construction

The first earthquake activity associated with dams was noticed at Hoover Dam, behind which the enormous reservoir called Lake Mead is backed up. The dam started impoundment of water in 1935. About 1 year later, a magnitude 5 earthquake occurred. For a number of years thereafter, small local earthquakes showed a close correlation in size and frequency with volume of water stored in the reservoir (Carder, 1945, 1970). It is believed that the tremendous weight of the reservoir water ($40 \times 10^9$ tons) acts on the earth's crust, causing some isostatic unbalance. Epicenter plots coincide with late Cenozoic basin and range faults, which appeared to have been reactivated and allow for settlement. It is also likely that the triggering of the earthquakes was facilitated by an increase in pore pressure in a rock system stressed by tectonic and reservoir weight forces.

At Koyna Reservoir, in eastern India, a magnitude 6.5 earthquake occurred on December 11, 1967 (Gupta, et al., 1969). It happened in an area of basaltic rocks, in one of the world's least seismically active areas. Additional smaller quakes had been noticed during the years right after filling occurred, between 1962 and 1967. The big earthquake happened just 5 years after the giant Koyna Dam was built. The epicenters of all the earthquakes were under the reservoir. The Koyna earthquake of 1967 killed 177 people and caused considerable loss of buildings within a 300 km radius. The dam itself was not destroyed, but did sustain damage. Seismic activity has continued to the present and correlates to changes in reservoir level.

Other dam earthquakes have been reported, similar to Lake Mead and Koyna Reservoir in that shallow focus earthquakes occurred shortly after construction. In 1966, at Kremasta, Greece, over 740 shocks occurred in a 6-month period (Gupta and Rastogi, 1976). A magnitude 6.2 earthquake caused one death and extensive damage. The large earthquake occurred the same year the Kremasta reservoir was impounded. In 1963, a magnitude 6.1 earthquake accompanied the filling of Kariba Reservoir, along the Rhodesia-Sambia border, in what was an aseismic area. The frequency of hundreds of smaller earthquakes at Kariba Reservoir correlates with the reservoir water level (Snow, 1982).

Table 11.4 is a list of reservoirs, where induced seismicity is believed to have occurred. One particularly hazardous situation is the Oroville Dam, completed in 1968 on the Feather River in California (see Fig. 8.25). On August 1, 1975, a magnitude 5.7 earthquake occurred near the dam. The horizontal ground acceleration was 0.13 g, with significant motion lasting 3 s. The Oroville dam problem has received considerable attention because of the large population that lives below the earthen dam. The recently completed Folsom Dam nearby has also generated some concern and considerable discussion by the Association of Engineering Geologists. The failure of this dam would send a wall of water into Sacramento.

On the positive side, it should be mentioned that, because earthquakes can be triggered by man, there is the possibility that injection wells may be deliberately used to set off small, harmless earthquakes along major faults (National Research Council, 1971). In other words, stress could be relieved before it builds up to the point where a catastrophic earthquake could occur. In theory, it may someday be possible to prevent catastrophic earthquakes such as might occur on California's San Andreas Fault by inducing gradual fault movement, releasing seismic energy

Table 11.4 Reservoirs with Induced Seismicity

| Location (dam–country) | Dam height (m) | Capacity ($m^3 \times 10^9$) | Basement geology | Date impounded | Date of first earthquake | Seismic effect |
|---|---|---|---|---|---|---|
| L'Oued Fodda, Algeria | 101 | 0.228 | Dolomitic marl | 1932 | 1933 | Felt |
| Hoover, USA | 221 | 38.3 | Granite, Precambrian shales | 1935 | 1936 | Noticeable (M = 5) |
| Talbingo, Australia | 162 | 0.92 | | 1971 | 1972 | Seismic (M < 3.5) |
| Hsinfengkiang, China | 105 | 11.5 | Granites | 1959 | | High activity (M = 6.1) |
| Grandval, France | 78 | 0.29 | | 1959 | 1961 | Intensity V in 1963 |
| Monteynard, France | 130 | 0.27 | Limestone | 1962 | 1963 | M = 4.9 |
| Kariba, Rhodesia | 128 | 160.0 | Archean gneiss and sediments | 1958 | 1961 | Seismic (M = 5.8) |
| Vogorno, Switzerland | 230 | 0.08 | | 1964 | 1965 | |
| Koyna, India | 103 | 2.78 | Basalt flows (Deccan Trap) | 1962 | 1963 | (M = 6.5) 177 people killed |
| Benmore, New Zealand | 110 | 2.04 | Greywackes, argillites | 1964 | 1965 | Significant (M = 5.0) |
| Kremasta, Greece | 160 | 4.75 | Flysch | 1965 | 1965 | Strong (M = 6.2) 1 death, 60 infuries |
| Nurak Tadzik, USSR | 317 | 10.5 | | 1972 | 1972 | Increased activity (M = 4.5) |
| Kurobe, Japan | 186 | 0.199 | | 1960 | 1961 | Seismic (M = 4.9) |
| Oroville, USA | 236 | 4.37 | Slate | 1968 | 1975 | M = 5.9 |

After Bolt et al., 1977.

in small doses (Raleigh, 1977). USGS geophysicists Barry Raleigh and James Deitrich proposed drilling three deep holes about 5 km deep and 500 m apart across the fault. Outer holes would be pumped out, effectively strengthening the surrounding rock. The middle row would be pumped into, increasing fluid pressure to the point of failure. Other scientists felt that such experiments should be undertaken with caution, lest they trigger a large quake. Control of the San Andreas Fault is still a long way off, and much more research needs to be done.

## 11.5 Planning for Earthquakes

People will continue to live in earthquake-prone areas. How can they best be protected? What building codes should be adopted? There are no universal answers to those questions. However, the following examples from California illustrate some ameliorating conditions (see also Chapter 13).

At any specific site, a first step in assessing earthquake hazards is the inspection of a general geological map (such as Fig. 11.25). Site-specific information is needed, such as bedrock type, topography, proximity to faults, landslide history, etc. (Data necessary to evaluate earthquake hazards in the United States are available in the environmental impact statements for large projects such as the construction of nuclear power plants.)

Some communities in California are cognizant of active faults. Recognizing that the San Andreas Fault is an active fault, for example, the city of Portola Valley, CA, required buildings set back from active fault traces. All new building construction is prohibited within a 31-m-wide zone, and structures with occupancies greater than a single family dwelling are required to be 38 m from an active fault trace (Nichols and Buchanan-Banks, 1974).

Many communities are less concerned. In 1981, Stanford University geologist Richard Jahns said:

> . . . There is an obvious risk in erecting homes or other structures athwart the most recent traces of past movements along active faults. Yet this is precisely what has been done, especially along the San Andreas fault southward from San Francisco and along the San Andreas and San Jacinto faults in the San Bernardino area of southern California. Moreover, this kind of "gamble in residence" is being taken more and more often, and in areas where the positions of faults are well known; it is particularly distressing to note the number of schools represented among installations that some day might serve as reference features for large-scale shear.

A second step is a detailed assessment of the subsurface conditions at the site in question. For example, earth materials such as saturated, loose, granular material (silt, sand) are subject to liquefaction, where earthquake deformation transform stable material into a fluidlike state (Nichols and Buchanan-Banks, 1974). Other materials are subject to landslides. Unconsolidated surficial deposits have greater wave amplitude than bedrock, and earthquake vibrations last longer in them. All surficial deposits should be regarded as suspect. During the 1951 earthquake near Bakersfield, CA, for example, a seismograph located on sedimentary fill recorded seismic waves with amplitudes four times greater than those recorded by nearby seismographs located on bedrock. Another example of a poor foundation is exemplified by the large apartment buildings in Nigata, Japan, which were tilted as much as 80° on their sides (with little structural damage) as a result of liquefaction of a shallow sand bed during an earthquake.

An assessment of earthquake hazards involves the assessment of all related geological hazards. For example, I was involved in the evaluation of the seismic safety of a large medical center in San Diego, CA. The parking lot and part of an attached wing to the structure rested on artificial fill at the edge of a ravine. Landslide scarps were beginning to appear in the fill. Even if an earthquake never occurred, the building was in a precarious position.

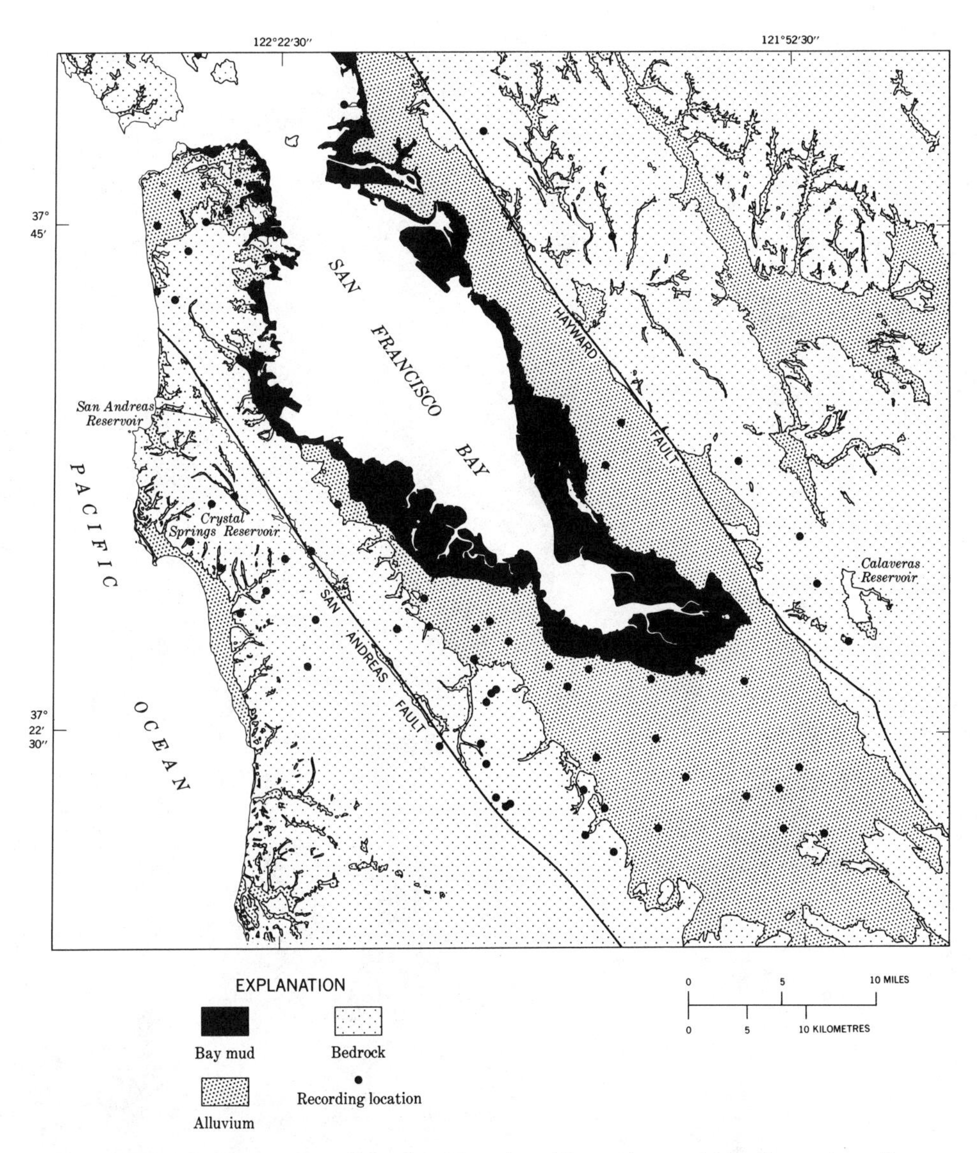


Figure 11.25 Geologic map of the San Francisco Bay region and locations where three components of ground motions generated by nuclear explosions were recorded (from Borcherdt et al., 1975).

The U.S. Geological Survey (Blair and Spangle, 1979) has initiated the seismic zoning of lands in the San Francisco Bay region. Figure 11.25 shows the geology of this region. Note the presence of Holocene Bay mud, and the location of the San Andreas Fault. A postulated 6.5 magnitude earthquake was used to develop probable zones of destruction in San Francisco (Fig. 11.26). The area within a kilometer of the fault is zoned very violent. Areas of Bay Mud such as the airport (lower right of Fig. 11.25) are zoned violent. Areas of bedrock are safer.

Recourse for the San Francisco seismic zonation map was made to the 1906

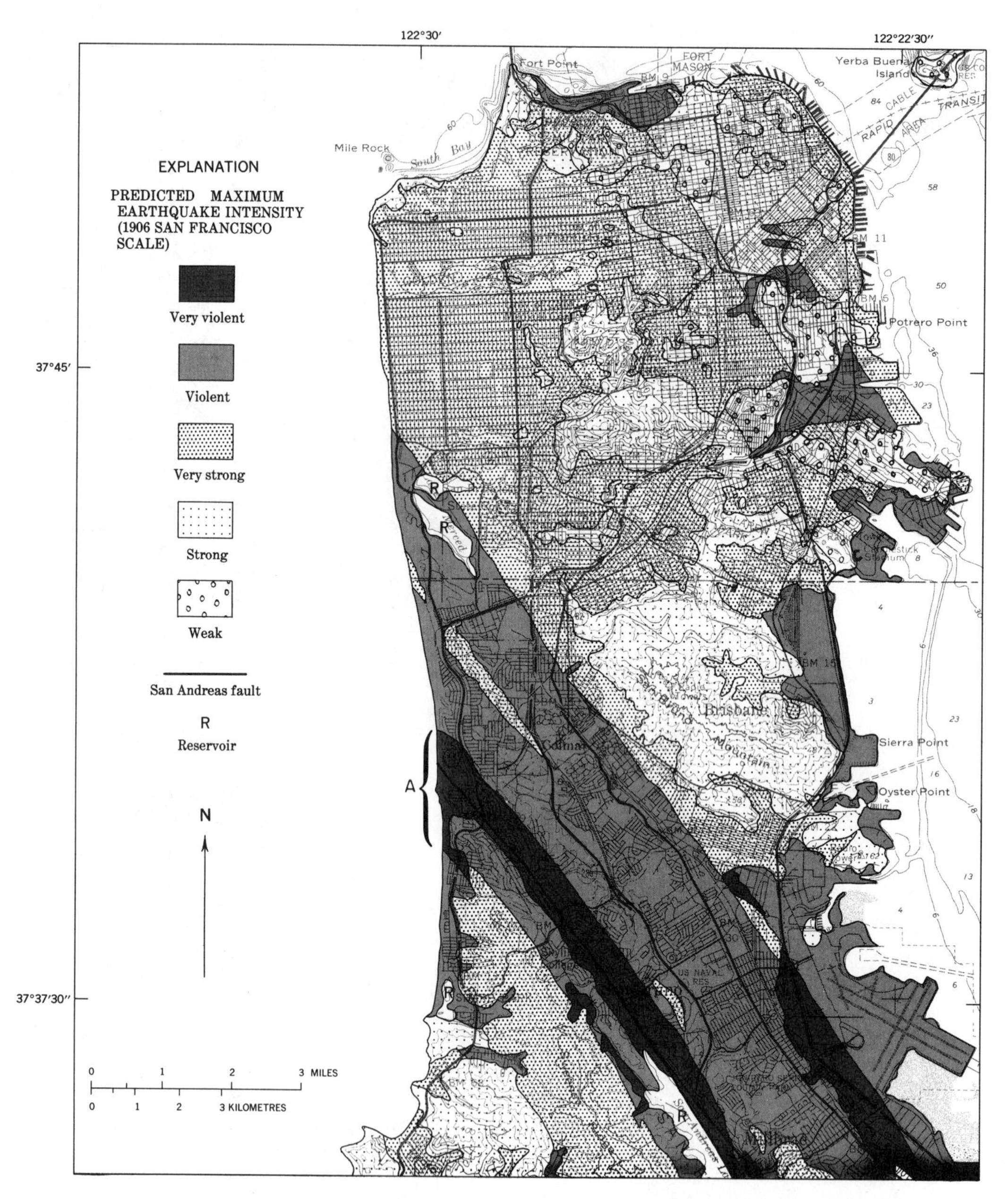


Figure 11.26 Maximum earthquake intensities predicted for San Francisco (from Borcherdt et al., 1975). Each value is the maximum of those predicted assuming a large earthquake of the San Andreas fault or the Hayward fault. The intensity values are predicted from the empirical relations based on only the good intensity data for the 1906 earthquake together with a generalized geological map. **A** indicates location of Figure 11.28.

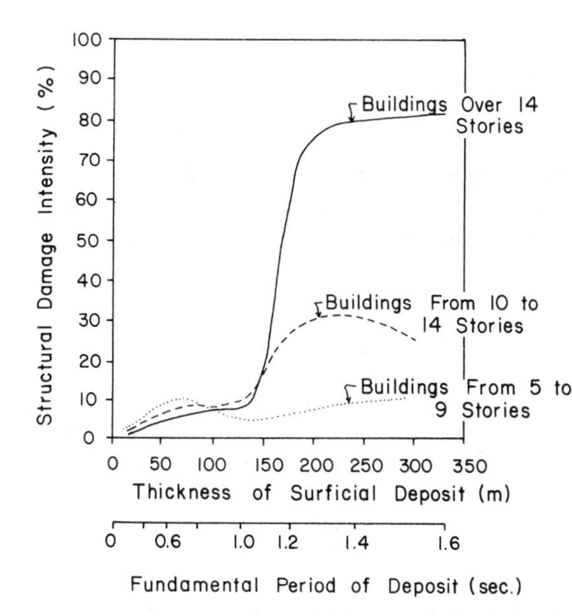


Figure 11.27 Graph showing the relationship between structural damage intensity to buildings of different heights, as related to soil depth and the computed fundamental period of vibration of soil deposits (after Nichols and Buchanan–Banks, 1974). Note that when the fundamental period of the underlying surficial deposit is short, buildings from 5–9 stories are damaged most; for longer periods, damage to higher structures increases.

earthquake (Borcherdt, 1975): " . . . in the Telegraph Hill area of San Francisco where rock is exposed at the surface, the effects of the earthquake were weak, with occasional fall of chimneys and damage to plaster partitions, plumbing and the like. But at a distance of less than one-fourth of a mile, in an area underlain by artificial fill and water-saturated mud, the effects of the earthquake were violent, with fairly general collapse of brick and frame structures. . . ." The report indicated that another 1906-type quake could be catastrophic.

There is no standard procedure for assigning the probable peak motions or duration of shaking of a future earthquake for any given site. Intensity can be attenuated from an earthquake source to a site by any number of intensity-attenuation charts. The art has been developing rapidly. According to Krinitzsky and Marcuson (1983, p. 264): "The safest general approach is to base one's selection of ground motions on a large catalogue of observed data. The trends in the data should be precisely projected where the data are insufficient and the values should be bracketed or encouraged in such a way that there will be as few surprises as possible were an earthquake to occur."

Many civil engineers emphasize that it is not earthquakes that kill people, but the failure of buildings. For example, five multistory buildings completely collapsed, during the 1967 Caracas, Venezuela, earthquake; 340 people lost their lives in this magnitude 6.3 earthquake. Structural engineers maintain the best earthquake strategy is not to try to control earthquakes, but simply to build sound structures on relatively safe sites. In California and Japan, where earthquake activity is an accepted way of life, some building codes include provisions for earthquake-resistant construction. In general, tall buildings should be avoided. For example, Figure 11.27 shows that in moderately deep soils buildings greater than 14 stories tall would suffer 80% structural damage, while 5–9 story buildings would suffer about 10% structural damage. One useful method of assessing potential building failure stems from knowledge of the fundamental period of the building relative to the thickness and properties of the surficial deposit upon which it rests. A damaging resonance commonly develops where the "fundamental period" of the building coincides with the natural resonance of the foundation material.

It has long been recognized that brick and/or adobe buildings have poor resistance to earthquakes (Page et al., 1975a). For example, Charles Darwin described the aftermath of the 1835 earthquake in Chile and the manner in which brick houses collapsed (Darwin, 1962):

> The next day I landed at Talcahuane, and afterwards rode to Concepcion. Both towns presented the most awful yet interesting spectacle I ever beheld. . . . In Concepcion, each house, or row of houses, stood by itself, a heap of ruins; but in Talcahuano, owing to the great (earthquake) wave, little more than one layer of bricks, tiles, timber, with here and there part of a wall left standing, could be distinguished. . . . Mr. Rouse, the English consul, told us that he was at breakfast when the first movement warned him to run out. He had scarcely reached the middle of the courtyard, when one side of the house came thundering down. He retained presence of mind to remember, that if he once got on top of that part which had already fallen, he would be safe. Not being able from the motion of the ground to stand, he crawled on his hands and knees; and no sooner had he ascended this little eminence, than the other side of the house fell in, the great beams sweeping close in front of his head."

As of 1982, structures of 16 stories or more must be built on the basis of a computerized dynamic analysis of their potential behavior during an earthquake, according to a Los Angeles building code. These properly designed tall buildings are safer than brick buildings, which are virtually certain to fall down, according to George Housner, Professor Emeritus of Engineering at the California Institute of Technology. There are some 8000 unreinforced brick buildings in Los Angeles County. Straps have been added to freeway overpass sections to prevent them from falling down as in the 1971 San Fernando earthquake. Nuclear power plants and dams are being designed to withstand the strongest of all possible earthquakes, but to design everything to nuclear reactor standards would be too expensive.

Steel skeleton buildings are superior to other types. Older masonry construction is hazardous; most were designed by masons with no knowledge of earthquake engineering. Daniel Schodek of the Harvard Univeristy Architectural Department stated that Boston's masonry buildings would crumble in an earthquake. Most American cities would face a similar dilemma.

Reinforced concrete, although behaving poorly in the 1972 Managua earthquake (Fig. 11.17), the 1967 Caracas, Venezuela, earthquake (Leveson, 1980) and elsewhere, can be designed to reduce damage. Metal hoops can be placed around reinforcing rods to keep them together. Steel buildings can be secured by more bolts or welds. Bearing walls, which act to resist twisting caused by lateral earth displacements, are being encouraged in seismic areas. Base isolation, large rubber shock absorbers placed between a structure and the foundation, has been used in several dozen new buildings, a number of bridges, and two French-designed nuclear power plants. The nine-story Oak Knoll Navy Hospital in Oakland, CA, is less than 1 km from the Hayward Fault; in 1985 buttresses were being added at the end of each wing.

The successful use of shear walls as a means of stiffening buildings and resisting earthquake forces has been observed in many earthquakes. A shear wall acts to stiffen the structure so that the deflection of upper floors is reduced. Shear walls typically include walls with no openings such as central elevator corridors or outer walls of buildings with no windows. Green (1978) presents theoretical and practical aspects of building design using shear walls.

The striking shape of San Francisco's Transamerica Building is an example of modern earthquake-resistant architecture and construction. Built in 1975, the triangular trusses and precast concrete-clad steel columns at the base of the 255 m high structure are designed to withstand more than twice the earthquake-induced stress anticipated, according to the city's building code. In a major quake, the building is expected to sway only 0.6 m, compared with the normal sway of a building of similar height of as much as 0.9 m. In many places building codes are formulated in terms of requiring resistance to acceleration of 0.01 g, or some other specific fraction (Lamar and Lamar, 1973).

Historic review of dam safety in earthquakes indicates dams are fairly stable. Damage to earthen dams, such as the Lower Van Norman Dam, California, during the 1971 earthquake, was due to liquefaction problems. The San Andreas Dam, California, built in the 1860s of gravel fill, survived the 1906 earthquake with 2.5 m of fault offset in its abutment. Dams can be engineered to resist acceleration (Wiegel, 1970). For example, the site of the proposed Auburn Dam, California, has been studied extensively by the U.S. Bureau of Reclamation. Originally designed as a concrete arch dam, it was redesigned in 1979 to withstand a fault displacement of 23 cm because it was observed that a 1975 earthquake at the nearby Oroville occurred along part of the presumed inactive fault system that passes beneath the Auburn site.

Despite the fact that earthquake damage can be reduced by proper design and construction, present building codes are insufficient in many places. Safety factors need to be increased on structures in earthquake-prone slopes and sensitive soils. For example, the October 10, 1980, earthquake (magnitude 7.3) in El Asman, Algeria, killed 20,000 people. Prior building codes had specified a 10% extra stiffening to resist earthquake stress. Yet the buildings failed. F. K. Farma, a civil engineer from Imperial College, London, stated that " . . . for that quake they needed 50 percent."

Despite the efforts of geologists, seismologists and engineers, cities such as San Francisco are essentially waiting for the catastrophe to happen. Scientists have repeatedly warned of the dangers, but they are largely ignored. Urban sprawl continues right across the San Andreas fault (Fig. 11.28), and homes are still being built at sea level on artificial fill in the Bay area. The reason is that in the San Francisco area, as in most parts of the country, planning has been primarily on the dollar basis. Many young people and recent immigrants to California are actually completely unaware of the hazards, or that they live, for example, in an unreinforced masonry building. April 18, 1906, has faded into history.

"Apathy reigns supreme . . .," says Philip Day, head of San Francisco's emergency services office; " . . . apathy reigns, and the only thing that will change it is a nice 7.5 or 7.0 earthquake."

Many readers of this book will probably live to hear about that earthquake.

Figure 11.28 Aerial photographs of San Francisco (from U.S. Dept. of Interior, 1969). **A**—photograph taken in 1956. **B**—photo taken in 1966.

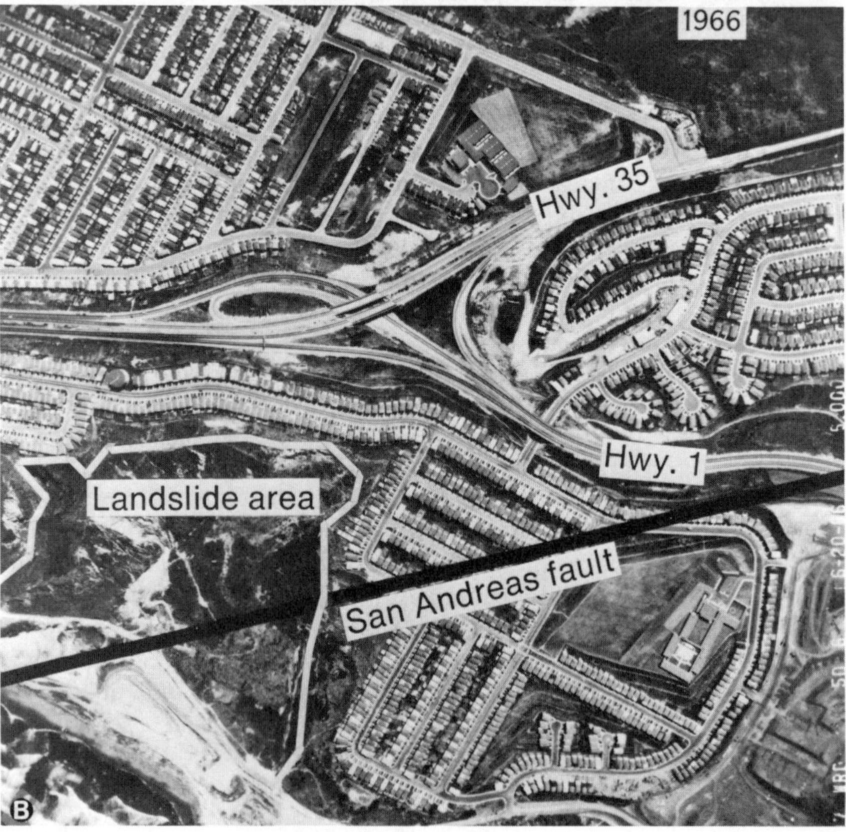

## Problems

1. Determine distance to epicenter using the three seismograms shown below (from Nehru et al., 1972). Using Fig. 11.7 to determine epicenter distance by using the interval between P and S waves. Plot the location of the epicenter on map. Determine latitude and longitude from the location of the following map.

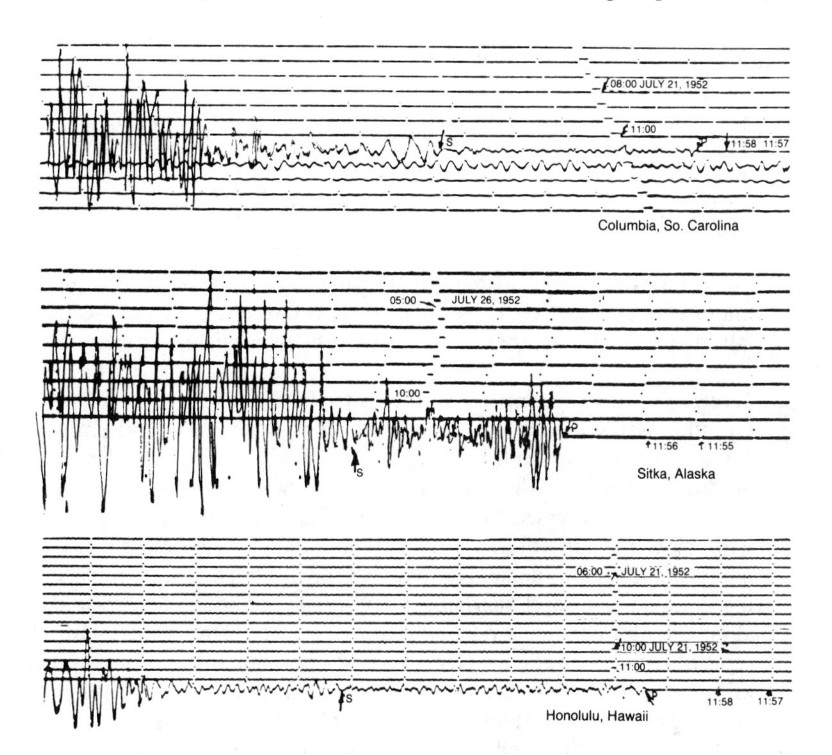


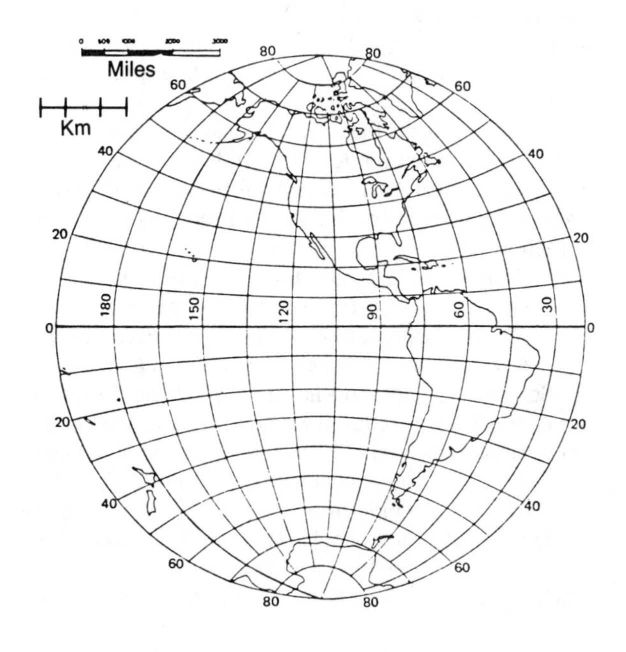

**2.** On arithmetic graph paper, plot world population (ordinate) versus time (abscissa) for 100 year increments from 1000 A.D. to present:

| Year | Population ($\times 10^6$) |
|---|---|
| 1000 | 275 |
| 1100 | 306 |
| 1200 | 348 |
| 1300 | 384 |
| 1500 | 446 |
| 1600 | 486 |
| 1700 | 600 |
| 1800 | 906 |
| 1900 | 1608 |
| 1980 | 4500 |

On same paper, plot a histogram showing the number of deaths (per 100 yr) due to major earthquakes (see Table 11.1). (Use two different vertical scales, one for population and one for deaths). Comment on the relationship between the two curves.

**3.** (Tongue-in-cheek example, modified after B. F. Howell, Jr.)
Grandpa Gries was sitting on the rear stoop of his farm 130 km east of Wretched City. He did not notice the shock. Several of his neighbors to the west did, however. So did his daughter who was a freshman at Southern Hills Institute of Technology. She was getting dressed to go to a dance, and when the shaking began she panicked with fright and ran out of the dorm without all her clothes, and caused more excitement than the earthquake. Her date, who commuted from Wretched City, was driving south on highway 17 and noticed nothing until he saw fissures and cracked ground along the west bank of Wretched River, about 15 km north of S. Hills Tech. Even the sight of fallen chimneys in Hermosa had not made him suspicious. Persons from 80 km south of S. Hills Tech generally had not noticed it either. Neither had people from Hangmans Hill or the Black and Blue Hills.

J. Martin in Wretched City first noticed the dishes rattling in the china closet. W. Roggenthen, who lived 16 km south of town, claimed his grandfather clock stopped. Reverend Redden from East Wretched reported a chandelier had swung in his chapel while he was praying. One of his parishioners 25 km east of east Wretched was busy feeding his cows and had not noticed it, although his wife who was sewing had felt it clearly.

At Thunderhead Falls, J. Mickelson noticed excess water flowing over the dam three times at 2½ min intervals, but did not notice the shock. The ticket cashier at the movies had felt it. Some patrons from the south end of town had heard creakings or even seen objects swing, but persons living north of the falls had, with few exceptions, not felt anything.

In North Wretched, old Mrs. Rich was awakened from her afternoon nap, but her cook thought it was just a truck rumbling along highway 71. Her husband noticed it, but probably because he was lying on his back under his car on the highway 40 km west of N. Wretched. The night watchman at Robert's Hardware Co. 32 km east of the School of Minds said he thought all the tinware was coming off the shelves, it rattled so. They were having a good time at the old barn dance down at the VFW 8 km west of the School of Minds, and not a soul noticed it except the couple who were "sitting" in a car outside; and they did not realize it was an earthquake at the time. Although many people in town observed it, no one was frightened and no damage was reported. Several people estimated the shaking lasted over 1 minute.

The most annoyed person was Sam Seism, who was an amateur seismologist at Blackhawk. In all his excitement on feeling the quake (only a few other people in town felt it), he put his seismogram into the fixer instead of the developer and ruined it, so he doesn't know if his new seismograph records anything. During the following month, Sam visited all the towns in the valley gathering information on the intensity of the shock using the modified Mercalli scale.

Since Sam heard you were taking a course in engineering geology, he gave you the above information and asked you to make a map showing isoseismal lines, the probable location of the epicenter, and approximate Richter magnitude.

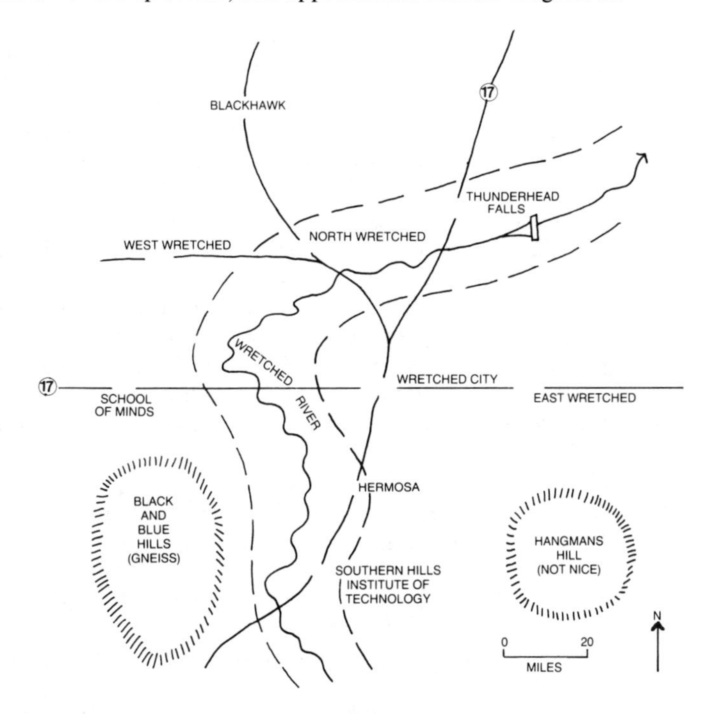


4. How much more energy does each integer on the Richter magnitude scale represent?

5. Consider Eq. (11.2). Calculate the energy released by the 1906 California, 1964 Alaska, 1971 San Fernando, and 1972 Managua earthquakes. Compare to the energy released in an atomic bomb where $E = mc^2$ and about 1 g of uranium is converted to energy.

6. Use Eq. (11.3) and Fig. 11.4 to determine approximate magnitude of the 1906 San Francisco earthquake.

7. Assume the seismogram for the 1939 Turkey earthquake (Fig. 11.6B) was recorded by a Wood-Anderson torsion seismometer 100 km from the epicenter. Determine the earthquake magnitude, assuming the seismogram has its original scale.

8. Use the Fig. 11.8 nomogram to determine the magnitude of an earthquake if the epicentral distance is 200 km and the amplitude is 10 mm.

9. Determine how much time would exist between the arrival of a seismic P wave and a tsunami at Los Angeles, CA, if a submarine earthquake occurred just off the coast of Kodiak Island, AK. (Hint: see Fig. 10.20A for tsunami velocities. Use a globe to obtain accurate measurements.)

**10.** Consider the slowest and fastest postulated slip rates for the San Andreas Fault.

**a.** Predict the date of the next large earthquake in the zone of the 1906 earthquake if stress builds up so that the fault displacement is the same as the displacement which caused the 1906 San Francisco earthquake. (Hint: solve for range of dates using maximum slip rates.)

**b.** What other earthquake hazards exist in San Francisco beside the specific event described in *a*?

**11.** In November 1981, the Nuclear Regulatory Commission voted to suspend the license granted to the Diablo Canyon nuclear plant near San Luis Obispo, CA. The reason was that construction diagrams for Unit 2 were mistakenly used in analysis for Unit 1, and the units are not quite the same distance from the Hosgri Fault, about 4 km offshore. the license is to be suspended until the $2.3 billion plant passed a series of seismic tests.

**a.** Use Eq. (11.4) to determine acceleration (in *g*'s) at the reactor for 1) a magnitude 6 earthquake at 4.0-km distance, 2) a magnitude 7 earthquake at 4.0-km distance, and 3) a magnitude 6 earthquake at 3-km distance.

**b.** Which parameter has a greater effect on acceleration at Diablo Canyon nuclear plant, small changes in distance, or small changes in earthquake magnitude?

**c.** Considering the answer to *b,* is it reasonable to be greatly concerned about small changes in the location of a nuclear reactor relative to an active fault when little is known about the magnitude of an earthquake which could be generated by this fault?

**12.** Consider 12 granite parallelepiped monuments having square cross sections of 1 m $\times$ 1 m, and heights of 1 m, 2 m, 3 m,. . . . 12 m. The monuments rest on a flat granite base where the coefficient of friction is 0.2. The density of the granite is 2.7 g/cm$^3$.

**a.** Determine which obelisks would slide on their bases due to an earthquake having a horizontal acceleration of 0.1 g.

**b.** Determine which obelisks would be tipped over by the same earthquake.

**13.** See Problem 8.12. Assume an earthquake causes a horizontal acceleration of 0.1 g at the dam.

**a.** Calculate the overturning moment about *A,* assuming the reservoir is empty. What is the F.S.?

**b.** If the reservoir is full, would the F.S. be increased or decreased?

# 12

# Geophysical Techniques

---

Buy yourselves stout shoes, get away to the mountains, search the valleys, the deserts, the shores of the sea, and the deep recesses of the earth ... In this way, and no other, will you arrive at a knowledge of things.

Peter Severinus, 1571

---

## 12.1 Engineering Seismology

Life is more complicated in the Twentieth Century than it was in 1571. Although few geologists would argue that stout shoes and field work are essential, there are sophisticated geophysical techniques that are also useful. Nowadays everyone has a black box.

The purpose of this chapter is to explain the basic theory and application of the most common kinds of geophysical techniques used in engineering geology. For further information the reader may consult Dobrin (1960, 1976), Trantina (1962), Heiland (1963), Parasnis (1966), Johnson (1966), Davis and DeWeist (1966), Morley (1970), Mooney (1973), Zohdy et al. (1974), Kesel (1976), Telford et al. (1976), Aki and Richards (1980), Griffiths and King (1981), and Fitch (1981).

### *12.1.1 Theory*

From Chapter 4, it was shown that there are a number of relationships between the elastic constants of rocks. Two other important relationships are the modulus of rigidity ($u$) and the bulk modulus ($k$). These parameters are referred to as Lamé's constants:

$$u = \frac{E}{2(1 + v)}, \tag{12.1}$$

$$k = \frac{E}{3(1 - 2v)}, \tag{12.2}$$

where:

$u$ = rigidity (shear) modulus ($N/m^2$),

$E$ = Young's modulus of elasticity ($N/m^2$),

$\upsilon$ = Poisson's ratio (dimensionless),

$k$ = bulk modulus ($N/m^2$).

The bulk modulus is defined as the stress imposed on a material divided by the corresponding proportional change in volume. Consider, for example, the change in volume for rock of density 3 $g/cm^3$ if it is buried at 10 km depth (from Howell, 1972). Each 1 $m^3$ of rock causes a stress of $(3 \times 10^3\ kg/m^3)\ (9.8\ m/s^2) = 2.94 \times 10^4\ N/m^2$ on the rock below. Therefore, the total stress at a depth of 10 km = $2.94 \times 10^8\ N/m^2$. If the bulk modulus is $10^6$ atmospheres (note: 1 atmosphere = $1.01 \times 10^5\ N/m^2$), then the change in volume of the material will be:

$$\frac{\Delta V}{V} = \frac{\Delta\ \text{stress}}{k},$$

$$= \frac{2.94 \times 10^8\ N/m^2}{1.01 \times 10^{11}\ N/m^2} = 2.91 \times 10^{-3}.$$

Young's modulus and the bulk modulus are not equal because they describe different properties. Young's modulus measures the change in length; the bulk modulus measures the change in volume.

As in the case of propagation of oceanic waves (Chapter 10), the wavelength ($\lambda$) of a seismic wave is related to velocity ($V$) and frequency ($n = 1/T$) as follows:

$$\lambda = \frac{V}{n}. \tag{12.3}$$

The theory of elasticity furnishes the following relationships among the constants describing the elastic behavior of an isotropic and homogeneous medium obeying Hooke's law (see Howell (1959) for derivation of formulas):

$$V_L = \left(\frac{k + \frac{4}{3}}{\rho}\right)^{1/2} = \left(\frac{E}{\rho} \cdot \frac{1 - \upsilon}{(1 + \upsilon)(1 - 2\upsilon)}\right)^{1/2}, \tag{12.4}$$

$$V_T = \left(\frac{\mu}{\rho}\right)^{1/2} = \left(\frac{E}{\rho} \cdot \frac{1 - \upsilon}{2(1 + \upsilon)}\right)^{1/2}, \tag{12.5}$$

where:

$V_L$ is the velocity of a longitudinal (P wave, or compressional-rarefaction wave),

$V_T$ is the velocity of a transverse (S wave, or shear wave),

$E$ is Young's modulus, and $\rho$ is rock density.

If Poisson's ratio is assumed to take a value of 0.33 (Attewell and Farmer, 1976), then Eq. (12.4) and (12.5) become:

$$V_L = 1.2\ (E/\rho)^{1/2}, \tag{12.6}$$

$$V_T = 0.61\ (E/\rho)^{1/2}. \tag{12.7}$$

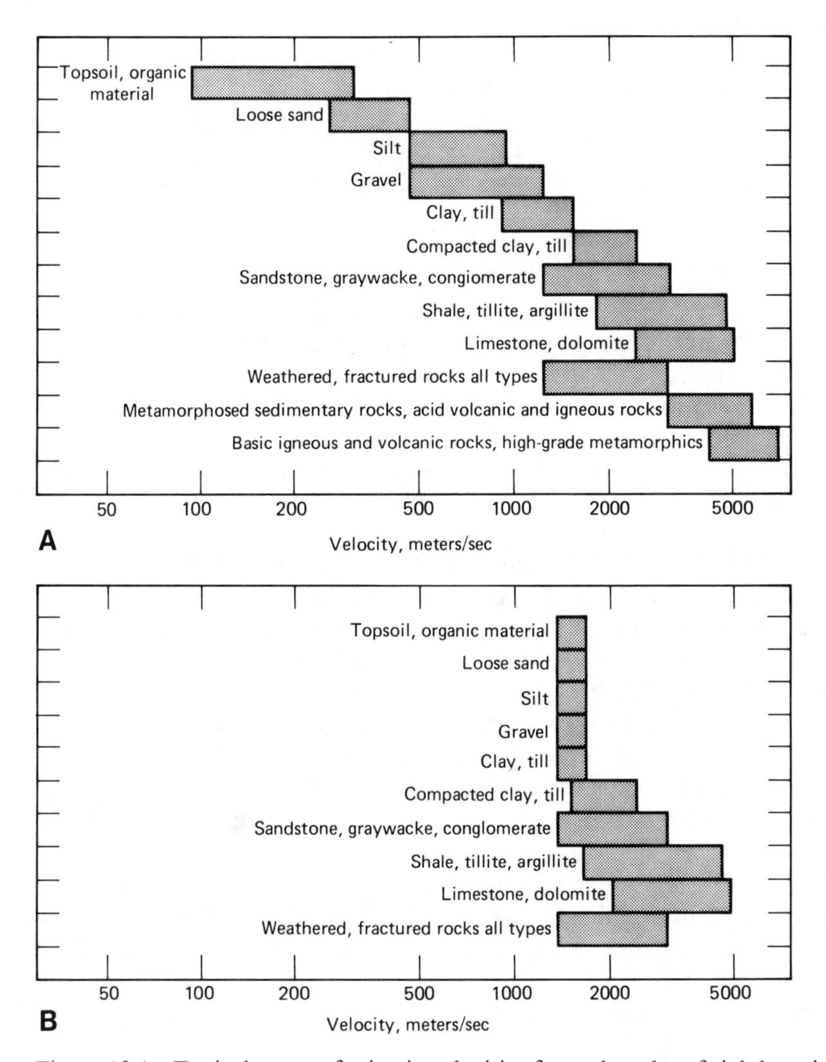


Figure 12.1 Typical range of seismic velocities for rock and surficial deposits (after Todd, 1980). **A**. Unsaturated material. **B**. Saturated material.

From these equations, it can be seen that the speed of a P wave is about twice as fast as the speed of an S wave. Equations (12.4) and (12.5) also show that the velocity of seismic waves is dependent upon rock density and Young's modulus. In general, these parameters increase with depth, and seismic wave velocity increases with depth (see Chapter 11).

Engineering geophysics generally relies on longitudinal wave data. Figure 12.1 shows typical longitudinal wave velocities. Since the strength of rock is related to its seismic wave velocity, Fig. 12.1 gives an indication of rock strength. In general, bulldozers are unable to rip apart materials whose seismic velocity exceeds 2,500 m/s.

From Eq. (12.4) and (12.5) it should be noted that Young's modulus of elasticity can be determined by observing the seismic velocities of rocks. This $E$ value is called the dynamic modulus of elasticity, as opposed to the conventional lab testing of stress versus strain to determine $E$. In general, the dynamic modulus of elasticity is higher for core-tested samples than the stress–strain determined modulus of elasticity values, because of the presence of small cracks which affect stress–strain loading curves more than seismic measurements (Jaeger, 1972).

## *12.1.2 Refraction Method*

The portable engineering seismograph is one of the most useful instruments for determining subsurface information. In the seismic method, a man-made seismic wave is created, and the times of arrival of P waves are sensed by regularly spaced geophones. Both refracted and reflected events can be detected from the same seismic shot.

Seismic methods are widely used in petroleum exploration where dynamite explosions are commonly used. In engineering geology, a simple blasting cap, sledge hammer, or small explosive charge can be used to generate the seismic wave. Hand-held sledge hammers rarely provide enough energy to determine subsurface information for depths greater than 10 m, however.

The theory of refraction stems from the fact that any ray bends (refracts) upon entering a different velocity medium. Consider the expanding hemispherical shells of P waves radiating from a source, simplified in two dimensions as rays (Fig. 12.2). The ray strikes the boundary at some angle of incidence ($i$), measured from a line perpendicular to the boundary. The angle of incidence changes to a refracted angle ($r$) when the wave travels through the new velocity medium. The larger the velocity differential between the two mediums, the larger angle $r$ becomes with respect to angle $i$.

The precise relationship between refraction and velocity differential can be established using simple trigonometry. Consider Fig. 12.2, where $V_1$ is the velocity of the first medium and $V_2$ is the velocity of the second medium. Two parallel rays are drawn, as well as orthogonals $AC$ and $DB$ from the two rays. Consider the right triangle $ACB$ and $ADB$:

$$BC = AB \sin i,$$
$$AD = AB \sin r.$$

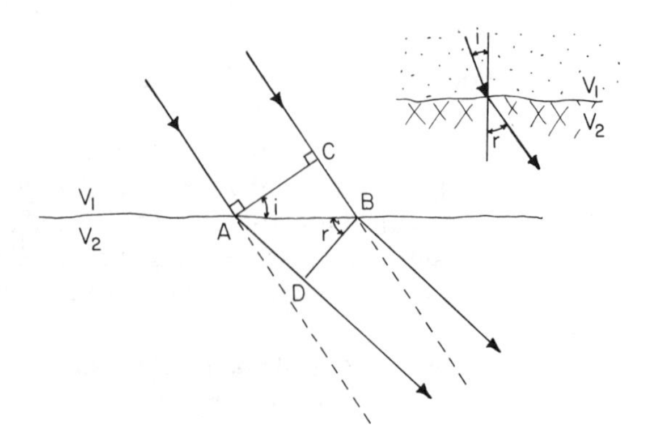


Figure 12.2 Refraction of elastic waves across an interface.

The time for a wave to travel distance $AD$ is the same as a wave traveling the distance $BC$:

$$t_{AD} = t_{CB}.$$

Since time equals $\dfrac{\text{distance}}{\text{velocity}}$,

$$\frac{AD}{V_1} = \frac{BC}{V_2}.$$

Substituting from above:

$$\frac{AB \sin r}{V_2} = \frac{AB \sin i}{V_1},$$

or

$$\frac{\sin i}{\sin r} = \frac{V_1}{V_2}. \tag{12.8}$$

Equation (12.8) is called Snell's law.

The depth ($z$) to some buried horizontal interface can be determined using Snell's law and the following theory. As a ray strikes the $V_1$ interface at increasingly smaller angles (i.e., angle $i$ increases), there eventually will be a refracted ray which travels along the interface (i.e., angle $r = 90°$). This angle of incidence is called the critical angle ($i_c$). Because $\sin 90° = 1$, from Eq. (12.8),

$$\sin i_c = \frac{V_1}{V_2}. \tag{12.9}$$

As waves travel along the boundary at velocity $V_2$, they continuously emit new disturbances into the upper medium, also at the critical angle. Consider geophones stationed at various distances ($x$) away from the seismic source (Fig. 12.3). A short distance away, geophones pick up both the direct wave and the refracted wave. Geophones nearest the seismic source receive the direct wave first, traveling at velocity $V_1$. But, because the refracted wave travels along the interface at the faster velocity $V_2$, there will be a critical distance ($x_c$) beyond which the refracted wave reaches geophones before the direct wave. The relationship between $x_c$ and depth ($z$) of a buried boundary can be derived as follows. The time-distance plot for the direct wave passes through the origin so that:

$$T = \frac{x}{V_1}.$$

The refracted wave must traverse three legs in order to reach the geophone: distance $AB$, $BC$, and $CD$. Three trigonometric relations involving these legs are necessary for the following derivation of the formula which determines the depth ($z$) to a buried interface:

$$\sin i_c = \frac{V_1}{V_2},$$

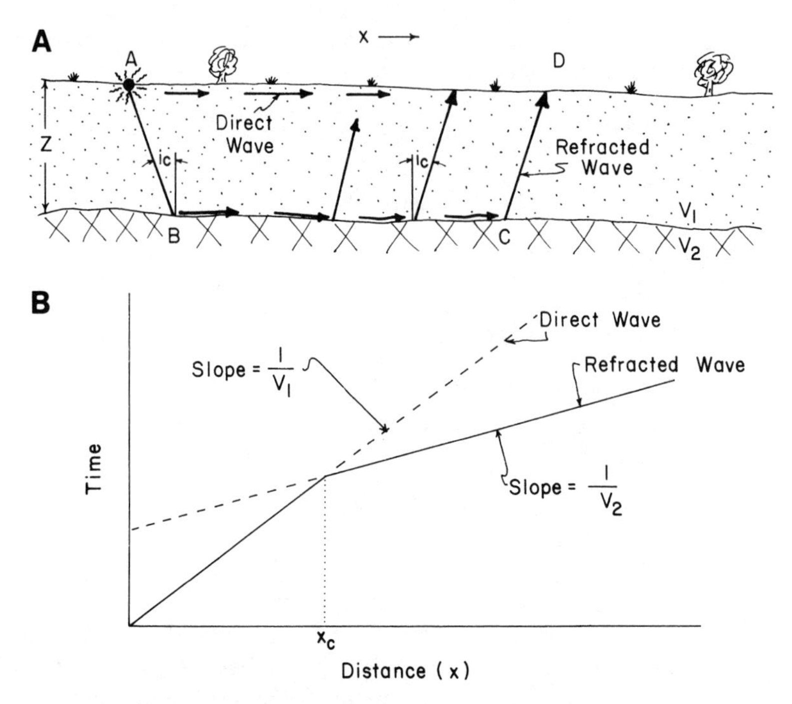


Figure 12.3 Ray paths for two layers separated by a horizontal interface. **A**. Geologic cross section. **B**. Time-distance plot.

$$\cos i_c = (1 - \sin^2 i_c)^{0.5} = \left(1 - \frac{V_1^2}{V_2^2}\right)^{0.5},$$

$$\tan i_c = \frac{\sin i_c}{\cos i_c} = \frac{V_1}{(V_2^2 - V_1^2)^{0.5}}.$$

The total time along refraction path *ABCD* is:

$$T = T_{AB} + T_{BC} + T_{CD}.$$

For a thickness $z$, since time $= \dfrac{\text{distance}}{\text{velocity}}$, the above equation can be written as:

$$T = \frac{AB}{V_1} + \frac{BC}{V_2} + \frac{CD}{V_1}.$$

From Fig. 12.3,

$$T = \frac{\dfrac{z}{\cos i_c}}{V_1} + \frac{x - 2(z \tan i_c)}{V_2} + \frac{\dfrac{z}{\cos i_c}}{V_1}.$$

Substituting trigonometric relationships yields:

$$T = \frac{2z}{V_1 \cos i_c} - \frac{2z \sin i_c/\cos i_c}{V_2} + \frac{x}{V_2},$$

$$
\begin{aligned}
T &= \frac{2z}{V_1 \cos i_c} \cdot (1 - \sin^2 i_c) + \frac{x}{V_2}, \\
T &= \frac{x}{V_2} + \frac{2z \cos i_c}{V_1}, \\
T &= \frac{x}{V_2} + \frac{2z(V_2^2 - V_1^2)^{0.5}}{V_1 V_2}.
\end{aligned}
\tag{12.10}
$$

At the critical distance, both direct and refracted waves arrive at the same time so that:

$$T_{\text{direct}} = T_{\text{refracted}},$$

$$\frac{x_c}{V_1} = \frac{x_c}{V_2} + \frac{2z(V_2^2 - V_1^2)^{0.5}}{V_1 V_2},$$

which simplifies to:

$$
\begin{aligned}
z &= \frac{1}{2} \frac{V_1 V_2 x_c}{(V_2^2 - V_1^2)^{0.5}} \cdot \left(\frac{1}{V_1} - \frac{1}{V_2}\right), \\
z &= \frac{x_c}{2} \left(\frac{V_2 - V_1}{V_2 + V_1}\right)^{0.5}.
\end{aligned}
\tag{12.11}
$$

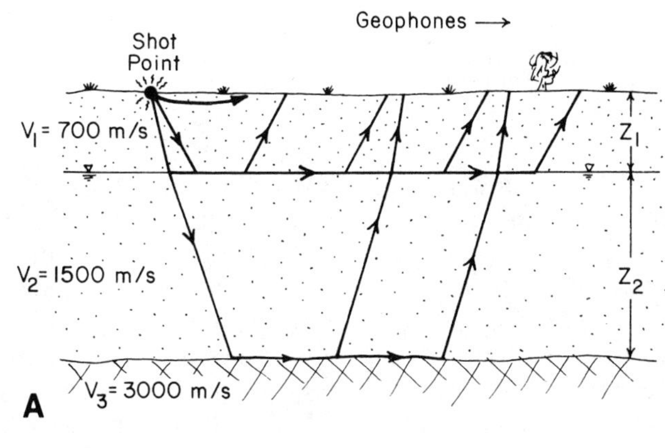


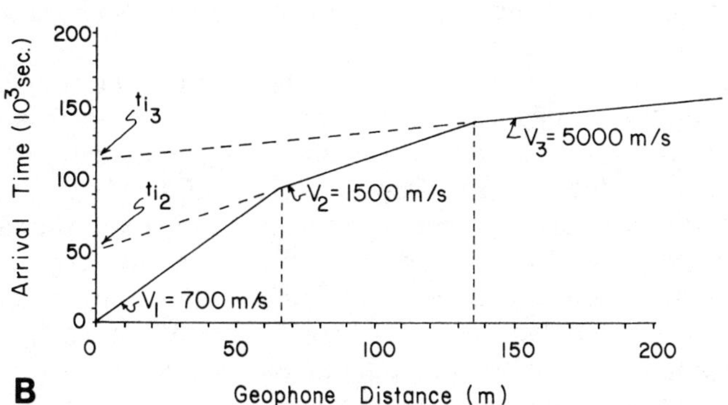


Figure 12.4 Refraction of seismic waves for a three-layered system (after Bouwer, 1978). **A**. Geologic cross section showing physical setup. **B**. Plot of time vs. distance for setup shown in **A**.

Equation (12.11) is used to determine the depth to a horizontal interface for a two-layer case. By using trigonometric methods, it is possible to derive formulas for depths to boundaries for a three-layer case or other configurations (see Dobrin, 1960, for derivations). For example, for a three-layer case (Fig. 12.4), the formula for the thickness $z_2$ of the second layer can be determined (after the upper layer thickness $z_1$ is established) as follows:

$$z_2 = \frac{x_2}{2}\left(\frac{V_3 - V_2}{V_3 + V_2}\right)^{0.5} - \frac{z_1}{6}, \tag{12.12}$$

where $x_2$ is the critical distance for the second break point in the distance–time data curve. For example, using Eq. (12.11) to solve for the thickness of the upper layer in Fig. 12.4, we obtain

$$\begin{aligned} z_1 &= \frac{x_1}{2}\left(\frac{V_2 - V_1}{V_2 + V_1}\right)^{0.5}, \\ &= \frac{67\ \text{m}}{2}\left(\frac{1500\ \text{m/s}^2 + 700\ \text{m/s}^2}{1500\ \text{m/s}^2 + 700\ \text{m/s}^2}\right)^{0.5}, \\ &= 20.2\ \text{m}. \end{aligned}$$

Using Eq. (12.12) to solve for the thickness of the next layer, we obtain:

$$\begin{aligned} z_2 &= \frac{x_2}{2}\left(\frac{V_3 - V_2}{V_3 + V_2}\right)^{0.5} - \frac{z_1}{6}, \\ &= \frac{135\ \text{m}}{2}\left(\frac{5000\ \text{m/s}^2 - 1500\ \text{m/s}^2}{5000\ \text{m/s}^2 + 1500\ \text{m/s}^2}\right)^{0.5} - \frac{20.2\ \text{m}}{6}, \\ &= 46.1\ \text{m}. \end{aligned}$$

The seismic refraction method is typically used where a single charge is recorded by numerous geophones. For practical application of a refraction survey, it is common practice to space the geophones at a regular interval in a straight line. The arrival of the first shock waves at each setup is recorded (Fig. 12.4B). After the wave arrival is collected from the geophone array, it is conventional practice to move the shot point to the other end of the string of geophones and set off another charge. This confirms interpretations, as well as providing information on dipping interfaces. In the case where only one geophone is available, it is necessary to set off a number of shots, moving the shot point out at regular intervals, such as every 5 m.

There are many engineering seismographs on the market. The simplest have digital printout of the time difference between the shot and wave arrival at the geophone. Others have a moving stylus which burns electrosensitive paper, making a permanent record of the distance–drawdown curve for the first arrivals, and all the other later wave arrivals. Another engineering seismograph uses an oscilloscope to portray wave arrivals. The waves, frozen on the screen, can be enhanced by repeated shots at the same point. Some oscilloscope-type engineering seismographs have capacity for multigeophone readouts and cameras mounted to photograph the oscilloscope readout. Hobson (1970) gives additional information on seismic instruments commonly used in North America.

A new method is called vibroseis, which sends continuous pulses into the earth (Fitch, 1981). The term vibroseis is a registered trademark of Continental Oil Co. The technique differs from methods using an impulse source such as dynamite, in which energy is input at one time. Vibroseis utilizes a technique whereby the entire frequency of vibrations is gradually altered so that the entire frequency range that is needed has been vibrated (Kirk, 1981).

## *12.1.3 Reflection Method*

The reflection method is used extensively for petroleum exploration. It gives a detailed picture of the subsurface structure and interfaces with an accuracy that is exceeded only by test holes. The depths are determined by observing the travel times of P waves generated near the surface which are reflected back from deep formations. The method is comparable to that of echo sounding used to determine water depths. A unique advantage of the reflection method is that it permits the mapping of many horizons from each shot.

If the seismic velocity of some overlying medium is a constant $V_1$, the wave will reflect from a horizontal interface at a depth $z$ as shown in Fig. 12.5. The total length ($L$) of the wave path to some distance ($x$) from the shot point is:

$$L = V_1 T,$$

where $T$ is the total travel time. From the Pythagoras theorem:

$$L = 2\left(\frac{x^2}{2} + z^2\right)^{0.5}.$$

Therefore,

$$z = ½((V_1 T)^2 - x^2)^{0.5}. \tag{12.13}$$

When $x = 0$,

$$z = \frac{V_1 T}{2}. \tag{12.14}$$

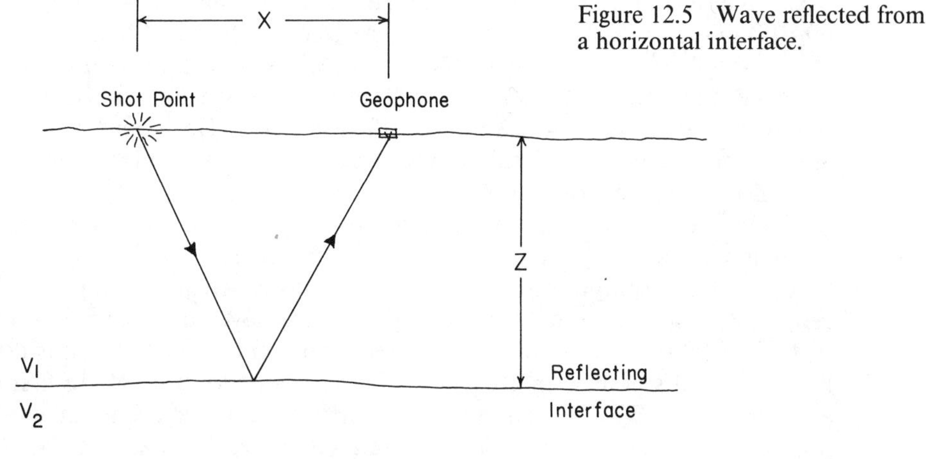


Figure 12.5 Wave reflected from a horizontal interface.

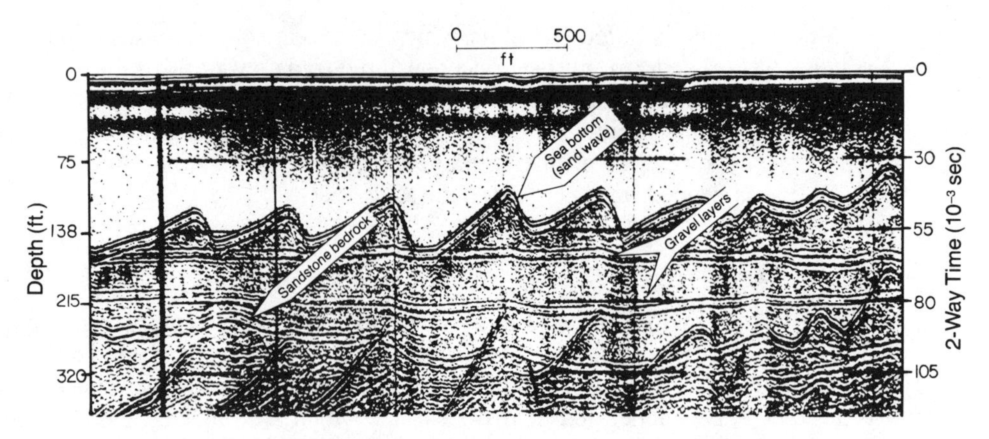


Figure 12.6 Reflection survey, offshore South Africa (from Oostdam, 1970).

The reflection method also makes possible the calculation of depths to dipping interfaces, as well as the angle of dip (see Dobrin, 1960, for derivation of dip formula and examples).

In petroleum prospecting, multiple geophones are often used, giving continuous printout on tapes. When several closely spaced geophones are placed along a line and the resulting seismograms held together for continuous viewing, the waves themselves line up on the seismograms in a manner that allows the viewer to see the interfaces as if the seismogams were a geologic cross section. This is because the waves corresponding to a common reflection will all line up across the record in such a way that the crests or troughs on adjacent traces appear to fit together. Recent advances in reflection methods using the finite difference or wave-equation migration are discussed by Claerbout (1971) and Hood (1981).

Figure 12.6 shows a continuous seismic profile used at sea. The seismic source used was a spark type explosive charge, which causes P waves to reflect off the sea floor as well as off geologic interfaces at greater depth. The profile in Fig. 12.6 shows the sea floor at about 50 m depth, with large sand waves (amplitude = 15 m) on the sea floor. Nearly horizontal strata (sandstone bedrock) are apparent at 60–100 m depth. Gravel layers are interpreted in Fig. 12.6. A ghost (repetition) of the sea floor appears at 70–120 m. In this study, geomorphologic features included drowned beaches and terraces, sea cliffs, drowned river beds, surf gullies, potholes, and basal gravels were identified at sea by Oostdam (1970), who was able to locate commercial submarine diamond and other placer deposits using the reflection seismograph.

The reflection technique has not received as much attention by engineering geologists as the refraction method. However, because the refraction technique is based on the assumption that deeper layers have increasingly higher velocities, refraction will not work where a high velocity layer overlies a low one. For instance, Hobson (1970) noted in Ontario that a hidden layer of till is detectable by a lineup of second arrivals with a different slope to those of the first or third arrivals (Fig. 12.7). Reflection profiling in permafrost areas is not affected by the high velocity permafrost whereas refraction techniques can be nullified completely.

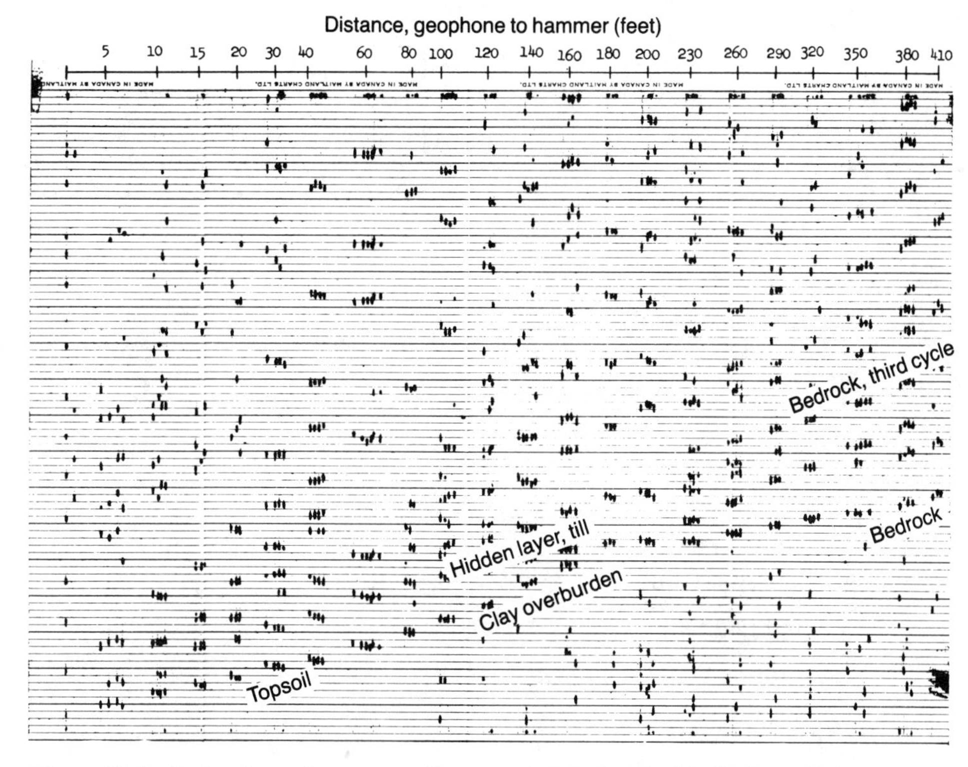


Figure 12.7 Reflection seismogram of area underlain by glacial till (from Hobson, 1970). The record was made by a moving stylus on electrochemically sensitive paper.

Seismic layers do not necessarily conform to major geologic strata or unconformities. For instance, at depth a dense lodgement till may overlie a preglacial, loose regolith on top of bedrock. Thus, a velocity interface detected at depth may not be the top of the bedrock, but may be the top of the lodgement till.

Seismic results must be correlated with drilling logs and geologic information. It must be emphasized that the seismic method is not the panacea to all geological problems. But it does give a view into the third dimension, which can assist the engineering geologist.

## 12.2 Electrical Resistivity

### *12.2.1 Theory*

If the total electrical resistance ($R$) of a conducting cylinder of length ($L$) and cross-sectional area ($A$) is known, then the (unit) resistivity ($\rho$) is defined as:

$$\rho = \frac{RA}{L}. \tag{12.15}$$

The unit of resistivity is an ohm-meter.

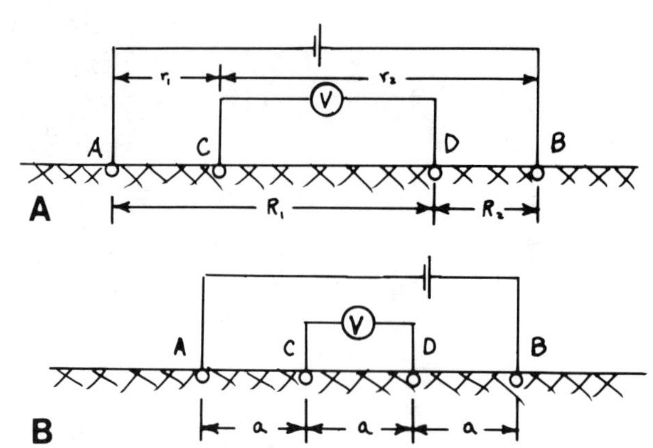


Figure 12.8 Electrical resistivity setup. **A**. Theoretical arrangement of electrodes. **B**. Wenner array.

From Ohm's law, the current ($I$) passing through this cylinder is related to the total resistance and the applied voltage ($V$) as:

$$I = V/R. \tag{12.16}$$

Substituting Eq. (12.15) into Eq. (12.16) and arranging terms, we obtain:

$$V = \frac{I\rho L}{A}. \tag{12.17}$$

For practical use, Dobrin (1960) shows the following analytical derivation of the electrical resistivity method. (An alternative derivation using calculus is given in Parasnis (1966).) The model assumes current is being sent into the earth from the surface (Fig. 12.8). It is assumed that the land surface is flat, and the rocks are isotropic and homogeneous. Consider an infinite section of land with uniform resistivity. Assume a current is introduced between points $A$ and $B$ on the surface as shown in Fig. 12.8A, and the voltage drop is measured between two other positions $C$ and $D$. The cross-sectional area through which current travels is $2\pi R$. The potential at $C$ is:

$$V_C = \frac{I\rho}{2\pi r_1} - \frac{I\rho}{2\pi r_2}. \tag{12.18}$$

Similarly, the potential at $D$ is:

$$V_D = \frac{I\rho}{2\pi R_1} - \frac{I\rho}{2\pi R_2}. \tag{12.19}$$

The voltage drop is obtained by subtracting Eq. (12.19) from Eq. (12.18). Solving for $\rho$, we have:

$$\rho = \frac{2\pi V}{I} \cdot (r_1 + r_2 - R_1 + R_2)^{-1}. \tag{12.20}$$

The most common setup for electrical resistivity apparatus used in North America is the Wenner array (Fig. 12.8B). In this case, all electrodes are spaced

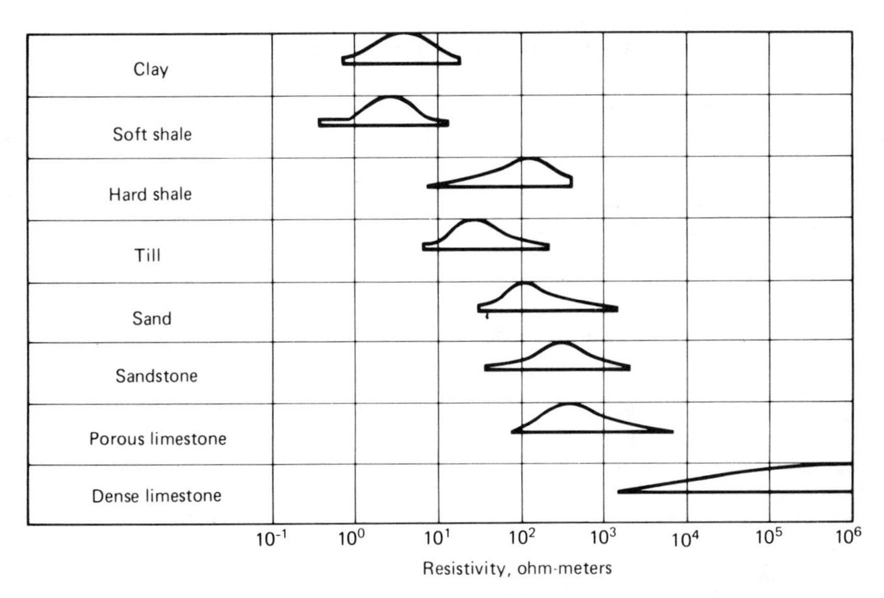


Figure 12.9 Representative ranges of electrical resistivities for various rocks and surficial deposits (from Todd, 1980).

at an even increment $a$. Therefore, Eq. (12.20) becomes:

$$\rho = 2\pi a \frac{V}{I}. \tag{12.21}$$

In actual practice, there are a number of different surface configurations for the electrical resistivity method (see Moore, 1952; Schwartz and McClymont, 1977; Carrington and Watson, 1981; or Larson and Hiegold, 1981, for examples). Koefoed (1979) gives a detailed theoretical treatment of electrical resistivity techniques, including the use of BASIC computer programs.

The range of resistivities ($\rho$) of natural rocks is enormous (Fig. 12.9). The amount of ground water and dissolved ions in the rocks and water are of great importance in a resistivity survey. Dry rock has virtually no electrical conductance. The resistivity of a water-bearing rock is a function of the amount of ground water present and the salinity. Because contaminated ground water typically has more dissolved solds than pure ground water, the electric resistivity method lends itself to ground-water contamination studies (Warner, 1969; Kean and Rogers, 1981; Yazicigil and Sendlein, 1982). Table 12.1 shows the resistivities of typical ground water samples. High chloride concentrations allow for easy passage of electricity (i.e., have low resistance). The resistivity of fine-grained sedimentary rocks tends to be lower than coarse-grained sedimentary rocks because of the greater porosity and hence water content of fine-grained materials (see Chapter 7 for porosity data). Verma et al. (1980) found productive wells associated with weathered metamorphic rocks whose low values of resistivity were obtained. Sound, unweathered rocks had "almost infinite resistivity values."

Coarse-grained materials (such as gravel) generally have better ground-water

Table 12.1 Resistivity Values ($\rho$) for Ground Water

| Rocks from which water samples were taken | Number of samples | Resistivity at 20°C, ohm-m | |
|---|---|---|---|
| | | Median | Range |
| Igneous rocks, Europe | 314 | 7.6 | 3–40 |
| Igneous rocks, South Africa | 175 | 11.0 | 0.5–80 |
| Metamorphic rocks, South Africa | 88 | 7.6 | 0.8–80 |
| Metamorphic rocks, Precambrian of Australia | 31 | 3.6 | 1.5–8.6 |
| Pleistocene to recent continental sediments, Europe | 610 | 3.9 | 1–27 |
| Pleistocene to recent sediments, Australia | 323 | 3.2 | 0.4–80 |
| Tertiary sedimentary rock, Europe | 993 | 1.4 | 0.7–3.5 |
| Tertiary sedimentary rock, Australia | 240 | 3.2 | 1.35–10 |
| Mesozoic sedimentary rock, Europe | 105 | 2.5 | 0.31–47 |
| Paleozoic sedimentary rock, Europe | 161 | 0.93 | 0.29–7.1 |
| Chloride waters from oil fields | 967 | 0.16 | 0.049–0.95 |
| Sulfate waters from oil fields | 256 | 1.2 | 0.43–5.0 |
| Bicarbonate waters from oil fields | 630 | 0.98 | 0.24–10 |

From Keller, 1970.

recharge and therefore tend to contain fresher water (i.e., better quality water because of less TDS) than fine-grained materials such as colluvium or till. It follows that the higher the resistivity (i.e., the lower the conductivity), the better the aquifer. In this way, aquifers can be located using resistivity surveys.

Much experimental work has been done by the petroleum industry correlating bulky (field) resistivity to water resistivity and to porosity. The simplest relationship is

$$\rho t = \frac{\rho_w}{\phi}, \tag{12.22}$$

where $\rho t$ is the bulk resistivity, $\rho_w$ is the resistivity of the water in the rock, and $\phi$ is the fractional porosity filled with water. This equation would describe the resistivity of a bundle of smooth tubes oriented in the direction of the applied voltage. The resistivity observed in the real rock will be larger than Eq. (12.22) predicts because of variability of cross section. Equation (12.22) is experimentally related as follows:

$$\rho t = \frac{A\rho_w}{\phi^m}, \tag{12.23}$$

where $A$ and $m$ are experimentally derived parameters (related to matrix properties such as packing and cementation) that fit a particular set of data. Equation (12.23) is known as Archie's Law. Keller (1970) reports that typical rock values of $A$ range from about 0.6 to 3.5, and values of $m$ range from 1.37 to 1.95, although Keys (1970) found $A$ = 0.62 and $m$ = 2.15 for granular rocks.

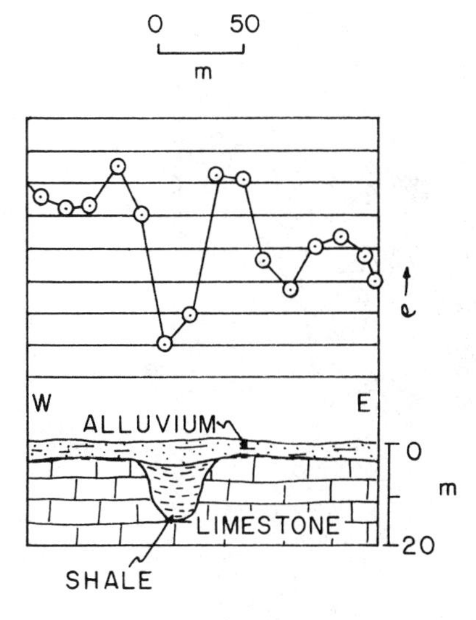


Figure 12.10 Observed resistivity profile over a sinkhole in Kansas (after Dobrin, 1960).

### *12.2.2 Practical Application*

Engineering geology aspects of electrical resistivity surveys are found in articles by Krynine and Judd (1957), Van Nostrand and Cooke (1966), Page (1968), Joiner et al. (1968), Keller (1970), and Keck et al. (1981). The field application of electrical resistivity to the investigation of the impact of waste disposal on shallow ground water has been documented by several studies (see Gilkeson and Cartwright, 1983, and references contained therein).

In practice, the Wenner array is used in two ways. 1) A survey is made over a wide area, using a constant "a" spacing. The data is plotted as a traverse line or directly on a map. The map data can be contoured, and anomalies detected. (2) A vertical survey (sounding) is made by increasing "a" spacings. This gives information concerning increasing depths. For a vertical survey it is generally assumed that all the current is passing through material at the same depth as the electrode spacing "a". (Problem 12.7 at the end of this chapter involves the utilization of several soundings along a strip of land.)

Figure 12.10 shows a traverse line using a Wenner setup in a karst terrain. The anomaly consists of a steep drop in the resistivity profile from about 350 to 150 ohm-m (Ω-m). The anomaly is due to a deep sinkhole that was not recognized at the surface. Caves and shallow underground mine tunnels have been located where the tunnel depth-to-diameter ratio is less than 10 (Peters and Burdick, 1983).

Figure 12.11A is a geologic map of an area in eastern South Dakota, where glacial deposits overlie Cretaceous sedimentary rocks. Figure 12.11B shows electrical resistivity contour highs (350–600 Ω-m) which correlate to glacial outwash deposits in the northwest part of the geologic map. (Boundaries of outwash aquifers such as these are often hard to determine in the field without extensive test drilling.) Another high anomaly in the center of the map (about 60 Ω-m) may indicate the bedrock surface (the Niobrara Formation) overlain by only a thin ($\sim$10 m) deposit of till.

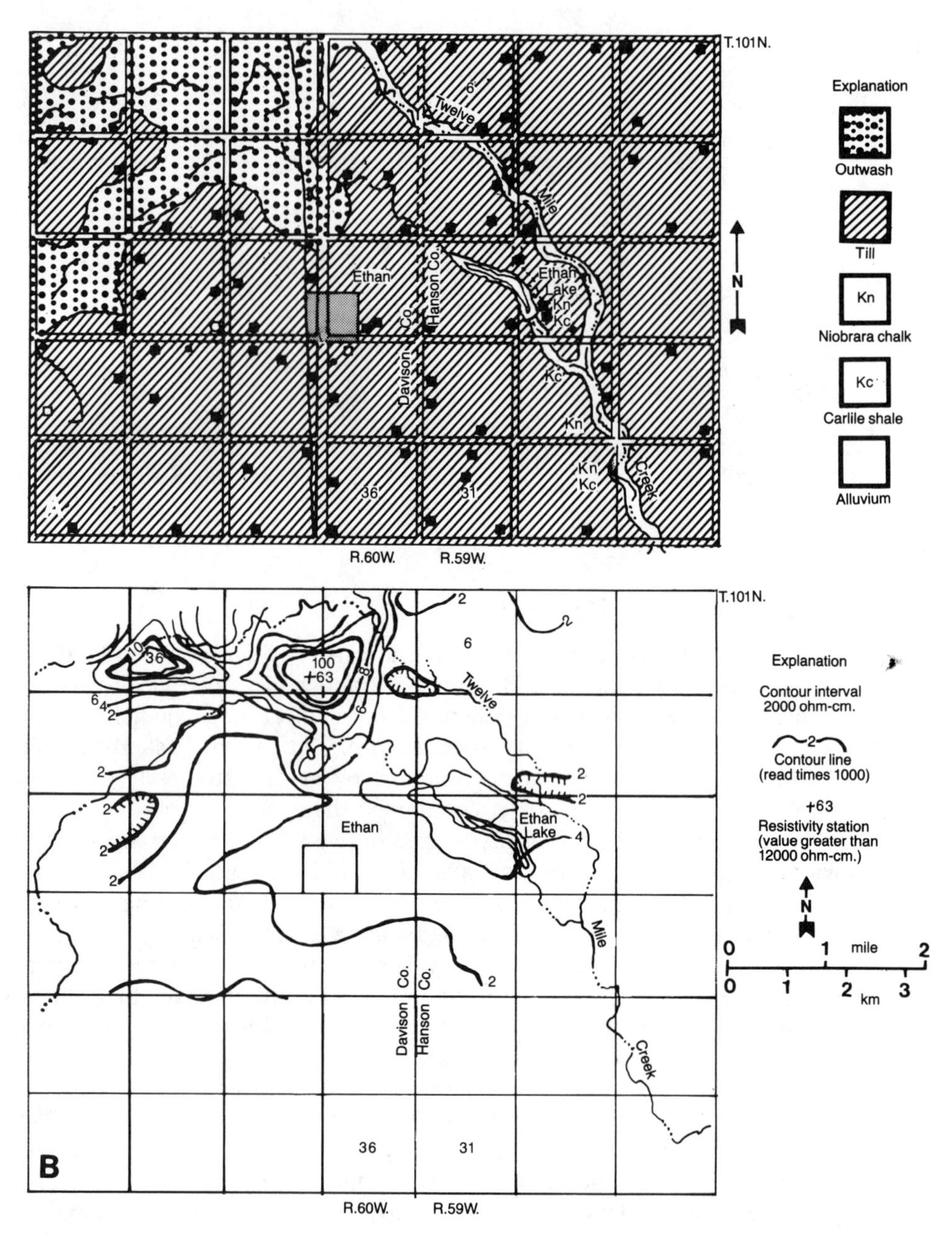


Figure 12.11 Maps of Ethan, SD area (from Lum, 1961). **A**. Geologic map. **B**. Iso-resistivity contour map.

Although anomalies may be detected such as illustrated above, it is difficult to obtain reliable quantitative resistivity results by comparing theoretical models and published resistivity values (Fig. 12.9) to the resistivity values obtained in a field survey (Sauck, 1981). For one thing, nature does not provide isotropic and homogeneous rocks. The saving grace of resistivity surveys is the tremendous

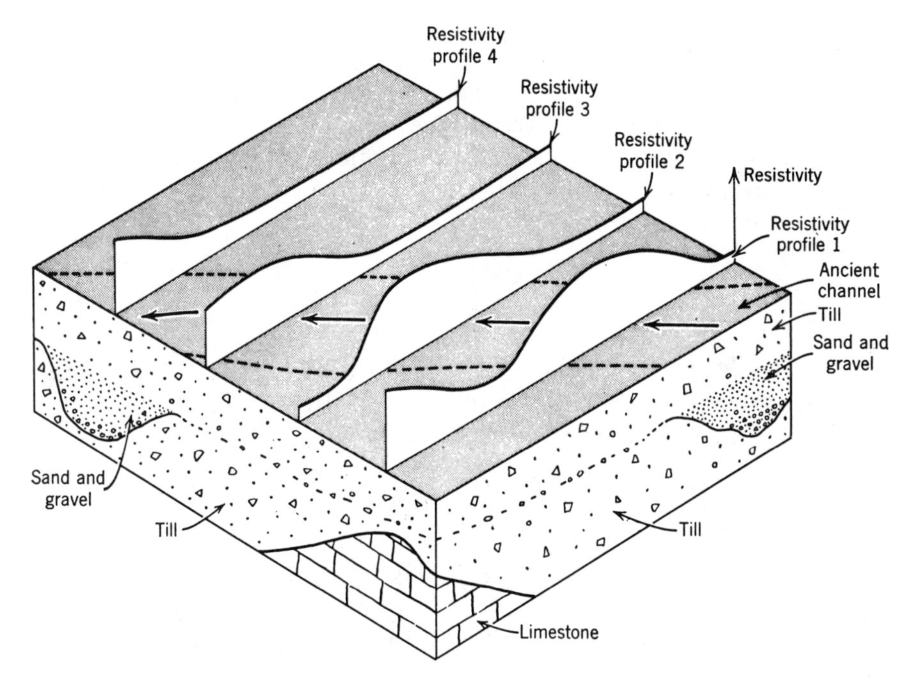


Figure 12.12 Sketch illustrating use of resistivity profiles (from Davis and DeWeist, 1966). The resistivity profiles are used to locate buried stream deposits sandwiched between two till sheets.

range in resistivities of natural materials, so that resistivity surveys do show profound anomalies. These anomalies can be used to make inferences about subsurface conditions.

Electrical resistivity profiles can be most effectively used when coordinated with test hole data. Figure 12.12 is an idealized sketch that shows how profiles may be used to detect buried alluvium sandwiched between till sheets. After initial discovery, subsequent exploration in this case is best made by profiles oriented perpendicular to the trend of the buried stream channel.

Electrical resistivity profiles may show a sharp change at the contact of the slip surface for a landslide (Hutchinson, 1983). For example, Pleistocene marine clays, originally deposited in seawater with an NaCl content of 35,000 mg/l commonly rest above sea level in Canada and Scandinavia. The leaching action of Holocene ground water has reduced the NaCl content, resulting in the formation of quick clays and occurrence of quick clay landslides. Test borings through these landslides may show a sharp discontinuity in the electrical profile where strongly leached clay exists in the landslide itself.

Electrical resistivity surveys have been used successfully to locate fault zones in potential tunnel alignments (Krynine and Judd, 1957). A critical factor in tunneling is roof stability, and faults are known to cause cave-ins and increased water inflow. In general, fault zones are wetter and have a lower electrical resistance than solid rock.

## 12.3 Gravity Methods

The gravity method is used to meaure variations in the earth's gravitational pull that are related to near-surface changes in rock density. The method has been used to detect large features such as intermountain basin fills or buried glacial valleys. To measure variations in gravity, extremely sensitive instruments, called gravimeters, have been developed.

### 12.3.1 *Theory*

The universal law of gravitational attraction developed by Newton states that two masses, $M_1$ and $M_2$ are attracted to each other by a force ($F$), which varies on the square of the distance ($R$) between them:

$$F = \gamma \frac{M_1 M_2}{R^2}. \tag{12.24}$$

The constant $\gamma$ is the universal gravitational constant,

$$\gamma = 6.670 \times 10^{-8} \frac{\text{dynes} \cdot \text{cm}^2}{\text{g}^2} = 6.670 \times 10^{-11} \frac{\text{N} \cdot \text{m}^2}{\text{kg}^2}.$$

If we consider earth as $M_1$, then the acceleration ($a$) of any mass $M_2$ on earth will be:

$$a = \frac{F}{M_2} = \frac{\dfrac{\gamma M_1 M_2}{R^2}}{M_2} = \gamma \frac{M_1}{R^2}. \tag{12.25}$$

Equation 12.25 indicates that the critical factors that determine gravitational acceleration on the earth's surface are the mass of the earth (including proximity of local deposits of rock of variable mass), and the distance to the earth's center. This equation also shows that the gravitational acceleration of any mass at a given location on the earth's surface is the same.

The units of acceleration are common $m/s^2$ or $cm/s^2$. The unit of $cm/s^2$ is called a gal in honor of Galileo. The average acceleration of gravity at the earth's surface is about 980 gal ($=980$ $cm/s^2$ or 9.8 $m/s^2$). A milligal (mgal), commonly used in gravity surveys, is $10^{-3}$ gal.

Gravity varies slightly on the earth's surface due to many factors. In order to isolate the effects of small differences in rock density, it is necesary to correct gravity measurements for latitude and topography. The reason for the latitude correction is because the earth is not a perfect sphere but is flattened at the poles. Thus the pull of gravity is greater at the poles because they are closer to the center of the earth than elsewhere on the earth's surface. Another factor counteracts this phenomenon, however; that is because the earth is rotating once every 24 hr, and in so doing produces a centrifugal force that reaches a maximum at the equator and zero at the poles. Latitude correction is extremely important in gravity surveys extending over large latitudinal distances. Correction tables are available in reference books on the subject.

Topographic correction involves the fact that a mountaintop is further from the earth's center than sea level. This free-air correction amounts to 0.308 mgal/m change in elevation. Further correction is necessary to account for the extra mass of the mountain itself. Assuming a 2.67 g/cm$^3$ density of the mountain, the Bouguer correction of 0.111 mgal/m change in elevation is used. Both topographic corrections (free-air less Bouguer) amount to 0.197 mgal/m change of elevation about sea level.

Other gravity factors such as correction for earth tides are explained in detail by Dobrin (1960).

## *12.3.2 Method*

In practice, a gravity survey is run in conjunction with an elevation survey. The gravimeter (Fig. 12.13) is set up at a station where the gravity is known, usually to 0.1 mgal accuracy. (In the United States, these stations are typically established at USGS bench marks.) The gravimeter is read by adjusting a null screw that reflects a ray of light off a weight supported by a quartz spring. Then the gravimeter is moved to a new location and read again. The difference in readings (corrected for the spring characteristics) allows the gravity at the new station to be determined. Most gravity measurements are made along railroads, at road intersections, or other bench mark locations where the elevation is known and latitude can be determined so that proper corrections can be made. The data is usually contoured and presented in map form as anomalies, i.e., differences between the theoretical and observed gravity.

Figure 12.13 Gravimeter being used in the field. After leveling the instrument, a dial is adjusted until a reflected light beam is nulled on a delicately suspended weight.

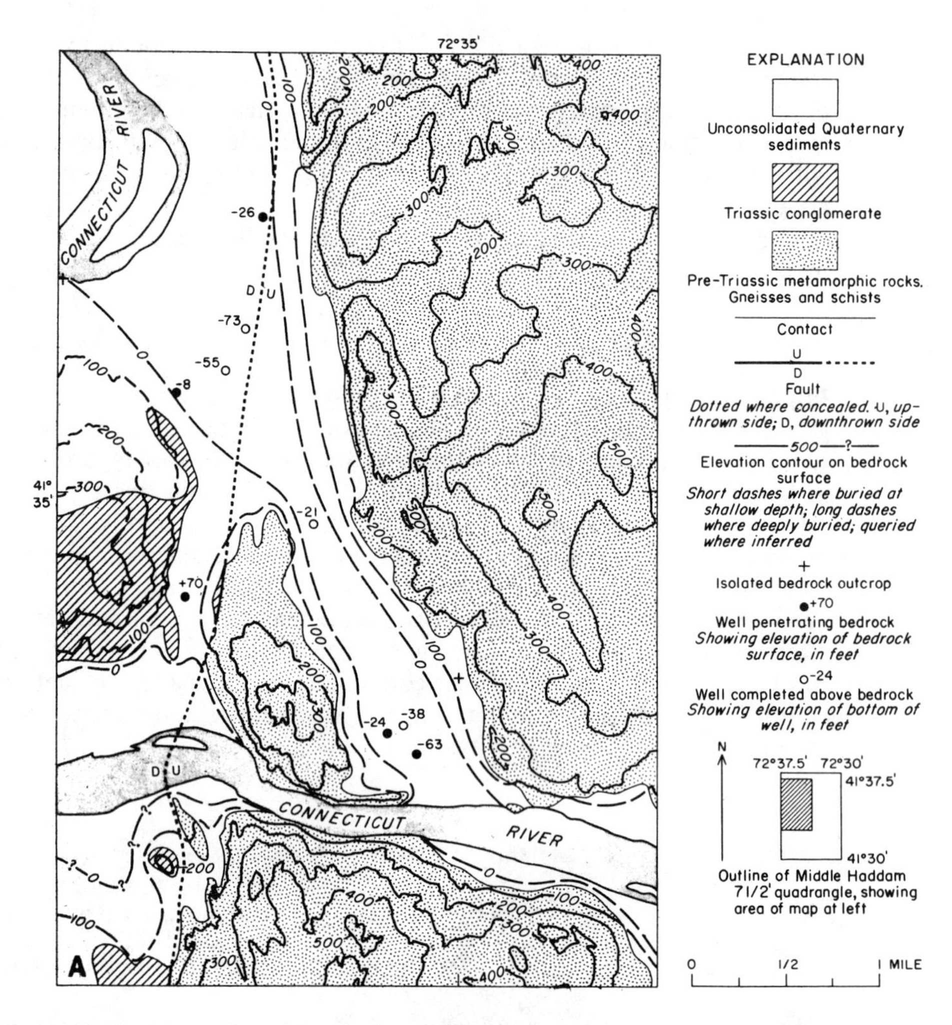


Figure 12.14 Maps of portion of Connecticut (from Eaton and Watkins, 1970). **A.** Generalized geological map of a portion of the Middle Haddam quadrangle, Connecticut, showing distribution of outcrops of bedrock and elevation contours

## *12.3.3 Examples*

For engineering geology, the most important use of the gravity method has been in the delineation of buried valleys or basin fills. For example, Fig. 12.14 shows a geologic and gravity map of a portion of Connecticut. Note the coincidence between the axis of a gravity anomaly and the position of the preglacial channel of the Connecticut River. A large gravel aquifer would be expected along the old channel.

Gravity surveys of Estevan, Saskatchewan (Hall and Hajnal, 1962; Robinson, 1963), northeastern Wisconsin (Stewart, 1980), and northwestern Pennsylvania (Rankine and Lavin, 1970), have delineated preglacial channels (gravel aquifers) that have been completely buried by till sheets. In order to make quantitative

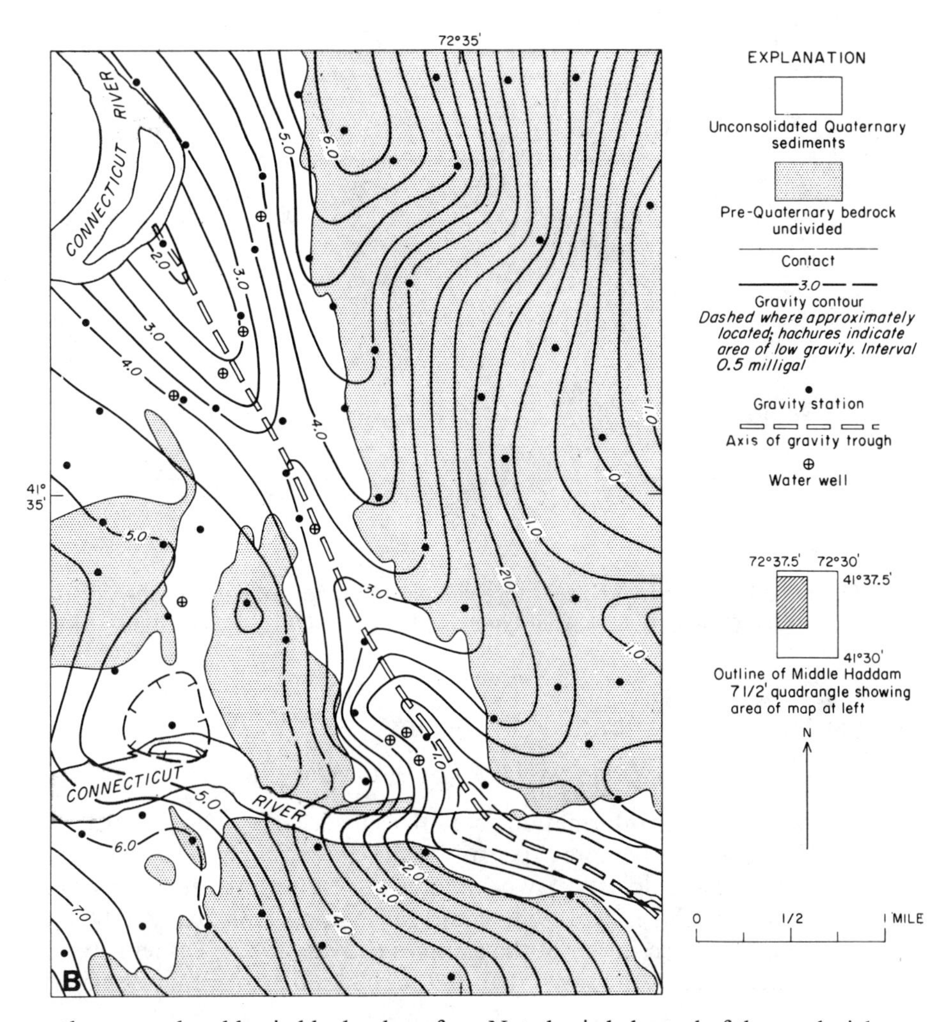


on the exposed and buried bedrock surface. Note buried channel of the preglacial Connecticut River. **B**. Complete Bouguer gravity map of the same area.

assessment of the thickness of gravel in these areas, it is necessary to know the surficial deposit density (see, for example, data in Lennox and Carlson, 1970), and to have simplified geometric models relating gravitational anomalies to density contrasts (see Dobrin, 1960, 1976).

Gravity anomalies correlated with the thickness of alluvial basin fill in the Basin and Range Province (Mabey, 1970; O'Brien and Stone, 1984). Figure 12.15 shows the gravity profile and a geologic cross section as interpreted from seismic profiling. Because gravity data can be obtained more easily than seismic data, Mabey stated that " . . . comparable information on the basement configuration could have been obtained with less expense from the few gravity observations."

Gravity methods have been used to locate large caves and soil-filled sinkholes (Omnes, 1977; Greenfield, 1979).

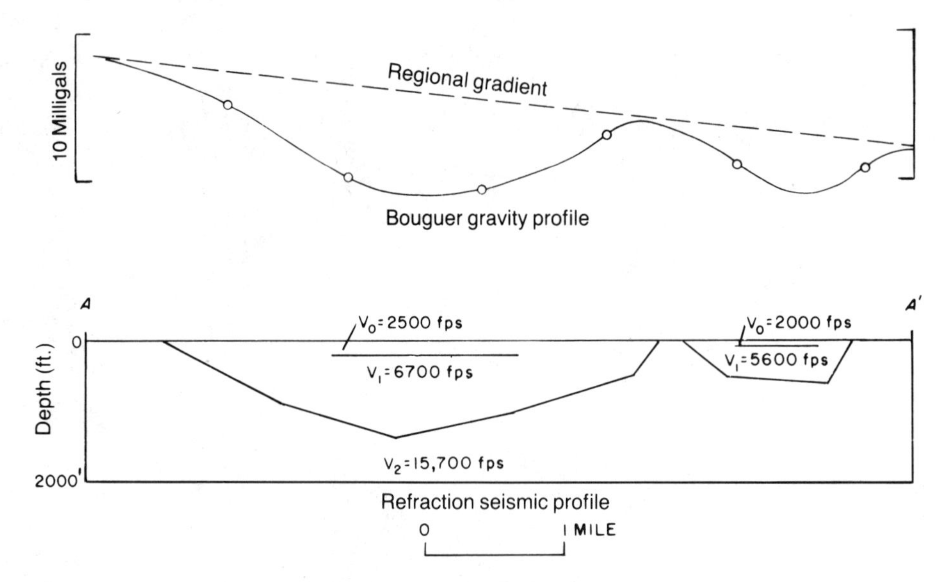


Figure 12.15 Gravity and seismic profile near Yuma, AZ (from Mabey, 1970). The interpreted geology is somewhat typical of the Basin and Range Province (see Figure 9.8).

# 12.4 Test Hole Drilling and Well Logging

## *12.4.1 Driller's Logs and Geologic Logs*

The most accurate information about the subsurface can be obtained by test drilling. This is generally more expensive than surface geophysical methods, however.

Engineering geologists may be involved in either soil test sampling and logging or drilling in hard rock. In either case, it is important that care be taken in the sampling and description of the samples (Boyce, 1982). Driller's logs tend to be incomplete (Fig. 12.16 and Table 12.2) or contain errors. Many engineering geol-

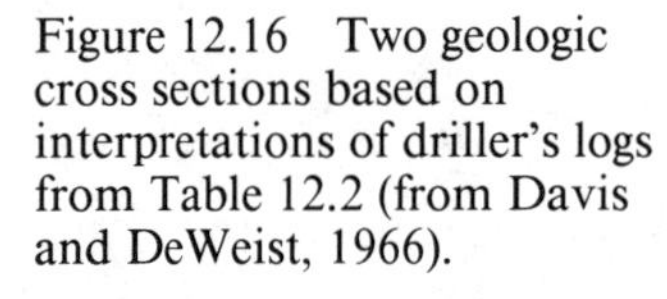

Figure 12.16 Two geologic cross sections based on interpretations of driller's logs from Table 12.2 (from Davis and DeWeist, 1966).

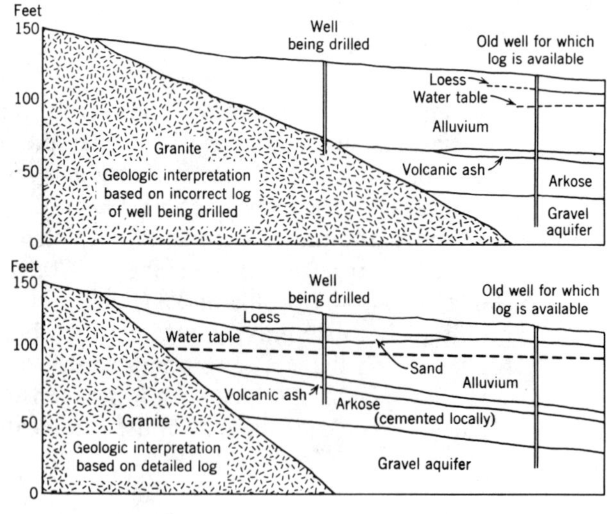

Table 12.2 Hypothetical Example of Good and Poor Driller's Logs

| Driller's log, poor | | Driller's log, good | | Geologic log, brief type | |
|---|---|---|---|---|---|
| Feet | | Feet | | Feet | |
| 0–10 | Soil | 0–10 | Silt, brown, soft | 0–2 | A and B horizons of modern soil. |
| | | | | 2–9 | Loess, light yellowish brown, friable, calcareous with gas-tropod remains. |
| 10–50 | Quicksand and water | 10–22 | Sand, tan, soft, some caving | 9–22 | Sand, medium to fine, well sorted, yellowish grey, friable. About ⅔ quartz and ⅓ feldspar. |
| | | 22–46 | Sand and silt with a little clay, tan, soft. Some water below 30 ft. Water stands in well at 29 ft. | 22–46 | Silt, sandy, poorly sorted, yellowish grey, friable. Small layers of clay less than 1 ft thick encountered every 1 to 2 ft. Clay is yellowish brown and compact. Sand, silt, and clay lack carbonates; composed of quartz, feldspar, and biotite. |
| | | 46–50 | Quicksand, white. Water rose in well, now stands at 25 ft. | 46–50 | Ash, rhyolitic, with about 10 per cent clean medium quartz, sand, light tan, friable. |
| 50–65 | Granite | 50–65 | Rock, hard, pink, looks like granite but drills easier than granite | 50–65 | Arkose, medium grained, light reddish brown, compact. |

[a]From Davis and DeWeist (1966).

ogists have stories about drillers who throw extra samples in the box at the end of the day, just to keep the box full! Shortcomings notwithstanding, driller's logs are often the main surce of information for test holes.

Drill cuttings obtained by sampling the drilling mud are often a mixture of material from the bottom of the hole, drilling mud, and material form higher layers that caved in. For these reasons, it is recommended that *changes* in sample composition are of greater importance than composition per se.

The penalties for careless or faulty logging in engineering are high, because the logs are often used as the basis for specifications available to bidders and become part of the legal contract. A practical, descriptive rock core log procedure for geotechnical investigation is given by Tepordei (1977). It includes a graphical boring log, written comments, and photographic records. Goodman (1976) describes

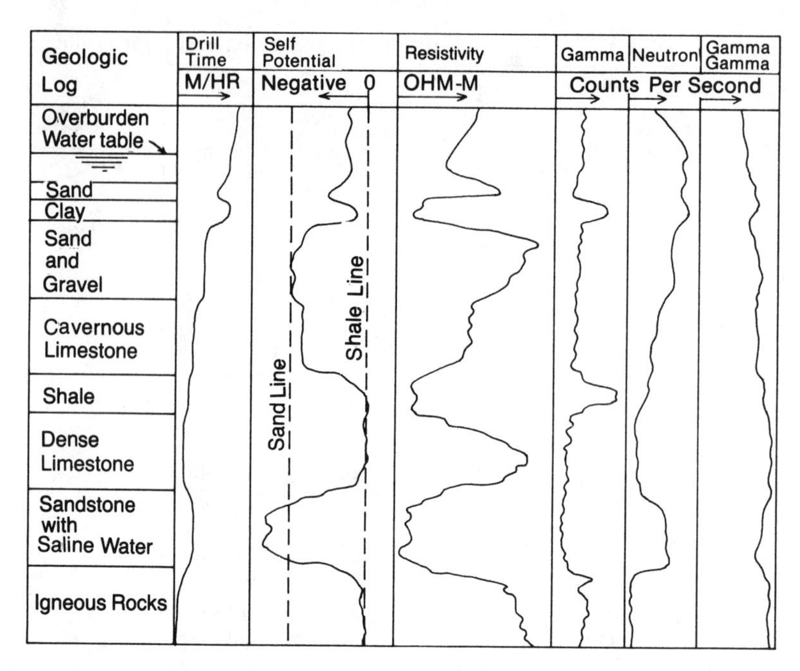


Figure 12.17 Hypothetical driller's logs and geophysical logs (from Bouwer, 1978).

methods of core logging. These logs may include the true attitudes of planar discontinuities. Absolute orientation of joints can be accomplished by downhole orientation measurements such as borehole periscopes or television cameras, paint marks, paleomagnetic techniques, or special core orienting devices.

There are numerous types of drilling log techniques besides geologic logs. A common technique is a log of the drilling rate, i.e., the record of distance drilled per unit time. Changes in drilling time may indicate discontinuities in lithology. Slow drilling rates normally are obtained in hard rock or clayey material. Figure 12.17 shows hypothetical driller's and geophysical logs. In addition to drilling time, there are numerous soil tests that can be utilized in connection with soil test borings (Figs. 5.10 and 5.11).

One problem with vertical boreholes is that they are not useful for analysis of vertical fractures. Attempts to use borehole data, for example, to locate optimum conditions from a hazardous liquid waste disposal well may not be adequate in an area where fractures are vertical.

## *12.4.2 Geophysical Logging*

Geophysical logs consist of the record obtained by lowering a probe into the uncased drill hole. Spontaneous potential and resistivity logging, collectively called electric logging, are among the most commonly used techniques. Electric logging was introduced in France by the Schlumberger brothers in 1927. Excellent references for geophysical logging include Schlumberger (1958), Davis and DeWeist (1966), Kelley (1969), and Keys (1970).

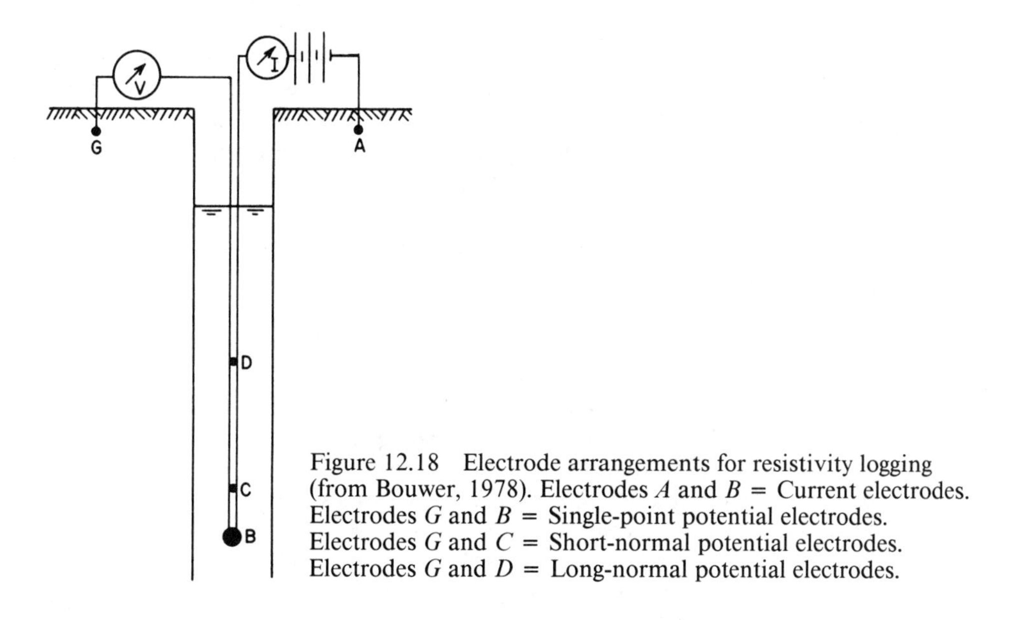


Figure 12.18 Electrode arrangements for resistivity logging (from Bouwer, 1978). Electrodes $A$ and $B$ = Current electrodes. Electrodes $G$ and $B$ = Single-point potential electrodes. Electrodes $G$ and $C$ = Short-normal potential electrodes. Electrodes $G$ and $D$ = Long-normal potential electrodes.

*Spontaneous potential (SP).* A log of the spontaneous (or self-) potential is obtained by recording the naturally occurring voltage difference between an electrode on the land surface and another electrode being lowered into the hole. The small variations in voltage differences are due to electrochemical or electrokinetic phenomena in the formations due to dissimilar fluids contained in the rocks penetrated and the bore holes. The equipment is basically very simple; it consists of two lead electrodes, one moving down the drill hole and the other stationary at the surface (or embedded in the pit of drilling mud, if used). A recorder plots millivolt changes in electrical potential as a function of depth.

The right (positive) side of the SP log is called the shale line and the left (negative) side the sand line (Fig. 12.17). Complicating factors such as salt water content make SP logs more difficult to interpret than idealized in Fig. 12.17 (Campbell and Lehr, 1973).

*Resistivity.* To obtain a resistivity log, a current is generated from the ground to a probe being lowered down the hole (Fig. 12.18). A voltage drop is determined between the surface and a probe located either at 1) the current electrode itself (single point device), 2) about 40 cm above the electrode (short normal device), or 3) about 100 cm above the electrode (long normal device). Single point and short-normal devices give best details of lithology (Bouwer, 1978). The single-point resistivity log, along with SP, are the two most widely used logging techniques in water wells (Keys, 1970).

The resistivity measured in a resistivity log is generally confined to the small spherical region around the electrode in the bore hole. The resistivity of the formation is heavily dependent upon the amount and salt content of water, as noted in the earlier part of this chapter. Clay and shale, typically having abundant porosity and containing highly mineralized water, offer little resistance to electricity. The resistivity is also affected by the drilling mud.

*Gamma ray.* All rocks and soils are naturally radioactive to some degree. The most common natural radioactive elements are $^{40}K$ and, to a lesser extent, the daughter products of uranium and thorium. Clays and shales have high potassium content in the form of clay materials. Gamma logs are used for lithologic interpretation and enhance the interpretation of electrical logs (Fig. 12.17).

*Neutron.* In this technique, a radioactive neutron source (such as 3 millicuries (mCi) of $^{241}Be$ which emits $8.62 \times 10^6$ neutrons per second) is lowered down into the hole. The intensity of attenuation (slowing) of these fast neutrons is a function of hydrogen present in the environs. A detector in the same probe measures the intensity of feedback of neutrons, which, for all extensive purposes, is related to water (or hydrocarbon) content.

In unsaturated deposits, neutron logging provides valuable information on water content (Wilson, 1981). Where saturated, the hydrogen content is related to porosity, so that a neutron log (Fig. 12.17) shows high readings in more porous rocks, such as shale. Keys (1970) noted that increased porosity, as evidenced by increased readings on a neutron log for a test hole in Texas, does not indicate that the effective porosity (interconnected pores) or permeability necessarily increases, because a gamma log showed these neutron log highs were merely due to an increase in clay content. When used in conjunction with a falling water table, neutron logging can determine the specific yield of an aquifer (Meyer, 1962).

Neutron-emitting loggers that have gamma ray detectors are presently being researched. The resulting neutron-gamma logs are sensitive to chloride content.

*Gamma-gamma.* These logs are obtained by lowering a gamma-emitting source (such as 10–35 mCi of $^{60}Co$) into the hole, and measuring the intensity of the backscattered and attenuated gamma rays with a detector in the same hole. This device is useful in determining the density of the rocks adjacent to the probe, so that bulk density and porosity data is obtainable (Fig. 12.17).

*Miscellaneous logging methods.* There are additional well-logging techniques being developed. Keys (1970) notes that the USGS uses, in addition to the logging techniques described above, the following: caliper, flow meter, radioactive tracer, fluid resistivity, and differential temperature. Qualitative log evaluation supply objective information on geometry, resistivity, bulk density, porosity, permeability, moisture content, and specific yield of rocks. Acoustic logs have been recently developed to measure the elastic wave properties of rocks by analysis of the compressional wave velocity. New advances in petroleum engineering include the digital recording of acoustic waveforms, which allow for interpretaion of both compressional and shear wave velocities (Timur, 1982). The acoustic images can be preserved on film, showing good resolution of fractures even when taken through a mud-filled hole (Nelson, 1982).

In summary, geophysical prospecting and logging techniques have proved to be an indispensable part of the petroleum industry. In the past few decades they have gained acceptance in the fields of ground water and engineering geology. Of course, there will always be a nonbeliever, a traditional geologist, who states: "I wouldn't trade a trunk load of seismic data for one good drill hole." In defense of geophysical logging, however, Keys (1970) responded that the USGS has 40 geophysical loggers, and the USGS uses ". . . . log wells when we need some *objective* information. I have very little use for driller's logs, which are more often than not misleading."

## Problems

**1.** Refer to the seismic refraction plot below:

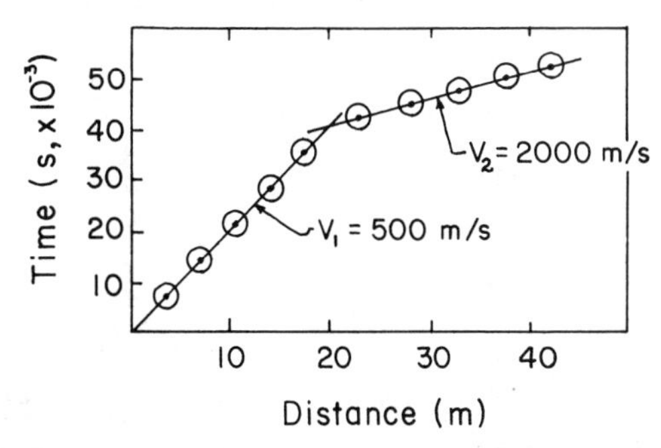


If $V_1$ is unsaturated alluvium (zone of aeration), and $V_2$ is saturated alluvium (zone of saturation), determine the depth to the water table.

**2.** Refer to Fig. 12.3B.

**a.** Use Snell's Law to predict the angle of incidence in order to have a wave bend so that it parallels the bedrock interface. Assume $V_1 = 1500$ m/s and $V_2 = 5000$ m/s.

**b.** Use a protractor on Fig. 12.3A and measure the angle of incidence. Is this angle equal to the angle calculated in a?

**3.** Use Eq. (12.4) to calculate $E$ for concrete if the longitudinal seismic velocity is 3570 m/s and the density is 2.18 g/cm$^3$. Assume Poisson's ratio is 0.21.

**4.** Determine the longitudinal and transverse seismic wave velocity for rock having a density of 2.65 g/cm$^3$, Young's modulus of $25 \times 10^9$ N/m$^2$, and Poisson's ratio of 0.25.

**5.** An explosion is detonated at the earth's surface at the top of a 500-m-thick layer where $V_1 = 2$ km/s. The underlying layer has a $V_2 = 4$ km/s.

**a.** Plot time versus distance on arithmetic paper for the direct and refracted wave from 0.5 to 6 km away. Use 0.5-km increments. Show first arrivals. Graphically determine the critical distance. Use Eq. (12.11) to confirm.

**b.** Which wave arrives first at a point 5 km away, the direct or refracted? How many seconds does it take?

**6.** Signals received from the lunar seismographs after impact of the abandoned Apollo 12 lunar module when it was intentionally crashed into the moon's surface brought surprising results. The signal reverberated for a long time, and the outer lunar material (regolith?) exhibited a low P-wave velocity, ranging from 1.84 to 1.75 km/s. A tongue-in-cheek comparison of these data with various kinds of cheeses (Schreiber and Anderson, 1970) shows that laboratory tests of compressional wave velocity of cheeses range from 2.12 to 1.54 km/s. Compare these two sets of P-wave data to data from terrestrial rocks (Fig. 12.1). What materials have seismic velocities similar to the moon's regolith?

**7.** The data below is from a resistivity survey at a golf course in Aberdeen, SD. The survey consists of 196 measurements (note: data are resistivity units, not ohm-m). The resistivity readings were made at five stations on a nearly level line. The stations were 61 m apart. Each station had six measurements with "a" spacing (Wenner array) progressively increased so that resistivity was being studied at depth increments of 3 m. (Note: this area is in the Lake Dakota plain, a very flat area which was once covered by a huge glacial lake. The surficial deposits consist of lacustrine silt containing gravel channels. The water table is very close to the land surface.)

| ←West | Station 1 | Station 2 | Station 3 | Station 4 | Station 5 | East→ |
|---|---|---|---|---|---|---|
| Depth 3 | 85 | 126 | 305 | 126 | 88 | |
| (m) 6 | 104 | 169 | 214 | 117 | 100 | |
| 9 | 89 | 128 | 140 | 112 | 100 | |
| 12 | 86 | 103 | 110 | 93 | 100 | |
| 15 | 79 | 85 | 95 | 88 | 95 | |
| 18 | 68 | 75 | 87 | 75 | 90 | |

**a.** Draw a cross-sectional profile from station 1 to 5, showing the resistivity data. Use arithmetic paper with vertical exaggeration equal to ×4. Draw contour lines of equal resistivity units, with 20 units as an interval.

**b.** Based on your interpretation of the data, where would be a good place to drill a water well? At what depth?

**c.** A year after the above geophysical survey was conducted, an out-of-state water well contractor was hired who located and drilled a production well near Station 5. A screen was set from 8.5 to 9.1 m depth. Despite the installation of a large pump, the well drew only small amounts of poor quality water, eventually becoming clogged with silt and calcite deposits. A few years later, the golf course went bankrupt mainly due to lack of irrigation water. Since the well driller had access to geophysical data, do you think the well driller was negligent in not paying any attention to the geophysical data?

**8.** The mass of the earth is about $598 \times 10^{25}$ g. Assume the distance from the center of the earth to the basement of the Mineral Industries Building at the South Dakota School of Mines and Technology is exactly 6370 km. The third floor is 6.92 m above the basement.

**a.** Calculate the theoretical difference in gravity between the basement and third floor based on 1) the free-air correction factor and 2) Newtonian physics [use Eq. (12.25)]. (Hint: use at least eight significant figures on calculations.)

**b.** Would this difference be discernible if a gravimeter reads to $10^{-3}$ mgal?

**c.** Using a gravity meter, it was determined that gravity was 980,268.21 mgal at the basement and 980,265.95 mgal on the third floor. What is the observed change in gravity?

# 13

# Mining and Energy

> The American Colossus was fiercely intent on appropriating and exploiting the riches of the richest of all continents—grasping with both hands, reaping where he had not sown, wasting what he thought would last forever.
>
> Gifford Pinchot

> Of all that was done in the past, you eat the fruit, either rotten or ripe.
>
> T. S. Eliot

> I suspect the energy crisis is over until we have our next energy crisis.
>
> James Schlesinger, 1982
> (former U.S. Department of Energy Secretary)

## 13.1 Mining

This chapter includes a discussion of mineral and energy resources, with emphasis on the environmental impacts caused by developing those resources.

In 1980, John B. Ivey, president of the Association of Engineering Geologists, made the following statement relative to the relationship between engineering technology and the mineral industry.

> Engineering geologists provide valuable data for the pre-mine design into the production phases of a mining operation. The work includes transportation networks (roads, railroads), dam sites and reservoirs (tailings, water supply, detention pads, etc.), site investigation, power transmission lines, soil and rock stability, hydrology (surface and ground water, water supply, waste-water disposal, etc.), geologic hazard determination and investigation, and many aspects of land reclamation.

### *13.1.1 Mineral Resources: Are We Running Out?*

The U.S. Geological Survey defines mineral resources as mineral concentrations "... discovered or only surmised, that are or might become economic sources of mineral raw materials." Mineral reserves "... are that portion of mineral resources that have actually been identified, and can be legally and economically extracted." The term "ore" is also used as synonymous with reserves of some minerals.

As a young nation, the United States had most of the resources it needed to be self-sufficient in lumber, iron, coal, copper, petroleum, etc. A few decades ago, oil was cheap because it was necessary only to figuratively scratch the surface to bring in a gusher. But now that easily obtained oil is gone, and the United States must use extremely expensive technology to obtain oil from great depths, from the continental shelves, and from nearly exhausted oil fields. In the past few decades, U.S. industries have become dependent upon foreign countries for a growing list of resources. Many minerals critical to well-being and defense such as cobalt and platinum are almost solely obtained from foreign sources thought to be unreliable. The U.S. mining and metals industry is also losing ground on a broad range of industries due, in part, to the fact that U.S. government regulations have added expenses not shared by many foreign competitors.

Experts differ on whether the mineral dependence of the U.S. should be cause for alarm. Some see the United States locked into a resource war that may ultimately lead to the equivalent of an oil embargo, so that the United States will be at the mercy of unstable and unfriendly suppliers of minerals. Others believe a reliance on international minerals is manageable.

In 1950, Congress established a National Defense stockpile. The stockpile contains 61 strategic materials, each with sufficient quantity to meet national security needs for three years. Over the past three decades, some of these materials have been sold off to help balance the federal budget.

In recent years, miners and environmentalists have become divided into two camps. On one extreme, there are preservationists (the "prophets of doom") and on the other, there are the hard-core miners (the "exploiters"). There has been much debate concerning the so-called "mineral crisis." Various opposing groups advocate either 1) the immediate expanded utilization of mineral deposits, 2) the preservation of mineral deposits for future generations, or 3) the cautious use of minerals.

Many of the proposals aimed at reducing U.S. dependence on unstable or unfriendly countries call for renewed efforts to exploit mineral resources in this country. There is no dearth of legislation, or lack of suggested remedies; many of them advocate easing restrictions or access to federal lands, relief from environmental regulation, improved stockpiling, import tariffs, fiscal incentives to U.S. industry, etc. Unfortunately, the public debate on U.S. mineral policy is stalemated due largely to the adversarial form of the pro- and anti-mining interests. "Thus, in arguing over the future of the public lands, the charge of 'locking up the nation's wealth' conventionally thrown at environmentalists is matched in silliness by the charge phrased by the opposite camp of 'ripping up the entire nation west of the Mississippi'" (Landsberg, 1982).

Facts can be stated in such a way as to bias a presentation. Consider Table 13.1, which is a list of mineral resources used in the United States, and a percentage of the mineral resources which are imported. Chromium, sheet mica, cobalt, etc., are mineral resources that, to a large degree, are not found in economic deposits in the United States. The data is alarming in that it shows how dependent the United States is on the mineral resources of other countries. To some people, the implication from Table 13.1 is that the United States needs to develop domestic mines for these short-supply minerals, or else be at the mercy of other nations in an emergency situation. On the other hand, Table 13.2 shows some other facts, in this case a list of the ratio of the per capita consumption of mineral resources in the United States versus other countries. It is clear that the United States consumes more (per person) than other countries. The implication from Table 13.2 is that the United State is a greedy nation, and in the future will probably have to cut back on its consumption as other nations become more affluent.

*Mining viewpoint.* There is an inherent optimism among most mining industries to the effect that American engineering can solve all the problems if only the government would get off their backs in terms of environmental and safety restrictions, and allow them freedom to mine in wilderness areas, in foreign countries, or wherever they choose. The 1981 Declaration of Policy of the American Mining Congress (AMC) states that "reserves of the more important minerals will be sufficient for several decades. Much of the major land areas and most of the ocean require further examination and improved methods of exploration, and improved techniques for extraction and processing . . . Over the last 20 years, the U.S. minerals processing industry has declined, in part as a consequence of the uncertainty deriving from restrictive environmental and safety regulations . . . The AMC seeks free access to foreign sources of minerals imports."

The optimistic enthusiasm of AMC is echoed by many people who feel that, although the percentage of a metal in ores keeps decreasing, American technology can keep refining lower and lower grade ores until, presumably, granite is mined for everything. (It is a well-known fact that every year lower grade ores are being mined. For example, the cutoff for uranium ore mined in 1950 was about 1% $U_3O_8$. In 1980, deposits containing only 0.025% $U_3O_8$ were being opened up.)

Promotion of the "mineral crisis" and its adverse effect on national security has been used by the mining industry to urge environmental cutbacks. In 1981, John W. Rold, president of the American Institute of Professional Geologists, said, "In the decade of the 1980s, the United States will see a minerals crisis which will make the energy crisis of the 1970s look like a Sunday School picnic . . . Our economy, the production of food, industrial products and energy, and our survival as a world leader depend upon the availability of the mineral materials." Rold advocates that government policy should encourage 1) exploration and development of our domestic sources of minerals, 2) discontinuation of future wilderness areas and implementation of periodic reviews of existing wilderness areas, and 3) the stockpiling of essential mineral materials by the government.

In 1982, M. O. Turner, President of the American Institute of Professional Geologists, urged that wilderness areas be opened for energy and mineral development:

> By denial of these vast expanses to the public for multiple use, these areas will become the private preserve of a new American aristocracy, the environmental groups whose current leadership is determined to seize control solely for the private use and privilege of the select few who are fortunate enough to have the time and money available for hiking and backpacking.

Table 13.1 United States Imports of Mineral Resources

| Minerals and metals | Net import reliance[a] as a percentage of apparent consumption[b] | Major foreign sources (1975–1978) |
|---|---|---|
| Columbium | 100 | Brazil, Canada, Thailand |
| Mica (sheet) | 100 | India, Brazil, Madagascar |
| Strontium | 100 | Mexico, Spain |
| Titanium (Rutile) | 100 | Australia, Japan, India |
| Manganese | 98 | South Africa, Gabon, Brazil, France |
| Tantalum | | Thailand, Canada, Malaysia, Brazil |
| Bauxite and alumina | 93 | Jamaica, Australia, Guinea, Surinam |
| Chromium | 90 | South Africa, U.S.S.R., S. Rhodesia, Turkey, Philippines |
| Cobalt | 90 | Zaire—Belg–Lux., Zambia, Finland, Canada |
| Platinum—group metals | 89 | South Africa, U.S.S.R., United Kingdom |
| Asbestos | 85 | Canada, South Africa |
| Tin | 81 | Malaysia, Thailand, Indonesia, Bolivia |
| Nickel | 77 | Canada, Norway, New Caledonia, Domin. Rep. |
| Cadmium | 66 | Canada, Australia, Mexico, Belg.–Lux |
| Potassium | 66 | Canada, Israel |
| Mercury | 62 | Algeria, Spain, Italy, Canada, Yugoslavia |
| Zinc | 62 | Canada, Mexico, Honduras, Spain |
| Tungsten | 59 | Canada, Bolivia, Rep. of Korea |

The mining industry has repeatedly urged the federal government to open wilderness areas. In 1982, John Goth, vice president of AMAX Corporation, said "It is totally impossible for Americans to achieve any kind of semblance of minerals self-sufficiency when our warehouses of known minerals have been padlocked. We must allow access to federal lands to explore and obtain information about our country's mineral resources."

The Heritage Foundation, a conservative think tank, urges relaxing environmental, health, and safety laws that are claimed to inhibit mining and mineral processing, and calls for tax laws favoring mining. In 1982, Secretary of the Interior James Watt said, "the best answer to mineral availability is domestic production." He proceeded to open up more public lands for this purpose.

*Environmental viewpoint.* The average American consumes 25 tons of raw material a year. More tonnage worldwide has been mined in the last 10 yr than all previous history combined. Raw materials, including fuels, are being produced

Table 13.1 (*Continued*)

| Minerals and metals | Net import reliance[a] as a percentage of apparent consumption[b] | Major foreign sources (1975–1978) |
|---|---|---|
| Gold | 56 | Canada, U.S.S.R., Switzerland |
| Titanium (ilmenite) | 46 | Australia, Canada |
| Silver | 45 | Canada, Mexico, Peru, United Kingdom |
| Antimony | 43 | South Africa, China Mainland, Mexico, Bolivia |
| Barium | 40 | Peru, Ireland, Mexico, Morocco |
| Selenium | 40 | Canada, Japan, Yugoslavia, Mexico |
| Gypsum | 33 | Canada, Mexico, Jamaica |
| Iron ore | 28 | Canada, Venezuela, Brazil, Liberia |
| Iron and steel scrap | (22) | |
| Vanadium | 25 | South Africa, Chile, U.S.S.R. |
| Copper | 13 | Canada, Chile, Zambia, Peru |
| Iron and steel products | 11 | Japan, Europe, Canada |
| Sulfur | 11 | Canada, Mexico |
| Cement | 10 | Canada, Mexico, Norway, Bahamas, |
| Salt | 9 | Canada, Mexico, Bahamas |
| Aluminum | 8 | Canada |
| Lead | 8 | Canada, Peru, Mexico, Honduras, Australia |
| Pumice and volcanic cinder | 4 | Greece, Italy |

[a]Net import reliance = imports-exports.
[b]Apparent consumption = U.S. Primary.
From U.S. Bureau of Mines

globally at a rate of some 15 billion tons annually. About a third of this production is consumed in the United States.

Environmentalists agree that there are shortages of some minerals, but that the "minerals crisis" is largely the result of mineral industry propaganda promoted to allow them to exploit public lands. Environmentalists point out that conservation and reuse of resources is a better alternative.

Despite the urgent warning of a mineral crisis by mining people, however, the National Research Council (1982a) of the National Academy of Sciences recently stated that humanity is not likely to run out of raw materials in spite of its voracious appetite. The report concluded that "non-fuel minerals are not likely to run short." It explains: "While current plus prospective reserves of some important minerals would probably be exhausted in a few decades at present relative prices, if there is no technological change and no change in recycling rates these conditions probably will not hold." The report asks rhetorically: "Can more abundant minerals be substituted for less abundant ones without substantial increases in

Table 13.2 Per Capita Consumption of Mineral Resources in 1969

| | United States | Rest of world |
|---|---|---|
| FUELS | | |
| Anthracite (tons) | 0.052 | 0.055 |
| Bituminous (tons) | 2.523 | 0.735 |
| Gas (cf) | 104,000 | 3.470 |
| Oil (bbl) | 19.41 | 3.16 |
| METALS (lbs) | | |
| Antimony | 0.4 | 0.015 |
| Bauxite | 158. | 19.4 |
| Beryllium | 0.1 | 0.0005 |
| Bismuth | 0.01 | 0.0017 |
| Cadmium | 0.08 | 0.005 |
| Chromite | 14.0 | 2.04 |
| Cobalt | 0.09 | 0.01 |
| Columbium–Tantalum | 0.03 | 0.005 |
| Copper | 20.0 | 2.35 |
| Gold (oz) | 0.34 | 0.011 |
| Ilmenite | 13.0 | 1.06 |
| Iron ore | 1400.0 | 303.0 |
| Lead | 7.8 | 1.6 |
| Magnesium | 1.0 | 0.05 |
| Manganese | 23.0 | 9.75 |
| Mercury | 0.03 | 0.004 |
| Molybdenum | 0.4 | 0.019 |
| Nickel | 1.3 | 0.21 |
| Platinum (oz) | 0.006 | 0.0007 |
| Silver (oz) | 0.8 | 0.033 |
| Thorium | 0.001 | 0.0004 |
| Tin | 0.6 | 0.095 |
| Tungsten | 0.06 | 0.016 |
| Uranium | 0.16 | 0.0045 |
| Vanadium | 0.07 | 0.005 |
| Zinc | 13.9 | 2.48 |
| NONMETALS (lbs) | | |
| Asbestos | 7.9 | 1.48 |
| Barite | 14. | 1.11 |

costs?" It concludes that, with few exceptions, "societies can turn to nearly inexhaustible minerals with little loss in welfare."

The U.S. Geological Survey recognizes that for some of the minerals in which U.S. production has been small for many years (tin, chrome, mercury, and the platinum metals), the United States does not have reserves. For several other minerals (copper, zinc, lead, manganese, nickel, tungsten, gold, and silver) mineral reserves plus resources are adequate until the end of the century. Only for a few minerals (iron, aluminum, molybdenum, and uranium) are resources apparently sufficient to meet demand for more than 100 yr.

Mineral reserve estimates are constantly changing. With the passage of time there are economic changes and a tendency toward technological innovation which result in a lowering of the grade of an ore. The U.S. National Commission on Materials Policy (1973) stated:

Table 13.2 *(Continued)*

| | United States | Rest of world |
|---|---|---|
| NONMETALS (lbs) | | |
| Borax | 8.2 | >0.337 |
| Cement | 760.0 | 280.0 |
| Clays | 560.0 | 120.0 |
| Corundum | n.a. | n.a. |
| Diamonds (industrial, carats) | 0.09 | 0.0037 |
| Diatomite | 4.8 | 0.73 |
| Feldspar | 7.2 | 0.81 |
| Fluorspar | 13.3 | 1.27 |
| Garnet | 0.2 | 0.022 |
| Gems (dollars) | $2.26 | $3.22 |
| Graphite | 0.5 | 0.23 |
| Gypsum | 155.0 | 21.7 |
| Kyanite | 1.0 | 0.14 |
| Lime | 197. | 26.6 |
| Lithium | 0.03 | 0.0011 |
| Mica (scrap) | 1.3 | 0.0168 |
| Mica (sheet) | 0.03 | 0.0039 |
| Olivine | n.a. | n.a. |
| Perlite | 4.0 | n.a. |
| Phosphate | 260.0 | 37.0 |
| Potash | 42.0 | 7.95 |
| Pumice | 30.0 | 6.4 |
| Quartz (crystal) | 0.001 | 0.00064 |
| Rutile | 1.7 | 0.145 |
| Salt | 462.0 | 45.0 |
| Saltcake | 18.0 | n.a. |
| Sand and gravel (tons) | 460.0 | 1.7 |
| Stone (tons) | 417.0 | 1.1 |
| Strontium | 0.07 | 0.0028 |
| Sulfur | 92.0 | 5.65 |
| Talc | 8.9 | 2.28 |
| Trona | 25.0 | n.a. |
| Vermiculite | 2.0 | 0.129 |

From Kruger, 1971.

Those unaware of the vast potential of our earth's finite crusts and seas fear increased consumption levels will deplete the earth's resources. Faulty comparisons of currently known mineral reserves with future demand, predicated upon exponential projections, serve to accentuate those fears.

Given the complexity of the task of establishing future resource availability, it is not surprising that widely different assessments have been made. Much depends on how optimistic the estimator is about the capability of technological innovation to redefine resources as mining and metallurgical processes improve, balanced against the rates of increased consumption. It is easy to show how inaccurate this method of estimation can be. For example, in 1950, the U.S. Bureau of Mines showed that world iron ore reserves would be exhausted by 1970. In 1970, however, enough proven reserves were inventoried to last 240 yr at the 1970 level of use.

*Relevance of free trade.* Mining crusaders see the struggle between the United States and the Soviet Union as a resource war. In the United States, environmental groups such as the Sierra Club or the Audubon Society feel that the resource war is being used as an excuse to relax pollution standards and open up wilderness lands to mining.

There are two sides of this complex issue. Is there a mineral crisis or not? Is it in the national interest to relax environmental, health, and safety regulations and grant new tax breaks and subsidies to mining companies? Would it be prudent to open public lands to indiscriminant mining? Some people say "yes" to these questions, and some say "no."

It is obvious that the geology of the world does not provide diversity so that every nation has just the right amount of all the ore deposits it needs. It logically follows that trade of minerals and other products is the sensible means to acquire mineral needs and to achieve the degree of technology that each nation desires. For example, a recent background document from the Commission of the European Communities encourages more trade, and points out that Japan, a relatively affluent nation, imports 90% of its raw materials. The Commission perceives mineral security not in terms of diplomacy designed to counter Soviet machinations, but in terms of a new level of cooperation with third-world nations.

History shows that over the past few hundred years, the major world powers have exploited the natural resources of their colonies. European colonization of America, Africa, and much of Asia enabled countries such as Spain, Germany, Portugal, Holland, Britain, France, and Belgium to enrich themselves by exploiting the food and mineral resources of vast territories. In the years leading up to World War II, Japan's desire for the resources of eastern Asia was a factor in its decision to use force. Since World War II, Russia has exploited East Germany, Manchuria, Poland, and other countries in its empire. Since World War II, the process of decolonization has, in general, reversed the tendency of powerful countries acquiring the possessions of others. England, for example, has recognized that decolonization has actually relieved the burden of administration and drain of maintaining military outposts. These factors, coupled with the colonies' desire for freedom, have resulted in a change in policy of the once-imperial powers, who have found it more profitable to trade with their former possessions.

But trade is fraught with political problems, and complex patterns of dependency emerge. Japan, for example, with few energy and mineral resources of its own, is in a potentially difficult position. Yet Japan's financial strength is so great that its mineral and energy suppliers find it attractive to offer long-term contracts. Brazil, on the other hand, has large resources but is heavily dependent upon foreign investment and technology to develop them. In developing mineral resources, technology and capital are often more important than the possession of a resource. Developing countries depend on multi-national companies for capital and technology. Because of these factors, there is no quick equilibration of fair trade among all nations.

To protect American steel plants, mining operations, and other industries, labor unions and industrial spokesmen often advocate that the United States should require tariffs against cheap imports. In the long run this action may hurt the U.S. economy because other nations would be tempted to reciprocate by imposing their own tariffs. Consider what happened in 1930 when President Hoover signed the highest of all tariffs. The Smoot-Hawley tariff put a protectionist wall around America. The loss of trade prevented prostrate Europe from paying its war debts. Within two years 25 countries had established retalitory tariffs. It opened the way for Hitler.

The general trend in the Twentieth Century has been away from imperialism and toward free trade among nations. The notable exception is the Organization of Petroleum Exporting Countries (OPEC), who have used their neomonopoly of petroleum supply to force other countries to pay heavily for oil.

*The future.* The future of mineral and energy industries depends on many factors. Kruger (1971) points out that: 1) the world population is growing, 2) most developing countries tend to imitate western affluent countries, and 3) the per capita consumption of minerals in newly industrialized countries has very high exponential growth. William Pecora, Undersecretary of the U.S. Department of the Interior, estimated in 1970 that the world's population would double in the next 35 yr, but the worldwide demand for minerals would increase more than fourfold. In 1971 William Pollard, director of Oak Ridge Associated Universities, warned, "Our present affluent society with its phenomenal standard of living has been created by abundant resources widely and cheaply available ... The removal of this resource base is certain to pull the rug out from under the affluent society ... The joyride we have been on in this country can last not much more than another 20 years. After that ... it will be simply impossible for us to continue to do simultaneously all the things we want to do."

The big question is: What level of affluence will the future have? Will everyone in the world of tomorrow be driving cars or peddling bicycles? The answer to this question lies not only in technology, but also in how soon we come to grips with population growth, education, environmental questions, and the recycling and conservation of resources. Perhaps the engineers of tomorrow can create a technical world where we will all drive big cars, yet the environment will remain pure. To do this will take a great deal of effort and even then may be impossible.

### *13.1.2 Environmental Impacts of Mining*

Engineering geologists more often become involved in environment-related problems associated with mining and milling than in the mining or metallurgical engineering per se.

Until about 100 yr ago, mining generally consisted of small scale operations which had a relatively minor impact on the environment. In fact, many old mining camps (ghost towns) in the western United States are tourist meccas today, and provide opportunity for recreation such as rock collecting. With the advent of hydraulic mining of gold in California around the turn of the century, mining impacts took a new dimension. (Hydraulic mining for placer gold in California at the turn of the century was devastating to the fluvial environment, and has been subsequently outlawed.)

In this century, mining has achieved an outstanding degree of efficiency of production from large volumes of low grade ores. Most U.S. mining takes place on relatively few, but enormous mines such as those producing gold at Lead, SD, copper at Butte, MT, Salt Lake City, UT, and Bisbee, AZ, silver at Coeur d'Alene, ID, lead and zinc at Joplin, MO, uranium near Gallup, NM, iron near Duluth, MN, phosphate near Tampa, FL, and coal in the Appalachians, Powder River Basin and elsewhere. In the United States, millions of hectares of land have been disturbed by mining, representing about 0.2% of the total U.S. land area. Most of the disturbance has been caused by coal (41%) and sand/gravel (26%), with the remainder mostly caused by stone, gold, clays, phosphate, and iron mining. About 40% of the disturbed area has been reclaimed. Many large open pit mines are clearly visible on Landsat imagery. For example, unreclaimed coal dumps in eastern Pennsylvania can be mapped using Landsat imagery (Short, 1982).

Environmental problems associated with underground mining include land subsidence (Chapter 9) and acid mine drainage from coal mines. Problems associated with surface mining include acid mine drainage and reclamation. Open-pit (surface) mining has all but replaced underground mining for metals (Church, 1981). Most veins of high-grade ore, which were best suited for underground mining, were discovered and played out long ago. Now, using mammoth earth-moving equipment, mining companies are tapping low-grade ore deposits. Nearly 90% of all metal ore mined in 1981 in the United States came from open-pit mines, which is up from 79% in 1959. About 60% of all coal is mined from open-pit mines.

Stream contamination as a result of mining and milling is well documented. An example is the mining of coal, particularly in Appalachia. The problem begins with the physical process of unearthing coal, which exposes pyrite ($FeS_2$), commonly associated with coal, to water and air. Pyrite reacts with oxygen and water to form ferrous sulfate and sulfuric acid (Dutcher et al., 1966). The following reactions are involved, resulting in the precipitation of ferric hydroxide:

$$
\begin{aligned}
&2FeS_2 + 2H_2O + 7O_2 \rightarrow 2FeSO_4 + 2H_2SO_4,\\
&4FeSO_4 + 2H_2SO_4 + O_2 \rightarrow 2Fe_2(SO_4)_3 + 2H_2O,\\
&2Fe_2(SO_4)_3 + 12H_2O \rightarrow 4Fe(OH)_3 + 6H_2SO_4.
\end{aligned}
\qquad (13.1)
$$

These reactions, whether within mines or waste piles, increase the concentrations of iron, sulfate, hardness, and total dissolved solids in water. Coal mines in particular have caused widespread acid mine drainage (Williams, 1975) and high sediment loads. For instance, in Kentucky, Collier et al., (1971) noted that streams draining coal strip-mine areas had a pH of 3–4, which is lethal to fish, and carried 12 times the dissolved solids and 75 times as much sediment load as streams draining unmined areas. Streams in West Virginia and Pennsylvania, where most coal production in Appalachia has occurred, have thousands of miles of adversely affected streams, many of which have a pH of less than 3 (Biesecker and George, 1972).

Western coals are low in pyrite but still have the potential for significant water degradation (Woessner et al., 1979). High sulfate and calcium concentrations are found in ground water associated with Powder River Basin spoils are primarily due to the solution of gypsum (Rahn, 1976).

Mining of sulfide-rich metallic ores can contaminate water resources. For example, lead and zinc were mined in the tristate (Kansas–Missouri–Oklahoma) area from the late nineteenth century until 1958. Pumping was required to keep the mines dewatered. Upon abandonment, ground water flooded the mines and in 1980 contaminated water overflowed into Tar Creek, OK. Because of the extent of contamination, the Tar Creek area was declared to be the nation's number one hazardous-waste site by the U.S. EPA.

### *13.1.3 Mine Reclamation*

Open-pit mining generally involves three steps: 1) the removal of unwanted overburden and dumping of this material as "spoils," 2) the mining of the mineral deposits, and 3) the reclamation of the disturbed land. Until about 1960, little reclamation was performed in the United States. Figure 13.1 illustrates the result of no reclamation in a coal strip-mined area near Sheridan, WY. The coal spoils were dragline-dumped about 1950. Almost no vegetation has grown on the piles in 35 yr. At nearby Decker, MT, active coal strip mining is in progress, but now state and federal laws regulate the mine reclamation. For example, about 30 m of overburden is removed by dragline or scrapers at Decker, and the 10 m thick subbituminous coal is then mined by bucket loader and power shovels. Spoils are dumped into the mined-out excavation and reclaimed by contouring, covering with topsoil, and seeding with grasses.

Reclamation is the process of artificially initiating and accelerating the natural recover and stabilization of a disturbed area. Two major aspects of reclamation include grading and revegetation. Adequate revegetation requires a suitable growth medium such as topsoil, although the growth medium need not be topsoil. For example, three products used to reclaim sterile tailings at the Urad molybdenum mine near Empire, CO, are waste rock, Denver sewage sludge, and wood chips from a nearby sawmill (Brown, 1982).

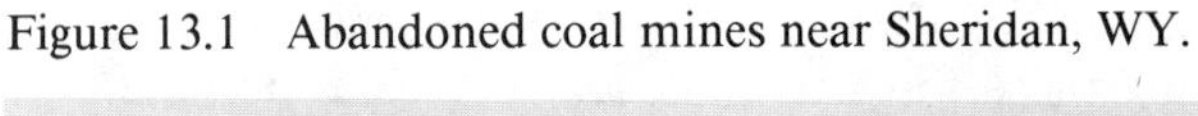

Figure 13.1 Abandoned coal mines near Sheridan, WY.

Reclamation is governed by climate, the physical and chemical composition of the mine wastes or spoils, the use of adjoining lands, and economics. The basic objectives of reclaiming mined lands are 1) stabilizing the spoils and open pit against erosion through grading and spoil replacement, 2) revegetating to help stabilize spoils and prevent erosion, and 3) reduction of mine drainage problems (Riley, 1973). Grading of spoils typically involves bulldozing to give the landscape a gently rolling appearance. In order to reduce erosion, some states require no slopes greater than 2°. Topsoil is usually stripped off and stored for use as a cover on the spoils. Then the topsoil is revegetated with fast-growing grass and other vegetation species, and possibly sowed with a bonding agent. Because mine wastes and tailings are usually infertile, fertilizers and mulch are often necessary to help promote initial plant growth. "Hydroseeding" is a simple operation in which the seed, wood fiber, soil seal, fertilizer, and water are all combined and sprayed on the topsoil (Ludeke, 1973).

Reclamation laws vary from state to state. There are also some federal regulations, such as the 1977 Surface Mine Control and Reclamation Act. There are, of course, questions as to the effectiveness of present reclamation practices. In 1980, mining company spokesman Leroy Balzer of Utah International said, "We in the business are committed to putting this land back the way we found it." Yet, environmental groups raise questions whether strip mining should be allowed at all in such fragile environments as the semi-arid west. Throughout arid areas of the world, millions of hectares of land are disturbed annually by mining (Day and Ludeke, 1979).

In some places, reclamation efforts include physical stabilization, such as adding clay or topsoil covers to these wastes, and chemical stabilization, such as by addition of petroleum by-products to form wind and water erosion-resistant crusts. Reclaimed land can be used for rangeland, row crops, or even orchards, depending upon climate and soil conditions. Surprisingly, in abandoned coal mines west of Joliet, IL, urban sprawl is making good use of unreclaimed open-pit coal mines. The old mine pits have filled with water and the spoils are forested over. Homesites among these forested hills sell for more money than the surrounding flat, undisturbed farmland. References on mine reclamation include Murray and Zahar (1973), Paone et al., (1974), Argall and Aplin (1976), Down and Stocks (1977), Coal Age (1978), and references contained therein.

One aspect of mining that should be emphasized is the sequential use of the land (Fig. 13.2). Once the land is urbanized, the resource (unless very deep) cannot be mined. Good planning, then, is of the essence. An example of sequential use of the land is the Denver area, which once had large sand and gravel quarries along the South Platte River. Once mined out, the quarries were used as a sanitary landfill site. The site was then reclaimed and today is the site of the Denver Coliseum (the "Mile-high Stadium"). Another example is Hackensack Meadowlands, New Jersey, long used as a dump, which was converted into DeKorte State Park and includes a football stadium and a ski slope.

Mined land can serve as a valuable community asset. For example, enormous underground limestone mines in Kansas City, Springfield, and Neosho, MO, have been converted into warehouses, cold storage facilities, offices, and manufacturing plants. These facilities offer advantages such as controlled humidity and temperature, freedom from street noises, lower fire risks, and decreased heating expenses. Today many underground mines in these horizontally bedded lime-

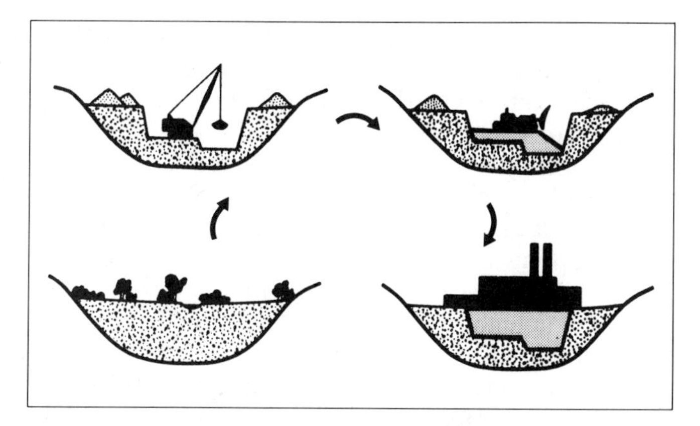

Figure 13.2 Sequential land use (from Robinson and Spieker, 1978). Undisturbed gravel-filled stream valley *(lower left)*; producing gravel pit *(upper left)*; pit filling and restoration of land surface *(upper right)*; use of filled-pit area—a factory site *(lower right)*.

stones are deliberately planned toward post-mining utilization. Originally the mines drifted in random pillar shape and arrangement. More recently, mine design has evolved to a strict geometric configuration (Hayes and Vineyard, 1969).

## 13.2 Metallurgical Processing

Most people are aware of the problem of smoke near metallurgical mills. Air pollution from smelters consists of particulates (smoke) and gases. The $SO_x$ group of pollutants is highly destructive. These gases are commonly caused by the roasting of sulfide ores and by burning pyrite-rich coal. Respiratory problems are caused at $SO_2$ concentrations of 4.5 ppm. Conifer trees are damaged at concentrations of 0.3 ppm. One of the largest smelters in the world, at Trail, British Columbia, was emitting 630 tonnes of $SO_2$ per month in 1930, and no conifers had survived nearer than 19 km (Down and Stocks, 1977). Similar incidents of denuded vegetation are known for a copper smelter at Ducktown, TN, a nickel smelter at Sudbury, Ontario, and a lead/silver/zinc smelter at Kellogg, ID. Let us briefly examine other environmental problems related to metallurgical processing with which engineering geologists are liable to become involved: water pollution and tailings dam construction.

### *13.2.1 Water Pollution*

Tailings (or tailing) are the residue left after the metal or mineral resource has been metallurgically extracted from the ore. Tailings differ from mine spoils in that the ore has been crushed, and usually chemicals of some kind have been added in order to extract the metal. Substances in tailings can be present in dissolved or mobilized form (including possible toxics), posing a greater hazard than mine spoils which consists only of broken rock.

The volume of tailings produced by a mill depends on the size of the operation and the type of metallurgical processes. At an iron mine, for example, where ore containing 30% by weight iron is enriched ("beneficiated") at a mill to pellets containing 60% iron, the tailings represent about 50% by weight of the ore that is mined. In a gold mine, ore containing perhaps only 10 ppm of gold is milled, and the gold extracted; the resulting tailings actually exceed 100% of the weight of the ore mined because chemicals and water are added in the milling process.

Several decades ago tailings disposal consisted simply of dumping into the nearest canyon, river, or lake. One of the worst U.S. mining horror stories is associated with the lead, silver, and zinc mines in northern Idaho. For decades, tailings laden with lead, cadmium, and other toxic chemicals were dumped into the pristine Coeur d'Alene River, one of the best trout streams in the United States. Government regulations in 1970 forced the mining companies to build tailings dams on mining company property. Today contaminants still seep into the Coeur d'Alene River to some degree, but the dams are a great improvement over the previous lamentable situation (Williams et al., 1971; 1979).

There are many other examples which show the historic lack of environmental concern by the mining industry. For example, at North America's largest gold mine, the Homestake in South Dakota, about 2500 tons/day of tailings containing arsenic, cyanide, and mercury were dumped directly into Whitewood Creek from about 1900 to 1978. In 1975, mercury was detected in fish as far away as the Missouri River, but Homestake Mining Company spokesmen said that if they had to build a tailings dam or stop using mercury to recover gold, the operation would be unprofitable and they would go out of business. The contamination problem was finally stopped by enforcement of the Federal Water Pollution Control Act of 1972, which prohibits disposal of industrial wastes into U.S. waters. In 1977, Homestake Mining Company spent $12 million to construct a tailings dam on low-permeability Precambrian rocks along Grizzly Gulch in the Black Hills (Carrigan and Shaddrick, 1977). At the same time, the price of gold jumped from about $35 to $570 per ounce, so the gold mine still made a profit.

An example of tailings problem and eventual cleanup is at Reserve Mining Company's taconite iron recovery mill at Silver Bay, Minnesota, where 67,000 tons/day of tailings were dumped directly into Lake Superior. The tailings moved as longshore drift toward Duluth, and soon Duluth's water supply (from Lake Superior) was found to contain asbestos fibers in tap water samples. Because asbestos is considered carcinogenic, in 1974 a U.S. District Court ordered Reserve Mining Company to stop dumping tailings into Lake Superior. (Note: There is debate among experts concerning the various types of asbestos fibers, and the toxicity of water-borne or atmospheric-borne asbestos in general. Rutstein (1982) discusses the toxicity of the various forms of asbestos minerals as well as cleavage fragments such as amphiboles, and concludes that the entire asbestos scare is an overreaction to a substance which has little health hazard. On the other hand, Reif (1981) points out that asbestos workers who smoke cigarettes incur an extremely heavy risk of developing lung cancer, presumably because prolonged irritation by asbestos increases susceptibility of the lung to inhaled carcinogens.) After much legal maneuvering, a new tailings repository was found about 11 km inland from Silver Bay, in an area underlain by low permeability glacial till over Precambrian volcanic rocks (Klohn and Dingeman, 1979). Today the tailings are slurried to the inland site and do not contaminate Lake Superior. The whole unpleasant problem would have been avoided if the lake was never viewed as a convenient dump in the first place.

Tailings dams can release contaminants to the environment. Seepage is fairly common. Uranium tailings have been dumped along the edge of the Colorado River and its tributaries. Uranium tailings typically contain uranium, selenium, arsenic, sulfate, and other contaminants. Precipitation entering the sandy tailings picks up contaminants such as radium and selenium and these solutions seep out into the river (Ford et al., 1977). Tailings such as these on flood plains can also be dispersed into the environment by floods. Uranium tailings at an active mill at Jeffrey City, WY, are discharged on top of arkosic sand, which is an alluvial-fan deposit. Tailings liquid from the mill seeps into the alluvium at a rate of about 80 liters per second (Wang and Williams, 1984) and a plume of arsenic-rich, contaminated ground water has been detected, moving downgradient toward Jeffrey City (D'Appolonia, 1977). In 1979, a uranium tailings dam burst completely and spread contaminants down the Little Colorado River.

### *13.2.2 Tailings Dam Safety*

The construction of a tailings dam follows traditional geotechnological principles for earth dam construction. The construction of safe tailings dams is important for environmental protection as well as public safety. With adequate construction, tailings can be practically isolated from the environment. The United States' most catastrophic tailings dam failure was the coal refuse dam in Buffalo Creek, WV, which was washed out February 26, 1972. The tailings dam retained coal refuse and tailings in a steep canyon. After heavy rain, the water overtopped the dam, and the dam failed, releasing $7 \times 10^5$ m$^3$ of water and tailings in a few minutes. More than 1500 homes were destroyed and 118 people killed. A law suit was initiated by more than 600 survivors who claimed some form of psychic impairment (Gleser et al., 1981). In Aberfan, South Wales, another coal tailings catastrophe involved a coal tip that failed in the form of a mud flow in 1966, (see Chapter 6). According to the National Research Council (1982c):

> In 1966, the failure of a coal tip at Aberfan, South Wales, killed 144 people. In 1972, a coal refuse disposal facility failure in Buffalo Creek, WV, resulted in a number of coal mining towns being wiped out, with the loss of about 100 lives. Investigation of the Buffalo Creek failure revealed that this and other (coal mining) disposal facilities were generally constructed with little thought given to planning and design. Geotechnical and hydraulic aspects of refuse materials and disposal were seldom considered. These factors are now being considered in constructing refuse disposal facilities, but much more needs to be done.

The safe construction of tailings dams is discussed in several publications, notably Williams (1975), Kuhnhausen (1979), Argall (1979), and Roberts (1981). The embankments may be constructed of earth, rock, or coarse coal refuse. Geotechnical analyses such as the method of slices (Chapter 6) should be conducted to design proper tailings dam slopes. Tailings fines are disposed upstream in the impoundment area. Typically, the fines are disposed in slurry form pumped directly from the plant.

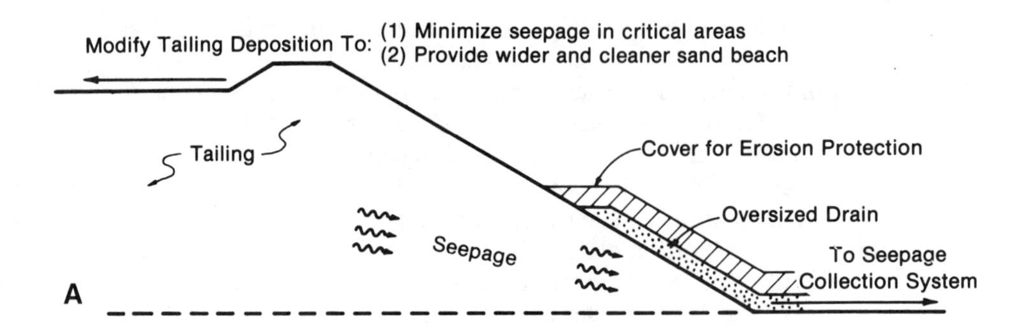


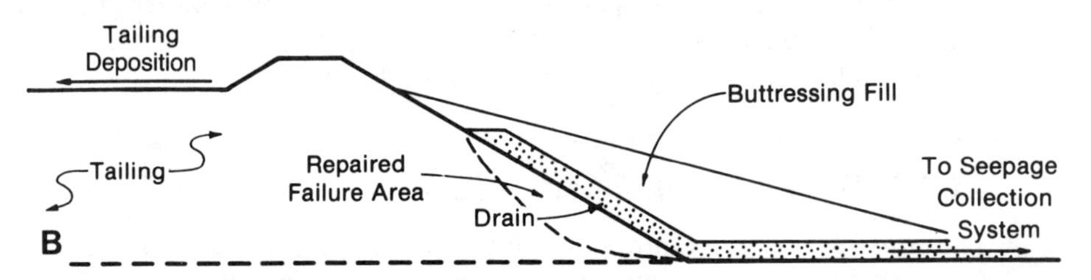


Figure 13.3 Tailings dam problems and correction methods (modified from Johnson, 1979). **A.** Seepage problem. **B.** Landslide problem.

Figure 13.3 illustrates two common types of problems that can lead to tailings dam failure. If the tailings have large permeability or large hydrostatic heads, seepage can occur. Seepage to the downstream side of a dam is especially critical because it can lead to piping or slumping, which can destroy the dam. Traditional remedial measures for a seepage problem are to cover the wet area with a coarse-grained drain material. Buttressing fills can be used to add support to the downstream face, underlain by proper drains (Fig. 13.3B).

## *13.2.3 Long-Term Environmental Impacts of Tailings*

Some tailings such as uranium tailings are environmentally hazardous for a long time after the mill is abandoned. Uranium is largely removed from the ore in the mill, but the tailings still contain radioactive elements such as thorium-230 and its daughter product, radium-226. Radon, a carcinogenic gas, is emitted from the decay of radium.

In recent years, a problem has arisen by the unregulated public usage of uranium tailings at Grand Junction, CO, and other places. The sandy tailings were hauled away by contractors and private individuals for use as sand in concrete, and as a backfill material around concrete floors and basements. Thousands of buildings, including schools, are now weakly radioactive and radon gas is continually released into the air inside the buildings (Bailey et al., 1978). Only removal of the radioactive material from the buildings will eliminate the hazard.

The large volumes and long half-life of thorium-230 (80,000 yr) makes uranium tailings as much of a long-term radioactive health hazard as the spent fuel

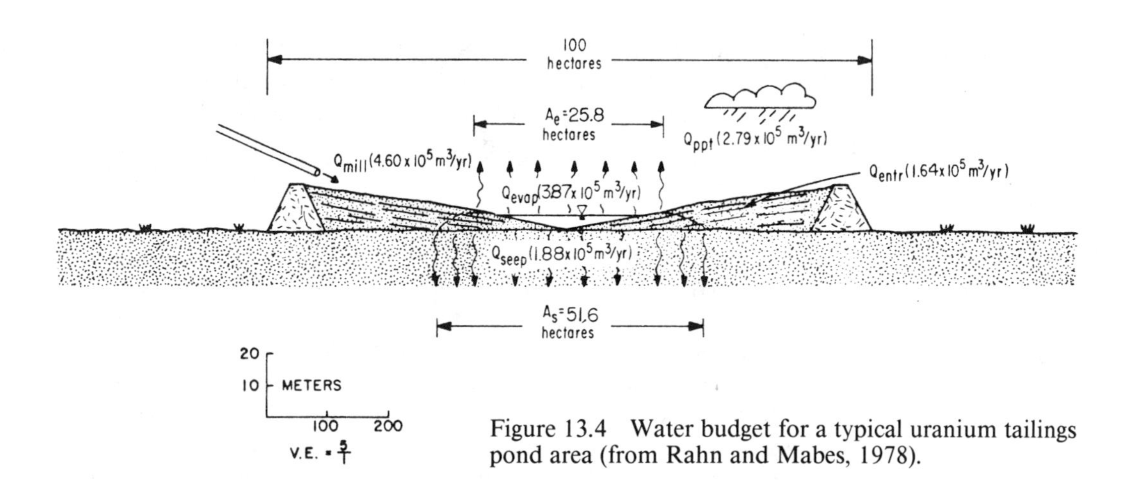


Figure 13.4 Water budget for a typical uranium tailings pond area (from Rahn and Mabes, 1978).

rods from nuclear reactors (Bredehoeft et al., 1978; Comey, 1983). Uranium tailings emit radon gas continuously. Field monitoring of the Anaconda uranium mill at Bluewater, NM, shows that radon gas concentration above the tailings diminishes to about 1% about 0.1 km downwind from the mill (Momeni et al., 1978). In 1978, the U.S. NRC regulatory policy required the covering of abandoned tailings with sufficient dirt to reduce the radon emanation rate to twice the emanation rate from the surrounding area (Scarano and Linehan, 1978). When a uranium mill has finished operation and is to be abandoned, the 1979 NRC mill licensing criteria require that tailings be covered with 3 m of earth so as to reduce radon emission to no more than 2 pCi/m$^2$/s. (Note: Mining company reaction to federal regulations concerned with abandoned uranium tailings has been negative. In response to the NRC regulations, three suits, brought by the American Mining Congress and uranium industries, challenge NRC's regulations as arbitrary and capricious.)

Uranium tailings can cause ground-water contamination for a long time after abandonment. Kaufman et al. (1976) documents elevated radium, selenium, nitrate, and other contaminants in water wells near uranium tailings in New Mexico. Figure 13.4 is a generalized model of a typical uranium tailings pond described by the NRC. In a semiarid climate such as New Mexico and Wyoming, an estimated $3.87 \times 10^5$ m$^3$/yr of tailings water evaporates, but some water containing toxic substances also seeps into the subsoil ($1.88 \times 10^5$ m$^3$/yr). Some contaminants such as thorium and radium may precipitate out a short distance into the aquifer but other contaminants such as selenium and uranium remain dissolved as constituents in ground water.

Most mining companies rarely acknowledge that any kind of tailings pond loses any water by seepage at all; they refer to the pond as an evaporation pond. However, a simple calculation of regional precipitation and evaporation rates (see Chapter 7) can be used to quantify seepage and evaporation rates. In some cases, where an active tailings pond lacks any water at all and is underlain by permeable substrata such as coarse alluvium, there can be no doubt what is happening. (Note: "Hypalon" and other synthetic liners may be used to reduce the adverse impacts of seepage. However, the Envrionmental Protection Agency (1980) has located over 24,000 mining waste water impoundments in the United States, not to mention tens of thousands of industrial, municipal, and agricultural impoundments and oil brine pits. Seventy percent of the impoundments were unlined, and only 5% were monitored for ground water quality.)

## 13.3 Energy

Engineering geologists will be called upon to contribute their talents toward solution of the energy crisis, and will become involved in such challenging and diverse problems as exploration for and development of coal and oil, as well as the environmental impacts and safety of related projects ranging from offshore drilling platforms to the disposal of high-level radioactive wastes. Issues relating to the development and use of energy have occupied a lot of public attention in the past decade. This section includes a brief discussion of some of the more important issues of energy resources and production.

### *13.3.1 Fossil Fuels*

Information representing U.S. energy consumption are shown in Fig. 13.5. The increase in energy consumption since 1930 has been substantial. Perhaps the most significant fact is the leveling off of U.S. oil and gas production, and the increasing amount of imported oil and gas.

Figure 13.6 shows the consumption of energy for U.S. electric utilities. In general, gas- and oil-fired electric plants are being phased out while increasing reliance for power is being made on coal and nuclear energy.

*Gas.* In the early part of this century, natural gas was generally considered a nuisance in U.S. oil fields and was deliberately burned off (flamed) at the site. (This is still practiced in the Middle East.) The economic value of natural gas was soon recognized in the United States, and today natural gas is a much sought-after commodity. Pipelines carry natural gas all over the country. Because of the seasonal demand for natural gas for residential heating, storage of gas must be practiced, including the use of natural features such as anticlines in Illinois and other places.

The use of gas in the near future depends on its price which is controlled by government regulations. The long-term use of gas (Fig. 13.7) indicates a very rapid decline in production due to diminishing reserves. Europe has a somewhat better future because in 1983 a 4500-km pipeline was completed from the Urengoi gas fields in northern Siberia to Czechoslovakia. The 1.4-m diameter pipeline will deliver up to 20 million $m^3$ of natural gas yearly to energy-hungry western Europe.

*Petroleum.* The primary use of petroleum (crude oil) is for the production of gasoline for automobiles. In the United States, the number of automobiles rose from 25 million to 125 million between 1940 and 1980. Industrial nations consume a lot of petroleum. The average per capita consumption of energy in the United States is the highest in the world. The United States has only 6% of the world's population, but uses 25% of the world's petroleum. As other countries try

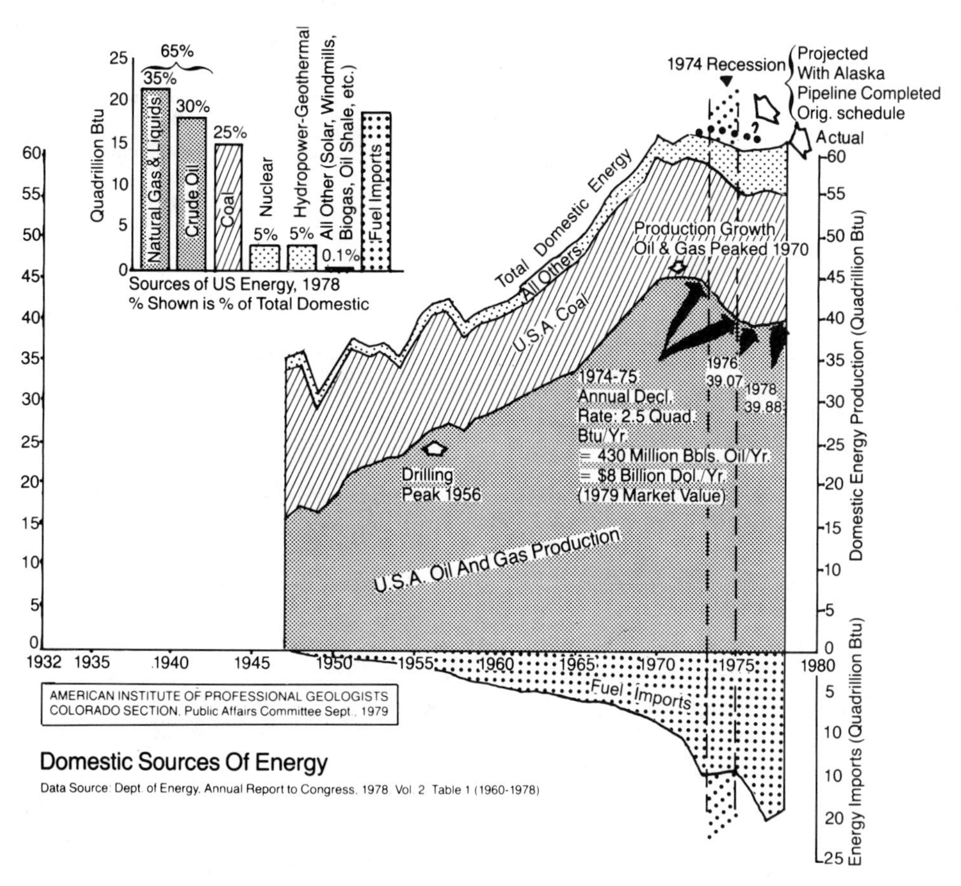


Figure 13.5 United States sources of energy (from American Institute of Professional Geologists, 1980).

Figure 13.6 Total United States energy consumption (from Energy Information Administration). (Note: 1 quad = $10^{15}$ BTU = $1.7 \times 10^{8}$ bbls of oil equivalent.)

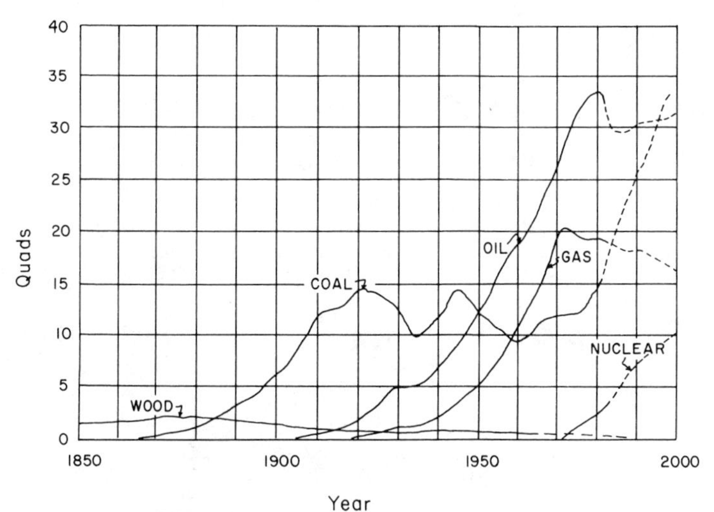

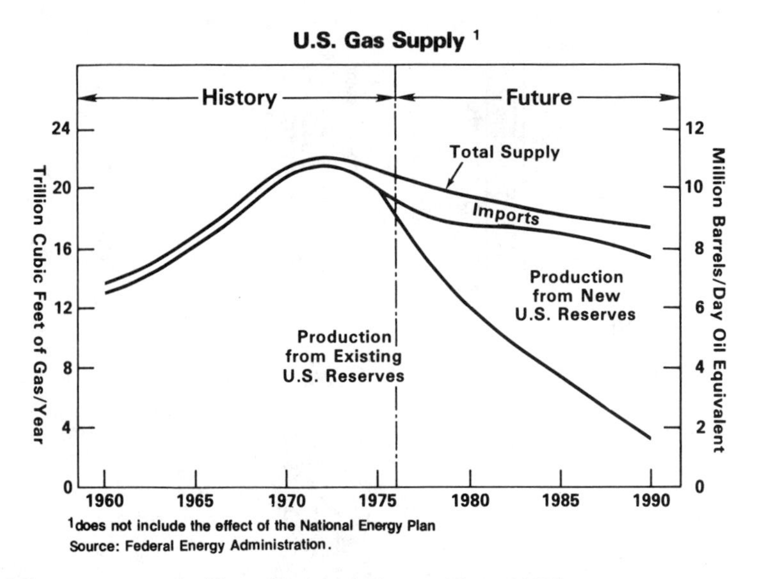


Figure 13.7 United States gas supply (from National Energy Plan, 1977).

to gain the affluence of developed countries, their need for oil skyrockets. The world price of crude oil rose from \$1.24/bbl in 1970 to \$41/bbl in 1980. Much of this was brought on by OPEC, capitalizing on their enviable position of abundant petroleum resources in a world whose population, cars, and attendant energy needs are skyrocketing.

Hubbert (1971, 1973) pointed out that United States oil production data follows a bell-shaped or normal (Gaussian) curve typical of the development, consumption, and eventual decline of any finite resource (Fig. 13.8). By 1983, it was clear that U.S. crude oil production had peaked in 1970, vindicating Hubbert's prediction. "There is no way to reverse the declining curve of oil production in the United States," stated Robert W. Baldwin, President of Gulf Oil Refining and Marketing Company, in 1979.

Figure 13.8 Hubbert's prediction (originally made in 1956) of the ultimate U.S. crude oil production (from Hubbert, 1973). The volume of one block in this graph is $25 \times 10^9$ bbls. A recent estimate of the ultimate total production is about $200 \times 10^9$ bbls.

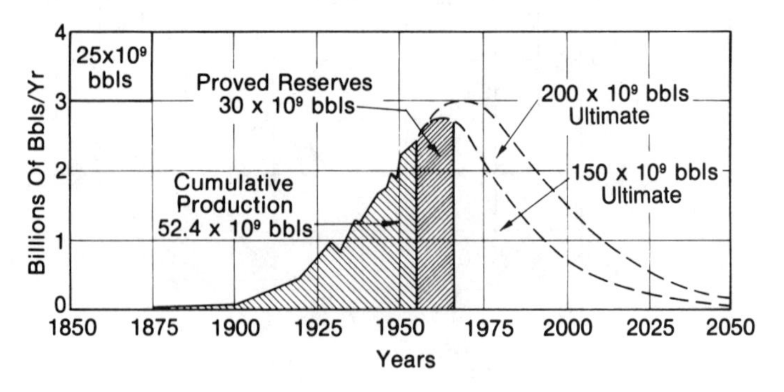

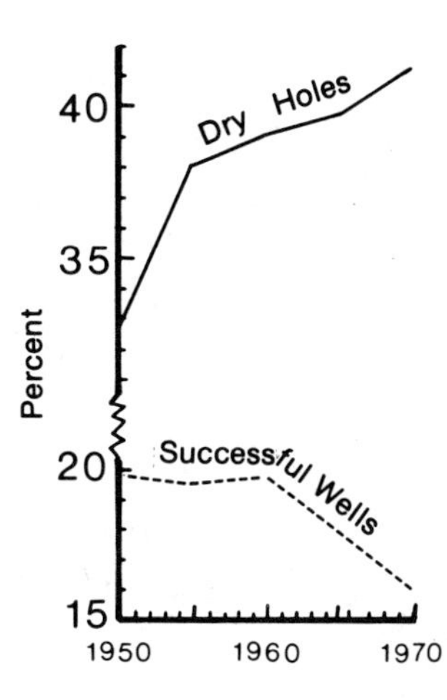


Figure 13.9 Percent of United States exploratory holes which were successful or dry holes (from Lindholm, 1980).

In the United States, petroleum companies are desperately coaxing the last oil and gas from the ground (Klins, 1979). The United States has been thoroughly drilled; it contains four times as many wells as the rest of the noncommunist countries combined. In 1980 almost 60,000 wells were drilled in the United States. Figure 13.9 illustrates that the percentage of exploratory oil drill holes which are succcessful is on the decline. It used to be that petroleum was plentiful (Fig. 13.10) and that a hole recovering hundreds of barrels of oil per day was only marginal. Today wells producing 3 bbl/day are kept in operation. High oil prices have caused a high level of drilling activity in the United States, but the new fields being found are small. The outlook: continuing decline in U.S. oil resources and production.

The development of an oil reservoir normally begins by primary production where gas drive is the predominant driving force. Secondary recovery, usually waterflooding, is a low-cost and highly predictable recovery method. In the United States, primary production results in about 4 million bbl/day of oil, whereas waterflood results in about 4.1 million bbl/day. Enhanced oil recovery such as steam recovery of heavy oil in California results in the production of another 0.4 million bbl/day (Ward, 1982). The Interstate Oil and Compact Commission (1982) contains numerous references detailing the state-of-the-art for enhanced recovery. Hydraulic fracturing and other types of enhanced recovery techniques have made a significant contribution to the petroleum industry as a method of increasing oil and gas production rates (Fayers, 1981). Introduced in 1949, about 40% of all currently drilled wells are hydraulically fractured. Different types of fracturing fluids are used. Proppants range from silica sand, the standard, to high-strength materials used in deep formations where fracture closure stresses exceed the strength of sand. The theory of fracture propigation (vertical versus horizontal fracture, geometry of expanding fracture zone, etc.) and economic considerations for hydraulic fracturing is discussed by Veatch (1983). The design of hydraulic fracturing systems still involves a considerable amount of judgement and experience, due to limited knowledge of in situ fracture shapes, fracture-induced permeabilities, and in situ rock properties and stress fields.

An example of the usefulness of enhanced oil recovery is given by the Jay Oil

Figure 13.10 Signal Hill oil field, Long Beach, CA (photo taken in 1930 by Fairchild Aerial Surveys). The density of oil wells on this anticline illustrates a negative facet of unregulated development of natural resources in that the oil could have been more judiciously obtained by fewer wells.

Field in northwestern Florida, the largest and most prolific oil field discovered in the lower 48 states in the past 25 yr. The field produces 45,000 bbl of oil and over a million $m^3$ of gas daily from the 5,000-m-deep Smackover Limestone. Secondary recovery techniques include waterflood, the pumping of fresh water into the producing formation. Tertiary recovery (or enhanced recovery) includes miscible flooding in which nitrogen gas is pumped into the producing formation (the nitrogen mixes with the oil and helps drive it up the producing wells). As a result of enhanced recovery techniques, the life of the Jay field is expected to increase from 1992 to 2004 A.D.

Despite enhanced recovery techniques such as waterflood, in situ combustion, and injection of surfactants, carbon dioxide, or other chemicals, these measures cannot be considered a solution to the energy crisis. According to Doscher (1981, p. 199):

> In view of the stark decline in producing capacity in the U.S. and the low probability of finding a significant number of giant and super-giant fields in the future, the enhanced recovery techniques developed to date can, at very best, be counted on only to slow down the decline in domestic crude oil supplies. Because of the national

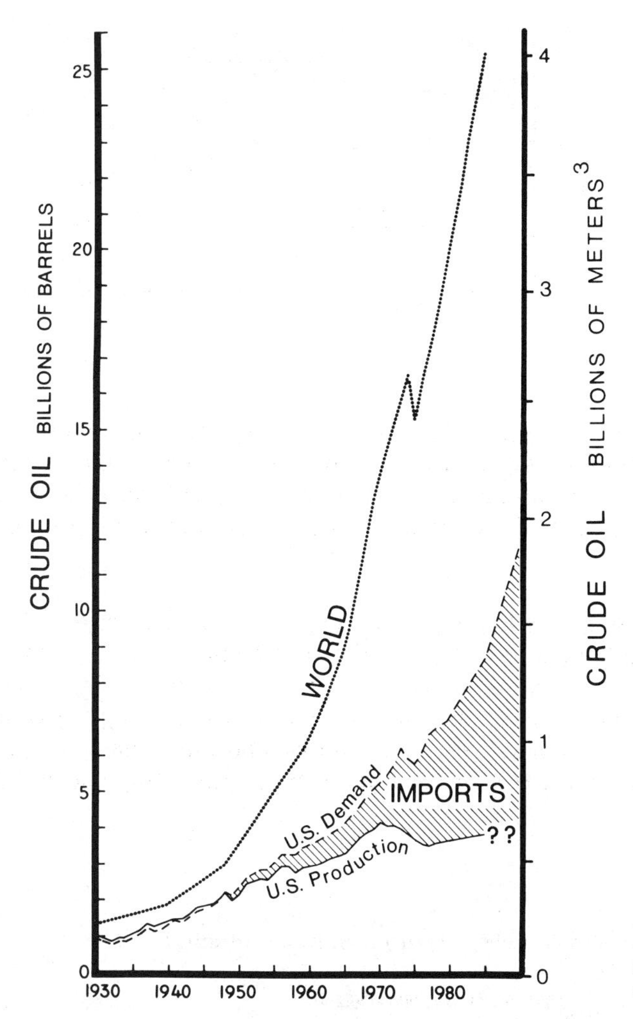


Figure 13.11 Worldwide annual crude oil production (from Lindholm, 1980).

need, continuing efforts are certainly called for, but they should not be substituted for the development of alternate energy sources.

Petroleum production worldwide is increasing (Fig. 13.11). But the production of oil in the world, like U.S. production, will eventually reach a maximum and decline. Figure 13.12 illustrates world sources of oil. In 1978, the United States was consuming 16 million bbl/day, of which 8 million bbl were imported. About 2 million bbl of the imported oil came from Saudi Arabia, which will probably continue to remain the chief source of new oil for the United States. Already the U.S. trade deficit, due mainly to oil expenditures, approaches $45 billion yearly. Continued reliance on imports puts the U.S. in a tenuous position and causes economic unbalance.

Between 1978 and 1983 a recession and the introduction of more fuel-efficient automobiles caused a reduction in U.S. oil consumption, from 18.9 million to 15.2 million bbl/day. This optimistic sign is the only good news in an otherwise

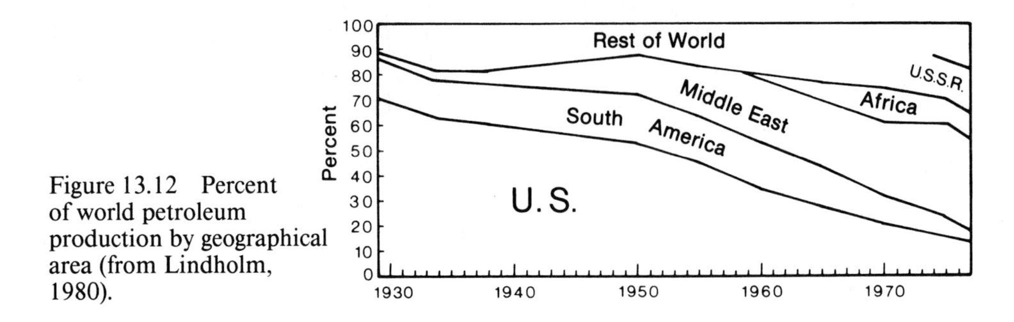


Figure 13.12 Percent of world petroleum production by geographical area (from Lindholm, 1980).

gloomy energy picture. There have been statements by some oil companies and politicians to the effect that the energy crisis could be eliminated if only the government would relax environmental restrictions and well-head price controls. The facts are, however, that the United States is simply running out of gas and oil.

*Coal.* The United States has vast coal deposits, and has been called the "Saudi Arabia of coal." Figure 13.13 shows the location of U.S. coal deposits. The largest deposits consist of anthracite coal in the folded Pennsylvanian strata of Pennsylvania, bituminous coal in the flat-lying Pennsylvanian strata of West Virginia, Kentucky, Ohio, Pennsylvania, and Illinois, and bituminous to sub-bituminous coal in the flat-lying Paleocene strata in Wyoming, Montana, and North Dakota.

Figure 13.13 United States coal deposits (from Gray and Bruhn, 1984).

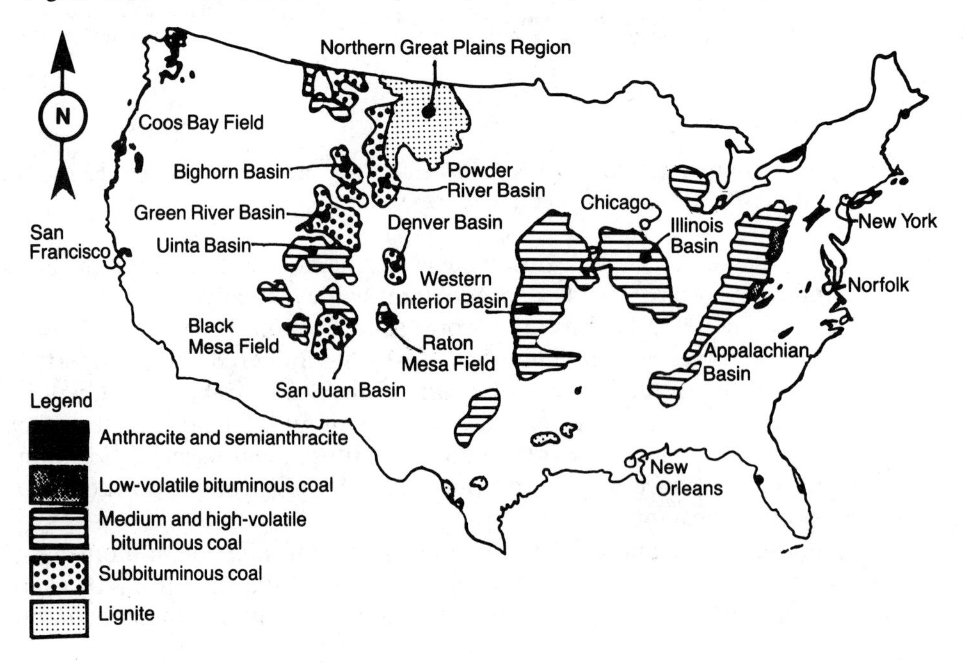

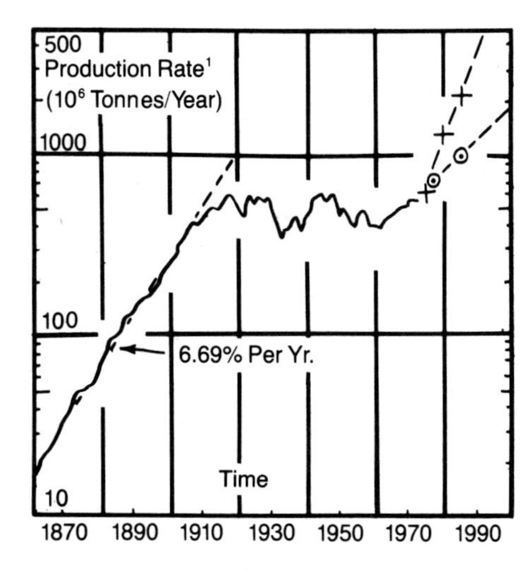


Figure 13.14 United States coal production (from Bartlett, 1980).

Coal is primarily used in the United States for electricity production and for residential heating. The future of coal for electric power is believed to be great (Fig. 13.6). A limiting factor in the use of coal is that it cannot be used for automobiles except by expensive synthetic conversion.

Figure 13.14 shows historic production of U.S. coal. From the close of the Civil War to about the year 1910, coal production grew steadily at 6.69%/yr. Production leveled off for 60 years as people switched from coal to fuel oil for residential heating because of oil's low price and because oil furnaces require less maintenance than coal furnaces. In the 1970s, renewed coal consumption occurred due to electrical power needs. The crosses in the steep dashed curve on the right side of Figure 13.14 show the coal production goals set by the Ford Administration, while the circles in the lower dashed line represent the production goal of the Carter Administration. The long dotted straight line at the left shows what would have happened if the growth of consumption had not stopped or slowed in 1910.

The ultimate depletion of a resource must take into account the reserves and rate of usage. (Note: It is difficult to quantify coal reserves, because coal beds dip into hillsides so that overburden gradually increases and the benefit/cost ratio for strip mining gradually decreases until the coal is no longer economical to mine.) The rate of coal consumption in the past has been exponential, and the future prediction of this resource can be determined using the simple exponential growth formula [Eq. (1.1)]. Figure 13.14 data follow an exponential equation of 6.69%/yr except for the period of coal stagnation as mentioned above. Using the 1976 production of 0.6 billion tons, a reserve base of $379 \times 10^9$ tons and a total identified resource of $1546 \times 10^9$ tons, the life expectancy of coal consumption at various rates of growth is shown in Table 13.3. At the present rate of consumption and zero annual growth, for example, the total identified coal resources would last 515 yr. However, at a continued 6.69% growth, they will last only 53 yr, and at 10% growth (1980 federal administration goal), the coal would last only 40 yr.

Conflicting statements have arisen due to misinterpretation of coal reserves in terms of exponential usage (Bartlett, 1980). A most common misinterpretation is

Table 13.3 U.S. Coal Production and Expiration Times

| Annual growth | Reserve base ($379 \times 10^9$ T) | Total identified resources (1546 $\times 10^9$ T) | Total including hypothetical ($3543 \times 10^9$ T) |
|---|---|---|---|
| Zero | 632 (126) | 2577 (515) | 5905 (1181) |
| 1 | 199 (82) | 329 (182) | 410 (255) |
| 2 | 131 (63) | 198 (121) | 239 (160) |
| 3 | 100 (52) | 145 (93) | 173 (120) |
| 4 | 82 (45) | 116 (77) | 137 (97) |
| 5 | 70 (40) | 97 (66) | 114 (82) |
| 6 | 61 (36) | 84 (58) | 98 (71) |
| 6.69 | 56 (34) | 77 (53) | 89 (66) |
| 7 | 54 (33) | 74 (52) | 86 (63) |
| 8 | 49 (30) | 67 (47) | 77 (57) |
| 9 | 45 (28) | 61 (43) | 70 (52) |
| 10 | 42 (26) | 56 (40) | 64 (48) |
| 11 | 39 (25) | 51 (37) | 59 (44) |
| 12 | 36 (23) | 48 (35) | 55 (41) |
| 13 | 34 (22) | 45 (32) | 51 (39) |
| 14 | 32 (21) | 42 (31) | 48 (37) |
| 15 | 30 (20) | 40 (29) | 45 (35) |

From Bartlett, 1980. The three columsn indicate the total tonnage of coal available. The data given is years until this tonnage will be consumed. The figures are based on the 1976 annual production rate of $0.6 \times 10^9$ tons, and they tell how many years U.S. coal will last if production grows at various rates. Each number in the table is the expiration time, which shows how long coal could satisfy U.S. requirements. Since coal supplies only about ⅕ of present U.S. energy needs, the figures in parenthesis are based in an annual production rate of $5 \times 0.6 \times 10^9$ tons per year, whereby coal could supply all our energy needs.

exemplified by the following statement, issued in 1975 by the American Electric Power Company: "We are sitting on half of the world's known supply of coal—enough for over 500 years." The 500-yr coal life expectancy figure has been published in many places. Clearly, however, as oil and gas are depleted and annual coal production increases, coal cannot support our needs for that length of time.

*Synfuels.* Many experts describe a transition between traditional fossil energy sources (such as gas, oil and coal) and the "unlimited" sources of fuel provided by 1) nuclear energy or 2) synfuels, the conversion of oil shale or coal into liquid fuel.

There is some reason to believe conversion of coal to liquid fuel can be done economically. During World War II, for example, the Germans ran many of their military trucks and tanks on synfuels. Presently, near Johannesburg, South Africa, three mammoth synfuel plants are being constructed. The United States, which imported $50 billion in crude oil in 1981, has huge coal and oil shale deposits, which are the fossil fuels most readily available for conversion. Sub-bituminous coal beds in Montana and bituminous coal beds in Illinois are considered by some to be the most favorable raw material. The richest oil shale deposits are found in the Green River Formation in western states, especially Colorado's Piceance Creek Basin. Oil shale is only 10% oil, but the deposit is capable of yielding about 57 liters of oil per ton of rock. The rocks underlie vast areas of land, and contain an estimated 2 trillion barrels of oil (or 10 times the total crude ever consumed in the U.S.). Oil shale, rich in kerogen, must be heated to a temperature of 500° C. in order to convert kerogen to "syncrude" (synthetic crude oil). The most developed method of processing oil shale is to strip-mine the shale, and process it at a surface facility. These facilities consume large amounts of water and yield tremendous quantities of waste rock. To eliminate mining and the waste rock disposal problem, in situ burning research is being conducted by the federal government.

Another synfuel process is the production of pipeline gas from coal. Governmental research has shown that pulverized coal, under pressure and temperature and with the proper solvents and catalysts, can be made into gas. There are no operating commercial gas plants in the U.S., but plans have been made to build giant plants near Fort Union coal deposits in North Dakota, Montana, and Wyoming. The "Great Plains" venture under construction near Beulah, ND, is the largest synfuels project in the U.S. In 1985 the project was having economic difficulties because the price of gas (which is regulated by the federal government and is tied to the price of fuel oil) declined slightly in the mid-1980s.

As of 1984, the U.S. has not produced any commercial synfuel. Problems of economics (large initial capital costs), mining and environmental considerations, and availability of water have proved to be great obstacles for private investors.

Another synfuel is called "gasohol," a blend of 90% gasoline and 10% alcohol. The alcohol (ethanol) is produced by fermenting and distilling agricultural crops such as corn, wheat, timber or sugar cane. There is, however, a problem with agricultural supply for production of alcohol. In 1980, the total corn crop processed in the U.S. amounted to $20 \times 10^6$ m$^3$, and about $3 \times 10^6$ m$^3$ went for gasohol production. If all $20 \times 10^6$ m$^3$ were diverted to alcohol production, it would make $5 \times 10^6$ m$^3$ of alcohol. This is about 1% of the present U.S. annual gasoline consumption. In other words, gasohol's impact is negligible. There is also a moral question to be asked in the use of gasohol. Is it right to use agricultural products for gas-guzzling automobiles when millions of people are dying of hunger in the world?

*Future of fossil fuels.* The fossil energy situation in the United States is not very bright. Gas and oil supplies in the United States will run out in several decades. Coal reserves are vast, but finite, and coal mining causes considerable environmental impacts.

In 1980, the National Academy of Science's report "Energy in Transition: 1985–2010" was published. The report stated that nuclear power (e.g., the breeder reactor) and coal were the keys to the nation's energy future. The immediate energy problem, however, is petroleum, and nuclear energy can provide little help there in the foreseeable future. (Nuclear reactors produce only electricity, and electrical generation consumes only 10% of the oil used in the United States.)

The energy crisis has not been as apparent in the 1980s as in the 1970s due to several factors. In 1980, governmental price controls on domestic oil were terminated. (The reasoning was to spur energy production by private industry in the hopes of finding new oil.) The governmental action resulted in increased fuel costs, which helped generate a slight recession, resulting in less usage of fuel. The increase in the price of gasoline also resulted in a shift from "gas-guzzlers" to more fuel-efficient cars, and hence reduced gasoline consumption. A worldwide economic slump in the early 1980s further reduced the impact of the energy crisis.

The United States must continue to rely on imported oil for some years to come. Because it requires energy to pump oil from the ground, Hall and Cleveland (1981) point out that it should be remembered that the time when petroleum will no longer be a viable resource " . . . is not when all the wells run dry, but rather at some point before that time, when the energy cost of obtaining a barrel of oil is the same as the energy in that barrel."

USGS geologist M. King Hubbert, in testimony before the U.S. Congressional Committee on Interior and Insular Affairs (July 5, 1974), said:

> "During the last two centuries of unbroken growth we have evolved what amounts to an exponential growth culture. . . . The exponential phase of the industrial growth which has dominated human activities for the last couple of centuries is now drawing to a close."

## *13.3.2 Nuclear Energy*

*Radioactivity.* In order to discuss engineering and environmental factors associated with the use of nuclear energy, it is necessary to understand some basic concepts of radioactivity and radioactive hazards.

Radionuclides are unstable nuclides which disintegrate naturally through various decay reactions which produce daughter elements. The basic unit of radioactivity is the roentgen, the quantity of x-ray or gamma radiation which will give rise to $2.08 \times 10^9$ ion pairs per $cm^3$ of dry air. A roentgen-equivalent-man, or "rem," is the absorption by the human body of radiation that produces an effect equivalent to the absorption of one roentgen of x-ray or gamma radiation. A maximum permissable dosage of 0.5 rem per year is recommended. Above 100 rem/year the red blood cell functions are impaired; death follows above 1000 rem for short-term exposure.

The quantity of radionuclides present in a sample is measured by the rate of decay reactions. A curie (Ci), originally defined as the radioactivity of 1 g of radium, is now defined as $3.7 \times 10^{10}$ disintegration per second. This unit is too large for most practical problems, so the common units of radioactivity are the millicurie (mCi) = $10^{-3}$ curie, the microcurie ($\mu$Ci) = $10^{-6}$ curie, and the micromicrocurie, or picocurie (pCi) = $10^{-12}$ curie.

Radiation hazards are not only due to direct radiation but can be caused indirectly by radiation-induced cancer which develops years after exposure. Ionizing radiation can also cause genetic mutations. Sources of information pertaining to radiation-induced cancer on human population is meager, but includes Japanese atomic bomb survivors, British patients given heavy x-ray dosages for arthritis, European uranium miners, and the 775 American women employed in painting radium numerals on watch dials between 1915 and 1935. In all these cases, cancer deaths in excess of normal populations were noted (Cohen, 1976).

Determination of acceptable risk for the general public for low-level radiation is very difficult. Radiation-induced cancers may take 30 years to develop. The establishment of health criteria standards is complicated by the fact that every radionuclide emits a different type of radiation, and at various rates. In addition, certain elements tend to accumulate in different parts of the body. Strontium-90, for instance, becomes incorporated into bones, and can cause bone cancer or leukemia. Iodine-129 accumulates in the thyroid gland. Tritium, on the other hand, is quickly passed through the body in the same manner as water. Thus, the drinking water limit for some radionuclides such as radium is low (5 pCi/l), but for others such as tritium, the limit is high (20,000 pCi/l).

It must be remembered that radioactivity in small concentration is ubiquitous in the lithosphere and hydrosphere. Elevated levels of radon gas are found in mines, caves and even poorly ventilated house basements. Rocks and soils containing potassium (such as granite or shale) have above-average radioactivity. (The radioactivity of an uranium mine was reportedly being discussed in the granite-walled Capitol in Washington, DC, when it was pointed out that the background radiation was greater in the chamber room than it was in the mine!) All water is radioactive to some degree. In fact, radium springs were once considered therapeutic.

*Nuclear reactors.* The radioactive hazards of nuclear energy result from either the mining and milling of uranium or the disposal of radioactive wastes from reactors and other sources. The danger of contamination of air and water from the "front end" of the nuclear cycle (mining and milling) is discussed in Sections 13.1 and 13.2. These hazards primarily result from inhalation of very low level radon near tailings, and ground-water contamination by radium and other toxics. Potential for severe contamination of the environment occurs at the "back end" of the nuclear cycle, i.e., the man-made wastes that come out of a reactor. About one-third of the uranium fuel in a commercial reactor is consumed and must be replaced every year. When fissionable materials such as $^{239}Pu$ or $^{235}U$ are subject to neutron bombardment, they form fission products plus additional neutrons and enormous amounts of energy. Of the fission products produced, $^{89}Sr$, $^{90}Sr$, $^{106}Ru$, $^{131}I$, $^{137}Cs$, $^{144}Ce$, and $^{237}Pu$ are probably the most important as far as biological hazards.

There are 73 operating commercial nuclear reactors in the United States; these reactors produce 12.5% of the nation's electricity. By 1990, an estimated 64 additional reactors may be operational. There are also research and military reactors. In addition, a large breeder reactor, which will convert uranium to fissionable plutonium, is being built in Tennessee.

The principle of the nuclear reactor is fairly simple. The most common type is the light water reactor (Nero, 1979), which uses small (1.5 cm $\times$ 1 cm diameter) ceramic fuel pellets of $UO_2$ enriched to about 3% $^{235}U$. About 200 pellets are lined up inside a zirconium alloy fuel rod (cladding) and an array of 49 fuel rods which make up a fuel bundle. About 800 of these bundles are emplaced between carbon control rods in a reactor. In the proper geometry, this arrangement permits a chain reaction in which a neutron striking a $^{235}U$ nucleus induces a fission reaction which releases neutrons, some of which induce other fission reactions. The fissioning creates heat, and the pressurized water is heated to about 300°C. The hot water is converted to steam in a boiling water reactor, and the steam drives a turbine and generator.

After fuel bundles are used up, they are carefully removed and temporarily stored in adjoining water pools. More than 30,000 spent fuel assemblies are stored in the commercial reactor sites in the U.S. The pools were built with limited capacities because it was originally expected that commercial spent fuel would be "reprocessed" to recover the remaining uranium and plutonium. Commercial reprocessing never became feasible, however. The nation's only commercial fuel reprocessing plant at West Valley, NY, closed in 1972. Another built at Morris, IL, was never operational.

*Radioactive wastes.* In addition to spent fuel rods and reactor wastes, the nuclear industry has created low-level and high-level wastes from the manufacture of weapons, medical uses, and research. In 1981, the national inventory of high-level liquid waste was about $3 \times 10^5$ $m^3$. High-level wastes are stored in tanks at four locations: the Hanford nuclear reservation at Hanford, WA ($2 \times 10^5$ $m^3$), the Savannah River plant near Aiken, SC ($10^5$ $m^3$), the Idaho National Engineering Lab near Idaho Falls, ID ($10^4$ $m^3$), and the defunct West Valley, NY, commercial plant ($2 \times 10^3$ $m^3$).

Man-made nuclear wastes began accumulating in 1943. Atomic bomb manufacture, test reactors, and nuclear submarines and ships have added to the inventory of fission products, mostly military. Beginning in 1957, civilian nuclear wastes (mostly from reactors) began to appear. It will be many years before civilian wastes overtake the military backlog. Most military wastes have been dissolved and at least partially reprocessed to recover plutonium, while civilian power-plant wastes are still in the form of undissolved spent fuel bundles. Most military wastes sit as liquids or slurries in large underground tanks, while civilian wastes reside in cooling ponds as zirconium-clad uranium oxide (Hammond, 1979).

Each of the licensed commercial nuclear reactors in the United States today generates about 1000 megawatts (1 GW) of electricity; in addition, each reactor produces about 340 $m^3$ waste/year. The U.S. Department of Energy estimates that by the year 2040 there will be 1.3 million cannisters (spent fuel bundles) of high-level waste, each weighing about 0.5 ton. There are also about 400 Navy nuclear reactors which produce spent fuel. In addition to the high-level waste storage areas mentioned above, there are landfills which contain low-level radioactive wastes such as Sheffield, IL, Barnwell, SC, Maxey Flats, KY, West Valley, NY, and Beatty, NV. The greatest volume and most highly radioactive waste area is at the U.S. defense facility at Hanford, WA. The cumulative waste from U.S. Defense Department (plutonium) atomic bomb manufacturing greatly exceeds all nuclear power plant wastes.

Ground water has been contaminated at some of these government facilities because the sites were not chosen on the basis of hydrogeology. For example, at Hanford, WA, the high-level radioactive liquid wastes are stored in about 150 underground tanks overlying basalt and gravel deposits. To date, there have been 20 leaks at Hanford. The most serious occurred in April 1973, when 450 $m^3$ of high-level waste soaked into the ground (Farney, 1974). This included 14,000 Ci of $^{90}Sr$ and 40,000 Ci of $^{137}Cs$. About 500,000 $m^3$ of low-level waste at Hanford has also been deliberately discharged into the earth by cribs and trenches. A large observation well system has been installed by government hydrogeologists to study the movement of contaminated ground water; $^{106}Ru$ and $^{3}H$ were observed to be the more mobile elements (Bouwer, 1978; Graham, 1981).

At the National Reactor Testing Station in southeastern Idaho, large volumes of radioactive water have been injected by a recharge well. Since 1952, radioactive wastes have been discharged into the Snake River aquifer, consisting of gravel beds between basalt flows. Tritium, 66 Ci of $^{90}$Sr, 120 Ci of $^{137}$Cs, 120 Ci of $^{60}$Co and other wastes have also been discharged into seepage pits since 1954 (Robertson, 1977). Dissolved $^{137}$Cs and $^{90}$Sr have mostly been absorbed by the sediment and basalt layers between the ponds and the water table, which is 137 m below the land surface (Bredehoeft et al., 1976). Routine monitoring of the aquifer by federal officials shows that tritium has traveled at least 12 km from the well and plutonium has traveled at least 2.5 km.

Industrial contamination also has occurred. For example, in 1974 radioactive wastes spilled into the Cedar River near Cedar Rapids, IA. Radioactive contamination of ground water has been observed at several low-level dumps, notably West Valley, NY, Maxey Flat, KY, Sheffield, IL, and Beatty, NV. At Oak Ridge National Laboratory, TN, and Savannah River Plant, GA, radioactive contamination of ground water has been documented (Duguid, 1974). In 1957, a fire at the Windscale facility in England released 33,000 Ci of radioactivity to the atmosphere, causing contamination of rangelands. In 1977, Windscale discharged 1.5 million $m^3$ of liquid wastes into the Irish Sea, as authorized by the English government. Low-level wastes were dumped into the sea by many countries prior to 1977. Nobody knows how many drums are lying, for example, on the bottom of the North Sea. The Paris-based Nuclear Energy Agency, made up of 20 Western countries plus Japan, set up a surveillance of ocean dumping and developed a policy which effectively has led to the end of ocean dumping by Western countries.

The first U.S. commercial site for the shallow land disposal of low-level radioactive wastes was in 1962. The disposal of wastes by burial in the ground using landfill methods seemed simple, safe, and inexpensive. Sites were selected in clay, glacial till, etc., with hydrogeologic characteristics which, it was believed, would ensure that radioactivity would not migrate from the sites. Recent studies (Carter et al., 1979) have shown, however, that under some circumstances, particularly in the humid eastern United States, disposal was not performing as anticipated. At several sites, trenches had filled with water and small quantities of radioactive contaminants had migrated tens of meters away from the trenches. Tritium was found to be moving at about 8 m/yr through glacial deposits at Sheffield, IL (Foster et al., 1984). Of concern is that, although stop-gap measures at leaking facilities can be made by on-going maintenance, perpetual care cannot be guaranteed.

The nature of radioactive ground-water contamination is witnessed by the following example in Illinois. Figure 13.15 shows an abandoned experimental reactor site south of Chicago. In 1943, this was the site of CP-2 and CP-3, the world's first heavy-water cooled and moderated reactors. The geology of this area consists of about 30 m of glacial till overlying a dolomite aquifer. Ground water has been contamined and shows up as seasonal pulses of high tritium concentrations (up to 14,000 pCi/l) in a shallow well at a county park (Red Gates Woods) about 1 km north of this site. The well is used by the public (the MPC for tritium is 20,000 pCi/l). The radioactivity concentration varies seasonably due to abundant meteoric water infiltrating into the site in the spring of the year. Because the abandoned reactor site and related bins contain exotic radionuclides as well as tritium, there is concern (Golchert and Sedlet, 1977, p. 25) that " ... following the tritiated

Figure 13.15 Abandoned reactor site in Illinois. In 1956, a 12-m pit was excavated next to the reactor containment shell and with the use of explosives, the reactor shell and miscellaneous contaminated items tumbled into the pit. The pit was then covered with dirt.

water, although at a considerably slower rate, should be the strontium . . ." and other radionuclides.

Atmospheric atomic bomb testing has led to worldwide radioactive fallout. During the testing of weapons in the late 1950s, radioactive elements were dispersed into the atmosphere and have been found in soils across the globe (Walton, 1963). A total of about 5000 kg of plutonium has been dispersed in atmospheric bomb tests. (As an example of the nature of the problem, Argonne National Laboratory, Illinois, contains atmospheric sensors to monitor the air near five experimental reactors. In 1963, the alarms were set off following fallout from Chinese H-bomb tests in the Pacific.) Tritium fallout from atmospheric H-bomb tests has allowed scientists to date ground water because of the worldwide recharge of $^{3}H$ during the late 1950s and early 1960s (Davis and Bentley, 1982). The concentration of tritium, a weak beta emitter with a half-life of 12.26 yr, is usually described in tritium units (TU), which are the number of $^{3}H$ atoms/$10^{18}$ atoms of H. The natural $^{3}H$ level in rainfall was about 10 TU before 1954, but thermonuclear explosions caused peak values of about 4000 TU in North America in 1963–64. Thus, near-zero tritium concentrations in ground water indicate the water consists entirely of pre-1954 water, and high tritium concentrations on the order of 100 TU indicate a mixture of post-1954 water (Smith, 1976).

In summary, there are numerous examples of radioactive contamination. This poor track record is due in part to the military urgency for nuclear weapons man-

Table 13.4 Factors Affecting Life Spans of the Average U.S. Citizen

| Factors decreasing lifetime | Decrease in lifetime |
|---|---|
| 100% nuclear power production | 1.2 hours |
| Overweight by 25% | 3.6 years |
| Male rather than female | 3.0 years |
| Smoking one pack per day | 7.0 years |
| City living | 5.0 years |

From Cohen, 1977.

ufacture, and in part to lack of concern. During the 1940s and 1950s, nuclear physicists were generally interested only in the excitement their young science afforded. The early developers of nuclear power knew much about radioactivity, but not enough about waste disposal or geology, and almost nothing about dealing with the public (Hammond, 1979). It is now clear, however, that if nuclear energy is to contribute to the solution of the energy crisis, careful waste disposal practices must be made.

*Reactor safety.* The safety of commercial nuclear reactors has been debated at great length. Antinuclear people worry about core meltdown or nuclear explosion, while nuclear advocates argue that radioactive releases due to operating reactors have been insignificant and the probability of accidents very remote.

Among the many surveys of probabilities and consequences of reactor accidents, the most famous is the 1975 report by Norman Rasmussen of MIT, the document called WASH-1400 (U.S. Nuclear Regulatory Commission, 1975). This report shows that the chances of a major nuclear disaster are inconsequentially small. For example, the WASH-1400 report predicts that an all-nuclear economy would produce 20 deaths/yr due to all forms of accidents and radiation-induced deaths. Thus the average American would lose 1.2 hours of his life (Cohen, 1976).

Other forms of life-shortening activities are shown in Table 13.4. For example, one pack of cigarettes per day reduces life expectancy 7 yr (which corresponds to about 12 min/cigarette). In addition, there are other wastes in the environment besides radioactive wastes which are toxic. For example, a lethal dose of arsenic oxide ($As_2O_3$) is 3 g. Most arsenic is used as a herbicide in regions where food is grown. Chlorine is a deadly gas, and yet it is stockpiled in giant cannisters in every city's water and sewage treatment plant. Figure 13.16 compares radioactive waste toxicity to other poisons. (It should be added that electric production from coal produces 750 deaths/yr from mining as well as perhaps 10,000 deaths annually from air pollution.)

There have been some nuclear reactor accidents. Perhaps the worst reactor accident in the free world occurred in 1957 at Britain's Windscale plutonium facility when radioactive iodine and polonium were released (Fuller, 1975). On March 28, 1979, there was an accident at Three-Mile Island (TMI) Nuclear Generating Station near Harrisburg, PA. A series of human errors and mechanical failures contributed to the mishap. No one was killed, but the plant owner, Gen-

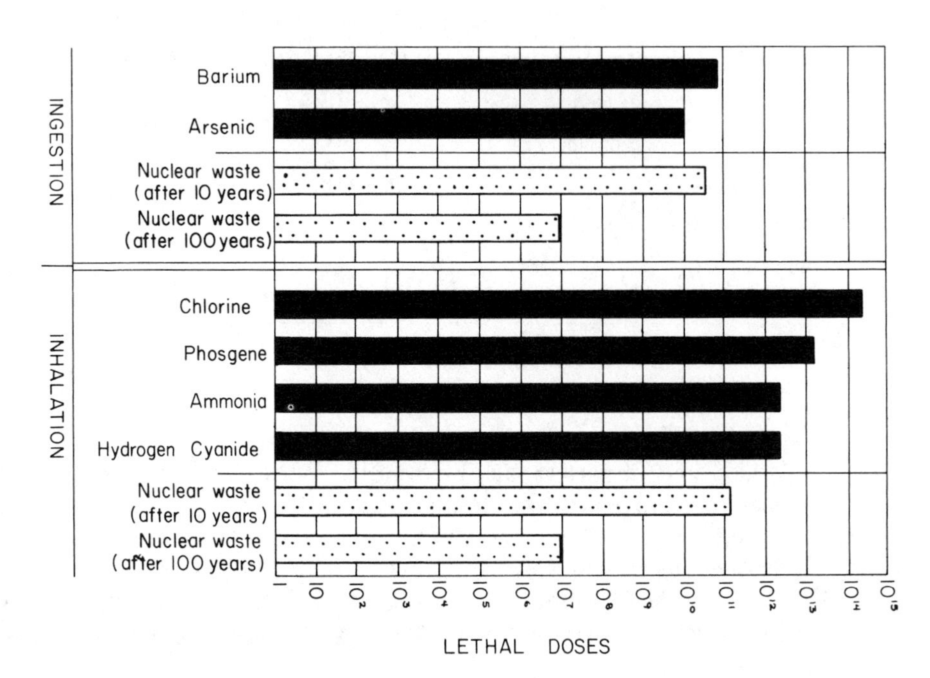


Figure 13.16 Comparison of health hazards presented by high-level radioactive wastes from nuclear reactors with those of other poisonous substances routinely used in large quantities in the United States (from Cohen, 1977). The graph demonstrates that there is nothing uniquely dangerous about nuclear wastes.

eral Public Utilities Corporation, has a $1 billion cleanup problem. The effects of TMI have been devastating to the U.S. nuclear industry. The orders for new reactors, which hit a high of 41 in 1973, declined dramatically. There have been no new nuclear reactor orders since 1978, and the orders for 25 were cancelled.

U.S. electric utilities in the 1980s have mostly given up on the idea of nuclear energy. The public is scared. The high cost of building and safely operating nuclear reactors in the U.S. has skyrocketed. Meanwhile, the rest of the world is pursuing nuclear energy with vigor. In 1984, 40 nations around the globe have 555 nuclear commercial generating plants either in operation (207), under construction (163), actively in planning (172), or on order (13).

*Radioactive waste disposal.* In 1982 there were about 4,000 metric tons of radioactive waste from commercial U.S. nuclear operations. They sit in temporary storage awaiting permanent disposition. By the year 2020, about 70,000 $m^3$ of solid high-level radioactive wastes, largely spent fuel assemblies, will have been produced by nuclear energy. It is imperative that a safe, permanent repository for high-level radioactive wastes be found. The wastes are accumulating rapidly and are with us now whether we build any more nuclear reactors or not. By the year 2000, Bredehoeft et al. (1978) estimate up to 476,000 spent fuel assemblies will be on hand.

Each spent fuel assembly weighs about 430 kg, 80% by weight core (spent fuel) and 20% "zircalloy" cladding and stainless steel hardware. The bundles are over 5 m long. Because the spent fuel rods are thermally hot (they still produce the equivalent heat after 10 yr of about twenty 100-watt light bulbs), as well as intensely radioactive, they will have to be aged (cooled) about 7 yr in deep pools of water prior to delivery to a permanent repository.

The overall average half-life of all the radioactivity in spent fuel is about 30 yr. Some radionuclides from spent fuels are short-lived, such as $^{95}Z$, which has a half-life of 65 days. Three major radionuclides are longer lived: $^{90}Sr$ and $^{137}Cs$ both have approximately 30-yr half-lives, and $^{237}Pu$ has a half-life of 24,000 yr. Longer containment is required for the transuranic isotopes of plutonium and americium which, along with their radioactive daughters, may require 500,000 yr of isolation from the environment (Winograd, 1974). de Marsily et al. (1977) discuss the confining ability of geologic formations for the three major elements which have long half-lives: $^{129}I$ (16 million years), $^{137}$ Np (2.13 million years), and $^{239}Pu$ (24,400 yr).

Various schemes have been proposed to get rid of radioactive wastes or to keep them isolated from the environment for thousands of years (McCarthy, 1979; Northrup, 1980; Davis, 1982). The repository must be leak-proof as well as a barrier for protection against both criminal actions and accidental encounter. The repository should be stable geologically (seismically inactive and sufficiently deep to escape future erosion). The principal danger from burial is that it might be contacted by ground water, be leached in solution and eventually make its way to the surface. The earliest, and perhaps the best scheme, was to put the wastes into a "repository" in deep salt mines.

In 1971, the Atomic Energy Commission (AEC) tentatively selected a salt mine near Lyons, KS, for experimentation, hopefully leading to design of a repository (Holden, 1971; Goebel, 1971). The Kansas Geological Survey noted the presence of unplugged oil and gas exploration wells in the vicinity, and questioned the integrity of the salt formation in view of the fact that ground water could possibly gain entry to the repository through the drill holes. Further, there was an active salt mine only 560 m from the repository where large quantities of water were lost during solution mining (the use of large quantities of water to extract salt from the formation) near the area. Because of these problems, the Lyons project was discontinued.

In 1974, the federal AEC was abolished, and two new agencies were founded: 1) the Nuclear Regulatory Commission (NRC) and 2) the Energy and Research Development Administration (ERDA), now the Department of Energy (DOE). Increased budgeting for commercial radioactive waste management was given. In 1977, a new Presidential policy was announced declaring that U.S. commercial nuclear power spent fuel would not be reprocessed for the indefinite future. This added impetus to finding a solution for the waste disposal problem. (The reason given for the 1977 U.S. policy not to reprocess fuel was for fear that enough plutonium could be stolen to make a bomb.)

Three basic types of geologic formations have been identified as theoretically suitable for the safe, permanent isolation of high-level radioactive waste (U.S. Department of Energy, 1977): salt (including both bedded and domed salt), crystalline rock (basalts and granite), and argillaceous rock (certain clays and shales). The U.S. DOE is studying these geologic environments as potential repository sites. Granite terrains of northern Wisconsin and the basalts in central Washington are being investigated as possible repository sites, as well as the Pierre Shale in western South Dakota and volcanic tuffs at the Nevada test site.

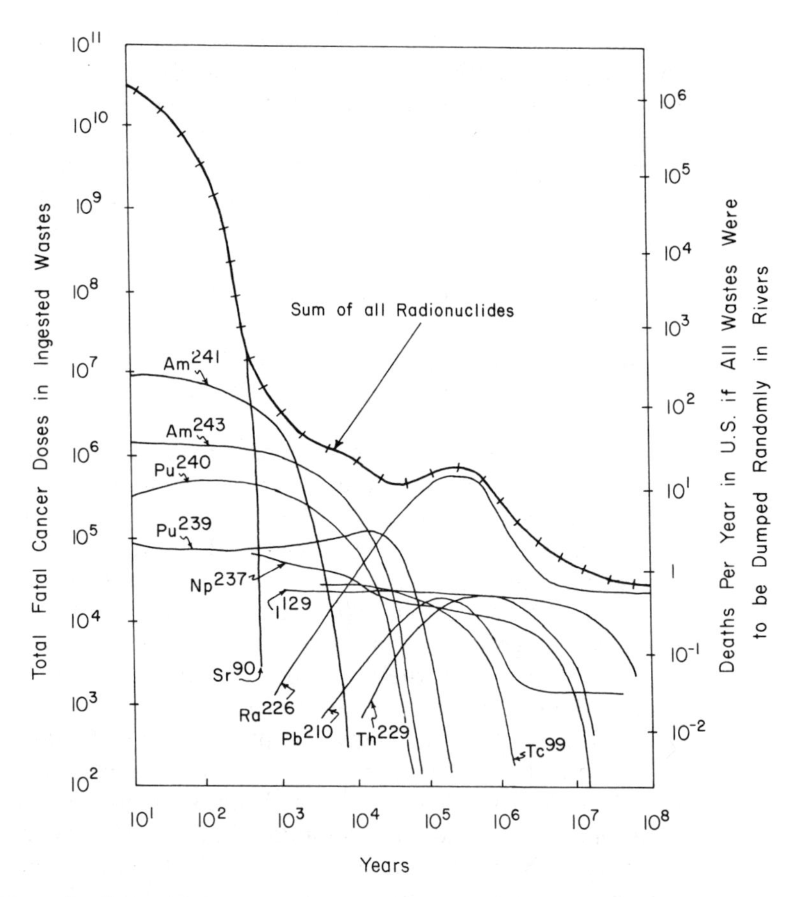


Figure 13.17 Hazards of spent fuel components (from Cohen, 1977). If all wastes were to be ingested, the biological effects on the human population of the United States would be considerable. As this graph shows, the number of cancer-causing doses in the United States is such that if all the wastes, after aging for 10,000 years, were to be converted into digestible form and fed to people, one would expect a million fatal cancers to ensue (scale at left). If, instead, wastes were to be converted into soluble form and immediately after reprocessing dumped at random into rivers throughout the United States, the result could again be a million fatalities (scale at right).

The general geologic areas which appear most promising for salt beds in the U.S. are the Permian Basin of Kansas, New Mexico, Oklahoma, and Texas, the salt domes of the Gulf Coast Embayment in Louisiana, Mississippi, and Texas, and the Salina Basin in New York, Ohio, Pennsylvania, and West Virginia. Of primary concern is how dry the rock tends to be. This is extremely important because ground water must not be contaminated by these long-lived wastes.

Due to the exponential decay of radionuclides present in spent fuel (Fig. 13.17), the first 1000 yr will be critical. During the first 200 yr, $^{90}Sr$ will account for more than 99% of the hazard potential. During a period of 1000–50,000 yr, plutonium dominates the hazard potential. The radioactivity hazards shown in Fig. 13.17 are units of deaths caused by ingestion, i.e., if the wastes were ground up and fed to the populace, a rather unlikely scenario, but at least a way to portray

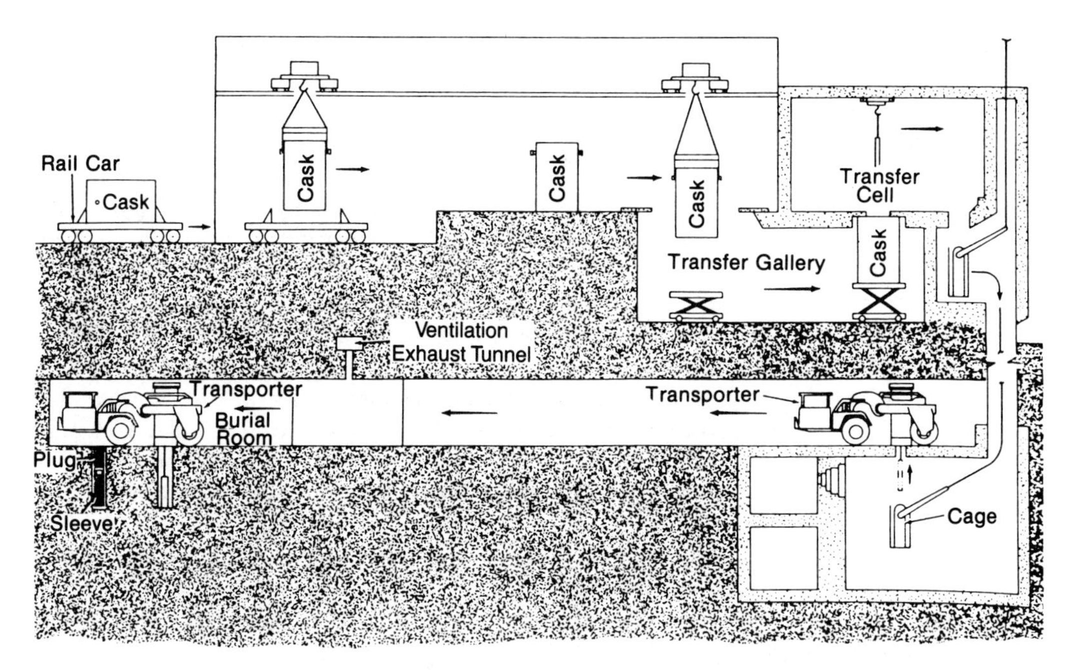


Figure 13.18 Model of high-level radioactive waste repository/bedded salt pilot plant handling schematic (from U.S. Atomic Energy Commission, 1974).

hazards. (It is interesting to note that on the same hazard scale as in Fig. 13.17, mill tailings would plot 1–2 orders of magnitude above all high-level waste elements (Bredehoeft et al., 1978). This is due to the tremendous volume of tailings. But, then, no one plans to eat tailings, either.)

The Nuclear Waste Policy Act of 1982 contains a deadline for finding the best possible answer to the nuclear waste problem. The repository is to be stable enough to keep the waste safe and dry for at least 10,000 yr. In 1984, the U.S. DOE narrowed its search to three potential sites: 1) bedded salt in Deaf Smith County, TX; 2) basalt near Richland, WA; and 3) volcanic tuff near Las Vegas, NV. Out of those three, one site is to be selected by March 1987, when the President is supposed to present a final recommendation to Congress.

The working model for a geologic repository in salt is described by the U.S. Department of Energy (1977) and Klinesberg and Duguid (1982). Figure 13.18 shows a schematic model of a proposed repository. Plans for a U.S. repository are not yet final, but the waste may be buried in salt beds about 600 m underground. U.S. DOE plans to have licensed, sometime in the 1980s, a Waste Isolation Pilot Plant (WIPP) in bedded salt near Carlsbad, NM (Bredehoeft et al., 1978). The facility would be located in bedded salt of Permian age in the Los Medanos area, about 40 km east of Carlsbad (Gonzales, 1982). The very existence of rock salt indicates the lack of circulating ground water, a major concern for any burial plan. In the possibility of contact with water by spent fuel, many of the radioactive materials would be absorbed by surrounding clays as the solution migrates through the evaporites and surrounding shale strata (see Chapter 7 for discussion of $K_d$, the distribution coefficient). One unresolved problem with salt beds is the presence of pockets of brines (Bredehoeft et al., 1978). The brine could migrate toward a waste cannister under the influence of the thermal field created by the emplaced wastes. Fluid inclusions and water-containing evaporite minerals such as gypsum could migrate toward waste cannisters (Kopp, 1982).

None of the repository sites still under consideration in 1984 seem completely ideal and Hunt (1983) suggests that other alternatives could include recycling, or storage in giant monuments impressive and durable as the Egyptian pyramids. Even more exotic methods of waste disposal have been proposed: in the seabed, in space, and in polar icecaps. The basic concept of seabed disposal is to emplace the spent fuel in a stable geologic formation under the sea floor. These may be deep muds or subducting trenches. Polar disposal would involve the waste being lowered into a 50-m-deep hole in Antarctic ice. The waste would theoretically melt down to the base of the glacier. The long-term accident risk of ice sheet disposal is the stability of the ice sheet itself. The space disposal alternative would require 110 shuttle launches by the year 2010 (U.S. Department of Energy, 1977). The Kennedy Space Center could handle only 20. The cost and safety of the extraterrestrial method make it unlikely. A 1981 environmental impact statement on commercial radioactive waste prepared by the Department of Energy found that ice sheet disposal, transmutation, pumping liquid wastes underground, and other exotic forms of disposal were not as desirable as burial in stable rock formations 600–900 m deep.

In addition to high-level radioactive waste, intermediate and low-level radioactive waste disposal is another problem. The United States generates about 90,000 $m^3$ of low-level radioactive waste annually from over 20,000 laboratories, hospitals, factories, and nuclear power plants. Low-level wastes must be stored safely for periods ranging from decades to centuries. Recognizing that the three national commercial disposal sites (Barnwell, SC, Beatty, NV, and Richland, WA) are overcrowded, Congress passed the Low-level Radioactive Waste Policy Act in 1980 which shifts the waste-disposal responsibility to the states.

Other countries are engaged in research pertaining to radioactive waste disposal. The Federal Republic of Germany has stored low and intermediate-level wastes at a radioactive waste repository in the Asse salt mine since 1967. Sweden and Canada are studying granitic masses.

In 1980, the Association of Engineering Geologists (AEG) adopted a policy statement on the disposal of nuclear wastes:

> It is the position of the Association of Engineering Geologists that radioactive nuclear wastes can be safely isolated and disposed of by deep underground burial in secure geological environments. The scientific and technical means to locate and define the boundaries of these environs and to achieve such safe disposal is well developed and increasing.

While the AEG statement is basically optimistic, it should be balanced by a USGS statement (Bredehoeft et al., 1978) warning that geologic phenomena prediction such as seismicity, climatic change, behavior of salt strength, and ground water flow and absorption phenomena are not credible millions of years into the future: "These uncertainties need to be faced candidly in public discussions of radioactive waste disposal."

Figure 13.19 Cartoon (from Kansas City Star).

*The nuclear dilemma.* If all exposures from radiation due to nuclear energy industry are added, and if all U.S. electric power were nuclear, it has been determined that the average American would receive 0.23 mrem/yr from routine emissions (Cohen, 1976). This is much less than the amount received from natural radioactivity or the exposure from medical and dental x-rays. An important variable in radiation is the dose received from cosmic radiation, which averages 28 mrem/yr at sea level but increases with altitude because of reduced atmospheric shielding. In Denver, for example, the annual dose from cosmic radiation is 50 mrem/year. Medical x-rays and therapy is the largest deliberate exposure of ionizing radiation to humans, averaging about 90 mrem/yr per person in the United States (Fig. 13.19). Living in a stone house adds 30 mrem/yr. Fallout of $^{90}Sr$ from pre-1963 weapons testing currently results in an average dose of about 4 mrem/yr.

If the small periodic releases of radioactivity from mining, reactors, transportation, etc., are determined, and a probability of life-shortening cancers determined, it results in a small reduction of the lives of very few individuals. A comparison of health hazards from nuclear hazards and other poisonous substances is shown in Table 13.4. It hardly makes sense for the average cigarette smoker, whose life can be statistically shown to be shortened 7 yr because of smoking, to worry about the hazards of nuclear energy.

In spite of these facts, the nuclear industry in the United States may be at a dead end. Whether the possible demise of nuclear power in the United States is a matter of rejoicing or despair is a complex issue and depends on personal values. Even if a new nuclear plant can produce electricity for less cost than coal and cause less environmental degradation such as acid rain, directors of utility companies avoid nuclear power plants because they embody political risks and financial uncertainties far beyond other energy sources.

The dream of many physicists and nuclear engineers in the 1950s was that in the near future nuclear-generated electricity would be "too cheap to meter." The fact is, however, that real and imaginary environmental concerns about nuclear energy have resulted in its apparent demise in the United States. To many scientists and engineers, this is tragic because today they are putting their best efforts into making nuclear energy as safe as possible.

### *13.3.3 Alternative Energy Sources*

In some places, unusually warm ground water may be used for residential house heating. For example, parts of Boise, ID, have been warmed by 77°C ground water since 1892. Over 400 buildings in Klamath Falls, OR, including the Oregon Institute of Technology (5000 students), are warmed by heat from the earth's interior. Ground water heat pumps make extraction of heat from large lakes or aquifers an economic investment in some areas (Parfit, 1983).

The United States leads the world in production of geothermal electricity. In 1982, 1200 megawatts were produced (Dick, 1983). This is about 0.2% of installed electrical capacity in the U.S. from all sources. The Phillipines and Italy each produced 450 megawatts, New Zealand, 225 megawatts, Mexico and Japan, each 180 megawatts. North of San Francisco, the Geysers geothermal project passed the 1000-megawatt level in 1982. By the mid-90s, it is estimated that this area should provide about 10% of California's electricity. Geothermal electrical power production will probably not be increased greatly in the foreseeable future because of the limited occurrence of natural geothermal areas and because of engineering and geochemical problems associated with the utilization of naturally occurring steam.

Water pumps, driven mechanically by the power of the wind, have been used for centuries as a method of lifting water from shallow wells to the surface. Thousands of "windmills" still dot the landscape in Holland or the open range country of the Sand Hills of Nebraska, for example. The conversion of wind to electrical power has been experimented with during the past few decades, but has met with only limited success. California has become the premier wind-farming state, where more than 4000 modern windmills are spinning in the windy mountain passes. Perched on ~25 m high poles, "wind farms" are fields full of windmills whose turbines spin electricity into the power grid of a regional utility company. In 1983, California had a total windmill capacity of 300 megawatts. The growth of wind farms is, to some degree, due to generous tax credits. In 1984, the U.S. Bureau of Reclamation built an enormous experimental windmill at Medicine Bow, WY. The single propeller blade, 100 m long, is capable of generating 2.5 million watts of electricity in a strong wind.

Solar energy offers the promise of a renewable, nonpolluting source of energy. There can be no doubt that passive solar heating of homes contributes to energy savings of the cost of heating a home. However, direct conversion of sunlight to electricity is limited. For example, if the entire roof of a typical house is covered by photovoltaic collectors (light-sensitive silicon cells which generate electricity when light strikes them), a typical house in the Chicago area could generate only enough electricity to run one 100-watt light bulb continuously. Future research may increase the efficiency of solar electricity collectors and storage units; theoretically enough sunlight falls on a 6 × 9 m solar panel to meet the electrical demands of the average house (Eaton, 1976).

Recently, solar energy projects have begun to produce electric power. In 1982, the world's largest photovoltaic installation (called "Solar One") was hooked up to the California Edison grid. A total of 1818 heliostats, each having 39 $m^2$ of movable mirrors, focus on a central receiver. Superheated steam drives a turbine to produce electricity; part of the steam is diverted to heat rocks held in a storage tank. The project can produce 10 megawatts of electric power for a period for 4 hours even on the shortest day of the year. By the early 90s, a series of 100-megawatt photovoltaic and solar thermal power plants may be a rapidly growing source of electricity for the state.

The production of hydropower in the U.S. has been nearly constant for the past few decades. Very few new dams are being built because the best dam sites have already been used. The percentage of U.S. energy produced by hydropower actually declined from 19% in 1960 to 12% in 1978. While the overall production of electricity by new dams may be bleak, there may be additional benefits of hydropower when used in conjunction with coal or nuclear. For example, "pumped storage" involves the construction of a reservoir in a tributary valley near a water supply. During off-hour electrical needs (at night), cheap electricity can be used to pump water up to the reservoir. During heavy electrical needs (during the day), electricity can be generated by releasing water from the reservoir.

Presently geothermal, wind, and solar energy do not contribute significantly to the overall need for electricity in the U.S. The future roles of alternative energy sources depends on economics. Unlike the multi-billion dollar investments and 15-year lead time necessary to build nuclear reactors, solar and wind farms can be financed and built by small power producer companies that presently enjoy special tax benefits. These renewable energy projects are smaller scale and more flexible, thus minimizing investment risks. Solar and geothermal developments in California signal that while nuclear power is presently failing the market test, renewable energy and energy-conservation investments are increasing. It remains to be seen if California is setting the pace for the rest of the nation, or whether California's favorable physical geography is the reason that it is in the enviable position of being able to develop wind, solar, and geothermal energy.

## Problems

**1.** Consider a typical subbituminous open-pit coal mine in the Gillette area of the Powder River Basin, WY (see cross-sectional sketch below. About 10 million tons/yr (in situ mass) are mined from a 10-m-thick coal bed having 15 m of overburden. The natural water table is at 10 m depth. Assume the natural terrain is flat and the strata beds are horizontal.

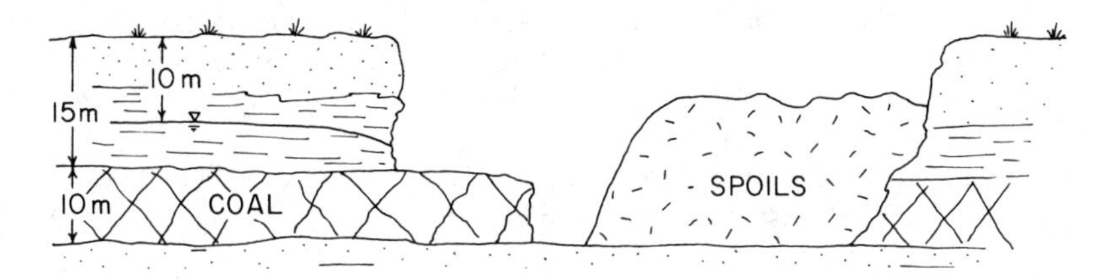

**a.** Upon abandonment, the water table will seek its original elevation. Assuming the dry density of the overburden (2.00 g/cm$^3$) becomes 1.80 g/cm$^3$ when dumped as spoils, will the top of the spoils be above or below the water table upon abandonment (i.e., will this be a lake or not)? By what amount?

**b.** How many hectares of land are consumed annually by this mine? (The in situ density of subbituminous coal is about 1.5 g/cm$^3$.)

**c.** The coal is being sent to Texas, where it will be burned to generate electricity. How many "unit trains" are necessary per week to transport this coal if each train consists of 100 cars, each carrying 100 tons?

**d.** If the coal is transported to Texas in a slurry pipeline consisting of 50% by weight coal and 50% water, what discharge (m$^3$/s) of water is necessary? Assume the mined coal has a bulk density of 1.0 g/cm$^3$. (Hint: Remember the in situ density of coal is 1.5 g/cm$^3$ so that it already has some water in it.)

**e.** Consult USGS Water-Supply Papers to determine nearest stream capable of sustaining this discharge.

**2.** Contaminated water seeps through the bottom of a uranium tailings pond (see sketch below, simplified from Fig. 13.4). Assume the vertical hydraulic conductivity ($K_v$) for the subsoil in the zone of aeration is $10^{-5}$ cm/s, the porosity ($n$) is 20%, and the bulk (dry) density ($\rho_b$) is 2.12 g/cm$^3$. The head of water in the tailings ponid is 4 m and the water table is a 25 m depth. Determine the velocity (m/yr) that radium will move if the distribution coefficient ($K_d$) is 0, 10, 100, and 1000 mg/g. [Hint: Refer to Chapter 7 for ground water equations. The following format may be used. First, determine the Darcy Law velocity ($V_d$) in the 25-m-thick unsaturated zone using Darcy's law (Eq. (7.3)), where

$$H/L = \frac{25 \text{ m} + 4 \text{ m}}{25 \text{ m}}.$$

Second, determine the true water velocity ($V_w$) from Eq. (7.6). Third, determine the velocity that radium moves ($V_{ra}$) by Eq. (7.27).]

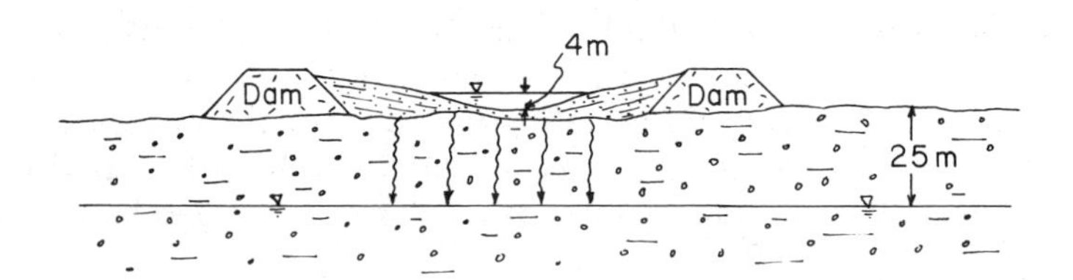


**3.** **a.** See Fig. 13.11. What has the annual growth rate of the world's oil production been over the past two decades?

**b.** If the entire planet earth were a spherical drop of oil, how many years would it last at this rate?

**4.** In 1979 the United States was using oil at an average rate of 18.4 million barrels a day (mbd), of which 8.2 mbd were imported. Natural gas usage averaged 10.4 mbd (in terms of the equivalent amount of oil) of which 0.6 mbd were imported.

**a.** At this rate, how many barrels of oil and gas does an average American consume in a 70-yr lifetime?

**b.** If 1 bbl of oil (42 gal) is 76 cm high and the daily U.S. consumption of barrels of oil and gas equivalent were lined end to end, how far (km) would the line extend from New York City in a westerly direction?

**5.** During the Three Mile Island nuclear reactor accident, release of radioactivity near the reactor reached about 0.05 rem/hr. How long would a person have to be exposed in order to get a year's maximum recommended dosage?

**6.** By the year 2000, U.S. nuclear reactors will have produced about $10^{10}$ Ci of $^{90}Sr$. If the drinking water MPC is 10 pCi/l, what is the ratio of water required to dilute this strontium to drinking water standards, compared to all the fresh lake water on earth? (See Chapter 7 for data on hydrosphere.) What does this indicate about using fresh water lakes as "dilution is the solution"?

**7.** Check the arithmetic presented in the last line in the following quotation (from Cohen, 1976). (The discussion relates to the potential time lost per average individual per lifetime due to U.S. nuclear power production. Assume that the U.S. population is $2 \times 10^8$, and the average lifespan is 70 yr.) Write one paragraph explaining the method used for determining the 1.2-hr figure. Include a critique of the rationale used.

> Estimates based on acceptance of WASH-1400 predict that an all-nuclear energy economy would result in about 20 deaths per year; critics of nuclear energy claim that the number is about 600. I shall now attempt to put these estimates in perspective. Since cancer is delayed by 15 to 45 years after exposure, the average loss of life expectancy per victim is 20 years. The loss of life expectancy for the average American from these 20 deaths per year is then ($20 \times 20$ man-years lost/$2 \times 10^8$ lived) $= 2 \times 10^{-6}$ (years) of a lifetime = 1.2 hours.

# 14

# Design with Nature

And God created man in His own image, in the image of God created he him; male and female created he them. And God blessed them: and God said unto them, be fruitful and multiply, and replenish the earth, and subdue it; and have dominion over the fish of the sea, and over the fowl of the air, and over every living thing that moveth upon the earth.

Genesis 1:27

The "control of nature" is a phrase conceived in arrogance, born of the Neanderthal age of biology and philosophy, when it was supposed that nature exists for the convenience of man.

Rachel Carson

Be mindful, when visiting a new city, of the air, the soils and the water.

Hippocrates

Man has indeed acted as a destructive geological agent in all too many instances in the past. But in the years ahead, he can be well guided by this experience from the past and remember that it is in his hands to see that cities of the future are indeed designed with nature and not in opposition to it.

Robert F. Legget

Like winds and sunsets, wild things were taken for granted until progress began to do away with them. Now we face the question whether a still higher "standard of living" is worth its cost in things natural, wild and free.

Aldo Leonard

Between the idea and the reality
Between the motion and the act
Falls the shadow.

T. S. Eliot

## 14.1 Geological Hazards

There is a distinction between an event, a hazard, and a disaster. A natural *event,* whether geological, climatological, etc., is simply a natural occurrence. A *hazard,* geological or otherwise, is the potential danger to human life or property. A *disaster* occurs when the hazard is realized. Tanaka (1981) makes the following analogy: "An earthquake in the wilderness is a seismic *event.* When humans clear the land and build by the fault and over a swamp, they have created a seismic *hazard.*

When the earthquake happens and buildings are destroyed and people are killed, a seismic *disaster* has taken place."

Earthquakes, floods, etc. are natural phenomena, and as such are not hazards at all. They become hazardous because man, in his ignorance or neglect, does not recognize areas where human activities are incompatible with natural events which are taking place or will in time take place. In 1981, Robert F. Legget, Canadian engineering geologist, said: "It is part of the order of nature that disasters such as volcanic eruptions and earthquakes take place. It is the high calling of geology to explain them, possibly one day to predict them, and always, when applied by the engineer, to minimize their effects upon the works of man." Recognition of geological hazards is the first step. The second step, frequently the most difficult, is to convince people in decision-making positions to do something about it.

The recognition, amelioration, and/or mitigation of geological hazards is an important job for the engineering geologist. Hopefully, in the future, people will be able to prevent geological events from becoming geological disasters. The track record to date has been much less than satisfactory. This chapter summarizes material presented in Chapters 6, 8, 9, 10, and 11, with emphasis on planning for the reduction of geological hazards. There are a number of fine general publications on this subject, notably Robinson and Spieker (1978, and references contained therein), and Hays (1981), as well as Flawn (1970), Nichols and Campbell (1971), McKenzie and Utgard (1972), Foose (1972), Geyer and McGlade (1972), Legget (1973), Cargo and Mallory (1974), Menard (1974), Spangle et al. (1976), Keller (1976), Howard and Remson (1978), Utgard et al. (1978), Costa and Baker (1981), Tank (1983a), Griggs and Gilchrist (1983), and Legget and Karrow (1983).

Geological processes which can be considered hazardous include 1) landslides and swelling soil, 2) floods, 3) coastal erosion and tsunamis, 4) earthquakes, 5) land subsidence, and 6) volcanic activity. The above list is arranged more or less in order of economic impact. Figure 14.1 shows annual economic loss and deaths in the United States due to geological hazards. According to a recent USGS report (Hays, 1981), geologic and hydrologic hazards now cause an estimated $8 billion loss annually in the United States.

There are ways to prevent or reduce these geological hazards. In general, they can be partially reduced or prevented either by 1) *amelioration,* i.e., reducing the impact by taking measures so the risk is minimized, or 2) *mitigation,* i.e., reducing the magnitude of the event itself. Flood hazards, for example, could be ameliorated by not building houses on the flood plain, or could be mitigated by building flood-control dams to reduce the size of floods.

### *14.1.1 Landslides and Expansive Soils*

Landslides and swelling soils cause great economic loss in the United States (Fig. 14.1). Mass-wasting processes, including landslides, are described in Chapter 6. Problems caused by expansive soils are more subtle than landslides but are, overall, more costly. Figure 14.2 is a generalized map of potential expansive soil areas in the United States. The plains province east of the Rocky Mountains is one of the highest shrink-swell areas. Swelling clay, e.g., montmorillonite, is the main

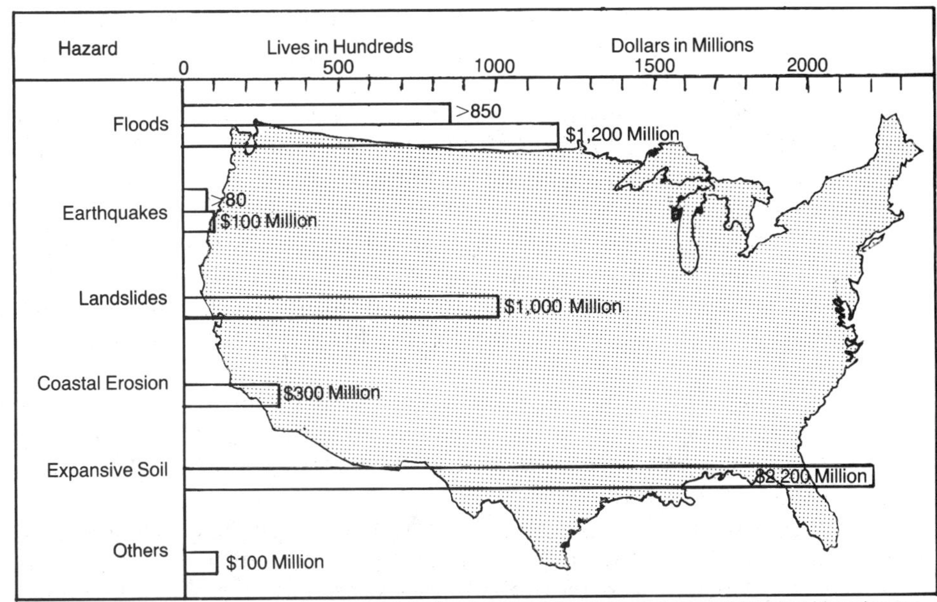


Figure 14.1 Mean annual United States losses due to geological processes (from Robinson and Spieker, 1978).

Figure 14.2 Map of swelling clay in near-surface rocks in the United States (from Robinson and Spieker, 1978).

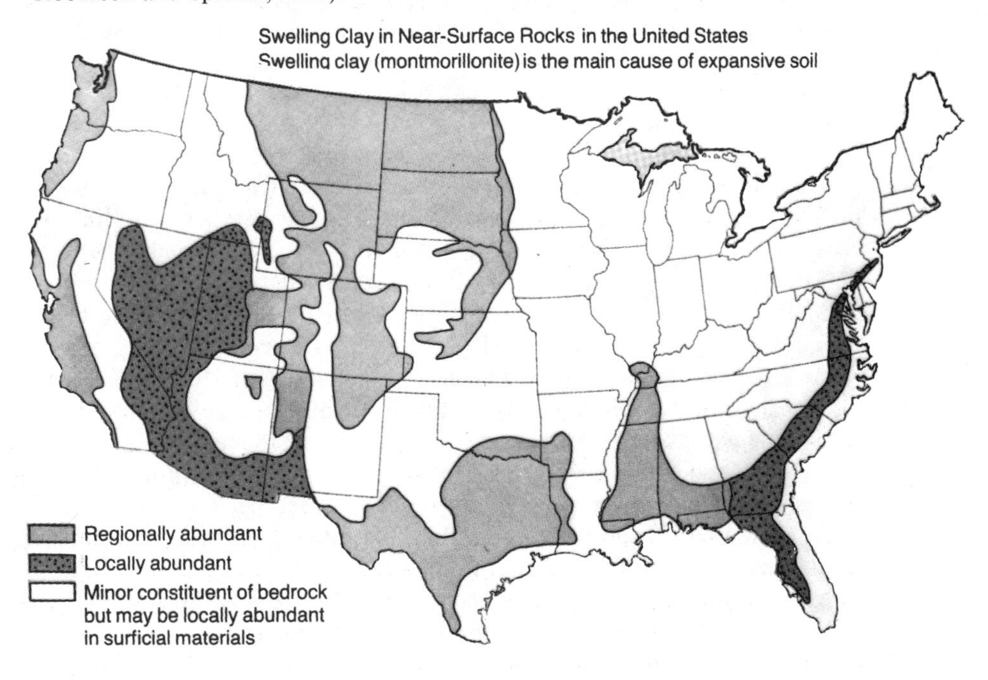

cause of expansive soil. Where the potential exists, cracked houses, roads, and bridges can result. Steeply sloping terrains underlain by expansive soils are also subject to landslides.

Engineering geologists can prepare generalized landslide and expansive soil-hazard maps, which convey information to nongeologists who hold planning or zoning decision-making positions. For example, the U.S. Geological Survey prepared potentially swelling soil maps for the Front Range Urban Corridor near Denver, CO (Robinson and Spieker, 1978). The hazard areas were ranked by engineering geologists and the maps colored accordingly. The basis of the ranking was primarily bedrock and/or surficial deposits, which contain greater or lesser amounts of clay. The Cretaceous Pierre Shale, for example, was pointed out as having a "very high swell potential."

Highway subgrade materials can be treated with hydrated lime, cement, or other compounds to help stabilize underlying swelling soils. Building up the lime content to about 10% is believed to allow calcium ions to exchange for sodium in the clay minerals.

Control of water is essential in order to reduce damage due to expansive soils. Drain spouts should carry rain water far away from house foundations. Runoff should be controlled by properly sloped ground or paving. A granular blanket about 1 m or more depth around a foundation will help control capillary water, maintaining a more uniform moisture content, thus reducing swelling problems (Bowles, 1982). Deep foundations, such as drilled piers, may be installed to depths below the seasonal moisture change.

Landslide hazards have increased due to population growth and man's habitation of less desirable sites. For example, in the Los Angeles, CA, area (Brown, 1962):

> Landslides have inflicted substantial damage to highways and railroads over the years. However, prior to the early 1950's, damage to homes by landslides was relatively minor because hillside homes were few, and generally in the best sites. The increasing demand for building sites, particularly in the large metropolitan areas, has accelerated residential development in the foothills and mountains.

The construction of deflection walls in southern California as a means of ameliorating landslide and debris flow damage is discussed by Hollingsworth and Kovacs (1981). The concrete walls, usually V-shaped in plan view, physically deflect the moving debris, helping to reduce inundation of property below.

Methods of stabilization of landslides for highways are given in Chapter 6. In urbanizing areas subject to landslides, land-use controls can mitigate future problems. For example, after heavy rains and landslides in 1952, aroused citizens of Los Angeles demanded adequate controls for hillslope development. The resulting ordinance enacted by Los Angeles is considered the nation's first grading code. The code has been modified through the years and now requires the submission of a geological report by a qualified engineering geologist.

As a member of the Rapid City, SD, Planning Commission, I participated in numerous decision-making situations relative to zoning and building permits. Local governmental boards such as this, often made up of unpaid lay people who are trying to do the best things, are under great pressure to allow contractors to develop the land in any way they see fit. The pressure to develop land is strengthened by the promise of added property tax base. However, to allow development in a geologically hazardous area is to be "penny wise and pound foolish," because these poor projects eventually will cost the taxpayers a lot of money to repair. Too often, geologists and engineers do not attend public planning meetings, and seem reluctant to speak out on issues. Because they cannot guarantee, for example, that a landslide will or will not occur on a certain hazardous hillslope, they remain silent. As a result, the board often makes a decision based on the advice of the developer's public relations person.

In order to convey geologic information to a local planning board, the geology must be reduced to very generalized maps and tables. A conventional geologic map is completely over the heads of nongeologists, who see it as rather abstract art, including a legend of strange names. Geologic hazard data must be presented in a simplistic manner, such as several colors showing relative hazards.

As an example of the successful utilization of hazard information for swelling soils, Robinson and Spieker (1978) show how the annexation and residential development of a tract of land in Boulder, CO, evolved. Originally the development was planned on high shrink-swell potential land below a leaking irrigation ditch. After the geologic hazards were studied and explained, the city council required that approval would be given only after the developer satisfied requirements to divert irrigation drainage, chemically stabilize the roadway pavement subbase, and use drilled piers and grade-beams for foundation construction. Thus, in this case, it was shown that it is possible to plan for swelling soil or landslide problems by mitigation (get rid of excess water), as well as amelioration (have drilled piers so that when expansion occurs it won't buckle the floor, as in the case of poured concrete slab floors).

### *14.1.2 Floods*

It is tragic that floods annually cause this nation losses of over $1.2 billion (Fig. 14.1) and 850 dead. Chapter 8 shows that there are ways to mitigate flood losses by engineering structures such as dams and levees. There is also the more natural way to ameliorate losses: simply avoid construction on flood plains.

Governmental agencies can ameliorate flood losses in the following ways:

1. adopt and enforce flood plain regulations
2. acquire high-risk flood areas so they can be devoted to open land use
3. make federal flood insurance, flood control measures, and financial assistance for flood disasters contingent upon land-use planning and implementation.

The Flood Disaster Protection Act of 1973 provided strong incentive for communities to participate in a flood protection program. In 1978, a special report by the Geological Society of America points out that national flood insurance has the sound goal of inducing local land-use planning in flood-prone areas and reducing governmental outlays for disaster relief. However, the insurance is being offered at highly subsidized rates; the Federal government pays a large part of the cost of protecting local interests. This means that the majority (88%) of the U.S. population who do not live in flood-risk zones are being forced to contribute to the welfare of the few (12%) who live in a hazardous area. Subsidized rates may actually encourage building on flood plains.

Many people in this country have a philosophy that freedom means people ought to be allowed to do as they want, and that any government regulation smacks of socialism ("self-serving bureaucrats" in Washington who are out to control the whole country). As a result of varying political climates, every state and county in the United States has regulations (or lack of regulations). Flood plain zonation (or lack of zonation) is a local issue, generally at a city or county level. Flood plain regulations are very diverse, and generally throughout much of this country flood plains continue to become urbanized with little regard for safety. Unfortunately, it is often the innocent who suffer. The developer retires to Arizona with his money leaving occupants of house trailers in eminent danger.

### *14.1.3 Coastal Erosion and Tsunamis*

Chapter 10 contains examples of coastal hazards and man's folly with respect to coastal erosion. Natural coastal processes can cause erosion of either coastal mainland areas (e.g., San Diego area) or barrier islands (e.g., Cape Hatteras). In some places, man has locally accelerated coastal erosion by engineering structures such as groins (e.g., Wisconsin, California, and Florida). The results of coastal erosion are ever too clear: houses destroyed by storm waves (Fig. 10.23), or ugly barricades constructed to temporarily hold off the forces of nature (Fig. 10.15).

Some enlightened coastal communities have taken steps to ameliorate coastal erosion. Figure 14.3 is a map of a portion of San Mateo County, California. The Pacific shoreline is zoned into three categories based on erosion rates and bedrock type. In general, Zone C includes areas of granite bedrock, not susceptible to much erosion, as evidenced by repetitive photography and map coverage. Zone A includes semiconsolidated Miocene sediments which are very susceptible to erosion. No structures at all are allowed along cliffs (labeled EX) in Zone A. In order for construction to take place in the area labeled DS of any of the three zones, a developer must submit a geologic report which demonstrates geotechnical stability of the planning structures. Howard and Remson (1978) describe other environmental constraints used in determining land-use planning maps for the San Mateo area. These include land slopes, landslide potential, vegetation communities, and existing land use.

San Mateo's planning zones may seem somewhat crude and simplistic, yet they are a positive step in ameliorating coastal erosion losses. There are examples of other coastal areas which have site-specific problems and hence site-specific solutions. For example, Dutch engineers have begun construction of a $5 billion effort to ameliorate flooding of the Dutch territory below sea level (Section 10.2). The Delta Works will consist of 63 giant sluice-gates, which will keep flood surges out, but allow normal tides through. The barrier will stop flooding and protect residents, but at the same time please environmentalists.

One partial solution of the urbanization of barrier islands (as discussed in Section 10.5) is to eliminate government subsidies in the form of construction funds, disaster relief, and reconstruction programs to barrier island communities.

As eustatic sea level continues to rise, artificially stabilized beaches become more hazardous. Short-sighted people will advocate larger sea walls. A more environmentally acceptable alternative for barrier island development is to establish a buffer zone and let the sea roll in. Given the certainty of erosion and landward

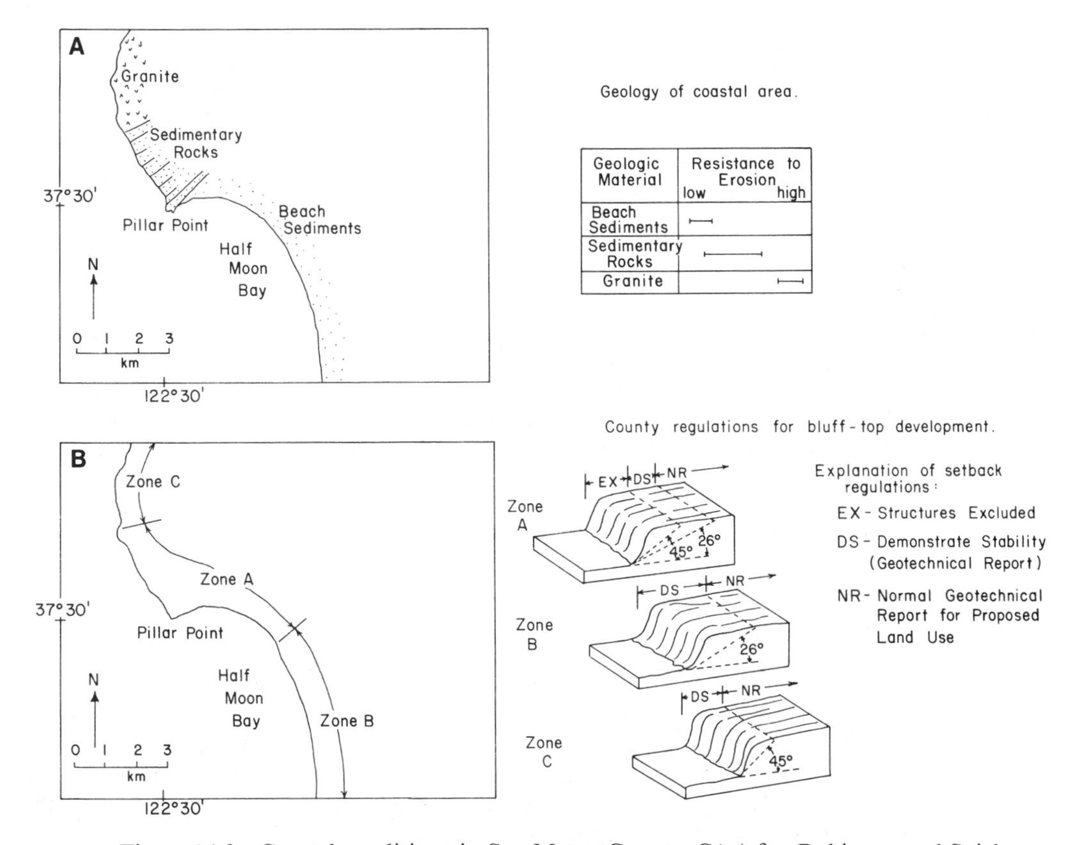


Figure 14.3 Coastal conditions in San Mateo County, CA (after Robinson and Spieker, 1978). **A**. Geology of coast line. **B**. County regulations for bluff-top development.

retreat of barrier islands, a static or permanently defined line cannot be considered as a long-range solution. It is encouraging in this respect that North Carolina is experimenting with a setback line of 30 times the annual erosion rate. Another ameliorating measure is the concept of a rolling setback line, one that by definition would periodically shift landward (Pilkney, 1981). Further ameliorating measures could be 1) the cessation of redevelopment funds (e.g., using tax money) after storms, and 2) real estate deeds containing a hazard document.

We are clearly at a point where decisions relating to shoreline management must be changed. The present absurdity of tax-promoted urbanization of ephemeral shorelines is due to short-sighted politicians, developers, and coastal engineers, among others, who through ignorance or haste, or in response to political pressure, fail to utilize the overwhelming engineering and scientific knowledge of shoreline processes.

## *14.1.4 Earthquakes*

When it comes to earthquakes, all eyes are on California. Most of the nation's seismic activity occurs in California, and, with its burgeoning population, the scene is being set for a first-class disaster. The California state government rec-

Table 14.1 Predicted Dollar Losses in California from1970 to 2000

| Geologic problem | Dollar losses to year 2000 (in millions of dollars) |
|---|---|
| Earthquake shaking | $21,000 |
| Tsunami | 40 |
| Fault displacement | 76 |
| Volcanic eruption | 49 |
| Flooding | 6500 |
| Landslides | 9850 |
| Subsidence | 26 |
| Expansive soil | 150 |
| Erosion activity | 565 |
| Loss of mineral resources used in urban development | 17,000 |
| Ground water depletion or degradation | 50 |
| Total | 55,306 |

After Bolt, et. al., 1977. One fatality is assessed at $360,000

ognizes earthquake hazards. Table 14.1 shows that earthquakes are expected to cause $21 billion in damage between 1970 and the year 2000, far exceeding damage due to all other geologic phenomena. It is interesting to note in this respect that because of the potential losses of a major quake, earthquake insurance is not vigorously sold by many reputable private companies. In Los Angeles, only 4% of all structures are insured against earthquakes.

The examples of disastrous earthquakes given in Chapter 11 illustrate that damage is not only caused by the magnitude of the earthquake and distance to epicenter, but also to the physical properties of the bedrock and surficial materials. Engineering geologists can help ameliorate earthquake damage by detailed geologic mapping of earthquake prone areas, and working to ensure that planning agencies develop regulations to minimize structural failures in high risk areas.

In terms of planning for earthquakes, California also leads the nation. Building codes for earthquake-prone areas vary from city to city (Berlin, 1980). The building codes utilize parameters such as bedrock type and proximity to active faults (see Chapter 11). Most small towns have no codes at all.

In general, new large buildings located in earthquake-prone areas (such as the Bank of America in Managua—see Section 11.2.4) are designed to harden the structure against damage. While the designs have proven to be moderately successful, the construction does little to protect the contents of the building. For example, an earthquake near Berkeley's Lawrence-Livermore Laboratory left the physical plant intact, but vibrations caused millions of dollars to lab equipment inside. A new technique, called base isolation has been used for a nuclear reactor in South Africa and a county courthouse complex in San Bernadino, CA. The California courthouse, built in 1983, will rest on a foundation of rubber-like shock absorbers, which will allow the building to move gently during a quake.

Despite the concern of many people, effective planning and zoning for earthquakes has not been too effective, even in California. In some high-risk areas of California, the same types of buildings are built in earthquake-prone areas as in

Figure 14.4 Suburbs of San Francisco built across the San Andreas Fault (photo by R. E. Wallace, USGS). Some of the suburbs in this view, looking southeast from above Daly City, lie across strands of the San Andreas Fault. The fault follows the distinct topographic alignment from the bottom center to the top center of the photo. The earthquake of 1906 caused as much as 2.4 m horizontal displacement along fault strands like these; much less movement would wreak havoc here. State laws enacted in 1972 require geologic investigations prior to construction of a public structure in the "special studies zone" and prohibit building of public structures across the fault strands. Site investigations may also be required by local governments before construction of new single family dwellings.

non-earthquake areas except that conventional engineering safety factors (for the design of reinforced concrete, I-beams, etc.) are raised slightly. Zoning an entire earthquake-prone area like San Francisco is exceedingly complex. Many housing developments are already built on top of faults (Fig. 14.4). It is too late to move the houses. But future construction can be regulated, thereby ameliorating the earthquake's effects. This regulation would include knowledge of fault proximity, and mapping of soils and rocks susceptible to pronounced shaking or susceptible to the thixotropic effect.

Based on the experience of building materials during earthquakes, it is obvious that some construction materials such as masonry, cinder block, and brick buildings are vastly inferior to steel framed buildings or buildings having reinforced concrete shear walls. The contrast in building materials was dramatically illustrated by the 6.5 magnitude earthquake which struck Coalinga, CA, on May 2, 1983. There were no fatalities, and the majority of newer structures survived very well with the exception of fireplace chimneys. The worst damage was in unreinforced brick and masonry buildings located in the downtown area, most of which were built during the 1920s and 1930s.

Following the 1971 San Fernando earthquake, the California State Capitol Building in Sacramento was carefully studied and found wanting. Finished in 1874, the building was constructed of thick, unreinforced brick. In 1972 the building was declared unsafe and most of its space vacated. Of special concern was the rotunda, which had virtually no lateral strength. To ameliorate the situation, structural engineers added reinforced concrete to the rotunda walls and steel framing beneath the upper dome. Interior brick walls were removed and replaced by reinforced concrete shear walls so that the entire building would behave like a rigid box.

Approval for the planning, design, and operation of nuclear reactors is handled by the U.S. Nuclear Regulatory Commission. In California and other places, the NRC has been very thorough in its evaluation of earthquake hazards.

In 1972, the State of California enacted a law which requires special geologic reports prior to construction of any public building within a 0.2-km strip along each side of faults recognized by the State Geologist as active (displacement in historic time) or potentially active (moved in the past 11,000 yr). Public structures within 15 m of an active fault would be prohibited. In this way, it is believed that damage to public structures by displacement along a fault would be reduced or eliminated for future buildings. However, excluded from the plans are 1) the many private structures which are not subject to the restrictions, 2) existing structures which are not "grandfathered" into the law, and 3) areas which are located farther than 15 m from an active fault.

Damage to structures built across faults can be ameliorated by special engineering techniques. For example, the active Hayward Fault near Berkeley, CA, was crossed underground in the Bay Area Rapid Transit 5-km-long Berkeley Hills tunnel. Extra clearance was made in the fault zone to accommodate up to 0.6 m horizontal and 0.3 m vertical displacement. Permanently operating warning systems were installed in the tunnel as a safety measure (Taylor and Conwell, 1981).

Actual displacement of the ground along faults and consequent shearing of structures in the sheared zone only accounts for a small fraction of the total damage. The foci of earthquakes are typically 5–10 km deep. Isoseismal plots of historic earthquakes show that intense and widespread damage is typically caused by ground shaking at some distance from the fault that caused the earthquake. So the California 15-m law may save a structure from being literally sheared into two pieces, but this is only part of the overall earthquake hazard. On the other hand, California officials cannot prohibit all construction within, say, 10 km of an active fault. Perhaps the most prudent ameliorating methods would include zonation based on combinations of 1) subsurface soil or bedrock shaking susceptibility data, 2) slope and land-slide potentials, and 3) proximity to active faults.

As in the case of many engineering geology situations, the lack of clear-cut answers to technical questions about earthquakes gives rise to engineering ambiguity or to conflicting testimony by geologists. In 1983, engineering geologist Henry H. Heel offered the following commentary:

> At about that time Frank (Morgan) was involved with the geological hassle that was taking place in connection with the location of the Corral Canyon nuclear power plant which the Department of Water and Power of the City of Los Angeles wanted to install in the mouth of Corral Canyon on the Malibu Coast, directly astride the Malibu Coast fault. This was perhaps a classic example of locating a plant such as this on the basis of engineering factors and then trying to make the geology fit the location. Before the thing was over with, there were perhaps 50 to 100 geologists involved, and the quality of the testimony of some of the geologists involved (no names mentioned) was such that it became obvious that something needed to be done about the ethics of the profession.

In 1980, California geologist Mason L. Hill also lamented that conflicting statements concerning earthquake hazards may be given by geologists:

> In the case of the Hosgri fault, offshore Central California, some of the 'experts' conclude that the nearby Diablo Canyon electrical generating facilities are in seismic jeopardy, and other "experts" think not . . . Why can conclusions regarding seismic risk be so divergent? Obviously, it is because we lack knowledge about the generation of earthquakes, which might provide a better basis for judgments regarding the seismic potential of specific faults. G. K. Gilbert, in 1909, knew practically as much about the origin and prediction of earthquakes as we do now.

Sooner or later, tough decisions regarding earthquake zonation must be made, hopefully utilizing engineering and geology expertise. The present land-use regulations such as those in parts of California are only a first step in ameliorating earthquake damage. While there is the available expertise to conduct these types of studies and to put them together in a meaningful way for planners, it will probably take a catastrophic earthquake before planners and politicians take proper earthquake planning seriously.

### *14.1.5 Land Subsidence*

Future land subsidence due to withdrawal of ground water can be mitigated by simply stopping the mining of ground water. In subsiding areas, it is, however, possible to ameliorate conditions to some degree by using telescoping casing for oil or water wells, building asphalt highways (which are easier to patch as cracks open), etc.

Land subsidence due to underground mining (Chapter 9) is almost impossible to mitigate because the cavities from abandoned mines already exist, and in time the land above will subside. The U.S. Bureau of Mines has experimented with the injection of slurry made of coal refuse and water. The slurry was pumped into abandoned coal mines at Wilkes-Barre, PA.

Urbanization of mine subsidence areas should be discouraged. However, in

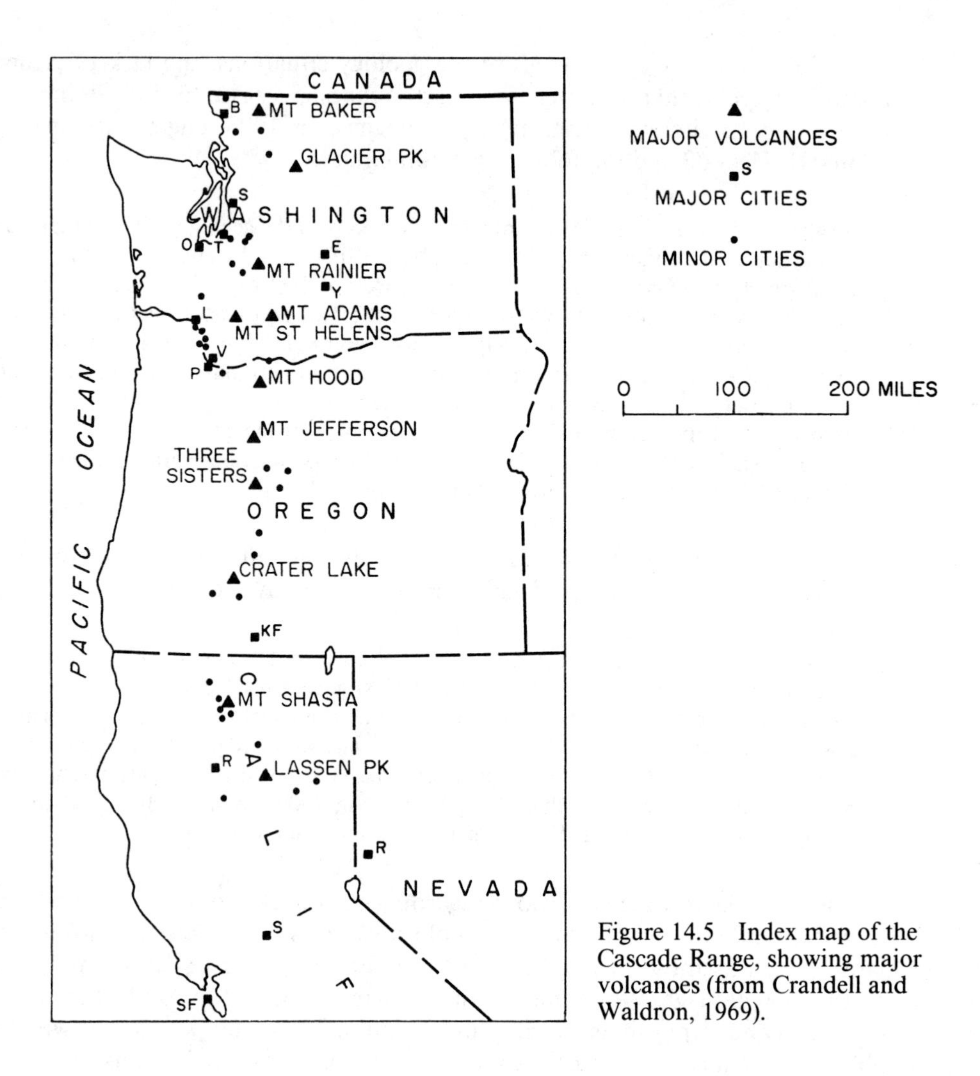


Figure 14.5 Index map of the Cascade Range, showing major volcanoes (from Crandell and Waldron, 1969).

many urban parts of the United States, it is infeasible to prevent construction over mines. Almost the entire towns of Butte, MT, and Wilkes-Barre, PA, for example, are built over underground mines.

Some old coal mines are on fire and cause air pollution as well as subsiding ground. Near Boulder, CO, maps prepared by the USGS show the land subsidence potential of these areas. The maps provided a basis for zoning decisions in Boulder County, including rejection of some development proposals (Robinson and Spieker, 1978). Amelioration of subsidence hazards over coal mines can be pursued by providing carbon monoxide detectors, etc., as described in Chapter 9.

## *14.1.6 Volcanic Activity*

*Volcanoes and volcanic eruptions.* The location of active and dormant volcanoes of the world are well known. The contiguous United States has 13 active or dormant volcanoes, all in the Cascade Range (Fig. 14.5).

Volcanic hazards result from lava flows, pyroclastic tephra (bombs, cinders, or ash), lahars (hot mud flows), nuée ardente (pyroclastic density current or glowing-cloud avalanche), or cataclysmic eruption. The reader may consult an excellent summary of these types of eruptions by Bolt et al. (1977).

Tephra falls are usually confined to the slopes and plains adjacent to an active volcano. For example, in 79 A.D., Mt. Vesuvius buried the cities of Pompeii and Herculaneum under about 3 m of hot ash lapilli and bombs.

Lava flows are common in Hawaii, particularly near Mauna Loa or Kilauea on the island of Hawaii. Basaltic eruptions are generally less violent than rhyolitic eruptions. On Hawaii, the basaltic lava flows (either aa or pahoehoe) are hazardous only to the degree that they slowly consume fields or roads. This commonly occurs in areas which have had a number of recent lava flows. On Hawaii the USGS has shown that the risk is almost wholly from lava flows, the majority of which erupt from fissures. Seismic activity usually precedes eruption. Statistical activity records and relative hazards can be compiled using dating and extent of previous flows.

Geologic evidence of Holocene lahars exists near volcanoes in the U.S. Coast Range. Crandell and Waldron (1969) show that 5000 and 500 yr ago, mud flows from Mt. Rainer extended over 50 km to the town of Tacoma, WA (Fig. 14.6). They would wreak extensive damage if they were to occur today. A lahar in 1877 traveled from Cotopaxi, the world's highest active stratovolcano in Equador, 270 km to the town of Esmeraldas on the Pacific coast.

Nuée ardente is the most hazardous of all volcanic activity. Hot ash, thrown into the air, is denser than air and rapidly flows down the side of the volcano. In 1902, the town of St. Pierre on the island of Martinique was covered by a 820°C nuée ardente from Mt Pelee. All 29,933 inhabitants perished except for a prisoner in a dungeon (Howell, 1959).

Cataclysmic eruptions are known from Krakatoa, an island near Java which blew up in 1883. The blast was heard 4800 km away, and the ash covered the world (Bolt et al., 1977). The 1815 eruption of Tambora Volcano, Indonesia, was an enormous explosion which dumped about 50 $km^3$ of rock into the air (Self et al., 1984). The eruption and related tsunami killed 90,000 people. Santorini, formerly the island of Thera, in the Mediterranean Sea, blew up about 1450 B.C. This caused the end of the Minoan (Cretian) civilization in Thera, and possibly resulted in the myth of the lost Atlantis (Doumas, 1981). Crater Lake, OR, formed about 7000 yr ago when "Mt. Mazama" blew up.

The eruption of Mexico's El Chichon volcano in 1982 caused a globe-encircling dust veil, which was higher and probably more massive than any produced this century. Tiny droplets of sulfuric acid were found at altitudes of 30 km.

*Mount St. Helens.* Mount St. Helens, Washington, provides another example of volcanic hazards. Until 1980, Mount St. Helens, about 50 km north of Portland, OR, was just another dormant volcano in the Cascade Range. Crandell and Mullineaux (1978) considered this the most dangerous volcano in the continental United States. They showed that 5000 yr ago Mount St. Helens had erupted violently, producing a widespread ash fallout. It blew up violently again at 8:32 A.M. on May 18, 1980, after a few months of small earthquakes and minor eruptions.

The eruption was initiated by a landslide on the north flank, which reduced

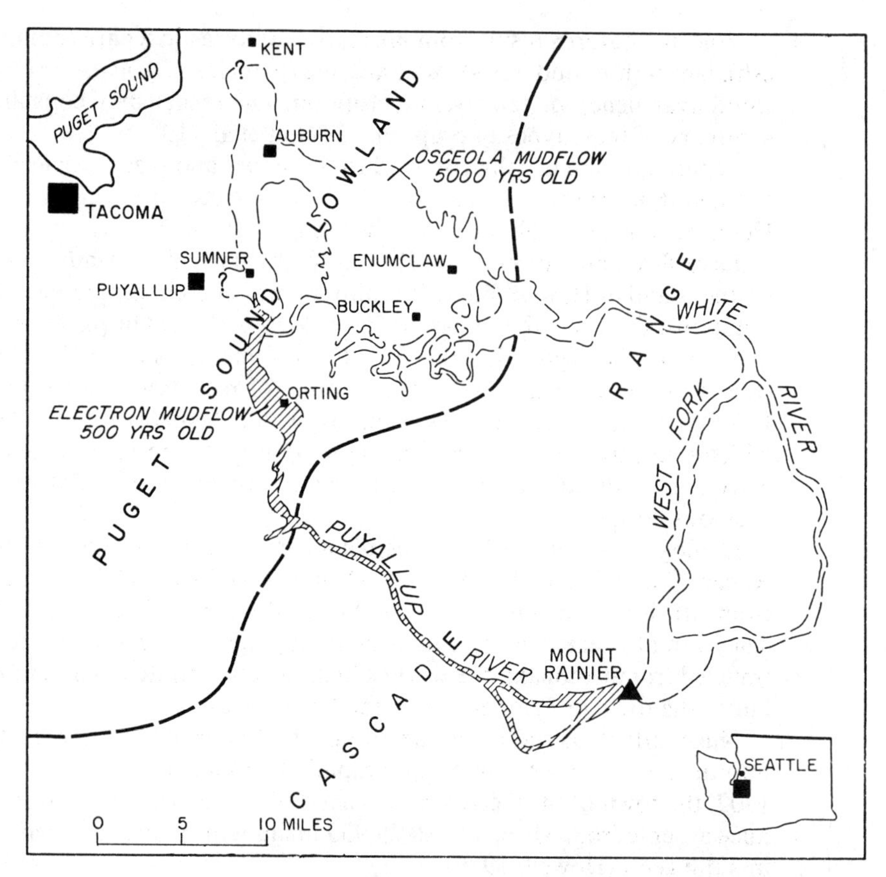


Figure 14.6 Map of Mt. Rainier and vicinity showing extent of the Osceola and Electron Mudflows (from Crandell and Waldron, 1969).

the pressure on the magma below, causing a gigantic explosion. (The scenario was predicted several weeks in advance by engineering geologist Barry Voight.) The landslide is described by geologist Keith Stoffel, who was viewing the crater from a private airplane: "Within a matter of seconds, perhaps 15 seconds, the whole north side of the summit crater began to move instantaneously." The detachment of the landslide reduced the pressure on super-heated ground water close to the magma, allowing the ground water to flash into steam, precipitating the eruption (Rosenfeld, 1980). About 400 km$^2$ of timber were blown flat by the base surge blast of tephra and hot volcanic gases. The nuée ardente was photographed by numerous people (see, for example, the January 1981 issue of *National Geographic*) as it sped over 320 km/hr. Studies of a sequence of photographs were used by Voight (1982) to document the velocity of the large landslide on the north slope at 50–70 m/s, and 50–100 m/s for the accompanying blast cloud.

Ash from the eruption of Mount St. Helens (Fig. 14.7) spread over several states (Shulters and Clifton, 1980). About 1.3 km$^3$ of ash fell on the State of Washington (Christman, 1983). The ash ranged from fine sand near Yakima (130 km downwind) to silt near Ritzville (310 km downwind). As much as 1 cm of ash accumulated 800 km away at Missoula, MT. Mount St. Helens, originally a stra-

Figure 14.7 Mount St. Helens, WA, ejecting ash on May 18, 1980 (USGS photograph). Ash from the eruption drifted eastward, whereas on March 30, the ash drifted south toward Mt. Hood, Oregon.

tovolcano 2951 m high, was reduced 400 m. Miraculously, only 71 people were killed in this catastrophe.

After the eruption, an area of about 2600 $km^2$ was completely devastated. The landslide mass, estimated at a volume of 3 $km^3$ swept 8 km down to Spirit Lake (Hammond, 1980). Up to 7 $km^3$ of mud flows (lahars) were generated. Mud flows traveled over 24 km down the north flank of the Toutle River, causing flooding. The lahar was steaming as it passed beneath a highway bridge about 20 km down valley. About 10 million $m^3$ of lahar were deposited near the mouth of the Cowlitz River and in the Columbia River, isolating ships because of the restricted shipping channel.

*Volcanic hazards.* Evaluation of volcanic risks can be accomplished by recognition of whether a volcano is 1) extinct (poses no risk); 2) dormant (no known historic record of eruption, but geologic evidence indicating return to activity could occur); or 3) active (erupted in historic time). Thermal imagery, seismic monitoring, and precise leveling can help monitor temporal changes in volcanic activity.

There is no way to stop a volcano from erupting. In this respect, reduction of volcanic hazards takes the form of amelioration, i.e., living with volcanoes as best possible.

One example of an attempt to ameliorate a volcanic lava flow occurred during the 1973 eruption of volcano Kirkjufell, which threatened the town of Vestmannaeyjar, Iceland. Many homes were buried under tephra, but a lava flow into the harbor area was slowed by a massive pumping effort. Cold sea water sprayed onto the flow caused the front to thicken and solidify prematurely, averting more damage (Williams and Moore, 1973).

Man's efforts don't always work, however. For example, the people on Sicily's east coast have always lived in fear of Mt. Etna, which has erupted 22 times this century. In March 1983, a ribbon of lava surged down the south face, destroying a dozen buildings. A team of volcanologists attempted to dynamite a section of old lava allowing for the hot lava to be redirected into a less habited area. However, the unrushing lava clogged the dynamite tubes so that the explosives were ineffective. Shortly thereafter the volcano showed contempt for man's feeble efforts: it released a poisonous cloud that swept workers from its slopes, showered gray ash on the town of Giarre, and sent a new river of lava toward the tourist center Rifugio di Sapienza. Finally, in May 1983, man-made earthen barriers successfully diverted some aa lava from Sapienza.

### *14.1.7 Alternatives*

In summary, natural hazards of one form or another exist everywhere on earth. If houses are built on hillslopes to avoid floods, they are subject to landslides, etc. No place is completely safe, even where mitigating measures are taken. There are only relative degrees of safety. It follows that habitation of this planet should be made in ways to ameliorate hazards as best as possible.

There are alternatives to the traditional development of geologically hazardous areas or to the construction of federal engineering projects funded by all American taxpayers to protect a few people who have the misfortune (or stupidity) to live in hazardous areas:

1. Community or state zoning.
2. A federal declaration that certain areas would be off limits for federal disaster relief.
3. Recognition that government insurance schemes should be completely self-sufficient and do not drain the U.S. treasury. (Or else leave the insurance business to private insurance companies.)

An Association of Engineering Geologists' position statement given by Ray Throckmorton in 1974 emphasizes, "It is more economical to properly design and construct in hazardous areas than it is to pour millions into disaster payments. By proper application of geology . . . the engineering geologist's expertise will assist in preventing misuse of our natural resources."

## 14.2 Environmental Abuses

Over the past several decades, a sense of alarm has spread across the United States and other industrialized nations. The alarm is over technological man's increasingly damaging impact on his natural environment and the quality of his own life. Some of the marvels of technology, such as nuclear power plants and large flood control structures, with all their benefits, have proven to be two-edged. Concern has given rise to new laws and new agencies whose responsibility is to control such things as air and water pollution, municipal and industrial dumps, and other forms of environmental degradation.

The remarkable photographs of earth taken from space are sobering reminders that, in contrast to other inhospitable planets and satellites, there is only one earth. All life forms are interrelated in this planet's giant ecosystem. Yet man has acquired the capacity to unbalance this system, on whose functioning his own long-term survival depends. Spaceship earth is in trouble with respect to its atmosphere, hydrosphere, and lithosphere.

### *14.2.1 Air Pollution*

The winds that move ceaselessly across the face of the globe carry with them an immense and growing burden of noxious wastes generated by man's industrial-urban life. Lead from automobile exhaust has been found in the Greenland ice cap. Sulfurous smoke from British factories blows with the prevailing winds and pollutes the fields and forests of Scandinavia. Insecticides and herbicides are spread all over the world.

Air pollution comes from many sources. In the United States, 100 million motor vehicles annually spew 66 million tons of carbon monoxide, 6 million tons of nitrogen oxides, and 1 million tons of sulfur oxides into the atmosphere. Many U.S. cities such as Los Angeles, Salt Lake City, and Denver have a chronic air pollution problem. In the winter, atmospheric inversion traps cool stagnant air on the ground, where carbon monoxide levels can reach levels dangerous to people with heart or respiratory ailments. In 1975, the U.S. EPA began requirements for emission controls on U.S. automobiles. Lead-free gasoline became required in new cars. These controls are helping to curb air pollution.

Sulfur oxides are among the largest source of air pollution in the United States, and are perhaps the most harmful of all in terms of public health. Each year, man injects a total of 650 million tons of sulfur into the air in the form of $SO_2$ (Lasaga, 1981). Coal burning contributes about 50%. The average content of the world's atmospheric $SO_2$ is about 0.2–1 $\mu g$ of sulfur/$m^3$. In downtown Los Angeles, the average is between 6 and 22 $\mu g/m^3$ reaching a peak of 150 around noon. Doctors advise emphysema and bronchitis patients to leave Los Angeles. A particularly severe London smog in 1952 killed about 12,000 people. In 1948, an atmospheric inversion in Donora, PA, made 6000 people ill, 20 of whom died. The effects of $SO_2$ on human health are not clear due to the complexity of the mixture of compounds in urban air, and the relatively greater importance of cigarette smoking to the incidence of lung cancer.

According to Fred Lipfert of Brookhaven National Laboratory (LaBastille, 1981), prevailing winds carry emissions from the industrial Ohio River Valley, where great quantities of coal are burned, eastward toward the Adirondack Mountains:

> The EPA in the 1970s permitted widely varying emission amounts from pre-existing smokestacks, with the tighter standards being applied in the heavily populated East.

> Some Midwest sources were allowed to disperse as much as 200 pounds of $SO_2$ per ton of coal.

The U.S. EPA now has regulations limiting the amount of sulfur oxides emitted from coal-burning electric plants. Attempts to meet these regulations are being made by using low-sulfur coal and by using smokestack scrubbers. Annual production levels of $SO_x$ remain high (unavoidably so according to the coal utilities, and unacceptably so according to environmentalists). Deliberate efforts to control the generation of $SO_x$ from coal-fired combustors are very expensive and remain only moderately successful (Reuther, 1982). Pollution from coal-burning power plants will be an increasing problem in industrial countries as the use of coal increases. Austrian foresters have found that a third of all the trees in the Alps are diseased. West Germany, the last major European country to take action to curb sulfur emissions, quickly changed its policy in 1983 when the government found that 8% of Germany's forests suffer from air pollution damage. Metal smelters in the United States capture about 60% of the sulfur emitted from mills, which is greater than those in some countries such as Australia, Africa, and Canada, but less than European smelters (Swan, 1982).

Acid rain is damaging to limestone and marble building stone and to outdoor works of art. Venice, Italy, loses an estimated 6% of the surfaces of its marble works of art each year due to acid rain caused by air pollution. Acid rain is blamed for the weathering of the 330-year-old Taj Mahal, India. Damage to the monument's white marble includes some areas etched gray by the rain. The Taj Mahal is located near two coal-fired power stations and 250 foundries.

Precipitation acidity, caused by $SO_2$ from fossil fuel combustion, has increased the acidity of certain lakes, primarily in the northeastern United States and in Scandinavia, so that aquatic ecosystems have been altered to the point that they no longer support some species of fish. Normal rain has an average pH of 5.6. Rain in the Adirondack Mountains, New York, has a pH of about 4.2, and one storm in West Virginia had a pH of 1.5 (Purdom and Anderson, 1983). In southern Norway, more than 1500 lakes have lost trout populations. Acid loading due to precipitation in eastern North America ranges between 25 and 45 kg of sulfate/ha per yr, compared to loads of 9–15 kg of sulfate/ha per yr in Scandinavian countries, where lakes have suffered severe loss of biota. In Sweden, acid rain, resulting in part from industries in Great Britain and West Germany, has reduced fresh water lake fish population because of the acid rain which falls on the crystalline bedrock which has little buffering capability. The rain in Switzerland is similar in acidity to that of Sweden, but little ecological damage results because of the abundant limestone in Switzerland.

In 1980, the New York State Department of Environmental Conservation declared that 264 Adirondack mountain lakes were dead, incapable of supporting life. Leaves that once decomposed in these lakes now rest on the bottom, because even the bacteria have disappeared. The State of New York has filed suit against EPA to prevent it from allowing Midwestern utilities to increase their sulfur dioxide emissions. About 50% of the sulfur deposited in eastern Canada originated in the United States. John Roberts, Canada's Minister of the Environment, said in 1982, "If we wait much longer, we will have lost our entire lake system in eastern Canada."

Coal-fired power plants cause more than just sulfur dioxide air pollution. Burning releases arsenic, nickel, and mercury, and the ash, if disposed of improperly, becomes a health hazard when water infiltrates through it and carries off leachates. The annual release from a typical 100-MWE coal-fired power plant includes over 1 Ci of radon gas as well as uranium and thorium contained in ash released to the atmosphere (Torrey, 1978).

Air pollution seems to have an effect on the weather. A century ago, there were 280–300 ppm of $CO_2$ in the world's air, compared to 315 ppm in 1958 and 339 in 1981. The concentration may rise to over 600 ppm in the next century, based on man's fossil fuel burning rate increase of 4%/yr. The world's temperature increased by about 0.5°C over the past century. Predictions are for even greater rises in temperature in the next century due to the greenhouse effect (Bailey et al., 1978). A 1979 National Academy of Science report stated that a doubling of atmospheric $CO_2$—expected within the next 50–100 yr—would most probably bring a global warming of 1.5–4.5°C. There are, however, many uncertainties in these predictions (National Research Council, 1982b). With the increased temperature, enough glacial ice could melt to inundate many low-lying areas of the world. In 1983, U.S. EPA associate director, Joseph Cannon, said "It appears that a rise in sea level within the next 100 years of as much as 8 feet cannot be ruled out." (Because there was actually a marked cooling on earth between 1940 and 1965, it is possible that the contribution of particulates (smoke) to the atmosphere and consequent reduction of solar energy (albedo) to the earth may have more than offset the increased temperature due to increased $CO_2$ emission during this time.)

### *14.2.2 Water Pollution*

Herodotus, the Greek historian, said of the Persians some 2600 yr ago, "They never defile a river with the secretions of their bodies, nor even wash their hands in one, nor will they allow others to do so, as they have a great reverence for rivers" (McGauhey and Eich, 1942). What a contrast to today's civilization!

Pollution of the world's streams and lakes has reached tragic proportions. The world's oceans have received all kinds of trash, including radioactive debris, oil spills, and chemicals. Widely used persistent compounds, of which DDT is the best known, concentrate in the bodies of aquatic organisms. There are more than 12,000 different toxic chemical compounds in industrial use today, with more than 500 new chemicals being developed every year. Little is known about the environmental and health impacts of the chemicals. In 1984, the U.S. National Academy of Science randomly sampled 675 chemicals produced by the chemical industry and found that adequate health hazard assessments could only be made for 10% of the pesticides and 18% of the drugs and ingredients mixed with them.

There are hundreds of examples of heavy metal pollution in streams, lakes, and near-shore marine areas (Forstner and Wittman, 1979). One of the worst examples was in the 1950s in Minamata, Japan. Fish and shellfish caught in sea water heavily polluted with industrial mercury were eaten. By 1965, 52 deaths and 160 cases of brain damage had been reported (Faust and Aly, 1981). There is growing international concern over the environmental impacts of dumping hazardous wastes, including radioactive wastes, in the ocean. In 1983, Great Britain dumped some 4000 tons of atomic waste 600 km northwest of the coast of Spain. In 1984 the U.S. Navy planned to dispose of 100 old nuclear submarines, perhaps by scuttling them, nuclear reactors and all, off the coast of Cape Hatteras, NC, and Cape Mendocino, CA.

In 1983, the U.S. Fish and Wildlife Service's Great Lakes Fishing Laboratory reported traces of 476 toxic substances found in fish in the Great Lakes. Of particular concern are pesticides, herbicides, toxic metals, and chlorinated hydrocarbons. Great Lake states now warn their residents not to eat large lake fish such as salmon and whitefish more than once a week. The state of Indiana urges its residents not to eat any Great Lakes fish.

In recent years, the U.S. EPA has identified hundreds of sites throughout the country where ground water has been polluted by hazardous wastes. A particularly tragic example is Love Canal, near Niagara Falls, NY, where a variety of discarded hazardous chemicals has been transported by ground water into a large area, including the basements of some houses (see Chapter 7). There are thousands of hazardous dump sites throughout the United States for which no environmental data has been collected. They are simply disasters waiting for public discovery. The National Research Council (1982c) concluded that, in general, geological studies for the safe disposal of industrial wastes have not been made prior to wastes being dumped upon or pumped into the earth's crust. A 1982 congressional committee survey revealed that 93% of all the hazardous wastes generated by America's 50 largest chemical firms during the past 30 yr has been disposed of by digging a hole out in the backyard of the manufacturer's site and dumping the stuff in. For example, the U.S. DOE's 800 $km^2$ Savannah River plant in South Carolina, manufacturer of raw materials for the nation's nuclear weapons program, has 59 seepage basins. In 1984, it was disclosed that about 350 tons of mercury were—as Defense Department officials put it—"lost to the environment" at the Y-12 nuclear weapons manufacturing facility at Oak Ridge, TN. For 30 years, effluent from the plant, stored in ponds, had been seeping into Bear Creek and East Poplar Creek.

The improper disposal of chemical wastes can contaminate soil as well as water. In 1983, significant traces of dioxin, a byproduct of herbacide and germacide manufacturing, was found in Times Beach and other sites in Missouri. Apparently the waste was distributed in the 1970s by one hauler who mixed it with waste oil and sprayed it over a broad variety of streets, parking lots, and horse arenas to keep the dust down. The U.S. EPA announced a plan to buy up homes and businesses in Times Beach, where two 1983 floods from the Meramec River compounded the problem.

In 1982, the Massachusetts Legislature was confronted with a list of 36 municipal wells and 29 private wells that were contaminated in the state. More than 8% of the communities in Massachusetts are affected by water supply contamination from organic chemicals (Bowly, 1983). The Massachusetts Department of Environmental Quality Engineering authorized a $10 million bond issue to clean up contamined water supplies and to provide emergency water supply augmentation.

In 1980, U.S. industry generated an estimated 56 million tons of hazardous waste. High-ranking EPA officials testified before a New York Special Grand Jury, "Over 90% of the hazardous waste generated in the U.S. today is handled improperly and may be or is causing detrimental effects to human health and the environment every day." Consider the following ironical situation at Woodbridge,

NJ, where very high levels of mercury were found all around an old WWI and WWII mercury processing plant (used for submarine ballast). It was suggested to excavate and reclaim the contaminated soil to prevent it from washing into Newark Bay. The commercial concern hired to do the cleanup was reluctant to undertake the project on the grounds that there would be so much recoverable mercury, estimated at 300 tons, that it would depress the world market price!

Ever since the 1979 discovery of widespread chemical contamination at the Love Canal area, the public has witnessed numerous episodes of communities being disrupted by the improper disposal of toxic substances. A series of chemical fires in Chester, PA; threatenea water supplies in Lathrop, CA, and Atlantic City, NJ; the discovery of PCBs in soil along State Highway 58 in North Carolina, and the dioxin poisoning of communities in Missouri are all examples of the consequences of the uncontrolled dumping of hazardous substances. After studying four hazardous waste landfills in New Jersey, Montague (1982) found that synethic (plastic) liners were not completely effective in containing chemical liquid wastes.

To help solve the problem of toxic waste, Congress passed the Comprehensive Environmental Response, Compensation, and Liability Act of 1980. Commonly known as the Superfund, this legislation provides for a $1.6 billion pool of money, derived primarily from taxes on manufacturers of certain chemicals, which will support cleanup efforts. Although the Superfund is definitely good news, the bad news is that cleanup costs for controlling a substantial fraction of the more than 15,000 hazardous waste sites identified by EPA will far exceed $1.6 billion. Congress's Office of Technology Assessment concludes that unless the government's regimen for managing and disposing of the millions of tons of hazardous wastes produced in the United States each year is strengthened, more dangerous dump-sites will be created in the next decade than will be cleaned up under the Superfund program.

Water and land pollution by industrial wastes is less severe in many countries where land is scarcer than in the United States. In Denmark, industry and citizens bring wastes to 200 collection centers, from which refuse is sent to a single treatment facility, where each year 50,000 tons are incincerated, detoxified, chemically neutralized, or solidified. The British mix cement with liquid wastes to imprison chemical wastes in concrete blocks used as subbase for highways. In the mid-1950s, the Thames River, England, was biologically dead. Since 1974, the Thames River Authority has carried out a $280 million cleaning project, and numerous forms of life have returned. Japan, where the poisoning of fish and people resulted from toxic waste disposal in the 1970s, has one of the toughest regulations concerning industrial dumping. Offending companies are required not only to pay stiff fines and clean up the mess, but also to pay medical damages and moving expenses for injured parties.

Garbage can be utilized for the betterment of the environment. In New Jersey, an ugly 815-acre landfill at Hackensack Meadowlands, containing 10 million tons of refuse, has been transformed into a 1-$km^2$ De Korte State Park. In San Francisco and in DuPage, IL, mountains of garbage have also been converted into recreational assets. "Mt. Trashmore," IL, is actually a 37 m deep layer of garbage that now serves as a slope for local skiers and tobogganers.

Land-use planning can reduce the impact of disposal of hazardous wastes. In eastern Connecticut, for example, sand and gravel pits have been used as the location of dump sites because of the ease of excavation. Unfortunately, these sand and gravel deposits (where saturated) are valuable aquifers. Proper land-use planning would identify thick till areas, where wastes could be better isolated. In Nashua, NH, over 1300 drums of highly hazardous wastes (solvents, plating water, organic alcohols), along with waste disposal by clandestine dumping, were dispensed into an abandoned gravel pit. Bulk chemicals had been poured directly into trenches. The illegal activities were discovered in the 1970s by local residents. A fast-moving contaminant plume measuring 450 m in length and 33 m in depth has been generated, moving at nearly 1 m/day and has reached a nearby stream (Morrison, 1983a). According to a Sierra Club book on hazardous waste (Epstein, 1982): "The ideal method for controlling hazardous waste contamination of groundwater is to end the production of hazardous waste. The next best method is to reduce the nature of hazardous wastes and to change our concept of land-use planning. It's time to stop siting landfills and lagoons in areas where vulnerable groundwater aquifers are located."

Indoor plumbing has been an environmental catastrophe for European and North American rivers. Even though more efficient sewage treatment plants are being built, the overall increase in population results in an increase in sewage wastes delivered to streams. These wastes not only degrade streams by causing noxious odors, but by supplying nutrients such as nitrates and phosphates which act as fertilizers and, hence, stimulate aquatic growth of algae and unwanted weeds. The biochemical oxygen demand (BOD) in streams increases due to the introduction of municipal waste; this can lower the dissolved oxygen in a stream to the point where fish or other aquatic organisms die. In Europe, many rivers cross national boundaries, so that little concern is given to downstream problems. The Rhine River, for example, leaves Germany and enters the Netherlands, where it exceeds Dutch water quality standards for sulfate, chloride, ammonia, cadmium, chromium, phosphate, mercury, biochemical oxygen demand, phenols, oil, and "smell" (Jansen et al., 1979).

The magnitude of the water problem is exemplified by the solid wastes produced in this country. In general, each person produces 2–3 kg of municipal waste and 3–4 kg of industrial waste daily. These wastes are usually buried in unlined pits where they can contaminate ground water. Solid waste disposal is a severe problem in some larger cities. For example, New York City disposes of solid and liquid wastes by dumping them in the Atlantic Ocean several kilometers southeast of the city. The total discharge of wastes from New York City is tremendous; it equals in weight the suspended sediment load of all rivers along the Atlantic coast from Maine to North Carolina (Costa and Baker, 1981).

Some innovative solutions to municipal waste disposal are worth mentioning. Recognizing the need for better sludge disposal, in 1970 Chicago began shipping sludge over 300 km to Fulton County to be used in reclamation of coal strip mines. The abandoned mines previously had had no reclamation, but, with the help of Chicago's sludge, the land is being suitably restored, eventually to become useful farmland.

In 1974, the Safe Drinking Water Act was passed, placing all of the 40,000 public water supplies in the United States under federal supervision. Passage of the law does not alter the fact that remarkably little is known about the toxicity of many substances, or that surface and ground waters continue to receive herbicides, pesticides, and fertilizers from agricultural lands, industrial effluents, mine drainage, and sewage from cities. The Federal Water Pollution Control Act of 1972 (PL 92-500) and the Clean Water Act of 1977 (PL 95-217) have helped reduce water pollution and have encouraged the development of land application of waste water.

There is little hope for termination of the pollution of surface and ground waters due to municipal sewage and other waste, but some mitigating steps may be taken (MacBerthoux and Rudd, 1977). Sewage treatment plants can be improved. One example utilized treated sewage water which is sprayed into fields and forest, allowing for renovation in soil zones, ultimately leading to aquifer recharge (Parizek, 1969). Sanitary and industrial landfills can be better engineered to protect ground water and allow for eventual reuse of the land.

On Long Island, NY, ground-water levels are declining due to pumpage and loss of infiltration due to urbanization. Ground water is also threatened by contaminants from sewage (Porter, 1980) and salt water intrusion. In the Queens section of New York City, excessive water-table lowering, salt water intrusion, and contamination of wells led to the abandonment of ground-water supply and the import of surface water from upstate New York. The use of recharge basins built for disposal of storm-water runoff, help maintain fresh water on Long Island and help prevent sea water intrusion.

Protecting the quality and preventing further contamination of the nation's ground-water resources, which now supply drinking water to almost half of the nation's population, are problems that have no easy answers or quick fixes, according to USGS hydrologist Gerald Meyer. In an article for the 1980 USGS *Yearbook,* Meyer said that the full magnitude of ground-water contamination in the United States is not known, and that complete control of contamination will never be economically or physically possible. Therefore, the task facing future hydrologists and planners lies in deciding what level of protection can be expected, implemented and maintained:

> Future preventive actions are much more promising than remedial reclamation measures. Once ground water has been contaminated, it has few uses, and practically speaking, deterioration in quality constitutes a permanent loss of water resources because treatment of the water or rehabilitiation of the aquifers is presently impractical.

McLindon (1981) points out that American business is now suffering because of its poor track record in terms of hazardous wastes:

> Emphasis on short-term, high-payoff programs had led to increased government regulation, many government programs, and an increased need for government to promulgate long-range policies for business. The degree to which government control is exercised depends in large measure upon the manner in which responsibilities are met. An industrialist being interviewed on television recently offered a reason for disposing of toxic wastes in a clay containment above a potable groundwater stratum the contention that he had to make a profit. All business must make a profit; that is the motivating force in free enterprise. But no business should make a profit at the risk of destroying potable water resources and possibly causing injury or death to users of that water. If the processes are such that the product is successful only if hidden subsidies have to be paid in the form of cost to governments to clean up, or the cost of damaged ecosystems or ill health or death for humans, then the business

> is not in free competition and does not belong in the marketplace. In effect what the industrialist said is that the company had not done sufficient research on the reason for toxic waste production or on the disposal of those wastes. They had not taxed their ingenuity or resourcefulness. They simply took their problems and transferred it to the community at large by burying it, literally and figuratively.

In 1981, Jay H. Lehr, editor of *Ground-Water Monitoring Review,* stated that while ground-water pollution is widespread, this polluted water constitutes less than 1% of America's total ground-water volume, and hence there is still time to protect the bulk of the nation's ground-water resources.

Hopefully the federal government will lead the way in recognizing the need to keep the country from becoming contaminated. Yet the federal government in the 1980s seems to have higher priority issues than environmental clean up measures. For example in 1982, Ann Gorsuch, administrator of the U.S. EPA, refused to enforce tight controls of toxic discharges at 2000 textile mills because such regulations "would have resulted in the closing of 9 mills and the loss of 1800 jobs." The EPA in 1983 loosened restrictions from ocean dumping and protection of ground water, contending that state officials should have the say over landfills, ground-water pumping, and water pollution issues.

### *14.2.3 Vanishing Lands*

The United States is blessed with rich soils and an agricultural abundance. Just as the world has come to depend heavily on the Middle East for oil, so now over 100 countries depend on the United States and Canada for grain shipments. Food exports are being looked upon as one solution to the multibillion-dollar U.S. deficit. This increased world demand for food has resulted in erosion problems as U.S. farmers abandon traditional crop rotational methods to concentrate on planting corn and other row crops. Losses in productivity can be masked for a time by an increased use of fertilizer, the use of which has gone from 14 million tons in 1950 to 113 million tons in 1980. The U.S. Department of Agriculture puts U.S. soil loss due to water erosion at 6 billion tons/yr. Every year, 3 billion tons of soil erodes from all cultivated U.S. cropland. The long-term (geologic) rate of erosion ranges between 0.22 and 2.2 tons/ha per yr, but intensely cultivated watersheds provide 445 tons/ha per yr (Toy, 1982). The Water Resources Council (1977) estimates that the average soil loss from U.S. croplands is 20 tons/ha/yr, twice the maximum tolerable loss required to sustain food production. Many farmers are actually mining the soil, knowing full well that every year a certain depth is eroded from their fields. In Iowa, for example, about 2.5 cm of topsoil is being lost every decade because the land is inadequately protected against water and wind erosion. Urbanization and flooding by reservoirs also takes it toll of land availability. Every year about 2 million hectares of the United States are used up by construction of everything from highways to new housing.

Currently, the United States has about $1.67 \times 10^6$ km$^2$ of cropland, but about 0.6% is lost annually by urban sprawl and reservoir inundation. Nevertheless, since 1950, U.S. food production has increased 50% largely due to the increased use of fertilizers, pesticides, irrigation, and mechanization, which all require large amounts of energy. Thus far, these techniques and the development of high-yield crops have offset America's shrinking farmland.

The world's land resources are finite and being consumed at an alarming rate. The total land area of the world is 149 million km$^2$, which is about 29% of the earth's surface area. The usable land area, which is not covered by glaciers or deserts, is 74 million km$^2$, or 14% of the earth's surface area. Of the usable land, only 13 million km$^2$ (or 3% of the earth's surface area) are arable, the remainder being forests or pastures. By the year 2000, the world's cities and built-up areas may be 1.8 million km$^2$, or 0.36% of the earth's surface area (Leveson, 1980). Building these cities will require vast quantities of raw materials. Each square kilometer which is urbanized is withdrawn from land formerly devoted to food production, forests, or recreation areas. The severity of urbanization or diminishing land is difficult to comprehend if you're from South Dakota. But consider Japan, where land in 1983 within 1 hour commuting distance of downtown Tokyo which sells for \$3,000/m$^2$ is considered a pretty good bargain.

Another form of vanishing lands is desertification, the spread of desert conditions beyond the boundaries of climatic deserts, materially assisted by man's activities. Biswas and Biswas (1981) show that desertification causes the annual loss of about 60,000 km$^2$ of productive land on earth. In the past half century, 650,000 km$^2$ of land have been swallowed up by the Sahara Desert on its southern fringe, in the area known as the Sahel. A United Nations study reveals that arable areas equal to 7% of the world's land surface have been turned into desert in the past 50 yr. The Near East is believed to be the scene of the beginnings of agriculture. Thousands of years ago the forested areas were cleared and cultivated and soil erosion set in. Today bare limestone slopes strewn with remnants of former terrace walls show that the battle against soil erosion was a losing one. Olson (1981) describes areas of western Turkey where "soils have been overgrazed by goats and sheep for centuries . . . Forests originally protected by the soils on the mountains, but the trees were cut for building construction and to provide firewood for the Roman baths. In a sense, the civilizations destroyed themselves because of the ecological and environmental degradation." In the United States, overgrazing, intensive farming, and drought combined to form the dust storms of the 1930s, which Lockeretz (1978) called "the worst man-made environmental problem the United States has ever seen." A related problem occurs where irrigation in arid areas has resulted in the salinization of the soil. Once fertile irrigated Mesopotamia and Ganges River valleys now are ruined because salts have evaporated, leaving the surface "as white as snow."

Fertile land is a precious resource. In overcrowded countries, where people are starving and competition for land is keen, erosion is often accelerated due to overgrazing. For example, millions of Ethiopians, struggling for survival, are scratching the surface of the land, cutting down trees for fuel and leaving the country denuded. One billion tons of topsoil flow from Ethiopia's highlands each year.

Protecting U.S. farmland from erosion is largely left to the individual farmer. The market does not recognize or reward him: a truckload of corn produced under careful soil stewardship brings no better price than a truckload produced on land being allowed to wash or blow away. In 1983, Senator Roger Jepsen of Iowa said "the agricultural production and potential of our land is our single greatest resource. . . . (Yet, after years of heavy erosion). . . . where we had eight to 12 inches, right now I can show you places with a quarter inch of topsoil left."

Because of the expanding population and the finite area of arable land on earth, there is a growing food problem. The development of new grain species (the "green revolution") during the past few decades has helped mitigate the vanishing farmland problem. However, high-yielding varieties of grain and rice only reach optimum production under ideal conditions such as plentiful water, large amounts of fertilizer, and chemical protection against disease. Fertilizers are in short supply throughout the world, and the increasing costs of pesticides, sprayers, and other mechanical equipment favor agri-corporation farming, which doesn't benefit the small farmer. In Mexico, for example, where the green revolution started, and where per capita wheat production has increased, most of the profits go to the owners of large, mechanized farms, and the produce itself goes for export.

Rand McNally (1979), summarizing the earth's resources, warns:

> It is impossible to exaggerate the severity of the global food problem. If this planet is soon to accommodate a massive increase in population—and this seems to be the one incontrovertible fact in a sea of uncertainty—it will not do so with a system of housekeeping that permits half the world to remain undernourished while the other half makes wasteful use of food. Without some radical attempt to redress the balance, we are faced with the distressing prospect that, in developing countries at least, population control in the future may be exercised by means of famine.

The Worldwatch Institute (Brown, 1981) warns that, at the present rate of soil erosion and loss of farmland, our civilization can't continue:

> A world that now has over 4 billion human inhabitants desperately needs a land ethic, a new reverence for land, and a better understanding of the need to use carefully a resource that is too often taken for granted. Civilization cannot survive a continuing erosion of the cropland base or the endless conversion of prime farmland to non-farm uses.

## 14.3 Environmental Impact Statements

The first step in solving environmental problems is to educate people so that they are cognizant of the problem. The second step is action.

### *14.3.1 National Environmental Policy Act of 1969*

In reponse to public recognition of environmental abuses to air, water, and earth, the U.S. Congress established the National Environmental Policy Act (NEPA) of 1969. The act established a presidential advisory group, the Council on Environmental Quality. NEPA also established a national policy requiring all federal agencies to give full consideration to environmental effects in planning their programs. Each federal agency must prepare an environmental impact statement (EIS) on every major federal action "significantly affecting the quality of the human environment." Private developments, not on public land and not using public money, are not required to publish an EIS. The statement must discuss alternatives to the proposed action and must be circulated for comment to other federal agencies, to state and local governments, and to the public (Canter, 1977).

Prior to 1969, federal agencies engaged in dam construction or other gigantic projects were not required to give full disclosure to the public of the specifics of the project. Many large dam projects were built, unmindful of environmental consequences. By the time the public found out, the project had been built. With NEPA, for the first time, the public could see what the government was going to do.

The EIS consists of two parts. A draft EIS (DEIS) is issued first. The agency is receptive for comments from agencies and individuals during a waiting period of at least 90 days. Then a final EIS (FEIS) is issued, which is supposed to consider and answer all questions raised by the DEIS. Public hearings are also held.

The contents of a DEIS consist of the following sections, which generally make chapter headings:

1. Description of the proposed action
2. Impact on the environment
3. Unavoidable adverse environmental effects
4. Alternatives to the proposed actions
5. Irreversible or irretrievable commitments of resources involved if the project is implemented.

In some respects, the EIS is an exercise in bureaucratic paperwork because, once an EIS is prepared, there is generally no way to stop a project except by separate litigation. However, termination of a project by federal initiative can be brought about by the President upon recommendation of the Council on Environmental Quality. This occurs rarely, but did happen in the case of the controversial Florida barge canal in 1973.

In spite of its shortcomings, NEPA does allow, for the first time, full disclosure to the public. Court action can then be taken by the people if they choose. Skillern (1981) presents a complete list of legal cases involving environmental impact statements. Lucchitta et al. (1981) state that, despite the drawbacks of environmental impact statements, they can identify and hopefully minimize problems:

> EIS's are not meant to be polemics. They are to be reasoned judgments of how proposed projects are likely to affect the future, and those judgments are to be built on the best information. Thus, despite their imperfections, EIS's are the best tool we have for making decisions and charting a safe course through our troubled times.

### *14.3.2 EIS Example: The Oahe Irrigation Project*

The following example of an EIS and ensuing legal battle illustrates some of the salient aspects of environmental issues in the United States.

The U.S. Bureau of Reclamation (USBR) is charged with the responsibility of developing irrigation in the western United States. From the first USBR project at Belle Fourche in western South Dakota in 1909, to the immense Glen Canyon Dam in Arizona, completed in 1960, the Bureau has achieved successful reclamation projects envisioned by its founder, John Wesley Powell. But the USBR is running out of good new projects. The USBR scheme for irrigating the James River lowland in eastern South Dakota originated after passage of the Flood Control Act of 1944, which allowed for the construction of large mainstem dams on the Missouri River. To compensate for the large fertile area of flood plain inundated by the reservoirs in South Dakota, the federal government promised South Dakota that a large irrigation project would be constructed. In the 1960s, the

James River Valley, about 160 km to the east and 60 m in elevation lower than Oahe Reservoir near Pierre, SD, was chosen. It was to be a gigantic project. When completed, 200,000 ha of land would be irrigated. One problem was the drainage of irrigation water. Because the project area is very flat and is underlain by about 12 m of low permeability glaciolacustrine silty clay from glacial "Lake Dakota," about 5000 km of underground drain pipes would be required to prevent waterlogging.

In 1974, the USBR issued an EIS describing the Oahe Irrigation project and began construction of an immense pumping station on Oahe Reservoir. At that time, a group of local ranchers filed suit against the USBR. The plaintiffs were farmers who lived in the project area or in areas whose farms were to be inundated by some of the holding reservoirs near the project area. The legal grounds of the suit (United Family Farmers versus Thomas S. Kleppe, South Dakota, 1976) was that the EIS was inadequate and did not fully explain the danger of waterlogging, soil salinization, return irrigation water, and the impact due to channelization of 195 km of the James River. The USBR argued that they had prepared a sufficiently accurate EIS.

In the course of the proceedings, it was discovered that the entire project was barely economic. It was pointed out that the drain pipes represented about 25% of the cost of the project, and the USBR had increased the spacing of expensive drain pipes from early estimates of 84 to 189 m apart. It appeared that the USBR had cut costs to make the project more economically attractive.

Nonetheless, a federal judge ruled in favor of the USBR, claiming that "The Final Environmental Impact Statement sufficiently alerts decision-makers of the possible adverse effects. . . . Disagreement among experts, however, as to the technical plans and engineering feasibility of proposed action will not invalidate a Final Environmental Statement."

However, because of public awareness of the environmental and economic problems of the Oahe project, most people, including South Dakota Congressmen, were opposed to the project after the trial. Backed by the USBR and local promoters, construction on the project continued until 1977 when President Carter decided to oppose the construction of 19 huge U.S. water projects on the grounds that they were simply pork-barrel projects. U.S. Congress backed the move, and financing for the Oahe Irrigation Project was dropped. The project was terminated.

The lesson learned here was that the United Family Farmers' stand against a questionable project was defeated in court, but because of the EIS and public exposure in the court proceedings, public opinion was changed and the government responded.

A controversial environmental issue was settled in a similar way in Pennsylvania. Recognizing the aesthetic value of the Delaware River valley near Delaware Water Gap, in 1970 the Delaware River Basin Commission reversed an earlier decision which instructed the Corps of Engineers to build the Tocks Island Dam. The dam, conceived following 1955 floods on the Delaware River, was opposed by 61 environmental organizations. Public exposure to the environmental consequences of Tocks Island Dam was spearheaded by Dr. Francis Trembley, a biology professor from Lehigh University. The dam most certainly would have been built prior to the days when governmental dam-building agencies were required to give full disclosure of their projects.

It must be remembered that prior to 1969, the public had almost no idea of the engineering, sociological, cultural, or environmental changes that could be brought about by a big federal project. So NEPA helped alert the public to potential hazards.

In some respects, however, NEPA is lacking. For instance, most EISs are prepared by the agency promoting the project, and so tend to downplay the environmental impacts. Another factor is that impact statements are generally too long and too technical for lay people. Most effort goes into making sure that all possible impacts, no matter how trivial, are mentioned in passing at least once. Lawyers recommend this sterile tactic to avoid a lawsuit challenging the complete omission of an impact. In this way, the EIS obscures rather than highlights project impacts. The purpose of the 1969 NEPA Act is to inform the public, with the intent, presumably, that the public can take a course of action if they have the will and resources to fight. Presently, the EIS seems to be more of a paper obstacle in the agency's way, which is cynically handled so as to confuse and overwhelm the public.

## 14.4 Pork Barrel Projects

It should be recognized that NEPA does not address the way in which this country decides upon and finances projects. The benefit/cost (B/C) ratio technique presently used is not very realistic, not only because the costs are estimates (see Table 8.3), but because no one knows what future inflation or benefits will be. If a Corps of Engineers project was authorized by 1968, the discount rate used in calculating annual benefits and costs was 3.25%. Projects falling into this category are said to have their discount rate "grandfathered"; i.e., 3.25% is used regardless of when construction is ultimately undertaken (Steinberg, 1982). The 1982 governmental rate of future payback (discount rate) is set at the low rate of 7.625%. When inflation (or cost of living) is 13%, as it was in 1981, then real prices rise accordingly and the discount rate is unrealistic. For instance, if $1 billion of federal tax money is used to build a dam, all to be repaid 30 years later as a lump payment computed at 7.625% interest, the payback fee, using Eq. (1.1), is $9.07 billion. Thus, the B/C ratio would be equal to unity if this payment were made. But if real inflation is 13% during those years, then $1 will only have an effective buying power of 2.6¢ in 30 years. Therefore, $9.07 billion is effectively worth only $232 million. Thus the B/C ratio is actually only 0.23. Using long backback times, low discount rates, and high inflation, one could manage a B/C $>$ 1 for the construction of a completely unproductive project such as an Egyptian pyramid. After several centuries, one tourist ticket for admission to see the pyramid would pay for the project.

Another factor that affects B/C ratios is the built-in bias in favor of construction. Economist Ronald Boster (1983), comments on the Central Arizona Project: "A common characteristic of pork-barrel projects is that benefits are highlighted while costs are de-emphasized. Yet, costs invariably exceed benefits."

Economic, engineering, and environmental analyses provided by an agency proposing a construction project are often very technical. Challenging these documents is difficult. Yet in several cases where independent challenges have been made, they have found B/C ratios greatly exaggerated. For example, in 1972, Governor Jimmy Carter of Georgia challenged the construction of a Corps of Engineers dam on the Flint River. It was found the Corps had omitted huge costs and had wildly inflated benefits; for example, the loss of $2 million annual income from forestry jobs as a result of drowning a forest had been ignored. In another example, the Paint Creek Dam in Kentucky was challenged in 1977 by a group of citizens. They found the Corps had valued coal deposits to be submerged by the reservoir at $48,000 when its value was probably worth millions. "The Corps doesn't want to pay what the coal is worth," said John Phipps, the county judge in Morgan County, where land will be flooded by the reservoir. "If they did, it would ruin their cost-benefit ratio and prove the dam isn't worth building."

Once Congress approves a project and appropriates money for it, there is, however, no longer any question of economics. It's like building B-1 bombers or rockets to the moon. When the pork barrel starts to roll—when Congressmen begin promising to vote for projects in each other's districts in return for votes on equally massive projects that will pump millions of federal dollars into their own districts—it becomes next to impossible to stop. The sad fate of the Little Tennessee River and the Teleco Dam is described in Chapter 8. The economics of this dreadful project was so bad that even the dam builder, the Tennessee Valley Authority, was against it. Yet Tennessee Congressmen were able to push it through.

A recent example of a political project is the Tennessee–Tombigbee Waterway, a 376 km canal completed in 1985. At 2 billion dollars it is the largest project ever undertaken by the Corps of Engineers. For barge operators in some Eastern states, it provides a shorter route to the Gulf of Mexico than the Mississippi River. However, the waterway has been condemned by critics as a classic example of government waste. Barge traffic the first two years was less than half what was expected. Joseph Carroll, an economist at Pennsylvania State University, says the waterway is "... primarily a political project" promoted by southern politicians in Congress.

Some projects, with $B/C < 1$, are built anyway. In Nebraska, the O'Neill Unit, including the proposed Norden Dam on the Niobrara River, was supposed to be built in 1963 for $72.5 million from the federal treasury. By 1982, the cost of the delayed project was estimated by the U.S. Bureau of Reclamation at $368 million. And of that amount, only $45 million will be recoverable from water users. The project has no sound financing. The 60-yr planned payback period interest charges alone could come to an estimated $8.3 million for every farm in the irrigated area. In addition, the land to be irrigated will produce corn which presently the U.S. government is attempting to discourage farmers from planting through the "payment in kind" program whereby farmers get paid for not growing something. In the face of boondoggles such as this, the 1864 words of Henry David Thoreau may reflect the sentiments of taxpayers who seek elimination of pork-barrel projects brought on by inept politicians: "If one were to judge these (legislators) wholly by the effects of their actions and not partly by their intentions, they would deserve to be classed with those mischievious persons who put obstructions on the railroad."

## 14.5 Planning Ahead

### *14.5.1 Land-use Planning*

As population pressure and urbanization continue, it is imperative that engineering geology and other disciplines are used to ensure that the best uses be made of limited resources. An excellent general reference for land-use planning is Ian

McHarg's (1969) book *Design with Nature.* McHarg, a geographer, shows how the topography, existing land use, soils, forests, mining, natural and cultural features, etc., all contribute to the compatibility or incompatibility of the land for habitation and use. He cites good examples as well as environmental horrors such as Glasgow, Scotland, and parts of Philadelphia, PA, where cities seem to trap people instead of providing a suitable habitat.

Legget (1973) provides numerous examples of geologic factors affecting the settling of cities in the world. The folowing passage stresses the role of geology to city planning:

> A new city can never be planned and designed in total disregard of its environment, upon which it cannot be just imposed. It must fit at least the topograhic limitations of the chosen site. It will be influenced inevitably by the geological conditions beneath the surface of the site. . . . Fundamentally, a new urban center must incorporate the landscape of its site as an integral part of its plan. Requirements of the regional plan must be reconciled with the main environmental features. The plan must take account of geological and foundation conditions, respecting the agricultural importance of some soils and protecting the original character of the landscape to the extent that is possible. Drainage of the site and the supply of water and building materials are all geological factors that must be given consideration. Assurance must be had that the plan will not interfere with the winning of building materials (especially sand and gravel) that may lie under part of the site, but rather allow for their extraction prior to use of this part of the area for building.

Land-use planning should take into account the changes in use that may occur over the course of time. A good example is Seaview Square Mall, a large shopping mall in Monmouth County, New Jersey, which was built in 1978. The site was formerly a large sanitary landfill. Engineering problems of compaction of waste, ground-water contamination, and methane gas production were overcome by intensive engineering geology studies and design. In this case, "useless" land was converted into useful land.

U.S. Geological Survey Professional Paper 950 by Robinson and Spieker (1978) provides illustrations of how geological factors such as unconsolidated sediments, depth of water table, slope of the land, depth to bedrock, etc., can be related into a general land-use map. For example, people from the town of East Granby, CT, a suburb of Hartford, recognized that a zoning map had to be developed that would maintain and enhance the rural character of the town and environs as well as protect the natural resources and ecology, and yet still provide for population, housing, commercial and industrial growth. A series of map overlays prepared by the USGS provided the town with information concerning which areas could be utilized for different intensities of urbanization. The map showed, for instance, where limited development would be desirable due to swampy areas, where moderate urban development could occur on till-covered hills, and where high or industrial development could occur in flat, sandy areas with deep water tables.

In eastern Connecticut, engineering map overlays (Hill and Thomas, 1972), include a series of maps which relate to the optimum location of sanitary landfill sites using the following criteria: 1) depth to water table, 2) flood-prone areas, 3) steep slopes, 4) valuable sand and gravel aquifers, 5) residential areas, and 6) thin glacial deposits over bedrock. Computer data base for maps showing suitability for solid waste disposal have been developed for a portion of Indiana (Hasan and West, 1982).

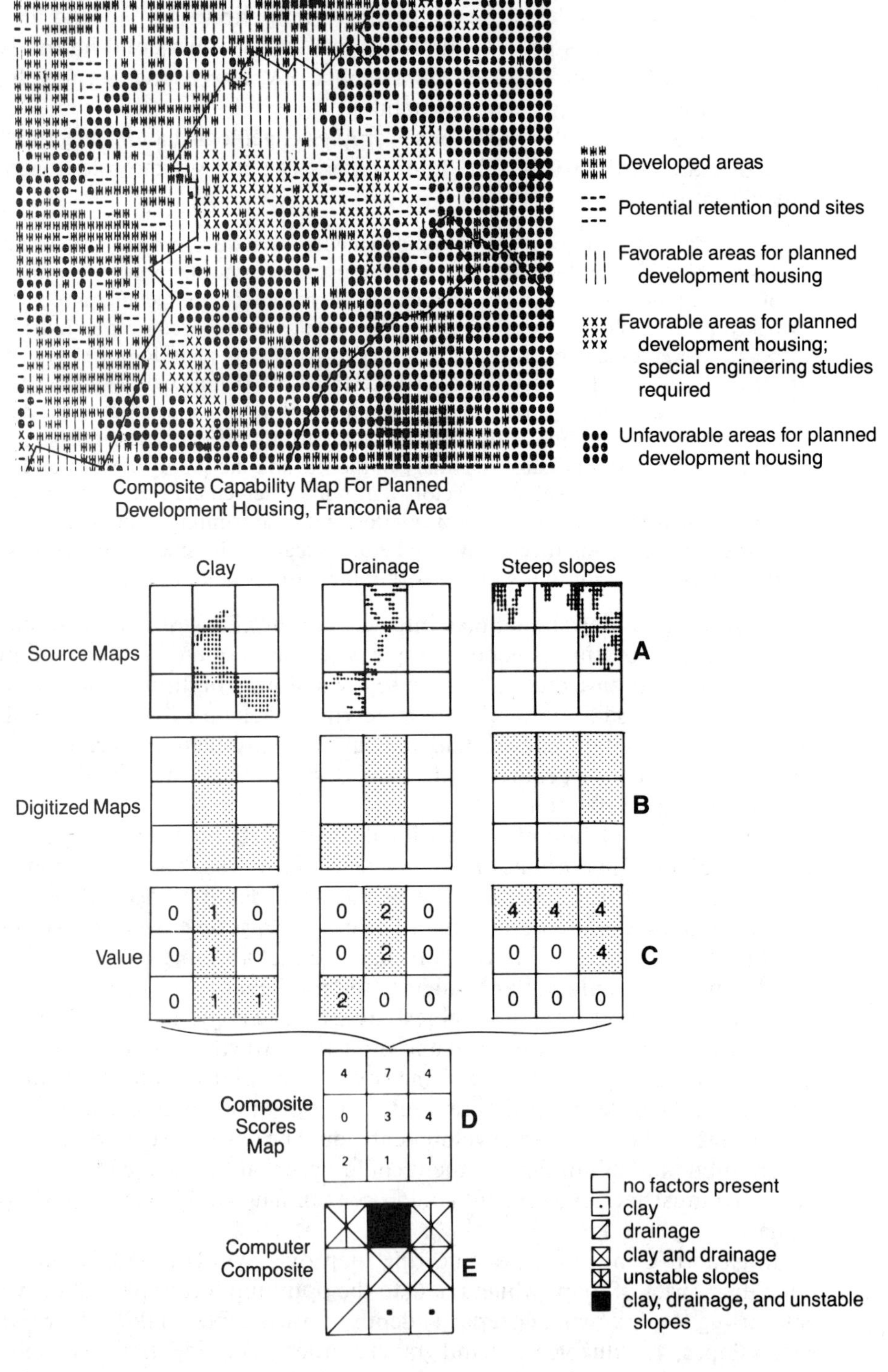


Figure 14.8 Computer composite mapping technique for a hypothetical area (from Robinson and Spieker, 1978).

Another example of land-use planning is a computerized printout of an area near Washington, DC (Figure 14.8). In this example, the Franconia area in Fairfax County (16 km southwest of Washington, DC) was about to be developed into an urban area. More than 400 ha of undeveloped land were to be transformed into a high-density new town. The USGS found that several factors could limit development in some places: storm-water runoff, expansive clay, and landslides. Base maps were prepared which ranked each hazard within a grid of the area. Basic geologic maps were used as a reference, emphasizing engineering geologic considerations. Figure 14.8 shows how "scores" were given to each criteria, and the criteria weighted according to the judgment of the engineering geologist. The end product is a computer printout of the land's capability for development. The land capability map shows how the complex interrelationships between the natural environment and man's use of the land can be dealt with in a consistent manner.

### *14.5.2 An Environmental Approach*

In summarizing his book *Cities and Geology,* Legget (1973) brings to focus the abandoned cities of Babylon and Timgad as well as other places—the wasted cedar forests of Lebanon, the Yellow River ("China's Sorrow," carrying great quantities of eroded silt to the sea), and the despoiling by industrial plants of the irreplaceable fruitlands of the Niagara Garden Belt in southwestern Ontario. Recognizing environmental problems such as these is one thing, but solving them is difficult. Consider, for example, the deforestation of western Turkey. The Spartan general Pausanias described the damage due to deforestation 2500 years ago. The problem is still not solved. The problem is rooted in the expanding population, the pressures of intensive farming, and the need to have wood for energy.

As long as the criteria for the use of land are strictly economic, the land will be utilized for a purpose that brings the most money. Many policymakers will not advocate land-use controls for fear of running against the spirit of free enterprise. Nevertheless, this strict economic approach is what contributes to the flood and earthquake disasters and other preventable geologic problems such as outlined in this book.

In the founding days of the United States, there seemed to be no limit to its resources, and the philosophy of manifest destiny meant that the colonists were free to conquer the wilderness as fast as possible. Perhaps the completely free enterprise system, which worked well in this country 100–200 yr ago in a new world of seemingly infinite resources, needs to be molded into a system wherein some recognition is given to the finiteness of resources as well as to the fragile ecosystem upon which all life depends. Lucchitta et al. (1981) put the idea of manifest destiny in this manner:

> North Americans, with their boundless optimism, have been particularly susceptible to the philosophy of the open system. As the pioneers marched westward, there were always other mountains and other valleys when the present ones were played out. But when the pioneers reached the Pacific—the western ocean—the system slammed shut. This closing was not percevied immediately, of course, because romantic ideas linger in death. Even today, many still believe that they live in the unfettered open system of the pioneers. But this is an illusion, the other place is here, the future time

Figure 14.9 Intensely cultivated hillside near Lauzanne, Switzerland.

is now. The new perception was made finally and poignantly clear by the beautiful Apollo photographs of the Earth—a bluish, ethereal and very finite sphere lost in the darkness of space.

There are many to whom unobstructed, unregulated growth is good. But with continued growth, this country's mountains and valleys will ultimately look like the crowded Ganges delta or industrial Europe, where cities are crowded and even mountain slopes are intensely cultivated (Fig. 14.9). Is this what we want? If it is not, and yet we are to continue to have better affluence with a growing population, it is up to scientists and engineers to become involved, because the good life requires not only that natural resources are used, but that they are developed in an environmentally acceptable manner. There must be balance. William Pecora, Undersecretary of the U.S. Department of the Interior, said in 1971:

> We must settle for conservation with controlled preservation rather than for whole preservation, and for a system which provides the best alternatives when no solution is a perfect one. This means management of resources to achieve the greatest good for the greatest number for the longest time. . . . Somewhere between the attitudes of unconcerned development and total preservation, there must be an acceptable point of balance. . . . This point of balance cannot be set by legislation. It must be located and kept in focus by continuing dialogue between those concerned primarily with supplying material needs and those concerned with maintaining pleasant surroundings.

Earth-science data can provide the foundation upon which land-use planners and developers can plan, design, and build in a manner harmonious with the natural environment. Future engineering geologists will be involved in this dialogue. In 1980, John P. Ivey, President of the Association of Engineering Geologists, urged:

> My message is that on issues where engineering geologists as a group are affected or have the potential to be affected, we must speak out, offer our support to those who may be in more of a front-line position than we, and take a stand on issues. Not to do so would be to forego our professional responsibility, thereby creating a vacuum which the unqualified tend to fill.

As a professional engineering geologist and as a private citizen, you should participate in the decision-making process. You live in a unique time on a unique earth. And there is nowhere else to go.

## Problems

**1.** Due to the burning of fossil fuels, the world's $CO_2$ atmosphere content is increasing 3% annually. Use Eq. (1.1) to determine how many years one requires for a doubling of the concentration at this rate.

**2.** In 1626, European settlers purchased Manhattan Island (New York) from the American Indians for the equivalent of $24. This sale is often called the worst real estate deal in history (at least from the Indians' point of view), because by 1984 the assessed value of taxable real estate reached $24 billion. But consider: if $24 had been put into a savings account at 6% annual compound interest, how much would the account be worth in 1984?

**3.** Until 1965, benefit/cost ratios for planned federal water projects were calculated on a discount rate of 3.125%.

- **a.** If a large dam project costing $1 billion was built in 1965 (with 1965 money) and was to be paid back 30 yr later in a lump payment determined by this discount rate, how much would the payment be in order for the project to have a B/C = 1?
- **b.** Assume inflation actually was 12% over the entire 30-yr period. If the original $1 billion were invested in a savings account earling 12%/yr interest, how much money would have accrued in 30 yr?
- **c.** Considering 12%/yr inflation, what is the value of the money returned as a "benefit" for a above?
- **d.** What does this exercise tell you about the B/C ratio as determined in a above?
- **e.** Do similar calculations using 1984 federal discount rates set at 7.25%.

**4.** When construction began in 1965, the Shoreham Nuclear Power Plant, NY, was supposed to cost $240 million. By 1984, the plant was still not finished even though Long Island Lighting Company had invested $4 billion into it.

- **a.** What is the ratio of actual cost to original estimated cost?
- **b.** If inflation averaged 8%/yr during this 20 yr period, what would the value of $240 million be worth in 1984?
- **c.** Is the cost overrun for this project mainly due to inflation, or is it mainly due to other factors?

**5.** In 1981, land in Switzerland sold for about $\$100/m^2$ in urban areas and $\$10/m^2$ in farming areas.

**a.** How much would a 10-ha farm cost in a farming area?

**b.** If money was borrowed at 18%/yr interest to pay for this farm, what are the yearly payments just to keep up with the interest (not applied toward reducing the principal)?

**c.** Is there any way someone can purchase a farm and make ends meet when population pressures force up land values?

Appendix A

# Metric–English Conversions

| Quantity | U.S. customary unit | SI equivalent |
|---|---|---|
| Acceleration | $ft/s^2$ | 0.3048 $m/s^2$ |
| Area | $ft^2$ | 0.09290 $m^2$ |
| | $in^2$ | 645.2 $mm^2$ |
| | $mi^2$ | 2.590 $km^2$ |
| | acre | 0.4047 hectare |
| Energy | ft·lb | 1.356 J = $1.356 \times 10^7$ ergs |
| Force | kip | 4.448 kN |
| | lb | 4.448 N = $4.448 \times 10^5$ dyne |
| Discharge | gpm | 5.446 $m^3/d$ |
| Hydraulic conductivity | $gpd/ft^2$ | 0.04070 m/d |
| Length | ft | 0.3048 m |
| | in | 2.540 cm |
| | mi | 1.609 km |
| Mass | oz (mass) | 28.35 g |
| | lb (mass) | 0.4536 kg |
| | slug | 14.59 kg |
| | ton | 907.2 kg |
| Moment of a force | lb·ft | 1.356 N·m |
| Moment of inertia | $in^4$ | $0.4163 \times 10^6$ $mm^4$ |
| Momentum | lb·s | 4.448 kg·m/s |
| Power | ft·lb/s | 1.356 W = 1.356 N·m/s |
| | hp | 745.8 W |
| Pressure or stress | $lb/ft^2$ | 47.88 $N/m^2$ |
| | psi | 6895 $N/m^2$ |
| Transmissivity | gpd/ft | 0.01240 $m^2/d$ |
| Velocity | ft/s | 0.3048 m/s |
| | mi/h (mph) | 1.609 km/h |
| Volume | $yd^3$ | 0.7646 $m^3$ |
| | $ft^3$ | 0.02832 $m^3$ |
| | $in^3$ | 16.39 $cm^3$ |
| | gal | 3.785 l = $3.785 \times 10^{-3}$ $m^3$ |
| | qt | 0.9463 l |
| | acre-ft | 1223 $m^3$ |
| Work | ft·lb | 1.356 J |

Appendix B

# Answers to Problems

## Chapter 1

**1.** 579 years (from 1985)

**2.** **(a)** 11:59 AM
**(b)** The last few minutes

**3.** 2027 A.D.

**4.** $6.26

**5.** $4.0 \times 10^9$ people/$m^2$ (by 1985)

**6.** 168 people/min (in 1985)

**7.** 10, 765 years

**8.** 9.4%

**9.** 2062 A.D.

**10.** 2400 ships/day

## Chapter 4

**1.** **(a)** 0.24
**(b)** 1.41

**2.** 1178 m

**3.** $c = 1.3 \times 10^7$ N/$m^2$
$\phi = 45°$

**4.** Block is moving. Acceleration = 1.04 m/$s^2$ downhill.

**5.** **(a)** $2 \times 10^8$ N/$m^2$
**(b)** none needed

6. **(a)** Wedge strikes N 44°W and plunges 69°NW
   **(b)** Cut should be no steeper than 74°

7. $\tau = 10^6\ N/m^2 + \sigma \tan 33°$

8. $3.2 \times 10^8\ N/m^2$

## Chapter 5

1. **(a)** 13%
   **(b)** 30%
   **(c)** 0.43
   **(d)** 80%
   **(e)** 1.86 $g/cm^3$

2. Conn. clay is more hazardous because it lies below the A line.

3. 8000 $N/m^2$

4. 8500 $N/m^2$

5. ~12 m

6. **(a)** ----------
   **(b)** gravelly sand
   **(c)** $C_u = 40$
   **(d)** Excellent

7. 0, 41°

8. $21 \times 10^4$ N per m of wall length, acting at 3.3 m above base at an angle of 30° from the horizontal

## Chapter 6

1. 32°

2. **(a)** $y = 1.1e^{0.22x}$
   **(b)** October 10
   **(c)** Yes

3. F.S. (ABCD) = 0.92; F.S. (AB′C′D′) = 1.04

4. **(a)** $\chi^2 = 40.7$, so there is a significant difference.
   **(b)** More landslides occur on north-facing slopes because it is probably wetter.

5. **(a)** F.S. = 2.09
   **(b)** F.S. = 0.84

**(c)** 1.11 rock bolts per m
**(d)** F.S. = 1.61

## Chapter 7

**1.** **(a)** 50 cm/yr
**(b)** 3.4%

**2.** 24.9 m

**3.** 2500 m

**4.** 2.4 cm/yr

**5.** **(a)** 38,000 $m^3/d$
**(b)** 800 d

**6.** 63,000 $m^3/d$

**7.** **(a)** Drained dam : $Q = 60\ m^3/d$
Undrained dam : $Q = 54\ m^3/d$
**(b)** Seepage area; could lead to slumping of downstream dam face.

**8.** $Q = 5100\ m^3/d$

**9.** 2.0 m, 3.6 m, and 5.1 m

**10.** 4.7 m

**11.** **(a)** $T = 25\ m^2/d$
$S = 2.3 \times 10^{-5}$
**(b)** 9.6 m

**12.** $T = 2900\ m^2/d$
$S = 0.0023$
$K = 162$ m/d
$K' = 0.023$ m/d

**13.** 45 m

## Chapter 8

**1.** **(a)** Flood losses are increasing.
**(b)** Dollar value and increasing population.
**(c)** No. Despite dam construction, flood losses continue to increase because of urbanization of flood-prone areas.

**2.** $L = 180\ Q^{1.9}$

**3.** **(a)** Battle Creek: 0.000,42 $m^3/s/km^2$
Castle Creek: 0.001,40 $m^3/s/km^2$
**(b)** Castle Creek because it does not tend to dry up.
**(c)** Battle Creek: 0.001,57 $m^3/s/km^2$
Castle Creek: 0.001,54 $m^3/s/km^2$
**(d)** The mean is larger than the median because a few very large floods have a great effect on the calculation of the mean discharge.
**(e)** Castle Creek drainage basin consists of permeable rocks (limestone) which yield abundant ground water (as base flow) to Castle Creek. Battle Creek drainage basin consists of rocks having very low permeability (granite and schist).

**4.** **(a)** Every 33 years
**(b)** ~200 $m^3/s$
**(c)** ~500 yr
**(d)** This magnitude flood should probably occur less frequently than once every 33 years, but probably more frequently than once every 500 years.

**5.** **(a)** 5.9 m
**(b)** 2.1 m
**(c)** For a given discharge, levee construction causes a higher stage than natural conditions.

**6.** **(a)** 85%
**(b)** 2532 A.D.

**7.** **(a)** 121 yr
**(b)** 126 yr
**(c)** decreased
**(d)** increased

**8.** **(a)** 53 mm
**(b)** 0.5%
**(c)** 0.5%

**9.** $V$ = 3.24 m/s
$Q$ = 45.3 $m^3/s$

**10.** $D$ = 1.04 m
$V$ = 1.24 m/s

**11.** **(a)** Bald Eagle Creek: 1.25 $m^3/s/km^2$
Juniata River: 0.67 $m^3/s/km^2$
Susquehanna River: 0.45 $m^3/s/km^2$
**(b)** The flood peaks occur at later times downstream.
**(c)** The peak discharge (per unit drainage basin area) is larger for the small tributary stream because the short duration of the storm probably coincided with the optimum time of concentration necessary to produce a

high discharge for the small watershed, and because the flood peak crest was subdued by bank storage on the flood plain of the large river.

**12.** **(a)** Yes. F.S. = 3.7
**(b)** No. F.S. = 0.75, so a keyway is necessary.

**13.** **(a)** 21 $m^3$/d per m
**(b)** 1.7 $m^3$/d per m
**(c)** Few open fractures are encountered at depth.

## Chapter 9

**1.** 55%

**2.** **(a)** 0.07
**(b)** No

**3.** **(a)** 0.17
**(b)** 0.033
**(c)** 0.24 m/yr

## Chapter 10

**1.** **(a)** 280 m/s
**(b)** 9.9 m/s

**2.** 1,800 $m^2$. This is a minimum value due to wave setup, wave runup, and the fact that the significant wave is often exceeded.

**3.** **(a)** 81 km/hr
**(b)** 1,950 km
**(c)** 10 m waves should have been produced.

## Chapter 11

**1.** 36°N Latitude, 118°W Longitude

**2.** Earthquake fatalities increase with increasing population, probably because people continue to live in areas of high earthquake hazards.

**3.** Epicenter at 15 km SSW of Hermosa; magnitude 6.3

**4.** 27.5

**5.** 1906: $1320 \times 10^{21}$ ergs
1964: $4170 \times 10^{21}$ ergs
1971: $5.55 \times 10^{21}$ ergs
1972: $1.47 \times 10^{21}$ ergs
A-Bomb: $0.9 \times 10^{21}$ ergs

**6.** ~8.3

**7.** 4.3

**8.** 4.5

**9.** 5.4 hours

**10.** **(a)** Slowest rate: 3186 A.D.
Fastest rate: 2013 A.D.
**(b)** There are other faults besides the San Andreas Fault, and other places on the San Andreas Fault near San Francisco where no movement has been detected in historic time.

**11.** **(a)** (1) 0.26 g
(2) 0.43 g
(3) 0.27 g
**(b)** Small changes in proximity to fault do not have as great effect on acceleration as magnitude of earthquake.
**(c)** No

**12.** **(a)** None
**(b)** The 11 and 12 m high monuments would fall over.

**13.** **(a)** F.S. = 12.3
**(b)** Decreased

## Chapter 12

**1.** 7.8 m

**2.** **(a)** 18°
**(b)** yes

**3.** $2.47 \times 10^{10}$ N/m$^2$

**4.** $V_L = 3360$ m/s
$V_T = 1680$ m/s

**5.** **(a)** $x_c = 1730$ m
**(b)** Refracted wave; 1.68 s.

**6.** Till, soft shale and sandstone, and cheese

**7.** **(a)** ---------
**(b)** Station #3, within ~7 m of surface
**(c)** Yes

**8.** **(a)** 2.14 mgal
**(b)** yes
**(c)** 2.26 mgal

## Chapter 13

**1.** **(a)** Water table will be 1.7 m below top of spoils.
**(b)** 67 ha/yr
**(c)** 19 trains/week
**(d)** 0.21 $m^3/s$
**(e)** Powder River or Tongue River

**2.** 18 m/yr, 0.17 m/yr, 0.017 m/yr, 0.0017 m/yr

**3.** **(a)** 5.8%/yr
**(b)** 467 years

**4.** **(a)** 3140 bbls
**(b)** 22,000 km (to Turkey)

**5.** 10 hours

**6.** 8000 $\times$ MPC. Dilution is not the solution.

**7.** Arithmetic is OK. A major drawback of this technique is to average lost time over everyone equally.

## Chapter 14

**1.** 23 years

**2.** $27.6 billion

**3.** **(a)** $2.517 billion
**(b)** $29.96 billion
**(c)** $0.084 billion
**(d)** $B/C \neq 1$, but $= 0.084$
**(e)** $B/C = 0.27$

**4.** **(a)** 17:1
**(b)** $1.12 billion
**(c)** Other factors

**5.** **(a)** $1 million
**(b)** $0.18 million
**(c)** No

Appendix C

# Addresses of State Geology Surveys

ALABAMA
Geological Survey of Alabama
P.O. Drawer O
University, AL 35486

ALASKA
Division of Geological and Geophysical Surveys
3001 Porcupine Drive
Anchorage, AK 99501

ARIZONA
Arizona Bureau of Mines
University of Arizona
Tucson, AZ 85721

ARKANSAS
Arkansas Geological Commission
Vardelle Parham Geological Center
Little Rock, AR 72201

CALIFORNIA
Division of Mines and Geology
California Department of Conservation
1416 9th Street—13th floor
Sacramento, CA 95814

COLORADO
Colorado Geological Survey
1845 Sherman Street
Denver, CO 80203

CONNECTICUT
Connecticut Geological and Natural History Survey
Box 128, Wesleyan Station
Middletown, CT 06457

DELAWARE
Delaware Geological Survey
University of Delaware
Newark, DE 19711

FLORIDA
Bureau of Geology
P.O. Drawer 631
Tallahassee, FL 32302

GEORGIA
Department of Natural Resources
19 Hunter Street, S.W.
Atlanta, GA 30334

HAWAII
Division of Water and Land Development
Department of Land and Natural Resources
P.O. Box 373
Honolulu, HI 96809

IDAHO
Idaho Bureau of Mines and Geology
Moscow, ID 83843

ILLINOIS
Illinois State Geological Survey
121 Natural Resources Building
Urbana, IL 61801

INDIANA
Department of Natural Resources
Indiana Geological Survey
611 North Walnut Grove
Bloomington, IN 47401

IOWA
Iowa Geological Survey
Geological Survey Building
16 West Jefferson Street
Iowa City, IA 52240

KANSAS
State Geological Survey of Kansas
The University of Kansas
Lawrence, KS 66044

KENTUCKY
Kentucky Geological Survey
University of Kentucky
307 Mineral Industries Building
Lexington, KY 40506

LOUISIANA
Louisiana Geological Survey
Box G. University Station
Baton Rouge, LA 70803

MAINE
Maine Geological Survey
Room 211, State Office Building
Augusta, ME 04330

MARYLAND
Maryland Geological Survey
214 Latrobe Hall
Johns Hopkins University
Baltimore, MD 21218

MASSACHUSETTS
Department of Public Works
93 Worcester Street
Wellesley Hills, MA 02181

MICHIGAN
Michigan Department of Conservation
Geological Survey Division
Stevens T. Mason Building
Lansing, MI 48926

MINNESOTA
Minnesota Geological Survey
1633 Eustis Street
St. Paul, MN 55108

MISSISSIPPI
Mississippi Geological, Economic, and Topographical Survey
P.O. Box 4915
Jackson, MS 39216

MISSOURI
Division of Geological Survey and Water Resources
P.O. Box 250
Rolla, MO 65401

MONTANA
Montana Bureau of Mines and Geology
Montana College of Mineral Science and Technology
Butte, MT 59701

NEBRASKA
Conservation and Survey Division
University of Nebraska
Lincoln, NB 68508

NEVADA
Nevada Bureau of Mines
University of Nevada
Reno, NV 89507

NEW HAMPSHIRE
Department of Resources and Economic Development
Office of State Geologist
University of New Hampshire
Durham, NH 03824

NEW JERSEY
New Jersey Bureau of Geology and Topography
John Fitch Plaza—Room 709
P.O. Box 1889
Trenton, NJ 08625

NEW MEXICO
State Bureau of Mines and Mineral Resources
Campus Station
Socorro, NM 87801

NEW YORK
New York State Museum and Science Service, Geological Survey
New York State Education Building
Albany, NY 12224

NORTH CAROLINA
Office of Earth Resources
Department of Natural and Economic Resources
P.O. Box 27687
Raleigh, NC 27611

NORTH DAKOTA
North Dakota Geological Survey
University Station
Grand Forks, ND 58201

OHIO
Ohio Division of Geological Survey
Ohio Department of Natural Resources
Fountain Square
Columbus, OH 43224

OKLAHOMA
Oklahoma Geological Survey
The University of Oklahoma
Norman, OK 73069

OREGON
State Department of Geology and
Mineral Industries
1069 State Office Building
1400 S.W. Fifth Avenue
Portland, OR 97201

PENNSYLVANIA
Bureau of Topographic and Geological
Survey
Department of Environmental Resources
Harrisburg, PA 17120

PUERTO RICO
Mineralogy and Geology Section
Economic Development Administration
of Puerto Rico
Industrial Laboratory
P.O. Box 38
Roosevelt, PR 00929

SOUTH CAROLINA
Division of Geology
P.O. Box 927
Columbia, SC 29202

SOUTH DAKOTA
South Dakota State Geological Survey
Science Center
University of South Dakota
Vermillion, SD 57069

TENNESSEE
Department of Conservation
Division of Geology
G-5 State Office Building
Nashville, TN 37219

TEXAS
Bureau of Economic Geology
The University of Texas
University Station, Box X
Austin, TX 78712

UTAH
Utah Geological and Mineralogical
Survey
103 Utah Geological Survey Building
University of Utah
Salt Lake City, UT 84112

VERMONT
Vermont Geological Survey
University of Vermont
Burlington, VT 05401

VIRGINIA
Virginia Division of Mineral Resources
P.O. Box 3667
Charlottesville, VA 22903

WASHINGTON
Washington Division of Mines and
Geology
335 General Administration Building
P.O. Box 168
Olympia, WA 98501

WEST VIRGINIA
West Virginia Geological and Economic
Survey
P.O Box 879
Morgantown, WV 26505

WISCONSIN
Wisconsin Geological and Natural
History Survey
University of Wisconsin
1815 University Avenue
Madison, WI 53706

WYOMING
Geological Survey of Wyoming
Box 3008, University Station
University of Wyoming
Laramie, WY 82071

# References Cited


Adams, W. M., 1973. Tsunami effects and risk at Kahuko Point, Oahu, Hawaii. Geological Society of America, Engineering Geology Case Histories No. 6, pp. 63–70.

Aki, K. and Richards, P. G., 1980. Quantitative seismology: theory and methods, Vol. I and II. Freeman, San Francisco, p. 373.

Alden, W. C., 1928, Landslide and flood at Gros Ventre, Wyoming. Transactions, American Institute of Mining and Metallurgical Engineers, 76:347–360.

Aley, T. J., Williams, J. H., and Massello, J. W., 1972. Ground water contamination and sinkhole collapse induced by leaky impoundments in soluble rock terrain. Missouri Geological Survey, Engineering Geol. Series No. 5.

Algermissen, S. T., and Perkins, D. M., 1976. A problematic estimate of maximum accelerations in rock in the contiguous United States. U.S. Geological Survey, Open-file report, No. 76-416.

Allen, S., 1969. Geologic settings of subsidence. In: Varnes, D. J., and Kiersch, G. (eds.), Reviews in engineering geology, Vol. II. Geological Society of America, pp. 305–342.

Allen, D. R., 1973. Subsidences, rebound and surface strain associated with oil producing operations, Long Beach, California. In: Moran, D. E., Slosson, J. E., Stone, R. D., and Yelverton, C. A. (eds.), Geology seismicity, and environmental impact: Association of Engineering Geologists, Special Publication, University Publishing, Los Angeles, pp. 101–111.

Althoff, W. F., Cleary, R. W., and Roux, P. H., 1981. Aquifer decontamination of volatile organics: a case history. Ground Water, 19:495–504.

American Geological Institute, 1951. Outstanding aerial photographs in North America. American Geological Institute, Committee on Education, Rep. No. 5.

American Geological Institute, 1972. Glossary of geology and related sciences, American Geological Institute, Washington, DC.

American Institute of Professional Geologists, 1980. Oil and gas resources position statement: Colorado Section, American Institute of Professional Geologists, Golden, CO.

American Institute of Steel Construction, 1956. Steel construction, 5th ed. American Institute of Steel Construction, New York.

American Nuclear Society, 1973. Nuclear power and the environment. Hinsdale, IL.

Anderson, D. L., 1971. The San Andreas fault. Sci. Am., 225(5):52–66.

Anderson, M. P., and Berkebile, C. A., 1977. Hydrogeology of the South Fork of Long Island, New York: discussion, Geol. Soc. Am. Bull., 88:895.

Anderson, R. E., Zoback, M. L., and Thompson, G. A., 1983. Implication of selected subsurface data on the structural form and evolution of some basins in the northern Basin and Range province, Nevada and Utah. Geol. Soc. Am. Bull. 94:1055–1072.

Andrews, J. T., 1961. The development of scree slopes in the English lake district and central Quebec–Labrador. Cahiers de Geogr. de Quebec, 5(10):219–230.

Andrews, E. D., 1983. Entrainment of gravel from naturally sorted riverbed material. Geol. Soc. Am. Bull., 94:1225–1231.

Anon., 1889. The Johnstown disaster. Eng. News Rec., 21(23):517–518.

Anon., 1928. Commission finds failure of St. Francis Dam due to defective foundation: Eng. News Rec. 100(14):553–555.

Argall, G. O., Jr., and Aplin, C. L., eds., 1976. Tailing disposal today. First International Tailing Symposium, 1976, Tucson, AR. Miller Freeman, San Francisco.

Argall, G. O., Jr., ed., 1979. Tailing disposal today. Second International Tailing Symposium, Denver, CO. Miller Freeman, San Francisco.

Asphalt Institute, 1964. Soils manual. Manual Series No. 10 (MS-10). College Park, MD.

Attewell, P. B., and Farmer, I. W., 1976. Principles of engineering geology. Chapman and Hall, London.

Back, W., and Baedecker, M. J., 1979. Hydrogeological processes and chemical reactions at a landfill. Ground Water, 17:429–437.

Back, W., and Hanshaw, B. B., 1970. Comparison of chemical hydrogeology of the carbonate peninsulas of Florida and Yucatan. J. of Hydrol. 10(4)L330–368.

Back, W., Hanshaw, B. B., Plummer, L. N., Rahn, P. H., Rightmire, C. T., and Rubin, M., 1983. Process and rate of dedolomitization: mass transfer and $^{14}C$ dating in a regional carbonate aquifer. Geol. Soc. Am. Bull. 94:1415–1429.

Bagnold, R. A., 1941. The physics of blown sands and desert dunes. Methuen, London, reprinted 1954.

Bailey, D. J., 1980. Land movement monitoring system. Assoc. Eng. Geol. Bull. 17:213–221.

Bailey, R. A., Clarke, H. M., Ferris, J. P., Krause, S., and Strong, R. L., 1978. Chemistry of the environment. Academic, New York.

Bain, G. W., 1931. Spontaneous rock expansion. J. Geol. 39:715–735.

Bair, E. S., and O'Donnell, T. P., 1983. Uses of numerical modeling in the design and licensing of dewatering and depressurizing systems. Ground Water, 21:411–420.

Baker, V. R., 1975. Urban geology of Boulder, Colorado: a progress report. Environ. Geol. 1:75–88.

Baker, V. R., 1977. Stream-channel response to floods, with examples from central Texas. Geol. Soc. Am. Bull. 88:1057–1071.

Baker, V. R., Kochel, R. C., and Patton, P. C., 1979. Long-term flood frequency analysis using geological data. Proc. Canberra Symp. Dec. 1978. Int. Assoc. Scientific Hydrol. Publ. 128.

Bakker, W. T., 1968. The dynamics of a coast with a groin systems. Proc. 11th Conf. on Coastal Eng. Am. Soc. of Civil Engineers, pp. 492–517.

Banerji, P. K., 1982. Laterization processes: challenges and opportunities: Episodes, 1982, no.(3):16–20.

Barnes, H. H., Jr. 1967. Roughness chararacteristics of natural channels. US Geological Survey, Water-Supply Paper 1849.

Baron, D. M., 1982. A well system can be designed to mimimize the incrustation tendency: Johnson Drillers J. 54(1):8–10.

Barth, T. F. W., 1961. Abundance of the elements, areal averages, and geochemical cycles. Geochimica Cosmochimica Acta, 23:1–8.

Bartlett, A. A., 1980. Forgotten fundamentals of the energy crisis. J. Geol. Ed. 28:4–35.

Baski, H. A., 1979. Ground water computer models—intellectual toys. Ground Water, 17:177–179.

Bataan, L. J., 1962. Cloud physics and cloud seeding. Anchor Books, Garden City, NY.

Bath, M., 1966. Earthquake seismology: Earth Sci. Rev., 1:69–86.

Baumgardner, R. W., Gustavson, T. C., and Hoadley, A. D., 1980. Salt blamed for new sink in W. Texas. Geotimes, 25(9):16–17.

Bazant, Z., 1979. Methods of foundation engineering. Elsevier, Amsterdam.

Bea, R. G., 1971. How sea-floor slides affect offshore structures. Oil and Gas J. 29:88–92.

Beaty, C. B., 1956. Landslides and slope exposure. J. Geol., 64:70–74.

Beer, F. P., and Johnston, E. R., Jr., 1977. Vector mechanics for engineers. McGraw-Hill, New York.

Behre, C. H., Jr., 1933. Talus behavior above timber in the Rocky Mountains. J. Geol. 41:633–635.

Bell, F. G., 1983. Fundamentals of engineering geology. Butterworths, London.

Belt, C. B., Jr., 1975. The 1973 flood and man's constriction of the Mississippi River. Science, 189:681–684.

Berkey, C. P., 1939. Geology in engineering. In: Frontiers of Geology. Geological Society of America, pp. 31–34.

Berlin, G. L., 1980. Earthquakes and the urban environment, vol. 30 CRC Press, Boca Raton, FL.

Biesecker, J. E., and George, J. R., 1972. Stream quality in Appalachia as related to coal-mine drainage, 1965. In: Pettijohn, W. A. (ed.), Water quality in a stressed environment. Burgess, Minneapolis, pp. 45–60.

Birdsall, L. E., 1973. Sources of geologic data: In: Moran, D. E., Slosson, J. E., Stone, R. O., and Yelverton, C. A., (eds.), Geology, seismicity, and environmental impact. Special Publication, Association of Engineering Geologist, University Publishing, Los Angeles, pp. 57–63.

Birkeland, P. W., 1974. Pedology, weathering, and geomorphological research. Oxford, New York.

Birkeland, P. W., 1982. Subdivision of Holocene glacial deposits, Ben Ohau Range, New Zealand, using relative-dating methods. Geol. Soc. Am. Bull., 93:433–449.

Birkeland, P. W., 1984. Soils and geomorphology. Oxford, New York.

Birkeland, P. S., Colman, S. M., Burke, R. M., Shrobo, R. R., and Meierding, T. C., 1979. Nomenclature of alpine glacial deposits, or, what's in a name? Geology, 7:532–536.

Bishop, A. W., Hutchinson, J. N., Penman, A. D. M., and Evans, H. E., 1969. Geotechnical investigations into the causes and circumstances of the disaster of 21 October 1966. In: A selection of technical reports submitted to the Aberfan Tribunal. Her Majesty's Stationary Office, London (2 vol.).

Biswas, M. R., and Biswas, A. K., ed., 1981, Associated case studies prepared for the United Nations Conference on Desertification. Pergamon, New York.

Bittinger, M. W., and Green, E. B., 1980. You won't miss the water till.... (the Ogallala Story). Water Resources Publications, Littleton, CO.

Bitton, G., Farrah, S. R., Ruskin, R. H., Butner, J., and Chou, Y. J., 1983. Survival of pathogenic and indicator organisms in ground water. Ground Water, 21:405–410.

Blackwelder, E., 1928. Mudflow as a geologic agent in semi-arid mountains. Geol. Soc. Am. Bull. 39:465–484.

Blackwelder, E., 1933. The insolation hypothesis of rock weathering. Am. J. Sci. 26(152):97–113.

Blackwelder, E., 1954. Geomorphic processes in the desert. In: Jahns, R. H. (ed.), Geology of Southern California. California Division of Mines, Bulletin 170, P. 5, pp. 11–20.

Blair, M. L., Spangle, W. E., and William Spangler and Associates, 1979. Seismic safety and land-use planning—selected examples from California. US GS, Prof. Paper 941-B.

Blankennagel, R. K., and Weir, J. E., Jr., 1973. Geohydrology of the eastern part of Pahute Mesa, Nevada Test Site, Nye County, Nevada: US GS, Prof. Paper 712-B.

Blaschke, T. O., 1964. Underground command center. Civil Eng. 34:36–39.

Block, J. W., Clement, R. C., Lew, L. R., and deBoer, J., 1979. Recent thrust faulting in southeastern Connecticut. Geology, 7:79–82.

Blom, R. G., Crippen, R. E., and Elachi, C., 1984. Detection of subsurface features in SEASAT radar images of Means Valley, Mojave Desert, California. Geology, 12:346–349.

Blyth, F. G. H., de Freitas, M. H., 1984. A geology for engineers. Elsevier, New York.

Boffey, P. M., 1976. Teton Dam collapse: was it a predictable disaster?: Science, 193:30–32.

Boegli, A., 1980. Karst hydrology and physical speleology. Springer-Verlag, New York.

Bolt, B. A., Horn, W. L., Macdonald, G. A., and Scott, R. F., 1977. Geological hazards. Springer-Verlag, New York.

Borcherdt, R. D. (ed.), 1975. Studies for seismic zonation of the San Francisco Bay region. USGS, Prof. Paper 941-A.

Borcherdt, R. D., Joyner, W. B., Warmick, R. E., and Gibbs, J. R., 1975. Response of local geologic units to ground shaking. In: Borcherdt, R. D. (ed.), Studies for seismic zonation of the San Francisco Bay region: US GS, Prof. Paper 941-A, pp. 52–74.

Boster, R. S., 1983. Protectionism and water. Ground Water, 21:402–403.

Boudreau, E. H., 1982. Warm Springs Dam (and other disasters) revisited. Ground Water, 20:752–753.

Boudreau, E. H., 1983. Reply to discussion of Reader's Forum—"Warm Springs Dam (and other disasters) revisited": Ground Water 21:362–364.

Bouwer, H., 1977. Land subsidence and cracking due to ground water depletion. Ground Water, 15:358–364.

Bouwer, H., 1978. Groundwater hydrology. McGraw-Hill, New York.

Bowen, N. L., 1922. The reaction principle in petrogenesis. J. Geol. 30:177–198.

Bowen, R., 1984. Geology in engineering. Elsevier, New York.

Bowles, J., 1978. Engineering properties of soils. McGraw-Hill, New York.

Bowles, J. E., 1982. Foundation analysis and design. McGraw-Hill, New York.

Bowley, D. R., 1983. The Massachusetts water-supply contamination correction protection. 3rd Nat. Symp. and Exposition on Aquifer Restoration and Ground water Monitoring. National Water Well Assoc., Worthington, OH, pp. 171–174.

Boyce, R. C., 1982. An overview of site investigations. Assoc. Eng. Geol. Bull. 19:167–171.

Brady, B. H. G., and Brown, E. T., 1985. Rock mechanics. Allen and Unwin, Inc., Winchester, MA.

Braun, B., and Nash, W. R., 1985. Ground freezing for construction. Civil Eng., 55(1):54–56.

Bredehoeft, J. D., Counts, H. B., Robson, S. G., and Robertson, J. B., 1976. Solute transport in ground water systems. In: Rodda, J. C. (ed.), Facets of hydrology. Wiley, New York, pp. 229–256.

Bredehoeft, J. D., England, A. W., Stewart, C. G., Trask, N. J., and Winograd, I. J., 1978. Geologic disposal of high-level radioactive wastes—earth science perspectives. US GS, Circular 779.

Bretschneider, C. L., 1957. Hurricane design wave practices: Proceedings, American Society of Civil Engineers. J. Waterways Harbor Div., W.W. 2:1238-1–33.

Bridge, J. S., 1981. Hydraulic interpretation of grain-size distributions using a physical model for bedload transport. J. Sediment. Petrol. 51:1109–1124.

Brink, A. B. A., 1979. Engineering geology of Southern Africa, vol. I. Building Publications, Pretoria.

Brooker, E. W., 1967. Strain energy and behavior of overconsolidated soils. Can. Geotech. J., 4(3):326–333.

Brooks, A. H., 1920. The use of geology on the Western Front. USGS, Shorter contributions to general geology, 1920, Prof. Paper 128-D, pp. 85–124.

Brown, D. R., 1962. Preface to preliminary editions: In: Building Codes and Related Matters Committee, 1965, Geology and Urban Development. Los Angeles Section, Assoc. Eng. Geol., Special Pub., October.

Brown, Larry F., 1982. Reclamation of topsoil use: Am. Mining Cong. J., 68(6):48–52.

Brown, L., 1981. Building a sustainable society. Worldwatch Institute, Norton, New York.

Brown, R. D., Jr., Ward, P. L., and Plafker, G., 1973. Geologic and seismologic aspects of the Managua, Nicaragua, earthquakes of December 23, 1972. US GS, Prof. Paper 838.

Browning, J. M., 1973. Catastrophic rock slide, Mt. Huascaran, north-central Peru, May 31, 1970. Am. Assoc. Petrol. Geol., 47:1335–1341.

Bruun, P., and Lackey, J. B., 1962. Engineering aspects of sediment transport. In: Fluhr, T., and Legget, R. F., (eds.). Reviews in engineering geology, V. 1. Geological Society of America, pp. 39–103.

Bryan, J. H., 1981. Chalk slope failure at Gainsville Lock, Alabama. Assoc. Engin. Geol. Bull., 18:55–69.

Bryan, K., 1923. Erosion and sedimentation in Papago County, Ariz. US GS, Bull. 730, pp. 19–20.

Bryan, Kirk, 1934. Geomorphic processes at high altitudes. Geogr. Rev. 25:655.

Bue, C. D., 1967. Flood information for flood-plain planning. USGS, Circular 539.

Burwell, E. B., Jr., and Moneymaker, B. C., 1950. Geology in dam construction. Geological Society of America, Engineering Geology (Berkey) Volume, pp. 11–43.

Building Research Advisory Board, 1968. Criteria for selection and design of residential slabs-on-ground. Federal Housing Administration, Nat. Acad. Sci., Rep. 33, Washington DC.

Calder, L. R., 1970. Mortgaging the old homestead. In: Rhodes, F. H. T., and Stone, R. O., (eds.), Language of the earth. Pergamon, New York, pp. 360–363.

Campbell, M. D., and Lehr, J. H., 1973. Water well technology. McGraw-Hill, New York.

Campbell, R. H., 1975. Soil slips, debris flows and rainstorms in the Santa Monica Mountains and vicinity, Southern California. USGS, Prof. Paper 851.

Campbell, T. N., 1941. Petrolyphs as a criteria for slope stability. Science, 93(2417):400.

Canter, L., 1977. Environmental impact assessment. McGraw-Hill, New York.

Cape, C. O., McGeary, S., and Thompson, G. A., 1983. Cenozoic normal faulting and the shallow structure of the Rio Grande rift near Socorro, New Mexico. Geol. Soc. Am. Bull. 94:3–14.

Capp, S. R., 1941. Observations on the rate of creep—Idaho. Am. J. Sci. 239:25–33.

Carder, D. S., 1945. Seismic investigations in the Boulder Dam area, 1940–1944, and the influence of reservoir loading on local earthquake activity. Seismol. Soc. Am. Bull. 35:175–192.

Carder, D. S., 1970. Reservoir loading and local earthquakes. Geol. Soc. Am., Engineering Geology Case Histories 8, pp. 51–61.

Cargo, D. N. and Mallory, B. F., 1974. Man and his geologic environment. Addison Wesley, Reading MA.

Carlson, R. L., 1982. Cenozoic convergence along the California coast: a qualitative test of the hot-spot approximation. Geology, 10:191–196.

Carrigan, M. C., and Shaddrick, D. R., 1977. Homestake's Grizzly Gulch tailings disposal project. Proc. 28th Annu. Highway Geol. Symp., Rapid City, South Dakota Dep. of Transportation, pp. 1–16.

Carrington, T. J., and Watson, D. A., 1981. Preliminary evaluation of an alternate electrode array for use in shallow-subsurface electrical resistivity studies. Ground Water, 19:48–57.

Carroll, D., 1970. Rock weathering. Geoscience. Plenum, New York.

Carson, R., 1962. Silent spring. Fawcett, Greenwich, CT.

Carter, M. W., Hoghissi, A. A., and Kahn, B. (eds.), 1979. Management of low-level radioactive wastes, Vol. II. Pergamon, New York.

Carter, W. D., and Rinker, J. N., 1976. Structural features related to earthquakes in Manqua, Nicaragua, and Cordoba, Mexico. USGS, Prof. Paper 929, pp. 123–128.

Carter, W. D., Rowan, L. C., and Huntington, J. F., 1980. Remote sensing and mineral exploration. Adv. Space Explorat. (COSPAR), vol. 10.

Casagrande, A., 1948. Classification and identification of soils. Am. Soc. Civil. Eng. Trans., 113:901–991.

Castle, R. O., and Youd, T. L., 1972. Discussion, the Houston fault problem. Assoc. Eng. Geol. Bull. 19(1):57–68.

Cattermole, J. M., 1969. Geologic map of the Rapid City West Quadrangle, Pennington County, South Dakota. USGS, Map GQ-828.

Cedergren, H. R., 1977. Seepage, drainage, and flow nets. John Wiley, New York.

Cederstrom, D. J., Trainer, F. W., and Waller, R. M., 1964. Geology and ground-water resources of the Anchorage area, Alaska. USGS Water-Supply Paper 1773.

Chadwick, W. L., et al., 1976. Summary and conclusions from report to U.S. Department of Interior and the State of Idaho on failure of the Teton Dam: by Independent Panel to Review Cause of Teton Dam Failure. US Dep. Interior.

Chandler, R. J., 1973. The inclination of talus, arctic talus terraces, and other slopes composed of granular materials. J. Geol. 81:1–14.

Cherkauer, D. S., 1980. The effect of flyash disposal on a shallow ground water system. Ground Water, 18:544–550.

Chorley, Richard J., Schumm, Stanley A., and Sugden, David E., 1984. Geomorphology. Methuen, London.

Chow, V., 1959. Open-channel hydraulics. McGraw-Hill, New York.

Chow, V., ed., 1964. Handbook of applied hydrology. McGraw-Hill, New York.

Christman, R. A., 1983. Ashfall in Washington–courtesy of Mount St. Helens. J. Geol. Ed., 30:313–314.

Church, H. K., 1981. Excavation handbook. McGraw-Hill, Mining Information Services, New York.

Claerbout, J. F., 1971. Toward a unified theory of reflector mapping. Geophysics, 36:467–481.

Clanton, U. S., and Amsbury, D. L., 1975. Active faults in southeastern Harris County, Texas. Environ. Geol., 1:149–154.

Clements, T., 1966. St. Francis Dam failure of 1928. In: Lung, R., and Proctor, R., Engineering geology in southern California. Assoc. Engin. Geol., Special Publ., Glendale, CA, pp. 88–92.

Cloud, P., 1980. Beyond plate tectonics. Am. Sci. 68:381–387.

Clough, G., 1977. Development of design procedures for stabilized soil support systems for soft ground tunneling: a report on the practice of chemical stabilization around soft ground tunnels in England, France, and Germany, Vol. I. Prepared for US Dep. Transportation, Nat. Tech. Info. Service, Springfield, VA.

Clough, G. W., 1981. Innovations in tunnel construction and support techniques. Assoc. Eng. Geol. Bull. 18:151–167.

Clough, R. W., and Chopra, A. K., 1977. Earthquake analysis of concrete dams. In: Hall, W. L., ed., Structural and geotechnical mechanics. Prentice-Hall, Englewood Cliffs, NJ, pp. 378–402.

Coal Age, 1978. Operating handbook of coal surface mining and reclamation. McGraw-Hill Mining Information Services, New York, Vol. 2.

Coates, D. F., 1967. Rock mechanics principles. Canada Dept. Energy, Mines and Resources, Mines Branch Monogr. 874, Queen's Printer, Ottawa.

Coates, D. R., 1976. Geomorphic mapping. In: Coates, D. R., Geomorphology and engineering. Dowden, Hutchinson and Ross, Stroudsburg, PA, pp. 3–21.

Cohen, B. L., 1976. Impacts of the nuclear energy industry on human health and safety. Am. Scient., 64:550–559.

Cohen, B. L., 1977. The disposal of radioactive wastes from fission reactors. Sci. Am., 236(6):21–31.

Collier, C. R., Pickering, R. J., and Musser, J. J., 1971. Influences of strip mining on the hydrologic environment of parts of Beaver Creek Basin, Kentucky, 1955–56. USGS, Prof. Paper 427-C, pp. 1–79.

Colwell, R. N., 1973. Remote sensing as an aid to the management of earth resources. Am. Sci., 61:174–183.

Colwell, R. N., ed., 1983. Manual of remote sensing. 2nd ed., 2 vol., American Society of Photogrammetry, Falls Church, VA.

Comey, D. D., 1983. The legacy of uranium tailings. In: Tank, R., Environmental Geology. Oxford, New York, pp. 382–386.

Conwell, F. R., and Fuqua, W. D., 1960. Engineering geology of debris flows along the southwest side of the San Joaquin Valley, California, 1957–59. California Department of Water Resources, Div. Design and Construction, Rep. E-3.

Cook, K. L., 1972. Earthquakes along the Wasatch Front, Utah—the record and outlook. In: Environmental Geology of the Wasatch Front, 1971. Utah Geological Association Publication 1, pt. 2, pp. H1–H29.

Cooley, M. E., 1976. Spring flow from pre-Pennsylvanian rocks in the southwestern part of the Navajo Indian Reservation, Arizona. USGS, Prof. Paper 521-F.

Cooley, R. L., Fiero, G. W., Lattman, L. H., and Mindling, A. L., 1973. Influence of surface and near-surface caliche distribution on infiltration characteristics and flooding, Las Vegas area, Nevada. Unviersity of Nevada, Desert Research Institute, Center for Water Resources Research, Reno, NV, Project Rep. 21.

Corte, A. E., 1969. Geocryology and engineering. In: Varnes, D. J., and Kiersch, G., (eds.), Reviews in engineering geology, Vol. II. Geological Society of America, pp. 119–185.

Costa, J. E., 1978. The dilemma of flood control in the United States. Environ. Manag. 2(4):313–322.

Costa, J. E., 1983. Paleohydraulic reconstruction of flash-flood peaks from boulder deposits in the Colorado Front Range. Geol. Soc. Am. Bull. 94:986–1004.

Costa, J. E., and Baker, V. R., 1981. Surficial geology, building with the earth. John Wiley, New York.

Costa, J. E., and Bilodeau, S. W., 1982. Geology of Denver, Colorado, United States of America. Assoc. Eng. Geol. Bull. 19:261–314.

Costa, J. E., and Jarrett, R. D., 1981. Debris flows in small mountain stream channels of Colorado and their hydrological implications. Assoc. Eng. Geol. Bull. 18:309–322.

Coulter, H. W., and Magliaccio, R. R., 1966. Effects of the earthquake of March 27, 1964, at Valdez, Alaska. USGS, Prof. Paper 542-C, pp. 1–36.

Council on Environmental Quality, 1980. The global 2000 report to the president, vol. 1. Washington DC.

Crandell, D. R., 1951. Landslides in Shale at Rapid City, South Dakota: USGS, Open File Rep.

Crandell, D. R., 1958. Geology of the Pierre area, South Dakota. USGS, Prof. Paper 307.

Crandell, D. R., Miller, C. D., Glicken, H. X., Christiansen, R. L., and Newhall, C. G., 1984. Castastrophic debris avalanche from ancestral Mount Shasta volcano, California. Geology, 12:143–146.

Crandell, D. R., and Mullineaux, D. R., 1978. Potential hazards from future eruptions of Mount St. Helens volcano, Washington, USGS, Bull. 1383-C.

Crandell, D. R., and Waldron, H. H., 1969. Volcanic hazards in the Cascade Range: Proc., Conf. Geologic Hazards and Public Problems, May 27–28, 1969, Office of Emergency Preparedness, U.S. Govt. Printing Office, pp. 5–18.

Craun, G. F., 1979. Waterborne disease—a status report emphasizing outbreaks in ground water systems. Ground Water, 17:183–191.

Cross, J. P., 1982. Slope stability program. Civil Eng., 52(10):41–74.

Crouse, C. F., 1979. Probability of earthquake ground accelerations in San Diego. In: Abbot, P. L., and Elliott, W. J., (eds.), Earthquakes and other perils, San Diego region. San Diego Association of Geologists, San Diego, CA, pp. 107–113.

Curry, R. R., 1966. Observations of Alpine mudflows in the Ten-Mile Range, Central Colorado. Geol. Soc. Am. Bull. 77:771–776.

Curtin, G., 1973. Collapsing soil and subsidence. In: Moran, D. E., Slosson, J. E., Stone, R. D., and Yelverton, C. A., (eds.), Geology seismicity, and environmental impact. Assoc. Eng. Geologists, Special Publ. University Publishing Co., Los Angeles, CA, pp. 89–100.

Curtis, W. F., Culbertson, J. K., and Chase, E. B., 1973. Fluvial-sediment discharge to the oceans from the conterminous United States. USGS, Circular 670.

Dalrymple, T., 1960. Flood-frequency analysis. USGS, Water Supply Paper 1543-A.

Dalstead, K. J., Best, R. G., Smith, J. R., Eidenshing, J. C., Schmer, F. A., Andrawis, A. S., and Rahn, P. H., 1977. Application of remote sensing in South Dakota. South Dakota State University, Remote Sensing Institute, SDSU-RS1-77-08.

Daly, R. A., Miller, W. G., and Rice, G. S., 1912. Report on the Committee to Investigate Turtle Mt., Frank, Alberta. Department of Mines, Geological Survey of Canada Mem. 27.

Danilevsky, A., 1983. Holland's fight against the sea. Civil Eng., 53(4):50–53.

D'Appolonia, 1977. Report 3, Environmental effect of present and proposed tailings disposal practices, Split Rock Mill, Jeffrey City, Wyoming. D'Appolonia Consulting Engineers, Project No. RM 77-419, for Western Nuclear, Inc., Vols. I and II, Oct., 1977.

D'Appolonia, E., Alperstein, R., and Appolonia, D. J., 1967. Behavior of a colluvial slope. Proc. Am. Soc. Civil Eng., vol. 93 (SM4), pp. 447–473.

Darton, N. H., 1909. Geology and underground water of South Dakota. USGS, Water Supply Paper 227.

Darton, N. H., 1947. Geologic map of the District of Columbia. USGS, Scale 1:31,680.

Darwin, C., 1962. The voyage of the Beagle. Anchor, Garden City, NY.

Das, B. M., 1983. Advanced soil mechanics. McGraw-Hill, New York.

Davis, A. H., 1979. Hydrogeology of the Belle Fourche, South Dakota water infiltration gallery area. Proc. South Dakota Acad. Sci., vol. 58, pp. 122–143.

Davis, K. S., and Day, J. A., 1961. Water, the mirror of science. Anchor, Garden City, NY.

Davis, S. N., 1982. Hydrogeology of radioactive waste isolation: the challenge of a rational assessment. Geol. Soc. Am., Special Paper 189, p. 389–396.

Davis, S. N., and Turk, L. J., 1964. Optimum depth of wells in crystalline rocks: Ground Water, 2(2):6–11.

Davis, S. N., and DeWeist, R. J. M., 1966. Hydrogeology. John Wiley, New York.

Davis, S. N., and Bentley, H. W., 1982. Dating ground water. In: Currie, L. A. (ed.), Nuclear and chemical dating techniques interpreting the environmental record. Am. Chemical Soc., Symp. Series no. 176, pp. 187–222.

Day, A. D., and Ludeke, K. L., 1979. Disturbed land reclamation in an oil environment. In: Argall, G., Jr. (ed.), Tailing disposal today, Vol. 2. Proc. 2nd Int. Tailing Symp., Denver, Colorado. Miller Freeman, San Francisco, pp. 437–453.

de Marsily, G., Ledoux, E., Barbreau, A., and Margot, J., 1977. Nuclear waste disposal: can the geologist guarantee isolation. Science, 197:517–527.

Dean, R. G., and Jones, D. F., 1974. Equilibrium profile as affected by seawalls. Am. Geophys. Union Trans. 55(4):322.

Denny, C. S., 1961. Landslides east of Funeral Mountains, near Death Valley, Junction, California: USGS Prof. Paper, 424-D, pp. 85–89.

Deutsch, M., 1976. Optical processing of ERTS data for determining extent of the 1973 Mississippi River flood. USGS, Prof. Paper 929, pp. 209–213.

Deutsch, M., and Ruggles, F. H., Jr., 1978. Hydrological applications of Landsat imagery used in the study of the 1973 Indus River flood, Pakistan. Water Resources Bull. 14:261–274.

DeWiest, R. J. M., 1965. Geohydrology. John Wiley, New York.

Dick, J. D., 1983. How to get into hot water without taking a bath. Interstate Oil Compact Commission Bull., 25(1):21–25.

Dils, R. E., 1957. A guide to the Coweeta hydrological laboratory. U.S. Forest Service, Southeast Forest Experimental Station, Ashville, NC.

Dobrin, M. B., 1960. Introduction to geophysical prospecting. McGraw-Hill, New York.

Dobrin, M. B., 1976. Introduction to geophysical prospecting. McGraw-Hill, New York.

Dolan, R., Godfrey, P. J., and Odum, W. E., 1973. Man's impact on the barrier islands of North Carolina. Am. Scient., 61:152–162.

Dolan, R., Hayden, B., and Lins, H., 1980a. Barrier islands. Am. Scient. 68:16–25.

Dolan, R., Lins, H., and Stewart, J., 1980b. Geographical analysis of Fenwick Island, a Middle Atlantic coast barrier island. USGS, Prof. Paper 1177-A.

Donath, F. A., 1968. Role of experimental rock deformation in dynamic structural geology. In: Riecker, R. E. (ed.), NSF advanced science seminar in rock mechanics. Air Force Cambridge Research Lab., Bedford, MA, pp. 355–437.

Donovan, N. C., 1973. A statistical evaluation of strong motion data including the Feb. 9, 1971, San Fernando earthquake. Proc., 5WCEE, Rome, Italy, vol. 1, pp. 1252–1261.

Doscher, T. M., 1981. Enhanced recovery of crude oil. Am. Scient., 69:193–199.

Douglas, I., 1967. Man, vegetation and sediment, yields of rivers. Nature, 215:925–928.

Doumas, C., ed., 1981. Thera and the Aegean World. Aris and Phillips, London.

Down, C. G., and Stocks, J., 1977. Environmental impact of mining. Applied Science, London.

Duguid, J. D., 1974. Ground water transport of radionuclides from buried waste: a case study at Oak Ridge national Laboratory. ORNL Rep. WASH-1332.

Dumble, J. P., and Cullen, K. T., 1983. The application of a microcomputer in the analysis of pumping test data in confined aquifers. Ground Water, 21:79–83.

Dunham, C. W., 1953. The theory and practice of reinforced concrete. McGraw-Hill, New York.

Dunrud, C. R., 1976. Some engineering geologic factors controlling coal mine subsidence in Utah and Colorado. USGS, Prof. Paper 969.

Dunrud, C. R., and Osterwald, F. W., 1978. Effects of coal mine subsidence in the western Powder River Basin, Wyoming. USGS, Open-file Rep. 78-473.

Dupree, H. K., Taucher, G. J., and Voight, B., 1979. Bighorn Reservoir slides, Montana, USA. In: Voight, B., Rockslides and avalanches, vol. 2, engineering sites. Elsevier, Amsterdam, pp. 247–267.

Dutcher, R. M., Jones, E. D., Lovell, H. L., Parizek, R. R., and Stefanko, R., 1966. Mine drainage, part 1: abatement, disposal, treatment. Mineral Industries, vol. 36, no. 3, Pennsylvania State University.

Dyhr-Nielson, M., and Sorenson, T., 1970. Some sand transport phenomena on coasts with bars. Proc. 12th Coastal Eng. Conf., Soc. Civil Eng., pp. 855–865.

Dzemeshko, J. B., and Parabutchev, I. A., 1979. On the sealing of karst in the foundation of the Euphrates Project Dam in Syria. Int. Assoc. Eng. Geol., Symp. Engineering Geology Problems in Hydrotechnical Construction, pp. 185–189.

Easton, W. H., 1973. Earthquakes, rain, and tides at Portuguese Bend Landslide, California. Assoc. Eng. Geol. Bull., 10:173–194.

Eaton, G. P., and Watkins, J. S., 1970. The use of seismic refraction and gravity methods in hydrogeological investigations. In: Morley, L. W. (ed.), Mining and ground water geophysics/1967. Geological Survey of Canada, Economic Geology Rep. no. 26, pp. 598–612.

Eaton, W. W., 1976. Solar energy. Energy Research and Development Administration, Washington, DC.

Eckel, E. B., ed., 1958. Landslides and engineering practices. Highway Research Board Special Paper 29.

Eckel, B., 1970. The Alaskan earthquake, March 27, 1974, lessons and conclusions. USGS, Prof. Paper 546.

Ege, J. R., 1979. Surface subsidence and collapse in relation to extraction of salt and other soluble evaporites. USGS, Open File Rep. 79-1666.

Ege, John R., 1984. Mechanisms of surface subsidence resulting from solution extraction of salt. In: Holzer, Thomas L. (ed.), Man-induced land subsidence. Geol. Soc. Am., Reviews in Engr. Geol., v. 6, p. 203–221.

Ehrlich, P. R., and Ehrlich, A. H., 1970. Population, resources, environment. W. H. Freeman, San Francisco.

Eiby, G. A., 1980. Earthquakes. Van Nostrand Reinhold, New York.

Eisbacher, G. H., 1982. Slope stability and land use in mountain valleys. Geosci. Canada v. 9(1):14–27.

Eisenhart, C., 1977. A test for the significance of lithological variation. In: Cubitt, J. M., and Henley, S., Statistical analysis in geology: Benchmark Papers—Geology/37. Dowden, Hutchinson and Ross, Stroudsburg, PA, pp. 21–29.

Elachi, C., 1982. Radar images of the Earth from space. Sci. Am. 247(6):54–61.

El-Ashry, M. T., ed., 1977. Air photography and coastal problems: Benchmark Papers—Geology/38. Dowden, Hutchinson and Ross, Stroudsburg, PA.

El-Ashry, M. T., 1980. Ground water salinity problems related to irrigation in the Colorado River Basin. Ground Water, 18:37–45.

Embleton, C., and King, C. A. M., 1968. Glacial and periglacial geomorphology. Saint Martin's, New York.

Emerson, J. W., 1971. Channelization: a case study. Science, 173:325–326.

Emery, K. O., and Kuhn, G. G., 1982. Sea cliffs: their processes, profiles, and classification. Geol. Soc. Am. Bull. 93:644–654.

Environmental Protection Agency, 1980. Ground water protection, a water quality management report. U.S. Environmental Protection Agency, Nov. 1980, WH-554.

Epstein, S. S., 1982. Hazardous waste in America. Sierra Club, San Francisco.

Ericksen, G. E., 1983. The Chilean nitrate deposits. Am. Sci., 71:366–374.

Erlin, B., 1969. Portland cement and concrete. In: Varnes, D. J., and Kiersch, G. (ed.), Reviews in engineering geology, Vol. II. Geol. Soc. Am., pp. 81–103.

Erskine, C. F., 1973. Landslides in the vicinity of Fort Randall Reservoir, South Dakota. USGS, Prof. Paper 675.

Evans, D. M., 1970. The Denver area earthquakes and the Rocky Mountain Arsenal disposal well. Geological Society of America, Engineering Geology, Case Histories, no. 8, pp. 25–32.

Evans, L. T., 1980. The natural history of crop yield. Am. Scient., 68:388–397.

Everett, A. G., 1979. Secondary permeability as a possible factor in the origin of debris avalanches associated with heavy rainfall. J. Hydrol. 43:347–354.

Fairbridge, R. W., 1960. The changing level of the sea. Sci. Am., 202(5):70–79.

Falconer, A., Deutsch, M., and Meyers, L., 1981. Lake Ontario dynamics and water quality observations using theratically enhanced Landsat data. Satellite Hydrology, 5th Annu. William T. Pecora Memorial Symp. on Remote Sensing, Am. Water Resources Assoc., pp. 655–661.

Farina, R. J., 1977. The Teton Dam failure in retrospect. The Professional Geologist, American Institute of Professional Geologists, Dec. 1977, pp. 13–15.

Farmer, I. W., 1968. Engineering properties of rocks. Chapman and Hall, London.

Farney, D., 1974. Ominous problems: what to do with radioactive waste. Smithsonian, Apr. 1974, pp. 20–27.

Faust, S. D., and Aly, O. M., 1981. Chemistry of natural waters. Ann Arbor Science, Ann Arbor, MI.

Fayers, F. J., ed., 1981. Third European Symposium on enhanced oil recovery. Elsevier, Amsterdam.

Ferrians, O. J., Jr., Kachadoorian, R., and Greene, G. W., 1969. Permafrost and related engineering problems in Alaska. USGS, Prof. Paper 678.

Ferris, J. G., Knowles, D. B., Brown, R. H., and Stallman, R. W., 1962. Theory of aquifer tests. USGS, Water-Supply Paper 1536-E.

Fetter, C. W., Jr., 1980. Applied Hydrogeology. Charles E. Merrill, Columbus.

Fetter, C. W., Jr., 1981. Determination of the direction of ground water flow. Ground Water Monitor. Rev. 1(3):28–31.

Findley, R., 1981. Mount St. Helens II, in the path of destruction. Nat. Geogr. 162(1):35–49.

Fine, J. C., 1980. Toxic waste dangers. Wter Spectrum, 13(1):23–30.

Fisher, J. S., and Dolan, R., ed., 1977. Beach processes and coastal hydrodynamics. Benchmark Papers—Geology/39. Dowden, Hutchinson, and Ross, Stroudsburg, PA.

Fisk, H. N., 1951. Mississippi River Valley geology relation to river regime. *Trans.* Am. Soc. Civil Eng., 117:667–682.

Fitch, A. A., ed., 1981. Developments in geophysical exploration methods—Two: Applied Science, London.

Flawn, P. T., 1970. Environmental geology: conservation, land-use planning, and resource management. Harper and Row, New York.

Fleming, R. W., Johnson, A. M., and Hough, J. E., 1981. Engineering geology of the Cincinnatti area. Geological Society of America, Field Trip 18, Annual Meeting, Cincinnati, pp. 543–570.

Flint, R. F., 1957. Glacial and Pleistocene geology. John Wiley, New York.

Foose, R. M., 1953. Ground water behavior in the Hershey Valley. Geol. Soc. Am. Bull. 64:623–645.

Foose, R. M., 1967. Sinkhole formation by ground water withdrawal, Far West Rand, South Africa. Science, 157: 1045–1048.

Foose, R. M., 1972. From town dump to sanitary landfill: a case history. Assoc. Eng. Geol. 9(1):1–16.

Foose, R. M., 1981. Sinking can be fast or slow. Geotimes, 26(8):21–24.

Ford, Bacon and Davis, Utah, Inc., 1977. A summary of the phase II–title I engineering assessment of inactive uranium mill tailings—Grand Junction Site, Grand Junction, Colorado. Prepared for US Dep. Energy, Grand Junction, CO.

Forstner, U., and Wittmann, G. T. W., 1979. Metal pollution in the aquatic environment. Springer-Verlag, Berlin.

Foster, K. E., and Wright, N. G., 1980. Jojoba, an alternative to the conflict between agricultural and municipal ground water refinements in the Tucson area, Arizona. Ground Water, 18:31–36.

Foster, J. B., Erickson, J. R., and Healy, R. W., 1984. Hydrogeology of a low-level radioactive-waste disposal site near Sheffield, Illinois. USGS, Water-Resources Investigations Rep. 83-4125.

Fowler, L. C., 1981. Economic consequences of land surface subsidence. Proc. Am. Soc. Civil Eng., J. Irrigation and Draining Div., pp. 151–159.

Fox, P. P., and McHuron, C. E., 1984. Engineering geology of the Boysen Tunnel, Wyoming. Assoc. Eng. Geol. Bull., 21:101–109.

Frankfort, D., 1968. Factors affecting the talus slopes in central Connecticut. M.S. thesis, University of Connecticut.

Freeze, R. A., and Cherry, J. A., 1979. Ground water. Prentice-Hall, Englewood Cliffs, NJ.

Fried, J. J., 1975. Ground water pollution: theory, methodology, modelling, and practical rules. Elsevier, Amsterdam.

Friedman, J. D., Wilberin, R. S., Jr., Palmason, G., and Miller, C. D., 1969. Infrared surveys in Iceland. USGS, Prof. Paper 650-C, pp. 89–105.

Fritsch, M., 1980. Man in the Holocene. Harcourt Brace Jovanovich, New York.

Fuller, J. G., 1975. We almost lost Detroit. Ballantine, New York.

Furneaux, R., 1963. Krakatoa. Prentice-Hall, Englewood Cliffs, NJ.

Fussell, J., 1980. Lineaments lead to oil in Wyoming. Geotimes, 25(2):19–20.

Fyfe, W. S., Price, N. J., and Thompson, A. B., 1978. Fluids in the earth's crust. Elsevier, Amsterdam.

Gabrysch, R. K., and Bonnet, C. W., 1975. Land-surface subsidence in the area of Burwett, Scott, and Crystal Bays near Bayton, Texas. USGS, N.T.I.S. Pub. No. PB 237-660 AS.

Galster, R. W., 1977. A system of engineering geology mapping symbols. Assoc. Eng. Geol. Bull. 14:39–47.

Gambolati, G., Gatto, P., and Freeze, R. A., 1974. Mathematical simulation of the subsidence of Venice, 2, results. Water Resources Research, 10n(3):563–577.

Gardner, C. H., and Tice, J. A., 1977. A study of the Forest City creep slide on the Oahe Reservoir. Proc., 28th Annu. Highway Geol. Symp., Rapid City, South Dakota Dep. Transportation, pp. 97–107.

Gardner, R. L., 1972. Origin of the Mormon Mesa caliche, Clark County, Nevada. Geol. Soc. Am. Bull. 83:143–156.

Gaskill, G., 1965. The night the Mountain fell. Reader's Digest, May, 1965, pp. 59–67.

Gass, T., 1981a. Sinkholes. Water Well J., 35(9):36–37.

Gass, T., 1981b. The safe-yield dilemma. Water Well J. 35(6):37.

Gastil, R. G., Kies, R., and Melins, D. J., 1979. Active and potentially active faults, San Diego County and northernmost Baja California. In: Abbott, P. L., and Elliott, W. J. (eds.), Earthquakes and other perils, San Diego region. San Diego Association of Geologists, San Diego, CA, pp. 47–60.

Gedney, D. S., and Weber, W. G., Jr., 1978. Design and construction of soil slopes. In: Schuster, R. L., and Krizek, R. J. (eds.), Landslides, analysis and control. Transportation Res. Board and Nat. Acad. Sci., Special Rep. 176, Washington, DC, pp. 172–191.

Geotz, A.F.H., Billingsley, F. C., Gillespie, A. R., Abrams, M. J., Squires, R. L., Shoemaker, E. M., Lucchitta, I., and Elston, D. P., 1975. Application of ERTS images and image processing to regional geologic problems and geologic mapping in northern Arizona. Jet Propul. Lab., Pasadena, CA. Tech. Rep. 32-1597.

Geraghty and Miller, 1975. Ground water pollution problems in the northwestern United States. Geraghty and Miller, Inc., proposed for U.S. Environmental Protection Agency, PB-242.

Geraghy and Miller, 1977. The prevalence of subsurface migration of hazardous chemical substances at selected industrial wasteland disposal sites. US Environmental Protection Agency, Solid Waste management Series, Rep. EPA/530/SW-634.

Gerba, C. P., Wallis, C., and Melnick, J. L., 1975. Fate of wastewater bacteria and viruses in soil. Am. Soc. Civil Eng., J. Irrigation and Drainage Division, 101:157–174.

Geyer, A. R., and McGlade, W. G., 1972. Environmental geology for land-use planning. Environ. Geol. Rep. 2, Pennsylvania Dep. Environ. Resources, Harrisburg.

Ghikas, I. Ch., Knuseman, G. P., Leontiadis, I. L., and Wozab, D. H., 1983. Flow pattern of a karst aquifer in the Molai area, Greece. Ground Water, 21:445–455.

Giardino, John R., and Vicks, Steven G., 1985. The engineering geology hazards of rock glaciers. Assoc. Engr. Geol. Bull. 22:201–216.

Gibbs, H. J., and Holland, W. Y., 1960. Petrographic and engineering properties of loess. US Bureau of Reclamation, Eng. Monogr. 28.

Gibbs, R. J., 1967. The geochemistry of the Amazon River system: 1. The factors that control the salinity and the composition and concentration of suspended solids. Geol. Soc. Am. Bull. 78:1203–1232.

Gibbs, R. J., Matthews, M. D., and Link, D. A., 1971. The relationship between sphere size and settling velocity. J. Sediment. Petrol. 41:7–18.

Gibson, U., and Singer, R. D., 1971. Water well manual. Premier Press, Berkeley.

Gilbert, G. K., 1885. The topographic features of lake shores. USGS, 5th Annu. Rep. pp. 75–123.

Gilbert, G. K., 1890. Lake Bonneville. USGS, Monogr. No. 1, pp. 259–318.

Gilbert, G. K., 1904. Domes and dome structures of the High Sierras. Geol. Soc. Am. Bull. 15:29–36.

Gilbert, G. K., 1909. Earthquake forecasts. Science, 29:121–138.

Gilbert, G. K., 1914. The transportation of debris by running water. USGS, Prof. Paper 86.

Gilbert, G. K., 1917. Hydraulic-mining debris in the Sierra Nevada. USGS, Prof. Paper 105.

Gile, L. H., Hawley, J. W., and Grossman, R. B., 1981. Soils and geomorphology in the basin and range area of southern New Mexico—guidebook to the desert project. Memoir 39, New Mexico Bureau of Mines and Mineral Resources.

Gilkeson, R. H., and Cartwright, K., 1983. The application of surface electrical and shallow geothermic methods in monitoring network design. Ground Water Monitor. Rev., 3(3):30–42.

Gillham, R. W., and Cherry, J. A., 1982. Contaminant migration in saturated unconsolidated geologic deposits. Geol. Soc. Am., Special Paper 189, pp. 31–62.

Gillott, J. E., 1968. Clay in engineering geology. Elsevier, Amsterdam.

Gilluly, J., Waters, A. C., and Woodford, A. O., 1968. Principles of geology. W. H. Freeman, San Francisco.

Glaccum, R. A., Benson, R. C., and Noel, M. R., 1982. Improving accuracy and cost-effectiveness of hazardous waste site investigations. Ground Water Monitor. Rev. 2(3):36–40.

Glass, F. R., Jr., 1977. Horizontal drains as an aid to slope stability on I-26, Polk County, North Carolina. Proc. 28th Annu. Highway Geol. Symp., Rapid City, South Dakota Dep. Transportation, pp. 119–136.

Gleser, G. C., Green, B., and Winget, C., 1981. Prolonged psychological effects of disaster: a study of Buffalo Creek. Academic, New York.

Godfrey, K. A., Jr., 1984. Retaining walls: competition or anarchy. Civil Eng. 54(12):48–52.

Goebel, E. D., 1971. Subsurface geology at the proposed site of the National Radioactive Waste Repository at Lyons, Kansas. Geol. Soc. Am. Bull., 3(3)L237.

Golchert, N. W., and Sedlet, J., 1977. Radiological survey of Site A and Plot M: Argonne National Laboratory, Argonne, IL. Prepared for US Dep. Energy, DOE/EV-0005/7.

Goldich, S. S., 1938. A study in rock-weathering. J. Geol. 46:17–58.

Gonzales, S., 1982. Host rocks for radioactive-waste disposal. Am. Scient. 70:191–200.

Goodman, R. E., 1976. methods of geological engineering. West, St. Paul, MN.

Goodman, R. E., 1980. Introduction to rock mechanics. John Wiley, New York.

Goudie, A., 1973. Duricrusts in tropical and subtropical landscapes. Oxford, London.

Graf, W. H., 1971. The hydraulics of sediment transport. McGraw-Hill, New York.

Graham, M. J., 1981. The radionuclide ground water monitoring program for the separations area, Hanford Site, Washington State. Ground Water Monitor. Rev., 1(2):52–56.

Grants, A., Plafker, G., and Kachadoorian, R., 1964. Alaska's Good Friday earthquake, March 27, 1964: a preliminary evaluation. USGS, Circular 491.

Gray, R. E., and Bruhn, R. W., 1984. Coal mine subsidence—eastern United States. In: Holzer, T. L., ed., Man-induced land subsidence. Geol. Soc. Am., Rev. Eng. Geol. 6:123–149.

Gray, R. E., and Meyers, J. F., 1970. Mine subsidence and support methods in Pittsburgh area. Am. Soc. Civil Eng., 96(SM4):1267–1287.

Gray, W. G., and Hoffman, J. L., 1983. A numerical model study of ground water contamination. Ground Water, 21:7–14.

Green, J. P., 1973. An approach to analyzing multiple causes of subsidence. In: Moran, D. E., Slosson, J. E., Stone, R. O., and Yelverton, C. A. (eds.), Geology, seismicity, and environmental impact. Assoc. of Eng. Geol. Special Publication. University Publishing, Los Angeles, pp. 79–87.

Green, N. B., 1978. Earthquake resistant building design and construction. Van Nostrand Reinhold, New York.

Greenfield, R. J., 1979. Review of geophysical approaches to the detection of karst. Assoc. Eng. Geol. Bull. 16:393–409.

Gries, J. P., Rahn, P. H., and Baker, R. K., 1976. A pump test in the Powder River Basin. Old West Regional Commission, Billings, MT.

Griffiths, D. H., and King, R. F., 1981. Applied geophysics for geologists and engineers. Pergamon, Elmsford, NY.

Griggs, D. T., 1936. The factor of fatigue in rock exfoliation. J. Geol. 44:783–796.

Griggs, G. B., and Gilchrist, J. A., 1983. Geologic hazards, resources, and environmental planning. Wadsworth, Belmont, CA.

Grimes, W. W., 1977. Geotechnical investigation for the Des Plaines River system tunnels and shafts. Proc. 28th Annu. Highway Geol. Symp., Rapid City, South Dakota Dep. Transportation, pp. 83–95.

Gupta, H., Narain, H., and Mohan, I., 1969. A study of the Koyna earthquake of December 10, 1967. Seismol. Soc. Am. Bull. 59:1149–1162.

Gupta, H. K., and Rastogi, B. K., 1976. Dams and earthquakes. Elsevier, New York.

Gustavson, T. C., 1975. Microrelief (gilgai) structures on expansive clays of the Texas Coastal Plain—their recognition and significance in engineering construction. Texas Bureau of Economic Geology, Circ. 75-7, Austin, TX.

Gutenberg, B., and Richter, C. F., 1956. Earthquake magnitude, intensity, energy and acceleration. Seismol. Soc. Am. Bull. 46:105–145.

Hack, J. T., 1957. Studies of longitudinal stream profiles—Virgnia and Maryland. USGS, Prof. Paper 294-B.

Hadley, J. B., 1964. Landslides and related phenomena accompanying the Hegben Lake earthquake of August 17, 1959. USGS, Prof. Paper 435, pp. 107–138.

Hálek, V., and Svec, J., 1978. Ground water hydraulics. Elsevier, Amsterdam.

Hall, C. A. S., and Cutler, J. C., 1981. Petroleum drilling and production in the United States: yield per effort and net energy analysis. Science, 211:576–579.

Hall, D. H., and Hajnal, Z., 1962. The gravimeter in studies of buried valleys. Geophysics, 27(pt. II):939–951.

Hamel, J. V., 1971. The slide at Brilliant Cut. Proc. 13th Symp. Rock Mechanics, Urbana, IL, pp. 487–510.

Hamilton, R. B., 1978. The status of earthquake prediction. In: Utgard, R. O., McKenzie, G. D., and Foley, D., (eds.), Geology and the urban environment. Burgess, Minneapolis, pp. 102–105.

Hammer, M. J., 1981. An assessment of current standards for selenium in drinking water. Ground Water, 19:366–369.

Hammer, M. J., and Thompson, O. B., 1966. Foundation clay shrinkage caused by large trees. Proc. Am. Soc. Civil Eng., 92:1–17.

Hammer, M. J., and Elser, G., 1980. Control of ground water salinity, Orange County, California. Ground Water, 18:536–540.

Hammond, P. D., 1980. Mount St. Helens. Assoc. Eng. Geol. Newsletter, 23(4):12–21.

Hammond, R. P., 1979. Nuclear wastes and public acceptance. Scient., 67:146–160.

Hanegan, G. L., 1973. Castaic dam: a case history of geologic engineering problems and their solution. In: Varnes, D. J., and Kiersch, G. (eds.), Reviews in engineering geology, Vol. II. Geological Society of America, pp. 201–211.

Hansen, W. R., Eckel, E. B., Schaem, W. E., Lyle, R. E., George, W., and Chance, G., 1966. The Alaskan earthquake, March 27, 1964, field investigation and reconstruction effort. USGS, Prof. Paper 541.

Hansen, W. R., 1965. Effects of the earthquake of March 27, 1964, at Anchorage, Alaska. USGS, Prof. Paper, 542-A, pp. 1–68.

Hansen, W. R., and Eckel, E. B., 1977. The Alaska Earthquake March 27, 1964, field investigation and reconstruction effort. In: Turk, R. (ed.), Focus on environmental geology. Oxford, London, pp. 69–88.

Hansmire, W. H., 1981. Tunneling and excavation in soft rock and soil. Assoc. Eng. Geol. Bull. 18:77–89.

Hantush, M. S., 1964. Hydraulics of wells. In: Chow, V. (ed.), Advances in hydrosciences, Vol. 1, Academic, New York, pp. 281–432.

Harrison, W., 1958. Marginal zones of vanished galciers reconstructed from the preconsolidation-pressure values of overriden silts. J. Geol. 66:72–95.

Harshbarger, J. W., Lewis, D. D., Skibitzke, H. E., Heckler, W. L., and Kister, L. R., 1966. Arizona water. USGS, Water-Supply Paper 1648.

Hartley, R. P., 1964. Effects of large structures on the Ohio shore of Lake Erie. Ohio Division of Geological Survey, Report of Invest. No. 53.

Hartshorn, J. H., 1958. Flowtill in southeastern Massachusetts. Geol. Soc. Am. Bull. 69:477–482.

Hasan, S. E., and West, T. R., 1982. Development of an environmental geology data base for land use planning. Assoc. Eng. Geologists Bull. 19:117–132.

Hatheway, A. W., 1980. Geology of the total urban development—report of a visit to the Republic of Singapore—The Engineering Geologist. Geol. Soc. Am., 15(4):4–10.

Hatheway, Allen W., 1981, Engineering geology: Geotimes, v. 26, n. 2, p. 23–24.

Hatheway, Allen W., and Cole R. McClure, Jr. (eds.), 1979, Geology in the siting of nuclear power plants: Geological Society of America, Reviews in engineering geology, v. 4, 264.

Hatheway, A. W., 1982. Significance of glacial till terminology in engineering construction. In: Farquhar, D. C. (ed.), Geotechnology in Massachusetts. U. Mass., Amherst, MA, pp. 193–195.

Haupt, A., and Kane, T. T., 1982. Population handbook. Population Reference Bureau, Washington, DC.

Hays, W. W., 1979. Program and plans of the U.S. Geological Survey for producing information needed in national seismic hazards and risk assessment, fiscal years 1980–84. USGS, Circular 816.

Hays, W. W., ed., 1981. Facing geologic and hydrologic hazards—earthquake consideration. USGS, Prof. Paper 1240-B.

Hayes, W. C., and Vineyard, J. D., 1969. Environmental geology in town and country. Missouri Geol. Survey, Education Series, No. 2.

Healy, J. H., 1968. Denver earthquakes, disposal of waste fluids by injection into deep well has triggered earthquakes. Science, 161:1301–1310.

Heath, R. C., and Trainer, F. W., 1981. Introduction to ground water hydrology. National Water Well Assoc., Worthington, OH.

Heiland, C. A., 1963. Geophysical exploration. Hafner, New York.

Heim, A., 1932. Bergsturz und Menschenleben. Fretz und Wasmuth, Zurich.

Heindl, L. A., and Feth, J. H., 1955. Symposium on land erosion—Piping—a discussion. Am. Geophys. Union Trans. 36(2):342–345.

Heinemann, H. G., 1981. A new sediment trap efficiency curve for small reservoirs. Water Resource Bull. 17:825–830.

Heilman, J. L., and Moore, D. G., 1981. Soil mositure applications of the heat capacity mission. Satellite Hydrology, 5th Annu. William T. Pecora Memorial Symp. on Remote Sensing. American Water Resources Assoc., pp. 371–376.

Henderson, F. M., 1966. Open channel flow. Macmillan, New York.

Hicks, B. A., and Lienkaemper, G. W., 1983. Movement of two active earthflows in the Middle Santam Basin, western Cascades, Oregon (Abs.). Assoc. Eng. Geol. Annu. Meeting, San Diego, p. 74.

Hileman, J. A., 1979. Seismicity of the San Diego Region. In: Abbott, P. L., and Elliott, W. J. (eds.), Earthquakes and other perils, San Diego region. San Diego Association of Geologists, San Diego, CA., pp. 11–20.

Hill, D. E., and Thomas, H. F., 1972. Use of natural resource data in land and water planning. Conn. Agriculture Experiment Station Bull. 733, New Haven.

Hill, M L., 1981. San Andreas fault—history of concepts. Geol. Soc. Am. Bull., 92:112–131.

Hjulstrom, F., 1939. The transportation of detritus by running water. In: Trask, P. D. (ed.), Recent marine sediments, a symposium. Am. Assoc. Petroleum Geologists, Tulsa, OK.

Hobson, G. D., 1970. Seismic methods in mining and ground water exploration. In: Morley, L. W. (ed.), Mining and ground water geophysics/1967. Geological Survey of Canada, Economic Geology Report no. 26, pp. 148–176.

Hoek, E., and Bray, J., 1974. Rock slope engineering. Institute of Mining and Metallurgy, London.

Hoek, E., and Bray, J., 1977. Rock slope engineering. Institute of Mining and Metallurgy, London.

Hoexter, D. F., Holzhausen, G., and Soto, A. E., 1978. A method of evaluating the relative stability of ground for hillslope development. Eng. Geol., 12:319–336.

Holden, Constance, 1971, Nuclear waste, Kansans riled by AEC plans for atom dump: Science, v. 172, p. 249–250.

Holtz, R. D., and Kovacs, W. D., 1981. An introduction to geotechnical engineering. Prentice-Hall, Englewood Cliffs, NJ.

Holzer, T. L., 1978. Documentation of potential for surface faulting related to ground water withdrawal in Las Vegas Valley, Nevada. USGS, Open-File Report 78–79.

Holzer, T. L., 1980. Reconnaissance maps of earth fissures and land subsidence, Bowie and Wilcox Counties, Arizona. USGS, Miscellaneous Field Studies Map MF-1156.

Holzer, Thomas L., 1984. Ground failure induced by ground-water withdrawal from unconsolidated sediments. In: Holzer, Thomas L. (ed.), Man-induced land subsidence. Geol. Soc. Am., Reviews in Engr. Geol., 6:67–105.

Holzer, T. L., and Bluntzer, R. L., 1984. land subsidence near oil and gas fields, Houston, Texas. Ground Water, 22:450–459.

Holzer, T. L., Davis, S. N., and Lofgren, B. E., 1979. Faulting caused by groundwater extraction in Southcentral Arizona. J. Geophys. Res., 84(B2):603–612.

Holzer, T. L., 1984. Ground failure induced by ground-water withdrawal from unconsolidated sediment. In Holzer, T. L., ed., Man-induced land subsidence. Geol. Soc. Am., Rev. Eng. Geol., 6:67–105.

Hollingsworth, R., and Kovacs, G. S., 1981. Soil slumps and debris flows: prediction and protection. Assoc. Eng. Geol. Bull., 18:17–28.

Hood, P., 1981. Migration. In: Fitch, A. A. (ed.), Developments in geophysical exploration methods—2. Applied Science, Ltd., London, pp. 151–230.

Hooghoudt, S. B., 1940. Bijdragen tot de kennis van eenige natururkundige grootheden van de grond: Verslagen Landbouwk, Onderz No. 46 (14B). Government Printing Office, The Hague, Netherlands, pp. 515–707.

Hooke, J. M., 1979. An analysis of the process of river bank erosion. J. Hydrol. 42:39–62.

Horikawa, K., and Sunamura, T., 1967. A study of erosion of coastal cliffs by using aerial photographs. Coastal Eng. Japan, 10:67–83.

Hough, B. K., 1957. Basic soils engineering. Ronald Press, New York.

Housner, G. W., and Jennings, P. C., 1977. The capacity of extreme earthquake motions to damage structures. In: Hall, W. J. (ed.), Structural and geotechnical mechanics. Prentice-Hall, Englewood Cliffs, NJ, pp. 102–116.

Howard, A. D., and Remson, I., 1978. Geology in environmental planning. McGraw-Hill, New York.

Howard, K., 1973. Avalanche mode of motion: implications from lunar examples. Science, 180:1052–1055.

Howard, K. A., et al., 1978. Preliminary map of young faults in the United States as a guide to possible fault activity. USGS, Map MF-916.

Howe, R. C., and Patrick, R. R., 1979. Landsat, land use mapping and the classroom. J. Geol. Ed., 27:55–58.

Howell, B. F., 1959. Introduction to geophysics. McGraw-Hill, New York.

Howell, B. F., Jr., 1972. Earth and universe. Charles E. Merril, Columbus, OH.

Howell, B. F., Jr., 1973. Average regional seismic hazard index (ARSHI) in the United States. In: Moran, D. E., Slosson, J. E., Stone, R. O., and Yelverton, C. A. (eds.), Geology, seismicity and environmental impact. Association of Engineering Geologists, Special Publication. University Publishing, Los Angeles, pp. 285–297.

Hoyt, W. G., and Langbeim, W. B., 1955. Floods. Princeton University Press, Princeton, NJ.

Hsu, K. J., 1975. Catastrophic debris streams (sturzstroms) generated by rock falls. Geol. Soc. Am. Bull., 86:129–140.

Huang, Y. H., 1983. Stability analysis of earth slopes. Van Nostrand Reinhold, New York.

Hubbert, M. K., 1940. The theory of ground water motion. J. Geol., 48:785–944.

Hubbert, M. K., 1971. Energy resources of the earth. Sci. Am., 224(3):60–70.

Hubbert, M. K., 1973. U.S. energy resources, a review as of 1972. Prepared at the request of Sen. Henry M. Jackson, Chairman, Committee on Interior and Insular Affairs, Serial 93-40 (92-75), Part I, U.S. Gov. Printing Office, Washington, DC.

Hubbert, M. K., and Rubey, W. W., 1959. Role of fluid pressure in mechanics of overthrust faulting. Geol. Soc. Am. Bull. 70:115–166.

Hughes, J. L., 1975. Evaluation of ground water degradation resulting from waste disposal to alluvium near Barstow, California. USGS, Prof. Paper 878.

Hunt, C. B., 1972. Geology of soils. W. H. Freeman, San Francisco.

Hunt, C. B., 1983. How safe are nuclear wastes. Geotimes, 28(7):21–22.

Hunt, Roy E., 1983. Geotechnical engineering investigation manual. McGraw-Hill, New York.

Hutchinson, J. N., 1983. Methods of locating slip surfaces in landslides. Assoc. Eng. Geol. Bull., 20:235–252.

Hyers, A. D., and Marcus, M. G., 1981. Land use and desert dust hazards in central Arizona. In: Péwé, T. L. (ed.), Desert dust: origin, characteristics, and effect on man. Geol. Soc. Am., Special Paper 186, pp. 267–280.

Inman, D. L., and Brush, B. M., 1973. The coastal challenge. Science, 181:20–32.

International Association on Land Subsidence, 1977. Land Subsidence proceedings of the 2nd Int. Symp. on Land Subsidence held at Anaheim, California, Dec. 13–17, 1976. Int. Assoc. of Hydro. Sci., Ann Arbor, MI.

Interstate Oil and Compact Commission, 1982. Improved oil recovery. Interstate Oil and Compact Commission, Oklahoma City, OK.

Ippen, A. T. (ed.), 1966. Estuary and coastline hydrodynamics. McGraw-Hill, New York.

Isaacs, R. M., and Codel, J. A., 1973. Permafrost: bank stability and pipeline construction, Mackenzie Valley, Northwest Territories, Canada. In: Moran, D. E., Slosson, J. E., Stone, R. O., and Yelverton, C. A. (eds.), Geology, seismicity, and environmental impact. Assoc. Eng. Geol., Special Publication, University Publishing, Los Angeles, pp. 229–241.

Ives, R. L., 1940. Rock glaciers in the Colorado Front Range. Geol. Soc. Am. Bull., 51:1271–1294.

Iyer, L. S., Ramakrishnan, V., Russell, J. E., and Rahn, P. H., 1975. Durability tests on some aggregates for concrete. Am. Soc. Civil Eng., Construction Div., vol. 101, no. CO3, Proc. Paper 11574, pp. 593–605.

Jachens, R. C., and Holzer, T. L., 1983. Differential compaction mechanism for earth fissures near Casa Grande, Arizona. Geol. Soc. Am. Bull. 93:998–1012.

Jacob, C. E., 1946. A generalized graphical method for evaluating formation constants and summarizing well-field history. Am. Geophys. Union Trans., 27:526–534.

Jaeger, C., 1972. Rock mechanics and engineering. Cambridge University Press, Cambridge.

Jahns, R. H., 1943. Sheet structure in granite: its origin and use as a measure of glacial erosion. J. Geol. 51(2):71–98.

Jahns, R. H., 1949. Desert Floods. Engineering and Science Monthly, vol. 18, California Institute of Technology, Pasadena. CA, pp. 10–15.

Jahns, R. H., 1958, Residential ills in the Heartbreak Hills of southern California. Engineering and Science Monthly, vol. 22, California Institute of Technology, Pasadena, CA, pp. 13–20.

Jahns, R. H., and Vander Linden, K., 1973. Space-time relationships on the southerly side of the Palos Hills, California. In: Moran, D. E., Slosson, J. E., Stone, R. O., and Yelverton, C. A. (eds.), Geology, seismicity, and environmental impact. Assoc. Eng. Geol., Special Publications. University Publishing, Los Angeles, pp. 123–138.

Jansen, P. Ph., van Bendegom, L., van den Berg, J., de Vries, M., and Zahen, A., eds., 1979. Principles of river engineering: the nontidal alluvial river. Pittman, London.

Jansen, R. B., 1980. Dams and public safety. U.S. Dep. Interior, Water and Power Resources Service.

Jansen, R. B., Dukleth, G. W., Gordon, B. B., James, L. B., and Shields, C. E., 1969. Earth movement at Baldwin Hills Reservoir. Conf. on Stability and Performance of Slopes and Embankments, Berkeley, California. Am. Soc. Civil Eng., pp. 607–631.

Jeffreys, E. J., 1932. Scree slopes. Geology Magazine, 69(8):383–384.

Jelenianski, C. P., 1978. Storm surges. In: Geophysical predictions. Nat. Acad. Sci., Washington, DC, pp. 185–192.

Jennings, J. N., 1983. Karst landforms. Am. Scient. 71:578–586.

Johnson, A. M., 1970. Physical processes in geology. Freeman, Cooper, San Francisco.

Johnson, A. M., and Rahn, P. H., 1970. Mobilization of debris flows. Zeitschrift fur Geomorphologie Suppl. 9:168–186.

Johnson, E. E., Inc., 1966. Ground water and wells. St. Paul, MN.

Johnson, J. W., 1957. The littoral drift problem at shoreline harbors. Am. Soc. Civil Eng., J. Waterways and Harbors Div., 83(WW1): Paper 1211.

Johnson, T. D., 1979. Failures, what have we learned? In: Argall, G. O., Jr. (ed.), Tailing disposal today, Vol. 2. Proc. 2nd Int. Tailing Symp., Denver, CO. Miller Freeman, San Francisco, pp. 327–337.

Johnson, T. M., Cartwright, K., and Schuller, R. M., 1981. Monitoring of leachate migration in the unsaturated zone in the vicinity of sanitary landfills. Ground Water Monitor. Rev., 1(3):55–63.

Johnston, A. C., 1982. The major earthquake zone on the Mississippi. Sci. Am. 246(4):60–68.

Joiner, T. J., Warnon, J. C., and Scarbrough, W. L., 1968. Saprolite in Alabama. Ground Water, 6(1):19–25.

Jones, B. L., Chinn, S. S. W., and Brice, J. C., 1984. Olokele rock avalanche, island of Kauai, Hawaii. Geology, 12:209–211.

Jones, D. E., Jr., and Holtz, W. G., 1973. Expansive soils—the hidden disaster. Civil Eng. 42(8):49–51.

Jorgensen, D. G., 1981. Geohydrologic models of the Houston District, Texas. Ground Water, 19:418–428.

Judd, W. R., ed., 1974. Seismic effects of reservoir impounding. Eng. Geol., 8:212.

Judson, S., 1968. Erosion of the land, or what's happening to our continents. Am. Scient., 56:356–374.

Judson, S., and Ritter, D. F., 1964. Rates of regional denudation in the United States. J. Geophys. Res., 69:3395–3401.

Kapp, M. S., 1969. Slurry-trench construction for basement wall of World Trade Center. Civil Eng., 39:36–40.

Kaufman, R. F., Eadie, G. G., and Russell, C. R., 1976. Effects of uranium mines and milling on ground water in the Grants Mineral Belt, New Mexico. Ground Water, 14:296–308.

Kautzman, R. R., and Cavaroc, V. V., 1973. Temporal relation between urbanization and reservoir sedimentation: a case study in the North Carolina piedmont. Assoc. Eng. Geol. Bull. 10:195–218.

Kean, W. F., and Rogers, R. B., 1981. Monitoring leachate in ground water by corrected resistivity methods. Assoc. Eng. Geol. Bull., 18:101–107.

Keck, W. G., Henry, G., Jr., and Minning, R. C., 1981. The Keck method of computing apparent resistivity. Ground Water Monitor. Rev. 1(3):64–68.

Keefer, D. K., 1984. Landslides caused by earthquakes. Geol. Soc. Am. Bull., 95:406–421.

Keefer, W. R., 1971. The geologic story of Yellowstone National Park. USGS, Bull. 1347.

Keller, E. A., 1976. Channelization: environmental, geomorphic, and engineering aspects. In: Coates, D. C. (ed.), Geomorphology and engineering. Dowden, Hutchinson and Ross, Stroudsburg, PA, pp. 115–140.

Keller, E. A., 1976. Environmental Geology. Charles E. Merrill, Columbus, OH.

Keller, E. A., Bonkowski, M. S, Korsch, R. J., and Shleman, R. J., 1982. Tectonic geomorphology of the San Andreas fault zone in the southern Indio Hills, Coachella Valley, California. Geol. Soc. Am. Bull., 93:46–56.

Keller, G. V., 1970. Application of resistivity methods in mineral and ground water exploration programs. In: Morley, L. W. (ed.), Mining and ground water Geophysics/1967. Geol. Survey of Canada, Economic Geol. Rep. 26, pp. 51–66.

Kelley, D. R., 1969. A summary of major geophysical logging methods. Penn. Geol. Survey, Bull. M61.

Kennedy, M. P., 1979. Sea-cliff erosion at Sunset Cliffs, San Diego. In: Abbot, P. L., and Elliott, W. J. (ed.), Earthquakes and other perils, San Diego region. San Diego Assoc. Geol., Field trip prepared for Geol. Soc. Am. Annu. Meeting, pp. 197–206.

Kennedy, N. T., 1969. Southern California's Trial by Mud and Water. Nat. Geogr., 150(10):552–573.

Kern, P. F., 1963. Quick clay landslides in Scandinavia and Canada. Sci. Am 209:132–142.

Kerr, A., Smith, B. J., Whalley, B., and McGreevy, J. P., 1984. Rock temperatures from southeast Morocco and their significance for experimental rock-weathering studies. Geology, 12:306–309.

Kesel, R. H., 1976. The use of the refraction-seismic techniques in geomorphology. Catena, 3:91–98.

Keys, W. S., 1970. Borehole geophysics as applied to ground water. In: Morley, L. W. (ed.), Mining and ground water geophysics/1967. Geological Survey of Canada. Econ. Geol. Rep. no. 26, pp. 598–612.

Kiersch, G. A., 1964. Vaiont Reservoir disaster. Civil Eng., 34(3):32–39.

Kilburg, J. A., Lorence, W. M., and Alvi, J. M., 1980. Geology and engineering for the realignment of the Pennsylvania Turnpike form mile post 123.5 to mile post 130 across the Allegheny Front (abs.). Assoc. Eng. Geol. Annu. Meeting, p. 47.

King, Cuchlaine A. M., ed., 1976. Periglacial processes: Benchmark Papers in Geology/27. Dowden, Hutchinson and Ross, Stroudsburg, PA.

Kirk, P., 1981. Vibroseis processing. In: Fitch, A. A. (ed.), Developments in geo physical exploration methods—2. Applied Science, London, pp. 37–52.

Kirkham, D. S., Toksoz, S., and van der Ploe, R. R., 1974. Steady flow to drains and wells. In: van Schilfgaarde, J. (ed.), Drainage for agriculture. Am. Soc. Agronomists, Agronomy Monogr. no. 17, pp. 203–244.

Klinesberg, C., and Duguid, J., 1982. Isolating radioactive wastes. Am. Scient., 70:182–190.

Klins, M. A., 1979. The squeeze is on in old oil fields. Earth and Mineral Sci., 48(9):65–70.

Klohn, E. J., and Dingeman, D., 1979. Tailings disposal system for Reserve Mining Company. In: Argall, G. O., Jr. (ed.), Tailing disposal today, Vol. 2. Proc. 2nd Int. Tailing Symp., Denver, CO. Miller Freeman, San Francisco, pp. 178–205.

Klosterman, M. T., Wolff, T. F., and Jahren, N. G., 1982. Quantitative relationship of grouting to the reduction of ground water flow through rock formations. Assoc. Eng. Geol. Bull. 19(1):15–24.

Knesel, T. R., 1976. Washington Metro's topless tunnels. In: Tobbins, R. J., and Conlon, R. J., (eds.), Proc. rapid excavation and tunneling conference. Port City Press, Baltimore, MD, pp. 296–310.

Knight, F. J., 1971. Geologic problems of urban growth in limestone terrains of Pennsylvania. Assoc. Eng. Geol. Bull., 8:94–95.

Koefoed, O., 1979. Geosounding principles, Vol. 1—resistivity sounding measurements. Elsevier, Amsterdam.

Kohout, F.A., Leve, G. W., Smith, F. T., and Manheim, F. T., 1977. Red Snapper Sink and ground water flow, offshore northeastern Florida. In: Tolson, J. S, and Doyle, F. L., (ed.), Karst hydrogeology. Proc. 12th Int. Cong., Int. Assoc. Hydrogeol., Huntsville, AL., pp. 193–194.

Kohout, F. A., Wiesnet, D. R., Deutsch, M., Shanton, J. A., and Kolipinski, M. C., 1981. Applications of aerospace data for detection of submarine springs in Jamaica. Satellite Hydrology, 5th Annu. William T. Pecora Memorial Symp. on Remote Sensing, American Water Resources Assoc., pp. 437–445.

Kolb, C. R., and Steinriede, W. B., Jr., 1967. Geological notes on Vicksburg and vicinity. Geol. Soc. Am., Annu. Meeting, pp. B-1–B-14.

Komar, P. D., 1981. The application of the Gibbs equation for grain settling velocities to conditions other than quartz grains in water. J. Sediment. Petrol., 51:1125–1132.

Konikow, L. F., 1977. Modeling chloride movement in the alluvial aquifer at the Rocky Mountain Arsenal, Colorado. USGS, Water-Supply Paper 2,044.

Koopersmith, C. A., 1980. Computer analysis of bank erosion on Lake Sharpe, South Dakota, M.S. Thesis, South Dakota School of Mines and Technology.

Kopp, O. C., 1982. Comment on "Problems in determination of the water content of rock-salt samples and its significance in nuclear waste storage siting." Geology, 10:330.

Kovach, R. L., Nur, A., Wesson, R. L., and Robinson, R., 1975. Water-level fluctuations and earthquakes on the San Andreas fault zone. Geology, 3:437–440.

Kozlovsky, Ye. A., 1984. The world's deepest well. Scient. Am., 251(6):98–104.

Krabill, W. B., and Brooks, R. L., 1981. Land subsidence measured by satellite radar imagery. Satellite Hydrology, 5th Annu. William T. Pecora Memorial Symp. on Remote Sensing, American Water Resources Assoc., pp. 683–688.

Kraft, J. C., 1980. Grove Karl Gilbert and the origin of barrier shorelines. Geol. Soc. Am., Special Paper 183, pp. 105–113.

Krank. K. K., and Watters, R. J., 1983. Geotechnical properties of weathered Sierra Nevada granodiorite. Assoc. Eng. Geol. Bull., 20:253–265.

Kreitler, C. W., 1977. Fault control of subsidence, Houston, Texas. Ground Water, 15:203–214.

Kresse, F. C., 1966. Baldwin Hills Reservoir. In: Lung, R., and Proctor, R., (eds.), Engineering geology in Southern California. Assoc. Eng. Geol., Glendale, CA, pp. 93–103.

Krinitzsky, E. L., and Marcuson, W. F., III, 1983. Principle for selecting earthquake motions in engineering design. Assoc. Eng. Geol. Bull., 20:253–265.

Krinitzsky, E. L., and Turnbull, W. J., 1967. Loess deposits of Mississippi. Geol. Soc. Am., Field Trip Guidebook Annu. Meeting, New Orleans, pp. C-1–C-45.

Kruger, F. C., 1971. Our mineral heritage—over indulgence or self denial. Mining Congr. J., 57(7):52–57.

Krynine, D. P., and Judd, W. R., 1957. Principles of engineering geology and geotechnics. McGraw-Hill, New York.

Krynine, P. D., 1960. On the antiquity of "sedimentation" and "hydrology." Geol. Soc. Am Bull. 71:1721–1726.

Kuesel, T. R., 1976. Washington Metro's topless tunnels. In: Tobbins, R. J., and Conlon, M. J., (eds.), Proceedings, rapid excavation and tunneling conference. Port City Press, Baltimore, MD, pp. 296–310.

Kuhn, G. G., and Shepard, F. P., 1979. Coastal erosion in San Diego County, California. In: Abbott, P. L., Elliott, W. J. (eds.), Earthquakes and other perils, San Diego Region.

San Diego Assoc. Geol., Field trip prepared for Geol. Soc. Am., Annu. Meeting, San Diego, pp. 207–216.

Kuhnhansen, G. H., 1979. Tailings disposal at the Bear Creek Uranium Project. In: Argall, G. O.,, Jr. (ed.), Tailing disposal today, Vol. 2: Proc. 2nd Int. Tailing Symp., Denver, CO. Miller Freeman, San Francisco, pp. 109–121.

La Bastille, A., 1981. Acid rain, how great a menace. Nat. Geogr. 160:652–681.

Lachenbruch, A. H., 1970. Some estimates of the thermal effects of a heated pipeline in permafrost. USGS, Circular 632.

Lahee, H., 1952. Field geology. McGraw-Hill, New York.

Lama, R. D., and Vutukuri, V. S., 1978. Handbook on mechanical properties of rocks, Vol. II. Trans. Tech. Publ., Switzerland.

Lamar, D. L., and Lamar, J. V., 1973. Elevation changes in the Whittier Fault area, Los Angeles Basin, California. In: Moran, D. E., Slosson, J. E., Stone, R. O., and Yelverton, C. A. (eds.), Geology, seismicity, and environmental impact. Assoc. Eng. Geol., Special Publication, University Publishing, Los Angeles, CA, pp. 71–87.

Lamar, D. L., Merifield, P. M., and Proctor, R. J., 1973. Earthquake recurrence intervals on major faults in Southern California. In: Varnes, D. J., and Kiersch, G. (eds.), Reviews in engineering geology, Vol. II. Geol. Soc. Am., pp. 265–276.

Lambe, T. W., and Whitman, R. V., 1969, Soil mechanics. John Wiley, New York.

Landsberg, H., 1982. Is a U.S. materials policy really the answer. Prof. Eng., 52(3):16–21.

Lane, E. W., 1955. Design of stable channels. Am. Soc. Civil Eng., Trans., 120:1234–1279.

Lane, E. W., 1956. The importance of fluvial morphology in hydraulic engineering. Am. Soc. Civil Eng., Proc., 81, Paper 795, pp. 1–17.

Lane, K. S., 1974. Stability of Reservoir Slopes. In: Voight, B. (ed.), Rock mechanics, the American Northwest. College of Earth and Mineral Sciences, Experiment Station, Penn. St. Univ., University Park, PA, p. 64.

Langbein, W. B., 1949. Annual runoff in the United States. USGS, Circular 52.

Langbein, W. B., and Schumm, S. A., 1958. Yield of sediment in relation to mean annual precipitation. Am. Geophys. Union, Trans., 39:1076–1084.

Larsen, E. S., 1948. Batholiths and associated rocks of Corena, Elsinere, and San Luis Bay quadrangles, Southern California. Geol. Soc. Am., Mem. 29, pp. 114–119.

Larson, T. H., and Heigold, P. C., 1981. Discussion of "Preliminary evaluation of an attenode electrode array for use in shallow-substance electrical resistivity studies." Ground Water, 19:433–434.

Lasaga, A. C., 1981. The sulfur cycle. Earth and Mineral Sciences, Penn. St. Univ., no. 1, pp. 1–11.

Lattman, L. H., 1960. Cross-section of a flood plain in a moist area of moderate relief. J. Sediment. Petrol., 30:275–282.

Lattman, L. H., 1965. Aerial photographs in field geology. Holt, Rinehart and Winston, New York.

Lattman, L. H., 1973. Calcium carbonate cementation of alluvial fans in southern Nevada. Geol. Soc. Am. Bull., 84:3013–3028.

Lattman, L. H., and Parizek, R. R., 1964. Relationship between fracture traces and the occurrence of ground water in carbonate rocks. J. Hydrol., 2:73–91.

Lawson, A. C., et al., 1908. The California earthquake of 1906. A. M. Robertson, San Francisco.

Leeds, D. J., 1973. The design earthquake. In: Moran, D. E., Slosson, J. E., Stone, R. O., and Yelverton, C. A. (eds.), Geology, seismicity, and environmental impact. Assoc. Eng. Geol., Specail Publication, University Publishers, Los Angeles, CA, pp. 337–347.

Legget, R. F., 1973. Cities and geology. McGraw-Hill, New York.

Legget, R. F., 1976. The role of geomorphology in planning. In: Coates, D. C. (ed.), Geomorphology and engineering. Dowden, Hutchinson and Ross, Stroudsburg, PA, pp. 315–328.

Legget, R. F., 1979. Geology and geotechnical engineering. Am. Soc. Civil Eng., J. Geotechnol. Div., 105(GT3):339–391.

Legget, R. F., and Karrow, P. F., 1983. Handbook of geology in civil engineering. McGraw-Hill, New York.

Lehr, Jay H., 1980. Another hidden reservoir. Water Well J., 34(3):8–9.

Leifeste, D. K., 1974. Dissolved solids USGS, discharge to the oceans from the conterminous United States. Circular 685.

Leliavsky, S., 1955. An introduction to fluvial hydraulics. Constable and Co., London.

Lennox, D. H., and Carson, V., 1970. Integration of geophysical methods for ground water exploration in the prairie provinces, Canada. In: Morley, L. W., (ed.), Mining and ground water geophysics/1967. Geol. Survey of Canada, Econ. Geol. Rep. No. 26, pp. 517–533.

Leopold, L. B., 1953. Downstream change of velocity in rivers. Am. J. Sci., 251:606–624.

Leopold, L. B., 1960. River meanders. Geol. Soc. Am. Bull., 71:769–794.

Leopold, L. B., 1972. Hydrology for urban land planning. In: McKenzie, G. D., and Utgard, R. O. (eds.), Man and his physical environment. Burgess, Minneapolis, MN, pp. 43–55.

Leopold, L. B., 1974. Water, a primer. W. H. Freeman, San Francisco.

Leopold, L. B., and Langbein, W. B., 1960. A primer on water. US Gov. Printing Office.

Leopold, L. B., and Maddock, T., Jr., 1953. The hydraulic geometry of stream channels and some physiographic implications. USGS Prof. Paper 252.

Leopold, L.B., and Miller, J. P., 1956. Ephemeral streams-hydraulic factors and their relation to the drainage net. USGS, Prof. Paper 282-A.

Leopold, L. B., and Skibitzke, H. E., 1967. Observations on unmeasured rivers: Geogr. Ann., 49A:247–255.

Leopold, L. B., Wolman, M. G., and Miller, J. P., 1964. Fluvial processes in geomorphology. W. H. Freeman, San Francisco.

Leveson, D., 1980. Geology and the urban environment. Oxford, New York.

Lifrieri, J. J., and Raghu, D., 1982. Development of a foundation quality index for foundations in solution-prone carbonate regions. Assoc. Eng. Geol. Bull., 19:35–47.

Lillesand, T., and Kieffer, R., 1979. Remote sensing and image interpretation. John Wiley, New York.

Lillevang, O. J., 1965. Groins and effects—minimizing liabilities. Am. Soc. Civil Eng., Coastal Eng. Specialty Conf., Santa Barbara, CA, 749–754.

Lindholm, R. C., 1980. The oil shortage—a story geologists should tell. J. Geol. Ed., 28:36–45.

Lins, H. F., Jr., 1981. Pattern and trends of land use and land cover on Atlantic and Gulf Coast barrier islands. USGS, Prof. Paper 1156.

Linsley, R. K., and Franzini, J. B., 1979. Water resources engineering. McGraw-Hill, New York.

Lippy, E. C., 1981. Waterborne disease, occurrence is on the upswing. J. Am. Water Works Assoc., 73(1):57–62.

Lockeretz, W., 1978. The lessons of the Dust Bowl. Am. Scient., 66:560–569.

Lofgren, B. E., 1963. Land subsidence in the Arvin-Maricopa area, California. USGS, Prof. Paper 475-B, pp. 171–175.

Lofgren, B. E., 1969. Land subsidence due to the application of water. In: Varnes, D. J., and Kiersch, G., (eds.), Reviews in engineering geology, Vol. II. Geol. Soc. Am., pp. 271–303.

London, J., 1945. "All Gold Canyon." In: "Best short stories of Jack London." Sun Dial Press, Garden City, NY, pp. 248–264.

Longwell, C. R., Flint, R. F., and Sanders, J. E., 1969. Physical geology. John Wiley, New York.

Lucchitta, B. K., 1978. A large landslide on Mars. Geol. Soc. Am., Bull. 89:1601–1609.

Lucchitta, I., Schleicher, D., and Cheney, P., 1981. Of price and prejudice: the importance of being earnest about environmental impact statements. Geol. Soc. Am. Bull., 9:590–591.

Ludeke, K. L., 1973. Vegetative stabilization of tailings disposal berms. Mining Cong. J., 59(1):32–39.

Lueder, D. R., 1959. Aerial photograph interpretation. McGraw-Hill, New York.

Lugeon, M., 1933. Dams and geology, Paris.

Lum, D., 1961. The resistivity method applied to ground water studies of glacial outwash deposits in eastern South Dakota. South Dakota Geol. Survey, Report of Invest., no. 89.

Luthin, J. N., 1978. Drainage engineering. John Wiley, New York.

Lutton, R. J., Banks, D. C., and Strom, W. E., Jr., 1979. Slides in Gaillard Cut, Panama Canal Zone. In: Voight, B. (ed.), Rockslides and avalanches, 2, engineering sites. Elsevier, Amsterdam, pp. 152–224.

Maasland, D. E. L., and Bittinger, 1963. Proc. of the Symp. on transient ground water hydraulics. Colo. St. Univ., Ft. Collins, CO.

Mabey, D. R., 1970. The role of geophysics in the development of the world's ground water resources. In: Morley, L. W. (ed.), Mining and ground water geophysics/1967. Geol. Survey of Canada, Econ. Geol. Rep., no. 26, pp. 267–271.

MacBerthouex, P., and Rudd, D. F., 1977. Strategy of pollution control. John Wiley, New York.

Machemehl, J. L., and Forman, W., 1980. Flood tidal delta in a new barrier island tidal inlet. Assoc. Eng. Geol. Bull., 17:163–191.

Mackin, J. H., 1948. Concept of the graded river. Geol. Soc. Am. Bull. 59:463–512.

Main, L., 1973. What starts a coal mine fire. Penn. Geol. Survey, 4(6):7–9.

Mark, R., 1978. Structural experimentation in Gothic Architecture. Am. Scient., 66:542–550.

Marin, J., and Sauer, J. A., 1954. Strengths of materials. Macmillan, New York.

Martinez, J. D., Kumar, M. B., Kolb, C. R., and Coleman, J. M., 1981. Catastrophic drawdown shapes floor. Geotimes, 26(3):14–16.

Mathes, P. E., 1930. Geologic history of the Yosemite Valley. USGS, Prof. Paper 160.

Mathewson, C. C., 1977. Beach washouts, an adverse effect of dune construction. Assoc. Eng. Geol. Bull. 14: 13–25.

Mathewson, C. C., 1981. Engineering geology. Charles E. Merrill, Columbus, OH.

Mathewson, C. C., 1982. Geology, foundation of the city. Assoc. Eng. Geol. Bull. 19:251–259.

Mathewson, C. C., and Dobson, B. M., 1982. The influence of geology and the expansive potential of soil profiles. Geol. Soc. Am. Bull., 3:565–571.

Mathewson, C. C., Dobson, B. M., Dyke, L. D., and Lytton, R. L., 1980. System interaction of expansive soils with light foundation. Assoc. Eng. Geol. Bull., 17:55–94.

Mathewson, C. C., and Minter, L. L., 1981. Impact of water resource development on the hydrology and semimentology of the Brazos River, Texas, with implications on shoreline erosion. Assoc. Eng. Geol. Bull., 18:39–53.

Mathieson, S. A., and Lajoie, K. R., 1983. Drastic increase in climate retreat rates due to a lack of geologic input into the design of the Half Moon Bay Breakwater: a problem enhanced by the storm of 1982–83, San Mateo County, California (Abs.). Assoc. Eng. Geol., Annu. Meeting, San Diego, CA, p. 85.

Mayuga, M. N., and Allen, D. R., 1966, Long Beach subsidence. In: Lung, Richard, and Proctor, Richard, ed., Engineering geology in southern California: Association of Engineering Geologists, Special Publication, Glendale, CA, p. 280–285.

McCarthy, G. J., ed., 1979. Scientific basis for nuclear waste isolation, Vol. I. Plenum, New York.

McConnell, D., Mielenz, R. C., Holland, W. Y., and Greene, K. T., 1950. Petrology of concrete affected cement-aggegate reaction. Geol. Soc. Am., Engineering Geology (Berkey) Vol., pp. 225–250.

McCray, K., 1982. The Ogallala—half full or half empty. Water Well J., 36(7):53–62.

McCullough, D., 1972. The great bridge. Simon and Schuster, New York.

McCullough, D., 1977. The path between the seas. Simon and Schuster, New York.

McGauhey, P. F., and Eich, H. F., 1942. Physical and chemical tests of water. Virginia Polytechnic Inst. Bull., 25(6):11.

McHarg, I. L., 1969. Design with nature. Natural History Press, Garden City, NY.

McKenzie, G. D., and Utgard, R. O., eds., 1972. Man and his physical environment. Burgess, Minneapolis, MN.

McLindon, G. J., 1981. Environment: the big picture. Water Spectrum, 13(4):2–7.

McSaveney, M. J., 1978. Sherman Glacier rock avalanche, Alaska, USA. In: Voight, B. (ed.), Rockslides and avalanches, 1, natural phenomena. Elsevier, Amsterdam, pp. 197–258.

Meadors, G. S., Jr., 1977. The geology and construction techniques of the second Hampton Roads crossing, Norfolk, Virginia. Proc. 28th Annu. Highway Geol. Symp., Rapid City, South Dakota Dep. Transportation, pp. 61–82.

Meadows, D. H., Meadows, D. L., Rauders, J., and Behrens, W. W., 1972. The limits to growth. New American Library, New York.

Megaw, T. M., and Bartlett, J. V., 1982. Tunnels: planning, design, construction, Vol. 2. Halstead Press, New York.

Meinzer, O. E., 1923. Outline of ground water hydrology with definitions. USGS, Water Supply Paper 494.

Meinzer, O. E., ed., 1942. Hydrology. Dover, New York.

Melosh, H. J., 1983. Acoustic fluidization. Am. Scient., 71:158–165.

Menard, H. W., 1974. Geology, resources and society. W. H. Freeman, San Francisco.

Mencl, V., 1966. Mechanics of landslides with special reference to the Vaiont slide. Geotechnique, 18: 329–337.

Merriam, R. H., 1960. Portuguese Bend landslide, Palos Verdes Hills, Calif. J. Geol., 68(2):140–153.

Meunier, T. K., 1980. Topographic mapmaking. J. Geol. Ed., 29:88–91.

Meyer, W. R., 1962. Use of a neutron moisture probe to determine the storage coefficient of an unconfined aquifer. USGS, Prof. Paper 450E, pp. 174–176.

Meyers, J. S., 1962. Evaporation from the 17 western states. USGS, Prof. Paper 272-D.

Mielenz, R. C., 1962. Petrography applied to Portland-Cement concrete. In: Fluhr, T., and Legget, R. F. (eds.), Reviews in engineering geology, Vol. 1. Geol. Soc. Am., pp. 1–38.

Mildner, W. F., 1982. Erosion and sediment. Assoc. Eng. Geol. Bull., 19:161–166.

Miller, C. H., and Miller, D. R., 1977. Orientation of explosion-induced surface fractures estimated from pre-explosion fractures and in situ measurements. Assoc. Eng. Geol. Bull., 14:27–37.

Miller, D. J., 1960. Giant waves in Lituya Bay. USGS, Prof. Paper 354-C, pp. 51–86.

Miller, J. P., 1958. Climate and man in the Southwest. Univ. Ariz., Bull. 28, no. 4.

Miller, R. D., and Dobrovolny, E., 1959. Surficial geology of Anchorage and vicinity of Alaska. USGS, Bull. 1093.

Miller, V. C., and Miller, C. F., 1961. Photogeology. McGraw-Hill, New York.

Mills, H. H., 1977. Basal till fabrics of modern alpine glaciers. Geol. Soc. Am. Bull., 88:824–828.

Milne, W. G., ed., 1976. Induced seismicity. Eng. Geol., 10 (Special Issue):83–389.

Mirsky, A., 1983. Influence of geologic factors on ancient Mesopotamian civilization. J. Geol. Ed., 30:294–299.

Moening, H. J., 1973. An overview of the Trans-Alaskan pipeline: engineering and environmental geology problems. In: Moran, D. E., James E. Slosson, Stone, R. O., and

Yelverton, C. A. (eds.), Geology, seismicity, and environmental impact. Asso. Eng. Geol., Special Publication, University Publishing, Co., Los Angeles, pp. 405–409.

Mollard, J. D., 1962. Photo analysis and interpretation in engineering-geology investigations: a review. In: Fluhr, T., and Legget, R. F. (eds.), Reviews in engineering geology, Vol. 1. Geol. Soc. Am., pp. 105–127.

Momeni, M. H., Kisieleski, W. E., Yuan, Y., and Roberts, C. J., 1978. Radiological and environmental studies at uranium mills. Proc., Seminar on management, stabilization and environmental impact of uranium mill tails, Albuquerque, Jul. 1978. Organization for Economic Cooperation and Development, Nuclear Energy Agency, Paris, pp. 85–102.

Mooney, H. M., 1973. Handbook of engineering geophysics. Bison Instruments, Minneapolis, MN.

Moore, G. K., and Deutsch, M., 1975. ERTS imagery for ground water investigations. Ground Water, 13:214–226.

Moore, R. W., 1952. Geophysical methods adapted to highway engineering problems. Geophysics, 17:197–202.

Montague, P., 1982. Hazardous waste landfills: some lessons from New Jersey. Civil Eng., 52(9):53–56.

Morgan, A. E., 1971. Dams and other disasters. Porter, Sargent, Boston.

Morgenstern, N. R., and Price, V. E., 1965. The analysis of the stability of general slip surfaces. Geotechnique, 15(1):79–93.

Morisawa, M., 1966. Talus slopes in high mountain areas: US Dep. of Army Grant DA-ARO-49-092-64-G52.

Morley, L. W., ed., 1970. Mining and ground water geophysics/1967. Geol. Survey of Canada, Econ. Geol. Rep. no. 26.

Morrison, A., 1983a. Arresting a toxic plume. Civil Eng., 53(8):35–37.

Morrison, A., 1983b. Arizona's water strategy: bring more in and restrict its use: Civil Eng., 53(4):46–49.

Morton, R. A., 1976. Effects of Hurricane Eloise on beach and coastal structures. Florida panhandle. Geology, 4:277–280.

Munfakh, G. A., 1985. Wharf stands on stone columns. Civil Eng., 55(1):44–47.

Munson, R. C., 1970. Relationship of effect of water flooding of the Rangely Oil Field on seismicity. Geol. Soc. Am., Eng. Geol. Case Histories, no. 8, pp. 39–49.

Murray, D. R., and Zahar, G., 1973. An approach to revegetation of mine wastes. Canada Dep. Energy, Mines, and Resources, Mines Branch. Int. Rep. 73–68, Ottawa, Canada.

Muskat, M., 1947. The flow of homogenous fluids through porous media. McGraw-Hill, New York.

Mustoe, G. E., 1982. The origin of honeycomb weathering. Geol. Soc. Am. Bull., 93:108–115.

NASA, 1976. Mission to earth, Landsat views the world. Nat. Aeronaut. and Space Admin., Washington, DC.

National Academy of Science, 1969. Seismology, responsibilities and requirements of a growing science. Nat. Acad. Sci., Washington, DC.

National Energy Plan, 1977. Executive Office of the President Energy Policy and Planning. US Superintendent of Documents, Washington, DC.

National Research Council, 1971. Earthquakes related to reservoir filling: Report by joint panel on problems concerning seismology and rock mechanics. Division of Earth Sciences, Dec. 1971.

National Research Council, 1981. Rock-mechanics research requirements for resource recovery, construction, and earthquake-hazard reduction: Panel on rock-mechanics research requirements. Nat. Acad. Press, Washington, DC.

National Research Council, 1982a. Outlook for science and technology, the next five years. National Academy of Sciences, W. H. Freeman, San Francisco.

National Research Council, 1982b. Carbon dioxide and climate, a second assessment: Nat. Acad. Press, Washington, DC.

National Research Council, 1982c. Geological aspects of industrial waste disposal. Nat. Acad. Sci., Nat. Acad. Press, Washington, DC.

National Research Council, 1984. Groundwater contamination. National Academy Press, Washington, DC.

Nehru, C. E., Shimer, J. A., and Stewart, J. C., 1972. Studies of the earth. Burgess, Minneapolis, MN.

Nelson, P. H., 1982. Advances in borehole geophysics for hydrology. Geol. Soc. Am., Paper 189, pp. 207–219.

Nero, A. V., Jr., 1979. A guidebook to nuclear reactors. Univ. Calif. Press, Berkeley.

Neumann, Frank, 1962. Engineering seismology. In: Fluhr, T., and Legget, R. F. (eds.). Reviews in engineering geology, V. 1. Geol. Soc. Am., pp. 161–186.

Neuzil, C. E., and Pollock, D. W., 1983. Erosional unloading and fluid pressures in hydraulically "tight" rocks. J. Geol., 91:179–193.

Newton, J. G., 1984. Sinkholes resulting from ground-water withdrawals in carbonate terranes—an overview. In: Holzer, Thomas L., ed., Man-induced land subsidence. Geol. Soc. Am., Reviews in Engr. Geol. 6:195–202.

Nichols, D. R., and Campbell, C. C., eds., 1971. Environmental planning and geology. US Gov. Printing Office, Washington, DC.

Nichols, D. R., and Buchanan-Banks, J. M., 1974. Seismic hazards and land-use planning. USGS, Circular 690.

Noble, C. C., 1976. The Mississippi River flood of 1973. In: Coates, D. R. (ed.), Geomorphology and engineering. Dowden, Hutchinson and Ross, Stroudsburg, PA, pp. 79–98.

Northrup, C. J. M., ed., 1980. Scientific basis for nuclear waste isolation, Vol. II. Plenum, New York.

Northwood, T. D., and Crawford, R., 1965. Blasting and building damage. Canada Nat. Res. Council, CBD 63, Ottawa, Canada.

Oakeshott, G. B., 1973. Patterns of ground ruptures in fault zones coincident with earthquakes: some case histories. In: Varnes, D. J., and Kiersch, G. (eds.), Reviews in engineering geology, Vol. II. Geolog. Soc. Am., pp. 287–312.

Obert, L., and Duvall, W. I., 1967. Rock Mechanics and the design of structures in Rock. John Wiley, New York.

O'Brien, K. M., and Stone, W. J., 1984. Role of geological and geophysical data in modeling a southwestern alluvial basin. Ground Water, 22:717–727.

Ollier, C. D., 1969. Weathering: Oliver and Boyd, Edinburgh.

Olson, G. W., 1981. Soils and the environment. Chapman and Hall, New York.

Omnes, G., 1977. High accuracy gravity applied to the detection of karstic cavities. In: Tolson, J., and Doyle, F. L., 1977. Karst hydrogeology, Proc. 12th Int. Cong. Int. Assoc. Hydrogeol., Huntsville, AL, pp. 273–284.

Oostdam, B. L., 1970, Exploration for marine placer deposits of diamond. In: Morley, L. W., (ed.), Mining and ground water geophysics/1967: Geological Survey of Canada, Econ. Geol. Rep. no. 26, pp. 447–461.

Outland, C. F., 1977. Man-made disaster—the story of St. Francis Dam. Arthur N. Clark, Glendale, CA.

Owens, L. B., and Watson, J. P., 1979. Landscape reduction by weathering in small Rhodesian watersheds. Geology. 7:281–284.

Page, L. M., 1968. Use of electrical resistivity method for investigating geologic and hydrologic conditions in Santa Clara County, Calif., Ground Water, 6(5):31–40.

Page, R. A., Blume, J. A., and Joyner, W. B., 1975a. Earthquake shaking and damage to buildings. Science, 189: 601–608.

Page, R. A., Boore, D. M., and Dieterich, J. H., 1975b. Estimation of bedrock motion at the ground surface. In: Borcherdt, R. D. (ed.), Studies for seismic zonation of the San Francisco Bay region. USGS, Prof. Paper 941-A, pp. A31–A38.

Paige-Green, P., 1981. Current techniques in ground water control applied to cut slopes. Geol. Soc. S. Afr., Trans., 84: (Special Issue): 161–167.

Paone, J., Morning, J. L., and Giorgetti, L., 1974. Land utilization and reclamation—the mining industry, 1930–71. US Bureau of Mines, IC 8642, Washington, DC, pp. 1930–1971.

Parasnis, D. S., 1966. Mining geophysics. Elsevier, Amsterdam.

Parfit, M., 1983. Ground water proves a hidden element whose time has come. Smithsonian, 13(12):50–61.

Pariseau, W. G., and Voight, B., 1979. Rockslides and avalanches: basic principles and perspectives in the realm of civil engineering and mining operations. In Voight, B. (ed.), Rockslides and avalanches, 2, engineering sites. Elsevier, Amsterdam, pp. 1–92.

Parizek, E. J., 1957. A clarification of definition of classification of soil creep. J. Geol. 65:653.

Parizek, R. R., 1969. An environmental approach to land use in a folded and faulted terrane. In: Nichols, D. R., and Campbell, C. C. (eds.), Environmental planning and geology: Proc. Symp. Eng. Geol. in the Urban Environ., Assoc. Eng. Geol., US Dep. Housing and Urban Devel., US Dep. Interior, pp. 122–143.

Parizek, R. R., 1971a. Land use problems in carbonate terranes. Geol. Soc. Am., Guidebook for 1971 Annu. Meeting, pp. 135–153.

Parizek, R. R., 1971b. An environmental approach to land use in a folded and faulted carbonate terrane. In: Nicholas, D. R., and Campbell, C. C. (eds.), Environmental planning and geology. USGS and US Dep. Housing and Urban Devel., pp. 122–143.

Parrish, D. K., and Senseny, P. E., 1981. Creep in welded tuff (Abs.): Geol. Soc. Am., Annu. Meeting, Rocky Mountain Section, 13(4):233.

Peck, R. B., 1968. Problems and opportunities—technology's legacy from the Quaternary. The Quatern. of Ill., Univ. Ill. Centennial Symp., pp. 138–144.

Peck, R. B., 1969. Deep excavation and tunneling in soft ground: state-of-the art volume. 7th Int. Conf. on Soil Mechanics and Foundation Eng., Mexico City, pp. 225–290.

Peck, R. B., Hanson, W. E., and Thornburn, T. H., 1974. Foundation engineering. John Wiley, New York.

Penick, J., Jr., 1976. The New Madrid earthquakes of 1811–1812. Univ. Missouri Press, Columbia.

Perloff, W. H., and Baron, W., 1976. Soil mechanics, principles and applications. John Wiley, New York.

Peters, V. R., and Burckick, R. G., 1983. Use of an automatic resistivity system for detecting abaondoned mine workings. Mining Eng., 35(1):55–59.

Peterson, F. L., Williams, J. A., and Wheatcraft, S. N., 1978. Waste injection into a two-phase flow field: sand box and Hele-Shaw models study. Ground Water, 16:410–416.

Peterson, R., 1958. Rebound in the Bearpaw Shale, western Canada. Geol. Soc. Am. Bull., 69:1113–1124.

Pettyjohn, W. A., 1982. Cause and effect of cyclic changes in ground water quality. Ground Water Monitor., Rev., 2(1)43–49.

Péwé, T. L., 1966. Permafrost and its effect on life in the North. Oregon St. Univ. Press, Corvallis.

Péwé, T. L., 1982. Desert dust: origin, characteristics, and effect on man. Geol. Soc. Am., Special Paper 186.

Phipps, R. L., 1974. The soil creep—curved tree fallacy USGS, J. Res., 2:371–377.

Phukan, A., and Andersland, O. B., 1978. Foundations for cold regions. In: Andersland, O. B., and Anderson, O. M. (eds.), Geotechnical engineering for cold regions. McGraw-Hill, New York, pp. 266–362.

Pierce, W. G., 1963. Roof Creek detachment fault, northwest Wyoming. Geol. Soc. Am Bull., 74:1225–1236.

Pilkney, O. H., 1981. Old solutions fail to solve beach problem. Geotimes, 21(12):18–22.

Pincus, Howard, ed., 1973, Geological factors in rapid excavation: Geol. Soc. Am., Engineering Geology Case History No 9, 81 p.

Piteau, D. R., and Peckover, F. L., 1978. Engineering of rock slopes. In: Schuster, R. L., and Krizek, R. J., Landslides, analysis and control. Transportation Research Board, Nat. Acad. Sci., Washington, DC, pp. 192–228.

Plafker, G., 1969. Tectonics of the March 27, 1964 Alaska earthquakes. USGS, Prof. Paper 543-I.

Plafker, G., and Erickesn, G. E., 1978. Navados Huascaran avalanches. Peru. In: Voight, B. (ed.), Rockslides and avalanches, Vol. 1, natural phenomena. Elsevier, Amsterdam, pp. 277–314.

Ploessel, M. R., 1973. Engineering geology along the southern California coastline. In: Varnes, D. J., and Kiersch, G. (eds.), Reviews in engineering geology, Vol. II. Geol. Soc. Am., pp. 365–366.

Poland, J. F., 1969. Land subsidence in western United States. In: Olson, R. A., and Wallace, M. W. (eds.), Geologic hazards and public problems: May 27–28, 1969, Conf. Proc. US Gov. Printing Office, pp. 77–96.

Poland, J. F., 1981. Subsidence in United States due to ground water withdrawal: Am. Soc. Civil Eng., Proc., J. Irrigation and Drainage Div., 107(IR2):115–135.

Poland, J. F., and Davis, G. H., 1969. Land subsidence due to withdrawal of fluids. In: Varnes, D. J., and Kiersch, G., (eds.), Reviews in engineering geology, Vol. II. Geol. Soc. Am., pp. 187–269.

Poland, J. F., and Evenson, R. E., 1966. Hydrogeology and land subsidence, Great Central Valley, Calif. Div. of Mines & Geology, Bull. 190: 239–247.

Poland, J. F., Lofgren, B. E., Ireland, R. L., and Pugh, R. G., 1975. Land subsidence in the San Joaquin Valley, California, as of 1972. USGS, Prof. Paper 437-H.

Population Reference Bureau, 1970. India: ready or not, here they come. Population Bull, 26(5):1–32.

Porter, K. S., 1980. An evaluation of sources of nitrogen as causes of ground water contamination in Nassau County, Long Island. Ground Water, 18:617–625.

Porter, S. C., and Orombelli, G., 1981. Alpine rockfall hazards. Am. Scient., 69(1):67–75.

Portland Cement Association, 1962. PCA soil primer. Chicago, IL.

Posey, C. J., 1974. Tests of scour protection for bridge piers. Am. Soc. Civil Eng., J. Hydraul. Div., 100(HY12):1772–1783.

Postma, G., 1984. Slumps and their deposits in fan delta front and slope. Geology, 12:27–30.

Potter, N., Jr., 1969. Tree-ring dating of snow avalanche tracks and the geomorphic activity of avalanches, northern Absaroka Mountains, Wyoming. Geol. Soc. Am., Special Paper 123, pp. 141–165.

Potter, N., Jr., 1974. Snow avalanches and rock glaciers, Sunlight mining district, northern Absaroka Mountains, Wyoming. In: Voight, B. (ed.), Rock mechanics, the American Northwest. Penn. St. Univ., pp. 51–55.

Potter, N., Jr., and Moss, J. H., 1968. Origin of the Blue Rocks block field and adjacent deposits, Berks County, Pennsylvania. Geol. Soc. Am. Bull., 74:255–262.

Poulos, H. G., and Davis, E. H., 1980. Pile foundation analysis and design. John Wiley, New York.

Powers, J. P., 1981. Construction dewatering. John Wiley, New York.

Pratt, H. R., and Voegele, M. D., 1984. In situ tests for site characterization, evaluation and design. Assoc. Eng. Geol. Bull., 21:3–220.

Prestegaard, K. L., 1983. Variables influencing water-surface slopes in gravelbed streams at bankfull stage. Geol. Soc. Am., Bull., 94:673–678.

Prickett, T. A., 1975. Modeling techniques for ground water evaluation. In: Chow, V. (ed.), Advances in hydroscience, Vol. 10. Academic, New York, pp. 1–143.

Prickett, T. A., 1979. Ground water computer models—state of the art. Ground Water, 17:167–173.

Proctor, R. J., 1973, Geology of urban tunnels—including a case history in Los Angeles. In: Varnes, D. J., and Kiersch, G. (eds.), Reviews in engineering geology, Vol. II. Geol. Soc. Am., pp. 187–193.

Proctor, R. J., Brooks, D. C., and Pentegoff, V. P., 1966. Exploration for 52 miles of tunnels for the Foothills Feeder. In: Lung, R., and Proctor, R. J. (eds.), Engineering geology in Southern California. Assoc. of Eng. Geol., Special Publications, Glendale, CA, pp. 81–87.

Prokopovich, N. P., 1983. Reconnaissance estimates of subsidence along Salt-Gila aqueduct, Arizona. Assoc. Eng. Geol. Bull., 20:297–315.

Prokopovich, N. P., and Marriott, M. J., 1983. Cost of subsidence to the Central Valley Project, California. Assoc. Eng. Geol. Bull., 20:325–332.

Purdom, P. W., and Anderson, S. H., 1983. Environmental Science. Charles E. Merril, Columbus, OH.

Radbruch-Hall, D., 1978. Gravitation creep of rock masses on slopes. In: Voight, B. (ed.), Rockslides and avalanches, I, natural phenomena. Elsevier, Amsterdam, pp. 607–657.

Rahn, P. H., 1966. Inselbergs and nickpoints in Southwestern Arizona. Zeitschrift fur Geomorphologie, 10(3):L217–225.

Rahn, P. H., 1967. Sheetfloods, streamfloods, and the formation of pediments. Ann. Assoc. Am. Geogr. 57(3):593–604.

Rahn, P. H., 1968. Future ground water supplies for Providence, Rhode Island. Proc. 4th Annu. Am. Water Resources Assoc. Conf., 1968 Annu. Meeting, New York, pp. 380–391.

Rahn, P. H., 1969. The relationship between natural forested slopes and angles of repose for sand and gravel. Assoc. Eng. Geol. Bull. 80:2123–2128.

Rahn, P. H., 1971a. The weathering of tombstones and its relationship to the topography of New England. J. Geol. Ed., 19:112–118.

Rahn, P. H., 1971b. The surficial geology of the Spring Hill Quadrangle, Connecticut. Conn. Geol. and Nat. Hist. Survey, Bull. 26.

Rahn, P. H., 1973. Effects of urbanization on stream runoff in Rapid City. South Dakota Acad. Sci., 52:179–188.

Rahn, P. H., 1975. Lessons learned from the June 9, 1972, flood in Rapid City, South Dakota. Assoc. Eng. Geol. Bull., 12:83–97.

Rahn, P. H., 1976. Potential of coal strip-mine spoils on aquifers in the Powder River Basin. Old West Regional Commission, Billings, Montana.

Rahn, P. H., 1977. Erosion below main stem dams on the Missouri River. Assoc. Eng. Geol. Bull., 14(3):157–181.

Rahn, P. H., 1978. An exhibit to illustrate the role of fluid pressure in fault movement. J. Geol. Ed., 26(3):110–111.

Rahn, P. H., 1984. Flood-plain management program in Rapid City, South Dakota. Geol. Soc. Am. Bull. 95:838–843.

Rahn, P. H., Bump, V. L., and Steece, F. V., 1981. Engineering geology of the central and northern Black Hills, South Dakota. In: Rich: F. J., (ed.), Geology of the Black Hills, South Dakota and Wyoming. Am. Geol. Inst., pp. 135–153.

Rahn, P. H., and Mabes, D. L., 1977. Seepage from uranium tailings ponds and its impact on ground water. Organization for Economic Cooperation and Development, Nuclear Energy Agency, Proc., Seminar on management, stabilization and environmental impact of uranium mill tailings, Albuquerque, NM, Ju. 1978, pp. 127–140.

Rahn, P. H., and Gries, J. P., 1973. Large springs in the Black Hills, South Dakota and Wyoming. South Dakota Geol. Survey. Rep. Invest., no. 107.

Rahn, P. H., and Moore, D. G., 1981. Landsat data for locating shallow glacial aquifers in eastern South Dakota. American Water Resource Association, 5th Annu. William T. Pecora Symp. on Remote Sensing, Sioux Falls, SD, pp. 398–406.

Rahn, P. H., Paul, H. A., 1975. Hydrogeology of a portion of the Sand Hills and Ogallala aquifer, South Dakota and Nebraska. Ground water, 13:428–437.

Raleigh, B., 1977. Can we control earthquakes? Earthquake Inform. Bull., 9(1):4–7.

Raleigh, C. B., Healy, J. H., and Bredehoeft, J. D., 1976. An experiment in earthquake control at Rangely, CO: Science, 191:1230–1237.

Rand McNally and Co., 1979. Our magnificent earth. Rand McNally, New York.

Randell, D. H., Reardon, J. B., Hileman, J. A., Matuschka, T., Liang, G. C., Khan, A. I., and Laviolette, J., 1983. Geology of the city of Long Beach, California, United States of America. Assoc. Eng. Geol. Bull., 20:9–94.

Rankine, W. E., and Lavin, P. M., 1970. Analysis of a reconnaissance gravity survey for drift-filled valleys in the Mercer Quadrangle, Pennsylvania. J. Hydrol., 10:418–435.

Rapp, A., 1962. Recent development of mountain slopes in Karkavagge and surrounding northern Scandinavia: Geogrofiska Ann., 142:71–200.

Rapp, A., 1976. Talus slopes and mountain walls at Tempelfjorden, Spitzbergen, a geomorphological study of the denudation of slopes in an arctic locality. In: King, C. A. M. (ed.), Periglacial processes: Benchmark Papers in Geology/27. Dowden, Hutchinson and Ross, Stroudsburg, PA, pp. 208–218.

Ray, R. G., 1960. Aerial photographs in geologic interpretation and mapping. USGS, Prof. Paper 373.

Reiche, P., 1950. A survey of weathering processes and products. Univ. New Mexico Publications in Geol. no. 3.

Reif, A. E., 1981. The causes of cancer. Am. Scient., 69:437–447.

Reitz, H. M., 1965. Foundations on fissured limestone—how they were selected. Civil Eng. 35:52–55.

Réméniéras, G., 1967. Statistical methods of flood frequency analysis. In United Nations, Assessment of the magnitude and frequency of flood flows: Inter-regional seminar on the assessment of the magnitude and frequency of flood flows. Water Resources Series no. 30, Bangkok, Thailand.

Retallack, G. J., 1983. A paleopedological approach to the interpretation of terrestrial sedimentary rocks: the mid-Tertiary fossil soils of Badlands National Park. Geol. Soc. Am. Bull. 94:823–840.

Reuss, H. S., Chmn., 1973. Stream channelization: Hearings between the Conservation and Natural Resources Subcommittee of the House Committee on Government Operations. US Govt. Printing Office, p. 5, pp. 2780–3288; p. 6, pp. 3789–3711.

Reuther, J. J., 1982. The fate of sulfur in coal combustion. Earth and Mineral Sciences, Penn. St. Univ., 51:49–53.

Richards, P. W., 1955. Geology of the Bighorn Canyon–Hardin Area, Montana and Wyoming. USGS, Bull. 1026.

Richardson, H. W., and Mayo, R. S., 1975. Practical tunnel driving. McGraw-Hill, New York.

Richter, C. F., 1935. An instrumental earthquake magnitude scale. Seismol. Soc. Am. Bull. 251:1–32.

Richter, C. F., 1958. Elementary Seismology. W. H. Freeman, San Francisco.

Riestenberg, M. M., and Sovonick-Dunford, S., 1983. The role of woody vegetation in stabilizing slopes in the Cincinnati area, Ohio. Geol. Soc. Am. Bull., 94:506–518.

Riley, C. V., 1973. Design criteria for mined land reclamation. Mining Eng. 25(3):41–45.

Ritter, D. F., 1978. Process geomorphology. William C. Brown, Dubuque, IA.

Robb, J. M., 1984. Spring sapping on the lower continental slope, offshore New Jersey. Geology, 12:278–282.

Roberts, A. F., 1981. Applied geotechnology. Pergamon, Oxford, England.

Robertson, J. B., 1977. Numerical modeling of subsurface radioactive solute transport from waste-seepage ponds at the Idaho National Engineering Laboratory. USGS, Open-File Rep. 76-717.

Robinson, W. J., 1963. The gravity method and its application to ground water studies in southern Saskatchewan. Saskatchewan Res. Council, Phys. Div. Rep. P-63-1.

Robinson, G. D., and Spieker, A. M., (eds.), 1978. Nature to be commanded. . . . earth-science maps applied to land and water management. USGS, Prof. Paper 960.

Robinson, G. M., and Peterson, D. E., 1962. Notes on earth fissures in southern Arizona. USGS, Circular 466.

Robinson, T. W., 1958. Phreatophytes. USGS. Water-Supply Paper 1423.

Rodda, J. C., 1976. Basin studies. In: Rodda, J. C. (ed.), Facets of hydrology. John Wiley, New York, pp. 257–297.

Rodgers, G. P., and Holland, H. D., 1979. Weathering products within microcracks in feldspar. Geology, 7:278–280.

Romeo, J. C., 1970. The movement of bacteria and viruses through porous media. Ground Water, 8:37–48.

Rosenbaum, J. G., 1976. Shoreline structures as a cause of shoreline erosion, a review. In: Tank, R., Focus on environment geology: Oxford, London, pp. 166–179.

Rosenfeld, C. L., 1980. Observations on the Mount St. Helens eruption. Am. Scient., 68:494–509.

Ross, D. S., 1976. Image modification for aiding interpretation and additive color display. In: CENTO Workshop on applications of remote sensing data and methods, Proc., Istanbul, Turkey. Central Treaty Organization, EC/25/M/D6, Oct. 22, 1976, pp. 170–212.

Ross, S. S., 1977. Teton Dam: a collapse of confidence. New Eng., 6(3):4.

Ross, S. S., 1981. Is technology getting a fair shake in the general press. Prof. Eng., 51(3):8–11.

Rothe, J. P., 1973. Summary: geophysical report. In: Ackerman, W. C. (ed.), Manmade lakes: their problems and environmental effects. Am. Geophys. Union, Monogr. 17, pp. 441–454.

Roux, P. H., and Althoff, W. F., 1980. Investigation of organic contamination of ground water in South Brunswick Township, New Jersey. Ground Water, 18:464–471.

Ross, S. S., 1981. Is technology getting a fair shake in the general press. Prof. Eng., 51(3):8–11.

Royster, D. L., 1973. Highland landslide problems along the Cumberland Plateau in Tennessee. Assoc. Eng. Geol. Bull. 10:255–287.

Royster, D. L., 1974. Construction of a reinforced earth fill along I-40 in Tennessee. Proc. 25th Highway Geol. Symp. pp. 76–93.

Royster, D. L., 1977. Some observations on the use of horizontal drains in the correction and prevention of landslides. Proc., 28th Annu. Highway Geol. Symp., Rapid City. South Dakota Dep. Transportation, pp. 137–166.

Rubey, W. W., 1938. The force required to move particles in a stream bed. USGS, Prof. Paper 189-E, pp. 121–141.

Ruhe, R. V., 1969. Quaternary landscapes in Iowa. Iowa St. Univ. Press, Ames.

Ruhe, R. V., 1970. Stream regime and man's manipulation. In: Coates, D. R. (ed.), Environmental geomorphology. NY St. Univ., Binghampton, pp. 9–23.

Ruhe, R. V., 1973. Background of model for loess-deprived soils in the upper Mississippi River Basin. Soil Sci. 115(3):250–253.

Rutstein, M. S., 1982. Asbestos—friend, foe or fraud. Geotimes, 27(4):23–27.

Ruxton, B. P., and Berry, L., 1957. Weathering of granite and associated erosional features in Hong Kong. Geol. Soc. Am. Bull., 68:1263–1292.

Sabins, F. F., Jr., 1973. Engineering geology application of remote sensing. In: Moran, D. E. (ed), Geology, seismicity and environmental impact. Assoc. Eng. Geol., Special Publication, pp. 141–155.

Sabins, F. F., Jr., 1978. Remote sensing, principles and interpretation. W. H. Freeman, San Francisco.

Sagan, C., Toon, O. B., and Pollack, J. B., 1980. Human impact on climate: of significance since the invention of fire. Science, 206:1323–1362.

Saines, M., 1981. Errors in interpretation of ground water level data. Ground Water Monitor. Rev., 1(1):56–61.

Sanborn, J. F., 1950. Engineering geology in the design and construction of tunnels. Geol. Soc. Am. Eng. Geol. (Berkey) Vol., pp. 45–81.

Sauck, W. A., 1981. Discussion of "Preliminary evaluation of an alternate electrode array for use in shallow-subsurface electrical studies." Ground Water, 19:554–556.

Savage, W. Z., and Chleborad, F., 1982. A model for creeping flow in landslides. Assoc. Eng. Geol. Bull. 19:333–338.

Scarano, R. A., and Linehan, J. J., 1978. Current U.S. Nuclear Regulatory Commission licensing review process: uranium mill tailings management. In: Organization for Economic Cooperation and Development, Nuclear Energy Agency, management, stabilization, and environmental impact of uranium mill tailings, Albuquerque, pp. 407–425.

Schaffer, R. J., 1933. The weathering of natural building stones: Special Rep. No. 18 of the Building Research Station, His Majesty's Stationery Office, London.

Scheidegger, A. E., 1961. Mathematical model of slope development. Geol. Soc. Am. Bull., 72:37–50.

Schlemon, Roy J., 1985. Application of soil-stratigraphic techniques to engineering geology. Assoc. Engr. Geol. Bull., 22:129–142.

Schlumberger, 1958. Introduction to Schlumberger well logging. Schlumberger Well Surveying Corp., Doc. no. 8, Houston.

Schmidt, K. D., 1979. The 208 planning approach to ground water protection—what is wrong and what can be done about it. Ground Water, 17:148–153.

Schmidt, P. W., and Pierce, K. L., 1976. Mapping of mountain soils west of Denver, Colorado, for land use planning. In: Coates, D. R. (ed.), Geomorphology and engineering. Dowden, Hutchinson and Ross, Stroudsburg, PA, pp. 43–53.

Schneider, R. R., and Dickey, W. L., 1980. Reinforced masonry design. Prentice-Hall, Englewood Cliffs, NJ.

Schneider, S. R., McGinnis, D. F., Jr., and Pritchard, J. A., 1981. Delineation of drainage and physiographic features in North and South Dakota. American Water Resources Association, Satellite Hydrology, 5th Annu. William T. Pecora Memorial Symp. on Remote Sensing, pp. 324–330.

Scholten, R., 1974. Deformational mechanisms in the Northern Rocky Mountains of the United States: the question of gigantic landslides. In: Voight, B. (ed.), Rock Mechanics, the American Northwest. Special Publication, Experiment Station, Penn. St. Univ., University Park, PA., pp. 181–185.

Schowengerdt, R. A., and Glass, C. E., 1983. Digitally processed topographic data for regional tectonic evaluation. Geol. Soc. Am. Bull., 94:549–556.

Schreiber, E., and Anderson, O. L., 1970. Properties and composition of lunar materials: earth analogies. Science, 168:1579–1580.

Schumann, H. H., and Poland, J. F., 1969. Land subsidence, earth fissures, and ground water withdrawal in south-central Arizona, USA. Proc. Tokyo Symp. on Land Subsidence: IASH-UNESCO, pp. 295–302.

Schumann, H. H., 1974. Land subsidence and earth fissures in alluvial deposits in the Phoenix area, Arizona. USGS, Map I-845-H.

Schumm, S. A., 1961. Effect of sediment characteristics on erosion and deposition in ephemeral-stream channels. USGS, Prof. Paper 352-C.

Schumm, S. A., 1971. Fluvial geomorphology, channel adjustment and river metamorphosis. In: Shen, A. W. (ed.), Fluvial geomorphology in river mechanics. Water Resources Publications, Fort Collins, CO pp. 4-1–5-22.

Schumm, S. A., ed., 1977. Drainage basin morphology: Benchmark Papers in Geology/41. Dowden, Hutchinson and Ross, Stroudsburg, PA.

Schumm, S. A., and Chorley, R. J., 1966. Talus weathering and scarp recession in the Colorado Plateau. Zeitshrift fur Geomorphologie, 10(1):11–36.

Schuster, R. L., 1983. Engineering aspects of the 1980 Mount St. Helens eruptions. Assoc. Eng. Geol. Bull. 20:125–143.

Schuster, R. L., and Krizek, R. L., 1979. Landslides, analysis and control. Nat. Acad. Sci., Special Rep. 176.

Schwartz, F. W., and McClymont, G. L., 1977. Applications of surface resistivity methods. Ground Water, 15:197–202.

Seed, H. B., and Wilson, S. D., 1969. The Turnagain Heights landslide, Anchorage, Alaska.

Conf. Stability and Performance of Slopes and Embankments, Berkeley, Ca., Am. Soc. Civil Eng., pp. 357–385.

Segall, P., and Pollard, D. O., 1983. Joint formation in granitic rock of the Sierra Nevada. Geol. Soc. Am. Bull., 92:563–575.

Self, S., Rampino, M. R., Newton, M. S., and Wolff, J. A. 1984. Volcanological study of the great Tambora eruption of 1815. Geology, 12:659–663.

Sharp, R. P., 1966. Kelso dunes, Mojave Desert, California. Geol. Soc. Am. Bull., 77:1045–1074.

Sharp, R. P., and Nobles, L. H., 1953. Mudflow of 1941 at Wrightwood, southern California. Geol. Soc. Am. Bull., 64:547–560.

Sharpe, C. F. S., 1938. Landslides and related phenomenon. Columbia Univ. Press, New York.

Sharpe, C. F. S., and Desch, E. F., 1942. Relation of soil-creep to earthflow in the Appalachian Plateau. J. Geomorphol., 5(4): 312–384.

Shelton, J. S., 1966. Geology illustrated. W. H. Freeman, San Francisco.

Shepard, F. P., 1963. Submarine geology. Harper and Row, New York.

Shepard, F. P., and Inman, D. L., 1951. Proc. 1st Conf. on Coastal Eng. Long Beach, Calif., Oct. 1980. The Engineering Foundation, Council on Wave Research, New York.

Shepard, F. P., and Wanless, H. R., 1970. Our changing coastlines. McGraw-Hill, New York.

Sheppard, T., 1912, The lost towns of the Yorkshire coast. A. Brown & Sons, London.

Sherard, J. L., Woodward, R. J., Gizienski, S. F., and Clevenger, W. A., 1963. Earth and earth-rock dams. John Wiley, New York.

Short, N. M., 1982. The Landsat tutorial workbook. National Aeronautics and Space Administration, Scientific and Technical Information Branch, NASA Ref. Publ. no. 1078, Washington, DC.

Short, N. M., Lowman, P. D., Jr., Greden, S. C., and Finch, W. A., Jr., 1976. Mission to earth, Landsat views the world. Nat. Aeronaut. Space Admin., Washington, DC.

Shreve, R. L., 1968. The Blackhawk landslide. Geol. Soc. Am., Special Paper 108.

Shulters, M. V., and Clifton, D. G., 1980. Mount St. Helens volcanic-ash fall in the Boell River watershed, Oregon, March–June, 1980. USGS, Circular 850-A.

Siefring, W., and Hart, W. L., 1982. Baltimore Harbor Tunnel: maintaining water quality. Water Spectrum, U.S. Army Corps of Engineers, 14(2):17–23.

Siegal, G. S., and Gillespie, A. R., (eds.), 1980. Remote sensing in geology: John Wiley, New York.

Sieh, K. E., and Jahns, R. H., 1984. Holocene activity of the San Andreas fault at Wallace Creek, Calif. Geol. Soc. Am. Bull., 95:883–896.

Simpson, R. H., and Lawrence, M. B., 1971. Atlantic hurricane frequencies along the US coastline. National Ocean and Atmospheric Admin., Tech. Memo. no. NWS-SR-58, Washington, DC.

Simons, D. B., and Richardson, E. V., 1963. Forms of bed roughness in alluvial channels. Am. Soc. Civil Eng., Trans. 128:284–302.

Sinton, L. W., 1980. Two antibiotic-resistant strains of Escherichia coli for tracing the movement of sewage in ground water. J. Hydrol. 19:119–129.

Skibitzke, H. E., 1961. Electronic computers as an aid to the analysis of hydrologic problems. Int. Assoc. of Scientific Hydrol. Publ. 52.

Skillern, F. F., 1981. Environmental protection: the legal framework. McGraw-Hill, New York.

Smith, D. B., 1976. Nuclear methods. In: Rodda, J. C. (ed.), Facets of hydrology. John Wiley, New York, pp. 61–83.

Smith, H. T. U., 1953. The Hickory Run Boulder Field, Carbon County, Pennsylvania. Am. J. Sci., 251:625–642.

Smith, N. D., and Smith, D. G., 1984. William River: an outstanding example of channel widening and braiding caused by bed-load addition. Geology, 12:78–82.

Smith, R. S. U., 1978. Guide to selected features of aeolian geomorphology in the Algodones dune chain, Imperial Valley, California. In: Greeley, R., Womer, M. B., Papson, R. P., and Spudis, P. D., (eds.), Aeolian features of southern California. Department of Geology and Center for Meteorite Studies, Ariz. St. Univ., Tempe, AZ, pp. 74–98.

Soderberg, A. D., 1979. Expect the unexpected: foundation for dams in karst. Assoc. Eng. Geol. Bull., 16:409–427.

Snow, D. T., 1982. Hydrogeology of induced seismicity and tectonism: case histories of Kariba and Koyna. Geol. Soc. Am., Special Paper 189, pp. 317–360.

Soil Conservation Service, 1979. Engineering field manual for conservation practices: US Dep. Agriculture, Washington, DC.

Soil Survey Staff, 1975. Soil taxonomy. US Dep. Agriculture Handbook, vol. 436.

Sowers, G. F., 1975. Failures in limestones in humid subtropics. Soc. Civil Eng., J. Geotechn. Eng. Div. 101(GT8):771–787.

Spangle, W., and Associates, Leighton, F. B., and Associates, and McDonald Baxter and Co., 1976. Earth science information in land-use planning, guidelines for earth scientists and planners. USGS, Circular 721.

Stall, J. B., and Lee, M. T., 1980. Reservoir sedimentation and its causes in Illinois. Water Resources Bull., 16:814–880.

Stauffer, T. P., Sr., 1978. Kansas City: a center for secondary use of mined-out space. In: Utgard, R. O., McKenzie, G. D., and Foley, D. (eds.), Geology in the urban environment. Burgess, Minneapolis, MN, pp. 176–186.

Steinberg, B., 1982. Planning versus implementation. US Army Corps of Engineers, Water Spectrum, 14(1):28–37.

Steinman, D. B., 1945. The builders of the bridge. Harcourt, Brace, New York.

Stephens, J. C., 1956. Subsidence of organic soils in the Florida everglades. Soil Sci. Soc. Am., Proc. 20(1):77–80.

Sterrett, R. J., and Edil, T. B., 1982. Ground water flow systems and stability of slopes. Ground Water, 20:5–11.

Stewart, M. T., 1980. Gravity survey of a deep buried valley. Ground Water, 18:24–30.

Stout, M. L., 1969. Radio-carbon dating of landslides in southern California and engineering geology implications. Geol. Soc. Am., Special Paper, 123, pp. 167–180.

Strand, R. I., and Pemberton, E. L., 1982. Reservoir sedimentation. US Bureau of Reclamation, Denver.

Streeter, V. L., and Wylie, E. B., 1979. Fluid mechanics. McGraw-Hill, New York.

Stringfield, V. T., LeGrand, H. E., and LaMoreaux, P. E., 1977. Development of karst and its effects on the permeability and circulation of water in carbonate rocks with special reference to the southeastern states. Geol. Survey of Alaska Bull., part G, n 94.

Stringfield, V. T., Rapp, J. R., and Anders, R. B., 1979. Effects of karst and geologic structure on the circulation of water and permeability in carbonate aquifers. J. Hydrol. 43:313–332.

Sundaram, P. N., Goodman, R. E., and Wang, C.-Y., 1976. Precursory and coseismic water-pressure variations in stick-slip experiments. Geology, 4:108–110.

Swam, David, 1982. Flexible environnmental policy needed for primary nonferrous smelters. Mining Cong. J., 68(1):24–28.

Swanston, D. N., and Swanson, F. J., 1976. Timber harvesting, mass erosion, and steepland forest geomorphology in the Pacific Northwest. In: Coates, D. C. (ed.), Geomorphology and engineering. Dowden, Hutchinson and Ross, Stroudsburg, PA, pp. 199–221.

Swift, D. J. P., and Palmer, H. D., eds., 1978. Coastal sedimentation: Benchmark Papers in Geology/42. Dowden, Hutchinson and Ross, Stroudsburg, PA.

Talma, A. S., 1981. Chemical changes in ground water and their reaction rates. Trans., Geol. Soc. S. Afr., 84(pt. 2) (special issue):99–105.

Tanaka, J. M. C., 1981. Letters to the editor. Geotimes, 26(12):12.

Tank, R. W., ed., 1983a. Environmental geology. Oxford, New York.

Tank, R. W., 1983b. Legal aspects of geology. Plenum, New York.

Taranik, J. V., 1982. Geological remote sensing and space shuttle. Mining Congr. J., 68(7):18–23.

Taranik, J. V., and Trautmein, C. M., 1976. Integration of geological remote sensing techniques, subsurface analysis. USGS, Open-file Rep. 76-402.

Taylor, C. L., and Conwell, F. R., 1981. BART—influence of geology on construction conditions and costs. Assoc. Eng. Geol. Bull., 18:195–205.

Telford, W. M., Geldart, L. P., Sheriff, R. E., Keys, D. A., 1976. Applied geophysics. Cambridge University Press, New York.

Tepordei, V. V., 1977. A new rock core log for geotechnical investigation. Assoc. Eng. Geol. Bull. 14:49–58.

Terzaghi, K., 1943. Theoretical soil mechanics. John Wiley, New York.

Terzaghi, K., 1950a. Geologic aspects of soft ground tunneling. In: Trask, P. (ed.), Applied sedimentology. John Wiley, New York, pp. 193–209.

Terzaghi, K., 1950b. Mechanism of landslides. Geol. Soc. Am. Eng. Geol. (Berkey) vol. pp. 83–123.

Terzaghi, K., 1962. Stability of steep slopes on hard unweathered rock. Harvard Soil Mechanics, Series 69.

Terzaghi, K., and Peck, R. B., 1948. Soil mechanics in engineering practice. John Wiley, New York.

Terzaghi, K., and Peck, R. B., 1967. Soil mechanics in engineering practice. John Wiley, New York.

Tharp, T. M., 1983. A field exercise in fluvial sediment transport: J. Geol. Ed., 31:375–378.

Theis, C. V., 1935. The relation between the lowering of the piezometric surface and the rate and duration at discharge of a well using ground water storage. Am. Geophys. Union, Trans., 16:519–524.

Thom, W. T., Jr., 1927. Subsidence and earth movements caused by oil extraction, or by drilling oil and gas wells. American Institute of Mining and Metallurgical Engineers, Technical Publication No. 17.

Thomas, D. M., and Benson, M. A., 1969. Generalizations of streamflow characteristics from drainage-basin characteristics: USGS, Water-Supply Paper 1975.

Thomas, E., 1962. Stabilization of rock by bolting: In: Fluhr, T., and Legget, R. F., (eds.), Reviews in engineering geology, Vol. I. Geological Society of America, pp. 257–280.

Thomas, G., and Witts, M. M., 1971. The San Francisco earthquake: Stein and Day, New York.

Thomas, M. P., 1966. Effect of glacial geology upon the time distribution of streamflow in eastern and southern Connecticut. USGS, Professional Paper 550, pp. 209–212.

Threet, R. L., 1979. Rose Canyon Fault, an alternative interpretation: In: Abbott, P. L., and Elliott, W. J. (eds.), Earthquakes and other perils, San Diego Region. San Diego Association of Geologists, San Diego, CA, pp. 61–71.

Till, Roger, 1974. Statistical methods for the earth scientist. Macmillan, London.

Tillman, S. M., 1982. Determination of subcoal strata strength using ultra sonic testing. Assoc. Eng. Geol. Bull., 19:77–86.

Timur, A., 1982. Advances in well-logging. J. Petrol. Techn., 34:1181–1185.

Todd, D. K., 1980. Ground water hydrology. Wiley, New York.

Toran, L., 1982. Isotopes in ground water investigations. Ground Water, 20:740–745.

Torrey, S., ed., 1978. Trace contaminants from coal. Noyes Data Corp., Park Ridge, NJ.

Tourtelot, H. A., 1974. Geologic origin and distribution of swelling clays. Assoc. Eng. Geol. Bull., 11:259–275.

Toy, T. J., 1982. Accelerated erosion: process, problems, and prognosis. Geology, 10:524–529.

Trantina, J. A., 1962. Investigation of landslides by seismic and electrical resistivity methods. In: Field testing of soils. American Society for Testing Materials, Special Publication no. 322, pp. 120–134.

Trautman, C. H., and F. H. Kulhawy, 1983. Data sources for engineering geologic studies. Assoc. Eng. Geol. Bull., 20:439–454.

Trefzger, R. E., 1966. Tecolote tunnel. In: Lung, R., and Proctor, R. (eds.), Engineering geology in southern California. Assoc. Eng. Geol. Special Publication, Glendale, CA, pp. 109–113.

Tributsch, H., 1982. When the snakes awake. MIT Press, Cambridge, MA.

Trujillo, L. F., 1982. Trap-efficiency study, Highland Creek flood-retarding reservoir near Kelseyville, California, water years 1966–77: USGS, Water-Supply Paper 2182.

U.S. Atomic Energy Commission, 1974. The nuclear industry. WASH-1174-73, UC-2.

U.S. Bureau of Reclamation, 1973. Design of small dams: U.S. Department of Interior. Water Resources Technical Publication, second ed.

U.S. Bureau of Reclamation, 1978. Drainage manual. U.S. Department of Interior, U.S. Govt. Printing Office, Washington, DC.

U.S. Department of Energy, 1977. Draft environmental impact statement, management of commercially operated radioactive waste. U.S. Department of Energy and Battelle Northwest Laboratory.

U.S. Department of Interior, 1969. Environmental planning and geology. U.S. Govt. Printing Office, Washington, D.C.

U.S. Department of Transportation, 1977. Rock slope engineering: workshops for planning, design, construction and maintenance of rock slopes for highways and railways. Federal Highway Administration, Washington, DC.

U.S. Geological Survey, 1971. The San Fernando, California earthquake of February 9, 1971. USGS, Professional Paper 733.

U.S. Geological Survey, 1982. Faulting arrested by control of ground-water withdrawal in Houston, Texas. USGS Yearbook, fiscal year 1982, US Govt. Printing Office, Washington, DC, pp. 51–53.

U.S. National Commission on Materials Policy, 1973. Material needs and the environment today and tomorrow. U.S. Govt. Printing Office, Washington, DC.

U.S. Nuclear Regulatory Commission, 1974. General site suitability criteria for nuclear power stations. Regulatory Guide 4.7.

U.S. Nuclear Regulatory Commission, 1975. Reactor safety study, an assessment of accident risks in US commercial nuclear power plants. Wash-1400 (NUREG 75/014), NTIS, Springfield, VA.

Utgard, R. O., McKenzie, G. D., and Foley, D. (eds.), 1978. Geology in the urban environment. Burgess, Minneapolis.

Vacher, H. L., 1974. Ground water hydrology of Bermuda. Bermuda Public Works Department.

Van Arsdale, R., 1982. Influence of calcrete on the geometry of arroyos near Buckeye, Arizona. Geol. Soc. Am. Bull. 93:20–26.

Van Burkelow, A., 1945. Angle of repose and angle of sliding friction. Geol. Soc. Am. Bull., 56:669–707.

van der Tak, J., Haub, C., and Murphy, E., 1979. Our population predicament, a new look. Pop. Bull. 34(5):2–000.

Van Dorn, W. G., 1965. Tsunamis. In: Chow, V., ed., Advances in hydroscience, Vol. II. Academic, New York, pp. 1–48.

Van Nostrand, R. G., and Cooke, K. L., 1966. Interpretation of resistivity data. USGS, Professional Paper 499.

Vanoni, V. A., 1975. Sedimentation engineering. American Society of Civil Engineers, New York.

Varnes, David J., 1958. Landslides types and processes. In Eckel, E. B. (ed.), Landslides and engineering practice. Highway Research Board Special Report 29, pp. 20–47.

Varnes, D. J., 1978. Slope movement types and processes. In: Schuster, R. L., and Krizek, R. J. (eds.), Landslides, analysis and control. Transportation Research Board, National Academy of Sciences, Special Report, no. 176, pp. 11–33.

Veatch, R. W., Jr., 1983. Overview of current hydraulic fracturing design and treatment technology—Part 1. J. Petrol. Techn., 35:677–687.

Verbeek, E. R., Ratzlaff, K. W., and Clanton, U. S., 1979. Faults in parts of north-central and western Houston metropolitan area, Texas. USGS, Miscellaneous Field Studies Map MF-1136.

Verma, R. K., Rao, M. K., and Rao, C. V., 1980. Resistivity investigations for ground water in metamorphic areas near Dhanbad, India. Ground Water, 18:46–55.

Verstappen, H. T., 1983. Applied geomorphology. Elsevier, Amsterdam.

Vick, S. G., 1981. Morphology and role of landsliding in formation of some rock glaciers in the Mosquito Range, Colorado. Geol. Soc. Am. Bull., 92:75–84.

Vidal, M. H., 1979. The development and future of reinforced earth. American Society of Civil Engineers, Proceedings, Symposium on placement and improvement of soils, Geotechnical Engineering Division, Annual Convention, Pittsburgh, pp. 1–61.

Visocky, A. P., 1982. Impact of Lake Michigan allocations on the Cambrian-Ordovician aquifer system. Ground Water, 20:668–674.

Voight, Barry, 1974. Anchor dam and reservoir, the unsolved problem of sealing a sieve. In: Voight, B., (ed.), Rock mechanics and the American northwest. Expedition Guide, Third Congress of the International Society of Rock Mechanics, The Pennsylvania State University, pp. 56–58.

Voight, B., 1978. Lower Gros Ventre Slide, Wyoming, U.S.A. In: Voight, B. (ed.), Rockslides and avalanches, Vol. 1, Natural Phenomena. Elsevier, Amsterdam, pp. 113–166.

Voight, Barry, 1982. Time scale for the first moments of the May 18 eruption. In: Lipman, P. W., and Mullineaux, D. R. (eds.), The 1980 eruptions of Mount St. Helens, Washington, USGS, Professional Paper 1250, pp. 69–86.

Voight, B., and Faust, C., 1982. Frictional heat and strength loss in some rapid landslides. Geotechnique, 32:43–54.

Waananen, A. O., et al., 1977. Flood-prone areas and land-use planning—selected examples from the San Francisco Bay Region, CA. USGS Professional Paper 942.

Wahler, W. A., 1973. Analysis of coal refuse dam failure, Middle Fork, Buffalo Creek, Saunders, West Virginia. Report to U.S. Bureau of Mines, Contract No. S0122084, 2 Vol.

Wahrhaftig, C., 1965. Stepped topography of the southern Sierra Nevada. Geol. Soc. Am. Bull., 75:1165–1189.

Wahrhaftig, C., and Cox, A., 1959. Rock glaciers in the Alaska Range. Geol. Soc. Am. Bull., 70:383–436.

Wallace, R. E., 1980. G. K. Gilbert's studies of faults, scarps, and earthquakes. Geol. Soc. Am., Special Paper 183, pp. 35–44.

Walling, D. E., 1977. Suspended sediment and solute response characteristics of the River Exe, Devon, England. In: Davidson-Arnott, R., and Nickling, W. (eds.), Research in fluvial geomorphology. Proceedings, Fifth Guelph Symposium on Geomorphology, Guelph, Ontario, pp. 159–197.

Walters, R. C. S., 1962. Dam geology. Butterworth, London.

Walton, G., 1959. Public health aspects of the contamination of ground water in South Platte River Basin in vicinity of Henderson, Colorado, August, 1959. U.S. Public Health Service, Nov. 2, 1959, mimeographed report.

Walton, W. C., 1960. Leaky artesian conditions in Illinois. Illinois State Water Survey, Report of Investigation 39.

Walton, W. C., 1962. Selected analytical methods for well and aquifer evaluation. Illin. State Water Sur. Bull. 49.

Walton, W. C., 1963. The distribution in soils of radioactivity from weapons tests. J. Geophys. Res. 68:1485–1496.

Walton, W. C., 1970. Ground water resource evaluation. McGraw-Hill New York.

Walton, W. C., Hills, D. L., and Grundeen, G. M., 1967. Recharge from induced streambed infiltration under varying ground water level and stream gage conditions. Water Resources Research Center, University of Minnesota, Bull. 6.

Wang, C. P., and Williams, R. E., 1984. Aquifer testing, mathematical modeling, and regulatory risk. Ground Water, 22:285–296.

Wang, H. F., and Anderson, M. P., 1982. Introduction to ground water modeling, finite difference and finite element methods. Freeman, San Francisco.

Ward, D. C., 1982. Enhanced oil recovery program and outlook. Interstate Oil Compact Commission Committee Bull., 24(2):6–25.

Warner, D. L., 1969. Preliminary field studies using earth resistivity measurements for delineating zones of contaminated ground water. Ground Water, 7:9–16.

Warnke, D. A., 1973. Beach changes in a low-energy coastal area in Florida which is in the pathway of urbanization. In: Moran, D. E., Slosson, J. E., Stone, R. O., and Yelverton, C. A. (eds.), Geology, seismicity, and environmental impact. Special Publication, Association of Engineering Geologists, University Publishing Co., Los Angeles, pp. 367–377.

Washburn, A. L., 1967. Instrumental observations of mass-wasting in the Mesters Vig District, N.E. Greenland. Moddelelser Greenland, 166(4)

Water Resources Council, 1977. Erosion and sedimentation and related resources considerations. U.S. Department of Agriculture for 1975 National Water Assessment, Washington DC.

Watson, R. A., and Wright, H. E., Jr. 1969. The Saidmarreh Landslide, Iran, Geol. Soc. Am., Special Paper 123, pp. 115–140.

Watts, G. M., Vallianos, L., and Jachowski, R. A., 1977. Means of controlling littoral drift to protect beaches, dunes, estuaries and harbour entrances. In: El-Ashry, M. T., ed., Air photography and coastal problems. Benchmark Papers in geology/38, Dowden, Hutchinson and Ross, Stroudsburg, PA, pp. 246–270.

Weaver, K. F., 1981. Our energy predicament. National Geographic Special Report, February, 1981, pp. 2–22.

Wesson, R. L., Helley, E. J., Lajoie, K. R., and Wentworth, C. M., 1975. Faults and future earthquakes. In: Borcherdt, R. D. (ed.), Studies for seismic zonation of the San Francisco Bay region. USGS, Professional Paper 941-A, pp. A5–A30.

Westing, A. H., 1981. A note on how many humans that have ever lived. Bioscience, 31(7):523–524.

Whitcomb, J. H., Garmany, J. D., and Anderson, D. L., 1973. Earthquake prediction: variations of seismic velocity before the San Fernando earthquake. Science, 180:632–635.

White, E. M., 1962. Volume changes in some clayey soils. Soil Science, 94:168–172.

Whitehouse, I. E., and Griffiths, G. C., 1983. Frequency and hazard of large rock avalanches in the central Southern Alps, New Zealand. Geology, 11:331–334.

Wiegel, R. L., 1969. Seismic sea waves. In: Olson, R. A., and Wallace, M. M. (ed.), Geologic hazards and public problems. Office of Emergency Preparations, Conference Proceedings, Region 7, U.S. Govt. Printing Office, Washington, DC, pp. 53–75.

Wiegel, R. L. (ed.), 1970. Earthquake engineering. Prentice-Hall, Englewood Cliffs, NJ.

Wilbur, J. B., and Norris, C. H., 1948. Elementary structural analysis. McGraw-Hill, New York.

Williams, A., and Caldwell, N., 1981. Coastal problems growing in Bulgaria. Geotimes, 26(5): pp. 24–26.

Williams, G. P., 1977. Washington, D.C.'s vanishing springs and waterways. USGS, Circular 752.

Williams, R. S., Jr., et al., 1974. Environmental studies of Iceland with ERTS-1 imagery. Proc. Ninth Symposium on Remote Sensing of Environment, Ann Arbor, Michigan, vol. 1, pp. 31–81.

Williams, R. S., Jr., and Carter, W. D. (eds.), 1976. ERTS-1, a new window on our planet. USGS, Professional Paper 929.

Williams, R. S., Jr., Moore, J. G., 1973. Iceland chills a lava flow. Geotimes, 18:14–17.

Williams, R. E., 1975. Waste production and disposal in mining, milling, and metallurgical industries. Miller Freeman, San Francisco.

Williams, R. E., Ortman, D., and Mabes, D. 1979. Seepage loss from a tailings pond: a case study. In: Argall, G. O., Jr. (ed.), Tailing disposal today, Vol. 2. Proc. Second International Tailing Symposium, Denver, CO. Miller Freeman, San Francisco, pp. 397–427.

Williams, R. E., Wallace, A. T., and Mink, L. L., 1971. Impact of a well managed tailings pond system on a stream. Mining Cong. J., 57(10):48–56.

Williams, G. P., and Wolman, M. G., 1984. Downstream effects of dams on alluvial rivers. USGS, Professional Paper 1286.

Wilson, L. G., 1981. Monitoring in the vadose zone, part I, storage changes. Ground Water Monit. Rev., 1(3):32–41.

Wilson, W. E., and Gerhart, J. M., 1979. Simulated changes in potentiometric levels resulting from ground water development for phosphate mines, west-central Florida. J. Hydrol. 43:491–515.

Windham, S. R., and Campbell, K. M., 1981. Sinkholes follow the pattern. Geotimes, 26(8):20–21.

Winkler, E. M., 1966. Important agents of weathering for building and stone. Eng. Geol., 1:381–400.

Winkler, E. M., ed., 1978. Decay and preservation of stone. Geol. Soc. Am., Engineering Geology Case History, no. 11.

Winograd, I. J., 1974. Radioactive waste storage in an arid zone. Am. Geophys. Union, Trans., 55:884–894.

Winograd, I. J., and Thordarson, W., 1975. Hydrogeologic and hydrochemical framework, South-Central Great Basin, Nevada-California, with special reference to the Nevada Test Site. USGS, Professional Paper 712-C.

Wischmeier, W., and Smith, D., 1978. Predicting rainfall erosion losses. U.S. Department of Agriculture Handbook No. 537.

Wise, D. U., 1963. Keystone faulting and gravity sliding driven by basement uplift of Owl Creek Mountains, Wyoming. Assoc. Am. Petrol. Geol., 47:586–598.

Woessner, W. W., Andrews, C. B., and Osborne, T. J., 1979. The impacts of coal strip mining in the hydrogeologic system of the Northern Great Plains: case study of potential impacts on the Northern Cheyenne Reservation. J. Hydrol., 43:445–467.

Wolman, M. G., and Schick, P. A., 1967. Effects of construction on fluvial sediment, urban and suburban areas of Maryland. Water Resources Res., 3(2):451–462.

Wood, W. W., and Ehrlich, G. G., 1978. Use of Baker's Yeast to trace microbial movement in ground water. Ground Water, 16:398–403.

Wooley, R. R., 1946. Cloudburst floods in Utah, 1857–1938. USGS, Water-Supply Paper 994.

Yazicigil, H., and Sendlein, V. A., 1982. Surface geophysical techniques in ground water monitoring, part II. Ground Water Monitor. Rev., 2:56–62.

Yeats, R. S., Clark, M. N., Keller, E. A., and Rockwell, T. K., 1981. Active fault hazard in southern California, ground rupture versus seismic shaking. Geol. Soc. Am. Bull., 92:189–196.

Yen, B. C., 1969. Stability of slopes undergoing creep deformation. Proceedings, American Society of Civil Engineers, J. Soil Mech. Found. Div., 95(SM4):1075–1096.

Yerkes, R. F., Ellsworth, W. L., and Tinsley, J. C., 1983. Triggered reverse fault and earthquake due to crustal unloading, northwest Transverse Ranges, California. Geology, 11:287–291.

Youd, T. L., and Hoose, S. N., 1978. Historic ground failures in northern California associated with earthquakes. USGS, Professional Paper 993.

Young, A., 1969. Present rate of land erosion. Nature, 224:851–852.

Yu, Y. S., and Coates, D. F., 1979. Canadian experience in simulating pit slopes by the finite element method. In: Voight, Barry (ed.), Rockslides and avalanches, v. 2, engineering sites. Elsevier Amsterdam, pp. 709–758.

Zall, Linda, and Michael, R., 1980. Space remote sensing systems and their application to engineering geology. Assoc. Eng. Geol. Bull., 27:101–152.

Zanbak, C., 1983. Design charts for rock slopes susceptible to toppling: American Society of Civil Engineers. J. Geotech. Eng. Div. 109:1039–1062.

Zeevaert, L., 1957. Foundation design and behavior of Tower Latino Americana in Mexico City. Geotechnique, 7:115–133.

Zienkiewicz, O. C., 1968. Continuum mechanics as an approach to rock mass problems. In: Staff, K. G., and Zienkienwicz, O. C. (eds.), Roch mechanics in engineering practice. Wiley, New York, pp. 237–273.

Zienkiewicz, O. C., and Stagg, K. G. (eds.), 1968. Rock mechanics in engineering practice. Wiley, New York.

Zohdy, A. A. R., Eaton, G. P., and Mabey, D. R., 1974. Application of surface geophysics to ground water investigation. Techniques of water-resources investigations of the U.S. Geological Survey, Chap. D1, U.S. Govt. Printing Office, N. 2401-02543, Washington, DC.

# Index

**T**



**U**

# Illustration and Table Credits

I wish to thank all the authors who have kindly supplied published and unpublished materials, photographs, and information, and have given permission to reproduce, incorporate, or quote these in the text. In addition, the following have given permission to reproduce published textual, pictorial, or tabular materials in the book.

Association of Engineering Geologists: Figure 4.33 from Clough, Assoc. of Engineering Geologists Bull., 18:151–167 (1981); Figure 8.13 and 9.9 from Rahn, Assoc. of Engin. Geologists Bull., 80:2123–2128; Figure 9.10, 9.22A from Lung and Proctor, Engineering Geology in southern California, special publication, 1966; Figure 9.10(b) from Randall et al., Assoc. of Engineering Geolog. 20:9–94 (1983); Figure 11.22 from Leeds, Geology, Seismicity, and environmental impact, special publication, 1973; Figure 9.16 from Parizek, in Environment Planning and Geology, Proceedings of Symp. Engineering Geology in the Urban Environment, 1969.

American Geological Institute: Figures 6.8, 6.13, 6.20 from Rahn et al., Engineering Geology of the central and northern Black Hills, SD, 1981.

American Geophysical Union: Figure 8.34, from Am. Geophy. Union Trans., 39:1076–1084 Langbein and Schumm. Figure 3.6 from Judson and Ritter, Journal of Geophys. Res., 69:3395–3401, 1964. Figure 8.35 from Wolman and Schick, Water Resources Res., 3(2):451–462 (1967).

American Society of Civil Engineering: Figure 6.3 modified from Zanbank, J. Geotech. Eng. Div., 109:1039-1062; Figure 9.10 from Poland, J. Irrig. Drainage Div., 107 (no. IR2):115–135 (1981).

Asphalt Institute: Figures 5.1, 5.4, 5.5 from Soils Manual, Series No. 10, 1964.

Burgess Publ. Co.: Figure 8.21 from Leopold, Man and his Physical Environment, 1972. Figure 11.7 from Nehru et al., Studies of the Earth, 1972.

Canada Dept. of Energy, Mines, and Resources: Figure 4.14 from Coates, Rock Mechanics Principles, Mines Branch Monograph 874, 1967.

Charles E. Merrill Publishing Co.: Figure 7.24 from Fetter, Applied Hydrogeology, 1980; Figures 10.20 and 10.25 from Mathewson, Engineering Geology, 1981.

Civil Engineering: Figures 17 a and b, Table 6.2 from Kiersch, 34(3) 32–39 (1964).

Connecticut Geological and Natural History Survey: Figure 3.14 from Rahn, 1971, The Surficial Geology of the Spring Hill Quadrangle, Bull. 26.

Council on Environmental Quality: Table 1.1 from the Global 2000 Report to the President, 1980.

Dover Pub., Inc.: Figure 7.22 from Meinzer, Hydrology, 1942.

The Engineering Foundation: Figure 10.6 from Shepard and Inman, Proced. 1st Conf. on Coastal Eng. Long Beach, CA, 1980.

Freeman, Cooper & Co.: Figures 4.1, 4.2, 4.3 from Johnson, Physical Processes in Geology, 1970.

Geotimes: Figure 9.13 from Windham and Campbell, Sinkholes follow the pattern. Geotimes 26(8):20–21 (1981).

Geological Survey of Canada: Figures 12.6, 12.7, 12.14, 12.15 from Economic Geology Report No. 26, 1970; Table 12.1.

Holt, Rinehart and Winston: Figure 2.5 from Lattman, Aerial Photographs in Field Geology, 1965.

John Wiley & Sons: Figure 4.13 and 5.11 from Lambe and Whitman, Soil Mechanics, 1969. Figure 4.19 from Peck et al., Foundation Engineering, 1974. Figure 5.10 and 5.14 from Hough, Basic Soils Engineering (Ronald Press, 1957). Figures 7.1, 7.2, 7.5–7.7, 7.11, 7.15, 7.32, 12.12, 12.16, Table 7.1 from Davis and DeWeist, Hydrogeology, 1966; Figures 7.9, 7.10, 7.14, 7.16, 7.33, 8.1, 8.2, 8.3, 8.4, 12.1, 12.9, Table 7.2 from Todd, Ground Water Hydrology, 1980; Figures 11.5, 11.6, 11.7 from Longwell et al., Physical Geology, 1969.

Kansas City Star: Figure 13.19 by Lynch.

MacMillan Co.: Figure 4.5, 4.7, Table 4.1 from Marin and Sauer, Strengths of Materials, 1954.

McGraw-Hill Book Co.: Figures 4.10, 4.11, 4.28, 4.29, 4.30, 4.32, 4.33, 8.9, 8.37, 10.10, from Krynine and Judd, Principles of Engineering Geology and Geotechnics, 1957; Figure 4.34 from Legget, Cities and Geology, 1973; Figure 5.8 from Das, Advanced Soil Mechanics, 1983; Figures 7.23, 7.31, 7.39, 12.4, 12.18, 12.19, Tables 7.1 and 7.4 from Bouwer, Groundwater Hydrology, 1978; Figures 7.2, 7.6, 8.26, Table 8.4 from Linsley and Franzini, Water Resources Engineering, 1979; Figure 9.2 from Walton, Ground Water Resource Evaluation, 1970; Figures 10.9, 10.12 from Ippen, Estuary and Coastline Hydrodynamics, 1966; Figure 10.19, Table 11.3 from Howell, Introduction to Geophysics, 1959; Figure 12.10 from Dobrin, Introduction to Geophysical Prospecting; Table 4.2 from Beer and Johnson, Vector Mechanics for Engineering, 1977.

Miller Freeman Publications, Inc.: Figure 13.3 from Johnson, Proceedings of the 2nd International Tailings Symposium, 1979.

Mining Congress Journal: Table 13.2 from Kruger, Vol. 57(7):52–57, 1971.

National Academy of Sciences: Figure 6.16 from Schuster and Krizek, Special Report No. 176.

National Association of Geology Teachers: Figures 13.9, 13.11, 13.12 from Lindholm, J. Geol. Ed., 28:36–45 (1980); Figure 13.14 and Table 13.3 from Bartlett, J. Geol. Ed., 28:4–35 (1980).

National Water Well Association: Figure 7.19 from Heath and Trainer, Introduction to Groundwater Hydrology, 1981.

Oxford University Press: Figures 10.13, 10.16, 10.17 from Rosenbaum (Tank, ed.), Focus on Environmental Geology, 1976. Figure 10.26 from Leveson, Geology and the Urban Environment, 1980.

Pennsylvania State University: Figures 4.24, 4.26 from Lane, in Rock Mechanics, the American Northwest (B. Voight, ed.), 1974.

Population Reference Bureau: Figures 1.1 and 1.3 from van der Tak, Population Bull., 34(5):2(1979); Figure 1.2 from Haupt and Kane, Population Handbook, 1982.

Portland Cement Association: Figures 5.2 and 5.6 from PCA Soil Primer, 1962.

Prentice-Hall, Inc.: Figures 7.25 and 7.27 from Freeze and Cherry, Ground Water, 1979.

San Diego Association of Geologists: Figure 11.18 from Hileman, in Earthquakes and other perils, San Diego Area (Abbott and Elliott, eds., 1979). Figures 11.19 and 11.20 from Crouse, Ibid.

Science Magazine: Figure 8.40 from Vol. 189:681–684 (1975).

South Dakota State University: Figure 2.6 from Dalstead et al., Application of Remote Sensing in South Dakota, Remote Sensing Institute Rep. No. SDSU-RS1-77-08; 1979.

South Dakota Geological Survey: Figure 12.12 from Lum, Rep. of Invest. No. 89.

Springer-Verlag: Figures 11.4, 11.16, Tables 11.1, 11.4, 14.1 from Bolt et al., Geological Hazards, 1977.

Trans Technical Publishing: Table 4.1 from Lama and Vutukuni, Handbook on Mechanical Properties of Rocks, Vol. II, 1978.

Wadsworth Publishing Co.: Figure 11.8 from Griggs and Gilchrist, Geologic Hazards, Resources, and Environmental Planning, 1983.

Water Resources Bulletin: Figure 8.38 from Stall and Lee, Vol. 16:814–880 (1980).

Water Resources Research Center, Univ. of Minnesota: Figure 7.34 from Walton et al., Bulletin No. 6, 1967.

W. H. Freeman & Co.: Figures 8.7, 8.10 from Leopold et al., Fluvial Processes in Geomorphology, 1964; Tables 3.1 and 8.2 from Leopold, Water: A Primer, 1974; Figure 10.20 from Shelton, Geology Illustrated, 1966.

William C. Brown Publications Co.: Figure 8.5 modified from Ritter, Process Geomorphology, 1978.

Zeitschrift fur Geomorphologie: Figure 6.24 from Johnson and Rahn, Suppl. 9, pp. 168–186, 1970.

Special thanks to the USGS and the GSA for supplying unpublished photographs and tables.